Principles of Gas–Solid Flows

Gas–solid flows are involved in numerous industrial processes and occur in various natural phenomena. This authoritative book addresses the fundamental principles that govern gas–solid flows and the application of these principles to various gas–solid flow systems.

The book is arranged in two parts: Part I deals with basic relationships and phenomena, including particle size and properties, collision mechanics of solids, momentum transfer and charge transfer, heat and mass transfer, basic equations, and intrinsic phenomena in gas–solid flows. Part II discusses the characteristics of selected gas–solid flow systems such as gas–solid separators, hopper and standpipe flows, dense-phase fluidized beds, circulating fluidized beds, pneumatic conveying systems, and heat and mass transfer in fluidization systems.

As a comprehensive information source on gas–solid flows, this text will be useful to a broad range of engineers and applied scientists – chemical, mechanical, agricultural, civil, environmental, aeronautical, and materials engineers, as well as atmospheric and meteorological scientists.

Principles of Gas–Solid Flows

LIANG-SHIH FAN
The Ohio State University

CHAO ZHU
The Ohio State University

CAMBRIDGE
UNIVERSITY PRESS

CAMBRIDGE UNIVERSITY PRESS
Cambridge, New York, Melbourne, Madrid, Cape Town, Singapore, São Paulo

Cambridge University Press
The Edinburgh Building, Cambridge CB2 2RU, UK

Published in the United States of America by Cambridge University Press, New York

www.cambridge.org
Information on this title: www.cambridge.org/9780521581486

First published 1998
This digitally printed first paperback version 2005

A catalogue record for this publication is available from the British Library

Library of Congress Cataloguing in Publication data

Fan, Liang-Shih.
Principles of gas–solid flows / Liang-Shih Fan, Chao Zhu.
p.   cm.
Includes bibliographical references (p.   ).
ISBN 0-521-58148-6
1. Two-phase flow.   2. Gas flow.   3. Bulk solids flow.   I. Zhu,
Chao, 1961–   .   II. Title.
TA357.5.M84F36   1997
531′.163 – dc20   96-41141
CIP

ISBN-13 978-0-521-58148-6 hardback
ISBN-10 0-521-58148-6 hardback

ISBN-13 978-0-521-02116-6 paperback
ISBN-10 0-521-02116-2 paperback

To

Lois Liang-Yen, Liang-Fu, and Jeannine Liang-Chi

and

Zhichi Zhu, Zhu Ruan, Baolan Shi, and Victor

Contents

Preface page xv

Part I Basic Relationships

1 Size and Properties of Particles 3
1.1 Introduction 3
1.2 Particle Size and Sizing Methods 3
1.2.1 Equivalent Diameters of a Nonspherical Particle 4
1.2.2 Particle Sizing Methods 10
1.3 Particle Size Distributions and Averaged Diameters 17
1.3.1 Density Functions 18
1.3.2 Typical Distributions 19
1.3.3 Averaged Diameters of a Particulate System 23
1.4 Material Properties of Solids 24
1.4.1 Physical Adsorption 25
1.4.2 Deformation and Fracture 28
1.4.3 Thermal Properties 32
1.4.4 Electrical Properties 35
1.4.5 Magnetic Properties 37
1.4.6 Material Densities 38
1.4.7 Optical Properties 39
Nomenclature 40
References 42
Problems 43

2 Collision Mechanics of Solids 46
2.1 Introduction 46
2.2 Stereomechanical Impact 47
2.2.1 Collinear Impact of Spheres 47
2.2.2 Planar Impact of Spheres 48
2.3 Theory of Elastic Contact of Solids 49
2.3.1 General Relations of Stresses in a Solid Medium in Equilibrium 50
2.3.2 Concentrated Force at a Point in an Infinite Solid Medium 52
2.3.3 Force on the Boundary of a Semiinfinite Solid Medium 53
2.3.4 Hertzian Theory for Frictionless Spheres in Contact 59
2.3.5 Theories for Frictional Spheres in Contact 63
2.4 Collision of Elastic Spheres 72
2.4.1 Normal Collision of Elastic Spheres 72
2.4.2 Collision of Frictional Elastic Spheres 74

2.5 Collision of Inelastic Spheres 78
2.5.1 Onset of Plastic Deformation 78
2.5.2 Restitution Coefficient 80
Nomenclature 83
References 85
Problems 85

3 Momentum Transfer and Charge Transfer 87
3.1 Introduction 87
3.2 Particle–Fluid Interactions 87
3.2.1 Drag Force 87
3.2.2 Basset Force 88
3.2.3 Saffman Force and Other Gradient-Related Forces 95
3.2.4 Magnus Effect and Force Due to Rotation of a Sphere 97
3.3 Interparticle Forces and Field Forces 101
3.3.1 Van der Waals Force 101
3.3.2 Electrostatic Force 103
3.3.3 Collisional Force 104
3.3.4 Field Forces 105
3.4 Motion of a Single Particle 107
3.4.1 Basset, Boussinesq, and Oseen (BBO) Equation 107
3.4.2 General Equation of Motion 108
3.5 Charge Generation and Charge Transfer 111
3.5.1 Static Electrification of Solids 111
3.5.2 Charge Transfer by Collision 119
Nomenclature 123
References 126
Problems 128

4 Basic Heat and Mass Transfer 130
4.1 Introduction 130
4.2 Heat Conduction 130
4.2.1 Heat Transfer of a Single Sphere in a Quiescent Fluid 131
4.2.2 Heat Conduction in a Collision of Elastic Spheres 133
4.3 Convective Heat Transfer 138
4.3.1 Dimensional Analysis of Forced Convection in a Single-Phase Flow 138
4.3.2 Heat Transfer of a Single Sphere in a Uniform Flow 138
4.3.3 Thermal Convection in Pseudocontinuum One-Phase Flow 141
4.4 Thermal Radiation 142
4.4.1 Single-Particle Scattering 143
4.4.2 Radiant Heating of a Particle 148
4.4.3 General Considerations in Radiation with a Particle Cloud 150
4.4.4 Radiation Through an Isothermal and Diffuse Scattering Medium 154
4.5 Mass Transfer 156
4.5.1 Diffusion and Convection 156
4.5.2 Mass and Heat Transfer Analogy 157

Nomenclature 159
References 161
Problems 162

5 Basic Equations 164
5.1 Introduction 164
5.1.1 Eulerian Continuum Approach 164
5.1.2 Lagrangian Trajectory Approach 165
5.1.3 Kinetic Theory Modeling for Interparticle Collisions 166
5.1.4 Ergun Equation 166
5.1.5 Summary 167
5.2 Modeling of Single-Phase Flows 167
5.2.1 General Transport Theorem and General Conservation 167
5.2.2 Governing Equations 169
5.2.3 Kinetic Theory and Transport Coefficients 170
5.2.4 Modeling for Turbulent Flows 174
5.2.5 Boundary Conditions 179
5.3 Continuum Modeling of Multiphase Flows 182
5.3.1 Averages and Averaging Theorems 182
5.3.2 Volume-Averaged Equations 189
5.3.3 Volume–Time-Averaged Equations 193
5.3.4 Transport Coefficients and Turbulence Models 196
5.3.5 Boundary Conditions of Particle Phase 205
5.4 Trajectory Modeling of Multiphase Flows 205
5.4.1 Deterministic Trajectory Models 206
5.4.2 Stochastic Trajectory Models 208
5.5 Kinetic Theory Modeling for Collision-Dominated Dense Suspensions 210
5.5.1 Dense-Phase Transport Theorem 211
5.5.2 Hydrodynamic Equations 213
5.5.3 Collisional Pair Distribution Function 215
5.5.4 Constitutive Relations 217
5.6 Equations for Flows Through Packed Beds 222
5.6.1 Darcy's Law 223
5.6.2 Straight Capillaric Model 224
5.6.3 Ergun's Equation 225
5.7 Dimensional Analysis and Similarity 230
5.7.1 Scaling Relationships for Pneumatic Transport of Dilute Suspensions 230
5.7.2 Scaling Relationships for Fluidized Beds 232
Nomenclature 236
References 239
Problems 242

6 Intrinsic Phenomena in a Gas–Solid Flow 244
6.1 Introduction 244
6.2 Erosion and Attrition 244
6.2.1 Ductile Erosion and Brittle Erosion 245

6.2.2 Locations of Erosive Wear 247
6.2.3 Mechanisms of Attrition 252
6.3 Thermodynamic Properties of a Gas–Solid Mixture 254
6.3.1 Density, Pressure, and Equation of State 254
6.3.2 Internal Energy and Specific Heats 257
6.3.3 Isentropic Change of State 258
6.4 Pressure Waves Through a Gas–Solid Suspension 259
6.4.1 Acoustic Wave 259
6.4.2 Normal Shock Wave 265
6.5 Instability 270
6.5.1 Wave Motions in Stratified Pipe Flows 270
6.5.2 Continuity Wave and Dynamic Wave 280
6.6 Particle–Turbulence Interaction 285
Nomenclature 288
References 292
Problems 293

Part II System Characteristics

7 Gas–Solid Separation 297
7.1 Introduction 297
7.2 Separation by Rotating Flow 297
7.2.1 Mechanism and Type of Rotary Flow Dust Separators 297
7.2.2 Flow Field in a Cyclone 300
7.2.3 Collection Efficiency of Cyclones 303
7.3 Electrostatic Precipitation 309
7.3.1 Mechanism of an Electrostatic Precipitator 310
7.3.2 Migration Velocity and Electric Wind 311
7.3.3 Collection Efficiency of Electrostatic Precipitators 312
7.4 Filtration 314
7.4.1 Mechanisms of Filtration and Types of Filters 314
7.4.2 Pressure Drop in a Filter 315
7.4.3 Collection Efficiency of Fabric Filters 319
7.5 Gravity Settling and Wet Scrubbing 320
7.5.1 Gravity Settling Chambers 321
7.5.2 Mechanisms of Scrubbing and Types of Scrubbers 323
7.5.3 Modeling for Scrubbing and Collection Efficiency 324
Nomenclature 328
References 330
Problems 331

8 Hopper and Standpipe Flows 333
8.1 Introduction 333
8.2 Powder Mechanics in Hopper Flows 333
8.2.1 Mohr Circle for Plane Stresses 334
8.2.2 Mohr–Coulomb Failure Criterion and Coulomb Powders 336
8.2.3 Static Stress Distributions in Standpipes and Hoppers 337

	8.2.4	Stress Distribution in a Steady Hopper Flow	340
	8.2.5	Flowability of Powders in Hopper Design	342
8.3		Hopper and Standpipe Flow Theories	346
	8.3.1	Moving Bed Flows in a Feed Hopper	346
	8.3.2	Standpipe Flows	350
	8.3.3	Hopper–Standpipe–Discharger Flow	354
	8.3.4	Multiplicity of Steady Standpipe Flows	357
	8.3.5	Leakage Flow of Gas in a Standpipe	359
8.4		Types of Standpipe Systems	361
	8.4.1	Overflow and Underflow Standpipes	361
	8.4.2	Inclined Standpipe and Nonmechanical Valves	364
Nomenclature			366
References			368
Problems			369
9		Dense-Phase Fluidized Beds	371
9.1		Introduction	371
9.2		Particle and Regime Classifications and Fluidized Bed Components	371
	9.2.1	Classification of Fluidized Particles	372
	9.2.2	Fluidization Regimes	374
	9.2.3	Components in a Dense-Phase Fluidized Bed	376
9.3		Minimum Fluidization and Particulate Fluidization	378
	9.3.1	Minimum Fluidization	378
	9.3.2	Particulate Fluidization	380
9.4		Bubbling Fluidization	381
	9.4.1	Onset of Bubbling	381
	9.4.2	Single Bubble in a Fluidized Bed	382
	9.4.3	Bubble/Jet Formation and Bubble Coalescence and Breakup	388
	9.4.4	Bubble/Jet Size and Bubble Rise Velocity	389
	9.4.5	Gas Flow Division and Bed Expansion	392
9.5		Turbulent Fluidization	396
	9.5.1	Regime Transition and Identification	396
	9.5.2	Determination of Transition Velocity	398
	9.5.3	Hydrodynamic Characteristics	399
9.6		Entrainment and Elutriation	400
	9.6.1	Mechanisms of Solids Ejection into the Freeboard	401
	9.6.2	Correlations and Modeling	402
9.7		Slugging	403
	9.7.1	Shapes and Rise Velocities of Single Slugs	403
	9.7.2	Continuous Slugging	404
9.8		Spouted Beds	406
	9.8.1	Onset of Spouting	407
	9.8.2	Maximum Spoutable Bed Depth and Spout Diameter	408
	9.8.3	Fountain Height	408
	9.8.4	Gas Flow Distribution	408
Nomenclature			409
References			411
Problems			415

10 Circulating Fluidized Beds 421
10.1 Introduction 421
10.2 System Configuration 422
10.3 Flow Regimes and Transitions Between Regimes 423
10.3.1 Flow Regimes and Regime Diagrams 423
10.3.2 Regime Transition Determination 425
10.3.3 Operable Fluidization Regimes 429
10.4 Hydrodynamic Behavior in a Macroscale 438
10.4.1 Axial Profiles of Cross-Sectional Averaged Voidage 438
10.4.2 Radial Profiles of Voidage and Solids Flux 440
10.4.3 Overall Solids Holdup 442
10.5 Local Solids Flow Structure 443
10.5.1 Transient Nature of Solids Flow 444
10.5.2 Characterization of Intermittent Solids Flow 445
10.6 Mathematical Models of Fast Fluidization 447
10.6.1 Models Based on the Concept of Cluster 447
10.6.2 Models Based on the Core-Annular Flow Structure 448
10.6.3 Models Based on the Axial Profiles of Solids Holdup 451
10.6.4 Two-Phase Flow Models and Computational Fluid Dynamics 451
Nomenclature 453
References 455
Problems 459

11 Pneumatic Conveying of Solids 461
11.1 Introduction 461
11.2 Classifications of Pneumatic Conveying Systems 461
11.2.1 Horizontal and Vertical Transport 461
11.2.2 Negative- and Positive-Pressure Pneumatic Conveyings 462
11.2.3 Dilute Flow Versus Dense Flow 464
11.2.4 Flow Regimes and Regime Transitions 464
11.3 Pressure Drop 466
11.3.1 General Pressure Drop in One-Dimensional Flow 467
11.3.2 Drag Reduction 469
11.3.3 Pressure Drop and Acceleration Length in Developing Regions 472
11.4 Critical Transport Velocities 474
11.4.1 Minimum Transport Velocity 475
11.4.2 Pick-up Velocity 476
11.5 Flows in Bends 478
11.5.1 Single-Phase Flow in a Curved Pipe 478
11.5.2 Particulate Flow in a Bend 481
11.6 Fully Developed Dilute Pipe Flows. 482
11.6.1 Basic Equations and Boundary Conditions 483
11.6.2 Characteristic Relations 487
11.6.3 Temperature Distributions of Phases 489
Nomenclature 494
References 496
Problems 498

12 Heat and Mass Transfer Phenomena in Fluidization Systems 499
12.1 Introduction 499
12.2 Suspension-to-Surface Heat Transfer 499
12.2.1 Heat Transfer Modes and Regimes 500
12.2.2 Film Model 502
12.2.3 Single-Particle Model 503
12.2.4 Emulsion Phase/Packet Model 506
12.3 Heat Transfer in Dense-Phase Fluidized Beds 512
12.3.1 Particle-to-Gas and Bed-to-Gas Heat Transfer 512
12.3.2 Bed-to-Surface Heat Transfer 513
12.3.3 Effect of Operating Conditions 518
12.4 Heat Transfer in Circulating Fluidized Beds 521
12.4.1 Mechanism and Modeling 521
12.4.2 Radial and Axial Distributions of Heat Transfer Coefficient 524
12.4.3 Effect of Operating Parameters 525
12.5 Heat Transfer in Spouted Beds 526
12.5.1 Gas-to-Particle Heat Transfer 526
12.5.2 Bed-to-Surface Heat Transfer 527
12.6 Mass Transfer in Multiparticle Gas–Solid Systems 527
12.6.1 Mass Transfer in Dense-Phase Fluidized Beds 527
12.6.2 Mass Transfer in Circulating Fluidized Beds 532
Nomenclature 532
References 535
Problems 537

Appendix: Summary of Scalar, Vector, and Tensor Notations 540

Index 545

Preface

Gas–solid flows are involved in numerous industrial processes and occur in various natural phenomena. For example, in solid fuel combustion, gas–solid flows are involved in pulverized coal combustion, solid waste incineration, and rocket propellant combustion. Gas–solid flows are encountered in pneumatic conveying of particulates commonly used in pharmaceutical, food, coal, and mineral powder processing. Fluidization is a common gas–solid flow operation with numerous important applications such as catalytic cracking of intermediate hydrocarbons, and Fischer–Tropsch synthesis for chemicals and liquid fuel production. Gas–solid flows occur in gas–particle separations, as exemplified by cyclones, electrostatic precipitators, gravity settling, and filtration operations. Fine powder–gas flows are closely associated with material processing, as in chemical vapor deposition for ceramics and silicon production, plasma coating, and xerography. In heat transfer applications, gas–solid flows are involved in nuclear reactor cooling and solar energy transport using graphite suspension flows. Solid dispersion flows are common in pigment sprays, dust explosions and settlement, and nozzle flows. The natural phenomena accompanied by gas–solid flows are typified by sand storms, moving sand dunes, aerodynamic ablation, and cosmic dusts. The optimum design of the industrial processes and accurate account of the natural phenomena that involve gas–solid flows as exemplified previously require a thorough knowledge of the principles governing these flows.

This book is intended to address basic principles and fundamental phenomena associated with gas–solid flows, as well as characteristics of selected gas–solid flow systems. It covers the typical range of particle sizes of interest to gas–solid flows, *i.e.*, 1 μm–10 cm, recognizing that flow characteristics for submicrometer particles are also of great industrial importance. The book features a systematic account of important theories or models concerning particle mechanics as well as fluid dynamics from their origins of the development. The physical interpretation and limitations in application of these theories or models are emphasized. Various intrinsic phenomena underlying the gas–solid flow systems are also illustrated. The book is aimed as a textbook for seniors and graduate students who are interested in general or specific topics of gas–solid flows. In addition, it can be used as a reference for researchers and practitioners who are interested in the general field of multiphase flow. It is written with multidisciplinary engineering readers in mind. Specifically, it will be of benefit to chemical and mechanical engineering readers as well as readers in other engineering disciplines, including agricultural, civil, environmental, pharmaceutical, aeronautical, mining, and atmospheric and meteorological sciences.

The book contains two parts; each part comprises six chapters. Part I deals with basic relationships and phenomena of gas–solid flows while Part II is concerned with the characteristics of selected gas–solid flow systems. Specifically, the geometric features (size and size distributions) and material properties of particles are presented in Chapter 1. Basic particle sizing techniques associated with various definitions of equivalent diameters of particles are also included in the chapter. In Chapter 2, the collisional mechanics of solids, based primarily on elastic deformation theories, is introduced. The contact time, area, and force of colliding particles are discussed using theories of elastic collision, which are important to the formulation of the momentum, heat, and charge transfer processes involving collisions of solids. Chapter 3 is devoted to the momentum and charge transfer of gas–solid flows. Various forces in gas–solid flows due to gas–particle interactions, particle–particle interactions, and external fields are delineated. Equations for single-particle motion, based on a force balance analysis, are derived. Basic mechanisms of charge generation in gas–solid flows are also introduced in the chapter, along with a detailed discussion of charge transfer mechanism by particle collisions. Chapter 4 deals with fundamental concepts and theories of heat and mass transfer in gas–solid flows. Highlights include thermal radiation of the particulate phase and heat conduction in collisions of elastic particles. Chapter 5 presents four basic modeling approaches of gas–solid flows, namely, continuum modeling of multiphase flows or multifluid modeling, trajectory modeling, kinetic theory modeling for collision-dominated dense suspensions, and the Ergun equation for flow through a packed bed of particles. In this chapter, the hydrodynamic equations of single-phase flows are first discussed. Here, basic concepts of kinetic theory of gas and turbulence models are introduced as a preamble to discussion of these basic modeling approaches. In contrast to the $k-\epsilon$ turbulence model for single-phase flows, the $k-\epsilon-k_{\mathrm{p}}$ model is introduced with the continuum approach of gas–solid flows to account for gas–solid turbulence interactions. Chapter 6 focuses on the discussion of intrinsic phenomena in gas–solid flows, such as erosion and attrition, acoustic wave and shock wave propagation through a gas–solid suspension flow, thermodynamic properties of a gas–solid mixture, flow instability, and gas–solid turbulence interactions.

Chapter 7 is concerned with gas–solid separations. The basic separation methods introduced in this chapter include cyclone, filtration, electrostatic precipitation, gravity settling, and wet scrubbing. Chapter 8 deals with hopper flows and standpipe flows, which are commonly encountered in the bulk solids handling and transport processes. In order to understand the fundamental hopper and standpipe flow characteristics, some basic concepts of powder mechanics are illustrated. Chapter 9 introduces the general concept of gas fluidization. Specifically, the chapter addresses dense-phase fluidization, which represents a gas–particle operation of enormous industrial importance. Various operating regimes including particulate fluidization, bubbling/slugging fluidization, and turbulent fluidization are discussed along with spouting. The fundamental properties of bubble, cloud, and wake and the intrinsic bubble coalescence and breakup and particle entrainment phenomena are illustrated. Chapter 10 continues the discussion of fluidization under higher-velocity conditions which are characterized by fast fluidization. Fast fluidization is conducted in a riser of a circulating fluidized bed system where solid particles are circulating in a loop. This chapter illustrates the interactive relationship of gas–solid flows in a loop situation by considering the flow behavior of the individual loop components and their effects on the overall gas–solid flow characteristics. Chapter 11 is concerned mainly with the dilute transport or pipe flow of gas–solid suspensions. Some pertinent phenomena such as drag reduction are discussed. Fully developed pipe flow and gas–solid flow in a bend are also illustrated. Chapter 12 describes transport phenomena underlying heat and mass transfer in fluidized systems. Transport models and empirical correlations are introduced to allow heat and mass transfer properties in various fluidized systems to be quantified. An appendix which summarizes the scalar, vector, and tensor notations presented in the text is provided. Throughout the text, unless otherwise noted, the correlation equations presented are given in SI units. Common notations used across the chapters such as superficial gas velocity and particle Reynolds number are unified. The solution manual for homework problems is available. Interested instructors are urged to contact the publisher about it.

The book is intended to be used in various ways depending on the specific information that the readers desire. While the material is described in a logical sequence chapter by chapter, each chapter is presented with extensive cross-references and maintains reasonable independence. Thus, readers who wish to have a quick grasp of a specific subject may go directly to the relevant chapters. It is important to note that gas–solid flow is a rapidly developing field of research, and that the physical phenomena of gas–solid flows are so complex that a comprehensive understanding of the phenomena is far from complete. Therefore, the present text is also intended to provide readers with ample fundamental concepts to allow them to follow through new developments in the field.

We would like to express our sincere thanks to the following colleagues who have reviewed the text and provided constructive suggestions and overviews: Professor R. S. Brodkey, Professor R. Clift, Professor J. F. Davidson, Dr. R. Davis, Professor N. Epstein, Professor J. R. Grace, Dr. K. Im, Professor B. G. Jones, Professor D. D. Joseph, Dr. C.-H. Lin, Dr. P. Nelson, Dr. S. L. Passman, Professor R. Pfeffer, Professor M. C. Roco, Professor S. L. Soo, Dr. B. L. Tarmy, Professor U. Tüzün, and Professor L.-X. Zhou. We are grateful to Dr. E. Abou-Zeida, Dr. P. Cai, Mr. S. Chauk, Dr. T. Hong, Dr. P.-J. Jiang, Professor J. Kadambi, Dr. T. M. Knowlton, Dr. S. Kumar, Dr. R. J. Lee, and Dr. J. Zhang for their valuable technical assistance in providing information which was incorporated in the text. Special thanks are due to Mr. R. Agnihotri, Dr. D.-R. Bai, Dr. H.-T. Bi, Dr. A. Ghosh-Dastidar, Mr. E.-S. Lee, Dr. S.-C. Liang, Mr. J. Lin, Mr. T. Lucht, Mr. X.-K. Luo, Dr. S. Mahuli, Mr. J. Reese, Mr. S.-H. Wei, Dr. J. Zhang, Mr. T.-J. Zhang, and Mr. J.-P. Zhang, who have read part of the text and have provided valuable comments. The outstanding editorial assistance of Dr. T. Hong and Dr. K. M. Russ is gratefully acknowledged. Thanks are also extended to Dr. E. Abou-Zeida and Mrs. Maysaa Barakat for their excellent drawing of the figures. The inquisitive students in the Chemical Engineering 801 course, Gas–Solid Flows, and the 815.15 course, Fluidization Engineering, taught by the senior author in the Department of Chemical Engineering at the Ohio State University have provided important feedback about the text. Their input is indeed extremely helpful. Financial assistance to this writing project provided by the members of the Ohio State University/Industry Consortium Program on Fluidization and Particulates Reaction Engineering, including Shell Development Co., E. I. duPont de Nemours & Co., Hydrocarbon Research Inc., Exxon Research & Engineering Co., Texaco Inc., and Mitsubishi Chemical Co., is deeply appreciated.

PART I

Basic Relationships

CHAPTER 1

Size and Properties of Particles

1.1 Introduction

The flow characteristics of solid particles in a gas–solid suspension vary significantly with the geometric and material properties of the particle. The geometric properties of particles include their size, size distribution, and shape. Particles in a gas–solid flow of practical interest are usually of nonspherical or irregular shapes and polydispersed sizes. The geometric properties of particles affect the particle flow behavior through an interaction with the gas medium as exhibited by the drag force, the distribution of the boundary layer on the particle surface, and the generation and dissipation of wake vortices. The material properties of particles include such characteristics as physical adsorption, elastic and plastic deformation, ductile and brittle fracturing, solid electrification, magnetization, heat conduction and thermal radiation, and optical transmission. The material properties affect the long- and short-range interparticle forces, and particle attrition and erosion behavior in gas–solid flows. The geometric and material properties of particles also represent the basic parameters affecting the flow regimes in gas–solid systems such as fluidized beds.

In this chapter, the basic definitions of the equivalent diameter for an individual particle of irregular shape and its corresponding particle sizing techniques are presented. Typical density functions characterizing the particle size distribution for polydispersed particle systems are introduced. Several formulae expressing the particle size averaging methods are given. Basic characteristics of various material properties are illustrated.

1.2 Particle Size and Sizing Methods

The particle size affects the dynamic behavior of a gas–solid flow [Dallavalle, 1948]. An illustration of the relative magnitudes of particle sizes in various multiphase systems is given in Fig. 1.1 [Soo, 1990]. It is seen in this figure that the typical range of particle sizes of interest to gas–solid flows is roughly from 1 μm to 10 cm. The particle shape affects the flowability of powders, their packing, and the covering power of pigments. Qualitative definitions for particle shapes are given in Table 1.1. The shape of particles is commonly expressed in terms of shape factors and shape coefficients [Allen, 1990].

Particles used in practice for gas–solid flows are usually nonspherical and polydispersed. For a nonspherical particle, several equivalent diameters, which are usually based on equivalences either in geometric parameters (*e.g.*, volume) or in flow dynamic characteristics (*e.g.*, terminal velocity), are defined. Thus, for a given nonspherical particle, more than one equivalent diameter can be defined, as exemplified by the particle shown in Fig. 1.2, in which three different equivalent diameters are defined for the given nonspherical particle. The selection of a desired definition is often based on the specific process application intended.

Table 1.1. *Definitions of Particle Shape*

Acicular	needle-shaped
Angular	sharp-edged or having roughly polyhedral shape
Crystalline	freely developed in a fluid medium of geometric shape
Dendritic	having a branched crystalline shape
Fibrous	regularly or irregularly thread-like
Flaky	plate-like
Granular	having approximately an equidimensional irregular shape
Irregular	lacking any symmetry
Modular	having rounded, irregular shape
Spherical	global shape

Source: T. Allen's *Particle Size Measurements*, Chapman & Hall, 1990.

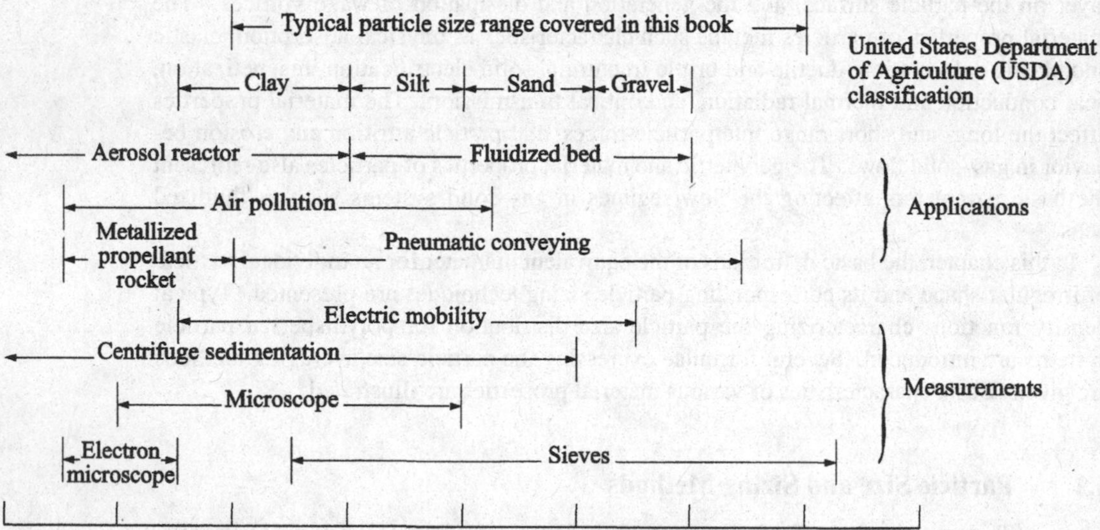

Figure 1.1. Magnitudes of particle sizes in gas–solid systems (after Soo, 1990).

1.2.1 *Equivalent Diameters of a Nonspherical Particle*

An equivalent diameter of a particle is usually defined in relation to à specific sizing method developed on the basis of a certain equivalency criterion. Several equivalent diameters of a spherical particle commonly employed are discussed in the following sections.

1.2.1.1 *Sieve Diameter*

A sieve diameter is defined as the width of the minimum square aperture through which the particle will pass. A common sizing device associated with this definition is a series of sieves with square woven meshes. Two sieve standards, *i.e.*, Tyler Standard and American

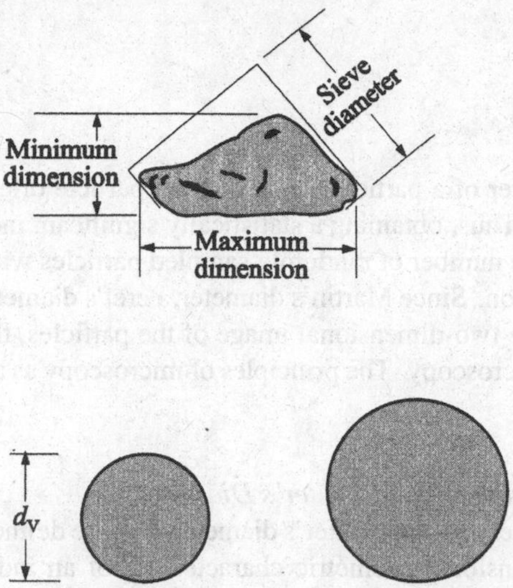

Figure 1.2. Schematic illustration of multidimensions of a particle and its equivalent volume diameter, surface diameter, and sieve diameter.

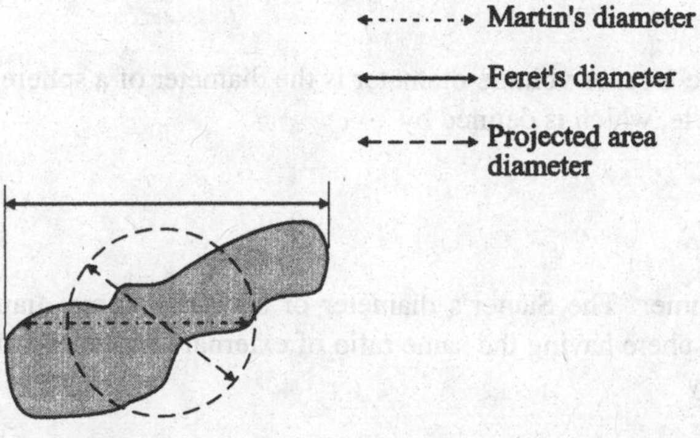

Figure 1.3. Schematic illustration of different particle diameters based on 2-D projected image.

Society for Testing and Materials (ASTM) Standard, are widely used; they are introduced in §1.2.2.1.

1.2.1.2 *Martin's Diameter, Feret's Diameter, and Projected Area Diameter*
Martin's diameter, Feret's diameter, and projected area diameter are three diameters defined on the basis of the projected image of a single particle. Specifically, Martin's diameter is defined as the averaged cord length of a particle which equally divides the projected area. Feret's diameter is the averaged distance between pairs of parallel tangents to the projected outline of the particle. The projected area diameter is the diameter of a sphere having the same projected area as the particle. These diameters are schematically represented in Fig. 1.3. The projected area diameter of a particle d_A can be related to the particle projected

area A by

$$d_A = \left(\frac{4A}{\pi}\right)^{1/2} \tag{1.1}$$

Martin's diameter and Feret's diameter of a particle depend on the particle orientation under which the measurement is made. Thus, obtaining a statistically significant measurement for these diameters requires a large number of randomly sampled particles which are measured in an arbitrarily fixed orientation. Since Martin's diameter, Feret's diameter, and projected area diameter are based on the two-dimensional image of the particles, they are generally used in optical and electron microscopy. The principles of microscopy as a sizing method are discussed in §1.2.2.2.

1.2.1.3 *Surface Diameter, Volume Diameter, and Sauter's Diameter*

The surface diameter, d_S, volume diameter, d_V, and Sauter's diameter, d_{32}, are defined such that each of them reflects a three-dimensional geometric characteristic of an individual particle. A surface diameter is given as the diameter of a sphere having the same surface area as the particle, which is expressed by

$$d_S = \sqrt{\frac{S}{\pi}} \tag{1.2}$$

where S is the particle surface area. A volume diameter is the diameter of a sphere having the same volume as the particle, which is defined by

$$d_V = \left(\frac{6V}{\pi}\right)^{1/3} \tag{1.3}$$

where V is the particle volume. The Sauter's diameter or surface-volume diameter is defined as the diameter of a sphere having the same ratio of external surface to volume as the particle, which is given by

$$d_{32} = \frac{6V}{S} = \frac{d_V^3}{d_S^2} \tag{1.4}$$

The concept of the surface diameter may be mostly used in the field of adsorption and reaction engineering, where the equivalent surface exposure area is important. The determination of the surface area depends on the method of measurements; for example, permeametry can give a much lower area than does gas adsorption. The latter often includes the contribution of pore surface area, which is accessible to the gas molecules. The determination of particle surface area by gas adsorption is given in §1.2.2.4. The fundamentals of gas adsorption are further covered in §1.4.1.

The volume diameter of a particle may be useful in applications where equivalent volume is of primary interest, such as in the estimation of solids holdup in a fluidized bed or in the calculation of buoyancy forces of the particles. The volume of a particle can be determined by using the weighing method. Sauter's diameter is widely used in the field of reacting gas–solid flows such as in studies of pulverized coal combustion, where the specific surface area is of most interest.

1.2.1.4 *Dynamic Diameter*

The dynamic response of a particle in gas–solid flows may be characterized by the settling or terminal velocity at which the drag force balances the gravitational force. The dynamic diameter is thus defined as the diameter of a sphere having the same density and the same terminal velocity as the particle in a fluid of the same density and viscosity. This definition leads to a mathematical expression of the dynamic diameter of a particle in a Newtonian fluid as

$$V(\rho_p - \rho)g = C_D \frac{\pi}{8}\frac{\mu^2}{\rho}\mathrm{Re}_t^2$$
$$d_t = \frac{\mathrm{Re}_t \mu}{\rho U_{pt}} \tag{1.5}$$

where Re_t is the particle Reynolds number at the terminal velocity; C_D is the drag coefficient, which is a function of Re_t; μ denotes the viscosity of the fluid; ρ and ρ_p represent the densities of the fluid and the particle, respectively; U_{pt} is the particle terminal velocity; g is the gravitational acceleration; and d_t is the equivalent dynamic diameter.

The relationship between C_D and Re_t for a sphere is given by Fig. 1.4 [Schlichting, 1979]. Mathematically, it can be expressed by

$$C_D = \frac{24}{\mathrm{Re}_t} \qquad \mathrm{Re}_t < 2$$
$$C_D = \frac{18.5}{\mathrm{Re}_t^{0.6}} \qquad 2 < \mathrm{Re}_t < 500 \tag{1.6}$$
$$C_D = 0.44 \qquad 500 < \mathrm{Re}_t < 2 \times 10^5$$

The three correlations in Eq. (1.6), in order from top to bottom, are known as Stokes's, Allen's, and Newton's equations, respectively. Combining these equations with Eq. (1.5),

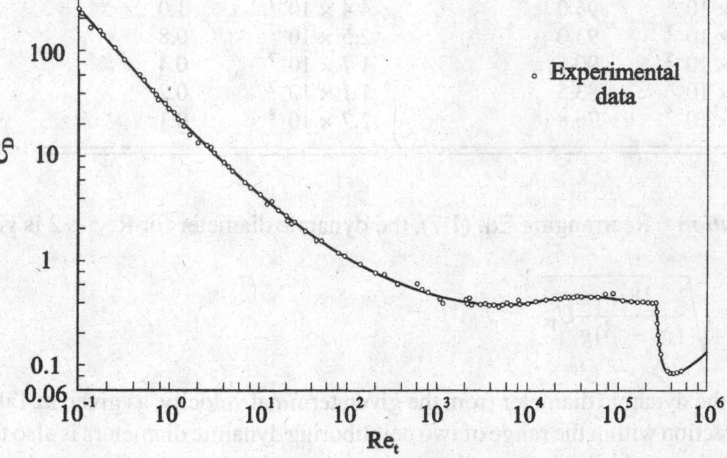

Figure 1.4. Drag coefficient for spheres as a function of Re_t (from Schlichting, 1979).

the terminal velocity of a sphere is related to its diameter by

$$
\begin{aligned}
U_{pt} &= \frac{d_t^2(\rho_p - \rho)g}{18\mu} && \mathrm{Re}_t < 2 \\
U_{pt}^{1.4} &= 0.072\frac{d_t^{1.6}(\rho_p - \rho)g}{\rho^{0.4}\mu^{0.6}} && 2 < \mathrm{Re}_t < 500 \\
U_{pt}^2 &= 3.03\frac{d_t(\rho_p - \rho)g}{\rho} && 500 < \mathrm{Re}_t < 2 \times 10^5
\end{aligned}
\tag{1.7}
$$

It is noted that in the laminar flow region, the particle moves in random orientation; however, outside this region it orients itself so as to give the maximum resistance to the motion. Thus, the dynamic diameter of an irregular-shaped particle in the intermediate region may be greater than that in the laminar flow region.

Example 1.1 One of the applications of using Stokes's law to determine the particle size is the Sedigraph particle analyzer. Table E1.1 shows the relationship between the cumulative weight percentage of particles and the corresponding particle terminal velocities for a powder sample. The densities of the particle and the dispersing liquid are 2,200 and 745 kg/m^3, respectively. The liquid viscosity is 1.156×10^{-3} kg/m·s. Find out the relationship of the mass fraction distribution to the equivalent dynamic diameter.

Table E1.1. *Cumulative Weight Percentage Versus Terminal Velocity*

U_{pt} (m/s)	Cumulative wt%	U_{pt} (m/s)	Cumulative wt%
4.4×10^{-3}	99.9	1.1×10^{-5}	65.6
2.5×10^{-3}	99.3	6.2×10^{-6}	47.2
1.7×10^{-3}	99.2	2.7×10^{-6}	21.2
6.2×10^{-4}	98.5	6.9×10^{-7}	1.2
1.5×10^{-4}	96.0	4.4×10^{-7}	1.0
6.9×10^{-5}	93.0	2.5×10^{-7}	0.8
4.4×10^{-5}	90.1	1.7×10^{-7}	0.4
2.5×10^{-5}	83.5	1.1×10^{-7}	0.2
1.7×10^{-5}	76.8	2.7×10^{-8}	0.1

Solution Rearranging Eq. (1.7), the dynamic diameter for $\mathrm{Re}_t < 2$ is given as

$$
d_t = \sqrt{\frac{18\mu}{(\rho_p - \rho)g}U_{pt}} \tag{E1.1}
$$

which yields the dynamic diameter from the given terminal velocity, as given in Table E1.2. The weight fraction within the range of two neighboring dynamic diameters is also tabulated in Table E1.2, from which the mass distribution versus the dynamic diameter is obtained, as shown in Fig. E1.1.

Table E1.2. *Mass Fraction (wt%) Versus Dynamic Diameter*

U_{pt} (m/s)	d_t (μm)	f_M (wt%)
4.4×10^{-3}	80	0.6
2.5×10^{-3}	60	0.1
1.7×10^{-3}	50	0.7
6.2×10^{-4}	30	2.5
1.5×10^{-4}	15	3.0
6.9×10^{-5}	10	2.9
4.4×10^{-5}	8	6.6
2.5×10^{-5}	6	6.7
1.7×10^{-5}	5	11.2
1.1×10^{-5}	4	18.4
6.2×10^{-6}	3	26.0
2.7×10^{-6}	2	20.0
6.9×10^{-7}	1	0.2
4.4×10^{-7}	0.8	0.2
2.5×10^{-7}	0.6	0.4
1.7×10^{-7}	0.5	0.2
1.1×10^{-7}	0.4	0.1
2.7×10^{-8}	0.2	

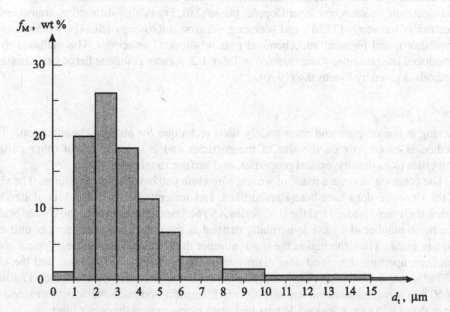

Figure E1.1. Mass fraction distribution based on data in Table E1.2.

Table 1.2. *Some Methods of Particle Size Measurement*

Method	Size range (μm)
Sieving	
Woven wire	37–5660
Electroformed	5–120
Punched plate	50–125,000
Microscopy	
Optical	0.8–150
Electron	0.001–5
Sedimentation	
Gravitational	5–100
Centrifugal	0.001–1,000
Fraunhofer diffraction	0.1–1,000
Doppler phase shift	1–10,000

1.2.2 *Particle Sizing Methods*

The sizing methods involve both classical and modern instrumentations, based on a broad spectrum of physical principles. The typical measuring systems may be classified according to their operation mechanisms, which include mechanical (sieving), optical and electronic (microscopy, laser Doppler phase shift, Fraunhofer diffraction, transmission electron miscroscopy [TEM], and scanning electron microscopy [SEM]), dynamic (sedimentation), and physical and chemical (gas adsorption) principles. The methods to be introduced later are briefly summarized in Table 1.2. A more complete list of particle sizing methods is given by Svarovsky (1990).

1.2.2.1 *Sieving*

Sieving is the simplest and most widely used technique for powder classification. This method is based only on the size of the particles and is independent of other particle properties (*e.g.*, density, optical properties, and surface roughness).

The common sieves are made of woven wire cloth and have square apertures. The sizes of the sieve openings have been standardized, and currently two different sets of standard series, the Tyler Standard and the U.S. Series ASTM Standard, are used in the United States. The mesh number of a sieve is normally defined as the number of apertures per unit area (square inch). Thus, the higher the mesh number the smaller the aperture. Typical mesh numbers, aperture sizes, and wire diameters are given for the Tyler sieves and the U.S. ASTM sieves in Table 1.3. Sieve analysis covers the approximate size range of 37 μm to 5,660 μm using standard woven wire sieves. Electroformed micromesh sieves extend the range down to 5 μm or less while punched plate sieves extend the upper limit.

It should be pointed out that longer sieving time can improve the recovery of a given particle size for a distribution; however, excessive sieving can lead to particle degradation due to attrition or mechanical wear. This effect can be especially pronounced for particles near the tail end of the size distribution. Unfortunately, neither good theories nor reliable empirical formulae are available to estimate the optimum sieving time under which a narrow error margin of the resulting size distribution can be ensured for a given sample.

Table 1.3. *Tyler Standard and U.S. ASTM Sieve Series*

Tyler standard			U.S. series ASTM standard		
Mesh no.	Size (μm)	Wire diameter (μm)	Mesh no.	Size (μm)	Wire diameter (μm)
$3\frac{1}{2}$	5,660	1,280–1,900	$3\frac{1}{2}$	5,613	1,650
4	4,760	1,140–1,680	4	4,699	1,650
5	4,000	1,000–1,470	5	3,962	1,120
6	3,360	870–1,320	6	3,327	914
7	2,830	800–1,200	7	2,794	833
8	2,380	740–1,100	8	2,362	813
10	2,000	680–1,000	9	1,981	838
12	1,680	620–900	10	1,651	889
14	1,410	560–800	12	1,397	711
16	1,190	500–700	14	1,168	635
18	1,000	430–620	16	991	597
20	840	380–550	20	833	437
25	710	330–480	24	701	358
30	590	290–420	28	589	318
35	500	260–370	32	495	300
40	420	230–330	35	417	310
45	350	200–290	42	351	254
50	297	170–253	48	295	234
60	250	149–220	60	246	179
70	210	130–187	65	208	183
80	177	114–154	80	175	142
100	149	96–125	100	147	107
120	125	79–103	115	124	97
140	105	63–87	150	104	66
170	88	54–73	170	88	61
200	74	45–61	200	74	53
230	62	39–52	250	61	41
270	53	35–46	270	53	41
325	44	31–40	325	43	36
400	37	23–35	400	38	25

1.2.2.2 *Microscopy*

Microscopy is often referred to as an absolute method for the determination of size and size distribution of small particles because it allows direct visualization and measurements of individual particles. Three commonly used types are optical microscopy, transmission electron microscopy (TEM), and scanning electron microscopy (SEM).

The optical microscope is one of the most basic instruments for particle sizing and is applicable to a typical size range of 0.8 μm to 150 μm. The lower limit is a result of the diffraction effects on the particle image as observed in a microscope. The limit of resolution of an optical microscope can be estimated by (Yamate and Stockham, 1977)

$$\delta = \frac{1.22\lambda}{2N_A} \tag{1.8}$$

where δ is the limit of resolution; λ is the wavelength of the light; and N_A is the numerical

Table 1.4. *The Maximum Useful Magnification and the Eyepiece Required for Different Objectives*

Objective					
Magnification	Focal length (mm)	Numerical aperture	Depth of focus (μm)	Maximum useful magnification	Eyepiece required
2.5	56	0.08	50	80	30
10	16	0.25	8	250	25
20	8	0.50	2	500	25
43	4	0.66	1	660	15
97	2	1.25	0.4	1,250	10

Source: A. G. Guy's *Essentials of Materials Science*, McGraw-Hill, 1976.

aperture of the objective. As an example, for visible light of $\lambda = 4{,}500$ Å and with an objective aperture having $N_A = 1.25$, the limit of resolution of the optical microscope can be calculated from Eq. (1.8) as 0.2 μm.

An appropriate selection of the maximum useful magnification of an optical microscope for a given sample is also important. The magnification of the microscope is the product of the objective-eyepiece combination. As a rule of thumb, the maximum useful magnification for the optical microscope is 1,000 times the numerical aperture. Table 1.4 summarizes the maximum useful magnification and the eyepiece required for different objectives.

The TEM and SEM are two advanced techniques which use electron beams for direct determination of the particle size and surface area. They are usually applied to measurement of particles in a size range of 0.001 μm to 5 μm. The TEM generates an image of a particle sample on a photographic plate by means of an electron beam, through the transmissibility of the electron beam on the sample. The SEM uses a fine beam of electrons of medium energy (5–50 keV) to scan across the sample in a series of parallel tracks. These scanning electrons produce secondary electron emission, back scattered electrons, light, and X-rays which can be detected. Both the TEM and SEM are extensively used in the determination of the pore structure and surface shape and area of the particle. The SEM is considerably faster and gives more three-dimensional information than the TEM. Details about the TEM and SEM are given by Kay (1965) and Hay and Sandberg (1967), respectively.

1.2.2.3 *Sedimentation*

The sedimentation techniques utilize the dependence of the terminal velocities of particles on their size either in a gravitational field or in a centrifugal field. The lower limit of the particle sizing by the gravitational sedimentation method is about 5 μm because of the effects of convection, diffusion, and Brownian motion as well as the long settling time involved. These effects can be overcome by centrifuging the suspension, which accelerates the settling process. Centrifugal sedimentation is mostly applied to the particle size range of 0.001 μm to 1 mm.

The sedimentation methods are normally used to measure the size of particles in a liquid medium because of the relatively high viscosity effects in liquids compared to gases. The particles in a liquid may become solvated, yielding increased weight and volume of the particle. Meanwhile, the buoyant effect on the solvated particle in the surrounding medium increases. In the determination of the overall driving force for sedimentation, these two effects are noted to cancel each other. Therefore, solvation usually has little effect on the particle sizing results when the sedimentation methods in liquids are used.

By analogy, the definition of dynamic diameter in a centrifugal field can be simply extended from Eq. (1.5) with the replacement of the gravitational acceleration, g, by the centrifugal acceleration, $\omega^2 r$, as

$$V(\rho_\mathrm{p} - \rho)\omega^2 r = C_\mathrm{D}\frac{\pi}{8}\frac{\mu^2}{\rho}\mathrm{Re}_\mathrm{t}^2 \qquad d_\mathrm{t} = \frac{\mathrm{Re}_\mathrm{t}\mu}{\rho U_\mathrm{pt}} \tag{1.9}$$

where ω is angular frequency and r is the radial distance from the center of the centrifugal field.

1.2.2.4 *Gas Adsorption*

As indicated earlier, the surface area of porous particles is an important variable in characterizing physical or chemical processes involving these particles. Porous particles commonly encountered in catalysis and reaction engineering include activated carbon, alumina, silica, and zeolites. For a given porous particle, the effective surface area is defined on the basis of the specific transport phenomenon of interest in a process system. For example, thermal radiation may be affected predominantly by the external surface area of the particle and the exposed surface area due to superficial cracks and fissures. On the other hand, for most chemical reactions and adsorption processes, the internal surface area provided by the interior pores of the particle may determine the overall rate process. A convenient classification of pores according to their width divides them into three categories: micropores, less than 20 angstrom (Å); mesopores, between 20 and 500 Å; and macropores, more than 500 Å. An exception of a large specific surface which is wholly external in nature is provided by a dispersed aerosol composed of fine particles free of cracks and indentations [Gregg and Sing, 1982].

The most common method used for the determination of surface area and pore size distribution is physical gas adsorption (also see §1.4.1). Nitrogen, krypton, and argon are some of the typically used adsorptives. The amount of gas adsorbed is generally determined by a volumetric technique. A gravimetric technique may be used if changes in the mass of the adsorbent itself need to be measured at the same time. The nature of the adsorption process and the shape of the equilibrium adsorption isotherm depend on the nature of the solid and its internal structure. The Brunauer–Emmett–Teller (BET) method is generally used for the analysis of the surface area based on monolayer coverage, and the Kelvin equation is used for calculation of pore size distribution.

It is noted that in the evaluation of the particle surface diameter and Sauter's diameter, as discussed in §1.2.1.3, only the external surface area of the particle is considered.

1.2.2.5 *Fraunhofer Diffraction*

The particle sizing technique using light scattering and diffraction possesses some advantages. It is nonintrusive and much faster than that using a mechanical means, requiring neither a conducting medium nor a large shearing force. The implementation of Mie theory

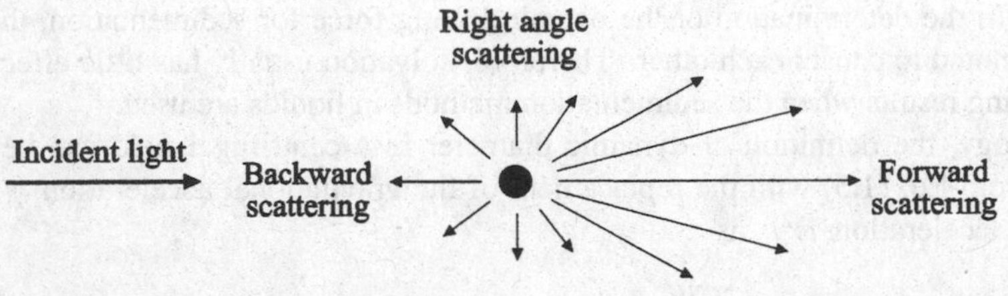

Figure 1.5. Illustration of the angular light intensity distribution of light scattered from a single particle.

with Fraunhofer diffraction and side scatter permits the measurement of particle sizes over a range of 0.1–1000 μm [Plantz, 1984].

From the Beer–Lambert law, the transmittance for a light beam through a sample of particles is given by

$$\frac{I_t}{I_i} = \exp(-nA_e l) \tag{1.10}$$

where I_t is the intensity of the transmitted beam; I_i is the intensity of the incident beam; n is the particle number concentration; A_e is the integrated cross section for extinction, which includes the effects of reflection, refraction, diffraction, and absorption; and l is the optical path length. The extinction cross section can be calculated from the Lorenz–Mie theory. A typical angular distribution of light scattered from a single particle is illustrated in Fig. 1.5. It shows that the most scattering is in the forward direction.

Although the Lorenz–Mie theory is exact, it does not lead to simple and analytical solutions relating the particle size to transmittance measurements. However, there are limiting cases where much simpler theories have been established. These limiting cases are the Rayleigh scattering for particles much smaller than the wavelength of light and the Fraunhofer diffraction for particles much larger than the wavelength of light. A criterion for discerning limiting cases is proposed by van de Hulst (1981) as

$$\begin{aligned} \kappa &< 0.3 && \text{Rayleigh scattering} \\ \kappa &\approx 1 && \text{Lorenz–Mie theory} \\ \kappa &\gg 30 && \text{Fraunhofer diffraction} \end{aligned} \tag{1.11}$$

The parameter κ is defined as

$$\kappa = \frac{2\pi d|n_i - 1|}{\lambda} \tag{1.12}$$

where n_i is the relative index of refraction of the particle.

In this book, particles larger than 1 μm are of primary interest, and thus, only the Fraunhofer diffraction method, which can account for particles larger than 2–3 μm, is discussed here. The Fraunhofer diffraction theory is derived from fundamental optical principles that are not concerned with scattering. To obtain the Fraunhofer diffraction, two basic requirements must be satisfied. First, the area of the particle or aperture must be much smaller than the product of the wavelength of light and the distance from the light source to the particle or aperture. Second, this area must also be smaller than the product

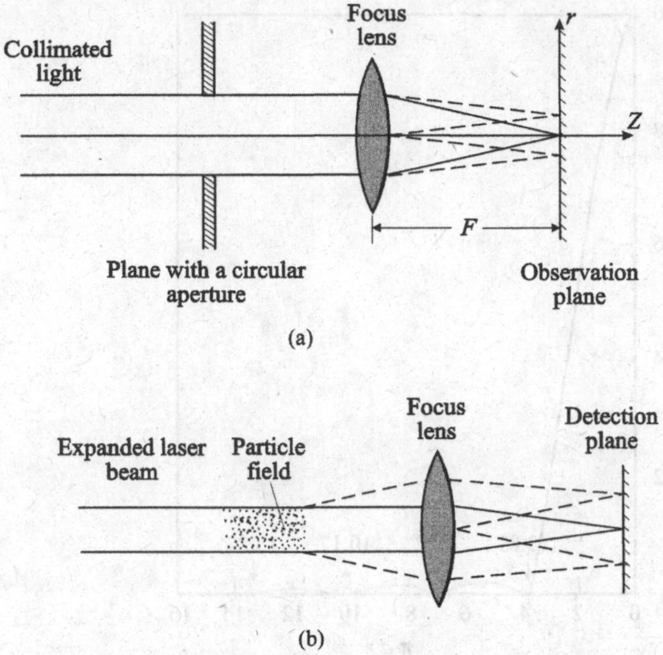

Figure 1.6. Fraunhofer diffraction system for particle size analysis: (a) Diffraction by a circular aperture; (b) Diffraction by a particle cloud.

of the wavelength and the distance from the particle or aperture to the observation plane. Therefore, Fraunhofer diffraction is known as far-field diffraction. A schematic diagram for the Fraunhofer diffraction of a single particle or aperture is illustrated in Fig. 1.6(a), whereas an optical schematic of a Fraunhofer diffraction instrument for the analysis of particle sizes in a gas–solid suspension system using a laser beam as the light source is shown in Fig. 1.6(b).

The transmittance of Fraunhofer diffraction for a circular aperture or spherical particles of diameter d can be expressed by

$$\frac{I_t}{I_i} = \left(\frac{2J_1(x)}{x}\right)^2 \tag{1.13}$$

where J_1 is the first-order spherical Bessel function and x is given by

$$x = \frac{\pi dr}{\lambda F} \tag{1.14}$$

where r is the radial distance in the observation plane as measured from the optical axis and F is the focal length of the lens. Thus, the Fraunhofer diffraction pattern for a circular aperture or spherical particles can be determined as shown in Fig. 1.7. Consequently, by measuring and analyzing the intensity distributions of the light beam over a finite area of the detector, the equivalent particle diameter can be obtained. More detailed information about the Fraunhofer diffraction method is given by Weiner (1984).

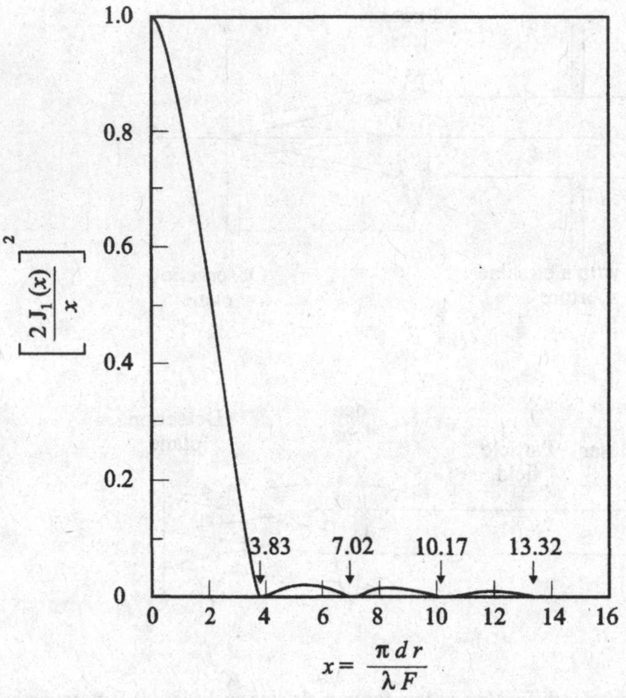

Figure 1.7. Fraunhofer diffraction pattern for circular aperture or opaque disk (from Weiner, 1984).

1.2.2.6 *Laser Doppler Phase Shift*

When a spherical particle enters the crossing volume of two laser beams, a Doppler effect occurs not only in frequency shift but also in phase shift of the scattered light. The frequency shift yields the velocity of the sphere, whereas the phase shift gives the particle size. The phase Doppler principle has been employed to measure the size and size distributions of spheres in addition to the particle velocity. The phase Doppler principle was first reported by Durst and Zaré (1975) and became a viable measurement tool one decade later [Bachalo and Houser, 1984].

The phase Doppler principle can be described as follows: When light is scattered by a small spherical particle traveling through a laser measurement volume, it yields frequency signals, which can be measured to obtain the particle velocity. This frequency is known as the Doppler shift frequency, which is identical in all spatial directions. When viewed from two separate spatial locations the scattered signals exhibit a phase shift whose magnitude depends on factors including the angle at which light is scattered to each photodetector, the index of refraction of the material of the spherical particle, and parameters such as the light wavelength and the beam intersection angle. When reflection is the dominant mode of scattering, the phase shift is independent of the index of refraction. The phase shift measured in the Doppler signal obtained from the same particle using two closely spaced photodetectors varies linearly with the particle diameter for spherical particles and hence provides a useful means for determining the spherical particle size. Evaluation of the relationship of the phase shifts from the signals received at each of the photodetector locations is complex but can be determined on the basis of Mie scattering theory [Bachalo and Houser, 1984]. In principle, the measurement of particle size requires that the particle entering the measurement volume be spherical, and the diameters of amorphous particles cannot be measured using the phase Doppler method.

Typically, the phase Doppler method is good for the measurement of particle sizes ranging from 1 μm to 10 mm with a variation by a factor of 40 at one instrument setting. As a rule of thumb, the maximum measurable concentration is 1,000 particles per cubic millimeter (mm^3). Commercial instruments using this technique are available, *e.g.*, the phase Doppler particle analyzer (PDPA) (Aerometrics) and the Dantec particle dynamics analyzer (DPDA) (Dantec Electronics).

1.2.2.7 *Coulter Principle*

The Coulter principle underlies a method for determining particle sizes and size distributions using an electrical sensing technique. The instrument based on the Coulter principle is known as the Coulter counter. In the Coulter counter, particles are first suspended in an electrolyte and then passed through a small orifice. The particle concentration is usually so low that particles traverse the orifice one at a time. The orifice has immersed electrodes. When a particle passes through the orifice, it displaces electrolyte within the orifice, which results in a change in impedance leading to a voltage pulse with an amplitude proportional to the volume of the particle. By regulating, sizing, and number counting of the pulses, the particle size and size distributions are obtained. The typical sizing range by the Coulter counter is from 1 to 50 μm.

1.2.2.8 *Cascade Impactor*

When particles are small enough, the sedimentation method becomes inefficient as a result of the impractically long settling time. An important design using the inertial technique is known as the cascade impactor, which samples and classifies particle sizes by their inertia. A cascade impactor consists of a series of collecting plates of the particle-laden gas flow, which is gradually increased in the form of a succession of jets. Thus, deflected by inertia, the particles are collected and graded on the series collecting plates. The extent of the particle deposition on each plate depends on the impact velocity of the gas stream. The intake velocity should be low enough to prevent any damage on the collecting plates. However, it should also be high enough to ensure sufficient inertia of the particles. The most commonly used cascade impactor is the one developed by May (1945). The May cascade impactor is capable of sampling airborne particles from 0.5 to 50 μm by using four or more collecting glass discs. The particle sizing range by cascade impactors is typically from 0.1 to 100 μm.

1.3 Particle Size Distributions and Averaged Diameters

For a system of polydispersed particles, various averaged diameters may be defined according to the diversity of needs in industrial applications. An averaged diameter depends not only on the type of particle size distribution but also on the selection of a weighing factor. A particle size density function can be defined in terms of either the number of particles or the mass of particles within a given size range. The number density function is interconvertible with its corresponding mass density function. Different weighing factors with their distinct physical significance may be imposed to yield various averaged diameters for particles in a polydispersed system.

1.3.1 *Density Functions*

A number density function, $f_{\mathrm{N}}(b)$, is defined so that $f_{\mathrm{N}}(b)\,\mathrm{d}b$ represents the particle number fraction in a size range from b to $b + \mathrm{d}b$. Thus,

$$\frac{\mathrm{d}N}{N_0} = f_{\mathrm{N}}(b)\,\mathrm{d}b \tag{1.15}$$

where $\mathrm{d}N$ is the number of particles within the size range of b to $b + \mathrm{d}b$ for a total number N_0 of the sample particles. Clearly, the preceding expression leads to a normalized condition

$$\int_0^\infty f_{\mathrm{N}}(b)\,\mathrm{d}b = 1 \tag{1.16}$$

Thus, over a range from d_1 to d_2, the fraction of the total sample N_0 of this size is obtained by

$$\frac{N_{12}}{N_0} = \int_{d_1}^{d_2} f_{\mathrm{N}}(b)\,\mathrm{d}b \tag{1.17}$$

A particle density function can also be defined in terms of the particle mass. A mass density function, $f_{\mathrm{M}}(b)$, represents the particle mass fraction in size by

$$\frac{\mathrm{d}M}{M_0} = f_{\mathrm{M}}(b)\,\mathrm{d}b \tag{1.18}$$

where $\mathrm{d}M$ is the mass of particles within the size range of b to $b + \mathrm{d}b$ for a total mass M_0 of the sample particles. Thus, the normalized condition for a mass density function is given by

$$\int_0^\infty f_{\mathrm{M}}(b)\,\mathrm{d}b = 1 \tag{1.19}$$

and, over a range from d_1 to d_2, the fraction of the total sample M_0 of the mass is found from

$$\frac{M_{12}}{M_0} = \int_{d_1}^{d_2} f_{\mathrm{M}}(b)\,\mathrm{d}b \tag{1.20}$$

It is noted that the mass of particles can be expressed in terms of the number of particles of the same size, or

$$\mathrm{d}M = m\,\mathrm{d}N \tag{1.21}$$

where m is the mass of a single particle of size b. For a spherical particle, m can be expressed by

$$m = \frac{\pi}{6}\rho_{\mathrm{p}} b^3 \tag{1.22}$$

From Eqs. (1.15), (1.18), and (1.21), the number density function is related to the mass density function by

$$f_{\mathrm{M}}(b) = \frac{N_0 m}{M_0} f_{\mathrm{N}}(b) \tag{1.23}$$

The number density function is usually obtained by using microscopy or other optical means such as Fraunhofer diffraction. The mass density function can be acquired by use of sieving or other methods which can easily weigh the sample of particles within a given size range.

1.3.2 *Typical Distributions*

In the applications of gas–solid flows, there are three typical distributions in particle size, namely, Gaussian distribution or normal distribution, log-normal distribution, and Rosin–Rammler distribution. These three size distribution functions are mostly used in the curve fitting of experimental data.

1.3.2.1 *Gaussian Distribution*

The Gaussian distribution, also known as the normal distribution, has the density function

$$f_{\mathrm{N}}(d) = A_{\mathrm{N}} \exp\left(-\frac{(d-d_0)^2}{2\sigma_{\mathrm{d}}^2}\right) \tag{1.24}$$

where A_{N} is the normalizing constant; d_0 is the arithmetic mean of d; and σ_{d} is the standard deviation of d. Therefore, as given in Fig. 1.8, $2\sqrt{2}\sigma_{\mathrm{d}}$ is the width of the distribution curve defined as the cord length of two points where

$$\frac{f_{\mathrm{N}}(d)}{f_{\mathrm{N}}(d_0)} = \frac{1}{e} \tag{1.25}$$

For a given sample, the particle size range is bounded by d_1 and d_2 as shown in Fig. 1.8. Thus, Eq. (1.16) becomes

$$\int_{d_1}^{d_2} A_{\mathrm{N}} \exp\left(-\frac{(b-d_0)^2}{2\sigma_{\mathrm{d}}^2}\right) db = 1 \tag{1.26}$$

from which A_{N} is obtained as

$$A_{\mathrm{N}} = \frac{1}{\sigma_{\mathrm{d}}}\sqrt{\frac{2}{\pi}}\left[\operatorname{erf}\left(\frac{d_2-d_0}{\sqrt{2}\sigma_{\mathrm{d}}}\right) + \operatorname{erf}\left(\frac{d_0-d_1}{\sqrt{2}\sigma_{\mathrm{d}}}\right)\right]^{-1} \tag{1.27}$$

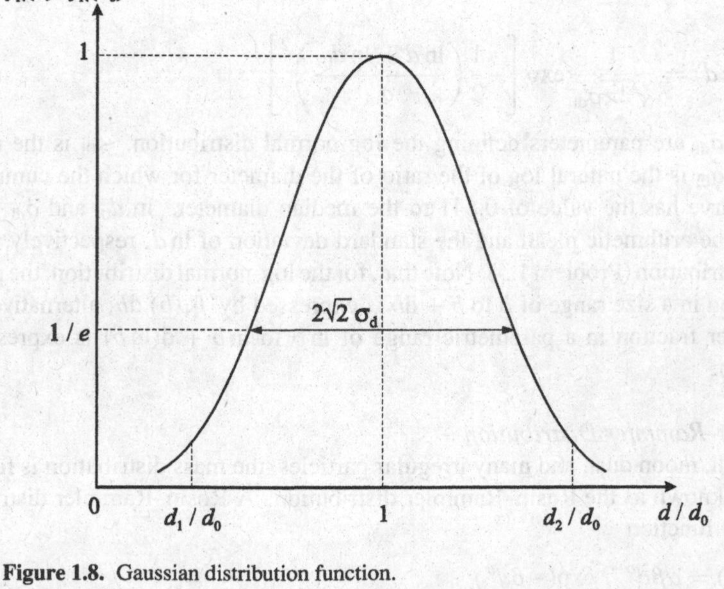

Figure 1.8. Gaussian distribution function.

Given the number density function of Eq. (1.24), the corresponding mass density function becomes

$$f_M(d) = A_M \frac{\pi}{6} \rho_p d^3 \exp\left(-\frac{(d - d_0)^2}{2\sigma_d^2}\right) \tag{1.28}$$

The normalizing constant A_M can be calculated from

$$\int_{d_1}^{d_2} A_M \frac{\pi}{6} \rho_p b^3 \exp\left(-\frac{(b - d_0)^2}{2\sigma_d^2}\right) db = 1 \tag{1.29}$$

There is no simple, exact, and explicit expression for A_M. However, for the case of a very narrow size distribution where $\sigma_d/d_0 \ll 1$, A_N and A_M are given by

$$A_N \approx \frac{1}{\sqrt{2\pi}\sigma_d} \tag{1.30}$$

and

$$\frac{1}{A_M} \approx \frac{(2\pi)^{3/2}}{6} \rho_p \frac{d_0}{\sigma_d} \left(\frac{3}{2} + \frac{d_0^2}{2\sigma_d^2}\right) \sigma_d^4 \tag{1.31}$$

For particle sizes following the Gaussian distribution described by Eqs. (1.24) and (1.30), 95 percent of the particles are of sizes between $(-2\sigma_d + d_0)$ and $(2\sigma_d + d_0)$.

1.3.2.2 *Log-Normal Distribution*

Most systems of fine particles have the log-normal type of particle size distribution. That is, with the logarithm of the particle size, the particle size distribution follows the normal or Gaussian distribution in semilog scales. Therefore, the density function for the log-normal distribution can be expressed by

$$f_N(d) = \frac{1}{\sqrt{2\pi}\sigma_{dl} d} \exp\left[-\frac{1}{2}\left(\frac{\ln d - \ln d_{0l}}{\sigma_{dl}}\right)^2\right] \tag{1.32a}$$

or

$$f_N(\ln d) = \frac{1}{\sqrt{2\pi}\sigma_{dl}} \exp\left[-\frac{1}{2}\left(\frac{\ln d - \ln d_{0l}}{\sigma_{dl}}\right)^2\right] \tag{1.32b}$$

Here, d_{0l} and σ_{dl} are parameters defining the log-normal distribution. d_{0l} is the median diameter, and σ_{dl} is the natural log of the ratio of the diameter for which the cumulative-distribution curve has the value of 0.841 to the median diameter. $\ln d_{0l}$ and σ_{dl} are not equivalent to the arithmetic mean and the standard deviation of $\ln d$, respectively, for the log-normal distribution (Problem 1.3). Note that, for the log-normal distribution, the particle number fraction in a size range of b to $b + db$ is expressed by $f_N(b)\, db$; alternatively, the particle number fraction in a parametric range of $\ln b$ to $\ln b + d(\ln b)$ is expressed by $f_N(\ln b) d(\ln b)$.

1.3.2.3 *Rosin–Rammler Distribution*

For broken coal, moon dust, and many irregular particles, the mass distribution is found to follow a form known as the Rosin–Rammler distribution. A Rosin–Rammler distribution has the density function

$$f_M(d) = \alpha\beta d^{\alpha-1} \exp(-\beta d^\alpha) \tag{1.33}$$

Table 1.5. *α and β for Some Materials*

Material	α	$\beta \times 10^3 (\mu m)^{-1}$
(a) Fine grinding		
Marlstone	0.675	33
Marlslate	0.839	33
Brown coal (lignite)	0.900	63
Feldspar	0.900	71
Cement clinker	1.000	29
Glass powder	1.111	25
Coal	1.192	21
(b) Coarse grinding		
Fullers clay	0.727	0.40
Coal, type 1	0.781	0.067
Coal, type 2	0.781	0.15
Limestone with 7% bitumen	0.781	0.13
Limestone, medium hardness	0.933	0.083
Limestone, hard	1.000	0.40
Clinker	1.036	0.50
Feldspar	1.111	0.50

Source: G. Herdan's *Small Particle Statistics*, Butterworths, London, 1960.

where α and β are constants. Integrating Eq. (1.33) yields the cumulative distribution function, F, as

$$F = \int_0^d f_M(b)\,db = 1 - \exp(-\beta d^\alpha) \tag{1.34}$$

However, the Rosin–Rammler distribution is often expressed in terms of R defined by

$$R = \int_d^\infty f_M(b)\,db = \exp(-\beta d^\alpha) \tag{1.35a}$$

Then, we have

$$\ln\left(\ln\frac{1}{R}\right) = \ln\beta + \alpha\ln d \tag{1.35b}$$

Equation (1.35b) shows that a linear relationship exists when $\ln[\ln(1/R)]$ is plotted against $\ln d$. From the slope and intercept of this straight line, α and β can be determined. α and β are typically obtained from the particle size distribution data based on sieve analyses. Table 1.5 provides a list of typical values of α and β for some materials for the Rosin–Rammler density function with d in the function having the unit micrometers (μm).

Example 1.2 A coarsely ground sample of corn kernel is analyzed for size distribution, as given in Table E1.3. Plot the density function curves for (1) normal or Gaussian distribution, (2) log-normal distribution, and (3) Rosin–Rammler distribution. Compare these distributions with the frequency distribution histogram based on the data and identify the distribution which best fits the data.

Table E1.3. *Data of Size Distribution*

Size range (mm)	Number of particles	Size range (mm)	Number of particles
0.05–0.10	1	0.50–0.55	3
0.10–0.15	5	0.55–0.60	1
0.15–0.20	6	0.60–0.65	2
0.20–0.25	7	0.65–0.70	0
0.25–0.30	8	0.70–0.75	1
0.30–0.35	6	0.75–0.80	0
0.35–0.40	4	0.80–0.85	1
0.40–0.45	4	0.85–0.90	0
0.45–0.50	4	0.90–0.95	1

Solution The data on numbers of particles in each particle range given in Table E1.3 can be converted to relative frequencies per unit of particle size as given in Table E1.4. The histogram for the relative frequency per unit of particle size for the data is plotted in Fig. E1.2; the histogram yields a total area of bars equal to unity. Superimposed on the histogram is the density function for the normal distribution based on Eqs. (1.24) and (1.30). For this distribution, the values for d_0 and σ_d are evaluated as 0.342 and 0.181, respectively. Also included in the figure is the density function for the log-normal distribution based on Eq. (1.32a). For this distribution, the values for $\ln d_{0l}$ and σ_{dl} are evaluated as -1.209 and 0.531, respectively.

Table E1.4. *Relative Frequency per Unit of Particle Size Data Given in Table E1.3*

Particle size (averaged, mm)	Number of particles	Relative frequency	Relative frequency per unit of particle size
0.075	1	0.019	0.370
0.125	5	0.093	1.852
0.175	6	0.111	2.222
0.225	7	0.130	2.593
0.275	8	0.148	2.963
0.325	6	0.111	2.222
0.375	4	0.074	1.481
0.425	4	0.074	1.481
0.475	4	0.074	1.481
0.525	3	0.056	1.111
0.575	1	0.019	0.370
0.625	2	0.037	0.741
0.675	0	0.000	0.000
0.725	1	0.019	0.370
0.775	0	0.000	0.000
0.825	1	0.019	0.370
0.875	0	0.000	0.000
0.925	1	0.019	0.370
Sum	54	1.000	20.00

For the Rosin–Rammler distribution, the distribution constants (α and β) are obtained from the particle mass distribution data. To obtain the mass density distribution, the data on the number density distribution given in Table E1.3 need to be converted by using Eq. (1.23). From the converted data and with the least-square fitting based on Eq. (1.35b), α and β can be obtained as $\alpha = 3.71$ and $\beta = 4.88\ \text{mm}^{-1}$. Note that the unit for d in Eq. (1.33) would be millimeters when these α and β values are used. Thus, the mass density function of the Rosin–Rammler distribution can be calculated. Converting the mass density function to the number density function, the results for the Rosin–Rammler distribution are plotted as shown in Fig. E1.2.

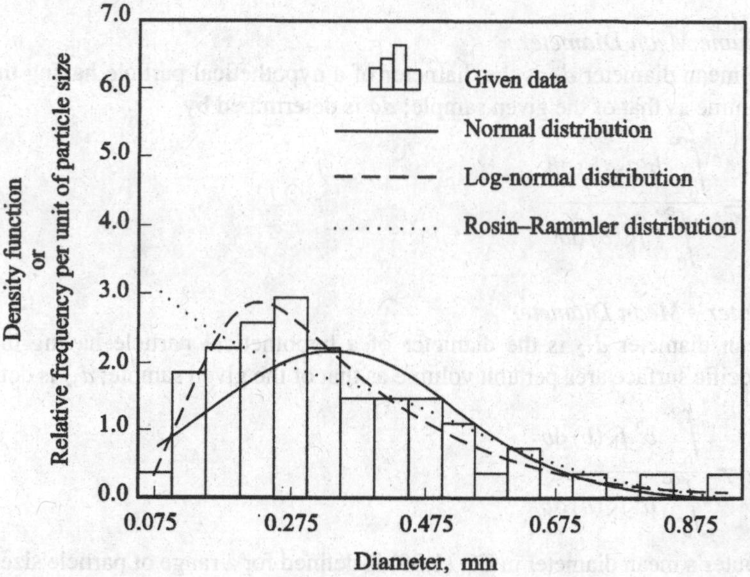

Figure E1.2. Comparisons of the relative frequency distribution based on the data with three density functions.

A graphical comparison of the three distributions with the given data shown in the figure reveals that the log-normal distribution best approximates the data.

1.3.3 *Averaged Diameters of a Particulate System*

For a given size distribution, various averaged diameters can be calculated, depending on the forms of weighing factors. The selection of an appropriate averaged diameter of a particle system depends on the specific needs of the application. For instance, in a pulverized coal combustion process, the surface area per unit volume may be important. In this case, Sauter's averaged diameter should be chosen.

1.3.3.1 *Arithmetic Mean Diameter*

The arithmetic mean diameter d_l is the averaged diameter based on the number density function of the sample; d_l is defined by

$$d_l = \frac{\int_0^\infty b f_N(b)\, db}{\int_0^\infty f_N(b)\, db} \tag{1.36}$$

1.3.3.2 *Surface Mean Diameter*

The surface mean diameter d_S is the diameter of a hypothetical particle having the same averaged surface area as that of the given sample; d_S is given by

$$d_S^2 = \frac{\int_0^\infty b^2 f_N(b)\,db}{\int_0^\infty f_N(b)\,db} \tag{1.37}$$

1.3.3.3 *Volume Mean Diameter*

The volume mean diameter d_V is the diameter of a hypothetical particle having the same averaged volume as that of the given sample; d_V is determined by

$$d_V^3 = \frac{\int_0^\infty b^3 f_N(b)\,db}{\int_0^\infty f_N(b)\,db} \tag{1.38}$$

1.3.3.4 *Sauter's Mean Diameter*

Sauter's mean diameter d_{32} is the diameter of a hypothetical particle having the same averaged specific surface area per unit volume as that of the given sample; d_{32} is defined by

$$d_{32} = \frac{\int_0^\infty b^3 f_N(b)\,db}{\int_0^\infty b^2 f_N(b)\,db} \tag{1.39}$$

Note that Sauter's mean diameter in Eq. (1.39) is defined for a range of particle size, which is different from Sauter's diameter in Eq. (1.4), defined for a single particle size.

1.3.3.5 *DeBroucker's Mean Diameter*

DeBroucker's mean diameter d_{43} is the averaged diameter based on the mass density function of the sample; d_{43} is evaluated by

$$d_{43} = \frac{\int_0^\infty b^4 f_N(b)\,db}{\int_0^\infty b^3 f_N(b)\,db} = \frac{\int_0^\infty b f_M(b)\,db}{\int_0^\infty f_M(b)\,db} \tag{1.40}$$

1.4 Material Properties of Solids

The material properties of solids are affected by a number of complex factors. In a gas–solid flow, the particles are subjected to adsorption, electrification, various types of deformation (elastic, plastic, elastoplastic, or fracture), thermal conduction and radiation, and stresses induced by gas–solid interactions and solid-solid collisions. In addition, the particles may also be subjected to various field forces such as magnetic, electrostatic, and gravitational forces, as well as short-range forces such as van der Waals forces, which may affect the motion of particles.

In this section, we briefly discuss several aspects of the material properties of solids that are of interest to gas–solid flow applications. They include physical adsorption, deformation and fracture, thermal properties, electrical properties, magnetic properties, and optical properties.

1.4.1 *Physical Adsorption*

The physical adsorption phenomenon (also see §1.2.2.4) was first described by de Saussure in 1814 [Bikerman, 1970]. In his experiment, activated carbon particles were put into a flask containing chlorine. The green color of the gas disappeared. With the heating of the flask, the greenish chlorine gas reappeared. This phenomenon, known as adsorption, is the accumulation of gas on a solid surface.

The outermost layer of the solid molecules on the surface is bound on only one side to the inner layer of molecules by atomic and molecular forces. To compensate for this imbalance of the binding force, an attraction force from the solid surface captures the surrounding gas, vapor, or liquid. This attraction can be either physical or chemical, depending on the temperature and the interacting forces between the solid and the surrounding fluid.

Physical adsorption, or van der Waals adsorption, results from a relatively weak interaction between the solid and the gas. The forces responsible for adsorption are "dispersion" forces (characterized by London; see §3.3.1) and/or electrostatic forces (Coulombic; see §3.3.2) if either the gas or the solid is polar in nature. Physical adsorption is reversible; hence all the gas adsorbed by physical adsorption can be desorbed by evacuation at the same temperature. Chemical adsorption is a result of a more energetic interaction between the solid and the gas than that of physical adsorption. Reversal of chemical adsorption using a vacuum requires elevated temperature, and even that may not be sufficient. Physical adsorption, being of more interest in gas–solid flows, is the focus of the following sections.

1.4.1.1 *Freundlich Isotherm*

The most common way to present adsorption data is an adsorption isotherm. The amount adsorbed is plotted as a function of the partial pressure of the adsorbate, p, at a constant temperature. The first empirical isotherm was proposed by Freundlich in 1906 [Freundlich, 1926], and since then, a number of empirical adsorption isotherms have been reported. The Freundlich isotherm is usually written in the form

$$V_a = \alpha p^\beta \tag{1.41}$$

where V_a is the volume of gas adsorbed while α and β are empirical constants derived from linear graphs of $\log(V_a)$ against $\log(p)$.

1.4.1.2 *Langmuir Isotherm*

Langmuir (1918) laid down a systematic and theoretically grounded adsorption isotherm, the well-known Langmuir isotherm, which has served as a basis for a number of future isotherms, such as the BET isotherm, described in §1.4.1.3. The basic assumptions of the Langmuir adsorption model include monolayer coverage, localized adsorption (*i.e.*, the specific adsorption sites exist and the interactions are between a specific molecule and the site), and homogeneous materials. In addition, the heat of adsorption is independent of coverage. He proposed that the rate of adsorption is proportional to the dimensionless pressure of the adsorbate, p/p_0 ($= p^*$), where p_0 is the saturated vapor pressure, and the percentage of unoccupied adsorption sites $(1 - \zeta)$, where ζ is the ratio of the number of occupied sites to the total number of adsorption sites on the solid surface. At equilibrium, the rate of adsorption balances the rate of desorption so that

$$K_a p^*(1 - \zeta) = K_d \zeta \tag{1.42}$$

where K_a is the adsorption rate constant and K_d is the desorption rate constant. When the entropy effects are neglected, the equilibrium constant, $K_e (= K_a / K_d)$, can be defined as

$$K_e = \frac{\zeta}{p^*(1 - \zeta)} = \exp\left(-\frac{\Delta H^\circ}{R_M^\circ T}\right) \tag{1.43}$$

where ΔH° is the heat of adsorption per mole at the standard temperature and pressure (STP) condition. R_M° is the universal gas constant, which is equal to $82.06 \times 10^{-3}\ \text{m}^3 \cdot \text{atm/kgmole} \cdot \text{K}$. The Langmuir isotherm is rearranged to the linearized form

$$\zeta^{-1} = 1 + (K_e p^*)^{-1} \tag{1.44}$$

Therefore, the plot of ζ^{-1} against p^{*-1} yields K_e from the linearized data.

1.4.1.3 *Brunauer–Emmett–Teller (BET) Isotherm*

The assumption of monolayer adsorption in the Langmuir isotherm model is unrealistic in most cases, and a modification to multilayer adsorption should be considered. In 1938, Brunauer, Emmett, and Teller modified the Langmuir approach of balancing the rates of adsorption and desorption for the various molecular layers [Brunauer *et al.*, 1938]. This approach is known as the BET method. The BET isotherm assumes that the adsorption of the first layer has a characteristic heat of adsorption ΔH_a and the adsorption and desorption on subsequent layers are controlled by the heat of condensation of the vapor, ΔH_c. The derivation of the BET equation is beyond the scope of this book; however, a common form of the BET equation is given as

$$\frac{p}{V_a(p_0 - p)} = \frac{1}{V_m C} + \frac{(C - 1)p}{V_m C p_0} \tag{1.45}$$

where V_m is the monolayer capacity at STP and C is given as

$$C = \exp\left(\frac{\Delta H_a - \Delta H_c}{R_M^\circ T}\right) \tag{1.46}$$

Most commercial instruments using gas adsorption for surface area and porosity determination are based on the BET isotherm. In Eq. (1.45), the monolayer capacity V_m can be used to calculate the surface area on the basis of the area occupied by each adsorbed gas molecule. According to Eq. (1.45), a plot of $p/[V_a(p_0 - p)]$ versus p/p_0 is linear. From the slope and the intercept, V_m can be obtained. Thus, the specific surface area S_V can be obtained as

$$S_V = \frac{V_m n_A s_m}{V_{gm} m} \tag{1.47}$$

where n_A is Avogadro's number (6.02×10^{23}); s_m is the area per molecule of the adsorbed gas; V_{gm} is the molar volume of the adsorbed gas; and m is the mass of the solid.

Example 1.3 From the data given in Table E1.5, estimate the specific surface area of 22.5 mg of $Ca(OH)_2$ powder. The data are for adsorption of nitrogen at a temperature of −195.8°C, with the specific surface area of the solid adsorbent $S_v = 4.35V_m/m$ (m^2/g).

Table E1.5. *BET Data for Specific Surface Area Calculation of $Ca(OH)_2$*

p/p_0	V_a (cm^3)
0.1	0.1855
0.2	0.2113
0.3	0.2472
0.35	0.2652

Solution The data given are plotted in the form as required by BET equation, *i.e.*, Eq. (1.45), as shown in Fig. E1.3, from which the intercept on the ordinate $I = 0.0122$ (cm^{-3}) and the slope of the curve $s = 5.77$ (cm^{-3}). The values of s and I are used to obtain V_m by

$$V_m = \frac{1}{(s+I)} = 0.173\,\text{cm}^3 \qquad \text{(E1.2)}$$

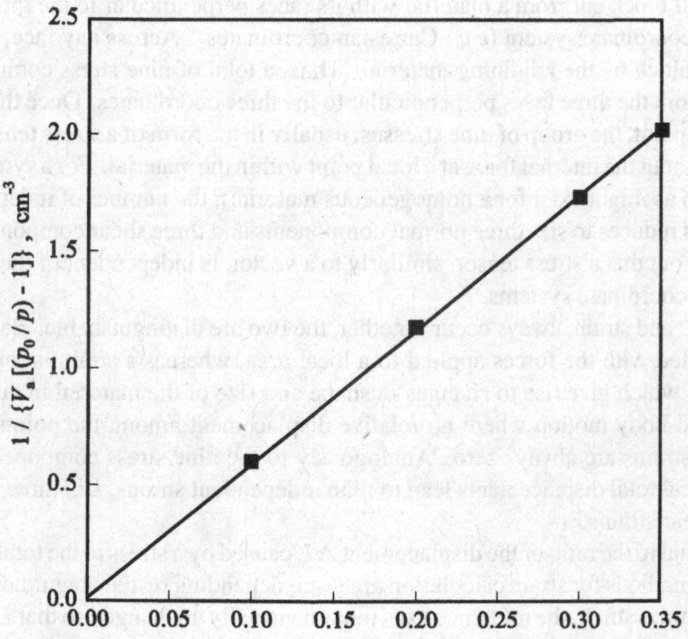

Figure E1.3. Plot of BET equation for data given in Table E1.5.

The specific surface area of $Ca(OH)_2$ can be obtained by substituting V_m and the weight of the solid in the expression for S_v.

$$S_v = 4.35 \times \frac{0.173}{0.0225} = 33.4\,\mathrm{m^2/g} \tag{E1.3}$$

1.4.2 *Deformation and Fracture*

All materials undergo a deformation, either noticeable or unnoticeable, whenever they are subjected to an external force. The deformation can be elastic, plastic, elastomeric, or viscoelastic. When an external force is applied on a body, the displacement of points of the body relative to neighboring points is measured as strain and the strength of the force applied to the local point is measured as stress. Elastic deformation such as rubber band stretch is recovered when the stress, or external force, is removed. Plastic deformation such as a dent on a metal car body is permanent and is not recoverable with the removal of the deforming stresses.

1.4.2.1 *Stress and Strain*

Deformation is caused by stress from either an external force or an imbalance of internal forces. Quantitatively, a stress σ on an area of a specimen is equal to the force applied per unit area. Since a force is a vector with three components, the stress component from the normal component of the force is called normal stress; it causes elongation or contraction of the material depending on the direction of the force. The stress components from the two tangential components of the force are called shear stresses; they are responsible for the shear deformation.

Imagine a small block cut from a material with its faces perpendicular to the three axes of an orthogonal coordinate system (*e.g.*, Cartesian coordinates). Across any face, a force is exerted on the block by the adjoining material. Thus, a total of nine stress components can be obtained from the three faces perpendicular to the three coordinates. Once the small block shrinks to a point, the group of nine stresses, usually in the form of a stress tensor, can give information about the internal force at a local point within the material. For a symmetric stress tensor (such as might exist for a homogeneous material), the number of independent stress components reduces to six, three normal components and three shear components. It should be pointed out that a stress tensor, similarly to a vector, is independent of the choice of the orthogonal coordinate systems.

Although stress and strain always occur together, the two are distinguishable. Basically, a stress is associated with the forces applied to a local area, whereas a strain is related to the displacements which give rise to changes in shape and size of the material in question. Therefore, in rigid-body motion where no relative displacement among the points in the body occurs, the strains are always zero. Analogously to the nine stress components at a local point, the local total displacements lead to nine independent strains, *i.e.*, three normal strains and six shear strains.

A strain ϵ is equal to the ratio of the displacement ΔL caused by a stress to the total length L. Two common methods for strain calculation are used, depending on the magnitude of the strain. When a strain is small, the reference axes remain virtually unchanged so that L essentially maintains its initial length L_0. Hence the strain, known as the nominal strain, is simply

$$\epsilon = \frac{\Delta L}{L_0} \tag{1.48}$$

In the cases where L changes appreciably from its initial value, the method for so-called true strain should be used instead of the nominal strain method. The true strain is calculated by

$$\epsilon = \int_{L_0}^{L} \frac{dl}{l} = \ln\left(\frac{L}{L_0}\right) \tag{1.49}$$

1.4.2.2 *Hooke's Law and Elastic Constants*

The mathematical relationship between the stress and the strain depends on material properties, temperature, and the rate of deformation. Many materials such as metals, ceramics, crystalline polymers, and wood behave elastically at small stresses. For tensile elastic deformation, the linear relation between the stress, σ, and strain, ϵ, is described by Hooke's law as

$$\epsilon = \frac{\sigma}{E} \tag{1.50}$$

where E is a material property known as Young's modulus. A material is anisotropic if the value of a given property depends on the direction in which testing is performed. Typical examples of anisotropic materials include sapphire crystal, whose atomic distribution varies with the testing directions, and wood, whose specific stiffness (defined as the ratio of Young's modulus to the material density) is much larger in the longitudinal direction (with the grain) than in the radial or tangential direction (against the grain). Polycrystalline materials such as copper are generally isotropic as a result of the random orientations of their crystals.

For isotropic materials, a general relationship correlating the three normal components of the stress tensor and strain is expressed as

$$\epsilon_1 = \frac{\sigma_1}{E} - (\sigma_2 + \sigma_3)\frac{\nu}{E} \tag{1.51}$$

where ϵ_1 is the tensile strain in the direction of σ_1, and σ_2 and σ_3 are the transverse stresses. This strain increases in proportion to the longitudinal stress σ_1 and decreases with the transverse stresses σ_2 and σ_3 as a result of the Poisson contraction. The degree of the Poisson contraction is determined by Poisson's ratio, ν, *i.e.*, the ratio of the contraction to the extension in the direction of a given applied stress. Young's modulus and Poisson's ratio are among the most important elastic properties of materials. Typical values of E and ν of some materials are listed in Table 1.6.

1.4.2.3 *Plastic Deformation*

Most solids are subjected to permanent deformation or breakup once the applied stresses exceed a certain limit. Hence, most solid particles may be classified into two categories: elastoplastic particles and elastic-brittle particles. Typical elastoplastic materials include metals and polymers, while typical elastic-brittle materials include coal, activated carbon, and ceramics. Materials that are elastoplastic at room temperature may become brittle at low temperatures and those that are brittle at room temperature may become plastic at high temperatures.

Some plastic materials have different tensile and compressive characteristics. For example, polystyrene is tough under compressive load but very brittle in tension. However, for most elastoplastic materials, the stress–strain curves in compression are the same as in tension. Hence, the deformation properties of these materials in tension may also be applied to those in compression, which is of great interest to gas–solid flows.

For elastoplastic materials, two stresses representing the yield strength and the tensile strength are important to consider. The yield strength is the stress at which appreciable

Table 1.6. *Elastic Properties of Some Solid Materials*

Material	Young's modulus E, 10^{10}N/m²	Poisson's ratio ν
Graphite	100	—
Al_2O_3 crystals		
[1010]	230	—
[1120]	125	—
[0001]	48	—
Boron	45	0.21
Sintered carbide (WC)	65	0.20
Glass-ceramic	10	0.25
Silica glass	8	0.24
Aluminum alloys	7	0.33
Steel	20	0.28
Tungsten	41	0.28
Wood (typical)		
Longitudinal	1	0.04
Radial	0.07	0.3
Tangential	0.06	0.5
Copper alloys	12	0.35
Nylon	0.3	0.48
Polyethylene	0.04	0.3

Source: A. G. Guy's *Essentials of Materials Science*, McGraw-Hill, 1976.

plastic deformation begins, whereas the tensile strength is the maximum nominal stress that a specimen can support before fracture. The yield strength is generally defined as the 0.2 percent offset yield stress, *i.e.*, the point at which the stress–strain curve crosses a line offset 0.2 percent along the strain axis parallel to the linear part of the stress–strain curve (see Fig. 1.9). Yield strength values of some typical materials are given in Table 1.7.

In contrast to the simplicity of elastic deformation, plastic deformation occurs in diverse ways. Figure 1.9 illustrates the stress–strain curves for two typical elastoplastic materials (hardened metal and polymer). Both materials show similar linear relationships between stress and strain for the elastic deformation (*i.e.*, before yield strength) but quite different correlations in the yielding processes before fracture.

1.4.2.4 *Fracture and Fatigue*

Besides deformation, fracture is the other response of materials to a stress. Fracture is the stress-induced breakup of a material. Two types of fracture are commonly defined. A brittle fracture is breakup which occurs abruptly without localized reduction in area. A ductile fracture is the failure of the material which is preceded by appreciable plastic deformation and localized reduction in area (necked region). The brittle fracture and ductile fracture are schematically illustrated in Fig. 1.10.

For brittle materials, the compressive strength is usually an order of magnitude greater than the tensile strength, and the fracture occurs by a shearing process [Guy, 1976]. For

Table 1.7. *Yield and Tensile Strength of Some Solid Materials*

Material	Yield strength (MPa)	Tensile strength (MPa)
(a) Metals and alloys		
Carbon steels	177–571	320–830
ASTM steels	204–340	306–680
Stainless steels	204–1,800	544–1,900
Cast aluminum alloys	55–435	35–70
Wrought aluminum alloys	27–618	68–666
Bronze	68–410	230–660
Brass	68–440	265–680
(b) Polymer		
Nylons	58–86	65–86
PVC–acrylic alloy	37–44	38–74
Polyacetals	60–68	60–68
Polystyrene	34–69	34–69
(c) Ceramics	(bending strength)	
Alumina	272–320	136–272
Electric porcelain	34–82	14–48
Polycrystalline glass	102–170	—
Mica	82–88	34–41
Graphite and carbon	4–27	5–108

Source: E. R. Parker's *Material Data Book*, McGraw-Hill, 1967 and J. F. Shackelford and W. Alexander's *CRC Materials Science and Engineering Handbook*, CRC Press, 1992.

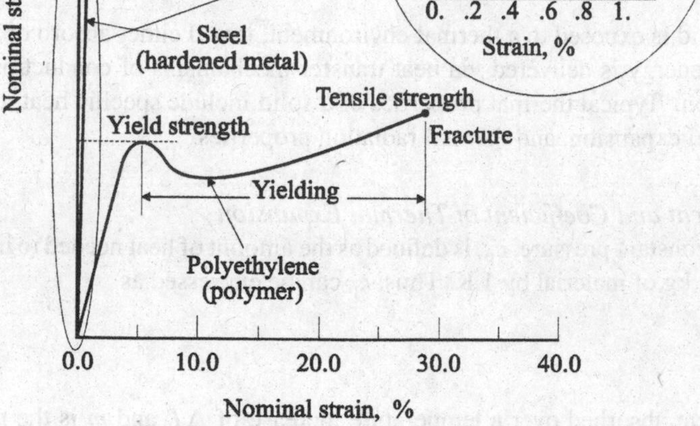

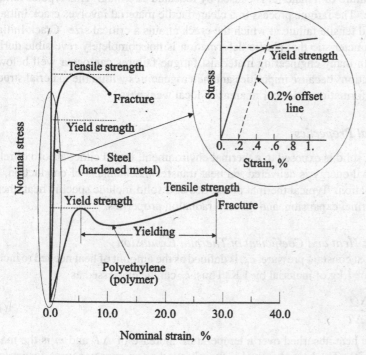

Figure 1.9. Deformation of typical elastoplastic materials (hardened metal and polymer) in the stress-strain diagram (after Guy, 1976).

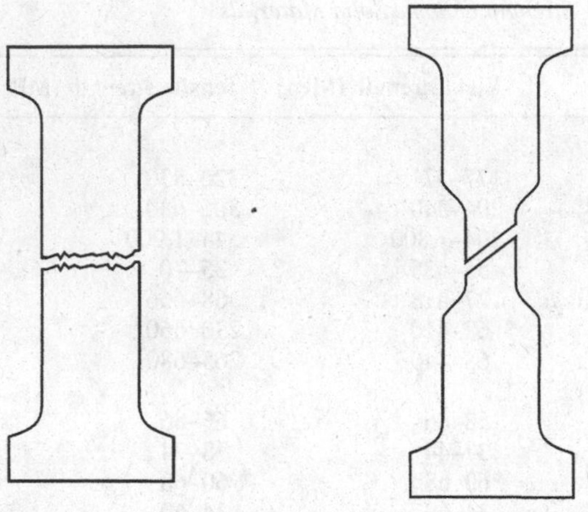

Figure 1.10. Brittle fracture and ductile fracture.

ductile materials, fracture requires a large amount of energy compared to that necessary for brittle fracture.

Fractures usually refer to a single cycle of either compressive or tensile loading. A more common failure of a material is caused by repeated stressing. This type of failure is known as fatigue. The fatigue process in a clean ductile material involves crack initiation, crack growth, and tensile failure at which the crack attains a critical size. Crack initiation is due to the characteristic that plastic deformation is not completely reversible during a cyclic loading. In many engineering materials, fatigue failures can occur well below the theoretical predictions because impurities and heterogeneities within the material structure permit plastic deformations to occur in areas of local weakness.

1.4.3 *Thermal Properties*

When a solid is exposed to a thermal environment, it will either absorb or release heat. This thermal energy is delivered via heat transfer mechanisms of conduction, convection, and radiation. Typical thermal properties of a solid include specific heat, thermal conductivity, thermal expansion, and thermal radiation properties.

1.4.3.1 *Specific Heat and Coefficient of Thermal Expansion*

The specific heat at constant pressure, c_p, is defined as the amount of heat needed to increase the temperature of 1 kg of material by 1 K. Thus, c_p can be expressed as

$$c_p = \frac{\Delta Q}{m \Delta T} \tag{1.52}$$

where ΔQ is the heat absorbed over a temperature increase of ΔT and m is the mass of the testing material. The heat absorbed is mainly used as an energy source for increased vibration of the crystal lattice. Since most solid materials exhibit similar lattice vibrations at room temperature, they have similar values of c_p. This value of c_p can be estimated as

$$c_p \approx 3R_M \tag{1.53}$$

where R_M is the gas constant (see Eq. (6.37)). At reduced temperatures, c_p initially decreases gradually but rapidly approaches zero as the temperature approaches 0 K.

When an object is heated or cooled, the characteristic length of the body is expanded or contracted by thermal expansion. Thermal expansion also changes stresses in the material and, in some cases, makes stresses large enough to cause fracture. The simplest form for a linear thermal expansion is given by

$$\frac{\Delta L}{L} = \alpha_T \Delta T \tag{1.54}$$

where α_T is the coefficient of thermal expansion and ΔL is the change in length.

1.4.3.2 *Thermal Conductivity*

When a temperature gradient exists in a material, energy in the form of heat is conducted from a high-temperature region to a low-temperature region through intermolecular and atomic impacts, lattice vibrations, and transport of electrons. This type of thermal energy transfer is called conductive heat transfer. The relation between heat flux induced by thermal conduction and temperature can be described by Fourier's law as

$$\boldsymbol{J}_q = -K\nabla T \tag{1.55}$$

where $\boldsymbol{J}_q$ is the heat flux vector driven by the temperature gradient ∇T, and the proportionality coefficient K is defined as thermal conductivity.

The thermal conductivity K mainly varies with temperature and type of the material. For solid materials, the quantum of lattice vibrational energy is known as a phonon. Heat conduction in solid material can be explained by the movement of phonons and electrons. The electrons acquire energy from a higher-energy phonon and carry it to a lower-temperature region, and then transfer the energy to a lower-energy phonon. For good metallic conductors, conduction of heat by electrons produces distinctly higher values of thermal conductivity than that from lattice vibration or phonon mechanism. This explains why a good thermal conductor, such as copper or silver, is usually a good electric conductor. For poor metallic conductors, such as a highly alloyed metal, phonons and electrons contribute comparably to the thermal conductivity. At ambient or higher temperatures, the thermal and electric conductivities of good metallic conductors can be described by the Wiedemann–Franz law, which has the following expression when the energy transfer by means of phonons is neglected

$$K = Lo\sigma T \tag{1.56}$$

where Lo is the Lorentz constant with a value of 2.45×10^{-8} W·Ω/K^2 and σ is the electrical conductivity, discussed in §1.4.4.1.

1.4.3.3 *Thermal Radiation Properties*

Radiative heat transfer of a solid involves both receiving and emitting radiant energy. Most solids in gas–solid flows can be regarded as thermally opaque. When a radiant heat flux strikes a solid surface, the incident energy is both reflected and absorbed with little heat transmitted. The fraction of absorbed radiant energy defines the thermal absorptivity of the solid, whereas the fraction reflected defines the thermal reflectivity of the material. Thus, we have

$$\alpha_R + \rho_R = 1 \tag{1.57}$$

where α_R is the absorptivity and ρ_R is the reflectivity.

The emissive power of a solid E_R is defined as the energy emitted by the body per unit area and per unit time. The body having a theoretical maximum emissive power at a given temperature is called a blackbody. The actual emissive power of a solid at a given temperature is less than or equal to that of the blackbody for the same temperature. Hence, we define the emissivity of a solid as the ratio of the emissive power of the solid to the emissive power of a blackbody at the same temperature, which is

$$\epsilon_R = \frac{E_R}{E_{Rb}} \tag{1.58}$$

For a blackbody, E_{Rb} is given by Stefan–Boltzmann's law

$$E_{Rb} = \sigma_b T^4 \tag{1.59}$$

where σ_b is the Stefan–Boltzmann constant, which has a value of 5.67×10^{-8} W/m$^2 \cdot$ K^4. From Kirchoff's law, we have

$$\alpha_R = \epsilon_R \tag{1.60}$$

In general, the emissivity of a solid is affected by the temperature as well as the wavelength of the radiation. The concept of monochromatic emissivity is related to the radiant emission by a solid at a specific wavelength. The monochromatic emissivity ϵ_λ is defined as the ratio of the monochromatic–emissive power of a solid E_λ to the monochromatic–emissive power of a blackbody $E_{b\lambda}$ at the same temperature and wavelength, *i.e.*,

$$\epsilon_\lambda = \frac{E_\lambda}{E_{b\lambda}} \tag{1.61}$$

Moreover, a solid is termed a gray body if its monochromatic emissivity ϵ_λ is independent of the wavelength. The monochromatic–emissive power of a blackbody at a specific temperature and wavelength is given by Planck's formula

$$E_{b\lambda} = \frac{C_1 \lambda^{-5}}{e^{C_2/\lambda T} - 1} \tag{1.62}$$

where λ is wavelength (μm); T is absolute temperature (K); $C_1 = 3.743 \times 10^8$ (W $\cdot$ μm^4/m^2); and $C_2 = 1.4387 \times 10^4$ μm $\cdot$ K. In addition, the wavelength λ_m at which the monochromatic–emissive power of a blackbody is maximum for a given temperature can be derived from Eq. (1.62) as

$$\lambda_m T = C_3 \tag{1.63}$$

where $C_3 = 2898$ μm $\cdot$ K. Equation (1.63) is known as Wien's displacement law.

The total emissivity of the solid is related to the monochromatic emissivity by

$$\epsilon = \frac{E}{E_b} = \frac{\int_0^\infty \epsilon_\lambda E_{b\lambda}\, d\lambda}{\sigma_b T^4} \tag{1.64}$$

The thermal properties of some solid materials are given in Table 1.8.

Table 1.8. *Thermal Properties of Solids*

Material	Thermal conductivity K, W/m · K	Coefficient of thermal expansion α_T, 10^{-6} K^{-1}
Copper	390	18
Beryllia (BeO)	250	10
Aluminum	200	25
Graphite	150	2
Sintered carbide (WC)	80	7
Steel	50	12
Boron nitride (BN)	28	8
Al_2O_3 ceramics	17	7
Pyrex glass	1.2	3
Kovar	15	5
Soda-lime glass	0.8	9
Polyethylene	0.3	300
Rubber	0.1	670
Urethane foam	0.05	90
Invar	11	1

Source: A. G. Guy's *Essentials of Materials Science*, McGraw-Hill, 1976.

1.4.4 *Electrical Properties*

Electrons are driven to move by electric force when an electrical potential difference exists in a material. The resistance R to this force depends on the cross-sectional area A, the length l, and the conductivity σ of an object as

$$R = \frac{l}{\sigma A} \tag{1.65}$$

1.4.4.1 *Electrical Conductivity*

The electrical conductivity σ is defined as the electrical current density or the amount of charges passing through a unit cross-sectional area per second in an electrical field with strength E of 1 V/m. The electrical conductivity can be determined from the particle number concentration n, the charge on a particle q, and the mobility of a particle μ by

$$\sigma = nq\mu \tag{1.66}$$

The mobility here is the average velocity of a particle that accelerates between successive collisions under unit electric driving force (*i.e.*, electric field strength E of 1 V/m).

For metallic materials, charges are conducted by electrons or holes because most of the energy bands of metals are partially filled. A common equation for the electrical conductivity of metals can be derived from the following simple model. Applying Newton's second law to an electron in a crystal, it yields

$$-eE = m_e \frac{d\nu_e}{dt} \tag{1.67}$$

where m_e is the effective mass of an electron and ν_e is the electron velocity. The electron will accelerate continuously under the driving force $-eE$ until it collides with impurities, lattice imperfections, *etc*. Assume that the velocity of the electron drops to zero at each collision and the average time between collisions is τ so that

$$eE\tau = m_e v_{em} \tag{1.68}$$

where v_{em} is the maximum velocity gained just before the next collision. From the definition of mobility, which equals the average velocity per unit electrical field strength,

$$\mu = \frac{v_{av}}{E} = \frac{v_{em}}{2E} = \frac{e\tau}{2m_e} \tag{1.69}$$

Therefore, the electrical conductivity of metals is obtained from Eqs. (1.66) and (1.69) as

$$\sigma = ne\mu = \frac{ne^2\tau}{2m_e} \tag{1.70}$$

For nonmetallic substances, the electrons cannot move as freely as in the case of metals because their energy bands are essentially completely full or empty. The electrical conductivity in nonmetallic materials is dominated by another mechanism, *i.e.*, the defect mechanism, instead of electron conduction. In ionic crystals such as salts (*e.g.*, sodium chloride), two types of ions, cations and anions, are driven to move by the electrical force qE once an electrical field is applied. The ions can move only by the defect mechanism; that is, they exchange position with a vacancy of the same type. At the room temperature, the fraction of vacancies for salt is very small (of the order of 10^{-17}) with low exchange frequency (of the order of 1 Hz) so that electrical conductivity is extremely low. Although impurities and high temperature can affect electrical conductivity by a large factor, nonmetallic materials generally have very low electrical conductivity and these substances are widely used as electrical insulators.

1.4.4.2 *Permittivity*

Permittivity ϵ is the most basic property of a dielectric material. To understand the concept of permittivity, consider first a parallel-plate capacitor with two metallic plates. When a voltage is applied across the metallic plates, charges are generated on the surface of each plate. The charge Q on the plate surface is given by

$$Q = \epsilon A \frac{V}{l} = CV \tag{1.71}$$

where A is the surface area of the plate; V is the voltage; l is the distance between the plates; ϵ is the permittivity of the medium; and C is the capacitance, which depends on the geometry of the capacitor and the medium with which it is filled.

In the case of a vacuum between the plates, the permittivity (ϵ_0) has a value of 8.854×10^{-12} F/m. When a dielectric material fills the space between the plates rather than a vacuum, the charge on the plate surface can be increased because of the nature of polarization of the dielectric as shown in Fig. 1.11. The ratio of ϵ to ϵ_0 is the relative permittivity or the dielectric coefficient of the dielectric.

In general, permittivity may be expressed in the complex form

$$\epsilon = \epsilon' - i\epsilon'' \tag{1.72}$$

where the imaginary part ϵ'' accounts for the electrical power loss or heating in the dielectric and the real part ϵ' is equivalent to the permittivity. Both ϵ' and ϵ'' vary with the frequency, which is a function of several mechanisms of polarization in a solid dielectric material.

Table 1.9. *Dielectric and Ferroelectric Materials*

Material	Dielectric coefficient ϵ_r, at 10^6 Hz and 25°C	Curie temperature °C
(a) Dielectrics		
Polyethylene	2.3	
Polyvinyl chloride	2.8	
Nylon	3.6	
Porcelain	5	
Mica	7	
Alumina ceramic	9.6	
(b) Ferroelectrics		
$BaTiO_3$ (ceramic)	1700	120
PZT-5A (ceramic)	1700	365
PZT-5H (ceramic)	3400	193

Source: A. G. Guy's *Essentials of Materials Science*, McGraw-Hill, 1976.

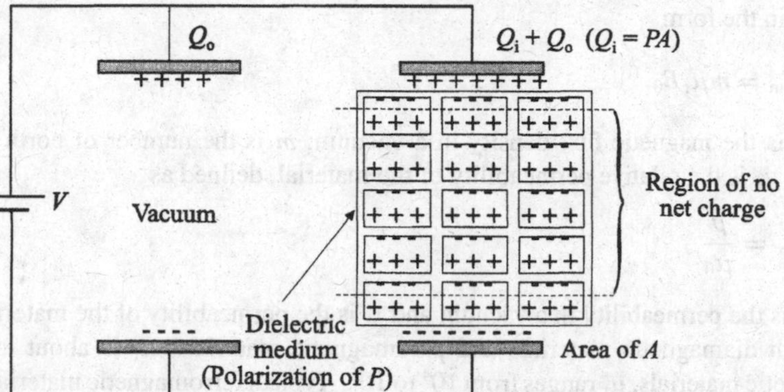

Figure 1.11. Illustration of induced charge by a dielectric medium (from Guy, 1976).

In comparison to ordinary dielectrics, the permittivities of the so-called ferroelectric materials are about 10^3 times larger. The ferroelectric material can be transformed into a new type of material called piezoelectric material by heating the ferroelectric above its Curie temperature and then cooling it in a powerful electric field. A piezoelectric crystal changes its polarization once subjected to a mechanical strain. As a result, it can deform mechanically under an electric field or produce electric impulses as a result of mechanical impulses. Currently, piezoelectric materials are widely used as force or pressure transducers with fast response times and very sensitive output. Permittivities of common dielectric and ferroelectric materials are given in Table 1.9.

1.4.5 *Magnetic Properties*

A few types of solids can be magnetized and therefore are responsive to the magnetic force of a magnetic field. Magnetic solids include ferromagnetic and ferrimagnetic materials. Most materials, however, are nonmagnetic. These nonmagnetic solids can be classified as either diamagnetic or paramagnetic.

Magnetic effects in solids originate in the tiny electrical currents associated either with electrons in atomic orbits or with the spin of electrons. In diamagnetic materials, the effects of those currents cancel out each other within an atom so that no net magnetic moment is generated. In paramagnetic materials, although each individual atom has net magnetic moments, the random orientations for a group of atoms known as the domain still result in zero net magnetization.

In ferromagnetic materials, the magnetic moments of all types are aligned within a single domain. In ferrimagnetic materials, the magnetic moments of each type are aligned within a domain. In the absence of an external magnetic field, the orientation of the net magnetic moment of a domain may be random so that the solid is unmagnetized. With a strong external magnetic field, both ferromagnetic and ferrimagnetic solids become magnetized because the domains in the material are aligned or nearly aligned. Once the external magnetic field is removed, thermal motion may cause most domains to return to a random arrangement. However, some domains retain a residual alignment which keeps the material partially magnetized. Figure 1.12 schematically illustrates the magnetic behavior of a typical material in each of the classes.

Analogously to the electrostatic force in an electric field, the magnetic force, F_m, is expressed in the form

$$F_m = m\mu_r B_0 \tag{1.73}$$

where B_0 is the magnetic flux density in a vacuum; m is the number of north magnetic poles; and μ_r is the relative permeability of the material, defined as

$$\mu_r = \frac{\mu}{\mu_0} \tag{1.74}$$

where μ_0 is the permeability in a vacuum and μ is the permeability of the material.

For both diamagnetic materials and paramagnetic materials, μ_r is about unity. For ferromagnetic materials, μ_r ranges from 10^4 to 10^5. Typical ferromagnetic materials include iron, nickel, cobalt, and six rare-earth elements, and their alloys.

1.4.6 *Material Densities*

One of the most important properties of a material is its density, for which there are several expressions, namely, bulk, particle, and skeletal densities. The bulk density of solids is the overall density of the material including the interparticle distance of separation. It is defined as the overall mass of the material per unit volume, which can be determined by simply pouring a preweighed sample of particles into a graduated cylinder and measuring the volume occupied. The material can become denser with time and settling, and its bulk density reaches a certain limiting value, known as the tapped or packed bulk density.

Particle density is the density of a particle including the pores or voids within the individual solids. It is defined as the weight of the particle divided by the volume occupied by the entire particle. Particle density is sometimes referred to as the material's apparent density. Direct measurement of particle density can be made by immersing a known quantity of the material in a nonwetting fluid, such as mercury, which does not penetrate into the pores. The volume of the particle is the volume change of the fluid.

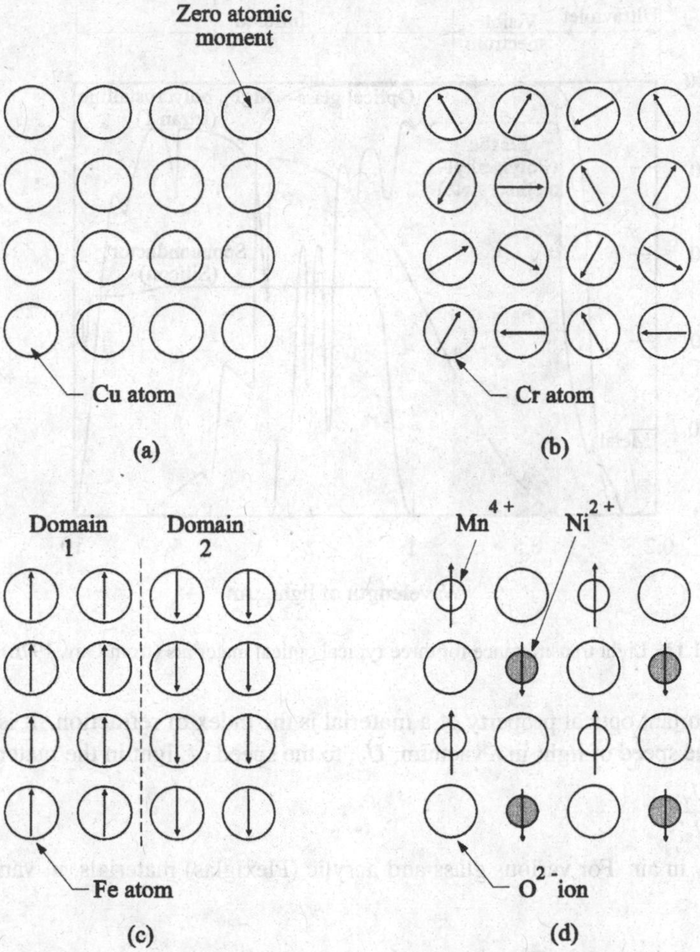

Figure 1.12. Schematic illustrations of the atomic origins of four types of magnetic behavior without external magnetic field (from Guy, 1976): (a) Diamagnetism (copper); (b) Paramagnetism (chromium); (c) Ferrimagnetism (iron); (d) Ferromagnetism ($NiMnO_2$).

The skeletal density, also called the true density, is defined as the density of a single particle excluding the pores. That is, it is the density of the skeleton of the particle if the particle is porous. For nonporous materials, skeletal and particle densities are equivalent. For porous particles, skeletal densities are higher than the particle density. Measurements of the skeletal density can be made by liquid or gas pycnometers.

1.4.7 *Optical Properties*

The most important optical property of a material is the ability to transmit light. The transmittance of a material is defined as the percentage of the light intensity of an initial light beam remaining after passing through 10 mm of the material. Thus, the transmittance of a material also depends on the wavelength of the light. Transmittances of some typical materials across the ultraviolet, visible, and infrared light spectra are illustrated in Fig. 1.13.

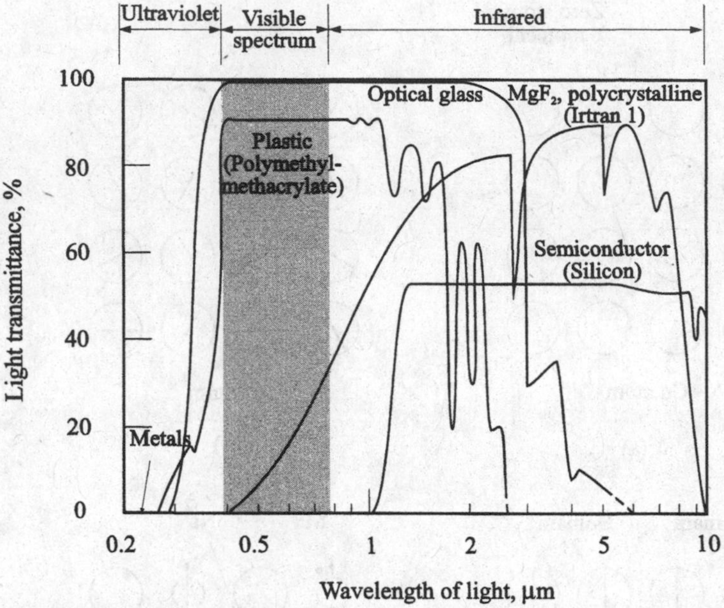

Figure 1.13. Light transmittance for three typical optical materials (from Guy, 1976).

Another important optical property of a material is the index of refraction. It is defined as the ratio of the speed of light in a vacuum, U_v, to the speed of light in the material U_m

$$n_i = \frac{U_v}{U_m} \tag{1.75}$$

n_i is about unity in air. For various glass and acrylic (Plexiglas) materials, n_i varies from 1.3 to 1.5.

Nomenclature

A	Cross-sectional area
A	Projected area
A_e	Integrated cross section for extinction
A_M	Normalized constant in the particle mass distribution
A_N	Normalized constant in the particle number distribution
B	Magnetic flux density
B_0	Magnetic flux density in a vacuum
b	Variable for particle diameter
C	Capacitance
C	Factor, defined by Eq. (1.46)
C_D	Drag coefficient
c_p	Specific heat at constant pressure
d	Particle diameter
d_0	Arithmetic mean diameter in the normal distribution
d_{01}	Parameter in the log-normal distribution
d_{32}	Sauter's diameter
d_{43}	DeBroucker's diameter
d_A	Projected area diameter
d_l	Arithmetic mean diameter
d_M	Martin's diameter
d_S	Surface diameter

d_t	Dynamic diameter
d_V	Volume diameter
E	Young's modulus
E	Electric field strength
E	Total emissive power of radiation
E_b	Total emissive power of blackbody radiation
E_R	Energy flux due to thermal radiation, or emissive power of radiation
E_{Rb}	Energy flux due to thermal radiation by a blackbody
E_λ	Monochromatic–emissive power of radiation
$E_{b\lambda}$	Monochromatic–emissive power of blackbody radiation
e	Charge carried by an electron
F	Force
F	Focal length
F	Cumulative distribution function
F_m	Magnetic force
f_M	Mass density function
f_N	Number density function
g	Gravitational acceleration
I	Intercept
I_i	Intensity of incident light beam
I_t	Intensity of transmitted beam
J_1	First-order spherical Bessel function
$\mathbf{J}_q$	Heat flux vector
K	Thermal conductivity
K_a	Adsorption rate constant
K_d	Desorption rate constant
K_e	Equilibrium constant
L	Length
L_0	Initial length
Lo	Lorentz constant
l	Optical path length
M	Mass of particles
M_0	Total mass of particles in a sample
m	Mass of a particle
m	Mass of the testing material
m	Number of north magnetic poles
m_e	Mass of an electron
N	Number of particles
N_0	Total number of particles in a sample
N_A	Numerical aperture
n	Number density, or number of particles per unit volume
n_A	Avogadro's number
n_i	Relative index of refraction
P	Polarization
p	Partial pressure of the adsorbate
p^*	Dimensionless pressure of adsorbate
p_0	Saturated vapor pressure
Q	Total charge
ΔQ	Heat absorbed by solids
q	Charge carried by a particle
R	Resistance
R	Rosin–Rammler distribution function
Re_t	Particle Reynolds number at the particle terminal velocity condition
R_M	Gas constant
R_M°	Universal gas constant
r	Radial distance
S	Surface area
S_V	Specific surface area of particles
s	Slope
s_m	Area per molecule of the absorbed gas
T	Absolute temperature
U_m	Speed of light in a material
U_{pt}	Particle terminal velocity
U_v	Speed of light in a vacuum
V	Particle volume
V_a	Volume of gas adsorbed at relative pressure p/p_0 of adsorbate
V_{gm}	Molar volume of the adsorbed gas

V_m	Monolayer capacity
v_e	Velocity of an electron
v_{em}	Maximum velocity of electron between collisions
x	Parameter, defined by Eq. (1.14)

Greek Symbols

α	Parameter, defined by Eqs. (1.33) and (1.41)
α_R	Absorptivity due to thermal radiation
α_T	Coefficient of thermal expansion
β	Parameter, defined by Eqs. (1.33) and (1.41)
ΔH°	Heat of adsorption
ΔH_a	Characteristic heat of adsorption
ΔH_c	Characteristic heat of condensation of vapor
δ	Limit of resolution of the optical microscope
ϵ	Normal strain
ϵ	Permittivity, or dielectric constant
ϵ	Total emissivity
ϵ_0	Permittivity in a vacuum
ϵ_R	Emissivity due to thermal radiation
ϵ_r	Dielectric coefficient
ϵ_λ	Monochromatic emissivity due to thermal radiation
ζ	Ratio of number of occupied sites to total number of sites on solid surface adsorption
κ	Parameter, defined by Eq. (1.12)
λ	Wavelength
λ_m	Wavelength at which the monochromatic–emissive power of a blackbody is maximum
μ	Mobility of a particle
μ	Permeability
μ	Dynamic viscosity of fluid
μ_r	Relative permeability
ν	Poisson's ratio
ν_{av}	Average electron velocity
ν_e	Electron velocity
ν_{em}	Maximum velocity gained by an electron just before the next collision
ρ	Density of fluid
ρ_p	Density of particles
ρ_R	Reflectivity due to thermal radiation
σ	Electrical conductivity
σ	Normal stress
σ_b	Stefan–Boltzmann's constant
σ_d	Root of mean square deviation in the normal distribution
σ_{dl}	Parameter in the log-normal distribution
τ	Average time between collisions
ω	Angular frequency

References

Allen, T. (1990). *Particle Size Measurement*, 4th ed. New York: Chapman & Hall.

Bachalo, W. D. and Houser, M. J. (1984). Phase/Doppler Spray Analyzer for Simultaneous Measurements of Drop Size and Velocity Distributions. *Optical Engineering*, **23**, 583.

Bikerman, J. J. (1970). *Physical Surfaces*. New York: Academic Press.

Brunauer, S., Emmett, P. H. and Teller, E. (1938). Adsorption of Gases in Multimolecular Layers. *J. Am. Chem. Soc.*, **60**, 309.

Cadle, R. D. (1965). *Particle Size*. New York: Reinhold.

Dallavalle, J. M. (1948). *Micrometrics: The Technology of Fine Particles*, 2nd ed. New York: Pitman.

Durst, F. and Zaré, M. (1975). Laser Doppler Measurements in Two-Phase Flows. *Proceedings of LDA Symposium*, Copenhagen.

Freundlich, H. (1926). *Colloid and Capillary Chemistry*. Ed. Trans. H. S. Hatfield. London: Methuen.

Ghadiri, M., Farhadpour, F. A., Clift, R. and Seville, J. P. K. (1991). Particle Characterization: Size and Morphology. In *Powder Metallurgy: An Overview*. Ed. Jenkins and Wood. London: Institute of Metals.

Gregg, S. J. and Sing, K. S. W. (1982). *Adsorption, Surface Area and Porosity*, 2nd ed. London: Academic Press.

Guy, A. G. (1976). *Essentials of Materials Science*. New York: McGraw-Hill.

Hay, W. and Sandberg, P. (1967). The Scanning Electron Microscope, a Major Breakthrough for Micropaleontology. *Micropaleontology*, **13**, 407.

Herdan, G. (1960). *Small Particle Statistics*. London: Butterworths.

Kay, D. H. (1965). *Techniques for Electron Microscopy*, 2nd ed. Oxford: Blackwell Scientific.

Langmuir, I. (1918). The Adsorption of Gases on Plane Surfaces of Glass, Mica, and Platinum. *J. Am. Chem. Soc.*, **40**, 1361.

May, K. R. (1945). The Cascade Impactor: An Instrument for Sampling Coarse Aerosols. *J. Sci. Instrum.*, **22**, 187.

Parker, E. R. (1967). *Material Data Book*. New York: McGraw-Hill.

Plantz, P. E. (1984). Particle Size Measurements from 0.1 to 1000 μm, Based on Light Scattering and Diffraction. In *Modern Methods of Particle Size Analysis*. Ed. H. G. Barth. New York: John Wiley & Sons.

Schlichting, H. (1979). *Boundary Layer Theory*, 7th ed. New York: McGraw-Hill.

Shackelford, J. F. and Alexander, W. (1992). *The CRC Materials Science and Engineering Handbook*. Boca Raton, Fla.: CRC Press.

Soo, S. L. (1990). *Multiphase Fluid Dynamics*. Beijing: Science Press; Brookfield USA: Gower Technical.

Svarovsky, L. (1990). Characterization of Powders. In *Principles of Powder Technology*. Ed. M. Rhodes. New York: John Wiley & Sons.

van de Hulst, H. C. (1981). *Light Scattering by Small Particles*. New York: Dover.

Weiner, B. B. (1984). Particle and Droplet Sizing Using Fraunhofer Diffraction. In *Modern Methods of Particle Size Analysis*. Ed. H. G. Barth. New York: John Wiley & Sons.

Yamate, G. and Stockham, J. D. (1977). Sizing Particles Using the Microscope. In *Particle Size Analysis*. Ed. Stockham and Fochtman. Ann Arbor, Mich.: Ann Arbor Science Publishers.

Problems

1.1 Two particles with an identical density of 2,000 kg/m^3 undergo free fall in an 80 percent glycerin/water solution at 20°C. Their terminal velocities, U_{pt}, are measured to be 0.007 and 0.114 m/s. What are the equivalent dynamic diameters of these particles? The viscosity and density of the solution are 62×10^{-3} kg/m · s and 1,208.5 kg/m^3, respectively.

1.2 A fly ash sample from a fluidized bed coal combustor is analyzed to obtain particle size data. Table P1.1 shows the distributions of the projected area equivalent diameter of the particle d_A obtained by the image analysis and the volume diameter of the particle d_V obtained by the electrozone technique [Ghadiri *et al.*, 1991].

Specify the type of density functions that can best represent the d_A and d_V distributions. How do they differ from the d_t distribution?

Table P.1.1. *Data for Cumulative Distribution of d_A and d_V for Fly Ash*

Volume % undersize	d_A µm	d_V µm	d_t (experimental) µm
2	0.8	0.50	—
6	1.2	0.75	—
10	1.6	0.90	—
20	2.0	1.05	1.00
30	2.2	1.40	—
40	2.8	1.80	—
50	3.0	2.00	2.10
60	3.5	2.20	3.00
70	4.0	2.80	—
80	4.2	3.20	4.05
90	5.0	4.05	5.20

1.3 The sizes of a powder sample are found to follow a log-normal distribution with $\sigma_{dl} = 0.411$ and $d_{01} = 0.445$ mm. Calculate (a) the arithmetic mean diameter, (b) the surface mean diameter, (c) the volume mean diameter, (d) the Sauter mean diameter, and (e) the DeBroucker mean diameter.

1.4 From the data given in Table P1.2, determine the specific surface area of 11.32 mg of CaO powder. The data are based on adsorption of nitrogen at −195.8°C, with $S_V = 4.35 V_m/m(\text{m}^2/\text{g})$.

Table P.1.2. *BET Data for Specific Surface Area Calculation of CaO*

p/p_0	$V_a(\text{cm}^3)$
0.1	0.080
0.2	0.086
0.3	0.105
0.35	0.115

Vapors in equilibrium with liquid in fine capillaries or pores will have depressed vapor pressure as a result of the Kelvin effect. In fact, if the pores are adequately small, the vapor will condense at pressures far below normal. By measuring the volume of nitrogen adsorbed at a relative pressure, *i.e.*, p/p_0, of 0.99 and with prior knowledge of the surface area, the average pore diameter can be calculated.

In an experiment with the given CaO powder it was found that at a relative pressure of 0.99, 1.13 cm³ of nitrogen gas was adsorbed. Assuming cylindrical pores for CaO, determine the average pore diameter of the given powder. To convert volume of gaseous nitrogen adsorbed to liquid nitrogen, V_{liq}, held inside the pores, use Eq. (P1.1),

$$V_{liq} = \frac{p_a V_a V_{mol}}{R_M^\circ T} \qquad \text{(P1.1)}$$

where p_a is the ambient pressure (1 atm); V_a is the volume of gaseous nitrogen adsorbed per gram of adsorbent; and V_{mol} is the molar volume of liquid nitrogen, 34.7 cm³/mole.

1.5 In a simple tensile test, a specimen of an aluminum alloy is elongated elastically by 0.18 mm. If the original length of the specimen was 25 cm, determine the axial stress in the specimen.

1.6 Prove that the Stefan–Boltzmann law for thermal radiation given in Eq. (1.59) can be derived by using Planck's formula given in Eq. (1.62). Also show that

$$\sigma = \left(\frac{\pi}{C_2}\right)^4 \frac{C_1}{15} \tag{P1.2}$$

$$\text{Hint}: \int_0^\infty \frac{x^3}{\exp(x) - 1} dx = \frac{\pi^4}{15} \tag{P1.3}$$

CHAPTER 2

Collision Mechanics of Solids

2.1 Introduction

In particulate multiphase flows, collision occurs between the particles or between the particle and the wall. In gas–solid flows, the collision frequency mainly depends on the particle concentration and particle size, although it is also influenced by the flow pattern and system configurations. Kinetic energy loss of the particles due to collisions may occur and can be in the form of frictional heat generation, wall surface erosion, particle breakage (attrition), particle deformation, particle agglomeration, or solid electrification. The total momentum of all colliding bodies obeys the law of conservation of momentum, namely, the total momentum before collision equals the total momentum after collision. However, as a result of the collision, the momentum of each individual particle may change.

A collision without permanent deformation or heat generation is called an elastic collision. Otherwise, the collision is inelastic and energy loss occurs mainly in the form of permanent deformation such as particle breakage and frictional heat loss. Generally, a force is exerted on the particle by its colliding counterpart during the collision. This force is expressed as an impulse, which is defined as the product of the force and the duration of the collision. From Newton's law, the impulse is equal to the momentum change of the particle during the collision. Thus, for the same change in momentum, a shorter duration requires a larger impact force. It is this impact force that causes the momentum transfer from one particle to other particles, particle deformation or breakage, and frictional heat loss.

The simplest theory of impact, known as stereomechanics, deals with the impact between rigid bodies using the impulse–momentum law. This approach yields a quick estimation of the velocity after collision and the corresponding kinetic energy loss. However, it does not yield transient stresses, collisional forces, impact duration, or collisional deformation of the colliding objects. Because of its simplicity, the stereomechanical impact theory has been extensively used in the treatment of collisional contributions in the particle momentum equations and in the particle velocity boundary conditions in connection with the computation of gas–solid flows.

In this chapter, two simple cases of stereomechanical collision of spheres are analyzed. The fundamentals of contact mechanics of solids are introduced to illustrate the interrelationship between the collisional forces and deformations of solids. Specifically, the general theories of stresses and strains inside a solid medium under the application of an external force are described. The intrinsic relations between the contact force and the corresponding elastic deformations of contacting bodies are discussed. In this connection, it is assumed that the deformations are processed at an infinitely small impact velocity and for an infinitely long period of contact. The normal impact of elastic bodies is modeled by the Hertzian theory [Hertz, 1881], and the oblique impact is delineated by Mindlin's theory [Mindlin, 1949]. In order to link the contact theories to collisional mechanics, it is assumed that the process of a dynamic impact of two solids can be regarded as quasi-static. This quasi-static approach is valid when the impact velocity is small compared to the speed of the elastic wave. Finally, collision of inelastic solids is addressed. The concepts, criteria, and modeling of plastic deformation and the coefficient of restitution for an inelastic collision are introduced and the rebounding velocities of colliding particles are illustrated. It should be noted that the effects of gas–solid interactions and field forces, such as gravitational and electrostatic forces, are usually insignificant during a collision process and are excluded in this chapter.

2.2 Stereomechanical Impact

In this section, collisions among rigid spherical particles are studied. Two simple cases, collinear and planar collisions, are described. For the general theory of stereomechanical impact of irregular-shaped rigid bodies in arbitrary motion, readers may refer to Goldsmith (1960).

2.2.1 *Collinear Impact of Spheres*

Consider a collinear collision of two rigid, frictionless, and nonrotating spheres, as shown in Fig. 2.1. Neither sphere has a tangential momentum component in this system. Therefore, conservation of the normal momentum component of the two-ball system yields

$$m_1 U_1 + m_2 U_2 = m_1 U_1' + m_2 U_2' \equiv U(m_1 + m_2) \tag{2.1}$$

where the superscript $'$ represents the state after impact and U is the normal velocity of the center of mass of the system. The normal impulses J and J' during the compression and rebounding processes of the collision, respectively, are given by

$$\begin{aligned} J &= m_1(U_1 - U) = m_2(U - U_2) \\ J' &= m_1(U - U_1') = m_2(U_2' - U) \end{aligned} \tag{2.2}$$

Now, define the coefficient of restitution, e, as

$$e = \frac{U_2' - U_1'}{U_1 - U_2} \tag{2.3}$$

Substituting Eqs. (2.1) and (2.2) into Eq. (2.3) shows

$$e = \frac{J'}{J} \tag{2.4}$$

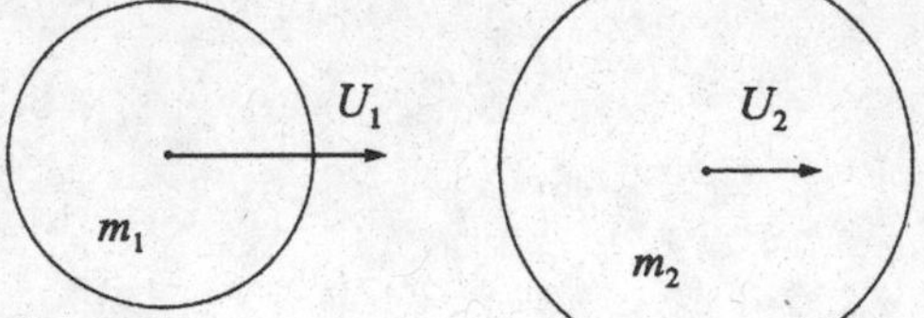

Figure 2.1. Normal collision of rigid and smooth spheres.

The rebounding velocities of the colliding spheres can be expressed in terms of the coefficient of restitution as

$$U_1' = U_1 - \frac{(1+e)m_2}{m_1+m_2}(U_1 - U_2)$$
$$U_2' = U_2 + \frac{(1+e)m_1}{m_1+m_2}(U_1 - U_2) \tag{2.5}$$

Therefore, the loss of the kinetic energy of the system is given by

$$\Delta E = \frac{1}{2}\frac{m_1 m_2}{m_1+m_2}(1-e^2)(U_1-U_2)^2 \tag{2.6}$$

It is clear that $e = 1$ for the normal impact of perfect elastic spheres, while $e = 0$ for the normal impact of perfect plastic spheres.

Equation (2.3) also provides a basis for the experimental determination of the coefficient of restitution. Consider the case where a ball at rest is dropped from a height h to a horizontal stationary massive rigid surface, rebounding back to a height of h'. If we label the ball with the subscript 1 and the massive plane with 2, Eq. (2.3) can be rearranged to

$$e^2 = \frac{h'}{h} \tag{2.7}$$

The coefficient of restitution in an impact depends not only on the material properties of the colliding objects but also on their relative impact velocity. More discussion of the coefficient of restitution is given in §2.5.2.

2.2.2 *Planar Impact of Spheres*

Consider a planar collision of a sphere of mass m_1 and radius a_1 against an initially stationary sphere of mass m_2 and radius a_2. Select the Cartesian coordinates so that the x- and y-axes are normal and tangential to the contact surfaces, respectively, as shown in Fig. 2.2. The conservation of linear momentum yields

$$m_1(U_1' - U_1) + m_2 U_2' = 0$$
$$m_1(V_1' - V_1) + m_2 V_2' = 0 \tag{2.8}$$

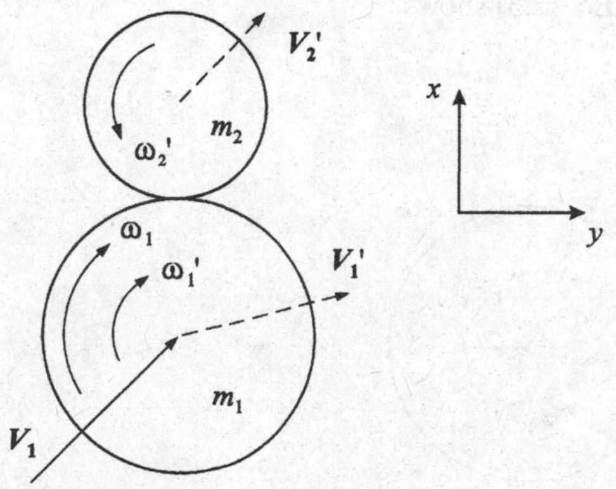

Figure 2.2. Planar collision of rigid spheres.

and the conservation of angular momentum with respect to the contact point gives

$$\begin{aligned} k_1^2(\omega_1' - \omega_1) - a_1(V_1' - V_1) &= 0 \\ k_2^2\omega_2' - a_2 V_2' &= 0 \end{aligned} \tag{2.9}$$

where k is the radius of gyration. For a sphere, k is evaluated by

$$k^2 \equiv \frac{I}{m} = \frac{2}{5}a^2 \tag{2.10}$$

where I is the moment of inertia.

In order to determine the unknown velocities after collision uniquely, two more relationships for impact velocities need to be specified. For certain values of the coefficient of friction and the coefficient of restitution, simple expressions of impact velocities can be obtained. In a collision of two completely rough and inelastic spheres, where the coefficient of friction f reaches infinity and the coefficient of restitution e is equal to zero, the relative velocities must vanish. Therefore, we have

$$\begin{aligned} U_1' &= U_2' \\ V_1' + \omega_1' a_1 &= V_2' + \omega_2' a_2 \end{aligned} \tag{2.11}$$

When two perfectly smooth and yet incompletely inelastic spheres collide, *i.e.*, $f = 0$ and $0 < e < 1$, no change in the tangential velocity of each colliding sphere occurs. Thus, it gives

$$\begin{aligned} V_1' &= V_1 \\ e &= \frac{U_2' - U_1'}{U_1} \end{aligned} \tag{2.12}$$

The condition $V_2' = V_2 = 0$ is automatically satisfied from Eq. (2.12) and Eq. (2.8). For an impact of two perfectly smooth and elastic spheres, we have

$$\begin{aligned} V_1' &= V_1 \\ m_1(U_1'^2 + k_1^2\omega_1'^2) + m_2(U_2'^2 + k_2^2\omega_2'^2) &= m_1(U_1^2 + k_1^2\omega_1^2) \end{aligned} \tag{2.13}$$

It can be easily shown that Eq. (2.12) and Eq. (2.13) are equivalent when $e = 1$.

The collisional impulse can be obtained by using the impulse–momentum equations [Goldsmith, 1960]. However, this stereomechanical approach does not yield the transient stresses, collisional forces, impact duration, or collisional deformation of the colliding objects. These quantities have to be obtained from the analysis of stresses and strains of the solids due to the impact, which is introduced later.

2.3 Theory of Elastic Contact of Solids

During a collision, the colliding solids undergo both elastic and inelastic (or plastic) deformations. These deformations are caused by the changes of stresses and strains, which depend on the material properties of the solids and the applied external forces. Theories on the elastic deformations of two elastic bodies in contact are introduced in the literature utilizing Hertzian theory for frictionless contact and Mindlin's approach for frictional contact. As for inelastic deformations, few theories have been developed and the available ones are usually based on elastic contact theories. Hence, an introduction to the theories on elastic contact of solids is essential.

In this section, an introduction of the general relations of stresses in equilibrium in an infinitely large solid medium is presented, followed by a special application where a concentrated force is acting on a point inside the solid. Also presented is the case of forces on the boundary of a semiinfinite solid medium, which is of importance to the contact of two solid objects. As consequences of the boundary compression, displacements due to the changes of stresses and strains in the region of contact can be linked to the contact force by the Hertzian theory for frictionless contacts and by Mindlin's theory for frictional contacts. For more details on the Hertzian theory for contact, interested readers may refer to books on elasticity [Goldsmith, 1960; Timoshenko and Goodier, 1970; Landau and Lifshitz, 1970].

2.3.1 *General Relations of Stresses in a Solid Medium in Equilibrium*

For an element in equilibrium with no body forces, the equations of equilibrium were obtained by Lamé and Clapeyron (1831). Consider the stresses in a cubic element in equilibrium as shown in Fig. 2.3. Denote T_{ij} as a component of the stress tensor $\boldsymbol{T}$ acting on a plane whose normal is in the direction of $\boldsymbol{e}_i$ and the resulting force is in the direction of $\boldsymbol{e}_j$. In the Cartesian coordinates in Fig. 2.3, the total force on the pair of element surfaces whose normal vectors are in the direction of $\boldsymbol{e}_x$ can be given by

$$\boldsymbol{F}_x = \left(\frac{\partial T_{xx}}{\partial x} \boldsymbol{e}_x + \frac{\partial T_{xy}}{\partial x} \boldsymbol{e}_y + \frac{\partial T_{xz}}{\partial x} \boldsymbol{e}_z \right) \Delta x \, \Delta y \, \Delta z \tag{2.14}$$

Similarly, the total forces on the other two pairs of element surfaces are given by

$$\boldsymbol{F}_y = \left(\frac{\partial T_{yx}}{\partial y} \boldsymbol{e}_x + \frac{\partial T_{yy}}{\partial y} \boldsymbol{e}_y + \frac{\partial T_{yz}}{\partial y} \boldsymbol{e}_z \right) \Delta x \, \Delta y \, \Delta z \tag{2.15}$$

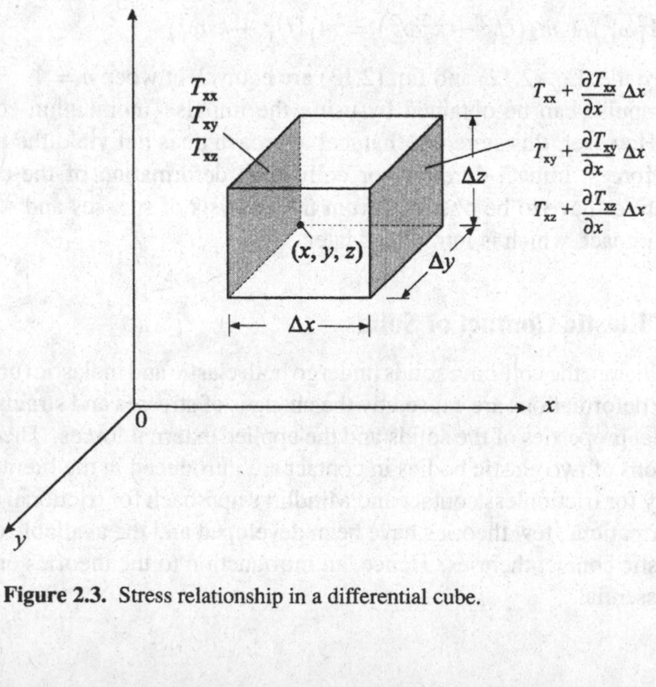

Figure 2.3. Stress relationship in a differential cube.

and

$$F_z = \left(\frac{\partial T_{zx}}{\partial z} e_x + \frac{\partial T_{zy}}{\partial z} e_y + \frac{\partial T_{zz}}{\partial z} e_z \right) \Delta x \, \Delta y \, \Delta z \tag{2.16}$$

From the condition of equilibrium, $\Sigma F_i = 0$, the equation of equilibrium is given as

$$\nabla \cdot T = 0 \tag{2.17}$$

For convenience in the following discussion, Eq. (2.17) is expressed in cylindrical coordinates. The symmetric stress tensor T is given as

$$T = \begin{bmatrix} \sigma_r & \tau_{r\theta} & \tau_{rz} \\ \tau_{r\theta} & \sigma_\theta & \tau_{\theta z} \\ \tau_{rz} & \tau_{\theta z} & \sigma_z \end{bmatrix} \tag{2.18}$$

where σ's are the normal stresses and τ's are the shear stresses. Thus, the equation of equilibrium takes the form

$$\begin{aligned} &\frac{\partial \sigma_r}{\partial r} + \frac{1}{r}\frac{\partial \tau_{r\theta}}{\partial \theta} + \frac{\partial \tau_{rz}}{\partial z} + \frac{\sigma_r - \sigma_\theta}{r} = 0 \\ &\frac{\partial \tau_{rz}}{\partial r} + \frac{1}{r}\frac{\partial \tau_{\theta z}}{\partial \theta} + \frac{\partial \sigma_z}{\partial z} + \frac{\tau_{rz}}{r} = 0 \\ &\frac{\partial \tau_{r\theta}}{\partial r} + \frac{1}{r}\frac{\partial \sigma_\theta}{\partial \theta} + \frac{\partial \tau_{\theta z}}{\partial z} + \frac{2\tau_{r\theta}}{r} = 0 \end{aligned} \tag{2.19}$$

The general relations between strains and stresses are represented by Hooke's law as

$$\begin{aligned} &\epsilon_r = \frac{1}{E}[\sigma_r - \nu(\sigma_\theta + \sigma_z)], \qquad \epsilon_\theta = \frac{1}{E}[\sigma_\theta - \nu(\sigma_r + \sigma_z)] \\ &\epsilon_z = \frac{1}{E}[\sigma_z - \nu(\sigma_\theta + \sigma_r)], \qquad \gamma_{rz} = \frac{2(1+\nu)}{E}\tau_{rz} \\ &\gamma_{r\theta} = \frac{2(1+\nu)}{E}\tau_{r\theta}, \qquad \gamma_{\theta z} = \frac{2(1+\nu)}{E}\tau_{\theta z} \end{aligned} \tag{2.20}$$

where ϵ's are the normal strains; γ's are the shear strains; E is Young's modulus of elasticity in tension and compression; and ν is Poisson's ratio. For most homogeneous solids, both E and ν are material properties whose values can be obtained from materials handbooks. Moreover, the displacements are related to the strains by

$$\begin{aligned} &\epsilon_r = \frac{\partial l_r}{\partial r}, \qquad \epsilon_\theta = \frac{l_r}{r} + \frac{1}{r}\frac{\partial l_\theta}{\partial \theta}, \qquad \epsilon_z = \frac{\partial l_z}{\partial z} \\ &\gamma_{rz} = \frac{\partial l_r}{\partial z} + \frac{\partial l_z}{\partial r}, \qquad \gamma_{r\theta} = \frac{1}{r}\frac{\partial l_r}{\partial \theta} + \frac{\partial l_\theta}{\partial r} - \frac{l_\theta}{r}, \qquad \gamma_{\theta z} = \frac{\partial l_\theta}{\partial z} + \frac{1}{r}\frac{\partial l_z}{\partial \theta} \end{aligned} \tag{2.21}$$

where l's are the displacements in the direction of the corresponding coordinates.

Most collisions with spherical particles can be regarded as axisymmetric, torsionless, and quasi-static. Therefore, Eq. (2.19) is simplified to

$$\begin{aligned} &\frac{\partial \sigma_r}{\partial r} + \frac{\partial \tau_{rz}}{\partial z} + \frac{\sigma_r - \sigma_\theta}{r} = 0 \\ &\frac{\partial \tau_{rz}}{\partial r} + \frac{\partial \sigma_z}{\partial z} + \frac{\tau_{rz}}{r} = 0 \end{aligned} \tag{2.22}$$

and the displacement–strain relationships become

$$\epsilon_{\mathrm{r}}=\frac{\partial l_{\mathrm{r}}}{\partial r},\qquad \epsilon_{\theta}=\frac{l_{\mathrm{r}}}{r},\qquad \epsilon_{\mathrm{z}}=\frac{\partial l_{\mathrm{z}}}{\partial z},\qquad \gamma_{\mathrm{rz}}=\frac{\partial l_{\mathrm{r}}}{\partial z}+\frac{\partial l_{\mathrm{z}}}{\partial r} \tag{2.23}$$

In order to solve Eq. (2.22), we introduce Love's stress function, ψ, which is defined [Love, 1944] by

$$\begin{aligned}\sigma_{\mathrm{r}}&=\frac{\partial}{\partial z}\left(\nu\nabla^2\psi-\frac{\partial^2\psi}{\partial r^2}\right),\qquad \sigma_{\theta}=\frac{\partial}{\partial z}\left(\nu\nabla^2\psi-\frac{1}{r}\frac{\partial\psi}{\partial r}\right)\\ \sigma_{\mathrm{z}}&=\frac{\partial}{\partial z}\left((2-\nu)\nabla^2\psi-\frac{\partial^2\psi}{\partial z^2}\right),\qquad \tau_{\mathrm{rz}}=\frac{\partial}{\partial r}\left((1-\nu)\nabla^2\psi-\frac{\partial^2\psi}{\partial z^2}\right)\end{aligned} \tag{2.24}$$

It can be shown that, with the substitution of Eq. (2.24), Eq. (2.22) becomes

$$\nabla^4\psi=0 \tag{2.25}$$

which is the equation of equilibrium of solids in terms of Love's stress function ψ in an axisymmetric and torsionless contact.

Hence, in principle, once ψ is found, all the stresses can be directly obtained from Eq. (2.24). Consequently, the corresponding strains and displacements can be evaluated from Eq. (2.20) and Eq. (2.23), respectively.

2.3.2 *Concentrated Force at a Point in an Infinite Solid Medium*

Consider the case of a concentrated force applied to a point in an infinite solid medium. To find the relationships between the point force and the resulting stresses, Love's stress function may be selected, from sets of solutions of Eq. (2.25), as [Timoshenko and Goodier, 1970]

$$\psi=A(r^2+z^2)^{\frac{1}{2}} \tag{2.26}$$

where A is a constant to be determined. By substituting this stress function into Eq. (2.24), the corresponding stresses are obtained as

$$\begin{aligned}\sigma_{\mathrm{r}}&=A\left[(1-2\nu)z(r^2+z^2)^{-\frac{3}{2}}-3r^2z(r^2+z^2)^{-\frac{5}{2}}\right]\\ \sigma_{\theta}&=A(1-2\nu)z(r^2+z^2)^{-\frac{3}{2}}\\ \sigma_{\mathrm{z}}&=-A\left[(1-2\nu)z(r^2+z^2)^{-\frac{3}{2}}+3z^3(r^2+z^2)^{-\frac{5}{2}}\right]\\ \tau_{\mathrm{rz}}&=-A\left[(1-2\nu)r(r^2+z^2)^{-\frac{3}{2}}+3rz^2(r^2+z^2)^{-\frac{5}{2}}\right]\end{aligned} \tag{2.27}$$

Note that all the stresses in Eq. (2.27) become singular at the origin where the point force is applied. To avoid this singularity, consider a small spherical cavity whose center is located at the origin of the coordinates as shown in Fig. 2.4. The coordinates are arranged such that the force is in the z-direction and is applied at the origin of the coordinates. Thus, the summation of the surface forces from the stresses in the direction of the z-axis balances the point force inside the solid medium.

Consider the ring-shaped element shown in Fig. 2.4. The z-component of the summation of surface stresses on the element, T_{z}, is given by

$$T_{\mathrm{z}}=-\tau_{\mathrm{rz}}\sin\phi-\sigma_{\mathrm{z}}\cos\phi=A\left(\frac{1-2\nu}{r^2+z^2}+\frac{3z^2}{(r^2+z^2)^2}\right) \tag{2.28}$$

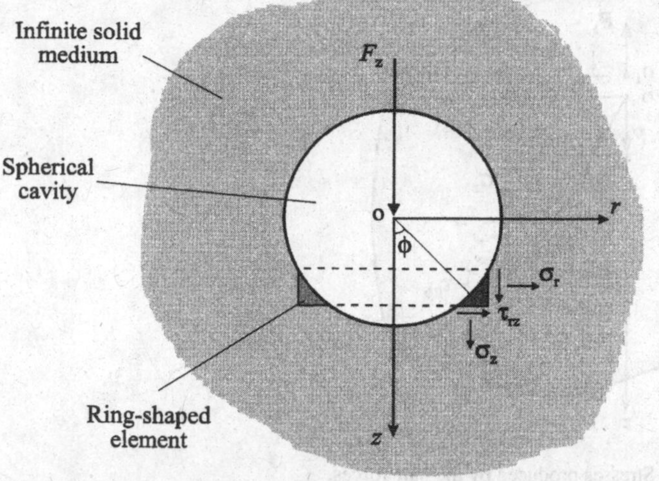

Figure 2.4. Surface stresses on a spherical cavity in an infinite solid medium.

Thus, the total surface force F_z is found by integrating the surface stresses over the entire surface as

$$F_z = 2\int_0^{\frac{\pi}{2}} 2\pi T_z r\sqrt{r^2+z^2}\,d\phi = 8\pi A(1-\nu) \tag{2.29}$$

Since F_z is equal to the applied point force, the constant A is obtained from the preceding equation as

$$A = \frac{F_z}{8\pi(1-\nu)} \tag{2.30}$$

The shear stress τ_{rz} on the plane $z = 0$ is then related to the point force F_z by Eq. (2.27) and Eq. (2.30) as

$$\tau_{rz} = -\frac{F_z(1-2\nu)}{8\pi(1-\nu)r^2} \tag{2.31}$$

2.3.3 *Force on the Boundary of a Semiinfinite Solid Medium*

In this section, the case of a semiinfinite solid with a concentrated force acting on the boundary is introduced. This case was originally solved by Boussinesq (1885). It should be noted that the only difference between this case and the case of a point force in an infinite solid medium is the boundary conditions. Shear stresses vanish on the boundary of the semiinfinite solid. In the following, the concept of a center of compression is introduced. The stress field in a semiinfinite solid with a boundary force can be obtained by superimposing the stress fields from a point force and a series of centers of compression. A center of compression is defined as the combination of three perpendicular pair forces.

2.3.3.1 *A Center of Compression*

Consider a case of two equal but opposite forces (in the z-direction) positioned a small distance δ apart in an infinite solid body, as shown in Fig. 2.5. At any point M, the stresses

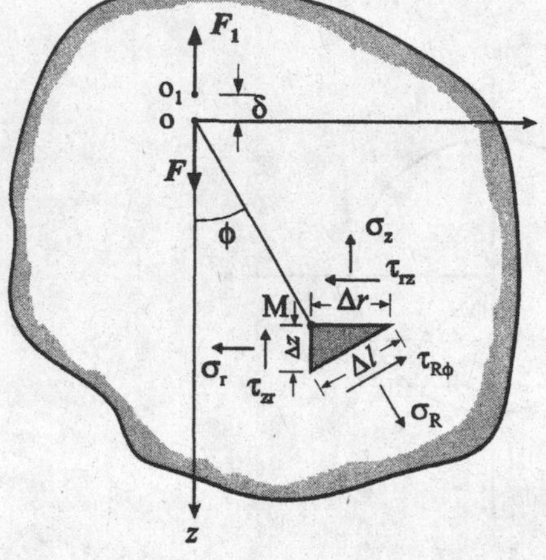

Figure 2.5. Stresses produced by the pair forces.

produced by force $\boldsymbol{F}$ at the origin may be given by Eq. (2.27). To arrive at an expression for the stresses produced by force $\boldsymbol{F}_1(\boldsymbol{F}_1 = -\boldsymbol{F})$ applied at O_1 in Fig. 2.5, noting that it is acting in the opposite direction and considering that δ is infinitely small, any term in Eq. (2.27) expressed in a general function $\Phi(r, z)$ can be replaced by $[-\Phi - (\partial\Phi/\partial z)\delta]$. The stresses produced by the pair forces can thus be given as

$$\begin{aligned}
\sigma_r &= -B\frac{\partial}{\partial z}\left[(1-2\nu)z(r^2+z^2)^{-\frac{3}{2}} - 3r^2z(r^2+z^2)^{-\frac{5}{2}}\right] \\
\sigma_\theta &= -B\frac{\partial}{\partial z}\left[(1-2\nu)z(r^2+z^2)^{-\frac{3}{2}}\right] \\
\sigma_z &= B\frac{\partial}{\partial z}\left[(1-2\nu)z(r^2+z^2)^{-\frac{3}{2}} + 3z^3(r^2+z^2)^{-\frac{5}{2}}\right] \\
\tau_{rz} &= B\frac{\partial}{\partial z}\left[(1-2\nu)r(r^2+z^2)^{-\frac{3}{2}} + 3rz^2(r^2+z^2)^{-\frac{5}{2}}\right]
\end{aligned} \tag{2.32}$$

where $B = A\delta$. Once in equilibrium, the relations of stresses on a triangular element at point M (shown in Fig. 2.5) can be expressed as

$$\begin{aligned}
\sigma_R \Delta l &= \sigma_r \Delta z \sin\phi + \sigma_z \Delta r \cos\phi + \tau_{rz}\Delta z \cos\phi + \tau_{rz}\Delta r \sin\phi \\
\tau_{R\phi}\Delta l &= \sigma_r \Delta z \cos\phi - \sigma_z \Delta r \sin\phi - \tau_{rz}\Delta z \sin\phi + \tau_{rz}\Delta r \cos\phi
\end{aligned} \tag{2.33}$$

or

$$\begin{aligned}
\sigma_R &= \sigma_r \sin^2\phi + \sigma_z \cos^2\phi + 2\tau_{rz}\cos\phi\sin\phi \\
\tau_{R\phi} &= (\sigma_r - \sigma_z)\sin\phi\cos\phi - \tau_{rz}(\sin^2\phi - \cos^2\phi)
\end{aligned} \tag{2.34}$$

Substituting Eq. (2.32) into Eq. (2.34) yields

$$\begin{aligned}
\sigma_R &= -\frac{2(1+\nu)B}{R^3}\left(-\sin^2\phi + \frac{2(2-\nu)}{1+\nu}\cos^2\phi\right) \\
\tau_{R\phi} &= -\frac{2(1+\nu)B}{R^3}\sin\phi\cos\phi
\end{aligned} \tag{2.35}$$

Thus, the stresses at any point due to the pair forces along the z-direction are given by Eq. (2.35). If the pair forces are relocated along the r-direction but still applied near the origin, the resulting stresses at any point M can be obtained by replacing ϕ by $(\phi - \pi/2)$ in Eq. (2.35) as

$$\sigma_R = -\frac{2(1+\nu)B}{R^3}\left(-\cos^2\phi + \frac{2(2-\nu)}{1+\nu}\sin^2\phi\right)$$
$$\tau_{R\phi} = \frac{2(1+\nu)B}{R^3}\sin\phi\cos\phi \tag{2.36}$$

Similarly, if the pair forces are relocated along the θ-direction and still applied near the origin, the stresses at M are obtained by setting $\phi = \pi/2$ in Eq. (2.35) as

$$\sigma_R = \frac{2(1+\nu)B}{R^3}; \qquad \tau_{R\phi} = 0 \tag{2.37}$$

Hence, by combining Eqs. (2.35), (2.36), and (2.37), the total stresses at M due to the three perpendicular pair forces are

$$\sigma_R = \frac{4(2\nu-1)B}{R^3}; \qquad \tau_{R\phi} = 0 \tag{2.38}$$

which indicates that the compression stress in the radial direction depends only on the distance from the center of compression and is inversely proportional to the cube of that distance.

To obtain an expression for the normal stress σ_ϕ in the tangential direction, consider the equilibrium of an element cut from a sphere by two concentric spherical surfaces of radii R and $R + dR$ and by a circular cone with a small angle $d\phi$, as shown in Fig. 2.6. The force balance along the radial direction gives

$$2\sigma_\phi\frac{\pi}{4}(d\phi)^2 R\,dR = \left(\sigma_R + \frac{d\sigma_R}{dR}dR\right)(R+dR)^2\frac{\pi}{4}(d\phi)^2 - \sigma_R R^2\frac{\pi}{4}(d\phi)^2 \tag{2.39}$$

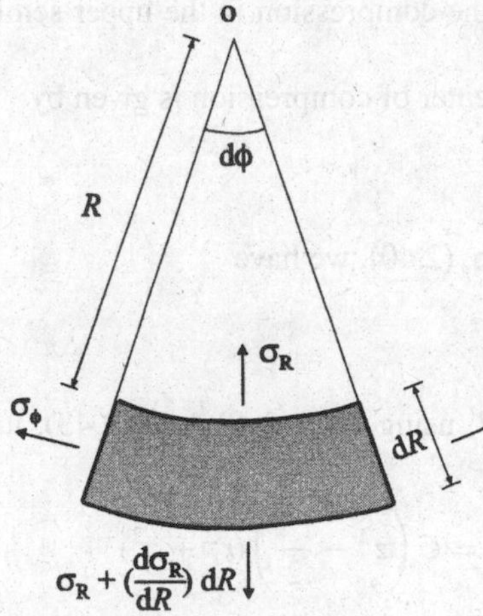

Figure 2.6. Equilibrium of an element in spherical coordinates.

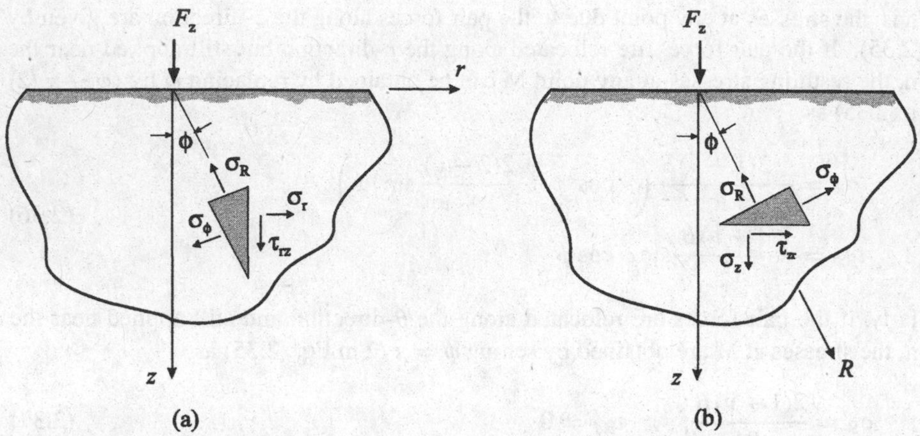

Figure 2.7. A point force on the boundary of a semiinfinite medium: (a) $r - z$ coordinates; (b) $R - \phi$ coordinates.

which yields

$$\sigma_\phi = \frac{R}{2}\frac{d\sigma_R}{dR} + \sigma_R \tag{2.40}$$

2.3.3.2 *Stresses Induced by a Boundary Point Force*

Consider the case where a point force is acting on the boundary of a semiinfinite body, as shown in Fig. 2.7. Let the boundary be in the plane $z = 0$. Then, the shear stress on the boundary would be given by Eq. (2.27) as

$$\tau_{rz} = -\frac{A(1-2\nu)}{r^2} \tag{2.41}$$

where A is a constant to be determined. Since shear stresses on the boundary cannot be produced by the concentrated force acting on the boundary, the shearing force in Eq. (2.41) has to be balanced by the stress resulting from the compression in the upper semiinfinite body.

The stress distribution corresponding to the center of compression is given by

$$\sigma_R = \frac{C}{R^3} \tag{2.42}$$

where C is a constant to be determined. From Eq. (2.40), we have

$$\sigma_\phi = \frac{d\sigma_R}{dR}\frac{R}{2} + \sigma_R = -\frac{1}{2}\frac{C}{R^3} \tag{2.43}$$

From the equilibrium of the element in Fig. 2.7, using Eqs. (2.42) and (2.43), it can be obtained that

$$\begin{aligned} \sigma_r &= C\left(r^2 - \frac{z^2}{2}\right)(r^2+z^2)^{-\frac{5}{2}}, \qquad \sigma_z = C\left(z^2 - \frac{r^2}{2}\right)(r^2+z^2)^{-\frac{5}{2}} \\ \tau_{rz} &= \frac{3}{2}Crz(r^2+z^2)^{-\frac{5}{2}}, \qquad \sigma_\theta = -\frac{1}{2}C(r^2+z^2)^{-\frac{3}{2}} \end{aligned} \tag{2.44}$$

Assume that the centers of pressure are uniformly distributed along the z-axis from $z = 0$ to $z = -\infty$. The stresses in an infinite solid medium can be obtained by integration from z to ∞ as

$$\sigma_r = \frac{C}{2r^2}\left[1 - z(r^2+z^2)^{-\frac{1}{2}} - zr^2(r^2+z^2)^{-\frac{3}{2}}\right], \qquad \sigma_z = \frac{C}{2}z(r^2+z^2)^{-\frac{3}{2}}$$

$$\tau_{rz} = \frac{C}{2}r(r^2+z^2)^{-\frac{3}{2}}, \qquad \sigma_\theta = -\frac{C}{2r^2}\left[1 - z(r^2+z^2)^{-\frac{1}{2}}\right] \tag{2.45}$$

Hence, on the plane $z = 0$, stresses are given by

$$\sigma_z = 0, \qquad \tau_{rz} = \frac{C}{2r^2} \tag{2.46}$$

To ensure that the boundary surface is free from stresses, the constants A and C in Eq. (2.41) and Eq. (2.46) can be selected such that the shearing forces on the boundary are eliminated. Thus, we have

$$C = 2A(1-2\nu) \tag{2.47}$$

The stresses in a semiinfinite body can be obtained by combining Eq. (2.27) with Eq. (2.45) as

$$\sigma_r = A\left\{\frac{1-2\nu}{r^2}\left[1 - z(r^2+z^2)^{-\frac{1}{2}}\right] - 3r^2z(r^2+z^2)^{-\frac{5}{2}}\right\}$$

$$\sigma_\theta = A\frac{1-2\nu}{r^2}\left[z(r^2+z^2)^{-\frac{1}{2}} - 1 + zr^2(r^2+z^2)^{-\frac{3}{2}}\right] \tag{2.48}$$

$$\sigma_z = -3Az^3(r^2+z^2)^{-\frac{5}{2}}, \qquad \tau_{rz} = -3Arz^2(r^2+z^2)^{-\frac{5}{2}}$$

To determine A, consider the equilibrium of the ring-shaped element shown in Fig. 2.4. The force on the element in the z-direction is

$$T_z = -(\tau_{rz}\sin\phi + \sigma_z\cos\phi) = 3Az^2(r^2+z^2)^{-2} \tag{2.49}$$

The balance between the point force and the integration of T_z over the hemispherical surface yields

$$A = \frac{F_z}{2\pi} \tag{2.50}$$

Therefore, with the substitution of Eq. (2.50) into Eq. (2.48), the stresses due to a normal force on the boundary of a semiinfinite solid are expressed in cylindrical coordinates by

$$\sigma_r = \frac{F_z}{2\pi}\left\{\frac{1-2\nu}{r^2}\left[1 - z(r^2+z^2)^{-\frac{1}{2}}\right] - 3r^2z(r^2+z^2)^{-\frac{5}{2}}\right\}$$

$$\sigma_\theta = \frac{F_z}{2\pi}\frac{1-2\nu}{r^2}\left[z(r^2+z^2)^{-\frac{1}{2}} - 1 + zr^2(r^2+z^2)^{-\frac{3}{2}}\right] \tag{2.51}$$

$$\sigma_z = -\frac{3F_z}{2\pi}z^3(r^2+z^2)^{-\frac{5}{2}}, \qquad \tau_{rz} = -\frac{3F_z}{2\pi}rz^2(r^2+z^2)^{-\frac{5}{2}}$$

2.3.3.3 *Displacements Induced by a Boundary Point Force*

As a result of applying a point force on the boundary, displacements in the semiinfinite solid are produced. These displacements can be calculated from Hooke's law and the displacement–strain relationships. The displacement in the r-direction, l_r, is given by Eq. (2.20) and Eq. (2.23) as

$$\begin{aligned} l_r = \epsilon_\theta r &= \frac{r}{E}[\sigma_\theta - \nu(\sigma_r + \sigma_z)] \\ &= \frac{(1-2\nu)(1+\nu)F_z}{2\pi E r}\left[z(r^2+z^2)^{-\frac{1}{2}} - 1 + \frac{r^2 z}{1-2\nu}(r^2+z^2)^{-\frac{3}{2}}\right] \end{aligned} \tag{2.52}$$

Similarly, the displacement in the z-direction, l_z, is given by

$$\begin{aligned} \frac{\partial l_z}{\partial z} &= \epsilon_z = \frac{1}{E}[\sigma_z - \nu(\sigma_r + \sigma_\theta)] \\ \frac{\partial l_z}{\partial r} &= \gamma_{rz} - \frac{\partial l_r}{\partial z} = \frac{2(1+\nu)}{E}\tau_{rz} - \frac{\partial l_r}{\partial z} \end{aligned} \tag{2.53}$$

which yields

$$l_z = \frac{F_z}{2\pi E}\left[(1+\nu)z^2(r^2+z^2)^{-\frac{3}{2}} + 2(1-\nu^2)(r^2+z^2)^{-\frac{1}{2}}\right] \tag{2.54}$$

where the integration constant is related to an axial rigid-body translation and is set to zero here for simplicity. Thus, the displacements on the boundary $z = 0$ are expressed as

$$l_{r0} = -\frac{(1-2\nu)(1+\nu)F_z}{2\pi E r}, \qquad l_{z0} = \frac{(1-\nu^2)F_z}{\pi E r} \tag{2.55}$$

The displacements given in Eq. (2.55) for a point force on a semiinfinite solid indicate that the displacements would be infinite when r approaches zero at the origin where the point force is applied. This singularity may be avoided by using a load distribution over a finite area to replace the point force.

Consider a pressure distribution p on a circular area A of radius r_c such that

$$\iint_A p \, \mathrm{d}A = F_z \tag{2.56}$$

The influence of this pressure distribution on the displacements in a neighboring area on the same boundary surface can be analyzed as shown in Fig. 2.8. Let M be an arbitrary point outside the circular area (Fig. 2.8(a)) or inside the circular area (Fig. 2.8(b)) with a distance r from the center of the circular area. A small element within the loaded zone is chosen as $s \, \mathrm{d}\beta \, \mathrm{d}s$, where s is the distance between this element and M. From the solution of l_z for a point force on a semiinfinite solid, the increment of the vertical displacement at location M under the influence of this element force would be

$$\mathrm{d}l_{z0} = \frac{(1-\nu^2)}{\pi E s} p s \, \mathrm{d}\beta \, \mathrm{d}s \tag{2.57}$$

Thus, the total vertical displacement on the boundary surface can be obtained by integrating over the whole circular area, which gives

$$l_{z0} = \frac{(1-\nu^2)}{\pi E} \iint p \, \mathrm{d}\beta \, \mathrm{d}s \tag{2.58}$$

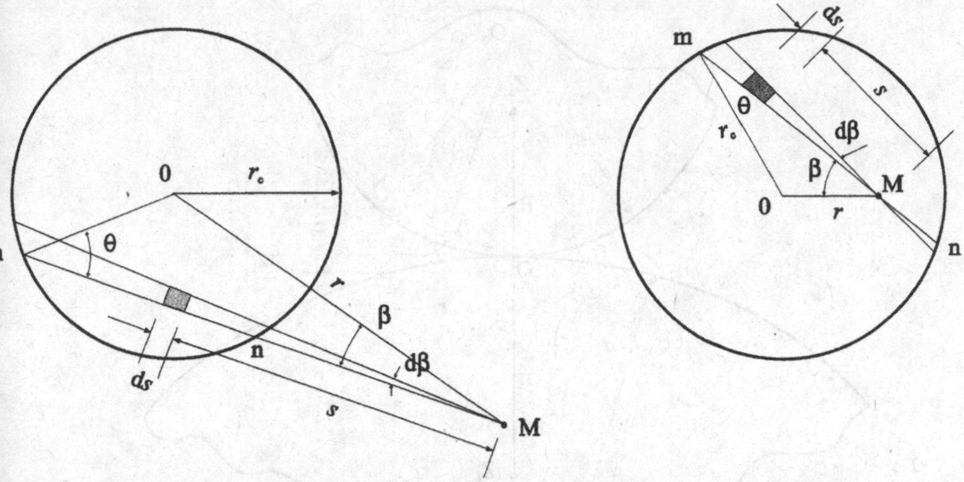

(a) (b)

Figure 2.8. Displacement under compression: (a) M is outside the compressional circle; (b) M is inside the compressional circle.

Similarly, the total radial displacement on the boundary surface is given by

$$l_{r0} = \frac{(1-2\nu)(1+\nu)}{\pi E} \iint p \, d\beta \, ds \tag{2.59}$$

Assuming uniform distribution of the pressure p, it can be shown that the maximum displacement at the center of the circle is $\pi/2$ times the displacement at the edge of the circle.

2.3.4 *Hertzian Theory for Frictionless Spheres in Contact*

When two elastic and frictionless spheres are brought into contact under compressional forces or pressures, deformation occurs. The maximum displacement and contact area depend not only on the compressional force but also on the elastic material properties and radii of the spheres. The contact between two elastic and frictionless spherical bodies under compression was first investigated by Hertz (1881) and is known as the Hertzian contact.

Consider two frictionless spheres in contact as shown in Fig. 2.9. M is a point on the surface of sphere 1 with a distance r from the Z_1-axis. N is the opposite point on the surface of sphere 2 with the same distance r from the Z_2-axis. O is the contact point and also serves as the origin of both Z_1-axis and Z_2-axis. N, M, and O are in the same plane. The distances from M or N to the tangential plane which is normal to the Z_1-axis and Z_2-axis are denoted by z_1 and z_2, respectively. In Fig. 2.9, the triangle O_2AO is similar to the triangle NBO. Hence, the theorem of similarity of triangles gives

$$\frac{2z_2}{\sqrt{z_2^2+r^2}} = \frac{\sqrt{z_2^2+r^2}}{a_2} \tag{2.60}$$

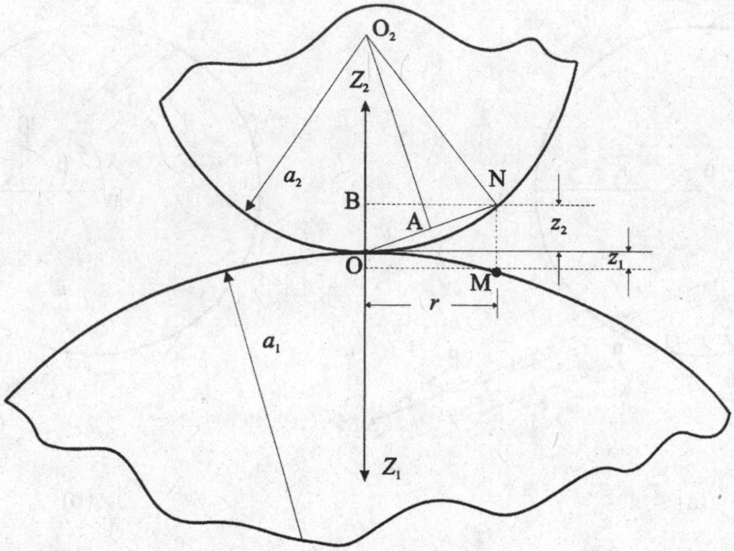

Figure 2.9. Hertzian contact.

which yields

$$z_2 = a_2\left[\frac{1}{2}\left(\frac{r^2}{a_2^2}\right) + O\left(\frac{r^4}{a_2^4}\right)\right] \tag{2.61}$$

When r/a_1 and r/a_2 are small enough, it can be obtained that

$$z_1 + z_2 = \frac{a_1 + a_2}{2a_1a_2}r^2 \tag{2.62}$$

During contact, location M tends to be moved in the direction of Z_1 by a vertical displacement l_{z01}. Similarly, N is going to be moved by a distance l_{z02} in the direction of Z_2. When the center of sphere 1 approaches the center of sphere 2 by a distance α (note that α here is an approaching distance rather than the distance between the two centers of the spheres), the change of distance between M and N is then given by $\alpha - (l_{z01} + l_{z02})$. Consider the case of deformation in which M and N are on the edge of the surface of contact as a result of local compression. We have

$$\alpha - (l_{z01} + l_{z02}) = z_1 + z_2 \tag{2.63}$$

where the vertical displacements l_{z01} and l_{z02} are given by Eq. (2.58) as

$$l_{z01} = \frac{(1 - \nu_1^2)}{\pi E_1}\iint p\,\mathrm{d}s\,\mathrm{d}\beta, \qquad l_{z02} = \frac{(1 - \nu_2^2)}{\pi E_2}\iint p\,\mathrm{d}s\,\mathrm{d}\beta \tag{2.64}$$

Hence, it gives

$$\frac{1}{\pi E^*}\iint p\,\mathrm{d}s\,\mathrm{d}\beta = \alpha - \frac{r^2}{2a^*} \tag{2.65}$$

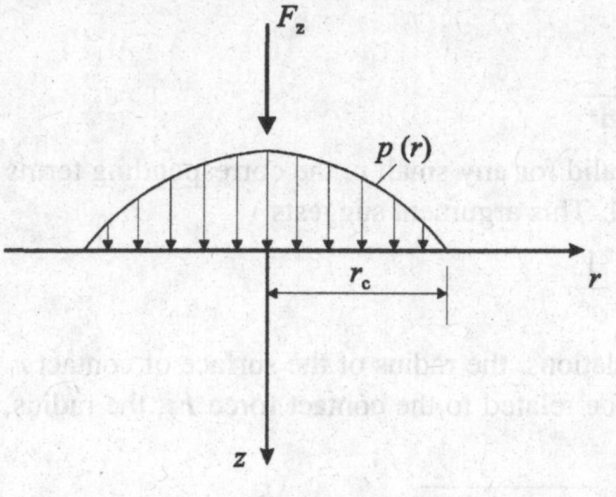

Figure 2.10. Hertz pressure distribution for frictionless contact.

where E^* is the contact modulus and a^* is the relative radius defined as

$$\frac{1}{E^*} = \frac{1-\nu_1^2}{E_1} + \frac{1-\nu_2^2}{E_2}; \qquad \frac{1}{a^*} = \frac{1}{a_1} + \frac{1}{a_2} \tag{2.66}$$

It is necessary to find an appropriate form for the pressure distribution p so that Eq. (2.65) can be satisfied for any r provided that r is small compared to the radii of spheres. One such pressure distribution is the Hertz pressure distribution in which the pressure at any location within the contact area is represented, as illustrated in Fig. 2.10, by

$$p(r) = C\left(r_c^2 - r^2\right)^{\frac{1}{2}} \tag{2.67}$$

It is noted that the total loading force F_z is equal to the pressure integrated over the contact area so that the constant C can be expressed by

$$C = \frac{3}{2}\frac{F_z}{\pi r_c^3} \tag{2.68}$$

Hence, the pressure distribution becomes

$$p(r) = \frac{3}{2}\frac{F_z}{\pi r_c^3}\left(r_c^2 - r^2\right)^{\frac{1}{2}} \tag{2.69}$$

It is shown in Eq. (2.69) that the maximum pressure p_o occurs at the center of the contact area and is given by

$$p_o = \frac{3}{2}\frac{F_z}{\pi r_c^2} \tag{2.70}$$

which is about 1.5 times the averaged pressure on the contact surface.

Using the Hertz pressure distribution, an integration along the cord mn in Fig. 2.8(b) gives

$$\int p\,\mathrm{d}s = \frac{p_o \pi}{2r_c}\left(r_c^2 - r^2\sin^2\beta\right) \tag{2.71}$$

Substituting Eq. (2.71) into Eq. (2.65) yields

$$\frac{p_o}{r_c E^*}\int_0^{\frac{\pi}{2}}\left(r_c^2 - r^2\sin^2\beta\right)\mathrm{d}\beta = \alpha - \frac{r^2}{2a^*} \tag{2.72}$$

Thus, we have

$$\frac{3F_z}{8E^*r_c^3}(2r_c^2 - r^2) = \alpha - \frac{r^2}{2a^*} \tag{2.73}$$

Since the preceding equation is valid for any small r, the corresponding terms on both sides of the equation should be equal. This argument suggests

$$\alpha = \frac{3F_z}{4E^*r_c}; \qquad \frac{3F_z}{4E^*r_c^3} = \frac{1}{a^*} \tag{2.74}$$

By rearranging the preceding two relations, the radius of the surface of contact r_c and the centers approaching distance α can be related to the contact force F_z, the radius, and the material properties of the spheres as

$$r_c = \sqrt[3]{\frac{3}{4}F_z\frac{a^*}{E^*}}; \qquad \alpha = \sqrt[3]{\left(\frac{3}{4}\frac{F_z}{E^*}\right)^2\frac{1}{a^*}} \tag{2.75}$$

The corresponding maximum pressure p_o at the center of the contact is thus obtained by substituting Eq. (2.75) into Eq. (2.70) as

$$p_o = \frac{3}{2\pi}\left(\frac{16}{9}F_z\frac{E^{*2}}{a^{*2}}\right)^{\frac{1}{3}} \tag{2.76}$$

Therefore, for Hertzian contact, using Eq. (2.75) and Eq. (2.76), the maximum radius of contact r_c, the maximum approaching distance α, and the corresponding maximum pressure can be calculated on the basis of the contact force, the elastic material properties of the spheres, and the radii of the spheres.

Example 2.1 Figure E2.1(a) shows a packed bed of height 1 m with monodispersed glass beads (silica glass) in a large container. Estimate the elastic displacement and contact area of a particle at the bottom of the bed. The density of the glass beads is 2,500 kg/m^3. The particle diameter is 1 cm.

Solution For a packed bed with monodispersed spheres in a large container, the volume fraction for solids is about 0.45. Thus, for particles under the bed height of 1 m, the average pressure caused by gravity is

$$p = \alpha_p \rho_p g H = 0.45 \times 2{,}500 \times 9.8 \times 1 = 1.1 \times 10^4 \text{ N/m}^2 \tag{E2.1}$$

As shown in Fig. E2.1(b), the effective area of a single particle in a horizontal layer is equivalent to that of an equal-sized triangle with side length of one particle diameter d_p. Thus, the vertical loading on a single sphere under the bed height of 1 m is given by

$$F = \frac{\sqrt{3}}{4}d_p^2 p = \frac{\sqrt{3}}{4} \times 10^{-4} \times 1.1 \times 10^4 = 0.48 \text{ N} \tag{E2.2}$$

Assume that the particles in consideration are not located in the bottom layer so that the contact arrangement among spheres may be represented as in Fig. E2.1(c). Hence, the compressional force between two neighboring spheres is given by

$$F_z = \frac{F}{\sqrt{6}} = \frac{0.48}{\sqrt{6}} = 0.20 \text{ N} \tag{E2.3}$$

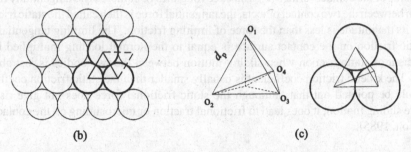

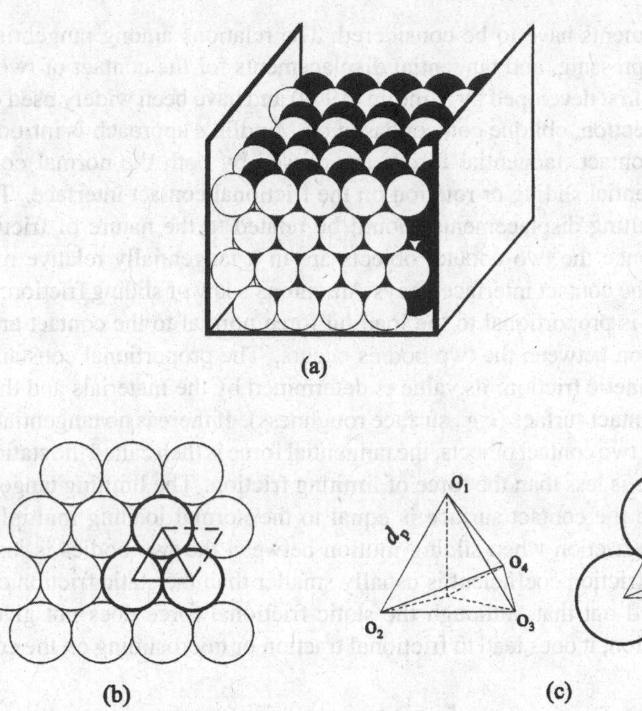

Figure E2.1. Particle contact arrangement in a packed bed: (a) A packed bed with monodispersed spheres; (b) Top view of a horizontal layer of monodispersed spheres; (c) Local contact mode of spheres.

The values for Young's modulus and Poisson's ratio for silica glass are found from Table 1.6 as

$$E = 8 \times 10^{10}\ \text{N/m}^2; \qquad \nu = 0.24 \tag{E2.4}$$

Hence, from Eq. (2.75), the radius of the contact circle of two contacting glass beads is

$$r_c = \left(\frac{3}{8}\frac{F_z d_p}{E}(1-\nu^2)\right)^{\frac{1}{3}} = \left(\frac{3}{8}\frac{0.2 \times 0.01}{8 \times 10^{10}}(1 - 0.24^2)\right)^{\frac{1}{3}} = 2 \times 10^{-5}\ \text{m} \tag{E2.5}$$

Note that the elastic displacement of a single sphere equals half of the total displacement of two contacting spheres. Therefore, from Eq. (2.74), we have

$$l_z = \frac{\alpha}{2} = \frac{3}{4}\frac{F_z(1-\nu^2)}{E r_c} = \frac{3}{4}\frac{0.2 \times (1 - 0.24^2)}{8 \times 10^{10} \times 2 \times 10^{-5}} = 8.8 \times 10^{-8}\ \text{m} \tag{E2.6}$$

2.3.5 *Theories for Frictional Spheres in Contact*

In the previous sections, only the normal contact of two elastic spheres with perfect smooth surfaces (*i.e.*, no tangential force) is considered. However, for oblique contact between two frictional spheres, tangential forces are encountered, and, consequently, tangential displacements have to be considered. The relations among tangential traction, normal traction or pressure, and tangential displacements for the contact of two frictional elastic bodies were first developed by Mindlin (1949) and have been widely used ever since. Therefore, in this section, oblique contact based on Mindlin's approach is introduced.

In an oblique contact, tangential forces are caused by both the normal compression and either the tangential sliding or rotation on the frictional contact interface. Thus, these forces and the resulting displacements should be related to the nature of friction on the contact surface. Once the two contact objects are in a tangentially relative motion, the frictional force on the contact interface obeys Amontons's law of sliding friction. That says the frictional force is proportional to the loading force normal to the contact area when a steady sliding motion between the two bodies occurs. The proportional constant is called the coefficient of kinetic friction; its value is determined by the materials and the physical conditions of the contact surface (*e.g.*, surface roughness). If there is no tangentially relative motion between the two contact objects, the tangential force is then called the static frictional force; its magnitude is less than the force of limiting friction. The limiting tangential force of static friction on the contact surface is equal to the normal loading multiplied by the coefficient of static friction when sliding motion between the two bodies is just about to occur. The kinetic friction coefficient is usually smaller than the static friction coefficient. It should be pointed out that, although the static frictional force does not give rise to a relative sliding motion, it does lead to frictional traction or microsliding on the contact area [Johnson, 1985].

2.3.5.1 *Sliding Contact of Spheres*

Assume that Amontons's law of sliding friction can be applied at each elementary area of the interface, such that

$$\frac{q(x,y)}{p(x,y)} = f_{\mathrm{k}} \tag{2.77}$$

where q is the tangential traction; p is the normal pressure; and f_{k} is the coefficient of kinetic friction, whose value depends on the materials and physical conditions of the contact interface only. Thus, the total tangential force F_{t} becomes

$$F_{\mathrm{t}} = f_{\mathrm{k}} F_{\mathrm{z}} \tag{2.78}$$

Assume that the normal pressure and displacements are not affected by the existence of the tangential traction and resulting displacements. Hence, the normal pressure and contact area can be determined by the Hertzian theory. For the sliding contact of spheres, substituting Eq. (2.69) into Eq. (2.77) gives rise to the tangential traction as

$$q(r) = \frac{3}{2}\frac{f_{\mathrm{k}} F_{\mathrm{z}}}{\pi r_{\mathrm{c}}^2}\left(1 - \frac{r^2}{r_{\mathrm{c}}^2}\right)^{\frac{1}{2}} \tag{2.79}$$

With the tangential traction parallel to the x-axis, the general displacements on the contact surface due to the tangential traction can be expressed by

$$\begin{aligned} l_{\mathrm{x}} &= \frac{1}{2\pi G}\iint_A q(\xi,\eta)\frac{(\xi - x)^2 + (1-\nu)(\eta - y)^2}{s^3}\,\mathrm{d}\xi\,\mathrm{d}\eta \\ l_{\mathrm{y}} &= \frac{1}{2\pi G}\iint_A q(\xi,\eta)\frac{\nu(\xi - x)(\eta - y)}{s^3}\,\mathrm{d}\xi\,\mathrm{d}\eta \end{aligned} \tag{2.80}$$

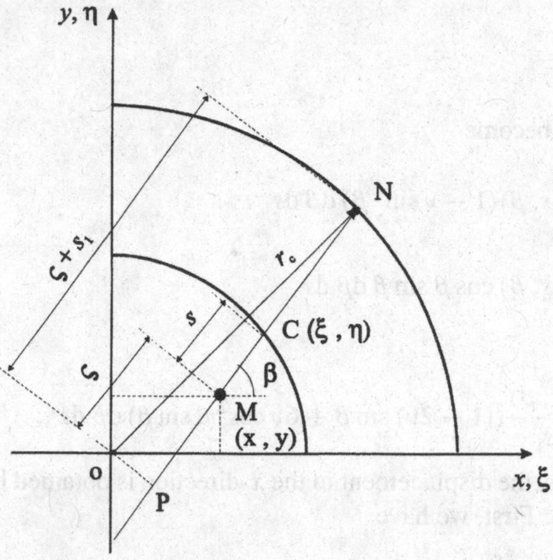

Figure 2.11. Coordinates conversion between (ξ, η) and (s, β).

where $s^2 = (\xi - x)^2 + (\eta - y)^2$, and G is the shear modulus, defined by $G = E/2(1+\nu)$. The corresponding surface shear stress τ_{xy} is

$$\tau_{xy} = \frac{1}{2\pi}\iint_A q(\xi, \eta)\left(\frac{(1-2\nu)(\eta - y)}{s^3} + \frac{6\nu(\xi - x)^2(\eta - y)}{s^5}\right) d\xi\, d\eta \qquad (2.81)$$

The detailed derivation of Eq. (2.80) and Eq. (2.81) is omitted here. Interested readers may refer to the work of Johnson (1985).

In order to perform the surface integration of Eq. (2.80) and Eq. (2.81), it is convenient to change the coordinates from (ξ, η) to (s, β), as shown in Fig. 2.11. Consider the displacements and stresses at a point M(x, y) when an element of tangential traction is applied at C(ξ, η). From the conversion of the coordinates, we have

$$\begin{aligned} \xi &= x + s\cos\beta \\ \eta &= y + s\sin\beta \end{aligned} \qquad (2.82)$$

By considering only the surface displacement within the loading circle of radius r_c, the limit of s should be on the edge of the circle. For a fixed β, this limiting point is represented by N in Fig. 2.11. Denote s_1 as the limiting length of s from M to N and ς as the distance between P and M. Thus, we have

$$\varsigma = x\cos\beta + y\sin\beta \qquad (2.83)$$

and

$$r_c^2 - (s_1 + \varsigma)^2 = x^2 + y^2 - \varsigma^2 \qquad (2.84)$$

which yields

$$s_1 = -\varsigma + \sqrt{\alpha^2 + \varsigma^2} \qquad (2.85)$$

where

$$\alpha = \sqrt{r_c^2 - (x^2 + y^2)} \tag{2.86}$$

Thus, Eq. (2.80) and Eq. (2.81) become

$$\begin{aligned} l_x &= \frac{1}{2\pi G}\int_0^{2\pi}\int_0^{s_1} q(s,\beta)(1 - \nu\sin^2\beta)\,d\beta\,ds \\ l_y &= \frac{\nu}{2\pi G}\int_0^{2\pi}\int_0^{s_1} q(s,\beta)\cos\beta\sin\beta\,d\beta\,ds \end{aligned} \tag{2.87}$$

and

$$\tau_{xy} = \frac{1}{2\pi}\int_0^{2\pi}\int_0^{s_1}\frac{q(s,\beta)}{s}((1-2\nu)\sin\beta + 6\nu\cos^2\beta\sin\beta)\,d\beta\,ds \tag{2.88}$$

For sliding contact of spheres, the displacement in the x-direction is obtained by substituting Eq. (2.79) into Eq. (2.87). First, we have

$$\begin{aligned} \int_0^{s_1} q(s,\beta)\,ds &= \frac{3}{2}\frac{f_k F_z}{\pi r_c^3}\int_0^{s_1}(\alpha^2 - 2\varsigma s - s^2)^{\frac{1}{2}}\,ds \\ &= \frac{3}{2}\frac{f_k F_z}{\pi r_c^3}\left[\frac{(\alpha^2+\varsigma^2)}{2}\left(\frac{\pi}{2} - \tan^{-1}\left(\frac{\varsigma}{\alpha}\right)\right) - \frac{1}{2}\alpha\varsigma\right] \end{aligned} \tag{2.89}$$

Note that, from Eq. (2.83), $\varsigma(\beta) = -\varsigma(\beta+\pi)$. Substituting Eq. (2.89) into Eq. (2.87) gives rise to

$$\begin{aligned} l_x &= \frac{3}{16\pi}\frac{f_k F_z}{r_c^3 G}\int_0^{2\pi}(\alpha^2+\varsigma^2)(1-\nu\sin^2\beta)\,d\beta \\ &= \frac{3}{64}\frac{f_k F_z}{r_c^3 G}(4(2-\nu)r_c^2 - (4-3\nu)x^2 - (4-\nu)y^2) \end{aligned} \tag{2.90}$$

Similarly, the displacement in the y-direction is

$$l_y = \frac{3}{32}\frac{f_k F_z}{r_c^3 G}\nu xy \tag{2.91}$$

It should be noted that the displacement in the z-direction due to tangential traction is usually not zero at the interface. Therefore, the actual interface with tangential force mismatches the flat surface caused by the normal loading [Johnson, 1985]. In this text, it is assumed that this mismatch of normal displacements can be neglected; this turns out to be an acceptable approximation for most cases.

2.3.5.2 *Nonsliding Contact of Spheres*

When a tangential force is less than the limiting tangential force of "static friction," no sliding motion is produced. However, tangential traction will still occur in this case. Consider the general case illustrated in Fig. 2.12. A_1 and A_2 represent two points on the contact surface which were coincident before a tangential force F_t is applied. With the action of F_t, A_1 and A_2 are shifted by tangential displacements l_{x1} and l_{x2} relative to the distant points B_1 and B_2 in the undeformed region of each sphere, and B_1 and B_2 are moved by rigid tangential displacements δ_{x1} and δ_{x2}, respectively. Denoting s_{x1} and s_{x2} as the absolute

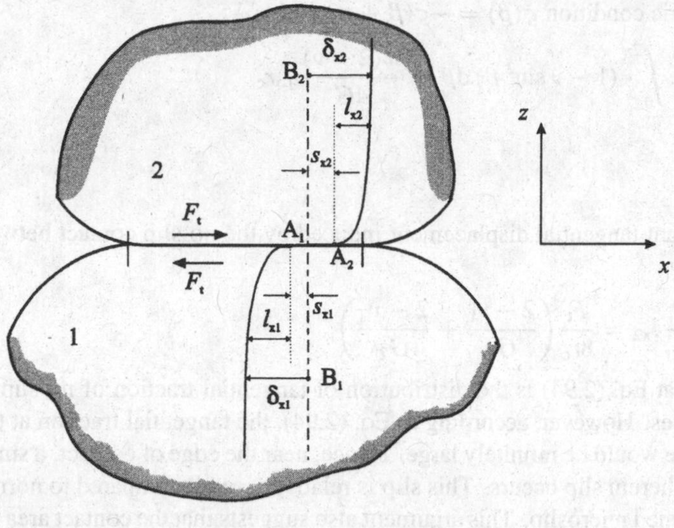

Figure 2.12. Tangential tractions in nonsliding contact.

tangential displacements of A_1 and A_2, respectively, the relative tangential slip s_x between A_1 and A_2 can be expressed by

$$s_x = s_{x1} - s_{x2} = (l_{x1} - l_{x2}) - (\delta_{x1} - \delta_{x2}) \tag{2.92}$$

A similar relation governs the tangential displacements in the y-direction. For the no-slip region, s_x becomes zero so that

$$\begin{aligned} l_{x1} - l_{x2} &= \delta_{x1} - \delta_{x2} = \delta_x \\ l_{y1} - l_{y2} &= \delta_{y1} - \delta_{y2} = \delta_y \end{aligned} \tag{2.93}$$

where δ_x and δ_y denote the relative but rigid tangential displacements between the two spheres so that both δ_x and δ_y are independent of the position of A_1 and A_2. Therefore, Eq. (2.93) leads to a conclusion that all surface points within the no-slip region undergo the same tangential displacement.

Consider the distribution of traction to be in the form of

$$q = q_o r_c \left(r_c^2 - r^2\right)^{-\frac{1}{2}} \tag{2.94}$$

This distribution leads to a uniform tangential displacement of the contact surface. q_o is the maximum tangential traction at the center of the contact surface. By integrating Eq. (2.94) over the contact surface, q_o is found to be

$$q_o = \frac{F_t}{2\pi r_c^2} \tag{2.95}$$

Substituting the distribution of traction in Eq. (2.94) into Eq. (2.87) gives

$$\int_0^{s_1} q(s, \beta)\, ds = q_o r_c \int_0^{s_1} (\alpha^2 - 2\varsigma s - s^2)^{-\frac{1}{2}}\, ds = q_o r_c \left[\frac{\pi}{2} - \tan^{-1}\left(\frac{\varsigma}{\alpha}\right)\right] \tag{2.96}$$

Using the symmetric condition $\varsigma(\beta) = -\varsigma(\beta + \pi)$ yields

$$l_{\mathrm{x}} = \frac{q_{\mathrm{o}} r_{\mathrm{c}}}{4G} \int_0^{2\pi} (1 - \nu \sin^2 \beta) \, \mathrm{d}\beta = \frac{\pi(2-\nu)}{4G} q_{\mathrm{o}} r_{\mathrm{c}} \tag{2.97}$$

Similarly, we have

$$l_{\mathrm{y}} = 0 \tag{2.98}$$

Therefore, a constant tangential displacement induced by the no-slip contact between two spheres is given by

$$\delta_{\mathrm{x}} = l_{\mathrm{x}1} - l_{\mathrm{x}2} = \frac{F_{\mathrm{t}}}{8 r_{\mathrm{c}}} \left(\frac{2 - \nu_1}{G_1} + \frac{2 - \nu_2}{G_2} \right) \tag{2.99}$$

which indicates that Eq. (2.94) is the distribution of tangential traction of no-slip contact between two spheres. However, according to Eq. (2.94), the tangential traction at the edge of the contact circle would be infinitely large. Hence, near the edge of contact, a small zone inevitably exists wherein slip occurs. This slip is relatively small compared to normal slip, and thus can be termed microslip. This argument also suggests that the contact area consists of a core region ($r \leq c$) for no-slip motion and an annulus region ($c \leq r \leq r_{\mathrm{c}}$) for microslip.

The distribution of traction for the microslip is similar to that of sliding contact. Thus,

$$q'(r) = \frac{f_{\mathrm{s}} p_{\mathrm{o}}}{r_{\mathrm{c}}} \sqrt{r_{\mathrm{c}}^2 - r^2} \tag{2.100}$$

where f_{s} is the coefficient of static friction. To match the no-slip condition in the core region, another distribution of traction should be obtained so that the superposition of this distribution with Eq. (2.100) can yield constant displacements in the nonslip region. To do so, consider a distribution of traction

$$q''(r) = -\frac{f_{\mathrm{s}} p_{\mathrm{o}}}{r_{\mathrm{c}}} \sqrt{c^2 - r^2} \tag{2.101}$$

Similarly to Eq. (2.90) and Eq. (2.91), the tangential displacements due to the distributions of tangential traction given by Eq. (2.100) and Eq. (2.101) are expressed, respectively, as

$$l'_{\mathrm{x}} = \frac{\pi f_{\mathrm{s}} p_{\mathrm{o}}}{32 r_{\mathrm{c}} G} \left[4(2-\nu) r_{\mathrm{c}}^2 - (4 - 3\nu) x^2 - (4-\nu) y^2 \right] \tag{2.102}$$

$$l'_{\mathrm{y}} = \frac{\pi f_{\mathrm{s}} p_{\mathrm{o}}}{16 r_{\mathrm{c}} G} \nu x y \tag{2.103}$$

and

$$l''_{\mathrm{x}} = -\frac{\pi f_{\mathrm{s}} p_{\mathrm{o}}}{32 r_{\mathrm{c}} G} \left[4(2-\nu) c^2 - (4 - 3\nu) x^2 - (4-\nu) y^2 \right] \tag{2.104}$$

$$l''_{\mathrm{y}} = -\frac{\pi f_{\mathrm{s}} p_{\mathrm{o}}}{16 r_{\mathrm{c}} G} \nu x y \tag{2.105}$$

Therefore, the tangential displacements in the core region under joint actions portrayed by Eq. (2.100) and Eq. (2.101) can be obtained as

$$l_{\mathrm{x}} = l'_{\mathrm{x}} + l''_{\mathrm{x}} = \frac{\pi f_{\mathrm{s}} p_{\mathrm{o}}}{8 r_{\mathrm{c}} G} (2-\nu) \left(r_{\mathrm{c}}^2 - c^2 \right) \tag{2.106}$$

$$l_{\mathrm{y}} = l'_{\mathrm{y}} + l''_{\mathrm{y}} = 0 \tag{2.107}$$

The displacements satisfy the condition for no-slip within the core region with the result that

$$\delta_x = \frac{3 f_s F_z}{16}\left(\frac{2-\nu_1}{G_1} + \frac{2-\nu_2}{G_2}\right)\frac{r_c^2 - c^2}{r_c^3} \tag{2.108}$$

The radius of the core region c can be determined from the given tangential force F_t by

$$F_t = \int_c^{r_c} q' 2\pi r\,dr + \int_0^c (q' + q'')2\pi r\,dr = f_s F_z\left(1 - \frac{c^3}{r_c^3}\right) \tag{2.109}$$

or

$$\frac{c}{r_c} = \left(1 - \frac{F_t}{f_s F_z}\right)^{\frac{1}{3}} \tag{2.110}$$

Thus, the area of the core region decreases with increasing tangential force while the normal force is maintained constant. Once the tangential force reaches the limiting static friction force ($F_t = f_s F_z$), the core region has shrunk to a single point so that the sliding motion is about to begin.

2.3.5.3 *Torsion of Elastic Spheres in Contact*

Besides the oblique contact, tangential displacements may also be produced in the contact of two elastic spheres under the actions of a compressional twist, as shown in Fig. 2.13. Since the torsional couple does not give rise to a displacement in the z-direction, the pressure distribution is not influenced by the twist and is thus given by the Hertzian contact theory.

If a torsional couple or twist moment M_z were applied to the two spheres in contact, the contact surface would undergo a rigid rotation relative to distant points in each body if there were no-slip motion at the interface. In this case, the tangential displacements could be expressed by

$$l_{\theta 1} = \beta_1 r, \qquad l_{\theta 2} = \beta_2 r \tag{2.111}$$

where β is the angle of twist in each body. To yield a uniform angle of twist throughout the whole contact area, the distribution of tangential traction may be in the form of [Johnson,

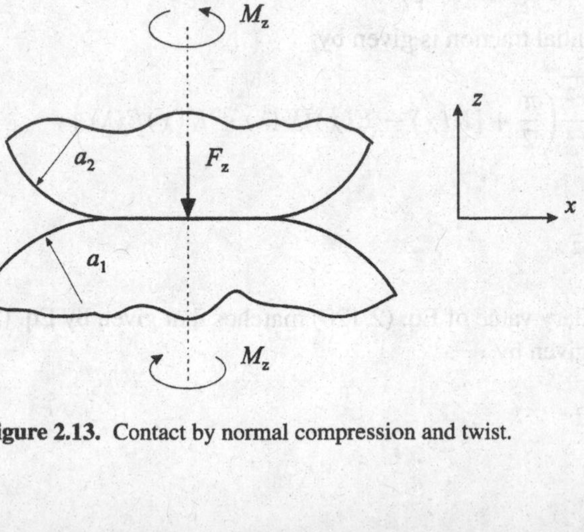

Figure 2.13. Contact by normal compression and twist.

1985]

$$q = \frac{3M_z r}{4\pi r_c^3}(r_c^2 - r^2)^{-\frac{1}{2}} \tag{2.112}$$

which leads to an infinite value of tangential traction at the edge of the contact circle. Hence, similarly to the argument for no-slip oblique contact, microslip has to occur in an annular region ($c \le r \le r_c$) while no-slip remains in the core region ($r \le c$).

The microslip motion is in the circumferential direction and the surface distribution of traction in this annular region takes the limiting value from Amontons's law as

$$q(r) = \frac{3}{2}\frac{f_s F_z}{\pi r_c^3}\sqrt{r_c^2 - r^2} \tag{2.113}$$

Several methods were presented by Lubkin (1951) for the determination of the radius of the no-slip core and the distribution of tangential traction in this region. Only the final results of Lubkin's work are introduced here.

The relation between the angle of twist and the penetration of microslip in sphere 1 is given by

$$\frac{G_1 r_c^2}{f_s F_z}\beta_1 = \frac{3}{4\pi}[K(\chi) - E(\chi)] \tag{2.114}$$

where χ is defined by

$$\chi^2 = 1 - (c/r_c)^2 \tag{2.115}$$

and $K(\chi)$ and $E(\chi)$ are the complete elliptic integrals of the first and second kinds of modulus of χ, respectively, defined by

$$K(\chi) = \int_0^{\frac{\pi}{2}} (1 - \chi^2 \sin^2\theta)^{-\frac{1}{2}}\,\mathrm{d}\theta \tag{2.116}$$

$$E(\chi) = \int_0^{\frac{\pi}{2}} (1 - \chi^2 \sin^2\theta)^{\frac{1}{2}}\,\mathrm{d}\theta \tag{2.117}$$

Thus, the total angle of twist is related to c by

$$\beta = \beta_1 + \beta_2 = \frac{3 f_s F_z}{4\pi r_c^2}\left(\frac{1}{G_1} + \frac{1}{G_2}\right)[K(\chi) - E(\chi)] \tag{2.118}$$

The distribution of tangential traction is given by

$$q = \frac{3 f_s F_z}{\pi^2 r_c^2}\sqrt{1 - \frac{r^2}{r_c^2}}\left(\frac{\pi}{2} + [K(\chi) - E(\chi)]K(\lambda) - K(\chi)E(\lambda)\right) \tag{2.119}$$

where λ is defined by

$$\lambda = \frac{c}{r_c} = \sqrt{1 - \chi^2} \tag{2.120}$$

It can be seen that the boundary value of Eq. (2.120) matches that given by Eq. (2.112).

The twisting moment is given by

$$M_z = 2\pi \int_0^{r_c} q r^2\,\mathrm{d}r \tag{2.121}$$

where $q(r)$ is obtained by Eq. (2.112) for $c \leq r \leq r_c$ and Eq. (2.119) for $r \leq c$. Thus, the relation between M_z and c is given by

$$\frac{M_z}{f_s F_z r_c} = \frac{3\pi}{16} + \frac{3\chi^2 K(\chi)}{4\pi}\left(2\lambda - \frac{\sin^{-1}\lambda}{\chi} + \int_0^{\frac{\pi}{2}} \frac{\sin^{-1}(\lambda \sin\alpha)\,d\alpha}{(1-\lambda^2\sin^2\alpha)^{\frac{3}{2}}}\right)$$
$$+ \frac{K(\chi) - E(\chi)}{4\pi}\left[\lambda(4\lambda^2 - 3) - 3\int_0^{\frac{\pi}{2}} \frac{\sin^{-1}(\lambda\sin\alpha)}{\sqrt{1-\lambda^2\sin^2\alpha}}\,d\alpha\right] \quad (2.122)$$

The relation between the twist moment M_z and the total angle of twist β is shown in Fig. 2.14(a), which is based on the numerical integration of Eq. (2.122). A typical distribution of the tangential traction $q(r)$ is illustrated in Fig. 2.14(b), which is obtained from Eqs. (2.112) and (2.119).

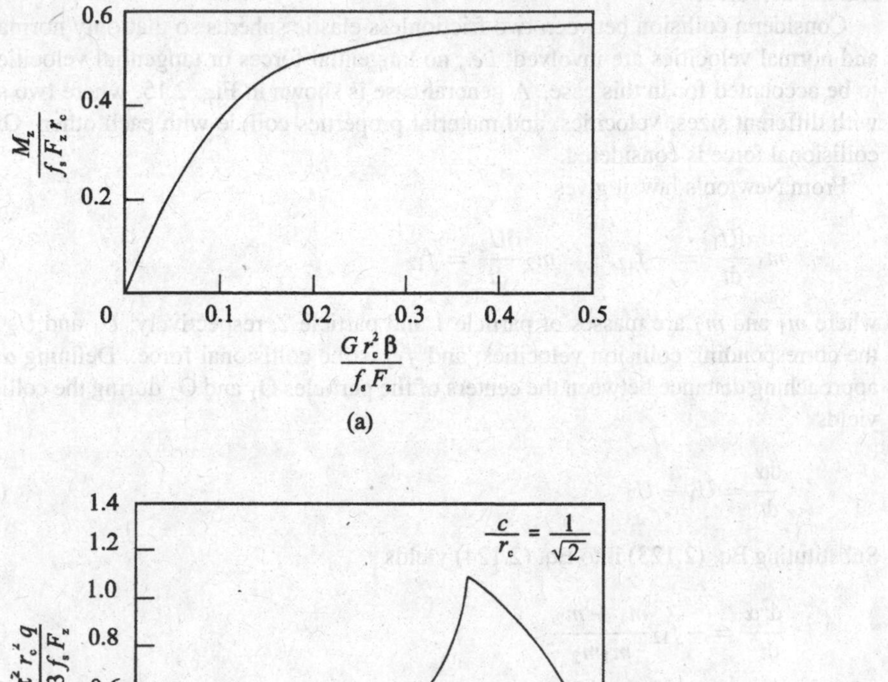

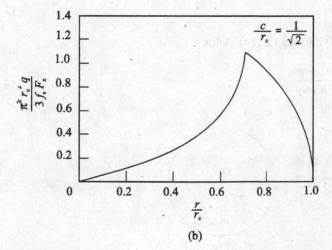

Figure 2.14. Twist angle and tangential traction by torsion (from Lubkin, 1951): (a) Twist moment M_z versus total twist angle β; (b) Typical distribution of tangential traction.

2.4 Collision of Elastic Spheres

The basic theories of elastic deformations associated with various contact forces under static contact conditions have been introduced in the last section. Assuming that an impact process of two solids can be regarded as quasi-static, the theories given in §2.3 are used directly to link the dynamic deformations of the colliding solids with the impact forces. In this section, the collisions of elastic spheres are described.

2.4.1 *Normal Collision of Elastic Spheres*

Collisions between particles with smooth surfaces may be reasonably approximated as elastic impact of frictionless spheres. Assume that the deformation process during a collision is quasi-static so that the Hertzian contact theory can be applied to establish the relations among impact velocities, material properties, impact duration, elastic deformation, and impact force.

Consider a collision between two frictionless elastic spheres so that only normal force and normal velocities are involved; *i.e.*, no tangential forces or tangential velocities need to be accounted for in this case. A general case is shown in Fig. 2.15, where two spheres with different sizes, velocities, and material properties collide with each other. Only the collisional force is considered.

From Newton's law, it gives

$$m_1 \frac{\mathrm{d}U_1}{\mathrm{d}t} = -f_{12}, \qquad m_2 \frac{\mathrm{d}U_2}{\mathrm{d}t} = f_{12} \tag{2.123}$$

where m_1 and m_2 are masses of particle 1 and particle 2, respectively; U_1 and U_2 denote the corresponding collision velocities; and f_{12} is the collisional force. Defining α as the approaching distance between the centers of the particles O_1 and O_2 during the collision, it yields

$$\frac{\mathrm{d}\alpha}{\mathrm{d}t} = U_1 - U_2 \tag{2.124}$$

Substituting Eq. (2.123) into Eq. (2.124) yields

$$\frac{\mathrm{d}^2\alpha}{\mathrm{d}t^2} = -f_{12}\frac{m_1 + m_2}{m_1 m_2} \tag{2.125}$$

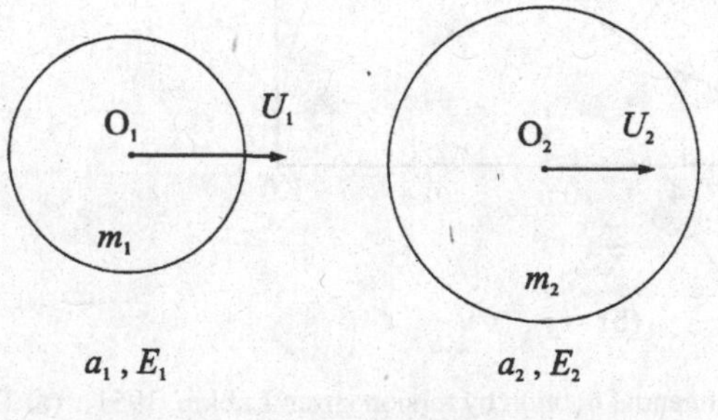

Figure 2.15. Normal collision of elastic spheres.

Note that f_{12} can be related to α by Eq. (2.75) as

$$f_{12} = \frac{4}{3}E^*\sqrt{a^*}\alpha^{\frac{3}{2}} \tag{2.126}$$

Hence, we have

$$\frac{d^2\alpha}{dt^2} = -\frac{4}{3}\frac{E^*}{m^*}\sqrt{a^*}\alpha^{\frac{3}{2}} \tag{2.127}$$

where m^* is the relative mass, defined by

$$\frac{1}{m^*} = \frac{1}{m_1} + \frac{1}{m_2} \tag{2.128}$$

The initial conditions are given by

$$\left(\frac{d\alpha}{dt}\right)_{t=0} = U_{10} - U_{20} = U_{12}; \quad \alpha_{t=0} = 0 \tag{2.129}$$

Multiplying both sides of Eq. (2.127) by $d\alpha/dt$ yields

$$\frac{d\alpha}{dt} = \left(U_{12}^2 - \frac{16}{15}\frac{E^*}{m^*}a^{*\frac{1}{2}}\alpha^{\frac{5}{2}}\right)^{\frac{1}{2}} \tag{2.130}$$

Since the maximum deformation occurs at $d\alpha/dt = 0$, we have

$$\alpha_m = \left(\frac{15}{16}\frac{m^*U_{12}^2}{E^*\sqrt{a^*}}\right)^{\frac{2}{5}} \tag{2.131}$$

From Eq. (2.126), the maximum collisional force f_m can be expressed by

$$f_m = (f_{12})_{max} = \frac{4}{3}E^*\sqrt{a^*}\left(\frac{15}{16}\frac{m^*U_{12}^2}{E^*\sqrt{a^*}}\right)^{\frac{3}{5}} \tag{2.132}$$

From Eq. (2.75), the corresponding maximum radius of collisional contact area can be given by

$$r_{cm} = \sqrt{a^*}\left(\frac{15}{16}\frac{m^*U_{12}^2}{E^*\sqrt{a^*}}\right)^{\frac{1}{5}} \tag{2.133}$$

The maximum pressure in a collision can also be given based on Eq. (2.70) as

$$p_{om} = \frac{3}{2}\left(\frac{f_{12}}{\pi r_c^2}\right)_{max} = \frac{2}{\pi}\frac{E^*}{a^*}r_{cm} \tag{2.134}$$

The collisional contact duration can be obtained by rearranging Eq. (2.130) to the form

$$dt = \left(U_{12}^2 - \frac{16}{15}\frac{E^*}{m^*}\sqrt{a^*}\alpha^{\frac{5}{2}}\right)^{-\frac{1}{2}} d\alpha = \frac{\alpha_m}{U_{12}}(1-\zeta^{\frac{5}{2}})^{-\frac{1}{2}} d\zeta \tag{2.135}$$

where ζ is defined as α/α_m. Hence, the total collisional time duration t_c can be obtained as

$$t_c = 2\frac{\alpha_m}{U_{12}}\int_0^1 (1-\zeta^{\frac{5}{2}})^{-\frac{1}{2}} d\zeta = \frac{2.94}{U_{12}}\left(\frac{15}{16}\frac{m^*U_{12}^2}{E^*\sqrt{a^*}}\right)^{\frac{2}{5}} \tag{2.136}$$

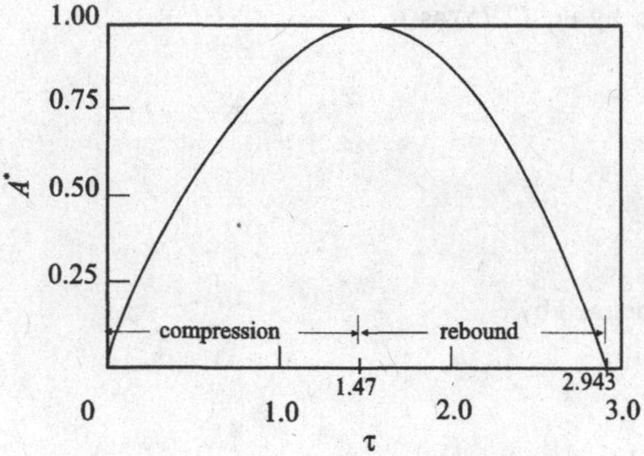

Figure 2.16. Change of contact area with time.

The dependence of the contact area A_c on the time of the compression process can be obtained from Eq. (2.130) and Eq. (2.75) as

$$\frac{dA_c}{dt} = \left((\pi U_{12} a^*)^2 - \frac{16}{15\sqrt{\pi}} \frac{E^*}{m^*} A_c^{\frac{5}{2}} \right)^{\frac{1}{2}} \tag{2.137}$$

Since the rebound process is identical to the compression process, the change of the contact area with time in a rebound process follows a path similar to that of the compression process, as illustrated in Fig. 2.16. In this figure, the dimensionless area A^* and dimensionless time τ are defined by

$$A^* = \frac{A_c}{A_{cm}} = \frac{r_c^2}{r_{cm}^2}$$
$$\tau = \left(\frac{16E^*}{15m^*} \right)^{\frac{2}{5}} (a^* U_{12})^{\frac{1}{5}} t \tag{2.138}$$

As an approximation, the change of A^* with τ in a collision may be fitted into a sinusoidal form by

$$A^* = \sin\left(\frac{\pi}{2.943} \tau \right) \tag{2.139}$$

2.4.2 *Collision of Frictional Elastic Spheres*

As noted before, the collision or dynamic contact from normal compression without any tangential forces involved is based on the Hertzian contact theory. Similarly, collisions with frictional forces can be approached using Mindlin's contact theory. To simplify the analysis, the effect of torsion is neglected. Consider the impact of two balls in a coplanar motion, as shown in Fig. 2.17. f_t is the frictional force, V_1 and V_2 are the tangential velocities, and ω_1 and ω_2 represent the angular velocities. The momentum equation in the tangential direction is given by

$$m_1 \frac{d}{dt}(V_1 + \omega_1 a_1) = f_t; \qquad m_2 \frac{d}{dt}(V_2 - \omega_2 a_2) = -f_t \tag{2.140}$$

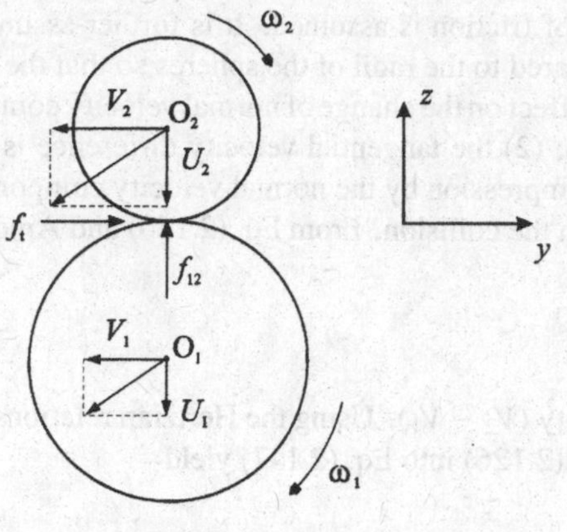

Figure 2.17. Two-dimensional collision with tangential friction and twist.

For each ball, the angular momentum about the center of contact O is conserved provided that the contact area is small and no external torque exists. Hence, we have

$$\frac{\mathrm{d}}{\mathrm{d}t}\left[m_1 V_1 a_1 + m_1\omega_1\left(a_1^2 + k_1^2\right)\right] = 0$$
$$\frac{\mathrm{d}}{\mathrm{d}t}\left[m_2 V_2 a_2 + m_2\omega_2\left(a_2^2 + k_2^2\right)\right] = 0 \tag{2.141}$$

where k_1 and k_2 are the radii of gyration of the balls about their centers of mass, which gives

$$\frac{\mathrm{d}\omega_1}{\mathrm{d}t} = -\frac{a_1}{a_1^2 + k_1^2}\frac{\mathrm{d}V_1}{\mathrm{d}t}; \qquad \frac{\mathrm{d}\omega_2}{\mathrm{d}t} = \frac{a_2}{a_2^2 + k_2^2}\frac{\mathrm{d}V_2}{\mathrm{d}t} \tag{2.142}$$

Thus, substituting the preceding equation into Eq. (2.140) yields

$$f_\mathrm{t} = \frac{m_1}{1 + a_1^2/k_1^2}\frac{\mathrm{d}V_1}{\mathrm{d}t} = -\frac{m_2}{1 + a_2^2/k_2^2}\frac{\mathrm{d}V_2}{\mathrm{d}t} \tag{2.143}$$

Let

$$\hat{m}_1 = \frac{m_1}{1 + a_1^2/k_1^2}; \qquad \hat{m}_2 = \frac{m_2}{1 + a_2^2/k_2^2}; \qquad \frac{1}{\hat{m}^*} = \frac{1}{\hat{m}_1} + \frac{1}{\hat{m}_2} \tag{2.144}$$

We have

$$\frac{\mathrm{d}}{\mathrm{d}t}(V_2 - V_1) = \frac{f_t}{\hat{m}^*} \tag{2.145}$$

or

$$\frac{\mathrm{d}^2\delta}{\mathrm{d}t^2} = \frac{f_\mathrm{t}}{\hat{m}^*} \tag{2.146}$$

where δ is the relative tangential displacement on the contact surface. The tangential deformation is affected by the existence of microslip, the magnitude of normal compressional force f_{12} (no matter whether f_{12} is constant or not), the coefficient of friction, and the loading history (i.e., loading, unloading, and reloading) [Johnson, 1985].

For the case in Fig. 2.17, the tangential velocities of the colliding spheres are not equal to each other before the impact, which would most likely lead to a sliding on the contact interface. Consider the simple case of two colliding spheres without relative rotation. To analyze this case, a constant coefficient of friction is assumed. It is further assumed that (1) the elastic deformation is small compared to the radii of the spheres so that the relative tangential shift of the two spheres has no effect on the change of normal velocity components throughout the whole course of collision; (2) the tangential velocity difference is not too high nor too low so that the maximum compression by the normal velocity component can be achieved while sliding is maintained in the collision. From Eq. (2.140) and Amontons's law, we have

$$\frac{dV_{12}}{dt} = -\frac{f_k f_{12}}{m^*} \tag{2.147}$$

where V_{12} is the relative tangential velocity $(V_2 - V_1)$. Using the Hertzian relations for the normal deformation and substituting Eq. (2.126) into Eq. (2.147) yield

$$\frac{dV_{12}}{dt} = -\frac{4E^* f_k \sqrt{a^*}}{3m^*} \alpha^{\frac{3}{2}} \tag{2.148}$$

The relation between α and t is obtained from Eq. (2.135) (where ζ is α/α_m) as

$$t = \frac{\alpha_m}{U_{12}} \int_0^{\zeta} \left(1 - \zeta^{\frac{5}{2}}\right)^{-\frac{1}{2}} d\zeta \tag{2.149}$$

for the compression process and

$$t_c - t = \frac{\alpha_m}{U_{12}} \int_0^{\zeta} \left(1 - \zeta^{\frac{5}{2}}\right)^{-\frac{1}{2}} d\zeta \tag{2.150}$$

for the rebound process. Thus, the change of relative tangential velocity after the collision can be calculated by

$$\Delta V_{12} = \frac{4E^* f_k \sqrt{a^*}}{3m^*} \alpha_m^{\frac{3}{2}} \int_0^{t_c} \zeta^{\frac{3}{2}} dt \tag{2.151}$$

Substitution of Eq. (2.131), Eq. (2.149), and Eq. (2.150) into Eq. (2.151) yields

$$\Delta V_{12} = 2 f_k U_{12} \tag{2.152}$$

which shows that the change of relative tangential velocity is related only to the surface friction coefficient and normal velocity component and is independent of other material properties. Equation (2.152) also indicates that, to ensure the sliding assumption, the relative incident angle should be larger than $\tan^{-1}(2 f_k)$.

It is clear that the collision between two elastic but frictional spheres is inelastic due to the inevitable sliding at contact which yields the kinetic energy loss by frictional work. Furthermore, the preceding analyses of both Hertzian collision and frictional collision can also be applied to the particle–wall collision, where the radius of the wall is simply set to be infinitely large.

Example 2.2 Consider an impact between a polyethylene particle ($d_p = 1$ cm) and a copper wall. The incident velocity is 2 m/s, and the incident angle is 30°. The friction coefficient of the interface is 0.2. The densities of polyethylene and copper are 950 and 8,900 kg/m^3, respectively. What is the contact time duration for the collision? Estimate the rebound velocity of the particle. Repeat the problem for a copper particle colliding with a polyethylene wall.

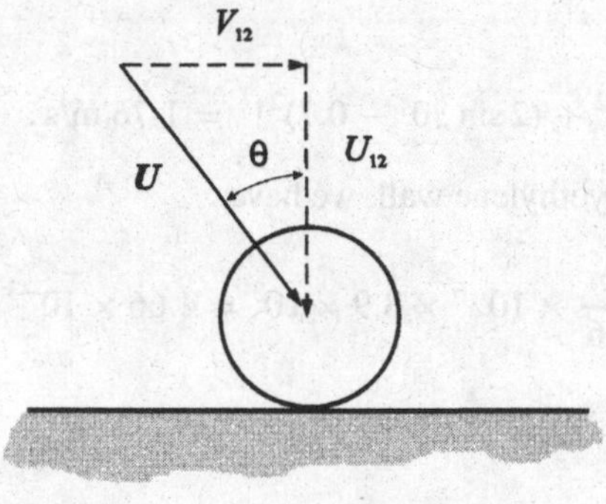

Figure E2.2. Collision of a sphere with a wall.

Solution As shown in Fig. E2.2, the normal velocity U_{12} is given by

$$U_{12} = U\cos\theta = 2\cos 30^\circ = 1.73 \text{ m/s} \tag{E2.7}$$

The wall can be regarded as a large sphere with an infinite radius. Thus, we have

$$a^* = \frac{d_p}{2}; \qquad m^* = m = \frac{\pi}{6}d_p^3\rho_p \tag{E2.8}$$

The following data are available from Table 1.6

$$E_1 = 0.04 \times 10^{10} \text{ N/m}^2; \qquad \nu_1 = 0.3 \tag{E2.9}$$

for polyethylene, and

$$E_2 = 12 \times 10^{10} \text{ N/m}^2; \qquad \nu_2 = 0.35 \tag{E2.10}$$

for copper.

Thus, it gives

$$E^* = \left(\frac{1-\nu_1^2}{E_1} + \frac{1-\nu_2^2}{E_2}\right)^{-1} = \left(\frac{1-0.3^2}{0.04\times 10^{10}} + \frac{1-0.35^2}{12\times 10^{10}}\right)^{-1} = 4.38 \times 10^8 \text{ N/m}^2 \tag{E2.11}$$

For a polyethylene particle hitting a copper wall, we have

$$a^* = \frac{d_p}{2} = 0.005 \text{ m}; \quad m^* = \frac{\pi}{6}d_p^3\rho_p = \frac{\pi}{6}\times 10^{-6} \times 0.95 \times 10^3 = 0.5 \times 10^{-3} \text{ kg} \tag{E2.12}$$

Hence, the collision time is calculated by Eq. (2.136)

$$t_c = \frac{2.94}{U_{12}}\left(\frac{15}{16}\frac{m^* U_{12}^2}{E^*\sqrt{a^*}}\right)^{\frac{2}{5}} = \frac{2.94}{1.73}\left(\frac{15}{16}\frac{0.5\times 10^{-3}\times 1.73^2}{4.38\times 10^8 \times \sqrt{0.005}}\right)^{\frac{2}{5}} = 1.2 \times 10^{-4}\text{s} \tag{E2.13}$$

and the change of tangential velocity is obtained by Eq. (2.152)

$$\Delta V_{12} = 2 f_k U_{12} = 2 \times 0.2 \times 1.75 = 0.7 \text{ m/s} \tag{E2.14}$$

Thus, the rebound velocity is

$$U' = \left[U_{12}^2 + (U\sin\theta - \Delta V_{12})^2\right]^{\frac{1}{2}} = [1.73^2 + (2\sin 30^\circ - 0.7)^2]^{\frac{1}{2}} = 1.76 \text{ m/s} \quad \text{(E2.15)}$$

Similarly, for a copper particle hitting a polyethylene wall, we have

$$a^* = \frac{d_p}{2} = 0.005 \text{ m}; \qquad m^* = \frac{\pi}{6} d_p^3 \rho_p = \frac{\pi}{6} \times 10^{-6} \times 8.9 \times 10^3 = 4.66 \times 10^{-3} \text{ kg} \quad \text{(E2.16)}$$

Hence, it yields

$$t_c = \frac{2.94}{U_{12}} \left(\frac{15}{16} \frac{m^* U_{12}^2}{E^* \sqrt{a^*}} \right)^{\frac{2}{5}} = \frac{2.94}{1.73} \left(\frac{15}{16} \frac{4.66 \times 10^{-3} \times 1.73^2}{4.38 \times 10^8 \times \sqrt{0.005}} \right)^{\frac{2}{5}} = 3.0 \times 10^{-4} \text{ s} \quad \text{(E2.17)}$$

while the rebound velocity remains the same as the previous case since the change in tangential velocity difference depends only on the friction coefficient and normal velocity.

2.5 Collision of Inelastic Spheres

Any solid material has its own upper limit of elastic deformation under either normal or tangential stresses. Once the stresses exceed this limit, plastic deformation will occur. In this section, collisions of inelastic spheres are presented. The degree of inelastic deformation is characterized by the restitution coefficient.

2.5.1 *Onset of Plastic Deformation*

In order to estimate the onset yield stress of the material, three common criteria are introduced. The Tresca criterion is based on maximum shear stress and is given as

$$\max(|\sigma_1 - \sigma_2|, |\sigma_2 - \sigma_3|, |\sigma_3 - \sigma_1|) = 2n = Y \quad (2.153)$$

where σ_1, σ_2, and σ_3 are the principal stresses; n is the yield stress in simple shear; and Y is the yield stress in simple compression or tension. The von Mises criterion considers the shear strain–energy, which is calculated by

$$(\sigma_1 - \sigma_2)^2 + (\sigma_2 - \sigma_3)^2 + (\sigma_3 - \sigma_1)^2 = 6n^2 = 2Y^2 \quad (2.154)$$

The third criterion is obtained from the maximum reduced stress and is expressed as

$$\max(|\sigma_1 - \sigma|, |\sigma_2 - \sigma|, |\sigma_3 - \sigma|) = n = \frac{2}{3}Y \quad (2.155)$$

where $\sigma = (\sigma_1 + \sigma_2 + \sigma_3)/3$.

No significant difference is found for the predictions of the yield stresses from these three criteria. However, Tresca's criterion is more widely used than the other two because of its simplicity. When two solid spherical particles are in contact, the principal stresses along the normal axis through the contact point can be obtained from the Hertzian elastic contact theory as [Johnson, 1985]

$$\sigma_r = \sigma_\theta = -p_o\left\{(1+\nu)\left[1-\frac{z}{r_c}\tan^{-1}\left(\frac{r_c}{z}\right)\right]-\frac{r_c^2}{2(r_c^2+z^2)}\right\}$$
$$\sigma_z = -\frac{p_o r_c^2}{r_c^2+z^2} \tag{2.156}$$

where z denotes the depth inside the sphere along the axis of symmetry; r_c is the radius of contact area; and p_o is the maximum pressure on the contact surface.

It can be proved from Eq. (2.156) that, for materials with Poisson's ratio of 0.3 (which is true for most solids), the maximum shear stress $|\sigma_z - \sigma_r|$ occurs at $z/r_c = 0.48$. Consequently, according to Tresca's criterion, the yield stress Y in a simple compression is $0.62\ p_o$. Therefore, when the hardness or the yield stress Y of the particle material is less than 0.62 times the maximum contact pressure, the sphere will, most likely, undergo plastic deformation. From the elastic collision of two solid spheres, the maximum contact pressure is given by Eq. (2.134). Thus, the relation between the critical normal collision velocity, U_{12Y}, and the yield stress is given by

$$U_{12Y} = 10.3\sqrt{\frac{a^{*3}Y^5}{m^*E^{*4}}} \tag{2.157}$$

Moreover, it can be shown from Eq. (2.156) that the maximum shear stress $|\sigma_z - \sigma_r|$ is insensitive to the variation of Poisson's ratio for solids, only deviating 5 percent when Poisson's ratio varies from 0.27 to 0.36 (see Problem 2.4). Thus, Eq. (2.157) can also be applied for solids with Poisson's ratio in the range of 0.27 to 0.36.

Example 2.3 For a normal impact between a polyethylene particle ($d_p = 1$ cm) and a copper wall, estimate the critical normal collision velocity of the particle, above which plastic deformation would occur. The yield strength for polyethylene is 2×10^7 N/m^2 and the yield strength for copper is 2.5×10^8 N/m^2. What would be the critical normal collision velocity for a copper sphere colliding with the copper wall?

Solution Since the yield strength for polyethylene is lower than that for copper, the polyethylene particle will undergo plastic deformation before the copper wall does in the collision. As obtained in Example 2.2, we have

$$m^* = 0.5 \times 10^{-3}\ \text{kg}; \quad a^* = 0.005\ \text{m}; \quad E^* = 4.38 \times 10^8\ \text{N/m}^2 \tag{E2.18}$$

Thus, from Eq. (2.157), the critical normal collision velocity is

$$U_{12Y} = 10.3\sqrt{\frac{a^{*3}Y^5}{m^*E^{*4}}} = 10.3\sqrt{\frac{0.005^3 \times (2\times 10^7)^5}{0.5\times 10^{-3}\times(4.38\times 10^8)^4}} = 1.5\ \text{m/s} \tag{E2.19}$$

If the particle is copper, we have

$$m^* = 4.66 \times 10^{-3}\ \text{kg}; \quad a^* = 0.005\ \text{m} \tag{E2.20}$$

$$E^* = \frac{E}{2(1-\nu^2)} = \frac{12\times 10^{10}}{2(1-0.35^2)} = 6.8 \times 10^{10}\ \text{N/m}^2 \tag{E2.21}$$

Hence, it yields

$$U_{12Y} = 10.3\sqrt{\frac{a^{*3}Y^5}{m^*E^{*4}}} = 10.3\sqrt{\frac{0.005^3 \times (2.5 \times 10^8)^5}{4.66 \times 10^{-3} \times (6.8 \times 10^{10})^4}} = 0.01 \text{ m/s} \tag{E2.22}$$

This example illustrates that even small impact velocities in collisions between metallic materials (or with hard solids) may lead to plastic deformation or breakage of colliding parts.

2.5.2 *Restitution Coefficient*

Once the relative impact velocity between two colliding spheres is higher than the critical yield velocity, plastic deformation must occur. Heat loss is another phenomenon often coupled with such collisions. Collisions with plastic deformation are referred to as inelastic collisions. All the energy transfer in the form of plastic deformation and heat loss in an inelastic collision is considered as a kinetic energy loss.

The recoverability or restitution of the kinetic energy during a normal collision between two objects can be represented by the coefficient of restitution defined by Eq. (2.3). Note that the coefficient of restitution cannot be used as a criterion to judge whether a collision is elastic or not unless the collision is solely considered as a normal collision. For example, the sliding at contact for the collision between two elastic spheres will make the collision inelastic while the value of the coefficient of restitution in this case is equal to 1.

A complete course of collision consists of the compression and rebound processes. The total work done during the compression process without heat loss can be calculated as

$$\frac{1}{2}m^*U_{12}^2 = \int_0^{\alpha_m} f_{12}\,d\alpha \tag{2.158}$$

and the total work done during the rebound process without heat loss is

$$\frac{1}{2}m^*U_{12}'^2 = \int_0^{\alpha_m'} f_{12}'\,d\alpha' \tag{2.159}$$

where α_m' is approximately equal to α_m when the ratio of the plastic deformation to the elastic deformation is negligibly small.

Prediction of the restitution coefficient has been a challenging research topic for decades. Unfortunately, no reliable and accurate prediction method has been found so far. However, some useful simplified models with certain limits have been developed. One of them is the elastic–plastic impact model in which the compression process is assumed to be plastic with part of the kinetic energy stored for later elastic rebounding, with the rebound process considered to be completely elastic [Johnson, 1985]. In this model, it is postulated that (1) during the plastic compression process, $\alpha = r_c^2/2a^*$; (2) during the compression process, the averaged contact pressure p_m is constant and is equal to $3Y$; and (3) the elastic rebound process starts when maximum deformation is reached. Therefore, the compressional force is

$$f_{12} = \pi r_c^2 p_m \tag{2.160}$$

so that the total work during the compression process is given by

$$\frac{1}{2}m^* U_{12}^2 = \int_0^{\alpha_m} f_{12}\, d\alpha = \frac{\pi r_{cm}^4 p_m}{4a^*} \tag{2.161}$$

Contrary to the rather straightforward calculation of the total work during the compression process, the rebound work is hard to estimate because of the difficulties in the formulation of the rebound force. In the recovery process, the repelling force is related to the radius of contact area by the elastic deformation. However, for the same degree of compression (*i.e.*, for the same α), the contact area in a fully plastic deformation is twice as big as that in an elastic case. Since the actual repelling force starts from the contact area by plastic deformation, the variation between the repelling force and the radius of the contact area in the rebounding process follows neither fully elastic nor fully plastic relations. Hence, an approximation has to be made to express the repelling force in terms of the radius of the contact area r_c or the approaching distance α.

First, for both elastic and plastic deformation, we have

$$\frac{f_{12}}{f_m} = \frac{r_c^3}{r_{cm}^3} \tag{2.162}$$

Furthermore, assume that Eq. (2.74), obtained for Hertzian contact, is valid with r_c being the actual radius of contact area so that

$$f_{12} = \frac{4 r_c E^*}{3}\alpha \tag{2.163}$$

which, combined with Eq. (2.162), yields the relation between α and r_c for the rebounding process as

$$\alpha = \frac{3 r_c^2}{4 E^* r_{cm}^3} f_m \tag{2.164}$$

Therefore, the rebounding velocity can be obtained by Eq. (2.159) as

$$\frac{1}{2}m^* U_{12}'^2 = \int_0^{\alpha_m} f_{12}\, d\alpha = \frac{3}{10}\frac{f_m^2}{E^* r_{cm}} = \frac{3\pi^2}{10}\frac{r_{cm}^3 p_m^2}{E^*} \tag{2.165}$$

The coefficient of restitution is then obtained by eliminating r_{cm} from Eq. (2.161) as

$$e = \left|\frac{U_{12}'}{U_{12}}\right| = 4.08\left(\frac{a^{*3} Y^5}{E^{*4} m^* U_{12}^2}\right)^{\frac{1}{8}} \tag{2.166}$$

The elastic-plastic model reveals that the restitution coefficient depends not only on the material properties but also on the relative impact velocity. Equation (2.166) also indicates that the restitution coefficient decreases with increasing impact velocity by an exponent of 1/4, which is supported by experimental findings, as shown in Fig. 2.18. For high relative impact velocities, the model prediction is reasonably good. However, for low relative impact velocities, the prediction may be poor because the deformation may not be in a fully plastic range as presumed.

Eliminating the yield stress in Eq. (2.166) by substitution of the critical yield velocity U_{12Y} from Eq. (2.157), the restitution coefficient becomes

$$e = 2.29\left(\frac{U_{12Y}}{U_{12}}\right)^{\frac{1}{4}} \tag{2.167}$$

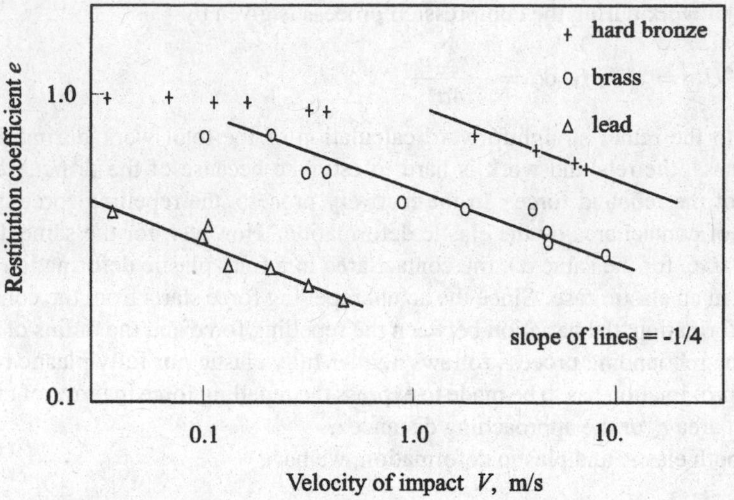

Figure 2.18. Measurements of the restitution coefficient of a steel ball on blocks of various materials (from Goldsmith, 1960).

The preceding equation is only valid for a certain range of impact velocities. The lower limit comes from the fact that the coefficient of restitution has to be less than unity so that

$$U_{12} > 27.5U_{12Y} \tag{2.168}$$

The upper limit of the impact velocity can be deduced from the fact that $r_{cm} < \min(a_1, a_2)$. For a collision between two spheres of the same diameter d_p, Eq. (2.161) yields

$$U_{12} < 1.09\sqrt{\frac{Yd_p^3}{m^*}} \tag{2.169}$$

Similarly, for the collision of a sphere with a solid wall, we have

$$U_{12} < 0.53\sqrt{\frac{Y}{\rho_p}} \tag{2.170}$$

where $Y = \min(Y_1, Y_2)$ and ρ_p is density of the sphere.

Example 2.4 A copper ball of 1 cm diameter normally collides with a stainless steel wall with an impact velocity of 0.5 m/s. Estimate the restitution coefficient using the elastic–plastic model. What is the rebound velocity of the ball? The yield strength of copper is 2.5×10^8 N/m^2. It can be assumed that the yield strength of the stainles steel is higher than that of copper.

Solution First, let us calculate the yield collisional velocity between the copper ball and the stainless steel wall. Assume the radius of the wall is infinitely large so that

$$a^* = \frac{d_p}{2} = 0.005\ \text{m}; \qquad m^* = \frac{\pi}{6}d_p^3\rho_p = \frac{\pi}{6} \times 10^{-6} \times 8.9 \times 10^3 = 4.66 \times 10^{-3}\text{kg} \tag{E2.23}$$

The following data are available

$$E_1 = 12 \times 10^{10}\ \mathrm{N/m^2}; \qquad \nu_1 = 0.35 \tag{E2.24}$$

for copper and

$$E_2 = 20 \times 10^{10}\ \mathrm{N/m^2}; \qquad \nu_2 = 0.28 \tag{E2.25}$$

for stainless steel. Thus, we have

$$E^* = \left(\frac{1-\nu_1^2}{E_1} + \frac{1-\nu_2^2}{E_2}\right)^{-1} = \left(\frac{1-0.35^2}{12\times 10^{10}} + \frac{1-0.28^2}{20\times 10^{10}}\right)^{-1} = 8.4\times 10^{10}\mathrm{N/m^2} \tag{E2.26}$$

The critical normal collision velocity is

$$U_{12\mathrm{Y}} = 10.3\sqrt{\frac{a^{*3}Y^5}{m^*E^{*4}}} = 10.3\sqrt{\frac{0.005^3\times(2.5\times 10^8)^5}{4.66\times 10^{-3}\times(8.4\times 10^{10})^4}} = 7.5\times 10^{-3}\ \mathrm{m/s} \tag{E2.27}$$

To examine the velocity range suitable for the elastic-plastic model, the lower limit is given by

$$U_{12} > 27.5U_{12\mathrm{Y}} = 27.5\times 0.0075 = 0.21\ \mathrm{m/s} \tag{E2.28}$$

and the upper limit is

$$U_{12} < 0.53\sqrt{\frac{Y}{\rho_\mathrm{p}}} = 0.53\times\sqrt{\frac{2.5\times 10^8}{8.9\times 10^3}} = 88.8\ \mathrm{m/s} \tag{E2.29}$$

Thus, $U_{12} = 0.5$ m/s is reasonable for the model. From Eq. (2.167), the restitution coefficient is obtained as

$$e = 2.29\left(\frac{U_{12\mathrm{Y}}}{U_{12}}\right)^{\frac{1}{4}} = 2.29\left(\frac{0.0075}{0.5}\right)^{\frac{1}{4}} = 0.80 \tag{E2.30}$$

Hence, the rebounding velocity is found from Eq. (2.166) as

$$U'_{12} = eU_{12} = 0.80\times 0.5 = 0.40\ \mathrm{m/s} \tag{E2.31}$$

Nomenclature

A	Constant, defined by Eq. (2.26)	C	Constant, defined by Eq. (2.42)
A_c	Contact area	C	Constant, defined by Eq. (2.67)
a	Particle radius	c	Radius of contact circle without microslip
a^*	Relative radius, defined by Eq. (2.66)	d_p	Particle diameter
B	Constant, defined by Eq. (2.32)	E	Total kinetic energy

E	Young's modulus
E	Complete elliptic integral of the second kind, defined by Eq. (2.117)
E^*	Contact modulus, defined by Eq. (2.66)
e	Coefficient of restitution
F	Force
F_t	Tangential force
F_z	Normal impact force (or force component)
f	Coefficient of friction
f_{12}	Collisional force between particles 1 and 2
f_k	Coefficient of kinetic friction
f_m	Maximum collisional force in a normal impact
f_s	Coefficient of static friction
G	Shear modulus
h	Height
I	Moment of inertia
J	Normal impulse during the compression process
J'	Normal impulse during the rebounding process
K	Complete elliptic integral of the first kind, defined by Eq. (2.116)
k	Radius of gyration
l	Displacement
M_z	Twist moment
m	Mass of a particle
m^*	Relative mass
n	Yield stress in simple shear
p	Pressure
p_m	Averaged contact pressure
q	Tangential traction
R	Spherical coordinate
r	Cylindrical coordinate
r_c	Radius of contact circle
s	Arbitrary coordinate
$\boldsymbol{T}$	Total stress tensor
t	Time
t_c	Contact time duration
U	Normal velocity component in a biparticle collision
U_{12Y}	Critical normal collision velocity
U_1	Normal velocity component for particle 1 in a biparticle collision
U_1'	Normal rebounding velocity component for particle 1 in a biparticle collision
U_2	Normal velocity component for particle 2 in a biparticle collision
U_2'	Normal rebounding velocity component for particle 2 in a biparticle collision
V	Tangential velocity component in a biparticle collision
x	Cartesian coordinate
Y	Yield stress in simple compression or tension
y	Cartesian coordinate
z	Cylindrical coordinate

Greek Symbols

α	Center approaching distance
α_m	Center approaching distance under maximum deformation
β	Arbitrary angle, defined by Fig. 2.8
β	Twist angle
γ	Shear strain
ϵ	Normal strain
θ	Cylindrical coordinate
θ	Spherical coordinate
δ	Distance between the points of application of pair forces
δ_x	Rigid displacement
ζ	Dimensionless center approaching distance
λ	Parameter for penetration of microslip, defined by Eq. (2.120)
ν	Poisson's ratio

ρ_p	Density of particles	χ	Parameter for penetration of microslip, defined by Eq. (2.115)
σ	Normal stress	ψ	Love's stress function
τ	Shear stress	ω	Angular velocity
ς	Length, defined by Eq. (2.83)		
ϕ	Spherical coordinate		

References

Boussinesq, J. (1885). *Application des Potentiels a l'Étude de l'Équlibre et du Mouvement des Solids Élastiques*. Paris: Gauthier-Villars.

Goldsmith, W. (1960). *Impact: The Theory and Physical Behavior of Colliding Solids*. London: Edward Arnold.

Hertz, H. (1881). Über die Berührung fester elastischer Körper. *J. Reine Angew. Math. (Crelle)*, **92**, 155.

Johnson, K. L. (1985). *Contact Mechanics*. Cambridge: Cambridge University Press.

Lamé, G. and Clapeyron, B. P. E. (1831). Mémore sur léquilibre intérieur des corps solides homogénes. *J. F. Math. (Crelle)*, **7**.

Landau, L. D. and Lifshitz, E. M. (1970). *Theory of Elasticity*, 2nd ed. New York: Pergamon Press.

Love, A. E. H. (1944). *A Treatise on the Mathematical Theory of Elasticity*, 4th ed. New York: Dover.

Lubkin, J. L. (1951). The Torsion of Elastic Sphere in Contact. *Trans. ASME, J. of Appl. Mech.*, **18**, 183.

Mindlin, R. D. (1949). Compliance of Elastic Bodies in Contact. *Trans. ASME, J. of Appl. Mech.*, **16**, 259.

Timoshenko, S. P. and Goodier, J. N. (1970). *Theory of Elasticity*. New York: McGraw-Hill.

Problems

2.1 Show that the selected set of stresses expressed by Love's stress function satisfy the equation of equilibrium for axisymmetric and torsionless deformation in a solid.

2.2 A spherical solid ball is in normal contact with a spherical seat, as shown in Fig. P2.1. Derive the expressions for the radius of the contact area, r_c, and the distance change between the centers of the contact bodies, α. The radii of the spherical seat and the solid ball are R_1 and R_2, respectively.

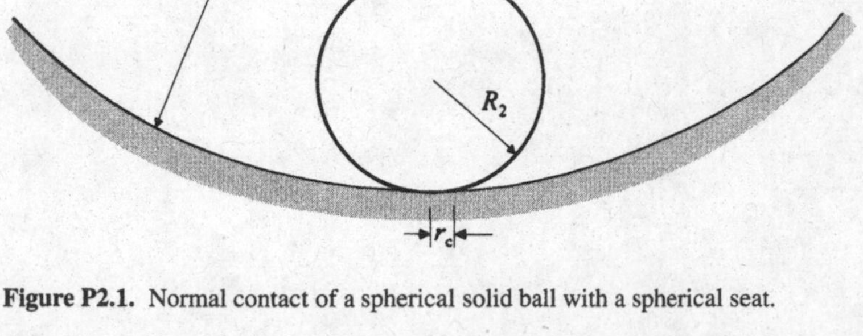

Figure P2.1. Normal contact of a spherical solid ball with a spherical seat.

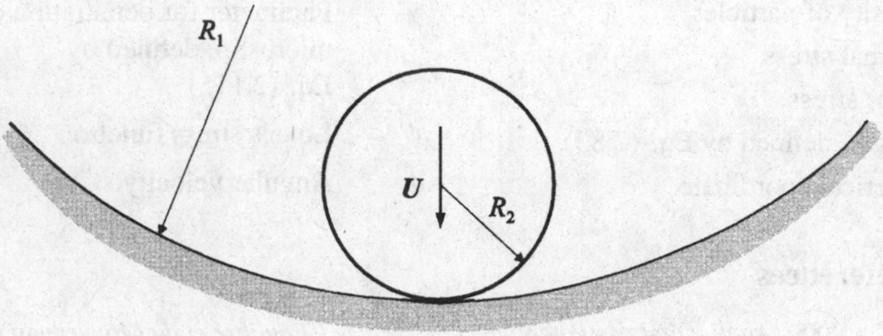

Figure P2.2. Collision between a solid ball and a spherical seat.

2.3 Derive the expressions for the maximum collision force and the collision time when a spherical solid ball collides with a spherical seat, as shown in Fig. P2.2, with a contact velocity U. This seat can be assumed unmovable.

2.4 Using the stresses given in Eq. (2.156), show that the yield stress in simple compression varies from 0.59 to 0.63 times the maximum pressure for materials with Poisson's ratio varying from 0.27 to 0.36.

2.5 Calculate the yield impact velocity for the collision between a stainless steel ball bearing and a brass ball. The diameter for each ball is 3 mm.

2.6 Verify that, for a collinear collision of two frictionless, nonspinning, rigid spherical particles, the total kinetic energy loss can be expressed by Eq. (2.6).

CHAPTER 3

Momentum Transfer and Charge Transfer

3.1 Introduction

In gas–solid flows, flow patterns of both phases depend not only on the initial conditions and physical boundaries of the system but also on the mechanisms of momentum transfer or the interacting forces between the phases. The forces controlling the motions of particles may be classified into three groups: (1) forces through the interface between fluid and particles, (2) forces due to the interactions between particles, and (3) forces imposed by external fields. Although interparticle forces and field forces do not directly change the course of the fluid motion, they may indirectly influence the motion via particle–fluid interactions.

Among the significant characteristics in a gas–solid flow are the charge transfer and charge accumulation among particles due to the interparticle collisions and/or particle–wall collisions. The charges carried by the particles may directly affect the particle motion and particle concentration distribution in the flow by the electrostatic forces, especially when an external electric field is applied. Even in the absence of an external electric field, the accumulation of particle charges may yield adverse effects such as arc discharge and dust explosion.

In this chapter, various forces affecting the motion of a gas–solid flow along with the momentum equations governing the motions of particles are delineated. The mechanisms of charge generation and charge transfer by collisions of particles are discussed.

3.2 Particle–Fluid Interactions

The motion of a single particle without short-range interactions such as the van der Waals force and collisional forces in the absence of external force fields may be generally described by a nonspherical particle accelerating and rotating in a nonuniform flow field. The complex movement of the particle may be logically perceived to comprise four types of simple motion of a sphere or an equivalent sphere assuming that they are noninteractive: (1) a particle moving with a constant velocity in a uniform flow field, (2) a particle accelerating in a uniform flow field, (3) a particle moving with a constant velocity in a nonuniform flow field, and (4) a particle rotating with a constant angular velocity in a uniform flow field. Four different forces, each representing one of the types of simple motion given, are the drag force, Basset force, Saffman force, and Magnus force, respectively.

3.2.1 *Drag Force*

In a particle–fluid flow, the particle velocity $\boldsymbol{U}_{\mathrm{p}}$ generally differs from the fluid velocity $\boldsymbol{U}$. The slip velocity, $\boldsymbol{U} - \boldsymbol{U}_{\mathrm{p}}$, leads to unbalanced pressure distribution as well as viscous stresses on the particle surface, which yield a resulting force known as the drag force. The drag force of a single particle in a uniform flow field can be generally expressed by

$$F_D = C_D A \frac{\rho}{2} |U - U_p| (U - U_p) \tag{3.1}$$

where A is the exposed frontal area of the particle to the direction of the incoming flow and C_D is the drag coefficient, which is a function of the particle Reynolds number, Re_p, and the local turbulent intensity of the fluid. Re_p is defined by

$$Re_p = \rho d_p \frac{|U - U_p|}{\mu} \tag{3.2}$$

At the two extremes, $Re_p \gg 1$ but not high enough to exceed that for the transition regime and $Re_p \ll 1$ for the creeping flow regime, C_D is well expressed in light of either experimental or analytical work. In the high Reynolds number regime, $C_D = 0.44$, originally determined from Newton's experiments in 1710. Newton's relation covers a range of Re_p from 700 to 10^5 over which the inertia effect is predominant. In the creeping flow regime where the viscous effect dictates the flow and the inertia is negligible, Stokes solved the simplified governing equations and revealed that $C_D = 24/Re_p$ in 1850.

An enormous amount of experimental data on the drag coefficient for a single sphere at various Re_p has been compiled into a standard curve [Schlichting, 1979], as shown in Fig. 1.4. As the figure illustrates, the sharp reduction in the drag coefficient at high Re_p (around 3×10^5) corresponds to the transition from a laminar to a turbulent boundary layer over the particle. This transition is due to the change of surface pressure distribution around the particle caused by the change of the wake structure behind the particle in the turbulent regime. It should be pointed out that the standard curve was obtained under uniform and undisturbed flow conditions. However, in most gas–solid flows, the fluid itself is usually turbulent even when the relative velocity and the resulting particle Reynolds number are small. The local turbulent intensity of the fluid causes deviations in C_D from the standard curve. Unfortunately, no general conclusive relations between C_D and the local turbulent intensity of the fluid are available at this stage. The drag coefficient of a particle in a multiphase system can be further affected by the existence and the movement of neighboring particles. Studies on the drag forces of interacting particles in multiphase flows [Rowe and Henwood, 1961; Lee, 1979; Tsuji *et al.*, 1982; Zhu *et al.*, 1994] suggest that the drag coefficient of an individual particle under the influence of a neighboring particle may be expressed by

$$\frac{C_D}{C_{D0}} = 1 - (1 - A) \exp\left(-B \frac{l}{d_p}\right) \tag{3.3}$$

where C_{D0} is the drag coefficient from the standard curve in Fig. 1.4; l is the distance between the two interacting particles; and A and B are empirical coefficients which may be a function of Re_p and the deflecting angle between the direction of the relative velocity and the line connecting the centers of the two interactive particles.

3.2.2 *Basset Force*

Once a particle is accelerating or decelerating in a fluid, a force known as the Basset force becomes important. The Basset force can be rigorously derived from the motion of a single accelerating sphere in the Stokes regime in a quiescent fluid. For particle Reynolds numbers beyond the Stokes regime, the original Basset term may be modified by introducing a correction factor [Odar and Hamilton, 1964; Clift *et al.*, 1978]. In this section, the original form of the Basset force is derived [Basset, 1888], along with a brief discussion of the importance of this force relative to the Stokes drag.

3.2.2.1 *Stream Function for Linear Motion of a Sphere*

For simplicity, we start with the derivation in cylindrical coordinates. The resultant equations are then transformed to polar coordinates in which the final form of the Basset force is obtained.

Consider a sphere of radius a moving along a straight line in an infinite stagnant fluid medium at a velocity V. The cylindrical coordinates are selected such that the z-axis coincides with the path of the sphere, and the origin is an arbitrarily fixed point on the path. Assuming that the fluid is incompressible and its motion is axisymmetric, the stream function ψ may be defined such that

$$u_r = -\frac{1}{r}\frac{\partial \psi}{\partial z}, \qquad u_z = \frac{1}{r}\frac{\partial \psi}{\partial r} \tag{3.4}$$

where u_r and u_z are the radial and axial velocity components of the fluid, respectively. Assume that the sphere moves in the Stokes regime where the inertia effects can be neglected. Thus, the Navier–Stokes equations reduce to

$$\frac{\partial u_r}{\partial t} = -\frac{1}{\rho}\frac{\partial p}{\partial r} + \nu\left(\nabla^2 u_r - \frac{u_r}{r^2}\right) \tag{3.5}$$

$$\frac{\partial u_z}{\partial t} = -\frac{1}{\rho}\frac{\partial p}{\partial z} + \nu\nabla^2 u_z \tag{3.6}$$

where ν is the kinematic viscosity. Defining an operator D as

$$D = \frac{\partial^2}{\partial z^2} + \frac{\partial^2}{\partial r^2} - \frac{1}{r}\frac{\partial}{\partial r} \tag{3.7}$$

Equations (3.4) through (3.6) yield

$$\begin{aligned} -\frac{1}{\rho}\frac{\partial p}{\partial z} &= \frac{1}{r}\frac{\partial}{\partial r}\left(\frac{\partial \psi}{\partial t} - \nu D\psi\right) \\ \frac{1}{\rho}\frac{\partial p}{\partial r} &= \frac{1}{r}\frac{\partial}{\partial z}\left(\frac{\partial \psi}{\partial t} - \nu D\psi\right) \end{aligned} \tag{3.8}$$

Eliminating the pressure term from the preceding equations leads to

$$D\left(D - \frac{1}{\nu}\frac{\partial}{\partial t}\right)\psi = 0 \tag{3.9}$$

The solution of this equation is

$$\psi = \psi_1' + \psi_2' \tag{3.10}$$

where ψ_1' and ψ_2', respectively, satisfy

$$D\psi_1' = 0; \qquad \left(D - \frac{1}{\nu}\frac{\partial}{\partial t}\right)\psi_2' = 0 \tag{3.11}$$

Consequently, the pressure field is determined by

$$\mathrm{d}p = \frac{\rho}{r}\frac{\partial}{\partial t}\left(\frac{\partial \psi_1'}{\partial z}\,\mathrm{d}r - \frac{\partial \psi_1'}{\partial r}\,\mathrm{d}z\right) \tag{3.12}$$

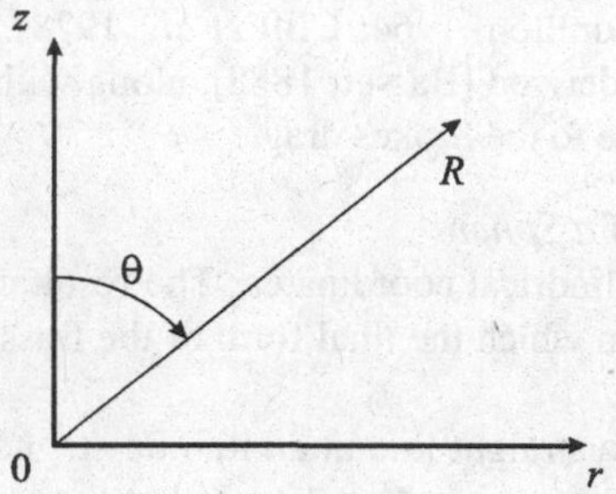

Figure 3.1. Transformation from (r, z) to (R, θ).

The preceding equations are obtained with the coordinate defined relative to a fixed origin. Now, consider a coordinate defined with the center of the sphere as the origin and denote ξ as the distance from a fixed point. We may express the axial component of velocity u_z as

$$u_z = f(z + \xi, r, t) \tag{3.13}$$

so that

$$\frac{du_z}{dt} = \frac{\partial f}{\partial t} + V\frac{\partial f}{\partial \xi} \tag{3.14}$$

The second term on the right-side of Eq. (3.14) is of the same order as the square of the velocity and so represents an inertia. Hence, this term can be omitted in Stokes flows. Therefore, it can be stated that both Eq. (3.5) and Eq. (3.6) are valid whether the origin is fixed or in motion.

We now transform the governing equations in cylindrical coordinates into polar coordinates. Since the motion is axisymmetric, the transformation from (r, z) to (R, θ), as shown in Fig. 3.1, is analogous to the transformation from Cartesian coordinates (x, y) to cylindrical coordinates (r, θ) in a two-dimensional domain. The stream function is related to the velocity components in polar coordinates by

$$u_R = \frac{1}{R^2 \sin\theta}\frac{\partial \psi}{\partial \theta}, \qquad u_\theta = -\frac{1}{R\sin\theta}\frac{\partial \psi}{\partial R} \tag{3.15}$$

The operator D, by analogy, is changed directly from Eq. (3.7) to

$$D = \frac{\partial^2}{\partial R^2} + \frac{1}{R}\frac{\partial}{\partial R} + \frac{1}{R^2}\frac{\partial^2}{\partial \theta^2} - \frac{1}{R\sin\theta}\left(\frac{\partial}{\partial r}\right)_z \tag{3.16}$$

Noting that

$$\left(\frac{\partial}{\partial r}\right)_z = \left(\frac{\partial R}{\partial r}\right)_z \frac{\partial}{\partial R} + \left(\frac{\partial \theta}{\partial r}\right)_z \frac{\partial}{\partial \theta} = \sin\theta\frac{\partial}{\partial R} + \frac{\cos\theta}{R}\frac{\partial}{\partial \theta} \tag{3.17}$$

Equation (3.16) becomes

$$D = \frac{\partial^2}{\partial R^2} + \frac{1}{R^2}\frac{\partial^2}{\partial \theta^2} - \frac{\cot\theta}{R^2}\frac{\partial}{\partial \theta} \tag{3.18}$$

Similarly, Eq. (3.12) can be expressed in polar coordinates as

$$dp = \frac{\rho}{\sin\theta}\frac{\partial}{\partial t}\left(\frac{\partial \psi_1'}{\partial R}\,d\theta - \frac{1}{R^2}\frac{\partial \psi_1'}{\partial \theta}\,dR\right) \tag{3.19}$$

The boundary conditions at the surface of the sphere are

$$\begin{aligned} u_R|_{R=a} &= V\cos\theta = \frac{1}{a^2\sin\theta}\frac{\partial \psi}{\partial \theta}\bigg|_{R=a} \\ u_\theta|_{R=a} &= -V\sin\theta = -\frac{1}{a\sin\theta}\frac{\partial \psi}{\partial R}\bigg|_{R=a} \end{aligned} \tag{3.20}$$

3.2.2.2 *Solution of Equations in Terms of Stream Function*

Equation (3.9) in polar coordinates can be satisfied if ψ is in a form

$$\psi = (\psi_1 + \psi_2)\sin^2\theta \tag{3.21}$$

where ψ_1 and ψ_2 are functions of R and t, which satisfy

$$\frac{\partial^2 \psi_1}{\partial R^2} - \frac{2\psi_1}{R^2} = 0 \tag{3.22}$$

$$\frac{\partial^2 \psi_2}{\partial R^2} - \frac{2\psi_2}{R^2} = \frac{1}{\nu}\frac{\partial \psi_2}{\partial t} \tag{3.23}$$

Solutions of Eq. (3.22) and Eq. (3.23) can be expressed by

$$\psi_1 = \frac{1}{2R}\sqrt{\frac{\pi}{\nu t}}\int_0^\infty \chi(\zeta)\exp\left(-\frac{\zeta^2}{4\nu t}\right)d\zeta \tag{3.24}$$

and

$$\psi_2 = \frac{R}{2}\sqrt{\frac{\pi}{\nu t}}\frac{\partial}{\partial R}\int_0^\infty \frac{\phi(\xi)}{R}\exp\left(-\frac{(R-a+\xi)^2}{4\nu t}\right)d\xi \tag{3.25}$$

where $\chi(\zeta)$ and $\phi(\xi)$ are functions to be determined by the boundary conditions given by Eq. (3.20).

Using Eq. (3.20) and assuming $\phi(0) = \phi'(0) = 0$ and $\phi(\xi)$ and $\phi'(\xi)$ are finite as $\xi \to \infty$, χ and ϕ can be determined as [Basset, 1888]

$$\chi(\zeta) = \frac{Va}{\pi}\left(\frac{3}{2}\zeta^2 + 3a\zeta + a^2\right) \tag{3.26}$$

and

$$\phi(\xi) = \frac{3}{2\pi}Va\xi^2 \tag{3.27}$$

Hence, ψ_1 and ψ_2 are obtained as

$$\psi_1 = \frac{Va}{2R}\left(3\nu t + 6a\sqrt{\frac{\nu t}{\pi}} + a^2\right) \tag{3.28}$$

and

$$\psi_2 = -\frac{3aV}{\sqrt{\pi}}\int_{\frac{R-a}{2\sqrt{\nu t}}}^\infty \left(\frac{1}{2R}(2\eta\sqrt{\nu t} - R + a)^2 + 2\eta\sqrt{\nu t} - R + a\right)e^{-\eta^2}\,d\eta = -Vf(t) \tag{3.29}$$

The preceding solution is for the case involving a particle moving at a constant velocity V. For the case involving an accelerating particle, in order to obtain the stream function, we introduce the following theorem first. Let $\mathrm{v_s}(t, R)$ be a solution of the equation

$$\Phi\left(\frac{\partial}{\partial R}\right)\mathrm{v_s} = \frac{\partial \mathrm{v_s}}{\partial t} \tag{3.30}$$

where Φ is an operator. Then, for the initial condition $\mathrm{v_s}(0, R) = 0$, w is also a solution of Eq. (3.30) if

$$w = \int_0^t G(t-\tau)\mathrm{v_s}(\tau, R)\,\mathrm{d}\tau \tag{3.31}$$

where $G(\tau)$ is an arbitrary function which is independent of R and t, and does not become infinite between the limits. Thus, replacing $G(\tau)$ in Eq. (3.31) with $F'(\tau)$ (=$\mathrm{d}F/\mathrm{d}\tau$) and noting that $f(0) = 0$ and $f(t)$ is a solution of Eq. (3.23), we may rewrite ψ_2 as

$$\psi_2 = -\int_0^t F'(t-\tau)f(\tau)\,\mathrm{d}\tau \tag{3.32}$$

where $F'(\tau)$ is a function to be determined. The new ψ_2 is also a solution of Eq. (3.23). Following the same procedure, a new ψ may be deduced from Eqs. (3.28) and (3.29) by changing t into τ and V into $F'(t-\tau)\,\mathrm{d}\tau$, and integrating the resulting equations from t to 0 as

$$\psi = \sin^2\theta \int_t^0 F'(t-\tau)\left[\frac{a}{2R}\left(3\nu\tau + 6a\sqrt{\frac{\nu\tau}{\pi}} + a^2\right) - f(\tau)\right]\mathrm{d}\tau \tag{3.33}$$

Equation (3.33) is also a solution of Eq. (3.9). If $F(0) = 0$, it can be found by substituting the new ψ into Eq. (3.20) that $F(t)$ is the velocity of the sphere from rest. Hence, Eq. (3.33) represents the stream function for the case of the accelerating motion of a particle.

3.2.2.3 *Basset Force and Carried Mass*

In evaluating the particle resistance due to relative motion to the fluid, it is convenient to begin with the constant velocity case and work to the time-dependent case. Consider the case of a sphere moving in a gravitational field and assume that the gravitational acceleration is in the same direction as the sphere motion. Denote F_z as the resistance. Since the total resistance is due to the surface shear and surface pressure, and the flow is axisymmetric, F_z can be postulated as

$$F_\mathrm{z} = 2\pi a \int_0^\pi \left(pa\cos\theta - \rho\frac{\partial\psi_2}{\partial t}\sin^2\theta\right)_a \sin\theta\,\mathrm{d}\theta \tag{3.34}$$

From Eq. (3.19) and taking the gravitational force into consideration, we have

$$\frac{\partial p}{\partial\theta} = \rho\sin\theta\frac{\partial^2\psi_1}{\partial t\,\partial R} - g\rho a\sin\theta \tag{3.35}$$

Substituting Eqs. (3.28), (3.29), and (3.35) into Eq. (3.34) yields

$$F_\mathrm{z} = \frac{M_\mathrm{a}}{a^2}\frac{\partial}{\partial t}\left[V\left(\frac{9}{2}\nu t + 9a\sqrt{\frac{\nu t}{\pi}} + \frac{a^2}{2}\right)\right] + M_\mathrm{a}g \tag{3.36}$$

where M_a is the mass of the fluid displaced. Changing t into τ and V into $F'(t-\tau)\,d\tau$ and then integrating the term in square brackets in Eq. (3.36) from t to 0, Eq. (3.36) becomes (Problem 3.3)

$$F_z = \frac{1}{2}M_a\frac{\partial v}{\partial t} + M_a g + \frac{9M_a}{2a^2}\nu v + \frac{9M_a}{2a}\sqrt{\frac{\nu}{\pi}}\int_0^t \frac{\frac{\partial v}{\partial \tau}}{\sqrt{t-\tau}}\,d\tau \tag{3.37}$$

where v is equal to $F(t)$ with $F(0) = 0$. The first and last terms on the right-hand side of Eq. (3.37) are the force due to the acceleration of the carried mass and the Basset force, respectively. The second term is the buoyancy force. The third term is the Stokes drag force. The carried mass is also termed the added mass or the virtual mass; it contributes to the resistance because acceleration of the particle is also accompanied by acceleration of the fluid; the mass of the carried fluid that has the same acceleration as the particle is noted as the carried mass. The volume of the carried mass, which generally depends on the particle geometry, is equal to one-half the particle volume when the particle is spherical. The Basset force or the Basset history integral accounts for the effect of past acceleration on the resistance. Here $(t-\tau)$ represents the time elapsed since past acceleration from 0 to t.

The Basset force can be substantial when the particle is accelerated at a high rate. The total force on a particle in acceleration can be many times that in a steady state [Hughes and Gilliland, 1952]. In a simple model with constant acceleration, the ratio of the Basset force to the Stokes drag, R_{BS}, was derived [Wallis, 1969] and rearranged to [Rudinger, 1980]

$$R_{BS} = \sqrt{\frac{18}{\pi}\frac{\rho}{\rho_p}\frac{\tau_S}{t}} \tag{3.38}$$

where τ_S is the Stokes relaxation time defined as

$$\tau_S = \frac{\rho_p d_p^2}{18\mu} \tag{3.39}$$

The Basset force may be negligible when the fluid–particle density ratio is small, *e.g.*, in most gas–solid suspensions, and the time change is much longer than the Stokes relaxation time or the acceleration rate is low.

Example 3.1 Consider the collision of two particles due to a wake attraction, as shown in Fig. E3.1. Some assumptions for the collision process are as follows:

(1) The motion of the leading particle is not affected by the approach of the trailing particle.
(2) Particles are equal-sized, rigid, and spherical.
(3) Initially, the velocities of the two particles differ only slightly and are close to their respective particle terminal velocities. The particles are separated by a characteristic distance l_0 which is within the wake-influencing distance.
(4) Particle acceleration is proportional to the relative velocity.

Develop a model to estimate the relative particle collision velocity due to the wake attraction.

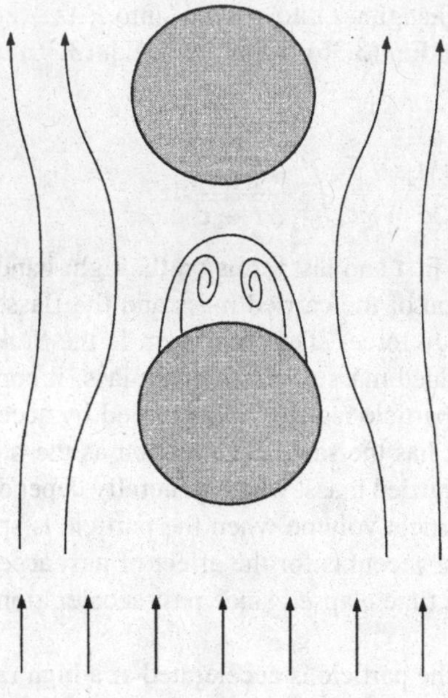

Figure E3.1. Schematic flow pattern of two interactive spheres due to wake attraction.

Solution A general momentum equation for the trailing particle takes the form

$$\left(\rho_p + \frac{1}{2}\rho\right) V \frac{dU}{dt} = (\rho_p - \rho) V g - F_D - F_B \tag{E3.1}$$

where V is the volume of the particle; U is the relative velocity of the trailing particle to the leading particle; and F_B is the Basset force. It is assumed that the expression for the Basset force given in the following can be extended beyond the Stokes regime

$$F_B = \frac{3}{2} d_p^2 \sqrt{\pi \rho \mu} \int_0^t \frac{\frac{dU}{d\tau}}{\sqrt{t - \tau}} d\tau \tag{E3.2}$$

On the basis of Assumption (3), the drag force of a single noninteracting particle F_{D0} is written as

$$F_{D0} = (\rho_p - \rho) V g \tag{E3.3}$$

The drag coefficient of the trailing particle can be given by Eq. (3.3). Substituting Eqs. (E3.2), (E3.3), (3.1), and (3.3) into Eq. (E3.1) yields

$$\frac{dU}{dt} = \frac{2(\rho_p - \rho)}{2\rho_p + \rho} g (1 - A) \exp\left\{ \frac{B}{d_p} \int_t^{t_c} U \, dt \right\} - \frac{18\sqrt{\pi \rho \mu}}{(2\rho_p + \rho)\pi d_p} \int_0^t \frac{\frac{dU}{d\tau}}{\sqrt{t - \tau}} d\tau \tag{E3.4}$$

where t_c is the elapsed time for particles to collide. From Assumption (4), we have

$$\frac{dU}{dt} = CU \tag{E3.5}$$

which leads to

$$U = U_c \exp[-C(t_c - t)] \tag{E3.6}$$

where U_c is the collision velocity and C is a constant. On the basis of Eq. (E3.6), the initial velocity may be expressed by

$$U_0 = U_c \exp(-Ct_c) \tag{E3.7}$$

Substituting Eqs. (E3.5) and (E3.6) into Eq. (E3.4) gives rise to

$$CU = \frac{2(\rho_p - \rho)}{2\rho_p + \rho} g(1 - A)\exp\left\{\frac{B}{d_p}\int_t^{t_c} U\,dt\right\} - \frac{18U}{(2\rho_p + \rho)d_p}\sqrt{C\rho\mu}\,\mathrm{erf}(\sqrt{Ct}) \tag{E3.8}$$

where $\mathrm{erf}(x)$ is the error function of x. At the moment of contact, Eq. (E3.8) reduces to

$$\frac{(U_c - U_0)}{l_0}U_c = \frac{2(\rho_p - \rho)}{2\rho_p + \rho} g(1 - A) - \frac{18U_c}{(2\rho_p + \rho)d_p}\sqrt{\frac{(U_c - U_0)}{l_0}\rho\mu}\,\mathrm{erf}\left[\sqrt{\ln\left(\frac{U_c}{U_0}\right)}\right] \tag{E3.9}$$

Equation (E3.9) expresses the collision velocity implicitly as a function of the given flow conditions and the properties of the fluid and particles.

3.2.3 *Saffman Force and Other Gradient-Related Forces*

When a sphere moves in a flow where a velocity gradient, pressure gradient, or temperature gradient exists, additional forces related to these gradients can be as important as the drag force.

3.2.3.1 *Saffman Force*

In a region where a velocity gradient exists, for instance, near a wall or in a high shear region, a sphere moving at a constant velocity is subjected to a lift force caused by the velocity gradient. This lift force is termed the Saffman force. The Saffman force was originally derived for the motion of a sphere at a constant velocity in a simple shear flow at low $\mathrm{Re_p}$ [Saffman, 1965, 1968]. Using asymptotic expansions and Fourier transforms of the velocity field, Saffman was able to deduce an expression for the lift force

$$F_S = \frac{K\mu}{4}|U - U_p|\,d_p^2\sqrt{\frac{1}{\nu}\left|\frac{\partial(U - U_p)}{\partial y}\right|} \tag{3.40}$$

where the constant K was determined to be 6.46 on the basis of a numerical integration for creeping flows at low shear rates. The ratio of the Saffman force to the Stokes drag, R_{SS}, is given as

$$R_{SS} = \frac{Kd_p}{12\pi}\sqrt{\frac{1}{\nu}\left|\frac{\partial(U - U_p)}{\partial y}\right|} \tag{3.41}$$

In a shear flow with a constant shear rate, R_{SS} can be estimated by

$$R_{SS} = \frac{K}{12\pi}\sqrt{\mathrm{Re_p}} \tag{3.42}$$

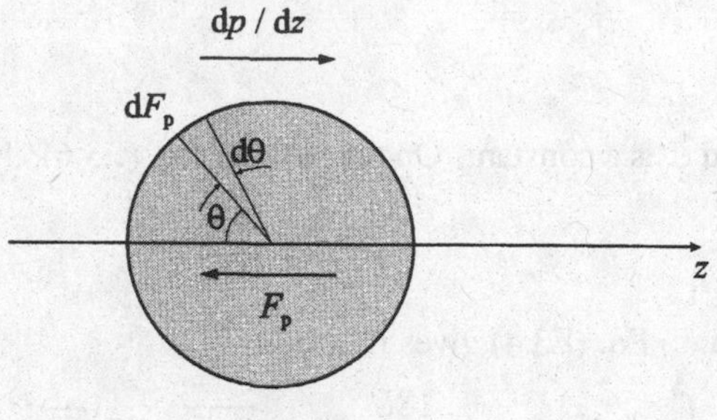

Figure 3.2. Force on a sphere due to pressure gradient.

which indicates that the Saffman force is negligible at very small shear rate or very low Re_p.

3.2.3.2 *Force Due to Pressure Gradient*

Besides the drag force, Basset force, and Saffman force, another force may act on the particle as a result of the existence of a pressure gradient in the fluid. Using the axisymmetric condition, the force on a differential element of a sphere in a pressure gradient field shown in Fig. 3.2 can be expressed by

$$dF_p = -2\pi a^3 \frac{\partial p}{\partial z} \sin\theta \cos^2\theta \, d\theta \tag{3.43}$$

Equation (3.43) gives

$$F_p = -2\pi a^3 \int_0^{\pi} \frac{\partial p}{\partial z} \sin\theta \cos^2\theta \, d\theta = -\frac{\partial p}{\partial z} \frac{\pi d_p^3}{6} \tag{3.44}$$

The minus sign indicates that the force is in the opposite direction of the pressure gradient [Tchen, 1947]. The significance of this force is evident, for example, when a shock wave propagates through a gas–solid suspension.

3.2.3.3 *Radiometric Force*

Forces acting on a particle as a result of a temperature gradient in a gas (thermophoresis) or nonuniform radiation (photophoresis) are known as radiometric forces. For a particle diameter much larger than the mean free path of a gas, the force due to the temperature gradient is given by [Hettner, 1926]

$$\boldsymbol{F}_T = -\frac{3\pi}{2} \mu^2 d_p \frac{R_M}{p} \nabla T_p \tag{3.45}$$

where R_M is the gas constant, and the temperature gradient inside the sphere, ∇T_p, is given by [Epstein, 1929]

$$\nabla T_p = \frac{3K}{2K + K_p} \nabla T \tag{3.46}$$

where K and K_p are the thermal conductivities of the gas and the solid, respectively. Equation (3.45) shows that the force is directed toward the lower-temperature region.

At room conditions and in air, the radiometric forces are important only for submicrometer particles. However, under conditions of high temperatures and/or large temperature gradients such as found in plasma coating, the thermophoresis effect may also be significant for larger particles.

3.2.4 *Magnus Effect and Force Due to Rotation of a Sphere*

Particle rotation may result from collisions of the particle with a solid wall or other particles, or from a velocity gradient in a nonuniform flow region. In low Reynolds number flows, particle rotation leads to fluid entrainment, resulting in an increase in the velocity on one side of the particle and a decrease in the velocity on the other side. Thus, a lift force is established which moves the particle toward the region of higher velocity. This phenomenon is known as the Magnus effect [Magnus, 1852] and the lift force is called the Magnus force.

For a sphere rotating in a uniform flow at low Reynolds number, the lift force can be obtained by solving the Navier–Stokes equations using an asymptotic expansion and a matching technique [Rubinow and Keller, 1961]. The solution procedure is noteworthy because of its uniqueness in yielding not only the lift force but also the Stokes drag and the Oseen drag of a spinning sphere. The Stokes stream function method cannot be applied to this situation because of the asymmetric nature of the flow around the rotating sphere.

Consider a steady motion of a sphere of radius a, velocity $\boldsymbol{U}_\mathrm{p}$, and angular velocity $\boldsymbol{\Omega}$ in an incompressible fluid at rest, as shown in Fig. 3.3. By selecting the Cartesian coordinates with the origin attached to the center of the sphere, with the x-axis pointing to the opposite direction of $\boldsymbol{U}_\mathrm{p}$, and with the x–y plane containing $\boldsymbol{\Omega}$, we have

$$\nabla \cdot \boldsymbol{U} = 0 \tag{3.47}$$

$$\boldsymbol{U} \cdot \nabla \boldsymbol{U} = -\frac{1}{\rho}\nabla p + \nu \nabla^2 \boldsymbol{U} \tag{3.48}$$

$$\boldsymbol{U}|_{r'=a} = \boldsymbol{\Omega} \times \boldsymbol{r}'; \qquad \boldsymbol{U}|_{r'=\infty} = (U_\mathrm{p}, 0, 0), \qquad p|_{r'=\infty} = 0 \tag{3.49}$$

where r' is the radial distance in the spherical coordinate. It is convenient to define dimensionless quantities $\overline{\boldsymbol{u}}$, $\overline{p}$, $\boldsymbol{\omega}$, and $\boldsymbol{r}$ by

$$\overline{\boldsymbol{u}} = \frac{\boldsymbol{U}}{U_\mathrm{p}}, \qquad \overline{p} = \frac{ap}{\mu U_\mathrm{p}}, \qquad \boldsymbol{\omega} = \frac{a}{U_\mathrm{p}}\boldsymbol{\Omega}, \qquad \boldsymbol{r} = \frac{\boldsymbol{r}'}{a} \tag{3.50}$$

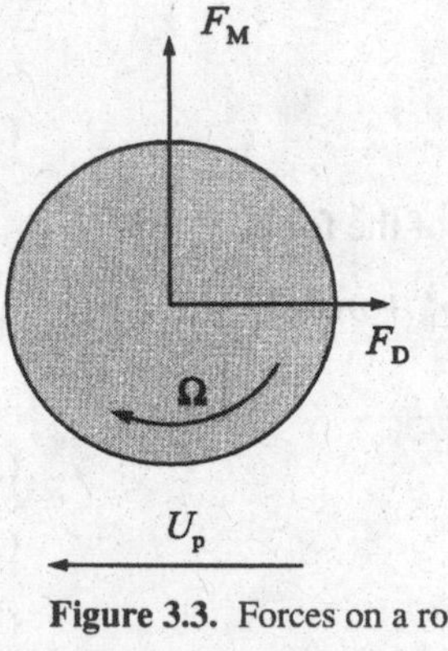

Figure 3.3. Forces on a rotating and moving sphere without wake separation.

so that Eqs. (3.47) through (3.49) become

$$\nabla \cdot \overline{\boldsymbol{u}} = 0 \tag{3.51}$$

$$\overline{\boldsymbol{u}} \cdot \nabla \overline{\boldsymbol{u}} = \frac{1}{\mathrm{Re_a}}(-\nabla \overline{p} + \nabla^2 \overline{\boldsymbol{u}}) \tag{3.52}$$

$$\overline{\boldsymbol{u}}|_{r=1} = \boldsymbol{\omega} \times \boldsymbol{r}; \qquad \overline{\boldsymbol{u}}|_{r=\infty} = (1, 0, 0), \qquad \overline{p}|_{r=\infty} = 0 \tag{3.53}$$

where $\mathrm{Re_a}$ is defined by

$$\mathrm{Re_a} = \frac{U_\mathrm{p} a}{\nu} \tag{3.54}$$

Assume that the solutions of $\overline{\boldsymbol{u}}$ and $\overline{p}$ can be expressed by the Stokes expansion in the form

$$\overline{\boldsymbol{u}} = \overline{\boldsymbol{u}}_0 + \mathrm{Re_a}\overline{\boldsymbol{u}}_1 + o(\mathrm{Re_a}); \qquad \overline{p} = \overline{p}_0 + \mathrm{Re_a}\overline{p}_1 + o(\mathrm{Re_a}) \tag{3.55}$$

The zero-order Stokes approximation thus satisfies

$$\nabla \cdot \overline{\boldsymbol{u}}_0 = 0 \tag{3.56}$$

$$\nabla^2 \overline{\boldsymbol{u}}_0 - \nabla \overline{p}_0 = 0 \tag{3.57}$$

$$\overline{\boldsymbol{u}}_0|_{r=1} = \boldsymbol{\omega} \times \boldsymbol{r} \tag{3.58}$$

and the first-order Stokes approximation satisfies

$$\nabla \cdot \overline{\boldsymbol{u}}_1 = 0 \tag{3.59}$$

$$\nabla^2 \overline{\boldsymbol{u}}_1 - \nabla \overline{p}_1 = \overline{\boldsymbol{u}}_0 \cdot \nabla \overline{\boldsymbol{u}}_0 \tag{3.60}$$

$$\overline{\boldsymbol{u}}_1|_{r=1} = 0 \tag{3.61}$$

It is noted that the Stokes expansion is not uniformly valid in the neighborhood of infinity. Therefore, another expansion, *i.e.*, the Oseen expansion, is introduced to satisfy the boundary conditions at infinity. By using the matching technique, the final results can be obtained.

The Oseen approximation satisfies Eqs. (3.51) through (3.53) in stretched coordinates

$$\nabla \cdot \boldsymbol{u}^* = 0 \tag{3.62}$$

$$\boldsymbol{u}^* \cdot \nabla \boldsymbol{u}^* = -\nabla p^* + \nabla^2 \boldsymbol{u}^* \tag{3.63}$$

$$\boldsymbol{u}^*|_{s=\mathrm{Re_a}} = \frac{1}{\mathrm{Re_a}} \boldsymbol{\omega} \times \boldsymbol{s}; \qquad \boldsymbol{u}^*|_{s=\infty} = (1, 0, 0), \qquad p^*|_{s=\infty} = 0 \tag{3.64}$$

where

$$\boldsymbol{u}^* = \overline{\boldsymbol{u}}, \qquad p^* = \frac{\overline{p}}{\mathrm{Re_a}}, \qquad \boldsymbol{s} = \mathrm{Re_a}\, \boldsymbol{r} \tag{3.65}$$

Similar to the Stokes expansion, the Oseen expansion is of the form

$$\boldsymbol{u}^* = \boldsymbol{u}_0^* + \mathrm{Re_a}\, \boldsymbol{u}_1^* + o(\mathrm{Re_a}), \qquad p^* = p_0^* + \mathrm{Re_a}\, p_1^* + o(\mathrm{Re_a}) \tag{3.66}$$

so that the zero-order Oseen approximation satisfies

$$\nabla \cdot \boldsymbol{u}_0^* = 0 \tag{3.67}$$

$$\nabla^2 \boldsymbol{u}_0^* - \nabla p_0^* = \boldsymbol{u}_0^* \cdot \nabla \boldsymbol{u}_0^* \tag{3.68}$$

$$\boldsymbol{u}_0^*\big|_{s=\infty} = (1, 0, 0), \qquad p_0^*\big|_{s=\infty} = 0 \tag{3.69}$$

The first-order Oseen approximation satisfies

$$\nabla \cdot \boldsymbol{u}_1^* = 0 \tag{3.70}$$

$$\nabla^2 \boldsymbol{u}_1^* - \nabla p_1^* = \boldsymbol{u}_0^* \cdot \nabla \boldsymbol{u}_1^* + \boldsymbol{u}_1^* \cdot \nabla \boldsymbol{u}_0^* \tag{3.71}$$

$$\boldsymbol{u}_1^*\big|_{s=\infty} = 0, \qquad p_1^*\big|_{s=\infty} = 0 \tag{3.72}$$

The matching principle requires that the Stokes expansion and the Oseen expansion must be asymptotically equal in their common domain of validity. The Stokes expansion is valid from the sphere out to some large distance, whereas the Oseen expansion is valid from infinity to some small radii in the stretched variables. Hence, the common domain is a spherical shell within which the unstretched radius is large but the stretched radius is small. In this domain, the matching principle yields

$$\overline{\boldsymbol{u}}_0|_{r=\infty} = \boldsymbol{u}_0^* + o(1), \qquad \overline{p}_0|_{r=\infty} = o(1) \tag{3.73}$$

Choose a particular solution for $\boldsymbol{u}_0^*$ and p_0^* as

$$\boldsymbol{u}_0^* = (1, 0, 0), \qquad p_0^* = 0 \tag{3.74}$$

On the basis of Eq. (3.74), $\boldsymbol{u}_1^*$ and p_1^* can be solved as [Rubinow and Keller, 1961]

$$\boldsymbol{u}_1^* = \frac{3\boldsymbol{s}}{2s^3} - \frac{3}{4s} e^{\frac{1}{2}(X-s)} \left(\boldsymbol{i} + \frac{\boldsymbol{s}}{s} - \frac{2\boldsymbol{s}}{s^2} \right), \qquad p_1^* = -\frac{3X}{2s^3} \tag{3.75}$$

where X is the x-component of $\boldsymbol{s}$, and $\boldsymbol{i}$ is the unit vector in the x-direction.

Using Eq. (3.73), $\overline{\boldsymbol{u}}_0$ and $\overline{p}_0$ are obtained by

$$\overline{\boldsymbol{u}}_0 = \left(1 - \frac{3}{4r} - \frac{1}{4r^3}\right)\boldsymbol{i} - \frac{3x}{4}\left(\frac{1}{r^3} - \frac{1}{r^5}\right)\boldsymbol{r} + \frac{1}{r^3}\boldsymbol{\omega} \times \boldsymbol{r}, \qquad \overline{p}_0 = -\frac{3x}{2r^3} \tag{3.76}$$

and $\overline{\boldsymbol{u}}_1$ and $\overline{p}_1$ are given by

$$\begin{aligned}
\overline{\boldsymbol{u}}_1 = {} & \frac{3}{32}\left[4 - \frac{3}{r} - \frac{1}{r^3} - x\left(\frac{4}{r} - \frac{3}{r^2} + \frac{1}{r^4} - \frac{2}{r^5}\right)\right]\boldsymbol{i} + \frac{x}{4}(\boldsymbol{\omega} \times \boldsymbol{r})\left(\frac{2}{r^3} - \frac{3}{r^4} + \frac{1}{r^5}\right) \\
& + \frac{3}{32}\left[\frac{2}{r} - \frac{3}{r^2} + \frac{1}{r^3} - \frac{1}{r^4} + \frac{1}{r^5} - 3x\left(\frac{1}{r^3} - \frac{1}{r^5}\right)\right. \\
& \left. + x^2\left(-\frac{2}{r^3} + \frac{6}{r^4} - \frac{3}{r^5} + \frac{4}{r^6} - \frac{5}{r^7}\right)\right]\boldsymbol{r} + \frac{\omega \sin\beta}{16}\left(\frac{6}{r} - \frac{6}{r^2} - \frac{1}{r^3} + \frac{1}{r^4}\right)\boldsymbol{k} \\
& - \frac{\omega z \sin\beta}{16}\left(\frac{2}{r^3} + \frac{3}{r^4} - \frac{7}{r^5} + \frac{2}{r^6}\right)\boldsymbol{r} + \frac{\boldsymbol{\omega} \times (\boldsymbol{\omega} \times \boldsymbol{r})}{4}\left(-\frac{1}{r^4} + \frac{1}{r^5}\right) \\
& + \frac{\omega^2}{8}\left(\frac{1}{r^3} - \frac{4}{r^4} + \frac{3}{r^5}\right)\boldsymbol{r} - \frac{(\boldsymbol{\omega} \cdot \boldsymbol{r})^2}{8}\left(\frac{3}{r^5} - \frac{8}{r^6} + \frac{5}{r^7}\right)\boldsymbol{r}
\end{aligned} \tag{3.77}$$

and

$$\overline{p}_1 = \frac{1}{16}\left[-\frac{9}{r^2}+\frac{7}{r^3}-\frac{3}{r^4}-\frac{1}{2r^6}-\frac{9x}{r^3}+x^2\left(\frac{18}{r^4}-\frac{21}{r^5}+\frac{18}{r^6}-\frac{3}{2r^8}\right)\right]$$
$$-\frac{\omega z\sin\beta}{8}\left(\frac{2}{r^3}+\frac{3}{r^4}-\frac{2}{r^6}\right)+\frac{\omega^2}{4}\left(\frac{1}{r^3}-\frac{2}{r^4}\right)-\frac{(\boldsymbol{\omega}\cdot\boldsymbol{r})^2}{4}\left(\frac{3}{r^5}-\frac{4}{r^6}\right) \tag{3.78}$$

where x and z are the two components of the Cartesian coordinate. β is the angle between $\boldsymbol{\Omega}$ and $\boldsymbol{u}_0^*$. The force per unit area exerted by the fluid on the surface of the sphere can be evaluated from $\overline{p}$ and $\overline{\boldsymbol{u}}$ as [Lamb, 1945]

$$\boldsymbol{f} = \frac{U_p\mu}{a}\left[-\frac{\boldsymbol{r}}{r}\overline{p}+\left(\frac{\partial}{\partial r}-\frac{1}{r}\right)\overline{\boldsymbol{u}}+\frac{1}{r}\nabla(\boldsymbol{r}\cdot\overline{\boldsymbol{u}})\right] \tag{3.79}$$

Thus, by integrating Eq. (3.79) over the entire surface of the sphere, the total force is obtained as

$$\boldsymbol{F} = -6\pi a\mu\boldsymbol{U}_p\left(1+\frac{3}{8}\mathrm{Re_a}\right)+\pi a\mu\,\mathrm{Re_a}\,\boldsymbol{\omega}\times\boldsymbol{U}_p+o(a\mu U_p\,\mathrm{Re_a}) \tag{3.80}$$

The first term on the right-hand side of Eq. (3.80) is the total drag force in the opposite direction of $\boldsymbol{U}_p$, including both Stokes drag and Oseen drag. The second term represents a lift force in the direction perpendicular to $\boldsymbol{U}_p$. Thus, the lift force or Magnus force for a spinning sphere in a uniform flow at low Reynolds numbers is obtained as

$$\boldsymbol{F}_M = \frac{\pi}{8}d_p^3\rho\boldsymbol{\Omega}\times\boldsymbol{U}_p \tag{3.81}$$

It is noted that $\boldsymbol{F}_M$ is independent of the viscosity μ and is similar to the Kutta–Joukowsky lift formula for two-dimensional potential flow over an airfoil.

The ratio of the Magnus force to the Stokes drag can be expressed by

$$R_{MS} = \frac{d_p^2}{24}\frac{\rho}{\mu}\Omega \tag{3.82}$$

Therefore, the lift force due to the particle spin is negligibly small compared to the drag force when the particle size is small or the spin velocity is low.

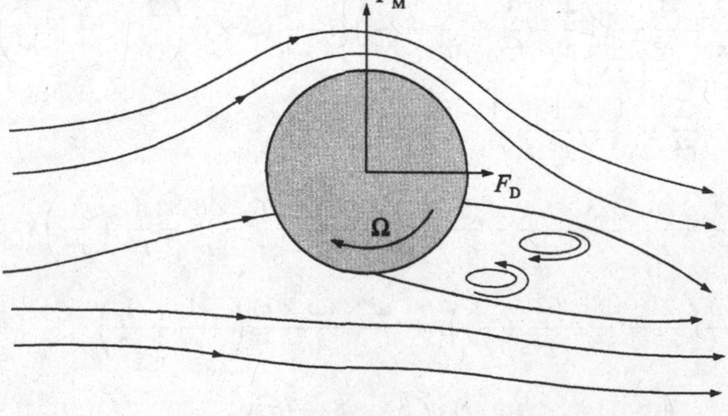

Figure 3.4. Forces on a rotating and moving sphere with wake separation.

At high Reynolds numbers, the rotation of the sphere yields an asymmetric wake, as shown in Fig. 3.4. In this case, the theoretical analysis of the Magnus force and the drag force becomes rather complex because of the difficulties in obtaining the expressions for the pressure and velocity distributions around the surface of the sphere. Thus, the determination of the lift force as well as the drag force relies mainly on the empirical approach.

3.3 Interparticle Forces and Field Forces

The motion of particles is affected by the short-range interparticle forces such as the van der Waals force, electrostatic force, and collision force. It is also affected by the long-range field forces such as the gravitational force, electric force, and magnetic force. This section discusses the basic relationships which quantify these interparticle and field forces.

3.3.1 *Van der Waals Force*

In 1873 van der Waals reported the nonideality of gases and attributed it to the existence of molecular or atomic interactions due to interacting dipoles. The van der Waals force is related to the dispersion effect, which is the interaction between the instantaneous dipoles formed in the atoms by their orbiting electrons [London, 1937]. The very rapidly changing dipole of one atom produces an electric field that acts upon the polarizability of a neighboring atom. The induced dipole of the neighboring atom tends to move in phase with the original dipole, producing an attractive atomic interaction known as the van der Waals force. The van der Waals force, F_{v}, can be expressed by

$$F_{\mathrm{v}} = \frac{\partial E}{\partial s} \tag{3.83}$$

where E is the energy of interaction and s is the separation distance between two atoms.

3.3.1.1 *Van der Waals Forces: Microscopic Approach*

For atoms in a vacuum, the induced dipole of the neighboring atom moves in phase with the original dipole. Thus, the energy of interaction between two atoms i and j separated by a distance s is obtained by [London, 1937]

$$E = -\frac{\lambda_{\mathrm{i,j}}}{s^6} \tag{3.84}$$

where $\lambda_{\mathrm{i,j}}$ is London's constant, which is a function of the interacting atoms. This energy of interaction is attractive provided the medium is not too optically dense such as in a vacuum or gas. Equation (3.84) is only valid at a distance less than the wavelength corresponding to the transition between the ground and the excited states of the atoms, *i.e.*, the absorption wavelength. At separations greater than the absorption wavelength, retardation effects become important and the attractive interaction energy is inversely proportional to s^7. Retardation effects result when the electromagnetic field has to travel farther at greater separations. By the time the field reaches the neighboring atom, the original atomic dipole has already changed its orientation. This effect leads to the interaction's being slightly out of phase. The interaction energy is reduced but still attractive.

The van der Waals force exists not only between individual atoms and molecules but also between solids. Equations for the van der Waals force between solids were deduced by use of the additivity concept, which allowed the force to be calculated in terms of the interaction between individual atoms in the solids [Hamaker, 1937]. The nonretarded energy of interaction between two materials, 1 and 2, of volumes V_1 and V_2 and atom density q_1 and q_2 is given by

$$E = -\int_{V_1}\int_{V_2} \frac{q_1 q_2 \lambda_{1,2}}{s^6}\, dV_2\, dV_1 \tag{3.85}$$

Thus, substituting Eq. (3.85) into Eq. (3.83) yields the van der Waals force between two solids. For two spheres of radii a_1 and a_2, F_v can be obtained as

$$F_v = \frac{A_{12}a^*}{6s^2} \tag{3.86}$$

Here s refers to the distance of separation between the two surfaces of the interacting materials. In Eq. (3.86), a^* is the relative radius defined by Eq. (2.66), and A_{12} is Hamaker's constant (or coefficient), given by

$$A_{12} = \pi^2 q_1 q_2 \lambda_{1,2} \tag{3.87}$$

Thus, for a sphere of radius a_1 and a plane surface where a_2 approaches ∞, Eq. (3.86) reduces to

$$F_v = \frac{A_{12}a_1}{6s^2} \tag{3.88}$$

For two plane surfaces, the solutions of Eqs. (3.83) and (3.85) give a van der Waals pressure, P_v, or the van der Waals force per unit area of contact by

$$P_v = \frac{A_{12}}{6\pi s^3} \tag{3.89}$$

Equations (3.86) and (3.88) give some examples of the nonretarded van der Waals forces for ideal contact geometries. For retarded interactions, the exponent for the distance of separation increases by 1 with the change of the corresponding numerical coefficients. The preceding theory, assuming complete additivity of forces between individual atoms, is known as the microscopic approach to the van der Waals forces.

3.3.1.2 *Van der Waals Forces: Macroscopic Approach*

The concept of additivity is unsatisfactory when applied to closely packed atoms in a condensed body. In this case, a new approach to the energy of interaction needs to be developed or a modification of Hamaker's constant is desirable.

According to Maxwell's equations, spontaneous electromagnetic fluctuations change the fluctuation field in the surrounding region, known as screening. A macroscopic theory of the van der Waals forces relating the nonadditivity to the thermodynamic fluctuations in the interior of a medium was developed by Lifshitz (1956). The macroscopic theory can eliminate the additivity problem by regarding solids as a continuum and deriving the forces in terms of bulk properties such as the dielectric constants and the refractive indices of the materials.

For a material, Hamaker's constant is expressed by

$$A = \frac{3}{4\pi} h\varpi \tag{3.90}$$

where $h\varpi$ is the Lifshitz–van der Waals constant, which depends only on the material and is independent of geometry. Therefore, Eqs. (3.86), (3.88), and (3.89) can be applied, in conjunction with Eq. (3.90), for various geometries in the evaluation of the van der Waals force or pressure.

The Lifshitz–van der Waals constant for the interaction between two bodies of materials 1 and 2 separated by a medium 3 is given by

$$h\varpi_{132} = h\int_0^{\infty}\left(\frac{\varepsilon_1 - \varepsilon_3}{\varepsilon_1 + \varepsilon_3}\right)\left(\frac{\varepsilon_2 - \varepsilon_3}{\varepsilon_2 + \varepsilon_3}\right)\mathrm{d}\zeta \tag{3.91}$$

where h is Planck's constant and ε_i is the dielectric constant of material i along the imaginary frequency axis ζ. This integral is a real function and decreases monotonously from the value ϵ_o, the static dielectric constant, when $\zeta = 0$, to 1 when $\zeta = \infty$. Equation (3.91) holds for two materials in a vacuum or gas by simply replacing ε_3 with 1.

In order to obtain the Lifshitz–van der Waals constant from Eq. (3.91), an expression for $\varepsilon(\zeta)$, related to the dielectric response, must be determined. The dielectric response of a material is represented by the complete dielectric spectra or the complex dielectric permeability, *i.e.*, $\varepsilon(\omega)$ versus ω. The real portion of the complex dielectric permeability measures the transmission of electromagnetic radiation of frequency ω through a body. The imaginary portion measures the absorption of electromagnetic radiation of frequency ω through the same body. This absorption determines the dissipation of energy in an electromagnetic wave propagated in the medium. Therefore, it is the macroscopic property of absorption that determines the strength of the van der Waals interaction between solid bodies.

Solving Eq. (3.91) for the Lifshitz–van der Waals constant or Hamaker's constant usually is not trivial. Many approximations for its calculation have been proposed for various cases [Landau and Lifshitz, 1960; Hough and White, 1980; Visser, 1989; Israelachvili, 1992]. For materials 1 and 2 interacting through material 3, Hamaker's constant becomes

$$A_{132} = c(A_{12} + A_{33} - A_{13} - A_{23}) \tag{3.92}$$

where c is a function of the materials interaction with an average value of about 1.7.

3.3.1.3 *Van der Waals Forces: Surface Effects*

Surface deformation increases the van der Waals forces as a result of the increasing area of contact [Krupp, 1967; Dahneke, 1972]. Thus, for deformable materials, the equations of the van der Waals force, which are valid for rigid materials, should be modified to account for the effect of the increased area of contact. Surface absorption also affects the van der Waals forces. The thickness of the adsorbed layer may increase the distance of separation between the interacting materials, thereby decreasing the van der Waals force [Krupp, 1967]. Moreover, if the thickness of the adsorbed layer is greater than the separation distance, the dielectric properties of the adsorbed layer will dominate the base material in the van der Waals interaction [Langbein, 1969].

3.3.2 *Electrostatic Force*

One of the important characteristics of gas–solid multiphase flows is concerned with the electrostatic effect. Particles can be charged by surface contact in a collision, by corona charging and scattering in an ionized gas, by thermionic emission in a high-temperature environment, and by other charging mechanisms such as colloidal propulsion for dielectric materials in an intense electric field [Soo, 1990]. The motion of a charged particle in a gas–solid flow is affected by the electrostatic force imposed by nearby charged particles.

Although the origin or basic mechanism of static electrification has not been fully understood, the well-known Coulomb's law, which quantifies the electrostatic force, was unraveled as early as two centuries ago. Coulomb's law states that for two charged objects that are much smaller than the distance between them, the force between them is proportional to the product of the charges and inversely proportional to the square of the distance of separation. The force is acting along a straight line from one charged object to the other. Thus, the electrostatic force can be expressed by

$$F_{\mathrm{e}} = \frac{1}{4\pi\epsilon}\frac{q_1 q_2}{r^2} \tag{3.93}$$

where q_{i} is the charge carried by object i; r is the distance between the two objects; and ϵ is the permittivity of the surrounding medium.

In a gas–solid suspension system, let n_{q} be the charge density and V be the effective domain. The electrostatic force of a particle carrying a charge q in the direction of $\boldsymbol{n}$ can be expressed by

$$F_{\mathrm{en}} = \frac{q}{4\pi\epsilon}\int_V \frac{n_{\mathrm{q}}}{r^3}(\boldsymbol{r}\cdot\boldsymbol{n})\,\mathrm{d}V \tag{3.94}$$

The origin of solid electrification and the mechanism of charge transfer are elaborated on in §3.5.

3.3.3 *Collisional Force*

The effect of the collisional force due to the impact of particles should be included when accounting for the motion of a particle except in a very dilute gas–solid flow situation. Basic mechanisms of collision between two particles or between a particle and a solid wall are discussed in Chapter 2. The collisional force between a particle and a group of neighboring particles in a shear suspension is discussed in §5.3.4.3. In a very dense system where particle collisions dominate the flow behavior, collisional forces can be described by using kinetic theory, as detailed in §5.5. The key equations derived in other chapters pertaining to the collisional forces can be summarized in the following.

For elastic spheres, the maximum collisional force in a collinear impact between two particles, F_{c} ($= f_{\mathrm{m}}$ in Eq. (2.132)), is given by

$$F_{\mathrm{c}} = \frac{4}{3}E^*\sqrt{a^*}\left(\frac{15}{16}\frac{m^*U^2}{E^*\sqrt{a^*}}\right)^{\frac{3}{5}} \tag{2.132}$$

where U ($= U_{12}$ in Eq. (2.132)) is the relative velocity, m^* is the relative mass defined by Eq. (2.128), a^* is the relative radius, and E^* is the contact modulus defined by Eq. (2.66).

The averaged collisional stress between a particle and a group of neighboring particles of the same diameter in a shear flow can be expressed by $-(\mu_{\mathrm{p}}\boldsymbol{\Delta}_{\mathrm{p}})$, where μ_{p} is given by

$$\mu_{\mathrm{p}} = \frac{1}{3}\alpha_{\mathrm{p}}^2\rho_{\mathrm{p}}d_{\mathrm{p}}^2(\Delta_{\mathrm{p}}:\Delta_{\mathrm{p}})^{\frac{1}{2}} \tag{5.220}$$

and Δ_{p} is evaluated by Eq. (5.217).

For a collision-dominated dense suspension, the collisional stress tensor $\boldsymbol{P}_c$ is given by

$$\boldsymbol{P}_c = \left(\frac{k}{d_p}\sqrt{\pi T_c} - \frac{3k}{5}\operatorname{tr}\boldsymbol{D}\right)\boldsymbol{I} - \frac{6k}{5}\boldsymbol{D} \tag{5.310}$$

where k is the thermal conductivity of collisional flux of fluctuation energy and is given by Eq. (5.309); $\boldsymbol{D}$ is a tensor defined by Eq. (5.311); T_c is the granular temperature; and the symbol tr denotes the trace.

3.3.4 *Field Forces*

The field forces, also known as the body forces, are long-range forces exerted by various fields outside the flow system. Typical field forces in a gas–solid system include the gravitational force, the electric force, and the magnetic force.

When an external electric field is applied to a gas–solid flow, the charged particles are subjected to an electric force, $\boldsymbol{F}_E$, which is expressed by

$$\boldsymbol{F}_E = q\boldsymbol{E} \tag{3.95}$$

where q is the charge carried by a particle and $\boldsymbol{E}$ is the electric field intensity. Driven by this force, particles carrying positive charges tend to move to the cathodic side while negatively charged particles move to the anodic side. Electrostatic precipitation is an example of the effect of the electric force.

If particles are magnetism-sensitive, they are subjected to a magnetic force once they are exposed to an electromagnetic field in a gas–solid flow. The magnetic force, F_m, is given by

$$F_m = m\mu_r B_0 \tag{1.73}$$

where B_0 is the magnetic flux density in a vacuum; m is the number of north magnetic poles; and μ_r is the relative permeability of the material. Most magnetic materials are highly conductive. Thus, the electrostatic effect in magnetic materials can usually be neglected. One example involving the magnetic force is magnetic fluidization (Problem 9.2).

In general, the forces and moments acting on a charged particle in an electromagnetic field comprise those due to the net charge, electric dipole (permanent or induced dipoles) in the electric field, and magnetic dipole in the induced magnetic field [Soo, 1964]. Neglecting the effects of magnetic dipoles, the force, known as Lorentz force, on a charged particle moving in an electromagnetic field is expressed by

$$\boldsymbol{F}_{em} = q(\boldsymbol{E} + \boldsymbol{U}_p \times \boldsymbol{B}) + \nabla(\boldsymbol{p}_d \cdot \boldsymbol{E}) \tag{3.96}$$

where $\boldsymbol{p}_d$ is the dipole moment, given for dielectric materials, by

$$\boldsymbol{p}_d = \frac{\pi}{2}\frac{(\epsilon_r - 1)}{(\epsilon_r + 2)} d_p^3\, \epsilon\, \boldsymbol{E} \tag{3.97}$$

where ϵ is permittivity of the material and ϵ_r is relative permittivity or dielectric coefficient of the dielectric material. Application of Eq. (3.97) to a dilute suspension of charged particles is illustrated in the following example [Soo, 1964].

Example 3.2 Consider a large number of uniformly charged solid particles initially kept in a spherical barrier of radius R_0 with a symmetric density distribution. When the barrier is suddenly removed, the particles start to emerge from that spherical domain. The viscous drag in the gas is assumed to be negligible. Find the ratio of the force due to dipole to that due to electrostatic repulsion and show that for dilute suspensions, the dipole effect due to self-field is negligible. Also discuss the spreading of the solid particles in this simple symmetric system.

Solution This is a one-dimensional problem. The equation of motion of a single particle in the R-direction is obtained from Eq. (3.96) as

$$m\frac{\mathrm{d}U_\mathrm{p}}{\mathrm{d}t} = qE + \frac{\mathrm{d}}{\mathrm{d}R}(p_\mathrm{d}E) \tag{E3.10}$$

where m is the mass of the particle. From Eq. (3.94), E can be given by

$$E = \frac{1}{4\pi\epsilon R^2}\int_0^R 4\pi r^2 n_\mathrm{q} q \, \mathrm{d}r = \frac{(q/m)}{4\pi\epsilon R^2}M_\mathrm{R} \tag{E3.11}$$

where

$$M_\mathrm{R} = \int_0^R 4\pi r^2 \rho_\mathrm{p}\alpha_\mathrm{p}\,\mathrm{d}r \tag{E3.12}$$

Expressing U_p by

$$U_\mathrm{p} = \frac{\mathrm{d}R}{\mathrm{d}t} \tag{E3.13}$$

and substituting Eq. (3.97) into Eq. (E3.10) give rise to

$$U_\mathrm{p}\frac{\mathrm{d}U_\mathrm{p}}{\mathrm{d}R} = \frac{1}{\epsilon}\left(\frac{q}{m}\right)^2\frac{M_0}{4\pi R^2} - \frac{12}{\rho_\mathrm{p}\epsilon}\frac{(\epsilon_\mathrm{r}-1)}{(\epsilon_\mathrm{r}+2)}\left(\frac{q}{m}\right)^2\frac{M_0^2}{(4\pi)^2R^5} \tag{E3.14}$$

where $M_0 = M_\mathrm{R}(R_0)$ at $t = 0$, where $U_\mathrm{p} = 0$. It is seen that the dipole effect is due to the gradient of electric field intensity, which, in this case, produces a deceleration. The ratio of the force due to dipole F_di to that due to electrostatic repulsion F_er is thus obtained by

$$\frac{F_\mathrm{di}}{F_\mathrm{er}} = \frac{12}{\rho_\mathrm{p}}\frac{(\epsilon_\mathrm{r}-1)}{(\epsilon_\mathrm{r}+2)}\frac{M_0}{4\pi R^3} \tag{E3.15}$$

Hence, for uniform distribution of solids within $R = R_0$, Eq. (E3.15) reduces to

$$\frac{F_\mathrm{di}}{F_\mathrm{er}} = 4\frac{(\epsilon_\mathrm{r}-1)}{(\epsilon_\mathrm{r}+2)}\alpha_\mathrm{p} \tag{E3.16}$$

Thus, for a dilute suspension of charged particles, the dipole effect due to self-field is negligible compared to that due to the electrostatic repulsion.

With the dipole force neglected, the direct integration of Eq. (E3.14) yields the particle velocity distribution as

$$\frac{2\pi\epsilon R_0}{M_0(q/m)^2}U_\mathrm{p}^2 = \left(1 - \frac{R_0}{R}\right) \tag{E3.17}$$

Using Eq. (E3.13), Eq. (E3.17) can be integrated with the initial condition of $R = R_0$ at $t = 0$ to yield

$$t\sqrt{\frac{M_0(q/m)^2}{2\pi\epsilon R_0^3}} = \sqrt{\frac{R}{R_0}\left(\frac{R}{R_0} - 1\right)} + \ln\left(\sqrt{\frac{R}{R_0}} + \sqrt{\frac{R}{R_0} - 1}\right) \tag{E3.18}$$

Thus, Eqs. (E3.17) and (E3.18) determine both the position and the velocity at a given time of each particle initially at radius R_0. In addition, each particle reaches an asymptotic velocity as R approaches infinity.

3.4 Motion of a Single Particle

The most convenient mathematical description of the motion of a particle in a flow field can be given using the Lagrangian coordinate which has its origin attached to the center of the moving particle. For the linear motion of a particle at very low particle Reynolds numbers, the particle motion is governed by the BBO equation, originated by Basset (1888), Boussinesq (1903), and Oseen (1927). To describe the general motion of particles subjected to various forces, the governing equation may be formulated by superposition of all applicable forces. This general formulation for particle motion forms the basis of the Lagrangian trajectory modeling for gas–solid flows, which is presented in this section. Examples of the application of the Lagrangian trajectory model in describing the particle motion in rotating and oscillating flow fields are also presented.

3.4.1 *Basset, Boussinesq, and Oseen (BBO) Equation*

The BBO equation delineates a sphere in linear motion at low Reynolds numbers [Basset, 1888; Boussinesq, 1903; Oseen, 1927]. For a spherical particle, the BBO equation, based upon Eq. (3.37) with the replacement of the buoyancy force with the pressure gradient force, can be expressed as [Soo, 1990]

$$\begin{aligned}\frac{\pi}{6}d_\mathrm{p}^3\rho_\mathrm{p}\frac{\mathrm{d}\boldsymbol{U}_\mathrm{p}}{\mathrm{d}t} &= 3\pi\mu d_\mathrm{p}(\boldsymbol{U} - \boldsymbol{U}_\mathrm{p}) - \frac{\pi}{6}d_\mathrm{p}^3\nabla p + \frac{\pi}{12}d_\mathrm{p}^3\rho\frac{\mathrm{d}}{\mathrm{d}t}(\boldsymbol{U} - \boldsymbol{U}_\mathrm{p})\\ &\quad + \frac{3}{2}d_\mathrm{p}^2\sqrt{\pi\rho\mu}\int_{t_0}^{t}\frac{\frac{\mathrm{d}}{\mathrm{d}\tau}(\boldsymbol{U} - \boldsymbol{U}_\mathrm{p})}{\sqrt{t-\tau}}\,\mathrm{d}\tau + \sum_\mathrm{i} \boldsymbol{f}_\mathrm{i}\end{aligned} \tag{3.98}$$

where $\mathrm{d}/\mathrm{d}t$ is the substantial derivative following the particle flow. The five terms on the right-hand side of Eq. (3.98) are, in order from left to right, the forces due to the Stokes drag, pressure gradient, added mass, Basset historic integral, and other external forces such as gravitational force and electrostatic force.

It is suggested that the pressure gradient may be estimated from the Navier–Stokes equation of a single-phase fluid by [Corrsin and Lumley, 1956]

$$-\nabla p = \rho\frac{\mathrm{D}\boldsymbol{U}}{\mathrm{D}t} - \mu\nabla^2\boldsymbol{U} \tag{3.99}$$

where D/Dt is the substantial derivative following the gas flow (see Eq. (5.7)). Substitution of Eq. (3.99) into Eq. (3.98) yields

$$\frac{d\boldsymbol{U}_p}{dt} = \frac{3\rho}{2\rho_p + \rho}\left(\frac{d\boldsymbol{U}}{dt} - \frac{2}{3}\nu\nabla^2\boldsymbol{U}\right) + \frac{2}{2\rho_p + \rho}\left[\frac{18\mu}{d_p^2}(\boldsymbol{U} - \boldsymbol{U}_p) + \rho(\boldsymbol{U} - \boldsymbol{U}_p)\cdot\nabla\boldsymbol{U}\right]$$
$$+ \frac{18}{(2\rho_p + \rho)d_p}\sqrt{\frac{\rho\mu}{\pi}}\int_{t_0}^{t}\frac{\frac{d}{d\tau}(\boldsymbol{U} - \boldsymbol{U}_p)}{\sqrt{t-\tau}}\,d\tau + \frac{12}{\pi d_p^3(2\rho_p + \rho)}\sum_i f_i \tag{3.100}$$

In gas–solid flows well beyond the Stokes regime, the effect of convective acceleration of the gas surrounding the particle is important. To incorporate this effect into the preceding formulation, modifications of the expressions for the Stokes drag, carried mass, and Basset force in the BBO equation are necessary [Odar and Hamilton, 1964]. The modified BBO equation takes the form [Hansell *et al.*, 1992]

$$\frac{d\boldsymbol{U}_p}{dt} = \frac{3C_D}{4d_p}\frac{\rho}{\rho_p}|\boldsymbol{U} - \boldsymbol{U}_p|(\boldsymbol{U} - \boldsymbol{U}_p) + \frac{1}{2}\frac{\rho}{\rho_p}C_I\frac{d}{dt}(\boldsymbol{U} - \boldsymbol{U}_p) + \frac{\rho}{\rho_p}\frac{D\boldsymbol{U}}{Dt}$$
$$+ \frac{9}{d_p}\frac{\rho}{\rho_p}\left(\frac{\nu}{\pi}\right)^{1/2} C_B\int_{t_0}^{t}\frac{\frac{d}{d\tau}(\boldsymbol{U} - \boldsymbol{U}_p)}{\sqrt{t-\tau}}\,d\tau + \left(\frac{\rho}{\rho_p} - 1\right)\boldsymbol{g} \tag{3.101}$$

where C_D, C_I, and C_B are the drag, carried mass, and Basset coefficients, respectively. In general, C_I and C_B are functions of the particle Reynolds number, Re_p, defined by Eq. (3.2) and the acceleration number, An, defined by [Faeth, 1983]

$$\mathrm{An} = \frac{\left|\frac{d(\boldsymbol{U} - \boldsymbol{U}_p)}{dt}\right|}{|\boldsymbol{U} - \boldsymbol{U}_p|^2}d_p \tag{3.102}$$

For a simple harmonic motion at $Re_p < 62$, correlations of C_I and C_B are suggested as [Odar and Hamilton, 1964]

$$C_I = 2.1 - \frac{0.132\,\mathrm{An}^2}{(1 + 0.12\,\mathrm{An}^2)};\qquad C_B = 0.48 + \frac{0.52\,\mathrm{An}^3}{(1 + \mathrm{An})^3} \tag{3.103}$$

3.4.2 *General Equation of Motion*

For general particle motion formulation, additional forces such as the Saffman force, Magnus force, and electrostatic force should be included. Assuming that all forces applied on the moving particle are additive, the equation of motion of a particle in an arbitrary flow can be expressed by

$$m\frac{d\boldsymbol{U}_p}{dt} = \boldsymbol{F}_D + \boldsymbol{F}_A + \boldsymbol{F}_B + \boldsymbol{F}_S + \boldsymbol{F}_M + \boldsymbol{F}_c + \cdots \tag{3.104}$$

where $\boldsymbol{F}_D$ is the drag force; $\boldsymbol{F}_A$ is the added (or carried) mass force; $\boldsymbol{F}_B$ is the Basset force; $\boldsymbol{F}_S$ is the Saffman force; $\boldsymbol{F}_M$ is the Magnus force; $\boldsymbol{F}_c$ is the collisional force; and "$\cdots$" denotes other forces such as the electrostatic force, the van der Waals force, the gravitational force, the magnetic force, and forces associated with various field gradients due to shock wave propagation, electrophoresis, thermophoresis, and photophoresis phenomena.

It should be noted that the forces in Eq. (3.104) are not generally linearly additive. The drag force, Basset force, Saffman force, and Magnus force all depend on the same flow field. The inertia term in the Navier–Stokes equation, $\nabla \cdot (\boldsymbol{UU})$, is not generally negligible, so the equation is highly nonlinear. Thus, linear superposition of the flow field for each flow mode becomes undesirable because of the nonlinearity of the equation. However, when the governing equations of fluids are linear, the forces on the solids may be additive; this relation has been partially proved by Passman (1986), using one type of the mixture theory which allows for separated phases.

Example 3.3 Consider a rotating gas flow in a cylindrical chamber with a small particle injected into the flow. Assume that the gas rotates as a rigid body with a constant angular velocity ω and the only driving force is the Stokes drag [Kriebel, 1961]. Initially, the relative particle velocity is normal to the flow. Develop the equations for the particle trajectory in this rotating flow and discuss the effect of particle sizes on the trajectory.

Solution Select a Cartesian coordinate system with the origin on the axis of rotation and the x-axis coinciding with the line connecting the origin and the initial location of the particle, as shown in Fig. E3.2. The velocity of gas is given by

$$\begin{aligned} U &= -\omega r \sin\phi = -\omega y \\ V &= \omega r \cos\phi = \omega x \end{aligned} \tag{E3.19}$$

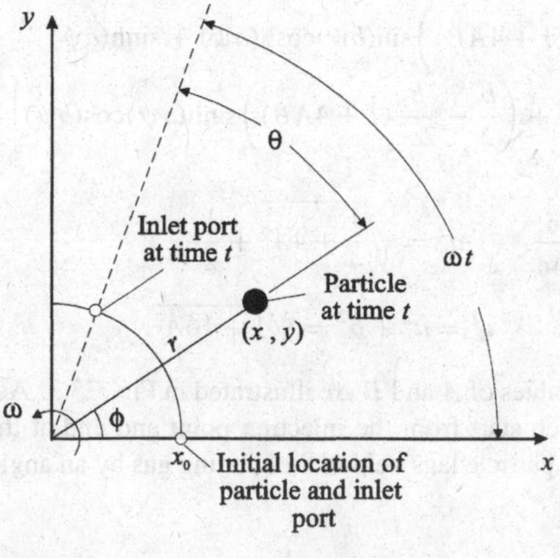

Figure E3.2. Coordinate system for particle motion in a rotating gas.

From Eq. (3.98), we have

$$\begin{aligned} \frac{dU_p}{dt} &= \frac{d^2x}{dt^2} = -\frac{1}{\tau_s}\left(\frac{dx}{dt} + \omega y\right) \\ \frac{dV_p}{dt} &= \frac{d^2y}{dt^2} = \frac{1}{\tau_s}\left(\omega x - \frac{dy}{dt}\right) \end{aligned} \tag{E3.20}$$

where τ_s is the Stokes relaxation time given by Eq. (3.39). Replacing d/dt by $\omega(d/d\phi)$,

Eq. (E3.20) becomes

$$\begin{aligned}\frac{d^2x}{d\phi^2} + \frac{1}{A}\frac{dx}{d\phi} + \frac{y}{A} = 0\\ \frac{d^2y}{d\phi^2} + \frac{1}{A}\frac{dy}{d\phi} - \frac{x}{A} = 0\end{aligned} \tag{E3.21}$$

where A is defined by

$$A \equiv \omega\tau_s \tag{E3.22}$$

The initial conditions can be expressed by

$$\begin{aligned}x|_{\phi=0} = x_0, \qquad y|_{\phi=0} = 0\\ \left.\frac{dx}{d\phi}\right|_{\phi=0} = \frac{\dot{x}_0}{\omega}, \qquad \left.\frac{dy}{d\phi}\right|_{\phi=0} = x_0\end{aligned} \tag{E3.23}$$

Using the Laplace transformation to solve Eq. (E3.21) with the initial conditions given by Eq. (E3.23), the particle trajectory can be obtained by

$$\begin{aligned}x = x_0 e^{-\psi}\left[\left(\cosh(a\psi) + \frac{a(1+2AB)+2Ab}{c^2}\sinh(a\psi)\right)\cos(b\psi)\right.\\ \left. + \left(\frac{b(1+2AB)-2Aa}{c^2}\cosh(a\psi)\right)\sin(b\psi)\right]\end{aligned} \tag{E3.24}$$

$$\begin{aligned}y = x_0 e^{-\psi}\left[\left(\frac{a}{2} + \frac{a}{2c^2}(1+4AB)\right)\sin(b\psi)\cosh(a\psi) + \sinh(a\psi)\right.\\ \left. \times \sin(b\psi) + \left(\frac{b}{2} - \frac{b}{2c^2}(1+4AB)\right)\sinh(a\psi)\cos(b\psi)\right]\end{aligned} \tag{E3.25}$$

where

$$\begin{aligned}\psi = \frac{\phi}{2A}, \qquad B = \frac{\dot{x}_0}{\omega x_0}, \qquad a^2 = \sqrt{\frac{1}{4} + 4A^2} + \frac{1}{2}\\ b^2 = \sqrt{\frac{1}{4} + 4A^2} - \frac{1}{2}, \qquad c^2 = a^2 + b^2 = \sqrt{1 + 16A^2}\end{aligned} \tag{E3.26}$$

The trajectories for different values of A and B are illustrated in Fig. E3.3. As expected, these trajectories are spirals which start from the injection point and end at the cylinder wall. The angular position of the particle lags behind the rotating gas by an angle θ, which is given by

$$\theta = \omega t - \tan^{-1}\left(\frac{y}{x}\right) \tag{E3.27}$$

For very small particles, $A \ll 1$. Equations (E3.24) and (E3.25) are reduced to

$$\begin{aligned}x \approx x_0(1 + AB)e^{A\phi}\cos(1 - 2A^2)\phi\\ y \approx x_0(1 + AB)e^{A\phi}\sin(1 - 2A^2)\phi\end{aligned} \tag{E3.28}$$

For large particles, $A \gg 1$. The particle trajectory is given by

$$\begin{aligned}x \approx x_0 + Bx_0\phi\\ y \approx x_0\phi\end{aligned} \tag{E3.29}$$

$$\omega = 7,200 \text{ rpm}$$
$$\rho_p = 3,000 \text{ kg} / \text{m}^3$$
$$\mu = 1.81 \times 10^{-5} \text{ kg} / \text{m·s}$$

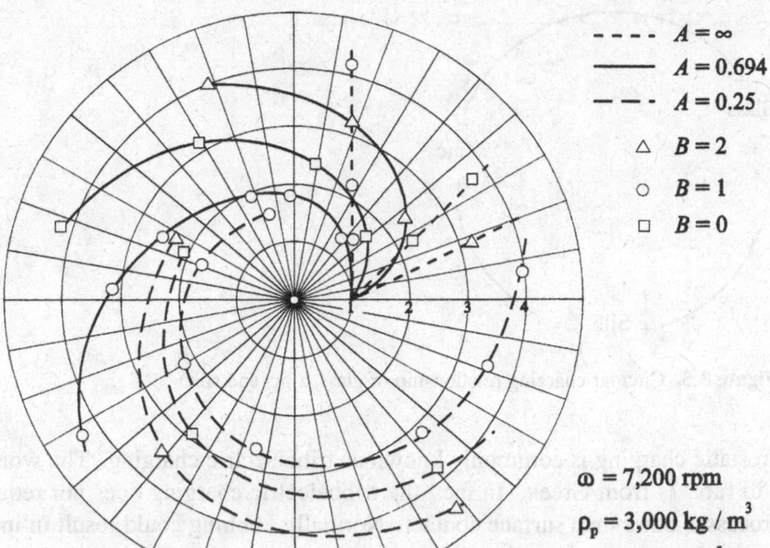

Figure E3.3. Calculated particles' trajectories (from Kriebel, 1961).

3.5 Charge Generation and Charge Transfer

Charge transfer may occur during the contact between particles or between particles and the solid surface. A gas–solid flow field can be significantly affected by the electrostatic forces induced by the charges carried by the particles. In this section, the origins of particle charges due to electrification involving surface contact and field charging are discussed. The mechanism of charge transfer during the collision of two particles is described. In terms of this mechanism, the electrostatic ball probe theory, which illustrates the interrelationship between the particle mass flux and the charging current on the ball probe in a gas–solid flow, is presented.

3.5.1 Static Electrification of Solids

The phenomena of the electrification of solids are complex. In gas–solid flows, surface contact by collisions, ion collection, and thermionic emission are known to be the major modes of particle electrification. Details of these three charging modes are introduced in the discussion that follows.

3.5.1.1 Charging by Surface Contact

The charge equilibrium between two objects in contact requires the Fermi energy level of the objects to be equalized. When two surfaces of dissimilar materials are brought into contact, charge transfer may occur through the contact surface as a result of the difference in the initial Fermi energy levels of these materials. Thus, the particles become electrostatically charged after their separation from the contact. One particle is positively charged because of the loss of electrons while the other is negatively charged because of the gain of electrons.

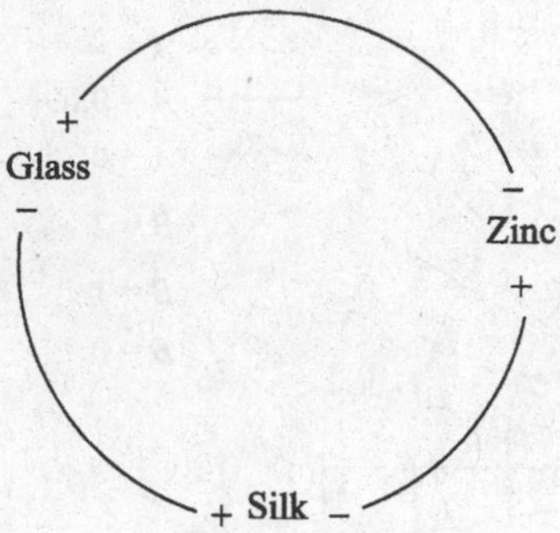

Figure 3.5. Circular-charging relationship of glass, zinc, and silk.

This electrostatic charging is commonly known as triboelectric charging. The word *tribo*, meaning "to rub," is from Greek. In fact, the triboelectric charging does not require any rubbing processes other than surface contact. Normally, rubbing could result in increased charge transfer due to an increased contact area.

Most solid materials can be arranged into a triboelectric series based on charge polarity in a simple contact and separation process. Several triboelectric series are given in Table 3.1. Although a triboelectric series can predict the charge polarity with reasonable accuracy, exceptions are not uncommon because of many influential factors, such as surface finish, preconditioning, and contamination. An overlapping circular polar relationship shown in Fig. 3.5 gives an example of these exceptions [McAteer, 1990].

It is interesting to estimate the percentage of atoms on a solid surface involved in a charge transfer process during a simple contact and separation [Cross, 1987]. A maximum accumulated charge on a surface is limited by the breakdown electric field strength of the surrounding medium. At atmospheric pressure with air as the medium, this breakdown strength is about 3×10^6 V/m. From Gauss's law, the electric field strength due to an assembly of charges can be estimated.

Gauss's law states that if an imaginary enclosed surface of area S is drawn around a uniform density of charge, the component of the electric field at the surface, which is perpendicular to the surface, is proportional to the total charge enclosed. Hence, Gauss's law is expressed as

$$E = \frac{\sum Q_i}{S\epsilon} \tag{3.105}$$

where $\sum Q_i$ is the sum of the charges enclosed by the surface. At the maximum or breakdown electric field strength in air, the maximum charge density on the surface is

$$q_{max} = \frac{\sum Q_i}{S} = E_{max}\epsilon_0 = 2.6 \times 10^{-5} \frac{\mathrm{C}}{\mathrm{m}^2} \tag{3.106}$$

Since the charge carried by a single electron is 1.6×10^{-19} C, the maximum number density of electrons on the charged surface becomes 1.6×10^{14} m^{-2}. The atomic density on a solid surface is approximately 2×10^{19} m^{-2}. It can therefore be estimated that fewer than 10 atoms per million atoms on the surface are involved in the charge transfer process. Such a small percentage of surface atoms participating in the charge transfer indicates the sensitivity of the effect of the surface impurity and contamination on the amount of charges transferred.

Table 3.1. *Triboelectric Series*

	Material	Polymer type	Source
(a) Montgomery (1959)			
+ Charge	Wool		
↑	Nylon		
	Viscose	Cellulose	
	Cotton		
	Silk		
	Acetate rayon	Cellulose	
	Lucite or Perspex	Polymethyl methacrylate	
	Polyvinyl alcohol		
	Dacro	Copolyester of ethylene glycol and terephthalic acid	
	Orlon	Polyacrylonitrile	
	PVC		
	Dynel	Copolymer acrylonitrile/ vinyl chloride	
	Velon	Copolymer vinylidene chloride/vinyl chloride	
↓	Polyethylene		
− Charge	Teflon	PTFE	
(b) Webers (1963)			
+ Charge	Polyox	Polyethylene oxide	Union Carbide
↑	Polyethylene amine		Chemirad
	Gelatin		
	Vinac	Polyninyl acetate	Colton Chemical
	Lucite 44	Polybutyl methacrylate	Du Pont
	Lucite 42	Polymethyl methacrylate	Du Pont
	Acryloid A101	Polymethyl methacrylate	Rohm and Haas
	Zelec DX	Polycation	Du Pont
	Polyacrylamide		Cyanamid
	Cellulose		Eastman
	Acysol	Polyacrylic acid	Rohm and Haas
	Carbopol	Polyacid	BF Goodrich
	Polyethylene		
↓	Polyvinyl butyral		
− Charge	Polyethylene		Du Pont
(c) Williams (1976)			
+ Charge	Lucite 2041	Methyl methacrylate	Du Pont
↑	Dapon	Daillyl phthalate	
	Lexan 105	Poly-bisphenol-A-carbonate	GE
	Formvar	Polyvinylformal	Monsanto
	Estane	Polyurethane	Goodrich
	Du Pont 49000	Polyester	Du Pont
	Durez	Phenol formaldehyde	Durez
	Ethocel 10	Ethyl cellulose	Hercules
	Polystyrene 8X	Polystyrene	Kopper
	Epolene C	Polyethylene	Eastman
	Polysulphone P-3500	A diphenyl sulphone	Union Carbide
	Hypalon 30	Chorosulphonated PE	Du Pont
	Cyclolac H-1000	Acrylonitrile-butadiene-styrene terpolymer	Borg Warner

(*continued on p. 114*)

Table 3.1 *(cont.)*

	Material	Polymer type	Source
	Uncoated iron		
	Cellulose acetate butyrate		
	Epon 828/V125	Epoxy amine curing agent	Shell/General
	Polysulphone P-1700		Union Carbide
↓	Cellulose nitrate		
– Charge	Kynar	Polyvinyldene fluoride	Penwalt

Source: J. A. Cross' *Electrostatics: Principles, Problems and Applications,* Adam Hilger, 1987.

To understand the basic mechanism of charge transfer, a brief introduction of various electron energy bands in a solid is in order. Each electron in a single isolated atom moves along a given orbit surrounding the nucleus of the atom. Each orbit represents a specific electron energy level. An electron can jump across certain energy levels by gaining or losing energy. In a solid particle, atoms interact with each other and, thus, the energy levels of atoms are influenced by neighboring atoms. In light of this interaction, the energy levels can be defined in terms of energy bands. The inner electrons are in narrow bands, the so-called core bands. All the valence electrons stay in the valence bands. The free electrons are in the conduction bands. An area of forbidden energy can lie between energy bands, the forbidden energy band. The Fermi energy level, E_f, is defined as the highest filled energy level at an absolute temperature of zero degree. At an absolute temperature above zero degree, the thermal energy allows some electrons to stay at the energy level above the Fermi energy. The probability of an electron state of energy in a metal can be described by the Fermi–Dirac distribution function as

$$f(E) = \left[1 + \exp\left(\frac{E - E_f}{kT}\right)\right]^{-1} \tag{3.107}$$

where E is the energy level, k is the Boltzmann constant, and T is the absolute temperature. Unless the temperature is very high, the Fermi energy can be used as a reasonable approximation of the energy of electrons to be freed for most ambient conditions.

The work function, W, is defined as the work required to remove an electron out of a metal material from the top of the energy distribution. The work function is related to the potential function ϕ, known as the thermionic work function, by

$$W = e\phi \tag{3.108}$$

where e is the charge of an electron.

An intrinsic potential difference exists between dissimilar dry metals in contact. This potential difference, known as the Volta potential or the contact potential, can be related to the work functions of the two dissimilar materials A and B as

$$V_A - V_B = \frac{W_A - W_B}{e} = \phi_A - \phi_B \tag{3.109}$$

The contact potential reflects the work needed to transfer an electron from one solid to the other. Thus, when two solids are brought into contact, the charge transfer results from the potential difference, which is related to the difference in the Fermi energy level. Upon separation of the two bodies, although some electrons may escape back as a result of the tunnel effect, a net charge results.

The actual mechanism for charging by surface contact is complicated, and many causes remain unresolved. However, it is clear that electron and/or ion transfers are principal factors responsible for the contact charging. More information is available in Montgomery (1959), Harper (1967), and Cross (1987).

3.5.1.2 *Charging by Ion Collection*

Various electric effects in multiphase flows have been summarized by Soo (1990). Typically, particles can be electrified in an electric field via three distinct charging mechanisms: field charging, diffusion charging, and corona charging. Field charging becomes predominant for particles larger than 1 μm where random motion of ions is insignificant and, thus, the diffusion charging mechanism can be ignored. For particles smaller than 0.2 μm, diffusion charging must be accounted for and the contribution of an external electric field becomes insignificant. When the electric field strength exceeds a certain limit, ionization of the surrounding gaseous medium occurs. By the bombardments of the ions and electrons, the solids are charged. Such a charging process is generally known as corona charging.

In the following discussion on these charging mechanisms, three basic assumptions are imposed: *i.e.*, all particles are spherical; particles of the same size are charged to the same degree; and interacting effects between particles are negligible.

A. FIELD CHARGING

When a conducting particle of diameter d_p is exposed to a uniform electric field E_0 in an air medium or in a vacuum with a charge density distribution q_0, the electric potential V is governed by the Poisson equation given by

$$\nabla^2 V = -\frac{q_0}{\epsilon_0} \tag{3.110}$$

where ϵ_0 is the particle permittivity in a vacuum. The relation between the electric field strength, $\boldsymbol{E}$, and the electric potential is given as

$$\boldsymbol{E} = -\nabla V \tag{3.111}$$

With the particle surface charge of Q_s, it can be derived that the electric field strength at any point on the sphere, E_s, satisfies

$$E_s = 3E_0 \cos\theta - \frac{Q_s}{\pi d_p^2 \epsilon_0} \tag{3.112}$$

where θ is the angle between a point on the sphere and the electric field. The second term on the right-hand side of Eq. (3.112) represents the contribution of the particle surface charge based on Gauss's law. Thus, the total charge current, i, entering the particle can be obtained by [Oglesby and Nichols, 1978]

$$i = \frac{dQ_s}{dt} = \int_0^{\theta_m} q_0 K E_s 2\pi a^2 \sin\theta \, d\theta \tag{3.113}$$

where K is the mobility of ions and θ_m is the limiting angle for the charge flow to the particle. θ_m is determined from Eq. (3.112) by setting E_s equal to zero, *i.e.*,

$$\cos\theta_m = \frac{Q_s}{3\pi\epsilon_0 d_p^2 E_0} \tag{3.114}$$

Thus, from Eqs. (3.112) through (3.114), the equation for the charging rate of a single particle in an electric field with unipolar ion density can be written as

$$\frac{dQ_s}{dt} = \frac{3\pi}{4} d_p^2 E_0 q_0 K \left(1 - \frac{Q_s}{Q_{sm}}\right)^2 \tag{3.115}$$

where Q_{sm} is the saturation charge, which is given by

$$Q_{sm} = 3\pi\epsilon_0 E_0 d_p^2 \tag{3.116}$$

The total charge acquired by an initially neutral particle within the charging time t is given by

$$Q_s = Q_{sm}\left(\frac{t}{t+\tau}\right) \tag{3.117}$$

where τ is the time constant of the saturation charge, expressed by

$$\tau = \frac{4\epsilon_0}{q_0 K} \tag{3.118}$$

For dielectric particles, the saturation charge can be given as

$$Q_{sm} = \frac{9\pi\epsilon_0\epsilon_r}{\epsilon_r + 2} E_0 d_p^2 \tag{3.119}$$

where ϵ_r is the relative permittivity of the dielectric materials.

B. Diffusion Charging

When a particle is placed in a field of unipolar ions with a charge density of q_0 and an electric field strength of E_0, the particle will be charged by ion striking as a result of the random motion of ions. The rate of diffusion charging of particles can be derived in terms of the kinetic theory of gases. Assuming that every ion striking the particle is captured, the charging rate of a particle in the absence of an external electric field is expressed by [White, 1963]

$$\frac{dQ_s}{dt} = \frac{\pi}{4} d_p^2 \mathrm{v} q_0 \exp\left(-\frac{Q_s e}{2\pi\epsilon_0 d_p k T}\right) \tag{3.120}$$

where k is the Boltzmann constant; T is the absolute temperature; and v is the mean thermal speed of the ions, which is expressed by

$$\mathrm{v} = \sqrt{\frac{8kT}{\pi m}} \tag{3.121}$$

where m is the mass of an ion. Hence, the charges acquired by the initially neutral particle within the charging time t can be obtained from Eq. (3.120) as

$$Q_s = \frac{2\pi\epsilon_0 d_p k T}{e} \ln\left(1 + \frac{t}{\tau_d}\right) \tag{3.122}$$

where τ_d is the time constant of the diffusion charging, expressed by

$$\tau_d = \frac{8\epsilon_0 kT}{d_p v q_0 e} \tag{3.123}$$

The diffusion charging of particles in the presence of an external electric field is available in Murphy (1956) and Liu and Yeh (1968).

C. Charging in an Ionized Gas

Consider an ideal case where a group of particles are uniformly suspended in an ionized gas. Assume that all the charges carried by ions and electrons arriving at the particle surface are completely attached to the particle surface. As the electrons have higher random velocities than the ions do because of the smaller mass of the electron, the particles would be negatively charged. Because of the complexity involved in treating the electric interaction among particles, we only consider the cases where the interparticle spacing is larger than the Debye shielding distance, λ_D, given by

$$\lambda_D = \sqrt{\frac{\epsilon_0 kT}{q_{e0} e}} \tag{3.124}$$

where q_{e0} is the electron charge density. This case limits the analysis given late, making it to be applicable only when the particle number density n is less than $(2\lambda_D)^{-3}$.

From the kinetic theory, the charging rate by electrons can be estimated by [Soo, 1990]

$$\frac{dQ_{es}}{dt} = \frac{\pi d_p^2 q_{e0}}{4}\sqrt{\frac{8kT}{\pi m_e}} e^{\alpha} \tag{3.125}$$

where α is the electrothermal number, defined as the ratio of the particle electrostatic energy to the thermal energy, and is expressed as

$$\alpha = \frac{Q_s e}{2\pi\epsilon_0 d_p kT} \tag{3.126}$$

α is negative for electron charging and positive for ion charging. Similarly, the charging rate by ions is

$$\frac{dQ_{is}}{dt} = \frac{\pi d_p^2 q_{i0}}{4}\sqrt{\frac{8kT}{\pi m_i}} e^{-\alpha} \tag{3.127}$$

At equilibrium, the charging rate by ions equals that by electrons. Thus, we have

$$\frac{q_{e0}}{q_{i0}} = \sqrt{\frac{m_e}{m_i}} e^{-2\alpha} \tag{3.128}$$

For any volume element, we have the charge balance as

$$Q_s n = q_{i0} - q_{e0} \tag{3.129}$$

Thus, the saturated surface charge of each particle is given as

$$Q_s = \frac{q_{i0}}{n}\left(1 - \sqrt{\frac{m_e}{m_i}} e^{-2\alpha}\right) \tag{3.130}$$

Example 3.4 Particles of 10 μm size flowing at a velocity of 20 m/s collect charges from unipolar ions in a corona discharge over a distance of 5 cm. Assume that the ions can be regarded as electrons. The average ion number density, n_e, in the corona discharge field is 10^{16} m^{-3}. The particle density is 1,400 kg/m^3. Estimate the charge-to-mass ratio acquired. Assume that the interparticle effects can be neglected. In addition, the simultaneous charging of positive ions in the corona is negligible. Initially, the particles are neutral. The environment is at 300 K.

Solution The charging rate of a single particle by electrons can be expressed by Eqs. (3.125) and (3.126). In this case, the particles are only charged by electrons, which can be regarded as negative ions. Thus, $Q_s = -Q_{es}$. From Eqs. (3.125) and (3.126), we have

$$\frac{dQ_{es}}{dt} = Ae^{-BQ_{es}} \tag{E3.30}$$

where

$$A = \frac{\pi d_p^2 n_e e}{4}\sqrt{\frac{8kT}{\pi m_e}} \tag{E3.31}$$

and

$$B = \frac{e}{2\pi\epsilon_0 d_p kT} \tag{E3.32}$$

With the initial condition of

$$Q_{es}|_{t=0} = 0 \tag{E3.33}$$

Equation (E3.30) yields

$$Q_{es} = \frac{1}{B}\ln(1 + ABt) \tag{E3.34}$$

Thus, for a particle passing through a corona over a distance of l at a velocity of U_p, the charge-to-mass ratio can be expressed by

$$\frac{Q_{es}}{m} = \frac{6}{\pi d_p^3 \rho_p B}\ln\left(1 + AB\frac{l}{U_p}\right) \tag{E3.35}$$

Substituting all the given data into Eq. (E3.35) gives

$$\begin{aligned}\frac{Q_{es}}{m} &= \frac{12\epsilon_0 kT}{ed_p^2\rho_p}\ln\left(1 + \frac{d_p n_e e^2 l}{4\epsilon_0 U_p}\sqrt{\frac{2}{\pi m_e kT}}\right)\\ &= \frac{12 \times 8.85 \times 10^{-12} \times 1.38 \times 10^{-23} \times 300}{1.6 \times 10^{-19} \times (10^{-5})^2 \times 1400}\\ &\quad \times \ln\left(1 + \frac{10^{-5} \times 10^{16} \times (1.6 \times 10^{-19})^2 \times 0.05}{4 \times 8.85 \times 10^{-12} \times 20}\right.\\ &\quad \left.\times\sqrt{\frac{2}{\pi \times 9.1 \times 10^{-31} \times 1.38 \times 10^{-23} \times 300}}\right) = 0.00029\ \frac{\text{C}}{\text{kg}}\end{aligned} \tag{E3.36}$$

3.5.1.3 *Charging by Thermionic Emission*

When solid particles are exposed to a high-temperature environment, typically for $T >$ 1,000 K, charging by thermal electrification becomes important. The electrons inside the solid can acquire the energy from the high-temperature field and be freed by overcoming the energy barrier or the work function. By losing electrons in such a thermionic emission process, the particles are thermally electrified.

From the Richardson–Dushman equation, the current density J_e by thermionic emission can be obtained as

$$J_e = A_0 T^2 \exp\left(-\frac{\phi}{kT}\right) \tag{3.131}$$

where ϕ is the work function of the solid and A_0 is an experimentally determined coefficient.

It is noted that the rate of electrification is not constant. Once the tendency is established for an electron to escape from the solid particle by thermionic emission, the charge buildup occurs on the particle, which then attempts to recapture the to-be-freed electron by the attracting Coulomb force. Therefore, the equilibrium of thermal electrification of solid particles in a finite space is possible. Details on the equilibrium and the rate of electrification concerning the thermionic emission are available in Soo (1990).

3.5.2 *Charge Transfer by Collision*

Charge transfer occurs when particles collide with each other or with a solid wall. For monodispersed dilute suspensions of gas–solid flows, Cheng and Soo (1970) presented a simple model for the charge transfer in a single scattering collision between two elastic particles. They developed an electrostatic theory based on this mechanism, to illustrate the interrelationship between the charging current on a ball probe and the particle mass flux in a dilute gas–solid suspension. This electrostatic ball probe theory was modified to account for the multiple scattering effect in a dense particle suspension [Zhu and Soo, 1992].

3.5.2.1 *Charge Transfer by Single Impact*

Consider the charge transfer between two spherical particles of diameters d_{p1} and d_{p2}. A direct analogy between the charge transfer by collisions and the heat transfer by convection appears to be in order. Thus, the current density through the contact area of these two particles, J_{e21}, is expressed as

$$J_{e21} = h_{e12}(V_2 - V_1) = \frac{\sigma_1}{d_{p1}}(V_c - V_1) = \frac{\sigma_2}{d_{p2}}(V_2 - V_c) \tag{3.132}$$

where h_{e12} is the charge transfer coefficient through the contact area A_{12}; V_1, V_2, and V_c are the electric potentials of particles 1 and 2 and the contact surface, respectively; and σ_1 and σ_2 are the electrical conductivities of the materials. By eliminating the contact potential V_c from the preceding equation, the charge transfer coefficient can be obtained as

$$h_{e12} = \frac{\sigma_1 \sigma_2}{\sigma_1 d_{p2} + \sigma_2 d_{p1}} \tag{3.133}$$

For each isolated particle, the charging current can be related to the particle capacity. Thus, we have

$$C_1 \frac{dV_1}{dt} = -C_2 \frac{dV_2}{dt} = A_{12} h_{e12}(V_2 - V_1) \tag{3.134}$$

where C_1 and C_2 are the capacitances of the particles.

For a homogeneous spherical particle, the capacitance is expressed by

$$C = 2\pi \epsilon d_{\mathrm{p}} \tag{3.135}$$

where ϵ is the permittivity of the material. The electric potential of the particle is

$$V = \phi + \frac{q}{2\pi \epsilon d_{\mathrm{p}}} \tag{3.136}$$

where ϕ is the work function and q is the charge carried by the particle. Therefore, the equation for the charge transfer can be expressed as

$$-\frac{\mathrm{d}(V_2 - V_1)}{\mathrm{d}t} = \frac{A_{12} h_{\mathrm{e}12}}{C^*}(V_2 - V_1) \tag{3.137}$$

where C^* is the relative capacitance, defined by

$$\frac{1}{C^*} = \frac{1}{C_1} + \frac{1}{C_2} \tag{3.138}$$

From Eq. (2.74), the relationship between the contact radius r_{c} and the approaching distance α for elastic collision of particles can be given by

$$r_{\mathrm{c}}^2 = \alpha a^* \tag{3.139}$$

Thus, the contact area A_{12} is given as

$$A_{12} = \pi r_{\mathrm{c}}^2 = \pi \alpha a^* \tag{3.140}$$

The charge transfer equation becomes

$$-\frac{\mathrm{d}[\ln(V_2 - V_1)]}{\mathrm{d}t} = \frac{\pi a^*}{C^*} h_{\mathrm{e}12} \alpha \tag{3.141}$$

From Eq. (2.135), we obtain the relation of the contact time t and the approaching distance α as

$$\mathrm{d}t = \left(U_{12}^2 - \frac{16E^*}{15m^*} \sqrt{a^*} \alpha^{\frac{5}{2}} \right)^{-\frac{1}{2}} \mathrm{d}\alpha \tag{3.142}$$

Let $x = \alpha / \alpha_{\mathrm{m}}$ and define α_{m} as

$$\alpha_{\mathrm{m}} = \left(\frac{15}{16} \frac{m^* U_{12}^2}{E^* \sqrt{a^*}} \right)^{\frac{2}{5}} \tag{3.143}$$

The rearranged form of the charge transfer equation is obtained as

$$-\mathrm{d}[\ln(V_2 - V_1)] = \frac{\alpha_{\mathrm{m}}^2 \pi a^* h_{\mathrm{e}12}}{U_{12} C^*} x \left(1 - x^{\frac{5}{2}}\right)^{-\frac{1}{2}} \mathrm{d}x \tag{3.144}$$

Integrating the preceding equation for a complete elastic impact process including the compression and the rebounding processes, we have

$$(V_{20} - V_2) - (V_{10} - V_1) = (V_{20} - V_{10})(1 - e^{-B}) \tag{3.145}$$

where subscript "0" denotes the initial state and B is a constant given by

$$B = 2 \frac{\alpha_{\mathrm{m}}^2 \pi a^* h_{\mathrm{e}12}}{U_{12} C^*} \int_0^1 x \left(1 - x^{\frac{5}{2}}\right)^{-\frac{1}{2}} \mathrm{d}x \tag{3.146}$$

The transferred charge Q_{12} relates to the loss or gain of the colliding particles as given by

$$Q_{12} = Q_{20} - Q_2 = Q_1 - Q_{10} \tag{3.147}$$

Thus, Q_{12} can be evaluated from

$$Q_{12} = C^*\left(\phi_2 - \phi_1 + \frac{Q_{20}}{C_2} - \frac{Q_{10}}{C_1}\right)(1 - e^{-B}) \tag{3.148}$$

3.5.2.2 *Theory of Electrostatic Ball Probe*

Consider a ball probe of a conductive sphere such as a stainless steel ball bearing which is connected to an electrometer. When the ball probe is placed in a flow system, the electric current induced by charge transfer between the ball and the flowing charged particles can be measured. The amount of current measured will vary with the particle velocity and particle mass flux.

Let us first consider the case where only single scattering is responsible for the charge transfer. The electric current, i_b, measured by the ball probe is proportional to the product of the collision frequency, f_c, and the averaged charge transfer per impact, Q_{bp}, as given by

$$i_b = f_c C^*(V_{p0} - V_{b0})(1 - e^{-B}) \tag{3.149}$$

where B is

$$B = 2.03\frac{\alpha_m^2 \pi a^* h_{ebp}}{U_p C^*} \tag{3.150}$$

Here, subscripts b and p refer to ball probe and particle, respectively. The collision frequency is expressed by

$$f_c = nU_p\frac{\pi}{4}(d_p + d_b)^2 \tag{3.151}$$

where n is the particle number density and U_p is the particle velocity. For $B \ll 1$, from Eqs. (3.149) and (3.151) and the Taylor expansion, i_b becomes

$$i_b = 4.76(d_b + d_p)^2(V_{p0} - V_{b0})h_{ebp}\left(\frac{m^*}{E^*}\right)^{\frac{4}{5}} a^{*\frac{3}{5}} nU_p^{\frac{8}{5}} \tag{3.152}$$

Note that the preceding equation is for ideal cases, in which the particles are monodispersed, spherical, and totally elastic, and the contact surface is clean. In practice, the particles are usually nonspherical and polydispersed; the collision could have involved some heat loss, plastic deformation, or even breakup; and the contact surface may have impurities or contaminants. In these cases, a correction factor η is introduced to account for the effects of these nonideal factors. The applicable form of the electric current through the ball probe is, thus, given by

$$i_b = AU_p^{3/5} J_p \tag{3.153}$$

where J_p $(= \alpha_p \rho_p U_p)$ is the particle mass flux and A is a constant related to the characteristic properties of the ball probe and particles. A can be expressed by

$$A = \frac{4.76\eta}{m}(d_b + d_p)^2(V_{p0} - V_{b0})h_{ebp}\left(\frac{m^*}{E^*}\right)^{\frac{4}{5}} a^{*\frac{3}{5}} \tag{3.154}$$

The theory given is applicable to a very dilute gas–solid suspension where the charge transfer is dominated by single scattering. When applied to concentrated gas–solid suspensions, the effects of multiple scattering, sliding, and charge saturation become significant and should be incorporated in the formulation of the theory.

Consider a monodispersed dense suspension. From the kinetic theory analogy, the number of simultaneous impacts by particles on a ball probe is given by

$$N_c = \frac{1}{\lambda_B n^{1/3}} \tag{3.155}$$

where λ_B is the mean free path of particle-to-ball probe collisions. λ_B can be expressed as

$$\lambda_B = \frac{4}{n\pi(d_b + d_p)^2} \tag{3.156}$$

The flow conditions ranging from single scattering to "boundary layer flow" of particles are shown in Fig. 3.6. It is noted that the single scattering is bounded by the condition of $d_b/\lambda_B = 1$. In terms of Eq. (3.156), the boundary of the single scattering is a function of the particle size, ball probe size, and particle concentration.

The effect of multiple scattering, sliding, and charge saturation on the ball probe current can be taken into account in a theory via the modification of the electric potential difference between the ball and each single colliding particle. Assume that the initial potential of a charged particle is the same for both single and multiple scattering. For the ball probe, we have

$$C_b \frac{dV_b}{dt} = -\sum_i^{N_c} C_i \frac{dV_i}{dt} = \sum_i^{N_c} A_{ib} h_{eib}(V_i - V_b) \tag{3.157}$$

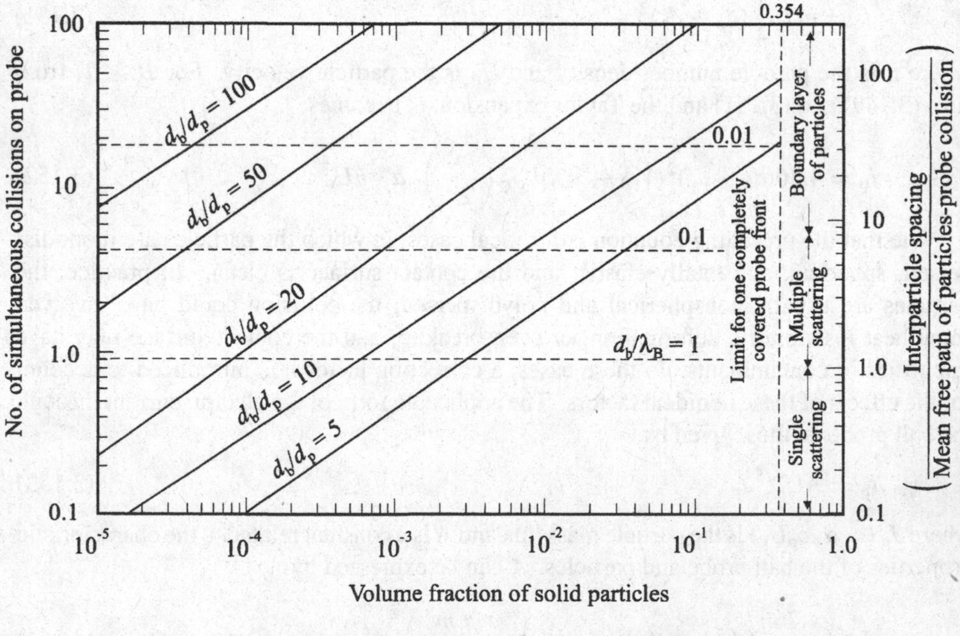

Figure 3.6. Particle flow regimes over an electrostatic probe (after Zhu and Soo, 1992).

Integration of the preceding equation yields

$$V_b - V_p = (V_{bs} - V_{ps}) \exp\left(-\sum_i^{N_c} \frac{A_{ib} h_{eib}}{C_b} t\right) \tag{3.158}$$

where subscript s denotes the single scattering and t can be treated as a time scale for the multiple scattering effect. It is desired to relate the time scale to the particle mass flux and the particle velocity so that the potential difference in dense suspensions is expressed as

$$V_b - V_p = (V_{bs} - V_{ps}) \exp\left(-C_c \frac{J_p}{U_p^2}\right) \tag{3.159}$$

where C_c is a dimensional concentration factor with a unit of $m^4/kg{\cdot}s$, which is determined experimentally. Thus, the ball probe current for dense suspensions takes the form

$$i_b = A U_p^{3/5} J_p \exp\left(-C_c \frac{J_p}{U_p^2}\right) \tag{3.160}$$

The preceding equation has been experimentally validated [Zhu and Soo, 1992]. By defining y as $i_b/U_p^{2.6}$ and x as J_p/U_p^2, the equation for the ball probe current can be arranged in a simple form as

$$y = A x e^{-C_c x} \tag{3.161}$$

Using the condition of $dy/dx = 0$, the upper limit of the particle mass flux that yields the univalent correspondence between the particle mass flux and the ball probe current can be obtained as given by:

$$J_p = \frac{U_p^2}{C_c} \tag{3.162}$$

Nomenclature

A	Exposed frontal area of a particle to the direction of an incoming flow
A	Hamaker's constant
A	Empirical coefficient, defined by Eq. (3.3)
A	Parameter, defined by Eq. (E3.22)
A	Parameter, defined by Eq. (3.154)
A_0	Constant, defined by Eq. (3.131)
An	Acceleration number
a	Particle radius
a^*	Relative radius, defined by Eq. (2.66)
$\mathbf{B}$	Magnetic field intensity vector
B	Empirical coefficient, defined by Eq. (3.3)
B	Parameter, defined by Eq. (E3.26)
B	Constant, defined by Eq. (3.146)
B_0	Magnetic flux density in a vacuum
C	Electric capacitance of a particle
C	Empirical constant, defined by Eq. (E3.5)
C_b	Electric capacitance of a ball probe
C_c	Concentration factor, defined by Eq. (3.159)
C_I	Carried mass coefficient
C_B	Basset coefficient
C_D	Drag coefficient
C^*	Relative capacitance, defined by Eq. (3.138)
c	Coefficient, defined by Eq. (3.92)
D	Operator, defined by Eq. (3.7)

d_p	Particle diameter
d_b	Ball probe diameter
E	Electric field strength
E	Energy of interaction between atoms
E	Energy level
E_0	Electric field strength in a vacuum
E_f	Fermi energy level
E_s	Electric field strength at surface
E^*	Contact modulus
$\mathbf{E}$	Electric field intensity vector
e	Charge of an electron
F_{di}	Force due to dipole
F_e	Electrostatic force
F_{er}	Electrostatic repulsion force
F_m	Magnetic force
F_p	Force due to pressure gradient
F_v	Van der Waals force
F_z	Total resistance to particle motion
$\mathbf{F}_A$	Carried mass force vector
$\mathbf{F}_B$	Basset force vector
$\mathbf{F}_c$	Collisional force vector
$\mathbf{F}_D$	Drag force vector
$\mathbf{F}_E$	Electric force vector
$\mathbf{F}_{em}$	Lorentz force vector
$\mathbf{F}_M$	Magnus force vector
$\mathbf{F}_S$	Saffman force vector
$\mathbf{F}_T$	Radiometric force vector
f	General function
f_c	Collision frequency
g	Gravitational acceleration
h	Planck's constant
h_e	Charge transfer coefficient
$h\varpi$	Lifshitz–van der Waals constant
i	Surface charge current
i	Total charge current
i_b	Electric current through a ball probe
J_e	Electric current density
J_p	Mass flux of particles
K	Constant, defined by Eq. (3.41)
K	Mobility of ions
K	Thermal conductivity of gas
K_p	Thermal conductivity of particles
k	Boltzmann constant
k	Collisional thermal conductivity
l	Interparticle distance
M_a	Mass of displaced fluid
m	Number of north magnetic poles
m	Mass of a particle
m	Mass of an ion
m^*	Relative mass
m_e	Mass of an electron
m_i	Mass of an ion
N_c	Number of simultaneous impacts by particles on a ball probe
n	Particle number density
n_q	Charge density
P_v	Van der Waals pressure
$\mathbf{P}_c$	Collisional stress tensor
p	Pressure
p^*	Dimensionless pressure for the Oseen expansion
$\overline{p}$	Dimensionless pressure for the Stokes expansion
p_f	Pressure of fluid
$\mathbf{p}_d$	Dipole moment vector
Q_{bp}	Averaged charge transfer per impact
Q_{es}	Surface charge of a particle carried by electrons
Q_{is}	Surface charge of a particle carried by ions
Q_s	Surface charge of a particle
Q_{sm}	Saturated surface charge of a particle
q	Charge carried by a particle
q	Atom density in the material
q_0	Charge density in a vacuum
q_{e0}	Electron charge density
q_{i0}	Ion charge density
R	Spherical coordinate

R_{BS} Ratio of Basset force to Stokes drag
R_M Gas constant
R_{MS} Ratio of Magnus force to Stokes drag
R_{SS} Ratio of Saffman force to Stokes drag
$\mathrm{Re_a}$ Particle Reynolds number, based on particle radius and relative velocity
$\mathrm{Re_p}$ Particle Reynolds number, based on particle diameter and relative velocity
r Cylindrical coordinate
r_c Contact radius
$\boldsymbol{r}$ Position vector
S Surface area
s Separation distance between two atoms
s Separation distance between two surfaces of interacting materials
s Stretched coordinate
T Absolute temperature
T_c Granular temperature
T_p Particle temperature
t Time
t_c Elapsed time for particles to collide
U Gas velocity in the x-direction, defined by Eq. (E3.19)
U Relative velocity of trailing particle to leading particle, defined by Eq. (E3.1)
U_p Particle velocity in the x-direction, defined by Eq. (E3.20)
$\boldsymbol{U}$ Gas velocity vector
$\boldsymbol{U}_p$ Particle velocity vector
u_R Gas velocity in the R-direction, defined by Eq. (3.15)
u_r Gas velocity in the r-direction, defined by Eq. (3.4)
u_z Gas velocity in the z-direction, defined by Eq. (3.4)
u_θ Gas velocity in the θ-direction, defined by Eq. (3.15)
$\boldsymbol{u}^*$ Dimensionless gas velocity vector for the Oseen expansion
$\overline{\boldsymbol{u}}$ Dimensionless gas velocity vector for the Stokes expansion
V Electric potential
V Volume of a particle
V Relative velocity of particle to gas, defined by Eq. (3.14)
V Gas velocity in the y-direction, defined by Eq. (E3.19)
V_p Particle velocity in the y-direction, defined by Eq. (E3.20)
v Mean thermal speed of ions
v Time-dependent relative velocity of particle to gas, defined by Eq. (3.37)
$\mathrm{v_s}$ A solution of Eq. (3.30)
W Work function
x Cartesian coordinate
y Cartesian coordinate
z Cartesian coordinate
z Cylindrical coordinate

Greek Symbols

α Approaching distance
α Electrothermal number, defined by Eq. (3.126)
ϵ Medium permittivity
ϵ_0 Particle permittivity in a vacuum
ϵ_r Relative permittivity of dielectric materials
ζ Imaginary frequency axis
η Correction factor
θ Cylindrical coordinate
θ Spherical coordinate
θ Angle between a point on the sphere and the electric field
λ_B Mean free path of particle-to-particle collisions

λ_D	Debye shielding distance	τ_d	Time constant of diffusion charging
$\lambda_{i,j}$	London's constant	τ_S	Stokes relaxation time
μ	Dynamic viscosity	ϕ	Potential function
μ_r	Relative permeability	ϕ	Thermionic work function
ν	Kinematic viscosity	ϕ	Rotating angle
ξ	Distance from a fixed point	ψ	Stream function
ρ	Density of gas	Ω	Angular velocity vector
ρ_p	Density of particles	ω	Angular velocity
σ	Electrical conductivity	$\boldsymbol{\omega}$	Dimensionless angular velocity vector
τ	Time constant of saturation charge		

References

Basset, A. B. (1888). *A Treatise on Hydrodynamics*, **2**. Cambridge: Deighton, Bell; (1961) New York: Dover.

Boussinesq, J. (1903). *Theorie Analytique de la Chaleur*, **2**. Paris: Gauthier-Villars.

Cheng, L. and Soo, S. L. (1970). Charging of Dust Particles by Impact. *J. Appl. Phys.*, **41**, 585.

Clift, R., Grace, J. R. and Weber, M. E. (1978). *Bubbles, Drops, and Particles*. New York: Academic Press.

Corrsin, S. and Lumley, J. (1956). On the Equation of Motion for a Particle in Turbulent Fluid. *Appl. Sci. Res.*, **6A**, 114.

Cross, J. A. (1987). *Electrostatics: Principles, Problems and Applications*. Philadelphia: Adam Hilger.

Dahneke, B. (1972). The Influence of Flattening on the Adhesion of Particles. *J. Colloid Interface Sci.*, **40**, 1.

Epstein, P. S. (1929). Zur Theorie des Radiometers. *ZS. f. Phys.*, **54**, 537.

Faeth, G. M. (1983). Evaporation and Combustion of Sprays. *Prog. Energy Combust. Sci.*, **9**, 1.

Hamaker, H. C. (1937). The London–Van der Waals Attraction Between Spherical Particles. *Physica IV*, **10**, 1058.

Hansell, D., Kennedy, I. M. and Kollmann, W. (1992). A Simulation of Particle Dispersion in a Turbulent Jet. *Int. J. Multiphase Flow*, **18**, 559.

Harper, W. R. (1967). *Contact and Frictional Electrification*. Oxford: Oxford University Press.

Hettner, G. (1926). Zur Theorie der Photophorese. *ZS. f. Phys.*, **37**, 179.

Hough, D. B. and White, L. R. (1980). The Calculation of Hamaker Constants from Lifshitz Theory with Applications to Wetting Phenomena. *Adv. Colloid Interface Sci.*, **14**, 3.

Hughes, R. R. and Gilliland, E. R. (1952). The Mechanics of Drops. *Chem. Eng. Prog.*, **48**, 497.

Israelachvili, J. N. (1992). *Intermolecular and Surface Forces*, 2nd ed. New York: Academic Press.

Kriebel, A. R. (1961). Particle Trajectories in a Gas Centrifuge. *Trans. ASME, J. Basic Eng.*, **83D**, 333.

Krupp, H. (1967). Particle Adhesion Theory and Experiment. *Adv. Colloid Interface Sci.*, **1**, 111.

Lamb, H. (1945). *Hydrodynamics*, 6th ed. New York: Dover.

Landau, L. D. and Lifshitz, E. M. (1960). *Electrodynamics of Continuous Media*. Oxford: Pergamon Press.

Langbein, D. (1969). Van der Waals Attraction Between Macroscopic Bodies. *J. Adhesion*, **1**, 237.

Lee, K. C. (1979). Aerodynamic Interactions Between Two Spheres at Reynolds Number Around 10^4. *Aerosp. Q.*, **30**, 371.

Lifshitz, E. M. (1956). The Theory of Molecular Attractive Force Between Solids. *Soviet Phys.*, **2**, 73.

Liu, B. Y. H. and Yeh, H. C. (1968). On the Theory of Charging of Aerosol Particles in an Electric Field. *J. Appl. Phys.*, **39**, 1396.

London, F. (1937). The General Theory of Molecular Forces. *Trans. Faraday Soc.*, **33**, 8.

Magnus, G. (1852). *Uber die Abweichung der geschosse, nebst einem Anhange: Uber eine auffallende Erscheinung bei rotirenden Korpern.* Berlin: F. Dummler.

Massimilla, L. and Donsi, G. (1976). Cohesive Forces Between Particles of Fluid-Bed Catalysis. *Powder Tech.*, **15**, 253.

McAteer, O. J. (1990). *Electrostatic Discharge Control.* New York: McGraw-Hill.

Montgomery, D. J. (1959). Static Electrification of Solids. *Solid State Phys.*, **9**, 139.

Murphy, A. T. (1956). *Charging of Fine Particles by Random Motion of Ions in an Electric Field.* Ph.D. Dissertation. Carnegie Institute of Technology.

Odar, F. and Hamilton, W. S. (1964). Forces on a Sphere Accelerating in a Viscous Fluid. *J. Fluid Mech.*, **18**, 302.

Oglesby, S. and Nichols, G. B. (1978). *Electrostatic Precipitation.* New York: Marcel Dekker.

Oseen, C. W. (1927). *Hydrodynamik.* Leipzig: Akademische Verlagsgescellschafe.

Passman, S. L. (1986). Forces on the Solid Constituent in a Multiphase Flow. *J. Rheology*, **30**, 1077.

Rowe, P. N. and Henwood, G. A. (1961). Drag Forces in a Hydraulic Model of a Fluidised Bed–Part I. *Trans. Instn. Chem. Engrs.*, **39**, 43.

Rubinow, S. I. and Keller, J. B. (1961). The Transverse Force on a Spinning Sphere Moving in a Viscous Fluid. *J. Fluid Mech.*, **11**, 447.

Rudinger, G. (1980) *Fundamentals of Gas–Particle Flow.* Amsterdam: Elsevier Scientific.

Saffman, P. G. (1965). The Lift on a Small Sphere in a Slow Shear Flow. *J. Fluid Mech.*, **22**, 385.

Saffman, P. G. (1968). Corrigendum. *J. Fluid Mech.*, **31**, 624.

Schlichting, H. (1979). *Boundary Layer Theory*, 7th ed. New York: McGraw-Hill.

Soo, S. L. (1964). Effect of Electrification on Dynamics of a Particulate System. *I&EC Fund.*, **3**, 75.

Soo, S. L. (1990). *Multiphase Fluid Dynamics.* Beijing: Science Press; Brookfield, Vt.: Gower Technical.

Tchen, C. M. (1947). *Mean Value and Correlation Problems Connected with the Motion of Small Particles in a Turbulent Field.* Ph.D. Dissertation. Delft University, Netherlands.

Tsuji, Y., Morikawa, Y. and Terashima, K. (1982). Fluid-Dynamic Interaction Between Two Spheres. *Int. J. Multiphase Flow*, **8**, 71.

Visser, J. (1989). Van der Waals and Other Cohesive Forces Affecting Powder Fluidization. *Powder Tech.*, **58**, 1.

Wallis, G. B. (1969). *One-Dimensional Two-Phase Flow.* New York: McGraw-Hill.

Webers, V. J. (1963). Measurement of Triboelectric Position. *J. Appl. Polymer Sci.*, **7**, 1317.

White, H. J. (1963). *Industrial Electrostatic Precipitation.* Reading, Mass.: Addison-Wesley.

Williams, M. W. (1976). The Dependence of Triboelectric Charging of Polymers on Their Chemical Composition. *J. Macromolecular Sci. Rev. Macromol. Chem.*, **C14**, 251.

Zhu, C., Liang, S.-C. and Fan, L.-S. (1994). Particle Wake Effects on the Drag Force of an Interactive Particle. *Int. J. Multiphase Flow*, **20**, 117.

Zhu, C. and Soo, S. L. (1992). A Modified Theory for Electrostatic Probe Measurements of Particle Mass Flows in Dense Gas–Solid Suspensions. *J. Appl. Phys.*, **72**, 2060.

Problems

3.1 Calculate the van der Waals force (newtons) between (a) two diamond spheres of radii $a_1 = a_2 = 10\ \mu$m; (b) two graphite spheres of radii $a_1 = 1\ \mu$m and $a_2 = 10\ \mu$m; and (c) a graphite plate and a diamond sphere of radius $a_1 = 5\ \mu$m. The Hamaker coefficients for diamond and graphite are 28.4×10^{-13} and 47.0×10^{-13} erg, respectively. For the van der Waals interaction between two different solid materials, the Hamaker coefficient can be approximated by the geometric mean of the individual coefficients [*e.g.*, Visser, 1989] as

$$A_{12} = \sqrt{A_{11}A_{22}} \tag{P3.1}$$

Assume that the separation distance is 4 Å.

3.2 Consider the collision of particles due to wake attraction, as shown in Fig. E3.1. It is assumed that (a) the motion of the leading particle is not affected by the approach of the trailing particle; (b) particles are equal-sized, rigid, and spherical; and (c) initially, the particles move nearly at their terminal velocities with a very small velocity difference and are separated by a characteristic distance l_0. An empirical relation can be used to describe the effects of the interparticle distance l and particle Reynolds number $\mathrm{Re_p}$ on the drag force of the trailing particle as

$$\frac{F_\mathrm{D}}{F_\mathrm{D0}} = 1 - (1 - A)\exp\left(B\frac{l}{d_\mathrm{p}}\right) \tag{P3.2}$$

where F_D0 is the drag force of a single noninteracting particle and the coefficients A and B, for $20 < \mathrm{Re_p} < 150$, are given by [Zhu *et al.*, 1994]

$$A = 1 - \exp\left(-0.483 + 3.45 \times 10^{-3}\mathrm{Re_p} - 1.07 \times 10^{-5}\mathrm{Re_p^2}\right) \tag{P3.3}$$

and

$$B = -0.115 - 8.75 \times 10^{-4}\mathrm{Re_p} + 10^{-7}\mathrm{Re_p^2} \tag{P3.4}$$

(a) Derive an expression for the particle contact velocity without considering the Basset effect. (b) Use the result of (a) and Eq. (E3.9), which considers the Basset effect (see Example 3.1), to predict the contact velocity under the conditions $\mathrm{Re_p} = 90$; $\mu = 0.057$ kg/m · s; $\rho_\mathrm{p} = 1{,}500$ kg/m^3; $\rho = 1{,}200$ kg/m^3; $d_\mathrm{p} = 0.019$ m; and $l_0 = 2d_\mathrm{p}$. The measured contact velocity is 0.13 m/s. Compare the calculated contact velocities with the measured velocity to show that the Basset force is important and cannot be neglected in the formulation.

3.3 Prove that the resistance on a sphere moving in a gravitational field with constant velocity V can be obtained in the form given in Eq. (3.36). If the sphere starts moving from rest with an accelerating velocity, v, show that the resistance can be expressed by Eq. (3.37).

3.4 Determine the initial acceleration of a free-falling glass bead with the density of 2,300 kg/m^3 in the air and that of a rising spherical air "bubble" in a fluidized bed. The fluidized bed can be regarded as a pseudocontinuum medium with a density of 600 kg/m^3. Discuss the significance of carried mass force for both cases.

3.5 For a fixed spherical particle in a fully developed laminar pipe flow, determine the Saffman force on the particle at various radial positions. Identify the location of the maximum Saffman force. Discuss the case if the flow is turbulent (using the 1/7 power law for the velocity profile).

3.6 In the transport of fine particles, it is important to understand interparticle force effects during the particle motion. On the basis of the data provided by Massimilla and Donsi (1976), *i.e.*, $h\varpi = 2$ eV; $s = 50$ Å; $\rho_\mathrm{p} = 1{,}000$–$2{,}000$ kg/m^3; $d_\mathrm{p} = 20 - 1{,}000\ \mu$m, discuss the relative significance of the gravitational force to the van der Waals force for disperse particles with these properties.

3.7 Consider a large number of uniformly charged solid particles initially kept in a spherical barrier of radius R_0 with a symmetric density distribution. When the barrier is suddenly removed, the particles start to emerge from that spherical domain. The viscous drag coefficient is assumed to be constant. It is also assumed that the dipole effect due to self-field can be negligible. Show that the relationship between the spreading velocity and the position of a charged particle initially located at R_0 can be given by

$$U_p = \exp\left(-\frac{AR}{2}\right)\left[\int_{R_0}^{R} \frac{B}{\xi^2} \exp(A\xi)\, \mathrm{d}\xi\right] \tag{P3.5}$$

where

$$A = \frac{3}{2}\frac{\rho C_D}{\rho_p d_p}; \qquad B = \frac{\left(\frac{q}{m}\right)^2 M_0}{2\pi\varepsilon}$$

(see Example 3.2 for notations).

3.8 An account of the behavior of acoustic wave propagation in a gas–solid suspension or particle movement in a turbulent eddy requires comprehensive knowledge of the dynamics of particle motion in an oscillating flow field. This oscillating flow can be analyzed in terms of one-dimensional simple harmonic oscillation represented by

$$U = U_0 \sin(\omega t) \tag{P3.6}$$

Assume that the only force imposed on the particle in the oscillating flow field is the Stokes drag. Derive expressions for the particle velocity and the phase lag between the particle velocity and the gas velocity for the condition when the particle is placed in the flow field for sufficiently long time (*i.e.*, $t \gg \tau_S$).

3.9 For an electron number density of 10^{15} m^{-3} and a mobility of 0.022 (m^2/V·s), calculate the time required to reach 99 percent of the saturation charge of an initially neutral particle by field charging. If the length of the charging chamber is 5 cm, what is the maximum particle velocity allowed to ensure 99 percent saturation charge?

3.10 For the ball–probe measurement of charge current in a dilute gas–solid suspension, prove that, for a given type of particle and a given mass flow ratio, the current measured is independent of the particle size for $d_b \gg d_p$. It is assumed that the charge transfer coefficient is constant.

CHAPTER 4

Basic Heat and Mass Transfer

4.1 Introduction

Gas–solid flows involving heat and mass transfer are common in many engineering operations including petroleum refining, nuclear reactor cooling, solid fuel combustion, rocket nozzle jetting, drying, and bulk material handling and transport. In the chemical process industries, reactors for polymerization and hydrogenation commonly require cooling or heating to maintain desired reaction temperatures. In nuclear reactor cooling or solar energy transportation, graphite suspension flows are used in the heat exchanger. Graphite suspension coolants can have high heat transfer coefficients in addition to other noted properties such as high thermal capacity, good temperature stability, and low pressurization characteristics [Boothroyd, 1971]. Combustion of pulverized coal, gas adsorption on a catalyst, and synthesis of chemicals are among many examples of mass transfer in gas–solid flows.

There are three basic modes of heat transfer, namely, thermal diffusion (heat conduction), convection, and radiation. All three modes may occur simultaneously, or one of them may dominate under certain conditions. The first two modes are analogous to their counterparts in momentum transfer in a fluid. Thermal radiation, a form of energy transfer through electromagnetic waves, is governed by rather different laws and can occur even through a perfect vacuum. It is also noted that an analogy can exist between mass transfer and heat transfer (excluding radiation) because of the similarity of their governing equations.

This chapter describes the fundamental principles of heat and mass transfer in gas–solid flows. For most gas–solid flow situations, the temperature inside the solid particle can be approximated to be uniform. The theoretical basis and relevant restrictions of this approximation are briefly presented. The conductive heat transfer due to an elastic collision is introduced. A simple convective heat transfer model, based on the pseudocontinuum assumption for the gas–solid mixture, as well as the limitations of the model applications are discussed. The chapter also describes heat transfer due to radiation of the particulate phase. Specifically, thermal radiation from a single particle, radiation from a particle cloud with multiple scattering effects, and the basic governing equation for general multiparticle radiations are discussed. The discussion of gas phase radiation is, however, excluded because of its complexity, as it is affected by the type of gas components, concentrations, and gas temperatures. Interested readers may refer to Özisik (1973) for the absorption (or emission) of radiation by gases. The last part of this chapter presents the fundamental principles of mass transfer in gas–solid flows.

4.2 Heat Conduction

Conductive heat transfer is the dominant mode of intraparticle heat transfer. Under low Reynolds number flow situations, conductive heat transfer is also an important mode for fluid heat transfer. This section analyzes the conductive heat transfer characteristics of a single spherical particle in a quiescent fluid and in an elastic collision with another spherical particle.

4.2.1 *Heat Transfer of a Single Sphere in a Quiescent Fluid*

For simplicity, it is assumed that the temperature distribution throughout the sphere is radially symmetric, no heat generation occurs inside the sphere, and no thermal radiation takes place between the sphere and the surrounding environment. Thus, the transient heat balance in the particle leads to

$$\frac{1}{R}\frac{\partial^2}{\partial R^2}(RT_\mathrm{p}) = \frac{\rho_\mathrm{p} c}{K_\mathrm{p}}\frac{\partial T_\mathrm{p}}{\partial t} \tag{4.1}$$

The boundary and initial conditions are

$$\begin{aligned} &\frac{\partial T_\mathrm{p}}{\partial R} = 0 && \text{at } R = 0 \\ &K_\mathrm{p}\frac{\partial T_\mathrm{p}}{\partial R} = h_\mathrm{p}(T_\infty - T_\mathrm{p}) && \text{at } R = a \\ &T_\mathrm{p} = T_\mathrm{pi}(R) && \text{for } t = 0 \end{aligned} \tag{4.2}$$

Define dimensionless variables as

$$\begin{aligned} &R^* = \frac{R}{a}, && T_\mathrm{p}^* = \frac{T_\mathrm{p} - T_\infty}{T_\mathrm{p0} - T_\infty} \\ &\mathrm{Bi} = \frac{h_\mathrm{p} a}{K_\mathrm{p}}, && \mathrm{Fo} = \frac{K_\mathrm{p} t}{\rho_\mathrm{p} c a^2} \end{aligned} \tag{4.3}$$

where Bi and Fo are the Biot number and Fourier number, respectively.

The Fo and Bi are two important dimensionless parameters in heat conduction. Fo can be interpreted as

$$\mathrm{Fo} = \frac{\dfrac{K_\mathrm{p}}{a^2}a^3}{\dfrac{\rho_\mathrm{p} c a^3}{t}} = \frac{\text{rate of heat conduction across } a^2 \text{ in } a^3}{\text{rate of heat storage in } a^3} \tag{4.4}$$

Bi can be interpreted as

$$\mathrm{Bi} = \frac{h_\mathrm{p}}{K_\mathrm{p}/a} = \frac{\text{convective heat transfer coefficient at sphere surface}}{\text{thermal conductance of solid across depth } a} \tag{4.5}$$

Equations (4.1) and (4.2) can be reduced to dimensionless forms as

$$\frac{1}{R^*}\frac{\partial^2}{\partial R^{*2}}\left(R^* T_\mathrm{p}^*\right) = \frac{\partial T_\mathrm{p}^*}{\partial \mathrm{Fo}} \tag{4.6}$$

with

$$\begin{aligned} &\frac{\partial T_\mathrm{p}^*}{\partial R^*} = 0 && \text{at } R^* = 0 \\ &\frac{\partial T_\mathrm{p}^*}{\partial R^*} + \mathrm{Bi}T_\mathrm{p}^* = 0 && \text{at } R^* = 1 \\ &T_\mathrm{p}^* = T_\mathrm{pi}^*(R^*) && \text{for } \mathrm{Fo} = 0 \end{aligned} \tag{4.7}$$

The solution for Eqs. (4.6) and (4.7) gives

$$T_p^* = \frac{2}{R^*} \sum_{n=1}^{\infty} e^{-\lambda_n^2 \mathrm{Fo}} \frac{\left[\lambda_n^2 + (\mathrm{Bi}-1)^2\right] \sin \lambda_n R^*}{\lambda_n^2 + (\mathrm{Bi}-1)^2 + \mathrm{Bi} - 1} \int_0^1 \xi T_{pi}^* \sin \lambda_n \xi \, d\xi \tag{4.8}$$

where λ_n's are the eigenvalues of

$$\lambda_n \cot \lambda_n + \mathrm{Bi} - 1 = 0 \tag{4.9}$$

Equation (4.8) indicates that the one-dimensional transient temperature distribution inside a solid sphere without internal heat generation varies with Fo and Bi. On the basis of the equation, the temperature distribution in the solid can be considered uniform with an error of less than 5 percent when $\mathrm{Bi} < 0.1$, which is the condition for most gas–solid flow systems. In transient heat transfer processes where the gas–solid contact time is very short, it also requires $\mathrm{Fo} \geq 0.1$ [Gel'Perin and Einstein, 1971] for the internal thermal resistance within the particles to be neglected. In the following, unless otherwise noted, it is assumed that the temperature inside a solid particle is uniform.

For a quasi-steady-state heat conduction between an isothermal sphere and an infinitely large and quiescent fluid, the temperature distribution in the fluid phase is governed by

$$\frac{1}{R^2} \frac{d}{dR} \left(R^2 K \frac{dT}{dR} \right) = 0 \tag{4.10}$$

with boundary conditions

$$\begin{aligned} T &= T_p \quad \text{at } R = a \\ T &= T_\infty \quad \text{at } R \to \infty \end{aligned} \tag{4.11}$$

which leads to the solution

$$T = \frac{a}{R}(T_p - T_\infty) + T_\infty \tag{4.12}$$

The corresponding heat flux at the surface of the sphere is

$$J_q = \frac{2K(T_p - T_\infty)}{d_p} \tag{4.13}$$

Note that the heat flux can also be expressed as

$$J_q = h_p(T_p - T_\infty) \tag{4.14}$$

Thus, the limiting Nusselt number at $\mathrm{Re}_p \to 0$ is expressed by

$$\mathrm{Nu}_p = \frac{h_p d_p}{K} = 2 \tag{4.15}$$

Equation (4.15) indicates that, for an isolated, isothermal solid sphere in an infinite fluid, Nu_p approaches 2 in the limit of negligible forced or natural convection. This relationship was confirmed experimentally by Ranz and Marshall (1952).

Consider the heat conduction through a spherical shell of an inner diameter d_1 and an outer diameter d_2. It can be shown that

$$\mathrm{Nu}_p = \frac{2}{1 - d_1/d_2} \tag{4.16}$$

Equation (4.16) can be reduced to Eq. (4.15) as d_2 becomes infinitely large. For a packed bed of monodispersed spheres arranged in a cubic particle array, the mean thickness of the "shell" of gas surrounding the particle can be approximated by

$$\delta \approx \frac{d_p^3 - \pi d_p^3/6}{\pi d_p^2} = 0.152 d_p \tag{4.17}$$

Thus, from Eqs. (4.16) and (4.17), $\mathrm{Nu_p} \approx 10$, indicating that the $\mathrm{Nu_p}$ of a single particle due to thermal conduction in a gas–solid suspension roughly varies from 2 to 10. A more detailed discussion on the variation of $\mathrm{Nu_p}$ with δ/d_p is given by Zabrodsky (1966).

4.2.2 *Heat Conduction in a Collision of Elastic Spheres*

In a collision between two spheres of different temperatures, heat conduction occurs at the interface. The contact area is usually negligibly small compared to the cross-sectional area of the spheres. Since the duration of the impact is also very short, the temperature change of the colliding particles is confined to a small region around the contact area. Therefore, the heat conduction between the two particles can be treated as that between two semiinfinite media. It is also assumed that there is no thermal resistance between the contact surfaces. Hence, the temperature and heat flux distributions are continuous across the contact area. The surfaces outside the contact area are assumed to be flat and insulated. For general information on collision mechanisms of solids, readers may refer to Chapter 2.

For simplicity, it is assumed that the impact is a Hertzian collision. Thus, no kinetic energy loss occurs during the impact. The problem of conductive heat transfer due to the elastic collision of solid spheres was defined and solved by Sun and Chen (1988). In this problem, considering the heat conduction through the contact surface as shown in Fig. 4.1, the change of the contact area or radius of the circular area of contact with respect to time is given by Eq. (2.139) or by Fig. 2.16. In cylindrical coordinates, the heat conduction between the colliding solids can be written by

$$\frac{1}{D_{ti}}\frac{\partial T_i}{\partial t} = \frac{1}{r}\frac{\partial}{\partial r}\left(r\frac{\partial T_i}{\partial r}\right) + \frac{\partial^2 T_i}{\partial z^2}; \qquad i = 1, 2 \tag{4.18}$$

The boundary and initial conditions are

$$\begin{aligned}
&T_1 = T_2; \qquad K_{p1}\frac{\partial T_1}{\partial z} = K_{p2}\frac{\partial T_2}{\partial z} && \text{at } z = 0 \text{ and } r \le r_c(t)\\
&\frac{\partial T_1}{\partial z} = \frac{\partial T_2}{\partial z} = 0 && \text{at } z = 0 \text{ and } r > r_c(t)\\
&\frac{\partial T_1}{\partial r} = \frac{\partial T_2}{\partial r} = 0 && \text{at } r = 0\\
&T_1 = T_{10}; \qquad T_2 = T_{20} && \text{at } t = 0
\end{aligned} \tag{4.19}$$

where D_t's are the thermal diffusivities of solids. D_t is defined by

$$D_t = \frac{K_p}{\rho_p c} \tag{4.20}$$

When the contact time is very short, the heat penetration by conduction from one medium to the other would not be very deep from the contact surface. Therefore, one-dimensional

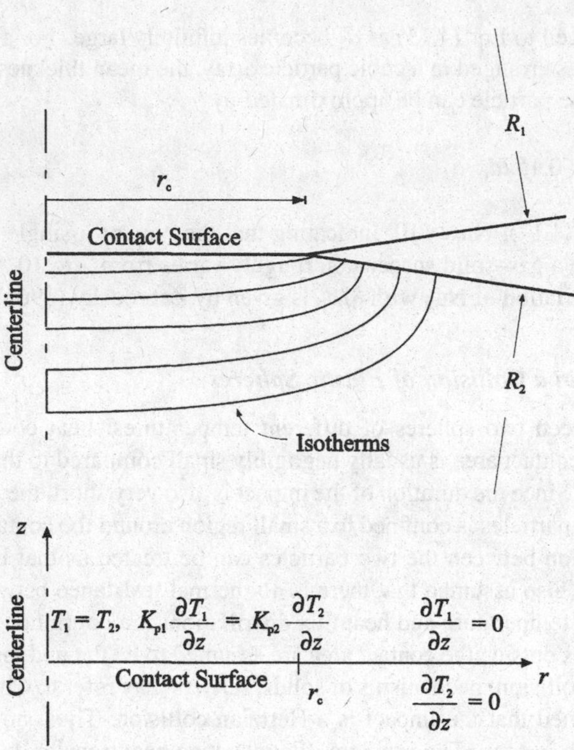

Figure 4.1. Conductive heat transfer by two-ball collision (from Sun and Chen, 1988).

approximation can be made to account for the heat transfer behavior. This behavior can usually be characterized by the Fourier number in terms of the total impact duration and maximum contact radius, as defined by

$$\mathrm{Fo_i} = \frac{D_{\mathrm{ti}} t_{\mathrm{c}}}{r_{\mathrm{cm}}^2}; \quad i = 1, 2 \tag{4.21}$$

Thus, for a small Fourier number, the one-dimensional heat transfer yields the following equation

$$\frac{1}{D_{\mathrm{ti}}} \frac{\partial T_{\mathrm{i}}}{\partial t} = \frac{\partial^2 T_{\mathrm{i}}}{\partial z^2}; \quad i = 1,\ 2 \tag{4.22}$$

with the boundary and initial conditions given by Eq. (4.19).

For constant properties, a general solution of the problem

$$\begin{aligned} &\frac{\partial T}{\partial t} = D_{\mathrm{t}} \frac{\partial^2 T}{\partial z^2} \\ &T|_{t=0} = 0; \qquad T|_{z=0} = \phi(t) \end{aligned} \tag{4.23}$$

is given by [Carslaw and Jaeger, 1959]

$$T = \frac{2}{\sqrt{\pi}} \int_{\frac{z}{2\sqrt{D_{\mathrm{t}} t}}}^{\infty} \phi\left(t - \frac{z^2}{4 D_{\mathrm{t}} \xi^2}\right) e^{-\xi^2}\,\mathrm{d}\xi \tag{4.24}$$

Thus, we have

$$\frac{\partial T}{\partial z} = -\frac{\phi(0)}{\sqrt{D_t \pi t}} e^{-\frac{z^2}{4D_t t}} - \int_{\frac{z}{2\sqrt{D_t t}}}^{\infty} \frac{z}{D_t \xi^2 \sqrt{\pi}} \frac{\partial \phi}{\partial z} e^{-\xi^2} \mathrm{d}\xi \tag{4.25}$$

At $z = 0$, the preceding equation reduces to

$$\left.\frac{\partial T}{\partial z}\right|_{z=0} = -\frac{\phi(0)}{\sqrt{D_t \pi t}} \tag{4.26}$$

Now, we can use the solutions, *i.e.*, Eqs. (4.24) and (4.26), for the problem imposed by Eq. (4.22) and its boundary and initial conditions. Using the continuity condition for the temperature across the interface, we have

$$\phi_1 = \phi_2 + (T_{20} - T_{10}) \tag{4.27}$$

Since the heat flux across the interface is continuous, we have

$$J_{qw} = \frac{K_1}{\sqrt{\pi D_{t1} t}} \phi_1(0) = -\frac{K_2}{\sqrt{\pi D_{t2} t}} \phi_2(0) \tag{4.28}$$

where the minus sign is due to the opposite outward-directed normals of the two media at the contact interface. The interface heat flux is thus obtained as

$$J_{qw} = \frac{T_{20} - T_{10}}{\sqrt{\pi t}} \left(\frac{\sqrt{D_{t1}}}{K_1} + \frac{\sqrt{D_{t2}}}{K_2} \right)^{-1} \tag{4.29}$$

Here t is measured from the moment of initial contact and, thus, can be expressed as a function of the radius of contact area r_c, as shown in Fig. 2.16. The total heat exchange per impact is

$$Q_c = 2\pi \int_0^{r_{cm}} \int_0^{t_c - 2t} J_{qw} \, \mathrm{d}t \, r \, \mathrm{d}r \tag{4.30}$$

with t related to r by either Eq. (2.137) or Eq. (2.139). The numerical integration of Eq. (4.30) gives

$$Q_{c0} = 2.73(T_{20} - T_{10}) r_{cm}^2 \sqrt{t_c} \left(\frac{\sqrt{D_{t1}}}{K_1} + \frac{\sqrt{D_{t2}}}{K_2} \right)^{-1} \tag{4.31}$$

where r_{cm} and t_c are given by Eqs. (2.133) and (2.136), respectively; and the subscript "0" for Q_c indicates that it is for the case in which the Fourier number approaches zero.

The preceding result of heat exchange is derived for heat conduction only in the z-direction when Fo is negligibly small. When heat conduction in the r-direction is not negligible, more heat would be transferred between the two colliding media. To quantify it, a correction factor C may be introduced in such a way that the total heat exchanged per impact becomes

$$Q_c = C Q_{c0} \tag{4.32}$$

Since there is no general analytical solution for Eq. (4.18) with the appropriate boundary and initial conditions, the factor C can only be evaluated by the numerical approach.

For most solid materials, the parameter $\rho_{p1}c_1/\rho_{p2}c_2$ varies from about 0.1 to 10 and the parameter K_1/K_2 varies from about 0.001 to 1. The Fourier number can be considered as a measure of the impact conditions. For example, a larger value of the Fourier number represents a smaller impact velocity. Within the range of interest, the solutions for the

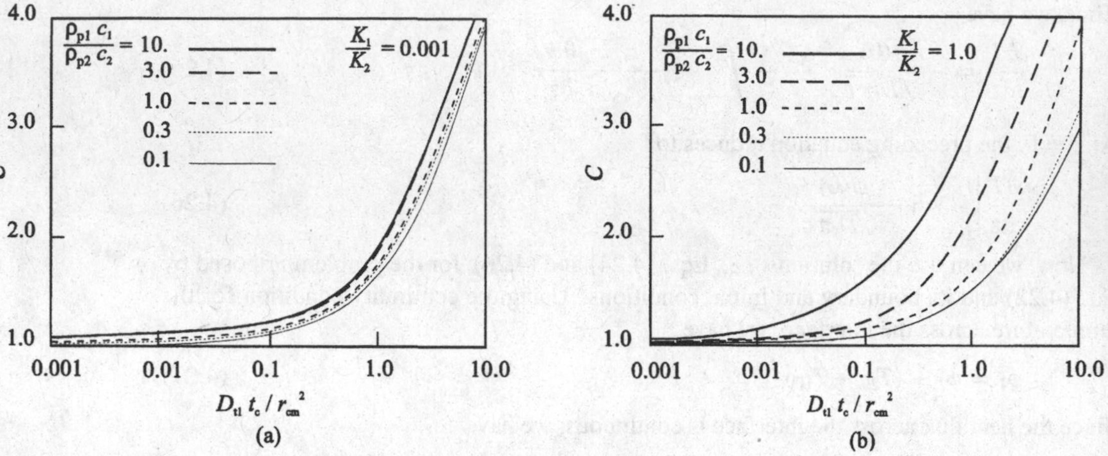

Figure 4.2. Relations of the correction factor with the Fourier number: (a) $K_1/K_2 = 0.001$; (b) $K_1/K_2 = 1.0$ (from Sun and Chen, 1988).

correction factor C are plotted in Figs. 4.2(a) and 4.2(b) for two limiting values of K_1/K_2. Each plot gives the variations of C with Fo for different $\rho_{p1}c_1/\rho_{p2}c_2$ values. It can be seen that, as Fo increases, the heat conduction in the radial direction becomes significant in the total heat conduction process.

In gas–solid suspensions and fluidized beds, the heat transfer between particles and the wall surface or between a particle at one temperature and a group of other particles at another temperature is largely due to particle impacts. Thus, the average rate of heat transfer may be expressed in terms of a collisional heat transfer coefficient h_c, which is defined by

$$h_c = \frac{n\Delta U_p Q_c}{T_{20} - T_{10}} \tag{4.33}$$

where ΔU_p is the relative impact velocity; n is the particle number density; and the product of n and ΔU_p is the rate of particle surface impact per unit area. Thus, from Eqs. (4.31), (4.32), and (4.33), the collisional heat transfer coefficient is given by

$$h_c = 2.73 n \Delta U_p C r_{cm}^2 \sqrt{t_c}\left(\frac{\sqrt{D_{t1}}}{K_1} + \frac{\sqrt{D_{t2}}}{K_2}\right)^{-1} \tag{4.34}$$

Example 4.1 Determine the collisional heat transfer coefficients under each of the following conditions: (1) collisions of a cloud of hot particles with a cold particle; (2) collisions of a cloud of cold particles with a hot wall. Assume the particles are in random motion with the average impact velocity of 0.1 m/s. All the particles are spherical and of the same diameter of 100 μm. The particles and wall are made of steel with $\nu = 0.3$, $E = 2 \times 10^5$ MPa, $\rho_p = 7{,}000$ kg/m^3, $K_p = 30$ W/m · K, and c = 500 J/kg · K. The particle volume fraction is 0.4.

Solution The maximum contact radius and contact time for an elastic sphere collision can be estimated from Eqs. (2.133) and (2.136), respectively. For collisions among particles, we have

$$r_{\rm cm} = \frac{d_{\rm p}}{2}\left(\frac{5\pi}{16}\frac{\rho_{\rm p}}{E}(\Delta U_{\rm p})^2(1-\nu^2)\right)^{1/5}$$

$$= \frac{10^{-4}}{2}\left(\frac{5\pi}{16}\frac{7\times 10^3}{2\times 10^{11}}0.1^2\times(1-0.3^2)\right)^{1/5} = 0.63\ \mu\text{m} \tag{E4.1}$$

and

$$t_{\rm c} = \frac{2.94\times 4r_{\rm cm}^2}{\Delta U_{\rm p} d_{\rm p}} = \frac{11.76\times(0.63\times 10^{-6})^2}{0.1\times 10^{-4}} = 4.7\times 10^{-7}\ \text{s} \tag{E4.2}$$

The Fo is obtained from Eq. (4.21) as

$$\text{Fo} = \frac{K_{\rm p} t_{\rm c}}{\rho_{\rm p} c r_{\rm cm}^2} = \frac{30\times 4.7\times 10^{-7}}{7{,}000\times 500\times(0.63\times 10^{-6})^2} = 10 \tag{E4.3}$$

Therefore, from Fig. 4.2, the correction factor C is found to be 5. The collisional heat transfer coefficient by interparticle collisions is thus given by Eq. (4.34) as

$$h_{\rm c} = \frac{8.19 K_{\rm p}\alpha_{\rm p}\Delta U_{\rm p} C r_{\rm cm} t_{\rm c}}{\pi d_{\rm p}^3\sqrt{\text{Fo}}}$$

$$= \frac{8.19\times 30\times 0.4\times 0.1\times 5\times 0.63\times 10^{-6}\times 4.7\times 10^{-7}}{\pi(10^{-4})^3\sqrt{10}} = 1.5\frac{\text{W}}{\text{m}^2\cdot\text{K}} \tag{E4.4}$$

Similarly, for collisions between a group of particles and the wall, we have

$$r_{\rm cm} = \frac{d_{\rm p}}{2}\left(\frac{5\pi}{2}\frac{\rho_{\rm p}}{E}(\Delta U_{\rm p})^2(1-\nu^2)\right)^{1/5}$$

$$= \frac{10^{-4}}{2}\left(\frac{5\pi}{2}\frac{7\times 10^3}{2\times 10^{11}}0.1^2(1-0.3^2)\right)^{1/5} = 0.95\ \mu\text{m} \tag{E4.5}$$

and

$$t_{\rm c} = \frac{2.94\times 2r_{\rm cm}^2}{\Delta U_{\rm p} d_{\rm p}} = \frac{5.88\times(0.95\times 10^{-6})^2}{0.1\times 10^{-4}} = 5.3\times 10^{-7}\ \text{s} \tag{E4.6}$$

The Fo is obtained by

$$\text{Fo} = \frac{K_{\rm p} t_{\rm c}}{\rho_{\rm p} c r_{\rm cm}^2} = \frac{30\times 5.3\times 10^{-7}}{7{,}000\times 500\times(0.95\times 10^{-6})^2} = 5 \tag{E4.7}$$

From Fig. 4.2, the correction factor C is about 4. The collisional heat transfer coefficient by interparticle collisions is given by

$$h_{\rm c} = \frac{8.19 K_{\rm p}\alpha_{\rm p}\Delta U_{\rm p} C r_{\rm cm} t_{\rm c}}{\pi d_{\rm p}^3\sqrt{\text{Fo}}}$$

$$= \frac{8.19\times 30\times 0.4\times 0.1\times 4\times 0.95\times 10^{-6}\times 5.3\times 10^{-7}}{\pi(10^{-4})^3\sqrt{5}} = 2.8\frac{\text{W}}{\text{m}^2\cdot\text{K}} \tag{E4.8}$$

This example indicates that impact of particles among themselves or with the wall is in general not an effective means of heat transfer in gas–solid flows.

4.3 Convective Heat Transfer

When a thermal surface is exposed to a flow field, the heat transfer from the surface is enhanced by thermal convection in addition to heat conduction. Although thermal convection in general consists of forced convection and natural convection, the effect due to natural convection in a gas–solid system is usually insignificant. This section presents the dimensional analysis which gives rise to the basic dimensionless groups accounting for forced convection in a single-phase flow. The thermal convection involving a single spherical particle in a uniform flow and a flow of gas and solids treated as a pseudocontinuum medium is discussed.

4.3.1 *Dimensional Analysis of Forced Convection in a Single-Phase Flow*

The influential parameters in a thermal convection system may be represented by the characteristic length l, velocity U, density ρ, viscosity μ, specific heat at constant pressure c_p, and thermal conductivity K. Thus, we have

$$h = f(l, U, \rho, \mu, c_p, K) \tag{4.35}$$

where h is the convective heat transfer coefficient. In terms of dimensional units, Eq. (4.35) takes the form

$$\frac{[\mathrm{M}]}{[\mathrm{T}]^3[\Theta]} = \mathrm{f}\left([\mathrm{L}], \frac{[\mathrm{L}]}{[\mathrm{T}]}, \frac{[\mathrm{M}]}{[\mathrm{L}]^3}, \frac{[\mathrm{M}]}{[\mathrm{L}][\mathrm{T}]}, \frac{[\mathrm{L}]^2}{[\mathrm{T}]^2[\Theta]}, \frac{[\mathrm{M}][\mathrm{L}]}{[\mathrm{T}]^3[\Theta]}\right) \tag{4.36}$$

where [M] denotes the mass unit; [L] represents the length unit; [T] stands for the time unit; and [Θ] is the thermal unit. From the Pi theorem, it yields

$$\mathrm{Nu} = \mathrm{f}(\mathrm{Re}, \mathrm{Pr}) \tag{4.37}$$

where Nu is the Nusselt number and Pr is the Prandtl number. The physical significance of Nu and Pr may be interpreted as

$$\mathrm{Nu} \equiv \frac{hl}{K} = \frac{hl^2 \Delta T}{Kl^2 \dfrac{\Delta T}{l}} = \frac{\text{convection}}{\text{conduction}} \tag{4.38}$$

and

$$\mathrm{Pr} \equiv \frac{\mu/\rho}{D_t} = \frac{\text{momentum diffusivity}}{\text{thermal diffusivity}} \tag{4.39}$$

4.3.2 *Heat Transfer of a Single Sphere in a Uniform Flow*

For forced convective heat transfer over a sphere in a uniform flow, a frequently used empirical relation was proposed by Ranz and Marshall (1952) as

$$\mathrm{Nu_p} = 2.0 + 0.6(\mathrm{Re_p})^{1/2}(\mathrm{Pr})^{1/3} \tag{4.40}$$

where the particle diameter is used as the characteristic length. In the equation, the first term matches the theoretical requirement of $\mathrm{Nu_p} = 2$ at $\mathrm{Re_p} = 0$, whereas the second term may be regarded as the enhancement factor due to forced convection. Equation (4.40) is applicable to a range of $\mathrm{Re_p}$ up to 10^4 and $\mathrm{Pr} \geq 0.7$, which covers most gas–solid flows. It is noted that although Eq. (4.40) was originally obtained from the study of liquid droplets, this relation was proved also to be valid for the case of rigid spheres [Hughmark, 1967].

For most gases, Pr is nearly constant (~0.7). Therefore, for convective heat transfer of an individual solid sphere in a gaseous medium, $\mathrm{Nu_p}$ is only a function of $\mathrm{Re_p}$. An alternative estimation of $\mathrm{Nu_p}$ for thermal convection of a sphere in a uniform flow was recommended by McAdams (1954) as

$$\mathrm{Nu_p} = 0.37\,\mathrm{Re_p^{0.6}} \quad \text{for } 17 < \mathrm{Re_p} < 70{,}000 \tag{4.41}$$

where all associated thermal properties are determined on the basis of the ambient temperature of the gas.

The analytical solution for convective heat transfer from an isolated particle in a Stokes flow can be obtained by using some unique perturbation methods, noting that the standard perturbation technique of expanding the temperature field into a power series of the Peclet number ($\mathrm{Pe} = \mathrm{Re_p Pr}$) fails to solve the problem [Kronig and Bruijsten, 1951; Brenner, 1963]. The $\mathrm{Nu_p}$ for the thermal convection of a sphere in a uniform Stokes flow is given by

$$\mathrm{Nu_p} = 2 + \frac{1}{2}(\mathrm{Re_p Pr}) + o(\mathrm{Re_p Pr}) \tag{4.42}$$

which is proved by the following example using the method developed by Kronig and Bruijsten (1951).

Example 4.2 Show that the convective heat transfer from a sphere in a flowing medium at low Reynolds number can be expressed by Eq. (4.42). It is assumed that the sphere is kept at a constant temperature T_p and the temperature inside the sphere is uniform. The flow is uniform with velocity of U_∞ and temperature of T_∞ at infinity. All the thermal properties are constant.

Solution The energy balance equation can be expressed by

$$\frac{K}{\rho c_\mathrm{p}}\nabla^2 T - U \cdot \nabla T = 0 \tag{E4.9}$$

with the boundary conditions (in polar coordinates)

$$T = T_\mathrm{p} \quad \text{at } R = a; \qquad T = T_\infty \quad \text{at } R \to \infty \tag{E4.10}$$

For simplicity, take the dimensionless form of the equation as

$$\nabla^{*2} T^* - \varepsilon U^* \cdot \nabla^* T^* = 0 \tag{E4.11}$$

where

$$T^* = \frac{T - T_\infty}{T_\mathrm{p} - T_\infty}, \qquad U^* = \frac{U}{U_\infty}, \qquad \nabla^* = a\nabla, \qquad r = \frac{R}{a}, \qquad \varepsilon = \frac{1}{2}\mathrm{Re_p\,Pr} \tag{E4.12}$$

and the boundary conditions become

$$T^* = 1 \quad \text{at } r = 1; \qquad T^* = 0 \quad \text{at } r \to \infty \tag{E4.13}$$

Upon taking the polar axis parallel to the streamlines at infinity, the velocity components of the Stokes flow can be given by

$$U_r^* = \left(1 - \frac{3}{2r} + \frac{1}{2r^3}\right)\cos\theta, \qquad U_\theta^* = -\left(1 - \frac{3}{4r} - \frac{1}{4r^3}\right)\sin\theta \tag{E4.14}$$

Substituting Eq. (E4.14) into Eq. (E4.11) and setting $\cos\theta = \mu$, Eq. (E4.11) becomes

$$(\nabla^{*2} + \varepsilon F + \varepsilon G)T^* = 0 \tag{E4.15}$$

with

$$F = -\mu\frac{\partial}{\partial r} - \frac{1}{r}(1-\mu^2)\frac{\partial}{\partial \mu},$$
$$G = \left(\frac{3}{2r} - \frac{1}{2r^3}\right)\mu\frac{\partial}{\partial r} + \left(\frac{3}{4r^2} + \frac{1}{4r^4}\right)(1-\mu^2)\frac{\partial}{\partial \mu} \tag{E4.16}$$

The term F represents the contribution from the uniform flow, whereas the term G accounts for the deviation from the parallel flow due to the presence of the sphere.

Now express T^* by a new variable Θ with the relation

$$T^* = \Theta \exp\left(-\frac{\varepsilon}{2}r(1-\mu)\right) \tag{E4.17}$$

The governing equation for Θ is obtained by substituting Eq. (E4.17) into Eq. (E4.15) as

$$\left[\nabla^{*2} + \varepsilon\left(-\frac{\partial}{\partial r} - \frac{1}{r} + G\right) + o(\varepsilon)\right]\Theta = 0 \tag{E4.18}$$

while Eq. (E4.13) gives

$$\Theta = \exp\left(\frac{\varepsilon}{2}(1-\mu)\right) \quad \text{for } r = 1 \tag{E4.19}$$

For Θ, we use the first-order approximation

$$\Theta = \Theta_0 + \varepsilon\Theta_1 + o(\varepsilon) \tag{E4.20}$$

Equation (E4.18) will be satisfied, provided

$$\nabla^{*2}\Theta_0 = 0, \qquad \nabla^{*2}\Theta_1 = \left(\frac{\partial}{\partial r} + \frac{1}{r} - G\right)\Theta_0 \tag{E4.21}$$

while Eq. (E4.19) leads to

$$\Theta_0 = 1, \qquad \Theta_1 = \frac{1}{2}(1-\mu), \quad \text{for } r = 1 \tag{E4.22}$$

Equations (E4.21) and (E4.22) are fulfilled by

$$\Theta_0 = \frac{1}{r}, \qquad \Theta_1 = \frac{1}{2r} + \left(-\frac{3}{4r} + \frac{3}{8r^2} - \frac{1}{8r^3}\right)\mu \tag{E4.23}$$

Thus, from Eqs. (E4.17), (E4.20), and (E4.23), we have

$$\left(\frac{\partial T^*}{\partial r}\right)_{r=1} = -\left(1 + \frac{1}{2}\varepsilon + o(\varepsilon)\right) \tag{E4.24}$$

which yields Eq. (4.42).

4.3.3 *Thermal Convection in Pseudocontinuum One-Phase Flow*

General considerations of the heat transfer in a gas–solid flow require separate energy equations for the two phases that are coupled through a local heat transfer coefficient as well as local phase velocities. These approaches are complex and are introduced in Chapter 5.

One of the simplified heat transfer models of two-phase flows is the pseudocontinuum one-phase flow model, in which it is assumed that (1) local thermal equilibrium between the two phases exists; (2) particles are evenly distributed; (3) flow is uniform; and (4) heat conduction is dominant in the cross-stream direction. Therefore, the heat balance leads to a single-phase energy equation which is based on effective gas–solid properties and averaged temperatures and velocities. For an axisymmetric flow heated by a cylindrical heating surface at T_w, the heat balance equation can be written as

$$\rho_m c_{pe} U \frac{\partial T}{\partial z} = \frac{K_e}{r}\frac{\partial}{\partial r}\left(r\frac{\partial T}{\partial r}\right) \tag{4.43}$$

with

$$\begin{aligned} T &= T_w \quad \text{at } r = R \\ T &= T_i \quad \text{at } z = 0 \end{aligned} \tag{4.44}$$

where z and r are in the streamwise and cross-stream directions, respectively; K_e is the effective thermal conductivity; and U is the pseudocontinuum one-phase velocity.

Using the method of separation of variables, the temperature distribution from Eqs. (4.43) and (4.44) is obtained as

$$T(r, z) = T_w + 2(T_i - T_w)\sum_{n=1}^{\infty}\frac{J_0(\lambda_n r)}{\lambda_n R J_1(\lambda_n R)}\exp(-\beta\lambda_n^2 z) \tag{4.45}$$

where J_0 and J_1 are the zeroth and first-order Bessel functions of the first kind, respectively, and λ_n's are the eigenvalues obtained from the eigenequation

$$J_0(\lambda_n R) = 0 \tag{4.46}$$

and β is a system-related constant given by

$$\beta = \frac{K_e}{U\rho_m c_{pe}} \tag{4.47}$$

The heat transfer coefficient is

$$h = \frac{2K_e}{R}\sum_{n=1}^{\infty}\exp(-\beta\lambda_n^2 z) \tag{4.48}$$

and the length-averaged Nusselt number is

$$\mathrm{Nu} = \frac{\overline{h}L}{K_e} = \frac{2L}{R\beta}\sum_{n=1}^{\infty}\frac{1-\exp(-\beta\lambda_n^2 L)}{\lambda_n^2} \tag{4.49}$$

The applicability of the preceding pseudocontinuum approach to convective heat transfer of gas–solid systems without heat sources depends not only on the validity of the phase continuum approximation but also on the appropriateness of the local thermal equilibrium assumption. The local thermal equilibrium may be assumed only if the particle-heating

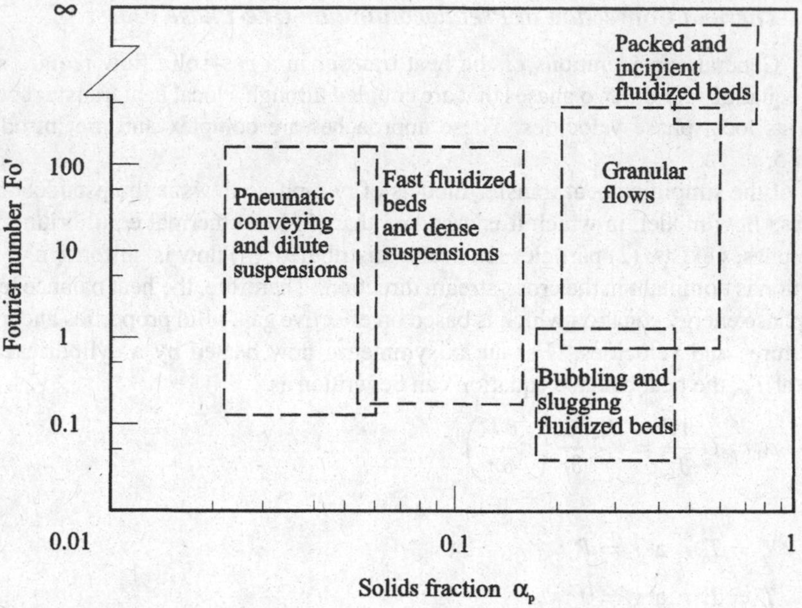

Figure 4.3. Regime map for gas–solid flows (from Hunt, 1989).

surface residence time, τ_{ps}, is much greater than the particle thermal diffusion time, τ_{pd}. Here, τ_{ps} and τ_{pd} may be estimated by

$$\tau_{pd} = \frac{d_p^2}{D_t}, \qquad \tau_{ps} = \frac{L}{U} \tag{4.50}$$

Furthermore, we may define a modified Fourier number as

$$Fo' = \frac{\tau_{ps}}{\tau_{pd}} = \frac{\tau_{ps} D_t}{d_p^2} \tag{4.51}$$

A regime map of Fo′ versus the solid volume fraction, α_p, for various gas–solid flows was presented by Hunt (1989), as shown in Fig. 4.3. Hunt (1989) suggested that except when $Fo' > 1$ and $\alpha_p > 0.1$, use of the pseudocontinuum model is inappropriate. Thus, from Fig. 4.3, it can be seen that the pseudocontinuum model is applicable to packed beds, incipient fluidized beds, and granular flows, whereas it is not applicable to pneumatic transport flows, dilute suspensions, bubbling beds, and slugging fluidized beds [Glicksman and Decker, 1982; Hunt, 1989].

4.4 Thermal Radiation

An element in a thermally radiative environment absorbs, reflects, refracts, diffracts, and transmits incoming radiative heat fluxes as well as emits its own radiative heat flux. Most solid materials in gas–solid flows, including particles and pipe walls, can be reasonably approximated as gray bodies so that absorption and emission can be readily calculated from Stefan–Boltzmann's law (Eq. (1.59)) for total thermal radiation or from Planck's formula (Eq. (1.62)) for monochromatic radiation. Other means of transport of radiative energy can be generally described by "scattering," which may encompass various effects due to reflection, refraction, diffraction, and transmission. In this section, we focus on the determination of scattering of radiative energy by solids in various gas–solid flows.

4.4.1 *Single-Particle Scattering*

When a beam of incident radiation strikes a particle, some of the radiation is absorbed by the surface while the remaining portion is scattered out of the surface. The reflection, refraction, diffraction, and transmission depend not only on the optical properties of the particle but also on the particle size d_p relative to the wavelength of the incident radiation λ.

The radiant energy scattered and absorbed by a particle can be characterized by the "cross section" of the particle for scattering and absorption, respectively. The cross section for scattering, C_s, is defined as the area on which the amount of incident energy equals the total energy scattered in all directions, and the cross section for absorption, C_a, is defined as the area on which the amount of incident energy equals the total energy absorbed into the particle. Similarly, we may also introduce the cross section for extinction, C_e, defined as the area on which the amount of incident energy equals the total energy removed from the original beam. From energy conservation, we have

$$C_e = C_s + C_a \tag{4.52}$$

It is convenient to introduce the efficiency factors for extinction, scattering, and absorption, which are defined by

$$Q_e = \frac{C_e}{C}, \qquad Q_s = \frac{C_s}{C}, \qquad Q_a = \frac{C_a}{C} \tag{4.53}$$

where C is the geometric cross section. From Eq. (4.52), we have

$$Q_e = Q_s + Q_a \tag{4.54}$$

In general, Q's are functions of the orientation of particles and the state of polarization of incident beams. However, for spherical and homogeneous particles, Q's are independent of both.

The most commonly used theory for single-particle scattering was originally developed by Mie (1908), who applied electromagnetic theory to derive the properties of the electromagnetic field with a plane monochromatic wave incident upon a sphere. Although the Mie scattering theory covers the entire range of particle diameters, general formulations for cross sections or efficiency factors are very complicated. The Mie scattering theory can be greatly simplified when $d_p \ll \lambda$ or $d_p \gg \lambda$, leading to Rayleigh scattering and geometric scattering, respectively. From Wien's displacement law (Eq. (1.63)), for the temperature range of most gas–solid flows, the wavelengths at maximum emissive power are less than 10 μm (see Example 4.3). Thus, particles for Rayleigh scattering are in the submicrometer range, which is beyond the particle size range of interest in this book. Therefore, discussions on Rayleigh scattering are excluded here. For Rayleigh scattering of submicrometer particles, interested readers may refer to Bohren and Huffman (1983).

Example 4.3 For gas–solid flows, the typical temperature range where thermal radiation should be considered is from 500 K to 2,000 K. Show that for Rayleigh scattering, the particle sizes of relevance are in the submicrometer range. The thermal radiation is assumed to be monochromatic, and the particles can be regarded as blackbodies.

Solution The wavelength at which the monochromatic-emissive power of a blackbody is maximum for a given temperature is given by Eq. (1.63) as

$$\lambda_m T_p = 2{,}898\ \mu\text{m}\cdot\text{K} \tag{E4.25}$$

For $500 < T_p < 2{,}000$ K, the range of λ_m is obtained as

$$1.4\ \mu\text{m} < \lambda_m < 5.8\ \mu\text{m} \tag{E4.26}$$

To fulfill the requirement of Rayleigh scattering (*i.e.*, $d_p \ll \lambda_m$), we may take the maximum particle size d_{pm} for Rayleigh scattering to be 10 times smaller than λ_m. Therefore, we have

$$0.14\ \mu\text{m} < d_{pm} < 0.58\ \mu\text{m} \tag{E4.27}$$

which is in the submicrometer range.

4.4.1.1 *Mie Scattering*

A typical single-particle scattering consists of an electromagnetic plane wave traveling through a medium and impinging upon a sphere. This problem was first solved by Mie in 1908 using Maxwell's equations, from which the efficiency factors for extinction, scattering, and absorption can be obtained [Özisik, 1973; Bohren and Huffman, 1983]. We have no intention of presenting the detailed derivation of the Mie theory here because of its complexity, but rather will focus on its applications to simple cases.

Considering an incident plane wave on a sphere of no net surface charge, the scattering and extinction efficiency factor for the field far away from the particle can be approximated by [Van de Hulst, 1957]

$$\begin{aligned} Q_s &= \frac{2}{\xi^2}\sum_{n=1}^{\infty}(2n+1)\{|A_n|^2 + |B_n|^2\} \\ Q_e &= \frac{2}{\xi^2}\sum_{n=1}^{\infty}(2n+1)\,\text{Re}(A_n + B_n) \end{aligned} \tag{4.55}$$

where $\text{Re}(z)$ denotes the real part of the complex z and $\xi = 2\pi a/\lambda$. The coefficients A_n and B_n are determined by

$$\begin{aligned} A_n &= \frac{\dfrac{d\psi_n(\eta)}{d\eta}\psi_n(\xi) - n_i\psi_n(\eta)\dfrac{d\psi_n(\xi)}{d\xi}}{\dfrac{d\psi_n(\eta)}{d\eta}\zeta_n(\xi) - n_i\psi_n(\eta)\dfrac{d\zeta_n(\xi)}{d\xi}} \\ B_n &= \frac{n_i\dfrac{d\psi_n(\eta)}{d\eta}\psi_n(\xi) - \psi_n(\eta)\dfrac{d\psi_n(\xi)}{d\xi}}{n_i\dfrac{d\psi_n(\eta)}{d\eta}\zeta_n(\xi) - \psi_n(\eta)\dfrac{d\zeta_n(\xi)}{d\xi}} \end{aligned} \tag{4.56}$$

where $\eta = 2\pi n_i a/\lambda$; a is the particle radius; n_i is the index of refraction; and functions $\psi_n(z)$ and $\zeta_n(z)$ are defined by

$$\psi_n(z) = \sqrt{\frac{\pi z}{2}} J_{n+1/2}(z), \qquad \zeta_n(z) = \sqrt{\frac{\pi z}{2}} H_{n+1/2}(z) \tag{4.57}$$

where J and H are the Bessel functions of the first and second kind of order $n + 1/2$, respectively.

The directional distribution of the scattering intensity can be described by phase functions. A phase function is defined as the ratio of scattering intensity in a direction to the scattering intensity in the same direction if the scattering is isotropic. Thus, it is a normalized function and is defined over all directions. Typical phase functions for small, intermediate, and large values of ξ and n_i are illustrated in Fig. 4.4, with the spheres assumed to be nonabsorbing [Tien and Drolen, 1987]. It is shown that the phase functions mainly vary

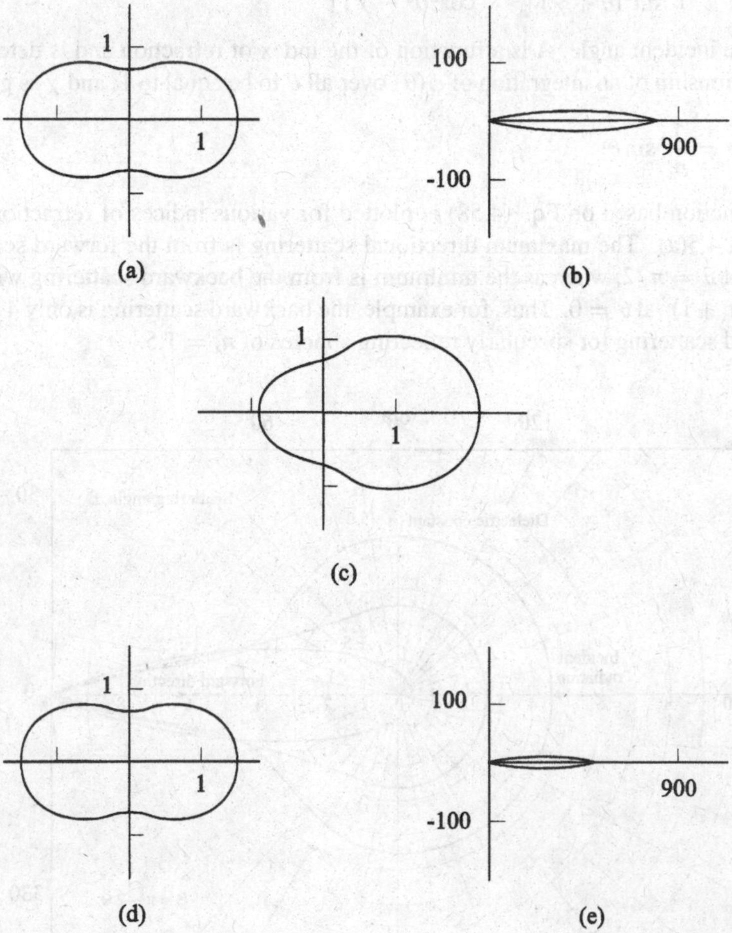

Figure 4.4. Scattering phase functions for small, intermediate, and large values of ζ, n_i (from Tien and Drolen, 1987): (a) $\zeta = 0.1$, $n_i = 1.05$; (b) $\zeta = 30.0$, $n_i = 1.05$; (c) $\zeta = 1.0$, $n_i = 1.5$; (d) $\zeta = 0.1$, $n_i = 3.0$; (e) $\zeta = 30.0$, $n_i = 3.0$.

with the change of particle size. With increasing particle size, the distribution becomes less uniform.

4.4.1.2 *Simple Scattering from Large Spheres*

For some typical modes of scattering from large spherical particles ($\xi > 5$), simple formulations of phase functions can be obtained. These modes include scattering from a specularly reflecting sphere, scattering from a diffuse reflection sphere, and scattering by diffraction from a sphere.

A. Specularly Reflecting Spheres

Consider an unpolarized incident radiation on a dielectric sphere. The phase function $\phi(\theta)$ can be expressed as [Siegel and Howell, 1981]

$$\phi(\theta) = \frac{A}{2}\frac{\sin^2(\theta - \chi)}{\sin^2(\theta + \chi)}\left[1 + \frac{\cos^2(\theta + \chi)}{\cos^2(\theta - \chi)}\right] \tag{4.58}$$

where θ is the incident angle; A is a function of the index of refraction and is determined from the relationship of an integration of $\phi(\theta)$ over all θ to be equal to 1; and χ is given by

$$\sin\chi = \frac{1}{n_i}\sin\theta \tag{4.59}$$

The phase function based on Eq. (4.58) is plotted for various indices of refraction n_i, as shown in Fig. 4.5(a). The maximum directional scattering is from the forward scattering with ϕ of A at $\theta = \pi/2$, whereas the minimum is from the backward scattering with ϕ of $A(n_i - 1)^2/(n_i + 1)^2$ at $\theta = 0$. Thus, for example, the backward scattering is only 4 percent of the forward scattering for specularly reflecting spheres of $n_i = 1.5$.

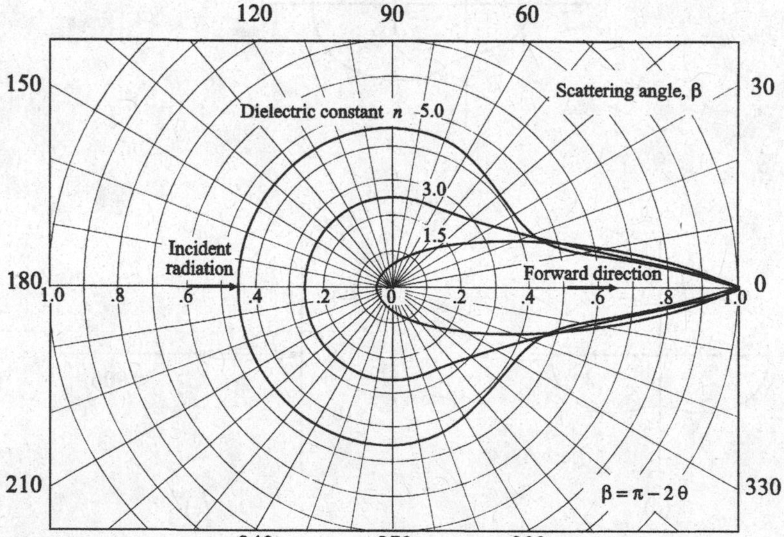

Figure 4.5a. Scattering diagram for a specularly reflecting sphere that is large compared with the wavelength of incident radiation (from Siegel and Howell, 1981).

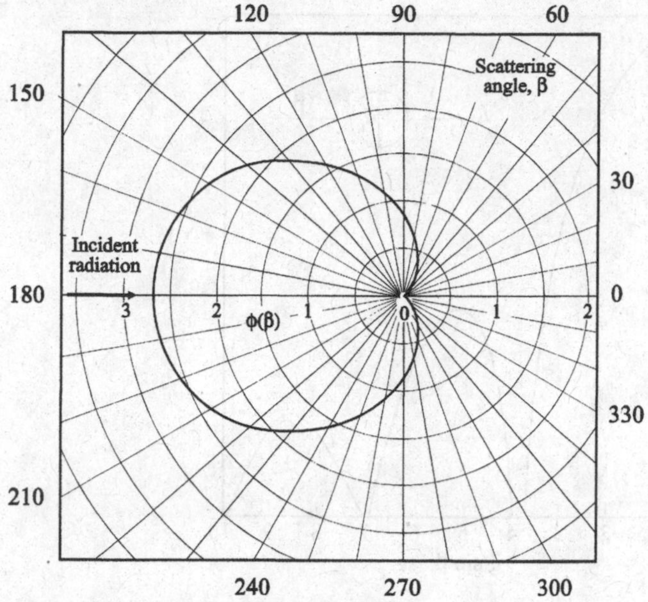

Figure 4.5b. Scattering phase function for a diffuse reflecting sphere which is large compared with the wavelength of incident radiation and with constant reflectivity (from Siegel and Howell, 1981).

B. Reflection from a Diffuse Sphere

For a diffuse sphere, each surface element that intercepts incident radiation will reflect the energy into the entire 2π solid angle above that element. Thus, the radiation scattered into a specified direction will arise from the entire region of the sphere that receives radiation and is also visible from this specified direction. Consequently, the phase function for a diffuse sphere can be obtained as [Siegel and Howell, 1981]

$$\phi(\beta) = \frac{8}{3\pi}(\sin\beta - \beta\cos\beta) \tag{4.60}$$

where β is the scattering angle between the observer and forward direction. From Eq. (4.60), $\phi(\beta)$ is plotted against various β, as shown in Fig. 4.5(b). In this case, the maximum scattering occurs from backward scattering where $\beta = \pi$, and the minimum is from forward scattering.

C. Diffraction from a Large Sphere

To account for the total directional scattering from a large sphere, effects of both diffraction and reflection must be considered. When a spherical particle is in the path of incident radiation, the diffracted intensity may be obtained from Babinet's principle, which states that the diffracted intensity is the same as that for a hole of the same diameter. The phase function for diffraction by a large sphere is given by [Van de Hulst, 1957]

$$\phi(\beta) = \frac{4\mathrm{J}_1^2(\xi\sin\beta)}{\sin^2\beta} \tag{4.61}$$

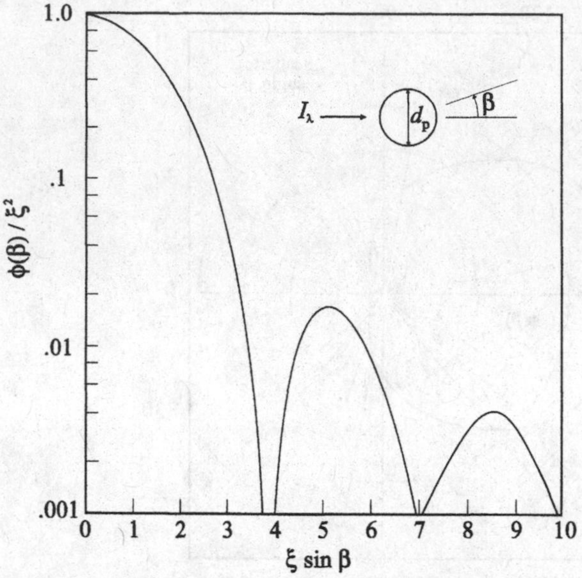

Figure 4.5c. Phase function for diffraction from a large sphere.

where J_1 is the Bessel function of the first kind of order one. Equation (4.61) is plotted against various β's, as shown in Fig. 4.5(c). Thus, for large ξ, the diffracted radiation lies within a narrow angular region in the forward scattering direction.

4.4.2 *Radiant Heating of a Particle*

Consider a situation in which a particle is suddenly exposed to a radiant flux from a thermal environment such as a heated wall or surrounding objects. For simplicity, it is assumed that the radiant flux is isotropic and the temperature distribution inside the particle is uniform. Absorption of radiation by the gas is negligible. Moreover, all the physical properties are constant.

The transient behavior of the particle temperature is defined by the energy balance

$$mc\frac{\mathrm{d}T_{\mathrm{p}}}{\mathrm{d}t} = hS(T_\infty - T_{\mathrm{p}}) + S\alpha_{\mathrm{R}}\sigma_{\mathrm{b}}\left(T_{\mathrm{w}}^4 - T_{\mathrm{p}}^4\right) \tag{4.62}$$

where S is surface area of the particle and h is the convective heat transfer coefficient. Equation (4.62) can be simplified to yield a simple solution as given later.

When the temperature difference between the particle and its thermal environment is small, we may linearize the radiation term by

$$T_{\mathrm{w}}^4 - T_{\mathrm{p}}^4 \approx 4T_{\mathrm{w}}^3(T_{\mathrm{w}} - T_{\mathrm{p}}) \tag{4.63}$$

Hence, Eq. (4.62) reduces to

$$mc\frac{\mathrm{d}T_{\mathrm{p}}}{\mathrm{d}t} = hS(T_\infty - T_{\mathrm{p}}) + 4S\alpha_{\mathrm{R}}\sigma_{\mathrm{b}}T_{\mathrm{w}}^3(T_{\mathrm{w}} - T_{\mathrm{p}}) \tag{4.64}$$

which yields a solution of the form

$$T_{\mathrm{p}} = T_{\mathrm{p}0}e^{-Bt} + A(1 - e^{-Bt}) \tag{4.65}$$

where the coefficients A and B are given by

$$A = \frac{hT_\infty + 4\alpha_R\sigma_b T_w^4}{h + 4\alpha_R\sigma_b T_w^3}, \qquad B = \frac{S}{mc}\left(h + 4\alpha_R\sigma_b T_w^3\right) \tag{4.66}$$

For particle heating with $T_w \gg T_p$, it is reasonable to neglect particle radiation to the surroundings. Thus, Eq. (4.62) becomes

$$mc\frac{dT_p}{dt} = hS(T_\infty - T_p) + 4S\alpha_R\sigma_b T_w^4 \tag{4.67}$$

which has a solution with the exact form of Eq. (4.65), but the coefficients A and B now are given by

$$A = T_\infty + \frac{4}{h}\alpha_R\sigma_b T_w^4, \qquad B = \frac{hS}{mc} \tag{4.68}$$

Example 4.4 Consider a solid sphere in a free fall passing through a heated chamber. The particle is a 500 μm alloy with a density of 4,000 kg/m^3, a thermal capacity of 150 J/kg · K, and an absorptivity of 0.6. The initial temperature of the particle is 300 K. Both the temperature of gas inside the chamber and the temperature of the wall are 1,000 K. Estimate the length of the chamber required for the particle to reach 700 K. The physical properties of gas are given as $\mu = 4 \times 10^{-5}$ kg/m · s, $\rho = 0.4$ kg/m^3, Pr = 0.7, and $K = 0.067$ W/m · K. The particle falls at its terminal velocity.

Solution The particle terminal velocity U_{pt} and the particle Reynolds number $\mathrm{Re_p}$ are obtained from Eq. (1.7) and Eq. (1.5), respectively, as

$$U_{pt} = \left(0.072\frac{d_p^{1.6}(\rho_p - \rho)g}{\rho^{0.4}\mu^{0.6}}\right)^{\frac{1}{1.4}} = \left(0.072\frac{(5\times 10^{-4})^{1.6}(4{,}000 - 0.4)9.8}{0.4^{0.4} \times (4\times 10^{-5})^{0.6}}\right)^{\frac{1}{1.4}} = 5\ \mathrm{m/s} \tag{E4.28}$$

and

$$\mathrm{Re_p} = \frac{\rho d_p U_{pt}}{\mu} = \frac{0.4 \times 5\times 10^{-4} \times 5}{4 \times 10^{-5}} = 25 \tag{E4.29}$$

In terms of Eq. (4.40), the convective heat transfer coefficient is obtained as

$$h = \frac{K}{d_p}\left(2 + 0.6\mathrm{Re_p^{\frac{1}{2}}}\mathrm{Pr^{\frac{1}{3}}}\right) = \frac{0.067}{5\times 10^{-4}}\left(2 + 0.6 \times 25^{\frac{1}{2}} \times 0.7^{\frac{1}{3}}\right) = 625\ \mathrm{W/m^2 \cdot K} \tag{E4.30}$$

From Eq. (4.68), the coefficients A and B are calculated as

$$A = T_\infty + \frac{4}{h}\alpha_R\sigma_b T_w^4 = 1{,}000 + \frac{4}{625}(0.6 \times 5.67 \times 10^{-8} \times 1{,}000^4) = 1{,}218\ \mathrm{K}$$
$$B = \frac{6h}{d_p\rho_p c} = \frac{6 \times 625}{5\times 10^{-4} \times 4{,}000 \times 150} = 12.5\ \mathrm{s^{-1}} \tag{E4.31}$$

Thus, the heating time required to reach 700 K can be estimated from Eq. (4.65) as

$$t = \frac{1}{B}\ln\left(\frac{A - T_{p0}}{A - T_p}\right) = \frac{1}{12.5}\ln\left(\frac{1{,}218 - 300}{1{,}218 - 700}\right) = 0.046\ \text{s} \tag{E4.32}$$

Thus, the required length of chamber is estimated by

$$l = U_{pt}t = 5 \times 0.046 = 0.23\ \text{m} \tag{E4.33}$$

4.4.3 *General Considerations in Radiation with a Particle Cloud*

The thermal radiation with a particle cloud may be conveniently studied by a simple model in which the energy is radiated from one heating surface to the other surface through a uniform particulate medium. The particulate medium is treated as a pseudocontinuum where particles absorb, emit, and scatter the radiative heat fluxes. Here, the scattering represents the combined effects of reflection, refraction, diffraction, and transmission of the radiation by the particles. In gas–solid flows involving thermal radiation, it is often assumed that the gas is transparent to the radiation. For radiation through a particle cloud, it is convenient to introduce the monochromatic coefficients for extinction, scattering, and absorption, which are defined by

$$\sigma_{e\lambda} = nC_{e\lambda} = \frac{3\alpha_p}{4a}Q_{e\lambda}; \qquad \sigma_{s\lambda} = nC_{s\lambda} = \frac{3\alpha_p}{4a}Q_{s\lambda}; \qquad \sigma_{a\lambda} = nC_{a\lambda} = \frac{3\alpha_p}{4a}Q_{a\lambda} \tag{4.69}$$

where α_p is the solid volume fraction. Thus, on the basis of Eq. (4.54), σ_e equals the sum of σ_s and σ_a.

4.4.3.1 *Single Versus Multiple Scattering*

If a single ray of radiative beam traversing a medium is scattered only once before leaving the medium, the medium is characterized by single scattering. If more than one scattering occurs per ray, the medium is characterized by multiple scattering.

In gas–solid flows, single scattering occurs in dilute suspensions where only a few particles exist in the domain of interest. In this case, the scattered intensity of the suspension is equal to the scattered intensity from a single particle multiplied by the number of particles. As the number of particles within the space increases, multiple scattering becomes increasingly significant. The classification between single and multiple scattering in a particulate medium may be based on the value of the characteristic optical path length l_s. One typical criterion for single scattering was suggested by Bayvel and Jones (1981) as

$$l_s = \int_0^L \sigma_s\, dl < 0.1 \tag{4.70}$$

where l is the characteristic length of the system and σ_s is the scattering coefficient.

4.4.3.2 *Independent Versus Dependent Scattering*

Independent scattering refers to the scattering from a single particle in a cloud which is not affected by the neighboring particles, whereas dependent scattering refers to the scattering which is affected by the neighboring particles. As suggested by Hottel *et al.* (1971), the general form of the efficiency factor for dependent scattering is

$$Q_{s,D} = f\left(\xi, n_i, \frac{c}{\lambda}, \alpha_p\right) \tag{4.71}$$

where c is the interparticle clearance and $\alpha_p = f(c/a)$. Similarly, the general form of the efficiency factor for independent scattering is

$$Q_{s,I} = f(\xi, n_i) \tag{4.72}$$

It is noted that the dependent effects are functions of α_p and c/λ instead of just the particle separation alone. Brewster and Tien (1982) suggested that the scattering is independent if $c/\lambda > 0.3$ for the range of $0.66 < \xi < 74$ and $0.01 < \alpha_p < 0.7$. A schematic diagram of independent and dependent scattering regimes for some particulate systems is given in Fig. 4.6. In this text, all scattering is assumed to be independent for simplicity.

4.4.3.3 *Radiation Transport Equation*

Consider the energy balance of radiant flux through a scattering medium of monodispersed spherical particles, in which absorption, emission, and scattering by spherical particles are included. As shown in Fig. 4.7, the equation for the change in monochromatic radiant

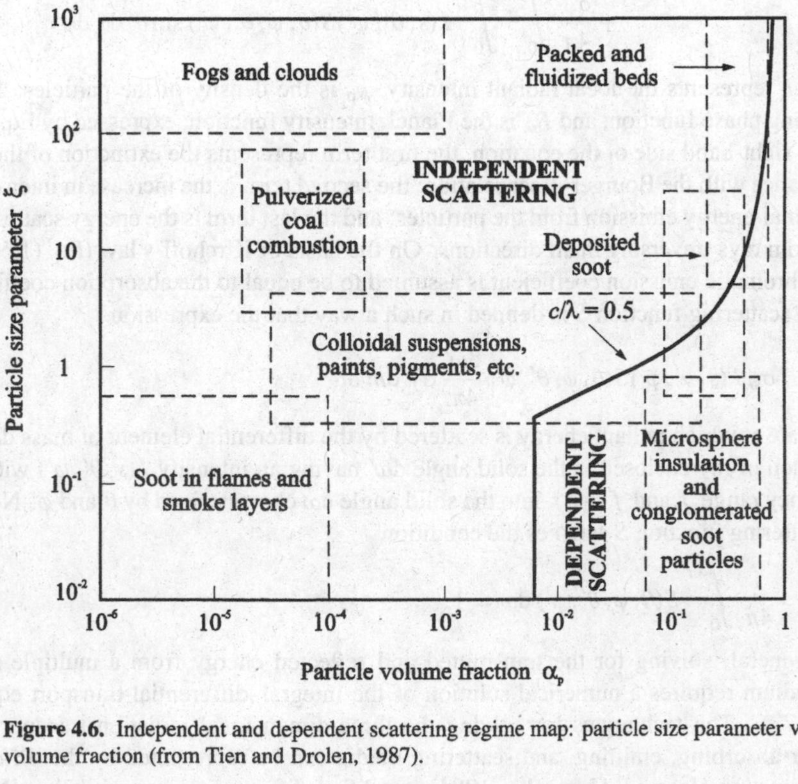

Figure 4.6. Independent and dependent scattering regime map: particle size parameter versus volume fraction (from Tien and Drolen, 1987).

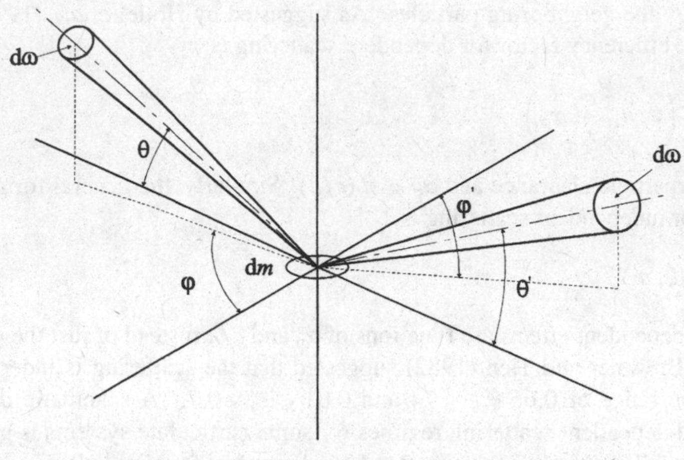

Figure 4.7. Coordinate system for scattering function.

intensity with respect to distance along a pencil of rays at s, θ, φ can be expressed by [Love, 1968]

$$\frac{\mathrm{d}I(s,\theta,\varphi)}{\mathrm{d}s} = -\sigma_{e\lambda} I(s,\theta,\varphi) + \sigma_{a\lambda} I_{b\lambda}(s) + \frac{\sigma_{s\lambda}}{4\pi}\int_0^{2\pi}\int_0^{\pi} I(s,\theta',\varphi') S(\theta,\varphi,\theta',\varphi') \sin\theta' \,\mathrm{d}\theta' \mathrm{d}\varphi' \tag{4.73}$$

where I represents the local radiant intensity, ρ_p is the density of the particles, S is the scattering phase function, and $I_{b\lambda}$ is the Planck intensity function, expressed by Eq. (1.62). On the right-hand side of the equation, the first term represents the extinction of the ray in accordance with the Bourger–Beer relation; the second term is the increase in intensity due to thermal-energy emission from the particles; and the last term is the energy scattered into θ, φ from rays traversing in all directions. On the basis of Kirchoff's law (Eq. (1.60)), the monochromatic emission coefficient is assumed to be equal to the absorption coefficient.

The scattering function S is defined in such a way that the expression

$$\sigma_{s\lambda} I(s,\theta',\varphi') S(\theta,\varphi,\theta',\varphi') \frac{\mathrm{d}\omega'}{4\pi} \mathrm{d}f \,\mathrm{d}m \,\mathrm{d}\omega \tag{4.74}$$

is the rate at which radiant energy is scattered by the differential element of mass dm from the pencil of ray enclosed in the solid angle dω' having an intensity $I(s,\theta',\varphi')$ within the frequency range f and $f + \mathrm{d}f$ into the solid angle dω characterized by θ and φ. Note that the scattering function S satisfies the condition

$$\frac{1}{4\pi}\int_0^{4\pi} S(\theta,\varphi,\theta',\varphi')\,\mathrm{d}\omega = 1 \tag{4.75}$$

In general, solving for the transmitted and reflected energy from a multiple scattering medium requires a numerical solution of the integral–differential transport equation, Eq. (4.73). The techniques for solving the three-dimensional radiation transport equation for absorbing, emitting, and scattering media can be represented by the differential approximation [Im and Ahluwalia, 1984], modified differential approximation [Modest,

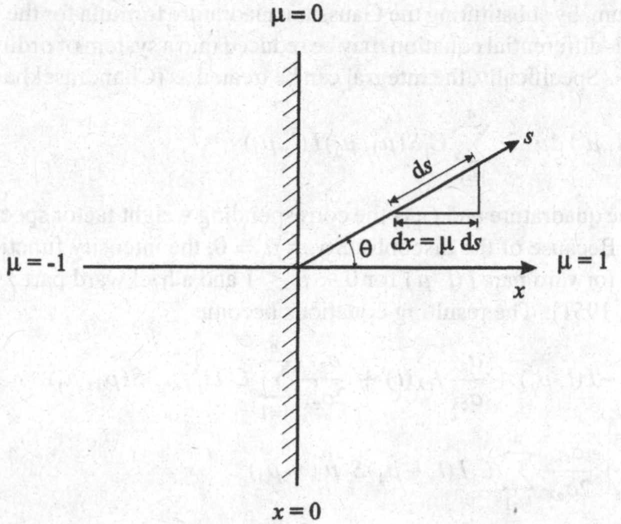

Figure 4.8. Coordinate scheme for axially symmetric case.

1989], discrete-ordinate approximation [Fiveland and Jamaluddin, 1991], hybrid method [Ahluwalia and Im, 1994], and exact Monte Carlo solution [Gupta *et al.*, 1983]. In the cases of single scattering, the equation can be greatly simplified since the last term in Eq. (4.73) vanishes.

4.4.3.4 *Approximation of Scattering Integral by Gaussian Quadrature*

In order to simplify the integral–differential equation of radiant transfer, certain geometric and optical restrictions have to be imposed. Consider a system which contains many uniformly sized spherical particles suspended in a transparent medium and bounded by parallel plates which emit and reflect radiation in a diffuse manner [Love and Grosh, 1965]. As shown in Fig. 4.8, x represents the component of s normal to surface 1. Because of symmetry, the radiant intensity varies only with respect to polar angle θ and is independent of the azimuthal angle φ. Express optical depth l as

$$l = \int_0^x \rho_p \sigma_{e\lambda}\, dx \tag{4.76}$$

Let

$$\mu \equiv \cos\theta \tag{4.77}$$

so that Eq. (4.73) becomes

$$\mu \frac{dI(l,\mu)}{dl} = -I(l,\mu) + \frac{\sigma_{a\lambda}}{\sigma_{e\lambda}} I_{b\lambda}(l) + \frac{\sigma_{s\lambda}}{2\sigma_{e\lambda}} \int_{-1}^{1} I(l,\mu') S(\mu,\mu')\, d\mu' \tag{4.78}$$

where

$$S(\mu,\mu') = \frac{1}{2\pi} \int_0^{2\pi} S(\theta,\varphi,\theta',\varphi')\, d\varphi' \tag{4.79}$$

For an isotropic medium, by substituting the Gaussian quadrature formula for the integral in Eq. (4.78), the integral–differential equation may be reduced into a system of ordinary linear differential equations. Specifically, the integral can be treated as [Chandrasekhar, 1960]

$$\int_0^1 S(\mu, \mu')I(l, \mu')\,\mathrm{d}\mu' \cong \sum_{j=1}^{n} C_j S(\mu_i, \mu_j) I(l, \mu_j) \tag{4.80}$$

where n is the order of the quadrature and C_j is the corresponding weight factor specified by the quadrature formula. Because of the discontinuity at $\mu = 0$, the intensity function may be further divided into a forward part $I(l, \mu)$ for $0 < \mu \leq 1$ and a backward part $I(l, -\mu)$ for $-1 \leq \mu < 0$ [Sykes, 1951]. The resulting equations become

$$\mu_i \frac{\mathrm{d}I(l, \mu_i)}{\mathrm{d}l} = -I(l, \mu_i) + \frac{\sigma_{a\lambda}}{\sigma_{e\lambda}} I_{b\lambda}(l) + \frac{\sigma_{s\lambda}}{2\sigma_{e\lambda}} \sum_{j=1}^{n} C_j I(l, \mu_j) S(\mu_i, \mu_j) + \frac{\sigma_{s\lambda}}{2\sigma_{e\lambda}} \sum_{j=1}^{n} C_j I(l, -\mu_j) S(\mu_i, -\mu_j) \tag{4.81}$$

for the forward radiant intensity, and

$$-\mu_i \frac{\mathrm{d}I(l, -\mu_i)}{\mathrm{d}l} = -I(l, -\mu_i) + \frac{\sigma_{a\lambda}}{\sigma_{e\lambda}} I_{b\lambda}(l) + \frac{\sigma_{s\lambda}}{2\sigma_{e\lambda}} \sum_{j=1}^{n} C_j I(l, \mu_j) S(-\mu_i, \mu_j) + \frac{\sigma_{s\lambda}}{2\sigma_{e\lambda}} \sum_{j=1}^{n} C_j I(l, -\mu_j)\, S(-\mu_i, -\mu_j) \tag{4.82}$$

for the backward radiant intensity.

4.4.4 *Radiation Through an Isothermal and Diffuse Scattering Medium*

Thermal radiation through an isothermal and diffuse scattering medium represents one of the simplest cases of radiation in a particulate medium. This problem was treated by Love and Grosh (1965).

For an isothermal medium at temperature T_p, the monochromatic radiant emission from particles $I_{b\lambda}$ is a function of T_p and λ only and is independent of l. In addition, we may assume that the solutions of Eqs. (4.81) and (4.82) are of the form

$$I(l, \mu_i) = x_i e^{\gamma l}, \qquad I(l, -\mu_i) = x_{(i+n)} e^{\gamma l} \tag{4.83}$$

Substitution of Eq. (4.83) into Eqs. (4.81) and (4.82) leads to 2n linear algebraic equations in x_i and $x_{(i+n)}$ with parameter γ. The matrix of the coefficients yields 2n eigenvalues γ_k as well as 2n sets of eigenvectors x_i and $x_{(i+n)}$ for each eigenvalue. Here, k is an iterative index. The solutions can be expressed as

$$\begin{aligned} I(l, \mu_i) &= \sum_{k=1}^{2n} D_k x_{i,k} \exp(\gamma_k l) + I_{b\lambda}(T_p) \\ I(l, -\mu_i) &= \sum_{k=1}^{2n} D_k x_{(i+n),k} \exp(\gamma_k l) + I_{b\lambda}(T_p) \end{aligned} \tag{4.84}$$

where D_k's are the 2n constants of integration.

The determination of D_k's depends on the boundary conditions defined. With the assumption of diffuse walls, the intensities at the wall may be written as

$$\begin{aligned} I(0, \mu_i) &= \epsilon_1 I_{b\lambda}(T_1) + 2\rho_1 \sum_{j=1}^{n} C_j \mu_j I(0, -\mu_j) \\ I(l_0, -\mu_i) &= \epsilon_2 I_{b\lambda}(T_1) + 2\rho_2 \sum_{j=1}^{n} C_j \mu_j I(l_0, \mu_j) \end{aligned} \tag{4.85}$$

where ϵ_1, ϵ_2, ρ_1, and ρ_2 represent the emissivity and reflectivity of wall surfaces 1 and 2, respectively. Substituting Eq. (4.85) into Eq. (4.84) gives

$$\begin{aligned} \sum_{k=1}^{2n} D_k \left[x_{i,k} - 2\rho_1 \sum_{j=1}^{n} C_j \mu_j x_{(j+n),k} \right] &= \epsilon_1 [I_{b\lambda}(T_1) - I_{b\lambda}(T_p)] \\ \sum_{k=1}^{2n} D_k \left[x_{(i+n),k} e^{\gamma_k l_0} - 2\rho_2 \sum_{j=1}^{n} C_j \mu_j x_{j,k} e^{\gamma_k l_0} \right] &= \epsilon_2 [I_{b\lambda}(T_2) - I_{b\lambda}(T_p)] \end{aligned} \tag{4.86}$$

Therefore, D_k's can be expressed in terms of $I_{b\lambda}(T_1)$, $I_{b\lambda}(T_2)$, and $I_{b\lambda}(T_p)$ by solving the preceding equations.

Now, substituting these D_k's from Eq. (4.86) into Eq. (4.84) leads to the solutions of radiant intensity inside the scattering medium in a form

$$\begin{aligned} I(l, \mu_i) &= E_1 I_{b\lambda}(T_1) + E_2 I_{b\lambda}(T_2) + E_p I_{b\lambda}(T_p) \\ I(l, -\mu_i) &= F_1 I_{b\lambda}(T_1) + F_2 I_{b\lambda}(T_2) + F_p I_{b\lambda}(T_p) \end{aligned} \tag{4.87}$$

where E's and F's are parameters independent of temperature and dependent on the nature of single scattering, emissivity of the bounding surfaces, optical spacing between the bounding surfaces, and optical depth of interest. Consequently, the net monochromatic flux $J_{r\lambda}(l)$ can be obtained by the integration of the intensity function as

$$\begin{aligned} J_{r\lambda}(l) &= 2\pi \int_{-1}^{1} \mu I(l, \mu)\, d\mu = 2\pi \sum_{j=1}^{n} C_j \mu_j [I(l, \mu_j) - I(l, -\mu_j)] \\ &= G_1 I_{b\lambda}(T_1) - G_2 I_{b\lambda}(T_2) - G_p I_{b\lambda}(T_p) \end{aligned} \tag{4.88}$$

where G's are parameters similar to E's and F's. As a result, the total radiant flux in the scattering medium is given by

$$J_r(l) = \int_0^\infty J_{r\lambda}\, df = \int_0^\infty [G_1 I_{b\lambda}(T_1) - G_2 I_{b\lambda}(T_2) - G_p I_{b\lambda}(T_p)]\, df \tag{4.89}$$

which generally requires numerical integration for the solution.

The significance of the scattering and emission by the particle cloud can be illustrated by the following numerical example [Soo, 1990]. Consider a gas–solid flow through a parallel-plate system. Wall 1 is at 1,111 K with reflectivity of 0.1, and wall 2 is at 278 K with reflectivity of 0.9. The gap between the two plates is 61 mm. The moving particles are 2 μm iron particles, and the temperature of the particles is maintained at 556 K. The particulate suspension density is 0.16 kg/m^3. It can be shown that the net heat flux from wall 1 is about 69.22 kW/m^2, whereas the net heat flux into wall 2 is about 5.17 kW/m^2.

The net heat input into the particle cloud is thus obtained as 64.05 kW/m^2. It is interesting to note that without particles the net heat flux between the two plates is only 8.54 kW/m^2.

4.5 Mass Transfer

The transfer of mass from one point to the other may take place by two different modes, namely, diffusion and convection. The basic mechanisms for these modes of mass transfer are similar to those for heat transfer discussed in §4.2 and §4.3. Specifically, the mechanism for mass convection is analogous to heat convection and that for mass diffusion is analogous to heat conduction.

4.5.1 *Diffusion and Convection*

Mass or molar diffusion is characterized by small-scale "random-walk" motion caused primarily by molecular collisions. The flux of diffusion follows the spatial gradients of a specific local driving force. If the driving force is concentration, the mass diffusion is termed ordinary diffusion. If the driving force is pressure (as in centrifugal separation) or temperature (as in thermophoresis), the mass diffusion is called pressure diffusion or thermal diffusion, respectively. In these cases, the direction of the flux is from the high to the low driving force. For diffusion of a species through a pore of a particle, ordinary diffusion occurs when the pore diameter is larger than the mean free path of the gas molecule. However, when the pore diameter is smaller than the mean free path of the gas molecule, molecules collide more frequently with the walls than with each other; this mode of molecular diffusion is known as Knudsen diffusion. In ordinary diffusion, the quantification of the mass flux is more complicated than that of the heat flux as the former requires accounting for the mixture properties such as mixture composition and individual species velocity in the mixture. As a phenomenological model equation, similar to Fourier's equation in heat conduction, Fick's equation expresses the flux of mass or molar diffusion for a species in a mixture at steady state, *i.e.*, constant temperature and pressure. For a binary mixture, the diffusive mass flux of species A, j_A, is linearly proportional to the local concentration gradient and is directed toward lower concentration, *i.e.*,

$$j_A = -D_{AB}\rho_m \nabla\omega_A \tag{4.90}$$

where D_{AB} is the binary diffusivity; ρ_m is the density of the mixture; and ω_A is the mass fraction of species A in the mixture. Fick's equation for the molar diffusive flux for species A, J_A, can be given by

$$J_A = -D_{AB}C_m \nabla X_A \tag{4.91}$$

where C_m is the overall molar concentration and X_A is the mole fraction of species A in the mixture.

Mass convection may be due to the bulk motion of the carrier gas or may be associated with the drift of the solute through the carrier gas as a result of net forces applied directly to the solute (*e.g.*, centrifugation due to centrifugal force). Here, we discuss only the convection due to bulk motion of the carrier gas.

For a binary mixture, the convective molar flux, J_c, for species A can be expressed as

$$J_{cA} = X_A(N_A + N_B) \tag{4.92}$$

where N_A and N_B are the molar flux for species A and B, respectively. Here, the molar flux N_A or N_B comprises both the diffusive flux and the convective flux. Thus, for species A, N_A can be expressed, from Eqs. (4.91) and (4.92), by

$$N_A = J_A + J_{cA} = -D_{AB}C_m \nabla X_A + X_A(N_A + N_B) \tag{4.93}$$

For multicomponent mixtures, the overall molar flux for species i in the mixture can be expressed by

$$N_i = -D_{im}C_m \nabla X_i + X_i \sum_{j=1}^{n} N_j \tag{4.94}$$

where D_{im} is the molecular diffusivity of species i in the mixture. ∇X_i can be expressed by the Stefan–Maxwell equation as [Taylor and Krishna, 1993]

$$\nabla X_i = \sum_{j=1}^{n} \frac{1}{C_m D_{ij}} (X_i N_j - X_j N_i) \tag{4.95}$$

The primary interest in the mass transfer occurring in gas–solid flow concerns the intraparticle diffusive mass transfer and interfacial convective mass transfer between the particle and the gas medium. In both cases, Fick's equation, *i.e.*, Eq. (4.91), or a more general form of the flux equation, *i.e.*, Eq. (4.93), can be applied. The basic formulation accounting for the mass transfer phenomenon for a single particle resembles that for heat transfer. Hence, the following discussion is focused on the analogy between the mass and heat transfer concerning a single particle.

4.5.2 *Mass and Heat Transfer Analogy*

In an analysis of mass (or molar) diffusion and mass (or molar) convection of a single particle, the resulting equations are in a form similar to those of heat conduction and thermal convection in a single particle given in §4.2 and §4.3. More precisely, the resulting equations for heat transfer between fluid and a particle given in §4.2 and §4.3 can be applied analogously to mass transfer by simply replacing the principal variables including the thermal diffusivity (see Eq. (4.20)), Prandtl number, and Nusselt number with the molecular diffusivity, Schmidt number, and Sherwood number, respectively. The analogy between heat and mass transfer for a single particle, however, is generally valid only under the following conditions:

(1) Constant physical properties
(2) Small rate of mass transfer
(3) No chemical reactions in the fluid
(4) No viscous dissipation
(5) No emission or absorption of radiant energy
(6) No pressure diffusion, thermal diffusion, or forced diffusion (*e.g.*, electrophoresis)

In flow situations, empirical analogies between mass and heat transfer are usually employed. For single-particle mass transfer, the boundary layer analysis for mass transfer is similar to that for heat transfer and thus is used for typical applications such as sublimation of a solid (*e.g.*, naphthalene ball) or evaporation of a liquid drop falling in air. For a single sphere of diameter d_p moving in a fluid, in terms of a boundary layer analysis analogous to heat transfer, the convective mass transfer coefficient, k_c, accounting for fluxes from the surface of a single sphere to the bulk, or vice versa, is given by Froessling (1938) as

$$\mathrm{Sh} = \frac{k_c d_p}{D_{AB}} = 2.0 + 0.6\,\mathrm{Re}_p^{1/2}\mathrm{Sc}^{1/3} \tag{4.96}$$

where Sc is the Schmidt number, defined as ν / D_{AB}. The equation provides a good correlation of data at low Re_p, but Sh calculated for high Re_p is lower than the values found experimentally. Equation (4.96) is practically the same as Eq. (4.40) when the heat and mass transfer analogy is applied, *i.e.*, Nu_p replaced with Sh and Pr replaced with Sc.

The conditions for negligible convection compared to diffusion for mass transfer analogous to heat transfer are [Rosner, 1986]

$$(\mathrm{Re}_p \mathrm{Sc})^{1/2} \ll 1 \tag{4.97}$$

and

$$(\mathrm{Ra}_m \mathrm{Sc})^{1/4} \ll 1 \tag{4.98}$$

where Ra_m is the Rayleigh number for mass transfer defined by

$$\mathrm{Ra}_m = \left[g\beta_w(\omega_{A,S} - \omega_{A,\infty})\frac{L^3}{\nu^2} \right] \left[\frac{\nu}{D_{AB}} \right] = \mathrm{Gr\,Sc} \tag{4.99}$$

where L is the characteristic length; Gr is the Grashof number; $\omega_{A,S}$ and $\omega_{A,\infty}$ are the mass fractions of species A on the solid surface and in the bulk, respectively; and β_W is the parameter defining the variation of local gas density with mass composition at constant temperature and pressure, *i.e.*,

$$\beta_w = -\frac{1}{\rho}\left(\frac{\partial \rho}{\partial \omega_A} \right)_{P,T} \tag{4.100}$$

Equation (4.97) is usually satisfied for mass transfer in the gas phase, such as transport of oxygen or hydrogen to the outer surface of a small catalyst particle which is stagnant relative to the surrounding gas. In this case, when Eq. (4.98) is also satisfied, Sh of 2 is a good approximation, which is analogous to the heat transfer counterpart given in Eq. (4.15). Equation (4.97) is especially important when there is simultaneous heat transfer and mass transfer.

These two conditions (Eqs. (4.97) and (4.98)) are usually sufficient for assuming the medium as quiescent in dilute systems in which both $\omega_{A,S}$ and $\omega_{A,\infty}$ are small. However, in nondilute or concentrated systems the mass transfer process can give rise to a convection normal to the surface, which is known as the Stefan flow [Taylor and Krishna, 1993]. Consider a chemical species A which is transferred from the solid surface to the bulk with a mass concentration $\omega_{A,\infty}$. When the surface concentration $\omega_{A,S}$ is high, and the carrier gas B does not penetrate the surface, then there must be a diffusion-induced Stefan convective outflux, which counterbalances the Fickian influx of species B. In such situations the additional condition for neglecting convection in mass transport systems is [Rosner, 1986]

$$K_m \ll 1 \tag{4.101}$$

where

$$K_m = \frac{\omega_{A,S} - \omega_{A,\infty}}{1 - \omega_{A,S}} \tag{4.102}$$

In very concentrated systems, the increased contribution of the Stefan flow leads the diffusional mass transfer rate corresponding to the Sh for quiescent systems to be modified as [Rosner, 1986]

$$\mathrm{Sh_s} = 2\frac{\ln(1 + K_m)}{K_m} \tag{4.103}$$

where $\mathrm{Sh_s}$ corresponds to the mass transfer coefficient with the Stefan flow. The Stefan reduction factor for mass transfer, S_{Fm}, is given as

$$S_{Fm} = \frac{\mathrm{Sh_s}}{\mathrm{Sh}} = \frac{\ln(1 + K_m)}{K_m} \tag{4.104}$$

The Stefan convection, however, does not alter the analogy between heat and mass transfer because the laws governing the change in the Nusselt number with the Stefan flow are identical to those governing $\mathrm{Sh_s}$.

Nomenclature

A	Constant, defined by Eq. (4.66) and Eq. (4.68)
A	Function of the index of refraction
A_n	Constant, defined by Eq. (4.56)
a	Particle radius
B	Constant, defined by Eq. (4.66) and Eq. (4.68)
Bi	Biot number
B_n	Constant, defined by Eq. (4.56)
C	Geometric cross section
C	Correction factor, defined by Eq. (4.32)
C_a	Cross section for absorption
C_e	Cross section for extinction
C_m	Overall molar concentration
C_s	Cross section for scattering
c	Specific heat of particles
c	Interparticle clearance
c_p	Specific heat of gas at constant pressure
c_{pe}	Effective thermal capacity of a mixture
D_{AB}	Molecular diffusivity of a species in a binary mixture
D_t	Thermal diffusivity
D_{im}	Molecular diffusivity of species i in a mixture
d_p	Particle diameter
Fo	Fourier number
Fo′	Modified Fourier number, defined by Eq. (4.51)
$\mathrm{Fo_i}$	Fourier number, based on contact time and maximum contact radius
Gr	Grashof number for mass transfer
h	Convective heat transfer coefficient
$\overline{h}$	Length-averaged heat transfer coefficient
h_c	Collisional heat transfer coefficient
h_p	Particle convective heat transfer coefficient
I	Radiant intensity
$I_{b\lambda}$	Planck intensity function
J_A	Diffusive molar flux of species A
j_A	Diffusive mass flux of species A
J_{cA}	Convective molar flux of species A

J_q	Heat flux
J_{qw}	Heat flux at the interface
J_r	Radiant heat flux
K	Thermal conductivity of gas
K_e	Effective thermal conductivity of a mixture
K_m	Parameter, defined by Eq. (4.101)
K_p	Thermal conductivity of particles
k_c	Convective mass transfer coefficient
L	Characteristic length
l	Optical depth
I	Characteristic length
l_s	Characteristic optical path length
m	Mass of a particle
N	Number of components in the mixture
N_A	Overall molar flux of species A
Nu	Nusselt number
$\mathrm{Nu_p}$	Particle Nusselt number
n_i	Index of refraction
n	Particle number density
Pe	Peclet number
Pr	Prandtl number
p	Pressure
Q_a	Efficiency factor for absorption
Q_c	Total heat exchange per impact
Q_e	Efficiency factor for extinction
Q_s	Efficiency factor for scattering
$Q_{s,D}$	Efficiency factor for dependent scattering
$Q_{s,I}$	Efficiency factor for independent scattering
R	Spherical coordinate
R	Radius of a cylindrical tube
R^*	Dimensionless spherical coordinate
$\mathrm{Ra_m}$	Rayleigh number for mass transfer
Re	Reynolds number
$\mathrm{Re_p}$	Particle Reynolds number, based on particle diameter and relative velocity
r	Cylindrical coordinate
r_c	Radius of contact area
r_{cm}	Maximum contact radius
S	Scattering phase function, defined by Eq. (4.74)
S	Surface area of a particle
S_{Fm}	Stefan reduction factor for mass transfer
Sc	Schmidt number
$\mathrm{Sh_s}$	Sherwood number in presence of Stefan flow
T	Absolute temperature
T_1	Temperature of colliding sphere 1
T_2	Temperature of colliding sphere 2
T_p	Temperature of particle
T_{p0}	Reference particle temperature
T_{pi}	Initial particle temperature distribution function
T_∞	Temperature of the surrounding environment when $R \to \infty$
t	Time
t_c	Contact time duration
U	Gas velocity
U	Pseudocontinuum one-phase velocity
U_p	Particle velocity
U_{pt}	Particle terminal velocity
X	Mole fraction

Greek Symbols

α	Volume fraction of the gas phase
α_p	Volume fraction of the particle phase
α_R	Absorptivity for thermal radiation
β	Scattering angle
β	Constant, defined by Eq. (4.47)
β_w	Parameter, defined by Eq. (4.100)
γ	Parameter, defined by Eq. (4.83)

δ	Thickness of gas surrounding a particle		absorption
ϵ	Emissivity of the surface	$\sigma_{e\lambda}$	Monochromatic coefficient for extinction
η	Constant, defined as $2\pi a n_i/\lambda$	$\sigma_{s\lambda}$	Monochromatic coefficient for scattering
θ	Incident angle		
λ	Wavelength	τ_{pd}	Particle thermal diffusion time, defined by Eq. (4.50)
λ_m	Wavelength at which the monochromatic emissive power of a blackbody is maximum	τ_{ps}	Particle–heating surface residence time, defined by Eq. (4.50)
λ_n	Eigenvalue		
μ	Dynamic viscosity of gas	ϕ	Phase function
μ	Parameter, defined by Eq. (4.77)	ϕ	Temperature distribution function, defined by Eq. (4.23)
ν	Kinematic viscosity		
ξ	Constant, defined as $2\pi a/\lambda$	χ	Angle, defined by Eq. (4.59)
ξ	Parameter, defined by Eq. (4.24)	ω_A	Mass fraction of species A
ρ	Density of gas	$\omega_{A,S}$	Mass fraction of species A at the solid surface
ρ	Reflectivity of the surface		
ρ_m	Density of a mixture	$\omega_{A,\infty}$	Mass fraction of species A in the bulk
ρ_p	Density of particles	ε	Parameter, defined by Eq. (E4.12)
σ_b	Stefan–Boltzmann constant		
$\sigma_{a\lambda}$	Monochromatic coefficient for absorption		

References

Ahluwalia, R. K. and Im, K. H. (1994). Spectral Radiative Heat-Transfer in Coal Furnaces Using a Hybrid Technique. *J. Institute of Energy*, **67**, 23.

Bayvel, L. P. and Jones, A. R. (1981). *Electromagnetic Scattering and Its Applications*. London: Applied Science.

Bohren, C. F. and Huffman, D. R. (1983). *Absorption and Scattering of Light by Small Particles*. New York: John Wiley & Sons.

Boothroyd, R. G. (1971). *Flowing Gas–Solids Suspensions*. London: Chapman & Hall.

Brenner, H. (1963). Forced Convection Heat and Mass Transfer at Small Peclet Numbers from a Particle of Arbitrary Shape. *Chem. Eng. Sci.*, **18**, 109.

Brewster, M. Q. and Tien, C. L. (1982). Radiative Transfer in Packed/Fluidized Beds: Dependent Versus Independent Scattering. *Trans. ASME, J. Heat Transfer*, **104**, 573.

Carslaw, H. S. and Jaeger, J. C. (1959). *Conduction of Heat in Solids*, 2nd ed. Oxford: Oxford University Press.

Chandrasekhar, S. (1960). *Radiative Transfer*. New York: Dover.

Fiveland, W. A. and Jamaluddin, A. S. (1991). Three-Dimensional Spectral Radiative Heat Transfer Solutions by the Discrete-Ordinate Method. *J. Thermophysics*, **5**, 335.

Froessling, N. (1938). The Evaporation of Falling Drops. *Gerlands Beitr. Geophys.*, **52**, 170.

Gel'Perin, N. I. and Einstein, V. G. (1971). Heat Transfer in Fluidized Beds. In *Fluidization*. Ed. Davidson and Harrison. New York: Academic Press.

Glicksman, L. R. and Decker, N. (1982). Heat Transfer from an Immersed Surface to Adjacent Particles in a Fluidized Bed: The Role of Radiation and Particle Packing. *Proceedings of the Seventh International Heat Transfer Conference*, München, Germany. Ed. Grigull, Hahne, Stephan, and Straub, **2**, 24.

Gupta, R. P., Wall, T. F. and Truelove, J. S. (1983). Radiative Scatter by Fly Ash in Pulverized-Coal-Fired Furnaces: Application of the Monte Carlo Method to Anisotropic Scatter. *Int. J. Heat & Mass Transfer*, **26**, 1649.

Hottel, H. C., Sarofim, A. F., Dalzell, W. H. and Vasalos, I. A. (1971). Optical Properties of Coatings: Effect of Pigment Concentration. *AIAA J.*, **9**, 129.

Hughmark, G. A. (1967). Mass and Heat Transfer from Rigid Spheres. *AIChE J.*, **13**, 1219.

Hunt, M. L. (1989). Comparison of Convective Heat Transfer in Packed Beds and Granular Flows. In *Annual Review of Heat Transfer*. Ed. C. L. Tien. Washington, D.C.: Hemisphere.

Im, K. H. and Ahluwalia, R. K. (1984). Combined Convection and Radiation in Rectangular Ducts. *Int. J. Heat & Mass Trans.*, **27**, 221.

Kronig, R. and Bruijsten, J. (1951). On the Theory of the Heat and Mass Transfer from a Sphere in a Flowing Medium at Low Values of Reynolds Number. *Appl. Sci. Res.*, **A2**, 439.

Love, T. J. (1968). *Radiative Heat Transfer*. Columbus, Ohio: Merrill.

Love, T. J. and Grosh, R. J. (1965). Radiative Heat Transfer in Absorbing, Emitting, and Scattering Media. *Trans. ASME, J. Heat Transfer*, **87c**, 161.

McAdams, W. A. (1954). *Heat Transmission*, 3rd ed. New York: McGraw-Hill.

Mie, G. (1908). Optics of Turbid Media. *Ann. Phys.*, **25**, 377.

Modest, M. F. (1989). Modified Differential Approximation for Radiative Transfer in General Three-Dimensional Media. *J. Thermophysics*, **3**, 283.

Özisik, M. N. (1973). *Radiative Transfer*. New York: John Wiley & Sons.

Ranz, W. E. and Marshall, W. R. (1952). Evaporation from Drops. *Chem. Eng. Prog.*, **48**, 141.

Rosner, D. E. (1986). *Transport Processes in Chemically Reacting Flow Systems*. Stoneham, Mass.: Butterworths.

Siegel, R. and Howell, J. R. (1981). *Radiative Heat Transfer*, 2nd ed. New York: McGraw-Hill.

Soo, S. L. (1990). *Multiphase Fluid Dynamics*. Beijing: Science Press; Brookfield, Vt.: Gower Technical.

Sun, J. and Chen, M. M. (1988). A Theoretical Analysis of Heat Transfer due to Particle Impact. *Int. J. Heat & Mass Transfer*, **31**, 969.

Sykes, J. B. (1951). Approximate Integration of the Equation of Transfer. *Monthly Notes Royal Astronomical Society*, **111**, 377.

Taylor, R. and Krishna, R. (1993). *Multicomponent Mass Transfer*. New York: John Wiley & Sons.

Tien, C. L. and Drolen, B. L. (1987). Thermal Radiation in Particulate Media with Dependent and Independent Scattering. In *Annual Review of Numerical Fluid Mechanics and Heat Transfer*. Ed. T. C. Chawla. Washington, D.C.: Hemisphere.

Van de Hulst, H. C. (1957). *Light Scattering by Small Particles*. New York: John Wiley & Sons.

Zabrodsky, S. S. (1966). *Hydrodynamics and Heat Transfer in Fluidized Beds*. Cambridge, Mass: MIT Press.

Problems

4.1 Solve the one-dimensional, unsteady conductive heat transfer equation for a homogeneous solid sphere. (Hint: (1) Let $\Theta = R^* T_p^*$ so that

$$\frac{\partial^2 \Theta}{\partial R^{*2}} = \frac{\partial \Theta}{\partial \mathrm{Fo}} \tag{P4.1}$$

with

$$\Theta = 0 \qquad \text{at } R^* = 0$$

$$\frac{\partial \Theta}{\partial R^*} + (\text{Bi} - 1)\Theta = 0 \qquad \text{at } R^* = 1 \qquad \text{(P4.2)}$$

$$\Theta = \Theta_i(R^*) \qquad \text{for Fo} = 0$$

(2) Use the separation of variables to obtain the solution.)

4.2 For a general forced convection system, prove that $\text{Nu} = f(\text{Re}, \text{Pr})$.

4.3 For a dilute suspension of particles in creeping flows ($\text{Re}_p \ll 1$), compare the difference in convective heat transfer coefficient in Eq. (4.40) and Eq. (4.42). It is assumed that $\text{Pr} = 0.7$.

4.4 Show that, using Eq. (4.58), the ratio of forward scattering to backward scattering is $(n_i - 1)^2/(n_i + 1)^2$ for specularly reflecting spheres.

4.5 Consider the cases where very hot particles are suddenly exposed to a cold environment. Derive an expression to account for the transient behavior of the particle temperature using Eq. (4.62). Assume that $T_p \gg T_w$ and the cooling is dominated by thermal radiation.

4.6 Consider a solid sphere in free fall passing through a heated chamber. The particle is a 500 μm alloy of density 4,000 kg/m^3, thermal capacity 150 J/kg · K, and absorptivity 0.6. The initial temperature of the particle is 300 K. The temperature of the gas inside the chamber is 900 K, and the temperature of the walls is 1,000 K. Determine the temperature of the particle at the chamber outlet if the length of the chamber is 0.3 m. Also, determine the ratio of the heat acquired by the particle as a result of thermal radiation to that due to heat convection. The physical properties of the gas are given as $\mu = 4 \times 10^{-5}$ kg/m · s, $\rho = 0.4$ kg/m^3, $\text{Pr} = 0.7$, and $K = 0.067$ W/m · K. The particle falls at its terminal velocity.

4.7 A spherical benzoic acid particle with a diameter of 2 mm is dropped into a quiescent air duct at ambient conditions (25°C). During the fall, sublimation of the particle occurs. Taking the molecular diffusivity of benzoic acid in air as 4×10^{-6} m^2/s, calculate the convective mass transfer coefficient when the particle is falling at its terminal velocity condition.

CHAPTER 5

Basic Equations

5.1 Introduction

Developing a mathematical model for general description of a particulate multiphase flow system has long been a challenging issue for researchers in the field. Although at this stage comprehensive constitutive relationships for modeling have not been fully established, several approaches which have proved useful are available for predicting the gas–solid flow behavior, particularly for engineering application purposes. Among them are the Eulerian continuum approach, Lagrangian trajectory approach, kinetic theory modeling for interparticle collisions, and Darcy's law and the Ergun equation for flows through packed beds. These approaches are the most commonly used and are introduced in this chapter.

The Eulerian continuum approach, based on a continuum assumption of phases, provides a "field" description of the dynamics of each phase. The Lagrangian trajectory approach, from the study of motions of individual particles, is able to yield historical trajectories of the particles. The kinetic theory modeling for interparticle collisions, extended from the kinetic theory of gases, can be applied to dense suspension systems where the transport in the particle phase is dominated by interparticle collisions. The Ergun equation provides important flow relationships, which are useful not only for packed bed systems, but also for some situations in fluidized bed systems.

5.1.1 *Eulerian Continuum Approach*

Evaluating the performance of a gas–solid transport system usually requires a means of macroscopic "field" description of the distribution of basic flow properties such as pressure, mass fluxes, concentrations, velocities, and temperatures of phases in the system. To conduct such an evaluation, the Eulerian continuum or multifluid approach is usually the best choice among the available approaches.

The Eulerian continuum approach is basically an extension of the mathematical formulation of the fluid dynamics for a single phase to a multiphase. However, since neither the fluid phase nor the particle phase is actually continuous throughout the system at any moment, ways to construct a continuum of each phase have to be established. The transport properties of each pseudocontinuous phase, or the turbulence models of each phase in the case of turbulent gas–solid flows, need to be determined. In addition, the phase interactions must be expressed in continuous forms.

The most fundamental step in constructing a continuum is to select a statistically valid "fluid element" of the continuum. The fluid element needs to be selected in such a way that it contains sufficient particles so that the fluid element possesses statistical thermodynamic properties such as temperature and density, as well as velocity. At the same time, the fluid element must be small enough compared to the characteristic length of the system so that, from a macroscopic point of view, the concept of a continuous phase of fluid can be adopted. Volume averaging is commonly applied to express the pseudocontinuum properties of each phase in a multiphase flow. In order that the volume averaging is of statistical significance and yet the control volume in the averaging is small enough to satisfy the pseudocontinuum condition, there exists a minimum control volume for the volume averaging. To account for the turbulence effects, time averaging needs to be adopted after the volume averaging [Soo, 1989]. Since the pseudocontinuum approach is based on the concept of volume averaging, the pseudocontinuum phase in a gas–solid flow should be regarded as compressible because of the nonuniform behavior of the volume fraction distribution of the phases even though the gas or particle itself may be incompressible.

The transport properties of gases in a single phase, laminar flow can be described in terms of the kinetic theory of gases. Particles in a gas–solid flow may be treated as "amplified molecules" so that a direct analogy of their behavior to gas molecules can be postulated. It is noted that, in the kinetic theory, the transport coefficients of gases are derived from the Maxwell–Boltzmann velocity distribution, which is based on Boltzmann statistics and the possible states governed by Schrödinger's wave equation. Since the interparticle forces in gas–solid systems can be quite different from the intermolecular forces in gas systems, the validity of the kinetic theory in describing the transport coefficients of particles needs to be further assured. For single-phase turbulent flows, the transport properties can be determined by turbulence models; these include the widely adopted mixing length model and k–ϵ model. In multiphase continuum modeling for gas–solid flows, these two models, with some modifications accounting for the compressibility of the pseudocontinuum phase and phase interactions, can be extended to illustrate the turbulence in the gas phase. In this chapter, the particle–particle interaction is described by using particle collision models while the particle–turbulence interaction is illustrated by using the Hinze–Tchen model and the k–ϵ–k_p model.

For the Eulerian continuum modeling discussed in this chapter, it is assumed that the basic form of the Navier–Stokes equation can be applied to all phases. For some dense suspension cases, the particle phase may behave as a non-Newtonian fluid. In these cases, a simple extension of the Navier–Stokes equation may not be appropriate.

5.1.2 *Lagrangian Trajectory Approach*

The Lagrangian approach, also known as trajectory modeling, provides a direct description of particulate flows by tracking the motion of individual particles. In trajectory modeling, the motion of a particle can be expressed by ordinary differential equations in Lagrangian coordinates. Consequently, the particle velocity and the corresponding particle trajectory can be solved by directly integrating the Lagrangian equation of the particle. Therefore, the Lagrangian approach is convenient to reveal the discrete and transient nature of the motion of particles.

Typical Lagrangian approaches include the deterministic trajectory method and the stochastic trajectory method. The deterministic trajectory method neglects all the turbulent transport processes of the particle phase, while the stochastic trajectory method takes into account the effect of gas turbulence on the particle motion by considering the instantaneous gas velocity in the formulation of the equation of motion of particles. To obtain the statistical average of the variables in the equations, statistical computations based on techniques such as the Monte Carlo method need to be performed.

To solve the Lagrangian equation for a moving particle, the dynamic behavior of the gas phase and other particles surrounding this moving particle should be predetermined. In this regard, the behavior of the gas phase can be simulated by the Eulerian method. Since the gas flow and the motion of all particles are coupled through gas–particle and particle–particle interactions, a large number of iterations are required in order to obtain converged solutions for all phases.

The development of the Lagrangian models has been limited mainly by the inherent need for large computing capacity to carry out statistical averaging and computation of the phase interactions. The Lagrangian models are particularly applicable to very dilute or discrete flow situations for which multifluid models are not appropriate, or to situations in which the historic tracking of particles is important (such as in pulverized coal combustion in a furnace or the tracking of radioactive particles in gas–solid flows).

5.1.3 *Kinetic Theory Modeling for Interparticle Collisions*

When the gas–solid flow in a multiphase system is dominated by the interparticle collisions, the stresses and other dynamic properties of the solid phase can be postulated to be analogous to those of gas molecules. Thus, the kinetic theory of gases is adopted in the modeling of dense gas–solid flows. In this model, it is assumed that collision among particles is the only mechanism for the transport of mass, momentum, and energy of the particles. The energy dissipation due to inelastic collisions is included in the model despite the elastic collision condition dictated by the theory.

The model yields a set of hydrodynamic equations for the solid phase. For equation closure, additional constitutive relations, which can be obtained by using the kinematic argument of the collision and by assuming the Maxwellian velocity distribution of the solids, are needed. Two examples are given to illustrate the applications of this model in this chapter.

5.1.4 *Ergun Equation*

The Ergun equation relates the pressure drop in a packed bed to the flow rate and the properties of particle and gas. However, the application of this equation has been extended beyond the limits of fixed bed systems since it was first formulated in 1951. Thus, a detailed account of the origin of this equation is necessary.

The general behavior of viscous flow through a homogeneous porous medium can be described by Darcy's law. In order to estimate the pressure drop across a porous medium, it is assumed that the general porous medium can be represented by an assemblage of channels of various cross sections with a fixed length. The viscous energy loss at low gas flow rates in this case is given by the Kozeny theory. When the gas flow rate increases, the effect of inertia becomes important. The kinetic energy loss at high gas flow rates can be described by the theory of Burke and Plummer. The general expression of the pressure drop for flow through a fixed bed, as suggested by Reynolds, may be presented in an algebraic form as the sum of the viscous and inertial contributions. The Ergun equation incorporates these two contributions in a semiempirical form. The equation may not be applicable to turbulent cases without further accounting for the turbulence effects.

5.1.5 *Summary*

In most engineering applications, the desired model is one that is simple, capable of providing a full flow field description, and yet reasonably accurate. The continuum model has such a simplicity feature and is widely used in spite of its crude continuum assumptions for all phases involved. When the particle concentrations are so high that the interparticle collisions become the dominant transport mechanism, the dynamic behavior of the particulate flow may be reasonably modeled on the basis of the kinetic theory. The kinetic theory can also be used in the formulation of the momentum transfer due to interparticle collisions in continuum modeling of the particle phase. For a very dilute multiphase suspension, the gas phase can be treated by the continuum approach while the motion of the particles may be described by the trajectory model. For gas flow through a packed bed or an unsuspended portion of a solids transport system, the Ergun equation can be applied. The Ergun equation is also able to describe the limiting situation of a dense suspension.

In this chapter, the continuum modeling of a single-phase flow is presented. The Eulerian approach of multiphase flows using the theories of volume averaging and time averaging is described. The Lagrangian method is introduced and the kinetic theory modeling for collision-dominated dense suspensions is delineated. Finally, the development of the Ergun equation is given.

5.2 Modeling of Single-Phase Flows

In this section, the governing equations and transport coefficients of a single-phase flow are presented. First, we derive the general transport theorem for conservation equations in continuum mechanics. The governing equations for mass, momentum, and energy transport are then obtained by applying the general transport theorem. The kinetic theory of gases, which provides the basis for the determination of transport coefficients, is described. In a turbulent flow, the transport coefficients are approximated by turbulence models. Hence, two commonly used turbulence models, the mixing length model and the k–ϵ model, are introduced. Various boundary conditions are also presented.

5.2.1 *General Transport Theorem and General Conservation*

All conservation equations in continuum mechanics can be derived from the general transport theorem. Define a variable $F(t)$ as a volume integral over an arbitrary volume $v(t)$ in an r-space

$$F(t) = \int_{v(t)} f(\boldsymbol{r}, t)\, \mathrm{d}v \tag{5.1}$$

where $f(\boldsymbol{r}, t)$ is an integrable function representing a parameter such as mass, momentum, or energy; t is time; and $\boldsymbol{r}$ is a position vector. In an r-space, $\boldsymbol{r} = (r_1, r_2, r_3)$. To calculate the rate of change of F, it is necessary to introduce a ξ-space where the volume is fixed with respect to time so that an interchange of differentiation and integration is possible for the derivative of F. A transformation of the volume from r-space to ξ-space can be expressed as

$$\mathrm{d}v = \mathrm{d}r_1\, \mathrm{d}r_2\, \mathrm{d}r_3 = J\, \mathrm{d}\xi_1\, \mathrm{d}\xi_2\, \mathrm{d}\xi_3 = J\, \mathrm{d}v_0 \tag{5.2}$$

where J is the Jacobian determinant, defined as

$$J = \frac{\partial(r_1, r_2, r_3)}{\partial(\xi_1, \xi_2, \xi_3)} \tag{5.3}$$

It can be proved that (Problem 5.1)

$$\frac{\mathrm{d}J}{\mathrm{d}t} = (\nabla \cdot \boldsymbol{U})J \tag{5.4}$$

where the velocity $\boldsymbol{U}$ is defined as

$$\boldsymbol{U} = \frac{\mathrm{d}\boldsymbol{r}}{\mathrm{d}t} \tag{5.5}$$

Therefore, we have

$$\begin{aligned} \frac{\mathrm{d}F}{\mathrm{d}t} &= \frac{\mathrm{d}}{\mathrm{d}t}\int_{v(t)} f(\boldsymbol{r}, t)\,\mathrm{d}v = \frac{\mathrm{d}}{\mathrm{d}t}\int_{v_0} f(\xi, t) J\,\mathrm{d}v_0 \\ &= \int_{v_0}\left(\frac{\mathrm{d}f}{\mathrm{d}t}J + f\frac{\mathrm{d}J}{\mathrm{d}t}\right)\mathrm{d}v_0 = \int_{v(t)}\left(\frac{\mathrm{d}f}{\mathrm{d}t} + f\nabla\cdot\boldsymbol{U}\right)\mathrm{d}v \end{aligned} \tag{5.6}$$

Note that $\mathrm{d}f/\mathrm{d}t$ is the substantial derivative, which can be expressed by

$$\frac{\mathrm{d}f}{\mathrm{d}t} = \frac{\partial f}{\partial t} + \boldsymbol{U} \cdot \nabla f \tag{5.7}$$

The equation describing the general transport theorem can be obtained by substituting Eq. (5.7) into Eq. (5.6) as

$$\frac{\mathrm{d}}{\mathrm{d}t}\int_{v(t)} f(\boldsymbol{r}, t)\,\mathrm{d}v = \int_{v(t)}\left(\frac{\partial f}{\partial t} + \nabla \cdot f\boldsymbol{U}\right)\mathrm{d}v \tag{5.8}$$

The rate of change of F can be found from the net flow of f across the closed surface of $v(t)$ and the generation of f inside $v(t)$. Denote ψ as the flux vector of f and Φ as the rate of generation of f per unit volume. Thus,

$$\frac{\mathrm{d}}{\mathrm{d}t}\int_v f\,\mathrm{d}v = -\int_A \boldsymbol{n}\cdot\psi\,\mathrm{d}A + \int_V \Phi\,\mathrm{d}v \tag{5.9}$$

where $\boldsymbol{n}$ is a unit normal vector directed outwardly so that the first term on the right-hand side represents the net inflow of f across the closed surface A. Note that, according to the Gauss theorem, a surface integral over a closed area can be transformed into a volume integral, *i.e.*,

$$\int_A \boldsymbol{n}\cdot\psi\,\mathrm{d}A = \int_v \nabla\cdot\psi\,\mathrm{d}v \tag{5.10}$$

Combining the general transport theorem, Eq. (5.8), with Eqs. (5.9) and (5.10) gives

$$\int_v \left[\frac{\partial f}{\partial t} + \nabla\cdot f\boldsymbol{U} + \nabla\cdot\psi - \Phi\right]\mathrm{d}v = 0 \tag{5.11}$$

Since the volume v is arbitrarily selected, the general conservation equation in a single phase of fluid can be written in the form

$$\frac{\partial f}{\partial t} + \nabla\cdot f\boldsymbol{U} + \nabla\cdot\psi - \Phi = 0 \tag{5.12}$$

5.2.2 *Governing Equations*

The conservation equations of mass, momentum, and energy of a single-phase flow can be obtained by using the general conservation equation derived previously.

5.2.2.1 *Continuity Equation (Conservation of Mass)*

For the total mass conservation of a single-phase fluid, f represents the fluid density ρ. ψ represents the diffusional flux of total mass, which is zero. For flow systems without chemical reactions, $\Phi = 0$. Therefore, from Eq. (5.12), we have the continuity equation as

$$\frac{\partial \rho}{\partial t} + \nabla \cdot (\rho \boldsymbol{U}) = 0 \tag{5.13}$$

5.2.2.2 *Momentum Equation (Newton's Law)*

For the momentum conservation of a single-phase fluid, the momentum per unit volume f is equal to the mass flux $\rho \boldsymbol{U}$. The momentum flux is thus expressed by the stress tensor $\psi = (p\boldsymbol{I} - \boldsymbol{\tau})$. Here p is the static pressure or equilibrium pressure; $\boldsymbol{I}$ is a unit tensor; and $\boldsymbol{\tau}$ is the shear stress tensor. Since $\Phi = -\rho \boldsymbol{f}$ where $\boldsymbol{f}$ is the field force per unit mass, Eq. (5.12) gives rise to the momentum equation as

$$\frac{\partial \rho \boldsymbol{U}}{\partial t} + \nabla \cdot (\rho \boldsymbol{U}\boldsymbol{U}) = -\nabla p + \nabla \cdot \boldsymbol{\tau} + \rho \boldsymbol{f} \tag{5.14}$$

where

$$\boldsymbol{\tau} = \mu \left(\frac{\partial U_{\mathrm{i}}}{\partial x_{\mathrm{j}}} + \frac{\partial U_{\mathrm{j}}}{\partial x_{\mathrm{i}}} \right) \boldsymbol{e}_{\mathrm{i}} \boldsymbol{e}_{\mathrm{j}} + \left(\mu' - \frac{2\mu}{3} \right) (\nabla \cdot \boldsymbol{U}) \delta_{\mathrm{ij}} \boldsymbol{e}_{\mathrm{i}} \boldsymbol{e}_{\mathrm{j}} \tag{5.15}$$

In Eq. (5.15), μ' is the second viscosity or bulk viscosity, which reflects the deviation of the averaged pressure from the pressure at equilibrium due to nonuniformity in the local velocity distribution. δ_{ij} is the Kronecker delta. μ' is given by

$$\mu' = -\frac{p_{\mathrm{m}} - p}{\nabla \cdot \boldsymbol{U}} \tag{5.16}$$

where p_{m} is the averaged pressure. Except for the strong nonequilibrium regions such as the shock wave region, it is assumed that $\mu' = 0$.

5.2.2.3 *Energy Equation (Conservation of Energy)*

For total energy conservation, f represents the total energy per unit volume ρE. The flux and the generation rate of the total energy per unit volume come from three sources: the heat flux vector $\boldsymbol{J}_{\mathrm{q}}$; the rate of heat generation per unit volume J_{e}, including Joule's heating and thermal radiation; and the rate of work done by the surface force and the field force. Thus, the energy equation can be given as

$$\frac{\partial \rho E}{\partial t} + \nabla \cdot (\rho \boldsymbol{U} E) = -\nabla \cdot \boldsymbol{U} p + \nabla \cdot (\boldsymbol{U} \cdot \boldsymbol{\tau}) + \rho \boldsymbol{U} \cdot \boldsymbol{f} - \nabla \cdot \boldsymbol{J}_{\mathrm{q}} + J_{\mathrm{e}} \tag{5.17}$$

where $E = e + 1/2\, \boldsymbol{U} \cdot \boldsymbol{U}$ and e is the internal energy per unit mass.

The energy equation expressed in terms of internal energy becomes

$$\frac{\partial \rho e}{\partial t} + \nabla \cdot (\rho \boldsymbol{U} e) = -p \nabla \cdot \boldsymbol{U} + \phi - \nabla \cdot \boldsymbol{J}_{\mathrm{q}} + J_{\mathrm{e}} \tag{5.18}$$

where ϕ is the dissipation function defined by

$$\phi = \nabla \cdot (\boldsymbol{U} \cdot \boldsymbol{\tau}) - \boldsymbol{U} \cdot (\nabla \cdot \boldsymbol{\tau}) \tag{5.19}$$

which represents the dissipation rate of energy per unit volume due to viscous effects. The energy is dissipated in the form of heat. ϕ can be proved to be always positive.

The energy equation expressed in terms of temperature is convenient for evaluating heat fluxes. Let $e = c_v T$ with c_v the specific heat at constant volume and T the absolute temperature of the fluid. Assuming the heat flux $\boldsymbol{J}_q$ obeys Fourier's law, Eq. (5.18) takes the form

$$\frac{\partial \rho c_v T}{\partial t} + \nabla \cdot (\rho \boldsymbol{U} c_v T) = -p \nabla \cdot \boldsymbol{U} + \phi + \nabla \cdot (K \nabla T) + J_e \tag{5.20}$$

where K is the thermal conductivity of the fluid.

5.2.3 *Kinetic Theory and Transport Coefficients*

A brief discussion of the kinetic theory of gases is essential to the understanding of the fundamental assumptions or conditions used in the derivation of the transport coefficients of the particle phase in the continuum modeling of gas–solid multiphase flows. When the conserved quantities such as the number of molecules, total momentum of molecules, and total kinetic energy of molecules (assuming elastic collisions) are initially nonuniformly distributed in an enclosed volume, a tendency for these quantities to be moving toward an equilibrium state is established. The rate of change of a conserved quantity toward the equilibrium state is characterized by the transport coefficient. Specifically, a transport coefficient is defined as the proportionality constant of the transport flux to the density gradient of the conserved quantity. The transport coefficients of a gas can be evaluated, from the kinetic theory, in terms of the microscopic properties of the fluid. The rigorous microscopic theory of transport processes can be represented by the Boltzmann transport equation. However, because of the complexity of this equation, it is omitted here. Instead, we adopt a phenomenological approach to derive the transport coefficients using the elementary kinetic theory and probabilistic concept.

In the kinetic theory of gases, the molecules are assumed to be smooth, rigid, and elastic spheres. The only kinetic energy considered is that from the translational motion of the molecules. In addition, the gas is assumed to be in an equilibrium state in a container where the gas molecules are uniformly distributed and all directions of the molecular motion are equally probable. Furthermore, velocities of the molecules are assumed to obey the Maxwell–Boltzmann distribution, which is described in the following section.

5.2.3.1 *Maxwell–Boltzmann Velocity Distribution*

Consider a system that comprises numerous identical, independent, and indistinguishable particles in equilibrium in a container at an absolute temperature T. If there is no limit on the number of particles occupying a given energy level, the probability of finding n_i particles with energy level ϵ_i in such a system can be described by Bose–Einstein statistics [Reif, 1965]. Denote g_i as the degeneracy of the energy level ϵ_i (*i.e.*, the number of the quantum states having the same or nearly the same energy). In Bose–Einstein statistics, the number of possible arrangements for the ith level of energy for any n_i, $W_{i,BE}$, is

$$W_{i,BE} = \frac{(n_i + g_i - 1)!}{(g_i - 1)!\, n_i!} \tag{5.21}$$

The total number of possible arrangements for a certain set of n_i in the system, W_{BE}, is therefore obtained as

$$W_{BE} = \prod \frac{(n_i + g_i - 1)!}{(g_i - 1)!n_i!} \tag{5.22}$$

In a special case where $g_i \gg n_i$, Eq. (5.22) can be reduced to the corrected Maxwell–Boltzmann statistics (note: the classical Maxwell–Boltzmann statistics is for distinguishable particles) as

$$W_{MB} = W_{BE} = \prod \frac{g_i^{n_i}}{n_i!} \tag{5.23}$$

When the gas in a container is in an equilibrium condition, it can be shown (Problem 5.2) that Eq. (5.23) can be expressed as

$$\frac{n_i}{n} = \frac{g_i e^{-\frac{\epsilon_i}{kT}}}{z} \tag{5.24}$$

where n is the total number of particles in the system; k is the Boltzmann constant; and z is the partition function, defined as

$$z = \sum g_i e^{-\frac{\epsilon_i}{kT}} \tag{5.25}$$

The possible energy levels are determined by Schrödinger's wave equation [Reif, 1965]. For translational motion of a particle, the wave equation takes the form

$$\nabla^2 \Psi = -\frac{8\pi^2 m}{h^2} \epsilon \Psi \tag{5.26}$$

where h is Planck's constant, m is the mass of the particle, and Ψ is the wave function. ϵ is the eigenvalue of Eq. (5.26), which represents the energy as expressed by

$$\epsilon = \frac{h^2}{8m}\left(\frac{k_x^2}{L_x^2} + \frac{k_y^2}{L_y^2} + \frac{k_z^2}{L_z^2}\right) \tag{5.27}$$

where k_x, k_y, and k_z are the integer quantum numbers; and L_x, L_y, and L_z are the geometric dimensions in the x-, y-, and z-directions, respectively. Moreover, the partition function z can be obtained as

$$z = \sum e^{-\frac{\epsilon}{kT}} = \frac{V}{h^3}(2\pi mkT)^{\frac{3}{2}} \tag{5.28}$$

where V is the volume of the container, which is equal to $L_x L_y L_z$.

If we consider a cubic system with a characteristic dimension L and assume that the energy distribution is continuous, we have

$$\epsilon = \frac{h^2 r^2}{8mL^2} \tag{5.29}$$

where r is the radius in a space of quantum numbers (note that only the first octant is meaningful since all the quantum numbers should be positive). As a result, the degeneracy $\mathrm{d}g$ between ϵ and $\epsilon + \mathrm{d}\epsilon$ can be expressed as

$$\mathrm{d}g = \frac{\pi}{2} r^2\,\mathrm{d}r = 2\pi\left(\frac{L}{h}\right)^3 (2m)^{\frac{3}{2}} \epsilon^{\frac{1}{2}}\,\mathrm{d}\epsilon \tag{5.30}$$

Substituting Eqs. (5.28) and (5.30) into Eq. (5.24) yields

$$\frac{dn}{n} = \frac{2}{\sqrt{\pi (kT)^3}} e^{-\frac{\epsilon}{kT}} \epsilon^{\frac{1}{2}}\, d\epsilon \tag{5.31}$$

With $\epsilon = 1/2\, mv^2$, we have finally obtained the Maxwell–Boltzmann velocity distribution as

$$dP = \frac{dn}{n} = \sqrt{\frac{2m^3}{\pi (kT)^3}} e^{-\frac{mv^2}{2kT}} v^2\, dv \tag{5.32}$$

where dP represents the probability of particles having velocity between v and $(v + dv)$.

5.2.3.2 *Collision Frequency*

The molecular collision is the basic mechanism that governs the transport processes in gases. Thus, it is necessary to examine the collision frequency and collision cross section before quantifying the transport coefficients. First consider the probability that a collision occurs between r and $r + dr$. Denote λ as the mean free path and $P(r)$ as the probability of no collisions within r. Hence, we have

$$P(r + dr) = P(r)\left(1 - \frac{dr}{\lambda}\right) \tag{5.33}$$

The Taylor series expansion of $P(r + dr)$ yields

$$\frac{dP}{dr} = -\frac{P}{\lambda} \tag{5.34}$$

With $P(0) = 1$, it is found that the probability of no collisions within the length of path r follows the Poisson distribution as

$$P(r) = e^{-\frac{r}{\lambda}} \tag{5.35}$$

Now, consider a gas with two different types of particles in a container. The radius of the sphere of influence for collision between the two different particles is

$$r_{12} = \frac{d_1 + d_2}{2} \tag{5.36}$$

where subscripts 1 and 2 denote the two different types of particles.

The average relative velocity between particles 1 and 2 can be evaluated by

$$\langle v_{12} \rangle = \int dv_1 \int dv_2 F_1 F_2 |v_1 - v_2| \tag{5.37}$$

where $\langle v_{12} \rangle$ is the averaged relative velocity and F is the velocity distribution function. Assume that the motion of each type of particle obeys the Maxwell–Boltzmann distribution. Then, $\langle v_{12} \rangle$ can be obtained as (Problem 5.3)

$$\langle v_{12} \rangle = \sqrt{\frac{8(m_1 + m_2)kT}{\pi m_1 m_2}} \tag{5.38}$$

The collision frequency per unit volume between different types of particles is obtained as

$$f_{12} = n_1 n_2 \pi r_{12}^2 \langle v_{12} \rangle = n_1 n_2 (d_1 + d_2)^2 \sqrt{\frac{\pi (m_1 + m_2)kT}{2m_1 m_2}} \tag{5.39}$$

which can be easily extended to yield the collision frequency among the same type of particles as

$$f_{11} = \frac{1}{2} n_1^2 \pi r_{11}^2 \langle v_{11} \rangle = 2 n_1^2 d_1^2 \sqrt{\frac{\pi k T}{m_1}} \tag{5.40}$$

where the factor 1/2 is to account for the overestimation of the collisions of identical particles.

5.2.3.3 *Transport Coefficients*

The coefficients of transport properties considered here include the viscosity, diffusivity, and thermal conductivity of a gas. The transport coefficients vary with gas properties if the flow is laminar. When the flow is turbulent, the transport coefficients become strongly dependent on the turbulence structure. Here we only deal with the laminar transport coefficients; the discussion of the turbulent transport coefficients is given in §5.2.4.

A. COEFFICIENT OF VISCOSITY

Consider a cubic unit volume containing n particles in a Cartesian coordinate system. On average, about $n/6$ particles move in the $+y$-direction, and the same number of particles move in the other five directions. Each particle stream has the same averaged velocity $\langle v \rangle$. Since particle collision is responsible for the momentum transport, the averaged x-component of the particle momentum transported in the y-direction may be reasonably estimated by

$$\tau_{xy} = \frac{1}{6} n \langle v \rangle [m U_x (y - \lambda) - m U_x (y + \lambda)] \tag{5.41}$$

where τ_{xy} is shear stress in the x–y-plane; λ is the mean free path; m is the mass of the particle; and $U_x(y{-}\lambda)$ represents the averaged x-component of velocity at the position $(y{-}\lambda)$ along the y-axis. Using the Taylor expansion, Eq. (5.41) becomes

$$\tau_{xy} = -\frac{1}{3} n \langle v \rangle m \lambda \frac{\partial U_x}{\partial y} \tag{5.42}$$

Thus, the transport coefficient of viscosity, denoted by μ, is

$$\mu = \frac{1}{3} n \langle v \rangle m \lambda \tag{5.43}$$

B. COEFFICIENT OF SELF-DIFFUSION (DIFFUSIVITY)

The diffusivity can be obtained by using the same method as that for the viscosity. The transported quantity in this case is the number of particles instead of the momentum of particles. The net flux of particles in the y-direction may be expressed as

$$J_y = \frac{1}{6} \langle v \rangle [n(y - \lambda) - n(y + \lambda)] \tag{5.44}$$

where $n(y{-}\lambda)$ represents the particle number density at the position $(y{-}\lambda)$ along the y-axis. From the Taylor expansion, Eq. (5.44) has the form

$$J_y = -\frac{1}{3} \langle v \rangle \lambda \frac{\partial n}{\partial y} \tag{5.45}$$

which gives the coefficient of self-diffusion D as

$$D = \frac{1}{3}\langle v\rangle\lambda \tag{5.46}$$

C. THERMAL CONDUCTIVITY

On the basis of the same approach, we can obtain an expression for the thermal conductivity based on the energy flux. The net energy transported in the y-direction can be estimated from

$$q_y = \frac{1}{6}n\langle v\rangle[\epsilon(y-\lambda) - \epsilon(y+\lambda)] = -\frac{1}{3}n\langle v\rangle\lambda\frac{\partial\epsilon}{\partial y} \tag{5.47}$$

where ϵ is the averaged energy of a particle. Since ϵ is related to the absolute temperature T, we have

$$\frac{\partial\epsilon}{\partial y} = \frac{\partial\epsilon}{\partial T}\frac{\partial T}{\partial y} = c\frac{\partial T}{\partial y} \tag{5.48}$$

where c is the specific heat of the particle. Therefore, the thermal conductivity K is in the form

$$K = \frac{1}{3}n\langle v\rangle c\lambda \tag{5.49}$$

D. TRANSPORT COEFFICIENTS BASED ON MAXWELL–BOLTZMANN DISTRIBUTION

The averaged velocity $\langle v\rangle$ can be expressed in terms of Eq. (5.32) as

$$\langle v\rangle = \int_0^\infty v\frac{dn}{n} = \left(\frac{2}{\pi}\right)^{\frac{1}{2}}\left(\frac{m}{kT}\right)^{\frac{3}{2}}\int_0^\infty v^3 e^{-\frac{mv^2}{2kT}}\,dv = \sqrt{\frac{8kT}{\pi m}} \tag{5.50}$$

Using Eq. (5.40), the mean free path can be estimated as

$$\lambda = \frac{1}{\sqrt{2}n\pi d^2} \tag{5.51}$$

Hence, the transport coefficients can be given as

$$\mu = \frac{2}{3\sqrt{\pi}}\frac{\sqrt{mkT}}{\pi d^2} \tag{5.52}$$

$$D = \frac{2}{3}\frac{1}{n\pi d^2}\sqrt{\frac{kT}{\pi m}} \tag{5.53}$$

$$K = \frac{2}{3\sqrt{\pi}}\frac{c}{\pi d^2}\sqrt{\frac{kT}{m}} \tag{5.54}$$

5.2.4 *Modeling for Turbulent Flows*

The most widely adopted method for the turbulent flow analysis is based on time-averaged equations using the Reynolds decomposition concept. In the following, we discuss the Reynolds decomposition and time-averaging method. There are other methods such as direct numerical simulation (DNS), large-eddy simulation (LES), and discrete-vortex simulation (DVS) that are being developed and are not included here.

In using the Reynolds decomposition, closure of the time-averaged Navier–Stokes equations cannot readily be realized because of the unknown correlation terms such as turbulent

Reynolds stresses generated by time averaging. Thus, additional equations are needed to correlate these terms with time-averaged quantities. These additional equations may come from turbulence models. The two most commonly used turbulence models, the mixing length model and the k–ϵ model, are introduced.

5.2.4.1 *Time-Averaged Navier–Stokes Equations*

It is assumed that the instantaneous Navier–Stokes equations for turbulent flows have the exact form of those for laminar flows. From the Reynolds decomposition, any instantaneous variable, ϕ, can be divided into a time-averaged quantity and a fluctuating part as

$$\phi(t) = \overline{\Phi} + \phi'(t) \tag{5.55}$$

where

$$\overline{\Phi} = \frac{1}{t_0}\int_0^{t_0} \phi(t)\,\mathrm{d}t \tag{5.56}$$

The integral time period t_0 should be short compared to the characteristic time scale of the system. At the same time, t_0 must also be long enough so that

$$\frac{1}{t_0}\int_0^{t_0} \phi'(t)\,\mathrm{d}t = 0 \tag{5.57}$$

Taking the time average over Eqs. (5.13) and (5.14), we can obtain the time-averaged continuity equation and time-averaged momentum equation as

$$\frac{\partial\overline{\rho}}{\partial t} + \frac{\partial}{\partial x_j}\left(\overline{\rho}\,\overline{U}_j + \overline{\rho' u'_j}\right) = 0 \tag{5.58}$$

$$\frac{\partial}{\partial t}\left(\overline{\rho}\,\overline{U}_i + \overline{\rho' u'_i}\right) + \frac{\partial}{\partial x_j}\left(\overline{\rho}\,\overline{U}_i\overline{U}_j + \overline{\rho}\overline{u'_i u'_j} + \overline{U}_i\overline{\rho' u'_j} + \overline{U}_j\overline{\rho' u'_i} + \overline{\rho' u'_i u'_j}\right)$$
$$= -\frac{\partial\overline{\sigma}_{ij}}{\partial x_j} + \sum_j F_{ji} \tag{5.59}$$

where $\overline{\rho}\overline{u'_i u'_j}$ is defined as the turbulent Reynolds stress and the time-averaged stress tensor $\overline{\sigma}_{ij}$ is

$$\overline{\sigma}_{ij} = \overline{p}\delta_{ij} - \overline{\mu}\left(\frac{\partial\overline{U}_i}{\partial x_j} + \frac{\partial\overline{U}_j}{\partial x_i}\right) + \frac{2}{3}\overline{\mu}\frac{\partial\overline{U}_m}{\partial x_m}\delta_{ij} + \frac{2}{3}\overline{\mu'\frac{\partial u'_m}{\partial x_m}}\delta_{ij} - \overline{\mu'\left(\frac{\partial u'_i}{\partial x_j} + \frac{\partial u'_j}{\partial x_i}\right)} \tag{5.60}$$

In Eq. (5.60), μ' denotes the viscosity fluctuation which may result from the temperature fluctuation. In Eq. (5.60) the quantities with the same subscript m in the third and fourth terms on the right-hand side of the equation indicate the Einstein summation.

For simplicity, consider the steady, incompressible, and isothermal turbulent flows where $\partial/\partial t = 0$; $\rho' = 0$; and $\mu' = 0$. The continuity equation is thus simplified from Eq. (5.58) to

$$\frac{\partial\overline{U}_j}{\partial x_j} = 0 \tag{5.61}$$

and the momentum equation is reduced from Eq. (5.59) to

$$\frac{\partial}{\partial x_j}\left(\overline{\rho}\,\overline{U}_i\overline{U}_j\right) = -\frac{\partial\overline{p}}{\partial x_i} + \frac{\partial}{\partial x_j}\left(\overline{\mu}\left[\frac{\partial\overline{U}_i}{\partial x_j} + \frac{\partial\overline{U}_j}{\partial x_i}\right] - \overline{\rho}\overline{u'_i u'_j}\right) + \sum_j F_{ji} \tag{5.62}$$

To render Eq. (5.62) solvable, it is necessary to provide an expression for the turbulent Reynolds stress. For isotropic turbulent flows, similar to the transport processes in laminar flows, a scalar turbulent viscosity μ_T is defined using the Boussinesq formulation

$$-\overline{\rho}\overline{u_i' u_j'} = \mu_T \left(\frac{\partial \overline{U}_i}{\partial x_j} + \frac{\partial \overline{U}_j}{\partial x_i} \right) \tag{5.63}$$

It should be noted that the turbulent viscosity is a function of the turbulence structure in addition to the physical properties of the gas. We may further introduce an effective turbulent viscosity μ_{eff} as

$$\mu_{eff} = \mu + \mu_T \tag{5.64}$$

Thus, Eq. (5.62) can be expressed as

$$\frac{\partial}{\partial x_j}(\overline{\rho}\overline{U}_i\overline{U}_j) = -\frac{\partial \overline{p}}{\partial x_i} + \frac{\partial}{\partial x_j}\left(\overline{\mu}_{eff} \left[\frac{\partial \overline{U}_i}{\partial x_j} + \frac{\partial \overline{U}_j}{\partial x_i} \right] \right) + \sum_j \overline{F}_{ji} \tag{5.65}$$

For nonisotropic turbulent flows such as strong swirling flows or buoyant flows, the turbulent viscosity is a tensor instead of a scalar. Therefore, all six components of the Reynolds stress need to be modeled individually.

To determine $\overline{u_i' u_j'}$ or μ_T, various types of turbulence models have been proposed. A turbulence model is characterized by a set of transport equations or constitutive relations to simulate higher-order correlations using lower-order correlations or time-averaged quantities. Currently, no general turbulence model is available since the physics of turbulence is not fully understood. However, for nonswirling and nonbuoyant flows (isotropic turbulence), the mixing length model and the k–ϵ model can be used and are introduced in the following. For nonisotropic turbulent flows, the second-order moment closure model in the form of algebraic expressions or differential transport equations is the simplest [Zhang *et al.*, 1992; Zhou, 1993]. The model in the form of differential transport equations is, however, too complex to be used for general engineering purposes. Specifically, it requires solving six equations instead of two in the k–ϵ model. In addition, eight empirical constants need to be determined instead of the five in the k–ϵ model. Furthermore, specifying the boundary conditions for each Reynolds stress component is difficult. Hence, in this text, we only discuss the isotropic turbulent flows.

5.2.4.2 *Mixing Length Model*

The mixing length model was originally proposed by Prandtl (1925). The model is based on two analogies: One is between the laminar viscosity and the turbulent viscosity of gas; the other is between the ratio of the averaged fluctuating velocity to the averaged fluctuating length and the strain from the averaged velocity. Prandtl proposed that

$$\mu_T = \overline{\rho} l_m \sqrt{\overline{u_i' u_i'}} \tag{5.66}$$

$$\frac{\sqrt{\overline{u_i' u_i'}}}{l_m} = \left| \frac{\partial \overline{U}_i}{\partial x_j} + \frac{\partial \overline{U}_j}{\partial x_i} \right| \tag{5.67}$$

Combining Eq. (5.67) with Eq. (5.66) yields

$$\mu_T = \overline{\rho} l_m^2 \left| \frac{\partial \overline{U}_i}{\partial x_j} + \frac{\partial \overline{U}_j}{\partial x_i} \right| \tag{5.68}$$

where l_m is the turbulent mixing length, which is determined empirically and depends on the type of flow.

The mixing length model is simple with μ_T defined by an algebraic expression. Hence, for simple flows such as jet flows, flows of the boundary layer type, and pipe flows, l_m can be easily estimated. For flows with recirculation regions or with complex geometries, the determination of l_m may be impossible. The selection of l_m for various simple flows can be found in many references [*e.g.*, Launder and Spalding, 1972]. However, the mixing length model indicates that the effect of turbulence will vanish where the strain, based on the averaged velocity, is zero (*e.g.*, near the end of a recirculation region or near a symmetric axis). This is not always true.

5.2.4.3 *k–ε Model*

If we assume that the local state of the turbulent flow depends on some turbulent quantities governed by the corresponding transport equations, how many independent scaling quantities are required? This question may be answered from the dimensional analysis.

To simulate the turbulent Reynolds stress, first we examine the dimensions of the Reynolds stress, the rate of strain, and the turbulent viscosity as follows

$$[\overline{\rho}\overline{u'_i u'_j}] = \frac{[M]}{[L][T]^2}; \qquad \left[\frac{\partial \overline{U}_i}{\partial x_j}\right] = \frac{1}{[T]}; \qquad [\mu_T] = \frac{[M]}{[L][T]} \tag{5.69}$$

where [M] denotes the mass unit; [L] represents the length unit; and [T] is the time unit. In order to correlate the Reynolds stress with the rate of strain, a length scale and a time scale or any two independent combinations of the length and time scales (with $\overline{\rho}$ as the mass scale) are needed to characterize the local turbulence. That is, two more governing equations are required to describe the transport of these two scaling quantities.

A common choice for the two scaling parameters is the kinetic energy of turbulence k and its dissipation rate ϵ, defined as

$$k = \frac{1}{2}\overline{u'_i u'_i}; \qquad \epsilon = \frac{\overline{\mu}}{\overline{\rho}}\overline{\frac{\partial u'_i}{\partial x_j}\frac{\partial u'_i}{\partial x_j}} \tag{5.70}$$

From the dimensional analysis, it can be shown that k and ϵ are the two independent scaling parameters of time and length scales

$$[T] = \frac{[k]}{[\epsilon]}; \qquad [L] = \frac{[k]^{\frac{3}{2}}}{[\epsilon]} \tag{5.71}$$

Thus, the turbulent viscosity μ_T may be related to k and ϵ in the form

$$\mu_T = C_\mu \overline{\rho}\frac{k^2}{\epsilon} \tag{5.72}$$

where C_μ is an empirical constant.

For steady-state, incompressible, and isothermal flows, the transport equation of the turbulent kinetic energy can be derived as (Problem 5.4)

$$\frac{\partial}{\partial x_j}(\overline{\rho}\overline{U}_j k) = -\overline{\rho}\overline{u'_i u'_j}\frac{\partial \overline{U}_i}{\partial x_j} - \overline{\rho}\epsilon + \frac{\partial}{\partial x_j}\left(\overline{\mu}\frac{\partial k}{\partial x_j} - \frac{1}{2}\overline{\rho}\overline{u'_i u'_i u'_j} - \overline{p' u'_j}\right) \tag{5.73}$$

The third term on the right-hand side of the equation represents the transport of k due to molecular and turbulent diffusion. Hence, with an analogy to the laminar transport, this term may be expressed as

$$\overline{\mu}\frac{\partial k}{\partial x_{\mathrm{j}}} - \frac{1}{2}\overline{\rho u'_{\mathrm{i}}u'_{\mathrm{i}}u'_{\mathrm{j}}} - \overline{p'u'_{\mathrm{j}}} = \frac{\mu_{\mathrm{eff}}}{\sigma_{\mathrm{k}}}\frac{\partial k}{\partial x_{\mathrm{j}}} \tag{5.74}$$

where σ_{k} is an empirical constant. Thus, Eq. (5.73) takes the form

$$\frac{\partial}{\partial x_{\mathrm{j}}}(\overline{\rho}\overline{U}_{\mathrm{j}}k) = \frac{\partial}{\partial x_{\mathrm{j}}}\left(\frac{\mu_{\mathrm{eff}}}{\sigma_{\mathrm{k}}}\frac{\partial k}{\partial x_{\mathrm{j}}}\right) + (\mu_{\mathrm{eff}} - \mu)\frac{\partial \overline{U}_{\mathrm{i}}}{\partial x_{\mathrm{j}}}\left(\frac{\partial \overline{U}_{\mathrm{i}}}{\partial x_{\mathrm{j}}} + \frac{\partial \overline{U}_{\mathrm{j}}}{\partial x_{\mathrm{i}}}\right) - \overline{\rho}\epsilon \tag{5.75}$$

Equation (5.75) is called the k-equation.

The transport equation for the rate of turbulent dissipation can also be derived as (Problem 5.5)

$$\begin{aligned}\frac{\partial}{\partial x_{\mathrm{j}}}(\overline{\rho}\epsilon\overline{U}_{\mathrm{j}}) = {}& \frac{\partial}{\partial x_{\mathrm{j}}}\left[\overline{\mu}\frac{\partial \epsilon}{\partial x_{\mathrm{j}}} - \overline{\rho u'_{\mathrm{j}}\epsilon'} - 2\overline{\upsilon}\overline{\frac{\partial p'}{\partial x_{\mathrm{l}}}\frac{\partial u'_{\mathrm{j}}}{\partial x_{\mathrm{l}}}}\right] - 2\overline{\mu}\left[\overline{\frac{\partial u'_{\mathrm{i}}}{\partial x_{\mathrm{l}}}\frac{\partial u'_{\mathrm{j}}}{\partial x_{\mathrm{l}}}} + \overline{\frac{\partial u'_{\mathrm{l}}}{\partial x_{\mathrm{i}}}\frac{\partial u'_{\mathrm{l}}}{\partial x_{\mathrm{j}}}}\right]\frac{\partial \overline{U}_{\mathrm{i}}}{\partial x_{\mathrm{j}}} \\ & - 2\overline{\mu}\overline{\frac{\partial u'_{\mathrm{i}}}{\partial x_{\mathrm{l}}}\frac{\partial u'_{\mathrm{j}}}{\partial x_{\mathrm{l}}}\frac{\partial u'_{\mathrm{i}}}{\partial x_{\mathrm{j}}}} - 2\overline{\mu}\,\overline{u'_{\mathrm{j}}\frac{\partial u'_{\mathrm{i}}}{\partial x_{\mathrm{l}}}}\frac{\partial^2 \overline{U}_{\mathrm{i}}}{\partial x_{\mathrm{j}}\partial x_{\mathrm{l}}} - 2\overline{\rho}\overline{\left(\upsilon\frac{\partial^2 u'_{\mathrm{i}}}{\partial x_{\mathrm{j}}\partial x_{\mathrm{l}}}\right)^2}\end{aligned} \tag{5.76}$$

Similarly to the approach used for Eq. (5.74), the first term on the right-hand side of Eq. (5.76) may be expressed as

$$\frac{\partial}{\partial x_{\mathrm{j}}}\left[\overline{\mu}\frac{\partial \epsilon}{\partial x_{\mathrm{j}}} - \overline{\rho u'_{\mathrm{j}}\epsilon'} - 2\overline{\upsilon}\overline{\frac{\partial p'}{\partial x_{\mathrm{l}}}\frac{\partial u'_{\mathrm{j}}}{\partial x_{\mathrm{l}}}}\right] = \frac{\partial}{\partial x_{\mathrm{j}}}\left(\frac{\mu_{\mathrm{eff}}}{\sigma_{\epsilon}}\frac{\partial \epsilon}{\partial x_{\mathrm{j}}}\right) \tag{5.77}$$

where σ_{ϵ} is an empirical constant. The next three terms on the right-hand side of Eq. (5.76) may be simplified to the form

$$\begin{aligned}& -2\overline{\mu}\left[\overline{\frac{\partial u'_{\mathrm{i}}}{\partial x_{\mathrm{l}}}\frac{\partial u'_{\mathrm{j}}}{\partial x_{\mathrm{l}}}} + \overline{\frac{\partial u'_{\mathrm{l}}}{\partial x_{\mathrm{i}}}\frac{\partial u'_{\mathrm{l}}}{\partial x_{\mathrm{j}}}}\right]\frac{\partial \overline{U}_{\mathrm{i}}}{\partial x_{\mathrm{j}}} - 2\overline{\mu}\overline{\frac{\partial u'_{\mathrm{i}}}{\partial x_{\mathrm{l}}}\frac{\partial u'_{\mathrm{j}}}{\partial x_{\mathrm{l}}}\frac{\partial u'_{\mathrm{i}}}{\partial x_{\mathrm{j}}}} - 2\overline{\mu}\overline{u'_{\mathrm{j}}\frac{\partial u'_{\mathrm{i}}}{\partial x_{\mathrm{l}}}}\frac{\partial^2 \overline{U}_{\mathrm{i}}}{\partial x_{\mathrm{j}}\partial x_{\mathrm{l}}} \\ & \approx -\mu_{\mathrm{T}}\frac{\overline{u'_{\mathrm{i}}u'_{\mathrm{j}}}}{[\mathrm{L}]^2}\frac{\partial \overline{U}_{\mathrm{i}}}{\partial x_{\mathrm{j}}} = -C_1\overline{\rho}\frac{\epsilon}{k}\overline{u'_{\mathrm{i}}u'_{\mathrm{j}}}\frac{\partial \overline{U}_{\mathrm{i}}}{\partial x_{\mathrm{j}}} = C_1\frac{\epsilon}{k}(\mu_{\mathrm{eff}} - \overline{\mu})\left(\frac{\partial \overline{U}_{\mathrm{i}}}{\partial x_{\mathrm{j}}} + \frac{\partial \overline{U}_{\mathrm{j}}}{\partial x_{\mathrm{i}}}\right)\frac{\partial \overline{U}_{\mathrm{i}}}{\partial x_{\mathrm{j}}}\end{aligned} \tag{5.78}$$

where C_1 is an empirical constant. The last term on the right-hand side of Eq. (5.76) may be related to k and ϵ by

$$-2\overline{\rho}\overline{\left(\upsilon\frac{\partial^2 u'_{\mathrm{i}}}{\partial x_{\mathrm{j}}\partial x_{\mathrm{l}}}\right)^2} \approx -\mu_{\mathrm{T}}\frac{\epsilon}{[\mathrm{L}]^2} = -C_2\overline{\rho}\frac{\epsilon^2}{k} \tag{5.79}$$

where C_2 is an empirical constant. The final form of Eq. (5.76) thus becomes

$$\frac{\partial}{\partial x_{\mathrm{j}}}(\overline{\rho}\epsilon\overline{U}_{\mathrm{j}}) = \frac{\partial}{\partial x_{\mathrm{j}}}\left(\frac{\mu_{\mathrm{eff}}}{\sigma_{\epsilon}}\frac{\partial \epsilon}{\partial x_{\mathrm{j}}}\right) + \frac{\epsilon}{k}\left[C_1(\mu_{\mathrm{eff}} - \overline{\mu})\left(\frac{\partial \overline{U}_{\mathrm{i}}}{\partial x_{\mathrm{j}}} + \frac{\partial \overline{U}_{\mathrm{j}}}{\partial x_{\mathrm{i}}}\right)\frac{\partial \overline{U}_{\mathrm{i}}}{\partial x_{\mathrm{j}}} - C_2\overline{\rho}\epsilon\right] \tag{5.80}$$

Equation (5.80) is called the ϵ-equation.

For the five empirical constants in the k–ϵ model, it is suggested by Launder and Spalding (1972) that $C_{\mu} = 0.09$; $C_1 = 1.44$; $C_2 = 1.92$; $\sigma_{\mathrm{k}} = 1.0$; and $\sigma_{\epsilon} = 1.22$.

5.2.4.4 *Closure Equations*

In order for a model to be closured, the total number of independent equations has to match the total number of independent variables. For a single-phase flow, the typical independent equations include the continuity equation, momentum equation, energy equation, equation of state for compressible flow, equations for turbulence characteristics in turbulent flows, and relations for laminar transport coefficients (*e.g.*, $\mu = f(T)$). The typical independent variables may include density, pressure, velocity, temperature, turbulence characteristics, and some laminar transport coefficients. Since the velocity of gas is a vector, the number of independent variables associated with the velocity depends on the number of components of the velocity in question. Similar consideration is also applied to the momentum equation, which is normally written in a vectorial form.

As an example, for steady, incompressible, and isothermal turbulent flows using the k–ϵ model, the independent equations are (1) the continuity equation, Eq. (5.61); (2) the momentum equation, Eq. (5.65); (3) the definition of the effective viscosity, μ_{eff} (combination of Eq. (5.64) and Eq. (5.72)); (4) the equation of turbulent kinetic energy, Eq. (5.75); and (5) the equation for the dissipation rate of turbulent kinetic energy, Eq. (5.80). Thus, for a three-dimensional model, the total number of independent equations is seven. The corresponding independent variables are (1) velocity (three components); (2) pressure; (3) effective viscosity; (4) turbulent kinetic energy; and (5) dissipation rate of turbulent kinetic energy. Thus, the total number of independent variables is also seven, and the model becomes solvable.

5.2.5 Boundary Conditions

The solutions of the governing equations are subjected to the boundary conditions specified. The number of boundary conditions required depends on the types of the governing equations, *e.g.*, elliptical, parabolic, or mixed. Because of the complexity of the nonlinear partial differential equations involved, there is no general form of boundary condition that can guarantee the solvability of the equations. For single-phase turbulent flows, the typical boundaries may include impermeable solid wall (the wall may be movable), free surface, axis of symmetry, and inlet and outlet conditions. The common independent variables whose boundary conditions are needed to obtain the solutions include velocity, temperature, turbulent kinetic energy, and dissipation rate of the turbulent kinetic energy (assuming the k–ϵ model is used for turbulence simulation). Since usually only the relative pressure field is of interest and it can be obtained by solving the momentum and continuity equations, the discussion of the boundary conditions concerning pressure is omitted here.

5.2.5.1 *Impermeable Solid Wall*

Assume that a no-slip condition of velocity at the solid wall boundary is valid. We have

$$\boldsymbol{U} = 0 \tag{5.81}$$

The boundary condition of temperature may be in one of the three following forms:
(1) Given temperature distribution

$$T = T_{\mathrm{w}}(\boldsymbol{r}_0) \tag{5.82}$$

where $\boldsymbol{r}_0$ is the position vector on the wall.

(2) Given heat flux distribution

$$-K\frac{\partial T}{\partial n} = q_{\mathrm{w}}(\boldsymbol{r}_0) \tag{5.83}$$

where n is a coordinate normal to the boundary.

(3) General balance of heat fluxes

$$-K\frac{\partial T}{\partial n} = h(T_{\infty} - T) + \sigma\epsilon\left(T_{\infty}^4 - T^4\right) \tag{5.84}$$

where ϵ is the wall emissivity.

For turbulent flows using the k–ϵ model, both k and ϵ near the wall must be determined. The most straightforward method is to apply experimentally established boundary conditions, such as the logarithmic law of the wall, at some point near the wall [Launder and Spalding, 1972]. For the near-wall region, the convection and diffusion of the turbulent kinetic energy are negligible. In most turbulent flows, the turbulent properties in the near-wall region are functions only of the normal coordinate. Thus, Eqs. (5.72) and (5.75) yield

$$k = \frac{1}{\rho\sqrt{C_{\mu}}}\left(-\mu_{\mathrm{T}}\frac{\mathrm{d}U}{\mathrm{d}n}\right)_{\mathrm{w}} = \frac{\tau_{\mathrm{w}}}{\rho\sqrt{C_{\mu}}} \tag{5.85}$$

where τ_{w} is the turbulent shear stress near the wall.

For the rate of turbulent energy dissipation near the wall, from Eq. (5.72) we have

$$\epsilon = C_{\mu}\rho\frac{k^2}{\mu_{\mathrm{T}}} = C_{\mu}\rho\frac{k^2}{\tau_{\mathrm{w}}}\left(\frac{\partial U}{\partial y}\right)_{\mathrm{w}} \tag{5.86}$$

In order to give an approximation of the velocity gradient near the wall, an empirical equation, known as the wall function, is introduced. The wall function is expressed as

$$U\sqrt{\frac{\rho}{\tau_{\mathrm{w}}}} = \frac{1}{\kappa}\ln\left(Ey\frac{\sqrt{\tau_{\mathrm{w}}\rho}}{\mu}\right) \tag{5.87}$$

where E and κ are empirical constants (κ is known as von Karman's constant). Thus, from Eq. (5.85) and Eq. (5.87), we obtain the boundary condition of ϵ near the solid wall, *i.e.*, $y = \delta_{\mathrm{w}}$, as

$$\epsilon = \frac{C_{\mu}^{\frac{3}{4}}k^{\frac{3}{2}}}{\kappa\,\delta_{\mathrm{w}}} \tag{5.88}$$

where δ_{w} is the normal distance of the point from the wall.

It is noted that turbulence may also affect the boundary conditions of velocity and heat flux near the wall. A detailed discussion using the k–ϵ model for these boundary conditions is given by Launder and Spalding (1974).

5.2.5.2 *Free Surface*

For the velocity at a free surface, we have

$$\frac{\partial U}{\partial n} = 0; \qquad U_{\mathrm{n}} = 0 \tag{5.89}$$

where U_{n} is the normal component of the velocity at the surface and n denotes the coordinate normal to the free surface.

For temperature, the same form of Eq. (5.84) may be applied. For both k and ϵ, it is assumed that

$$\frac{\partial k}{\partial n} = 0 \tag{5.90}$$

and

$$\frac{\partial \epsilon}{\partial n} = 0 \tag{5.91}$$

5.2.5.3 *Axis of Symmetry*

On an axis of symmetry, we have the symmetric condition

$$\frac{\partial \phi}{\partial r} = 0 \tag{5.92}$$

where ϕ can be $\boldsymbol{U}$, T, k, or ϵ, and r is the radial coordinate. The radial component of the velocity is equal to zero.

5.2.5.4 *Inlet Conditions*

The inlet velocity profile and temperature profile are given as

$$\boldsymbol{U} = \boldsymbol{U}_{\mathrm{i}} \tag{5.93}$$

$$T = T_{\mathrm{i}} \tag{5.94}$$

The inlet profiles of k and ϵ are generally assumed.

5.2.5.5 *Outlet Conditions*

Several types of boundary conditions at the outlet can be specified.

A. FULLY DEVELOPED FLOW

If the flow becomes fully developed at the outlet, the partial derivatives for $\boldsymbol{U}$, T, k, and ϵ along the outlet direction are zero at the outlet.

B. LARGE PECLET NUMBER NEAR THE OUTLET

The Peclet number is defined as the product of the Reynolds number with the Prandtl number. For a large Peclet number near the outlet, there is no upstream influence because the flow is dominated by the downstream convection. In this case, no boundary condition information is needed for the outflow boundary [Patankar, 1980].

C. FREE STREAM APPROXIMATION

The boundary conditions for k and ϵ at the outlet may be represented by the limiting cases of the k and ϵ equations in a free stream [Launder and Spalding, 1972]. In a free stream, all terms containing derivatives with respect to the coordinates perpendicular to the streamline are zero. Therefore, from the k-equation, we have

$$U_{\mathrm{i}} \frac{\partial k}{\partial x} = -C_{\mu} \rho \frac{k^2}{\mu_{\mathrm{Ti}}} \tag{5.95}$$

where μ_{Ti} is the turbulent viscosity at the outlet, and x is the coordinate along the streamline. Similarly, for ϵ at the outlet, we have

$$U_{\mathrm{i}} \frac{\partial \epsilon}{\partial x} = -C_{2} \rho \frac{k\epsilon}{\mu_{\mathrm{Ti}}} \tag{5.96}$$

5.3 Continuum Modeling of Multiphase Flows

Since a multiphase flow usually takes place in a confined volume, the desire to have a mathematical description based on a fixed domain renders the Eulerian method an ideal one to describe the flow field. The Eulerian approach requires that the transport quantities of all phases be continuous throughout the computational domain. As mentioned before, in reality, each phase is time-dependent and may be discretely distributed. Hence, averaging theorems need to be applied to construct a continuum for each phase so that the existing Eulerian description of a single-phase flow may be extended to a multiphase flow.

5.3.1 *Averages and Averaging Theorems*

To construct a continuum for each phase, volume averaging and time averaging are essential to the analysis of a multiphase flow where each phase has discrete spatial and temporal distributions. Although the continuum for each phase may be constructed either by volume averaging [Slattery, 1967a, 1967b; Whitaker, 1969; Delhaye and Archard, 1976] in terms of the volume fraction or by time averaging in terms of the fractional residence time [Ishii, 1975], the dynamic and thermodynamic properties of a mixture are cumulative with the volume fraction, but not with the fractional residence time [Soo, 1989]. Moreover, a priori time averaging based on the fractional residence time may eliminate the identity of different dynamic phases because of the different corresponding velocities. Hence, volume averaging is preferred for constructing a continuum for each phase while time averaging may be carried out after volume averaging to account for the high-frequency fluctuations [Soo, 1989].

5.3.1.1 *Phase Average and Intrinsic Average*

Using the subscript k to denote the transport quantities of phase k, for a control volume V as shown in Fig. 5.1, the volume average of any quantity of phase k can be defined as

$$\langle \psi_k \rangle = \frac{1}{V} \int_{V_k} \psi_k \, dV \tag{5.97}$$

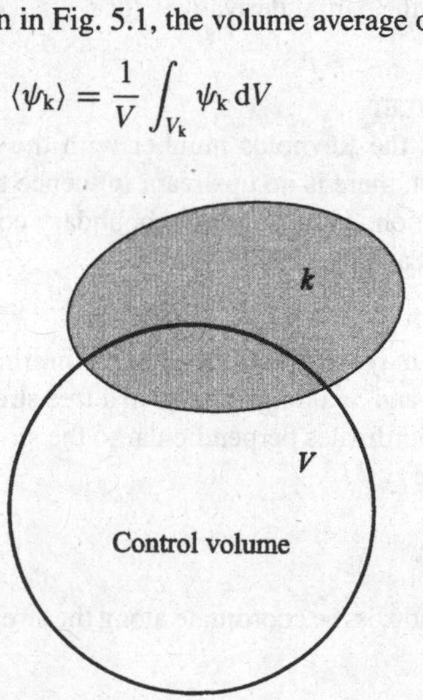

Figure 5.1. Concept of volume average.

where ψ_k is a quantity of phase k (ψ_k can be a scalar, vector, or tensor) and V_k is the volume occupied by phase k inside the control volume V. $\langle\psi_k\rangle$ is termed the phase average of ψ_k.

Besides the phase average, it is also important to introduce the intrinsic average, which is defined by

$$ {}^i\langle\psi_k\rangle = \frac{1}{V_k}\int_{V_k} \psi_k \, dV \tag{5.98}$$

Hence, the phase average can be related to the intrinsic average by the volume fraction of phase k, α_k, as

$$ \langle\psi_k\rangle = \frac{V_k}{V}\, {}^i\langle\psi_k\rangle = \alpha_k\, {}^i\langle\psi_k\rangle \tag{5.99}$$

Usually, the intrinsic average reflects the real physical property or quantity such as density and velocity, while the phase average gives a pseudoproperty or quantity based on the selection of control volume. Phase averages are used to construct the continuum of each phase to which Eulerian description can be applied.

Note that volume averaging can only be applied to quantities per unit volume, such as density, momentum per unit volume, and energy per unit volume. The intrinsic average of the velocity of phase k may thus be defined as

$$ {}^i\langle \boldsymbol{U}_k\rangle = \left[\frac{1}{V_k}\int_{V_k} \rho_k \boldsymbol{U}_k \, dV\right]\frac{1}{{}^i\langle\rho_k\rangle} = \frac{\langle\rho_k \boldsymbol{U}_k\rangle}{\langle\rho_k\rangle} \tag{5.100}$$

and the intrinsic averaged internal energy of phase k is

$$ {}^i\langle e_k\rangle = \frac{1}{{}^i\langle\rho_k\rangle}\left[\frac{1}{V_k}\int_{V_k} \rho_k e_k \, dV\right] = \frac{\langle\rho_k e_k\rangle}{\langle\rho_k\rangle} \tag{5.101}$$

5.3.1.2 *Minimum Control Volume of Phase Average*

As mentioned, the pseudocontinuum approach can be used for the multiphase flow analysis only when the control volume contains enough particles that the volume average is of statistical significance, and at the same time, the control volume is small enough to ensure the validity of the continuum assumption. Thus, in order to have a statistically meaningful phase average, there exists a minimum averaging volume.

The minimum averaging volume of a statistically meaningful phase average can be obtained from the study of the effects of control volumes on the computed volume fractions of a particulate phase [Celmiņš, 1988]. Consider a three-dimensional hexagonal model as an example to illustrate the method of analysis. The spherical particles are arranged with a regular repeating pattern, as shown in Fig. 5.2. For convenience, a series of cocentered spherical control volumes are used for volume averaging. In principle, the center of these spherical control volumes can be at any arbitrary location inside a typical three-dimensional hexagonal cell. Two special locations of the center, however, can adequately represent the limiting cases of the volume average, *i.e.*, one in the center of a sphere and the other in the symmetric center of four neighboring spheres whose centers form a pyramid, as shown in Fig. 5.2.

The computed volume fraction of particles, α_p^c, for a given control volume in general includes the volumes of fully and partially enclosed solid spheres inside the enclosure, as indicated by the shaded area in Fig. 5.3. With increasing control volumes, the volume fractions of solids obtained from different centered control volumes are close to the same

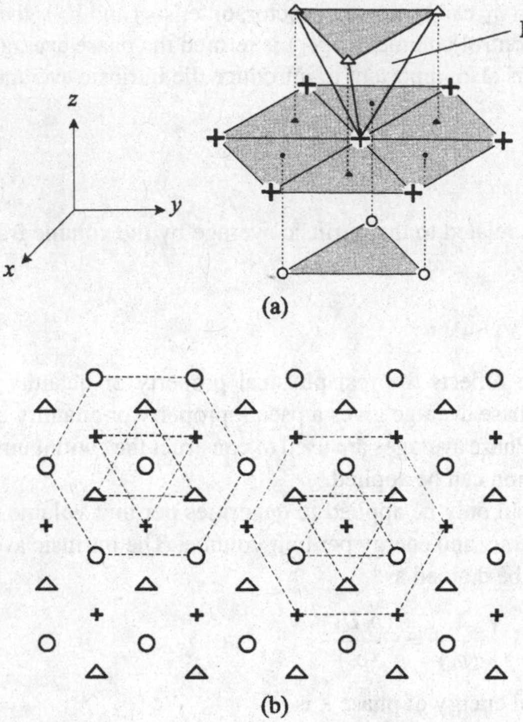

Figure 5.2. Three-dimensional hexagonal structure of spherical particles: (a) Three-dimensional view; (b) Top view.

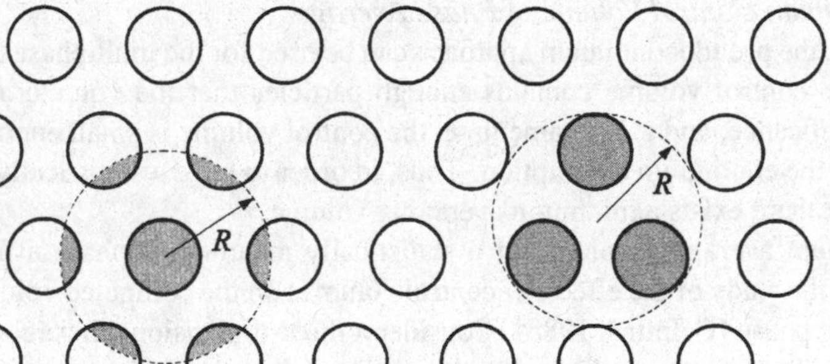

Figure 5.3. Top view of volume averaging over a hexagonal arrangement of spheres: (a) Cocentrated with a sphere; (b) Centered in a pyramid of spheres.

asymptotic value α_p, as shown in Fig. 5.4(a). This is the true ambient volume fraction of solids. Consequently, given an error margin of relative deviation of the calculated volume fraction of particles, δ (defined as $|\alpha_p^c - \alpha_p|/\alpha_p$), the minimum radius of averaging volume for statistically valid phase volume averages can be defined, as illustrated in Fig. 5.4(b). It can be realized that in the previous computation, the volume fraction of solids for a given type of particle arrangement depends on two geometric characteristic lengths, *e.g.*, diameter

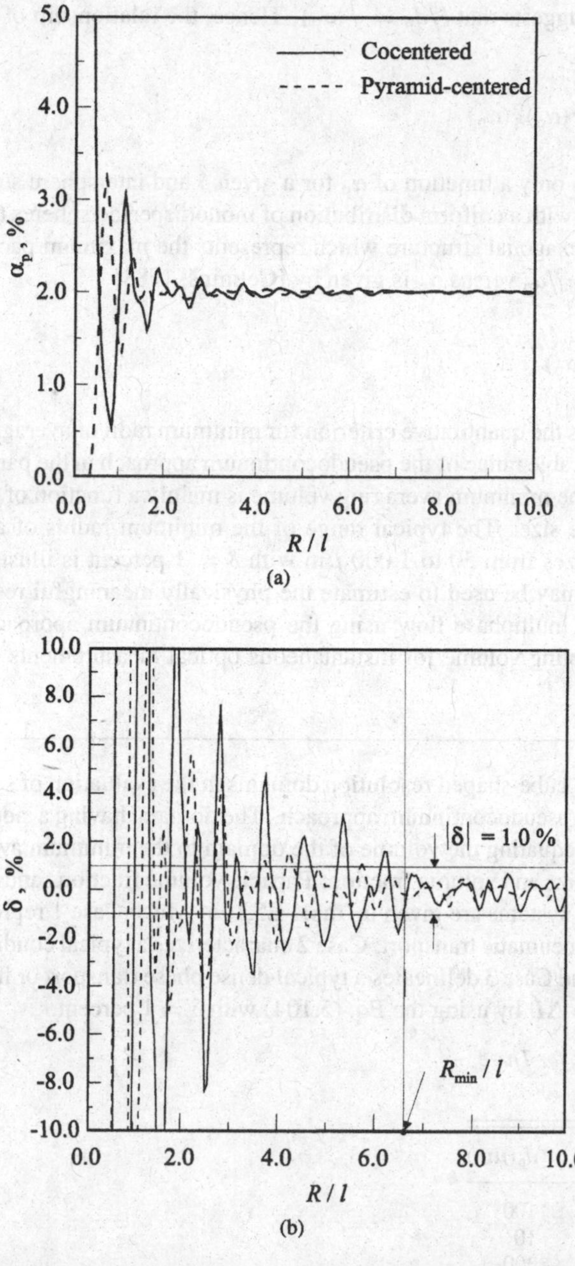

Figure 5.4. Determination of minimum radius of control volume for two different control volume centers: (a) Effect of control volume on calculated particle volume fraction; (b) Effect of control volume on relative deviation of calculated particle volume fraction.

of a solid sphere d_p and center-to-center distance of two neighboring particles l. The average interparticle distance can be further estimated from the particle number density, n, by

$$l \approx n^{-\frac{1}{3}} = \left(\frac{\pi}{6} \frac{d_p^3}{\alpha_p} \right)^{\frac{1}{3}} \tag{5.102}$$

Moreover, this expression suggests that $l/d_p = f(\alpha_p)$. Hence, the relationship of $R/l = g(\alpha_p)$ can be written as

$$\frac{R}{d_p} = \frac{l}{d_p} g(\alpha_p) = f(\alpha_p) g(\alpha_p) \tag{5.103}$$

which means that R_{min}/d_p is only a function of α_p for a given δ and intersphere structure. Thus, for a particulate phase with a uniform distribution of monodispersed spheres (including the three-dimensional hexagonal structure which represents the maximum packing of monodispersed spheres), R_{min}/d_p versus α_p is given by [Celmiņš, 1988]

$$\frac{R_{min}}{d_p} = \frac{\alpha_p^{-\frac{1}{3}}}{\sqrt{2\delta}}(1 - \alpha_p) \tag{5.104}$$

Equation (5.104) provides the quantitative criterion for minimum radii of averaging volumes to demarcate the applicable range of the pseudocontinuum approach in the particulate multiphase flow analysis. The minimum averaging volume is mainly a function of particle volume fraction and particle size. The typical range of the minimum radius of an averaging volume for particle sizes from 50 to 1,000 μm with $\delta = 1$ percent is illustrated in Fig. 5.5. Equation (5.104) may be used to estimate the physically meaningful resolution for phase distributions of a multiphase flow using the pseudocontinuum approach or to estimate the minimum sampling volume for instantaneous optical measurements of local particle concentrations.

Example 5.1 Consider the cube-shaped resolution domains in the evaluation of solutions for gas–solid flows using the pseudocontinuum approach. The domain, having a side length of ΔL, can be estimated by equating the volume of the domain to the minimum averaging volume for a given particle size and volume fraction. Particle volume fractions and particle diameters in three gas–solid systems are given in Table E5.1, in which Case 1 represents a typical condition for dilute pneumatic transport; Case 2 characterizes a typical condition for dispersion of dusty cloud; and Case 3 delineates a typical dense phase transport or fluidized bed. For each case, estimate ΔL by using the Eq. (5.104) with $\delta = 1$ percent.

Table E5.1. α_p *and* d_p *for Three Typical Cases*

Case	α_p(%)	d_p(μm)
1	1	100
2	0.1	10
3	40	300

Solution By equating the volume of the resolution domain to the minimum averaging volume for a given particle size and volume fraction, ΔL is related to R_{min} by

$$\Delta L = \left(\frac{4\pi}{3}\right)^{\frac{1}{3}} R_{min} = 1.6 R_{min} \tag{E5.1}$$

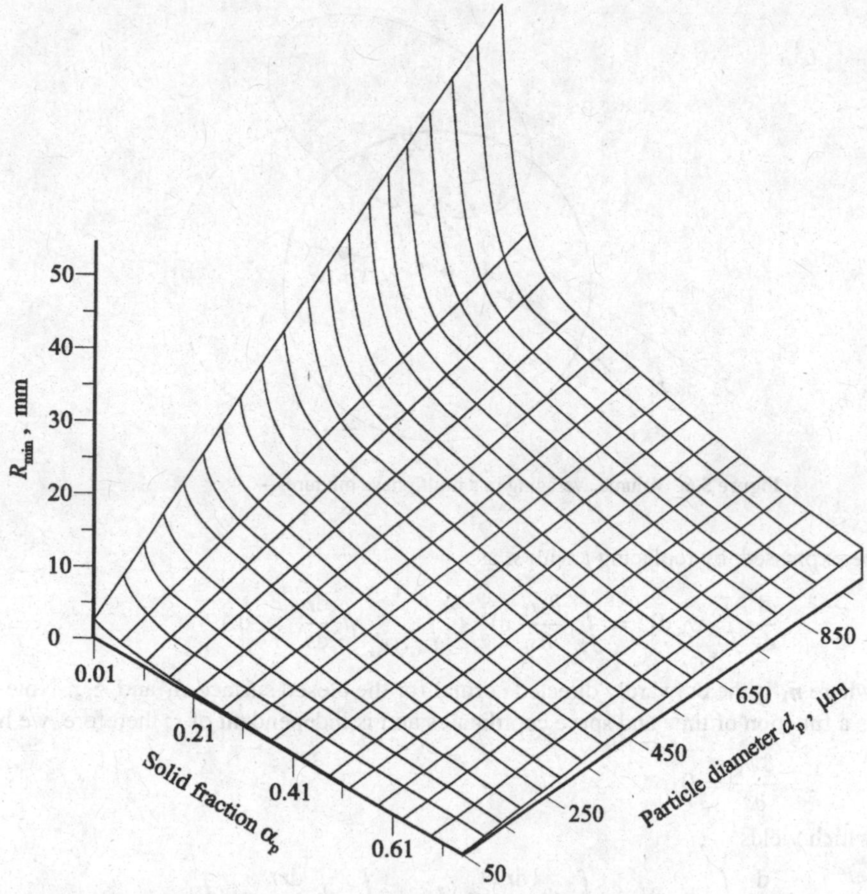

Figure 5.5. Typical range of minimum radius of control volume.

If R_{min} is evaluated from Eq. (5.104), the side lengths of the resolution domain for the numerical simulation of the cases given in Table E5.1 are calculated to be 5.2 mm, 1.1 mm, and 1.7 mm, respectively.

5.3.1.3 *Volume-Averaging Theorems*

Volume-averaging theorems are derived to express the volume averages over integrals or derivatives in terms of the integrals or derivatives of the volume averages. For simplicity, let us consider a point on an arbitrary continuous curve s in a multiphase mixture, as shown in Fig. 5.6. With each point on the curve as its center, we can have a sphere over which the volume averaging may be applied. The portion occupied by phase k is denoted as V_{k}. V_{k} is bounded by an interfacial area A_{k} and a surface area A_{ke}, which is on the sphere centered at the point s and is interfaced by phase k. The closed surface A_{c} $(=A_{\mathrm{k}} + A_{\mathrm{ke}})$ is considered to be moving by translation (without rotation) along this arbitrary curve. From the general transport theorem, *i.e.*, Eq. (5.8), the change of the integral over V_{k} as a function of s may

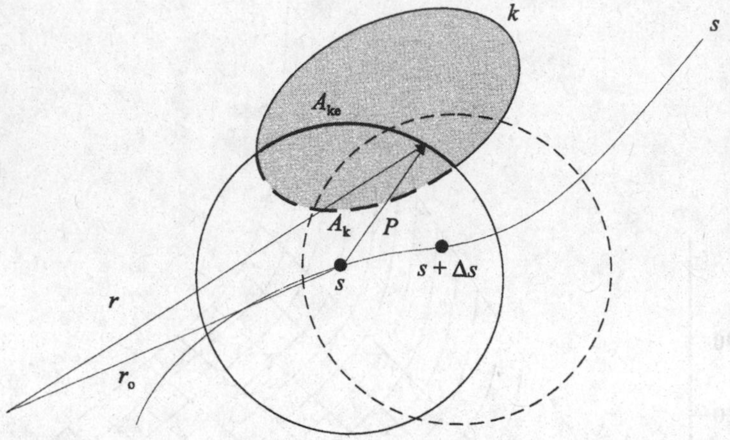

Figure 5.6. Volume averaging in a multiphase mixture.

be expressed, by replacing t with s, as

$$\frac{d}{ds}\int_{V_k}\psi_k\,dV = \int_{V_k}\frac{\partial\psi_k}{\partial s}\,dV + \int_{A_k\cup A_{ke}}\psi_k\frac{d\boldsymbol{r}}{ds}\cdot\boldsymbol{n}_k\,dA \tag{5.105}$$

where $\boldsymbol{n}_k$ is the outwardly directed normal for the closed surface A_k and A_{ke}. Note that ψ_k is a function of time and space coordinates and is independent of s; therefore, we have

$$\frac{\partial\psi_k}{\partial s} = 0 \tag{5.106}$$

which yields

$$\frac{d}{ds}\int_{V_k}\psi_k\,dV = \int_{A_k}\psi_k\frac{d\boldsymbol{r}}{ds}\cdot\boldsymbol{n}_k\,dA + \int_{A_{ke}}\psi_k\frac{d\boldsymbol{r}}{ds}\cdot\boldsymbol{n}_{ke}\,dA \tag{5.107}$$

The interfacial area A_k is assumed to be smooth and continuous when the center of the integral sphere shifts from s to $s + ds$ so that $d\boldsymbol{r}/ds$ is a tangential vector on $A_k(s)$. Thus, the surface integral over A_k in Eq. (5.107) becomes

$$\frac{d\boldsymbol{r}}{ds}\cdot\boldsymbol{n}_k = 0 \tag{5.108}$$

For the surface integral over A_{ke} in Eq. (5.107), it is noted that A_{ke} is not continuous when the center of the sphere shifts from s to $s + ds$; in other words, $A_{ke}(s + ds)$ can be a completely different surface from $A_{ke}(s)$. Hence, $d\boldsymbol{r}/ds$ is not a tangential vector on $A_{ke}(s)$ or $d\boldsymbol{r}/ds\cdot\boldsymbol{n}_{ke}$ is not zero. However, if $\boldsymbol{r}(s)$ is divided into $\boldsymbol{r}_0(s)$ and $\boldsymbol{P}(s)$, where $\boldsymbol{r}_0$ is the position vector of the center of the sphere and $\boldsymbol{P}$ is a position vector relative to the center s, and since the closed surface is translated, we have

$$\frac{d\boldsymbol{P}}{ds} = 0 \tag{5.109}$$

So Eq. (5.107) can be written in the form

$$\frac{d}{ds}\int_{V_k}\psi_k\,dV = \int_{A_{ke}}\psi_k\frac{d\boldsymbol{r}_0}{ds}\cdot\boldsymbol{n}_{ke}\,dA \tag{5.110}$$

The derivative d/ds is expressed in terms of $\boldsymbol{r}_0(s)$ as [Whitaker, 1969]

$$\frac{\mathrm{d}}{\mathrm{d}s} = \frac{\mathrm{d}x_i}{\mathrm{d}s}\frac{\partial}{\partial x_i} = \frac{\mathrm{d}\boldsymbol{r}}{\mathrm{d}s}\cdot\nabla = \frac{\mathrm{d}\boldsymbol{r}_0}{\mathrm{d}s}\cdot\nabla \tag{5.111}$$

Equation (5.110) finally takes the form

$$\frac{\mathrm{d}\boldsymbol{r}_0}{\mathrm{d}s}\cdot\left(\nabla\int_{V_k}\psi_k\,\mathrm{d}V - \int_{A_{ke}}\psi_k\boldsymbol{n}_{ke}\,\mathrm{d}A\right) = 0 \tag{5.112}$$

Since $\boldsymbol{r}_0$ is arbitrary, the preceding equation can be reduced to

$$\nabla\int_{V_k}\psi_k\,\mathrm{d}V = \int_{A_{ke}}\psi_k\boldsymbol{n}_{ke}\,\mathrm{d}A \tag{5.113}$$

From the Gauss theorem, we have

$$\int_{A_{ke}}\psi_k\boldsymbol{n}_{ke}\,\mathrm{d}A = \int_{V_k}\nabla\psi_k\,\mathrm{d}V - \int_{A_k}\psi_k\boldsymbol{n}_k\,\mathrm{d}A \tag{5.114}$$

Combining Eq. (5.114) with Eq. (5.113), we have an averaging theorem as

$$\int_{V_k}\nabla\psi_k\,\mathrm{d}V = \nabla\int_{V_k}\psi_k\,\mathrm{d}V + \int_{A_k}\psi_k\boldsymbol{n}_k\,\mathrm{d}A \tag{5.115}$$

or

$$\langle\nabla\psi_k\rangle = \nabla\langle\psi_k\rangle + \frac{1}{V}\int_{A_k}\psi_k\boldsymbol{n}_k\,\mathrm{d}A \tag{5.116}$$

Similarly, it can be proved that

$$\langle\nabla\cdot\boldsymbol{\psi}_k\rangle = \nabla\cdot\langle\boldsymbol{\psi}_k\rangle + \frac{1}{V}\int_{A_k}\boldsymbol{\psi}_k\cdot\boldsymbol{n}_k\,\mathrm{d}A \tag{5.117}$$

The averaging theorem for the time derivative can be derived directly from the general transport theorem. Consider the phase k in Fig. 5.7 to be fixed in space while A_k varies as a function of time as a result of phase change. Applying Eq. (5.8) to the system in Fig. 5.7, with $\mathrm{d}/\mathrm{d}t = \partial/\partial t$, we have

$$\frac{\partial}{\partial t}\int_{V_k}\psi_k\,\mathrm{d}V = \int_{V_k}\frac{\partial\psi_k}{\partial t}\,\mathrm{d}V + \int_{A_k}\psi_k\boldsymbol{U}_s\cdot\boldsymbol{n}_k\,\mathrm{d}A \tag{5.118}$$

where $\boldsymbol{U}_s$ is the speed of displacement of the interface. Let the control volume of the mixture be constant and larger than V_k. From the definition of the phase average, we obtain

$$\left\langle\frac{\partial}{\partial t}\psi_k\right\rangle = \frac{\partial}{\partial t}\langle\psi_k\rangle - \frac{1}{V}\int_{A_k}\psi_k\boldsymbol{U}_s\cdot\boldsymbol{n}_k\,\mathrm{d}A \tag{5.119}$$

Equations (5.116), (5.117), and (5.119) characterize the volume-averaging theorems of derivatives [Slattery, 1967b; Whitaker, 1969].

5.3.2 *Volume-Averaged Equations*

For a dispersed multiphase flow, each phase may be regarded as a single phase flowing over other phases. Thus, the volume-averaged equations for phase k in a dispersed multiphase flow system can be obtained by applying the averaging theorem to the governing equations of single-phase flows.

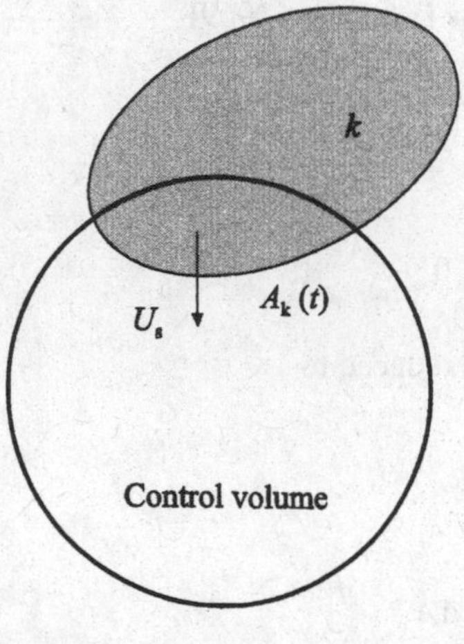

Figure 5.7. Expansion of phase k in a fixed control volume.

The volume-averaged continuity equation for dispersed laminar multiphase flows is given by applying the volume-averaging theorems to Eq. (5.13) as

$$\frac{\partial}{\partial t}\langle\rho_k\rangle + \nabla\cdot\langle\rho_k \boldsymbol{U}_k\rangle = \Gamma_k \tag{5.120}$$

where Γ_k is given by

$$\Gamma_k = -\frac{1}{V}\int_{A_k}\rho_k(\boldsymbol{U}_k - \boldsymbol{U}_s)\cdot\boldsymbol{n}_k\,\mathrm{d}A \tag{5.121}$$

Γ_k represents the rate of mass generation of phase k per unit volume. It is noted that the unit "volume" in this context is that of the gas–solid system. If a phase is defined in a physical sense such as the solid phase or the gas phase, Γ_k may be caused by chemical reactions or phase changes. On the other hand, when a phase is defined in a dynamic sense [Soo, 1965], Γ_k may result from the size change due to attrition or agglomeration in addition to the chemical reaction or phase change. From the mass balance of the mixture, we have

$$\sum_k \Gamma_k = 0 \tag{5.122}$$

Applying the averaging theorems to Eq. (5.14), the volume-averaged momentum equation of phase k is obtained as

$$\frac{\partial}{\partial t}\langle\rho_k\boldsymbol{U}_k\rangle + \nabla\cdot\langle\rho_k\boldsymbol{U}_k\boldsymbol{U}_k\rangle = -\nabla\langle p_k\rangle + \nabla\cdot\langle\boldsymbol{\tau}_k\rangle + \langle\rho_k\boldsymbol{f}\rangle + \boldsymbol{F}_{Ak} + \boldsymbol{F}_{\Gamma k} \tag{5.123}$$

where $\boldsymbol{F}_{Ak}$ is given as

$$\boldsymbol{F}_{Ak} = \frac{1}{V}\int_{A_k}(-p_k\boldsymbol{I} + \boldsymbol{\tau}_k)\cdot\boldsymbol{n}_k\,\mathrm{d}A \tag{5.124}$$

which accounts for the transfer of pressure and viscous stresses across the interface per unit volume. $\boldsymbol{F}_{Ak}$ includes the drag force, Saffman force, and Magnus force of the phase. $\boldsymbol{F}_{\Gamma k}$

in Eq. (5.123) takes the form

$$F_{\Gamma k} = -\frac{1}{V}\int_{A_k} \rho_k U_k (U_k - U_s) \cdot n_k \, dA \tag{5.125}$$

which represents the momentum transfer across the interface per volume due to the mass generation of the phase. The field force per unit mass f is taken to be a constant. For gas–solid suspensions, f may include the gravitational force, electromagnetic force, electrostatic force, and other material-related forces such as the van der Waals force.

The volume-averaged equation of energy conservation in terms of internal energy is given by averaging Eq. (5.18) as

$$\frac{\partial \langle \rho_k e_k \rangle}{\partial t} + \nabla \cdot \langle \rho_k U_k e_k \rangle = -\langle p_k \nabla \cdot U_k \rangle - \nabla \cdot \langle J_{qk} \rangle + \langle J_{ek} \rangle + \langle \phi_k \rangle + Q_{Ak} + Q_{\Gamma k} \tag{5.126}$$

where Q_{Ak} accounts for the heat transfer across the interface as

$$Q_{Ak} = -\frac{1}{V}\int_{A_k} J_{qk} \cdot n_k \, dA \tag{5.127}$$

and $Q_{\Gamma k}$ is the internal energy transferred by the mass generation as a result of phase change and is expressed by

$$\langle Q_{\Gamma k} \rangle = -\frac{1}{V}\int_{A_k} \rho_k e_k (U_k - U_s) \cdot n_k \, dA \tag{5.128}$$

The term $\langle \phi_k \rangle$ is the volume-averaged dissipation function for the energy dissipated by the viscous force, which is irreversible dissipation of mechanical work into thermal energy or heat. For the solid–particle phase, the kinetic energy loss by attrition or inelastic collision may be included in this term.

The volume-averaged products in the preceding volume-averaged equations can be further expressed by the products of the volume averages. From Eq. (5.100), the volume-averaged mass flux of phase k is given by

$$\langle \rho_k U_k \rangle = \langle \rho_k \rangle {}^i\langle U_k \rangle = \alpha_k {}^i\langle \rho_k \rangle {}^i\langle U_k \rangle \tag{5.129}$$

Similarly, the momentum flux of phase k can be denoted by

$$\langle \rho_k U_k U_k \rangle = \langle \rho_k \rangle {}^i\langle U_k U_k \rangle = \alpha_k {}^i\langle \rho_k \rangle {}^i\langle U_k U_k \rangle \tag{5.130}$$

Furthermore, the intrinsic average of velocity correlation may be approximated by a product of intrinsic averages of velocities as

$${}^i\langle U_k U_k \rangle \approx {}^i\langle U_k \rangle {}^i\langle U_k \rangle \tag{5.131}$$

From Eqs. (5.99) and (5.101), the internal energy per unit volume can be expressed by

$$\langle \rho_k e_k \rangle = \langle \rho_k \rangle {}^i\langle e_k \rangle = \alpha_k {}^i\langle \rho_k \rangle {}^i\langle e_k \rangle \tag{5.132}$$

and

$${}^i\langle \rho_k e_k \rangle = {}^i\langle \rho_k \rangle {}^i\langle e_k \rangle \tag{5.133}$$

Thus, the internal energy carried by the mass flux of phase k can be expressed by

$$\langle \rho_k \boldsymbol{U}_k e_k \rangle \approx \langle \rho_k e_k \rangle \, {}^i\langle \boldsymbol{U}_k \rangle = \alpha_k \, {}^i\langle \rho_k \rangle \, {}^i\langle e_k \rangle \, {}^i\langle \boldsymbol{U}_k \rangle \tag{5.134}$$

The approximation in the preceding equation is to circumvent any conflict with Eq. (5.100), which defines the intrinsic average of velocity. This approximate relation becomes an equivalent one under the isothermal condition of phase k with the expansion work estimated by

$$\langle p_k \nabla \cdot \boldsymbol{U}_k \rangle \approx \langle p_k \rangle \nabla \cdot {}^i\langle \boldsymbol{U}_k \rangle \tag{5.135}$$

For cases without the phase change between gas and solid phases, Γ_k, $\boldsymbol{F}_{\Gamma k}$, and $Q_{\Gamma k}$ vanish. Thus, assuming constant material densities and using Eq. (5.129), the volume-averaged continuity equation, Eq. (5.120), becomes

$$\frac{\partial \alpha_k}{\partial t} + \nabla \cdot \left(\alpha_k \, {}^i\langle \boldsymbol{U}_k \rangle \right) = 0 \tag{5.136}$$

Similarly, the volume-averaged momentum equation, Eq. (5.123), reduces to

$$\langle \rho_k \rangle \left(\frac{\partial \, {}^i\langle \boldsymbol{U}_k \rangle}{\partial t} + {}^i\langle \boldsymbol{U}_k \rangle \cdot \nabla \, {}^i\langle \boldsymbol{U}_k \rangle \right) = \nabla \cdot \langle \boldsymbol{T}_k \rangle + \langle \rho_k \rangle \boldsymbol{f} + \boldsymbol{F}_{Ak} \tag{5.137}$$

Moreover, the volume-averaged equation of energy conservation in terms of internal energy, given in Eq. (5.126), becomes

$$\langle \rho_k \rangle \left(\frac{\partial \, {}^i\langle e_k \rangle}{\partial t} + {}^i\langle \boldsymbol{U}_k \rangle \cdot \nabla \, {}^i\langle e_k \rangle \right) = -\langle p_k \rangle \nabla \cdot {}^i\langle \boldsymbol{U}_k \rangle - \nabla \cdot \langle \boldsymbol{J}_{qk} \rangle + \langle J_{ek} \rangle + \langle \phi_k \rangle + Q_{Ak} \tag{5.138}$$

For an isothermal gas–solid two-phase flow without the phase change, Eq. (5.136) and Eq. (5.137) can be expressed by

$$\frac{\partial \alpha}{\partial t} + \nabla \cdot (\alpha \, {}^i\langle \boldsymbol{U} \rangle) = 0 \tag{5.139}$$

$$-\frac{\partial \alpha}{\partial t} + \nabla \cdot ((1 - \alpha) \, {}^i\langle \boldsymbol{U}_p \rangle) = 0 \tag{5.140}$$

$$\rho \alpha \left(\frac{\partial \, {}^i\langle \boldsymbol{U} \rangle}{\partial t} + {}^i\langle \boldsymbol{U} \rangle \cdot \nabla \, {}^i\langle \boldsymbol{U} \rangle \right) = \nabla \cdot \langle \boldsymbol{T} \rangle + \rho \alpha \boldsymbol{f} + \boldsymbol{F}_A \tag{5.141}$$

$$\rho_p (1 - \alpha) \left(\frac{\partial \, {}^i\langle \boldsymbol{U}_p \rangle}{\partial t} + {}^i\langle \boldsymbol{U}_p \rangle \cdot \nabla \, {}^i\langle \boldsymbol{U}_p \rangle \right) = \nabla \cdot \langle \boldsymbol{T}_p \rangle + \rho_p (1 - \alpha) \boldsymbol{f} - \boldsymbol{F}_A \tag{5.142}$$

Equations (5.139) to (5.142) are the basic equations for a gas–solid flow. More detailed information on both the fluid–particle interacting force $\boldsymbol{F}_A$ and the total stresses $\boldsymbol{T}$ and $\boldsymbol{T}_p$ must be specified before these equations can be solved. One approach to formulate the fluid-particle interacting force $\boldsymbol{F}_A$ is to decompose the total stress into a component $\boldsymbol{E}$ representing the "macroscopic" variations in the fluid stress tensor on a scale that is large compared to the particle spacing, and a component ϵ representing the effect of detailed variations of the point stress tensor as the fluid flows around the particle [Anderson and

Jackson, 1967]. Under this condition, we may express the total tensor $\boldsymbol{T}$ of the fluid phase by

$$\boldsymbol{T} = \boldsymbol{E} + \boldsymbol{\epsilon} \tag{5.143}$$

and

$$^{i}\langle \boldsymbol{T} \rangle = {}^{i}\langle \boldsymbol{E} \rangle \tag{5.144}$$

The interacting force imposed by the particles on the fluid phase, $\boldsymbol{F}_A$, is given by

$$\begin{aligned} \boldsymbol{F}_A &= \frac{1}{V}\int_{A_p} \boldsymbol{T}\cdot\boldsymbol{n}\,\mathrm{d}A = -\frac{1}{V}\int_{A_p} \boldsymbol{T}\cdot\boldsymbol{n}_p\,\mathrm{d}A = -\frac{1}{V}\int_{A_p} (\boldsymbol{E}+\boldsymbol{\epsilon})\cdot\boldsymbol{n}_p\,\mathrm{d}A \\ &= -\frac{1}{V}\int_{V_p} \nabla\cdot\boldsymbol{E}\,\mathrm{d}V - \boldsymbol{F}_p = -(1-\alpha)\nabla\cdot{}^{i}\langle\boldsymbol{E}\rangle - \boldsymbol{F}_p \end{aligned} \tag{5.145}$$

where $\boldsymbol{F}_p$ is defined by

$$\boldsymbol{F}_p = \frac{1}{V}\int_{A_p} \boldsymbol{\epsilon}\cdot\boldsymbol{n}_p\,\mathrm{d}A \tag{5.146}$$

Thus, Eq. (5.141) and Eq. (5.142), respectively, become

$$\rho\alpha\left[\frac{\partial\,{}^{i}\langle\boldsymbol{U}\rangle}{\partial t} + {}^{i}\langle\boldsymbol{U}\rangle\cdot\nabla\,{}^{i}\langle\boldsymbol{U}\rangle\right] = \alpha\nabla\cdot{}^{i}\langle\boldsymbol{E}\rangle + \alpha\rho\boldsymbol{f} - \boldsymbol{F}_p \tag{5.147}$$

$$\rho_p(1-\alpha)\left(\frac{\partial\,{}^{i}\langle\boldsymbol{U}_p\rangle}{\partial t} + {}^{i}\langle\boldsymbol{U}_p\rangle\cdot\nabla\,{}^{i}\langle\boldsymbol{U}_p\rangle\right) = \nabla\cdot{}^{i}\langle\boldsymbol{T}_p\rangle + (1-\alpha)\nabla\cdot{}^{i}\langle\boldsymbol{E}\rangle + (1-\alpha)\rho_p\boldsymbol{f} + \boldsymbol{F}_p \tag{5.148}$$

In the preceding equations, $\boldsymbol{F}_p$ can be expressed as a combination of local averaged drag force and virtual mass force [Anderson and Jackson, 1967].

5.3.3 *Volume–Time-Averaged Equations*

The purpose of the time averaging after the volume averaging is to express averages of products in terms of products of averages and to account for turbulent fluctuations and high-frequency fluctuations [Soo, 1989]. The volume–time averaging is presented here in a similar way to that of the Reynolds analysis of single-phase turbulent flow.

The averaging time duration should be chosen in such a way that $\tau_{HF} \ll T \ll \tau_{LF}$. The characteristic time of the high-frequency component τ_{HF} may be estimated from the reciprocal of the characteristic spectral frequency of the fluctuation, while the characteristic time of the low-frequency component τ_{LF} may be determined from the time required to travel the characteristic dimension of the physical system at the local characteristic low-frequency speed. Thus, a time averaging after volume averaging can be defined as

$$\overline{\langle\psi_k\rangle} = \frac{1}{T}\int_0^T \langle\psi_k\rangle\,\mathrm{d}t \tag{5.149}$$

An instantaneous volume-averaged quantity is therefore expressed by two parts

$$\langle\psi_k\rangle = \overline{\langle\psi_k\rangle} + \langle\psi_k'\rangle \tag{5.150}$$

where the prime denotes the high-frequency component and

$$\overline{\langle \psi_k' \rangle} = 0 \tag{5.151}$$

Further, we have

$$\overline{\langle \psi_k \rangle} = \frac{1}{T}\int_0^T \left[\frac{1}{V}\int_V (\overline{\psi_k} + \psi_k')\,\mathrm{d}V\right]\mathrm{d}t = \frac{1}{V}\int_V \left[\frac{1}{T}\int_0^T (\overline{\psi_k} + \psi_k')\,\mathrm{d}t\right]\mathrm{d}V = \langle \overline{\psi_k} \rangle \tag{5.152}$$

which indicates that the time averaging over a volume-averaged quantity equals the volume averaging over a time-averaged quantity. This equivalence is valid when the spatial integral and temporal integral become interchangeable with a proper choice of T (*i.e.*, $\tau_{HF} \ll T \ll \tau_{LF}$). When the time averaging is applied to the mass flux of phase k, we obtain

$$\overline{\langle \rho_k \boldsymbol{U}_k \rangle} = \overline{\langle \rho_k \rangle\, {}^i\langle \boldsymbol{U}_k \rangle} = \overline{\langle \rho_k \rangle}\, \overline{{}^i\langle \boldsymbol{U}_k \rangle} + \overline{\langle \rho_k' \rangle\, {}^i\langle \boldsymbol{U}_k' \rangle} \tag{5.153}$$

The correlation of two fluctuating components may be best formulated using Boussinesq's approach [Boussinesq, 1877] by introducing a transport coefficient. Thus, the last term in Eq. (5.153) can be expressed by

$$\overline{\langle \rho_k' \rangle\, {}^i\langle \boldsymbol{U}_k' \rangle} = -D_k \nabla \overline{\langle \rho_k \rangle} \tag{5.154}$$

where D_k is the eddy diffusivity of phase k in the mixture.

Assuming the material density or the intrinsic averaged density of phase k to be constant (as is true for the solids and almost true for the fluid when the effect of temperature variation on the material density can be neglected), we have

$$\langle \rho_k \rangle = \alpha_k\, {}^i\langle \rho_k \rangle = \alpha_k \rho_{km} \tag{5.155}$$

where ρ_{km} is the material density of phase k.

The expression for the momentum flux of phase k in terms of the product of averages can be derived from the definition of the volume-averaged momentum flux of phase k. From Eq. (5.130), we obtain

$$\begin{aligned}\langle \rho_k \boldsymbol{U}_k \boldsymbol{U}_k \rangle &= \langle \rho_k \rangle\, {}^i\langle \boldsymbol{U}_k \boldsymbol{U}_k \rangle = \langle \overline{\rho_k} + \rho_k' \rangle\, {}^i\langle (\overline{\boldsymbol{U}_k} + \boldsymbol{U}_k')(\overline{\boldsymbol{U}_k} + \boldsymbol{U}_k') \rangle \\ &= (\langle \overline{\rho_k} \rangle + \langle \rho_k' \rangle)({}^i\langle \overline{\boldsymbol{U}_k}\, \overline{\boldsymbol{U}_k} \rangle + {}^i\langle \boldsymbol{U}_k' \overline{\boldsymbol{U}_k} \rangle + {}^i\langle \overline{\boldsymbol{U}_k} \boldsymbol{U}_k' \rangle + {}^i\langle \boldsymbol{U}_k' \boldsymbol{U}_k' \rangle)\end{aligned} \tag{5.156}$$

If the local low-frequency component of a phase quantity is assumed to be constant throughout the volume and time averaging, we can further obtain

$${}^i\langle \overline{\boldsymbol{U}_k}\, \overline{\boldsymbol{U}_k} \rangle = {}^i\langle \overline{\boldsymbol{U}_k} \rangle\, {}^i\langle \overline{\boldsymbol{U}_k} \rangle \tag{5.157}$$

which leads to Eq. (5.156) in the form of

$$\begin{aligned}\overline{\langle \rho_k \boldsymbol{U}_k \boldsymbol{U}_k \rangle} &= \langle \overline{\rho_k} \rangle\, {}^i\langle \overline{\boldsymbol{U}_k} \rangle\, {}^i\langle \overline{\boldsymbol{U}_k} \rangle + \left(\langle \overline{\rho_k} \rangle\, \overline{{}^i\langle \boldsymbol{U}_k' \boldsymbol{U}_k' \rangle} + \overline{\langle \rho_k' \rangle\, {}^i\langle \boldsymbol{U}_k' \boldsymbol{U}_k' \rangle}\right) \\ &\quad - {}^i\langle \overline{\boldsymbol{U}_k} \rangle (D_k \nabla \langle \overline{\rho_k} \rangle) - (D_k \nabla \langle \overline{\rho_k} \rangle)\, {}^i\langle \overline{\boldsymbol{U}_k} \rangle\end{aligned} \tag{5.158}$$

where the second term on the right-hand side of the equation represents the Reynolds stress and the last two terms account for the momentum flux arising from the eddy mass diffusion.

A similar procedure can be applied to the energy transport terms. When the volume-averaged energy equation is expressed in terms of the internal energy and time averaging is applied to the equation, the local volume-averaged internal energy flux becomes

$$\overline{\langle\rho_k U_k e_k\rangle} = \langle\overline{\rho_k}\rangle\ \overline{{}^i\langle U_k\rangle}\ {}^i\langle\overline{e_k}\rangle + \langle\overline{\rho_k}\rangle\ {}^i\overline{\langle e_k' U_k'\rangle} + {}^i\langle\overline{e_k}\rangle\overline{\langle\rho_k' U_k'\rangle} + {}^i\langle\overline{U_k}\rangle\overline{\langle\rho_k' e_k'\rangle} + \overline{\langle\rho_k' U_k' e_k'\rangle} \tag{5.159}$$

where the second and last terms on the right-hand side represent the thermal transport due to the fluctuations in velocity and internal energy (or temperature) and can be expressed by

$$\langle\overline{\rho_k}\rangle\ {}^i\overline{\langle U_k' e_k'\rangle} + \overline{\langle\rho_k' U_k' e_k'\rangle} = -\langle\overline{\rho_k}\rangle D_{Tk}\nabla\, {}^i\langle\overline{e_k}\rangle \tag{5.160}$$

where D_{Tk} is the eddy thermal diffusivity. The third term on the right-hand side of Eq. (5.159) accounts for the energy transport by diffusion, which may be expressed by

$${}^i\langle\overline{e_k}\rangle\overline{\langle\rho_k' U_k'\rangle} = -\,{}^i\langle\overline{e_k}\rangle D_k\nabla\langle\overline{\rho_k}\rangle \tag{5.161}$$

The fourth term on the right-hand side of Eq. (5.159) is the energy transport by the fluctuations in density and internal energy (or temperature), which is linked to the averaged quantities by

$${}^i\langle\overline{U_k}\rangle\overline{\langle\rho_k' e_k'\rangle} \approx {}^i\langle\overline{U_k}\rangle[\mathrm{L}]^2\nabla\langle\overline{\rho_k}\rangle\cdot\nabla\,{}^i\langle\overline{e_k}\rangle = {}^i\langle\overline{U_k}\rangle C_{ek}\frac{k^3}{\epsilon^2}\nabla\langle\overline{\rho_k}\rangle\cdot\nabla\,{}^i\langle\overline{e_k}\rangle \tag{5.162}$$

where C_{ek} is an empirical constant associated with the energy transport by fluctuations in density and temperature. In addition, the volume–time-averaged expansion work is estimated by

$$\langle\overline{p}_k\nabla\cdot\overline{U_k}\rangle + \overline{\langle p_k'\nabla\cdot U_k'\rangle} \approx \langle\overline{p}_k\rangle\nabla\cdot\,{}^i\langle\overline{U_k}\rangle \tag{5.163}$$

The volume–time-averaged equations can be summarized as follows:
(1) Continuity equation

$$\frac{\partial\overline{\alpha}_k}{\partial t} + \nabla\cdot\left(\overline{\alpha_k}\,{}^i\langle\overline{U_k}\rangle\right) = \nabla\cdot\left(D_k\nabla\overline{\alpha_k}\right) + \frac{\overline{\Gamma}_k}{\rho_{km}} \tag{5.164}$$

(2) Momentum equation

$$\begin{aligned}\frac{\partial}{\partial t}&\left(\overline{\alpha_k}\,{}^i\langle\overline{U_k}\rangle - D_k\nabla\overline{\alpha_k}\right) + \nabla\cdot\left(\overline{\alpha_k}\,{}^i\langle\overline{U_k}\rangle\,{}^i\langle\overline{U_k}\rangle\right)\\ &= \nabla\cdot\left[D_k(\nabla\overline{\alpha_k})\,{}^i\langle\overline{U_k}\rangle + D_k\,{}^i\langle\overline{U_k}\rangle\nabla\overline{\alpha_k}\right] - \frac{1}{\rho_{km}}\nabla\langle\overline{p}_k\rangle\\ &\quad + \frac{1}{\rho_{km}}\nabla\cdot\left(\langle\overline{\tau_k}\rangle + \overline{\tau}_k^{\mathrm{T}}\right) + \langle\overline{\alpha_k f}\rangle + \frac{\overline{F}_{Ak}}{\rho_{km}} + \frac{\overline{F}_{\Gamma k}}{\rho_{km}}\end{aligned} \tag{5.165}$$

where the Reynolds stress of phase k is defined as

$$\overline{\tau}_k^{\mathrm{T}} = -\left(\langle\overline{\rho_k}\rangle\,{}^i\overline{\langle U_k' U_k'\rangle} + \overline{\langle\rho_k'\rangle\,{}^i\langle U_k' U_k'\rangle}\right) \tag{5.166}$$

which needs to be modeled by the turbulence models of phase k.

(3) Energy conservation equation

$$\frac{\partial}{\partial t}\langle\overline{\alpha}_{\rm k}\rho_{\rm km}\overline{e}_{\rm k}\rangle + \nabla\cdot\langle\overline{\alpha}_{\rm k}\rho_{\rm km}\,{}^{i}\langle\overline{\boldsymbol{U}_{\rm k}}\rangle\overline{e}_{\rm k}\rangle$$
$$= \nabla\cdot(\overline{\alpha}_{\rm k}\rho_{\rm km}D_{\rm Tk}\nabla\,{}^{i}\langle\overline{e}_{\rm k}\rangle)$$
$$+\nabla\cdot({}^{i}\langle\overline{e}_{\rm k}\rangle D_{\rm k}\nabla\langle\overline{\alpha}_{\rm k}\rho_{\rm km}\rangle) + \nabla\cdot\left(C_{\rm ek}\frac{k^3}{\epsilon^2}\rho_{\rm km}\,{}^{i}\langle\overline{\boldsymbol{U}_{\rm k}}\rangle\nabla\overline{\alpha}_{\rm k}\cdot\nabla\,{}^{i}\langle\overline{e}_{\rm k}\rangle\right)$$
$$-\langle\overline{p}_{\rm k}\rangle\nabla\cdot{}^{i}\langle\overline{\boldsymbol{U}_{\rm k}}\rangle + \langle\overline{J}_{\rm ek}\rangle + \langle\overline{\phi}_{\rm k}\rangle + \overline{Q}_{\rm Ak} + \overline{Q}_{\Gamma\rm k} \tag{5.167}$$

For convenience, we remove the ⟨ ⟩ signs and time-averaging bars from the preceding equations. Thus, the volume–time-averaged equations can be simplified to the form

$$\frac{\partial\alpha_{\rm k}}{\partial t} + \nabla\cdot(\alpha_{\rm k}\boldsymbol{U}_{\rm k}) = \nabla\cdot(D_{\rm k}\nabla\alpha_{\rm k}) + \frac{\Gamma_{\rm k}}{\rho_{\rm km}} \tag{5.168}$$

with

$$\sum_{\rm k}\alpha_{\rm k} = 1 \tag{5.169}$$

$$\frac{\partial}{\partial t}(\alpha_{\rm k}\boldsymbol{U}_{\rm k} - D_{\rm k}\nabla\alpha_{\rm k}) + \nabla\cdot(\alpha_{\rm k}\boldsymbol{U}_{\rm k}\boldsymbol{U}_{\rm k})$$
$$= \nabla\cdot[D_{\rm k}(\nabla\alpha_{\rm k})\boldsymbol{U}_{\rm k} + D_{\rm k}\boldsymbol{U}_{\rm k}\nabla\alpha_{\rm k}]$$
$$-\frac{1}{\rho_{\rm km}}\nabla p_{\rm k} + \frac{1}{\rho_{\rm km}}\nabla\cdot\left(\boldsymbol{\tau}_{\rm k} + \boldsymbol{\tau}_{\rm k}^{\rm T}\right) + \alpha_{\rm k}\boldsymbol{f} + \frac{\boldsymbol{F}_{\rm Ak}}{\rho_{\rm km}} + \frac{\boldsymbol{F}_{\Gamma\rm k}}{\rho_{\rm km}} \tag{5.170}$$

$$\frac{\partial}{\partial t}(\alpha_{\rm k}e_{\rm k}) + \nabla\cdot(\alpha_{\rm k}\boldsymbol{U}_{\rm k}e_{\rm k})$$
$$= \nabla\cdot(\alpha_{\rm k}D_{\rm Tk}\nabla e_{\rm k}) + \nabla\cdot(e_{\rm k}D_{\rm k}\nabla\alpha_{\rm k}) + \nabla\cdot\left(C_{\rm ek}\frac{k^3}{\epsilon^2}\boldsymbol{U}_{\rm k}\nabla\alpha_{\rm k}\cdot\nabla e_{\rm k}\right)$$
$$+\frac{1}{\rho_{\rm km}}(-p_{\rm k}\nabla\cdot\boldsymbol{U}_{\rm k} + J_{\rm ek} + \phi_{\rm k} + Q_{\rm Ak} + Q_{\Gamma\rm k}) \tag{5.171}$$

Assuming that the effects of turbulence can be reasonably modeled such that the correlations of turbulent fluctuations can be linked to the averaged quantities, the dependent variables for a n-phase flow are $\alpha_{\rm k}$, $\boldsymbol{U}_{\rm k}$, $p_{\rm k}$, and $T_{\rm k}$, where k varies from 1 to n. Thus, the total number of dependent variables is $6n$. Although Eqs. (5.168) through (5.171) constitute $5n+1$ equations, additional $n-1$ equations ($p_{\rm k} = p_{\rm k}(\alpha_{\rm k}, T_{\rm k})$) are needed to reach a closure. The development of these $n-1$ equations requires the mechanistic understanding of the pseudofluid nature of each phase.

5.3.4 *Transport Coefficients and Turbulence Models*

The fluxes of mass, momentum, and energy of phase k transported in a laminar or turbulent multiphase flow can be expressed in terms of the local gradients and the transport coefficients. In a gas–solid multiphase flow, the transport coefficients of the gas phase may be reasonably represented by those in a single-phase flow although certain modifications may be needed to account for the effects of the solid phase, especially when the solids concentration is high.

In this section, we discuss the transport coefficients of the solid phase. Specifically, the basic mechanisms for the transport processes of solids suspended in a gas medium may be understood from the gas–particle interactions and the particle–particle collisions. In turbulent gas–particle flows, the diffusion of particle clouds by turbulent eddies was modeled by Tchen and Hinze [Tchen, 1947; Hinze, 1959]. The Hinze–Tchen model considers the motion of a small single particle tracking a turbulent eddy. The model is simple and always yields a particle turbulent intensity less than the gas turbulent intensity. A more advanced turbulence model for gas–solid flows is known as the k–ϵ–k_p model [Zhou, 1993]. This model accounts for the effect of gas–particle interactions on both the gas turbulence and particle velocity fluctuation in terms of the particle turbulent kinetic energy k_p. In addition, the k–ϵ–k_p model uses a transport equation of k_p instead of an algebraic relation as in Hinze–Tchen's model for the simulation of particle phase turbulence. Both the Hinze–Tchen model and the k–ϵ–k_p model are for dilute gas–solid flows where the effect of interparticle collision can be neglected. When the solids concentration is high, the momentum and energy transfers due to single scattering of the collisions among particles become significant. In this section, an expression for the viscosity of solids from a simple shear flow model for particle collisions with single scattering and specular reflection is derived [Soo, 1989]. Further increase in the solids loading leads to a dense suspension in which the dynamics of the solid phase is dominated by particle collisions. In this case, the viscosity of the solids may be simulated by using the kinetic theory of gases [Savage, 1983; Jenkins and Savage, 1983; Gidaspow, 1993], which is discussed in §5.5.

5.3.4.1 *Hinze–Tchen's Model*

Here we present the derivation of the transport coefficients due to diffusion of discrete particles in a homogeneous turbulent flow. Although the Hinze–Tchen model was developed in a more general form for general particle–fluid multiphase flows, we introduce this model only for the cases of gas–solid flows. Some assumptions for this model are the following:

(1) The turbulence of the fluid is homogeneous and steady.
(2) The domain of turbulence is infinite.
(3) The particle is spherical and follows Stokes's law.
(4) The particle is small compared to the smallest wavelength in turbulence.
(5) The particle is always entrapped inside the same turbulent eddy.
(6) For gas–solid suspensions, $\rho_p/\rho \gg 1$.

To simplify the derivation, we may consider only the one-dimensional case. Denote $s(t_0+t)$ as the distance traveled by a marked fluid element with starting time t_0 during the time interval t. Let u be the instantaneous velocity and u' be the fluctuating velocity. In the following derivation, the time-averaged fluid velocity is taken as zero. Thus, we have

$$s(t_0+t) = \int_0^t u'(t_0+\tau)\,d\tau \tag{5.172}$$

It can be shown that [Taylor, 1921]

$$\overline{s^2(t)} = 2\overline{u'^2}\int_0^t dt' \int_0^{t'} R(\tau)\,d\tau \tag{5.173}$$

where $R(\tau)$ is a dimensionless autocorrelation function or the Lagrangian correlation coefficient given by

$$R(\tau) = \frac{\overline{u'(t)u'(t+\tau)}}{\overline{u'^2}} \tag{5.174}$$

The transport rate of a scalar quantity γ can be expressed in the form suggested by Boussinesq (1877) as

$$\overline{u'\gamma'} = \frac{1}{T}\int_0^T \gamma'(\tau)u'(\tau)\,\mathrm{d}\tau = -D_\gamma \frac{\mathrm{d}\Gamma}{\mathrm{d}x} \tag{5.175}$$

where D_γ is the coefficient of diffusion of γ and Γ is the corresponding mean value of the quantity. Assuming a linear variation of Γ with respect to x, we have

$$\gamma'(t) = -s(t)\frac{\mathrm{d}\Gamma}{\mathrm{d}x} \tag{5.176}$$

Since

$$u'(t) = \frac{\mathrm{d}s}{\mathrm{d}t} \tag{5.177}$$

from Eq. (5.175), we obtain

$$\overline{u'\gamma'} = -\frac{\mathrm{d}\Gamma}{\mathrm{d}x}\frac{1}{T}\int_0^T s(\tau,t)\frac{\mathrm{d}s(\tau,t)}{\mathrm{d}t}\,\mathrm{d}\tau = -\frac{1}{2}\frac{\mathrm{d}\Gamma}{\mathrm{d}x}\left(\frac{\mathrm{d}}{\mathrm{d}t}\overline{s^2(t)}\right) \tag{5.178}$$

Hence, from Eq. (5.172) and Eq. (5.173), the coefficient of diffusion of γ is

$$D_\gamma = \frac{1}{2}\frac{\mathrm{d}\overline{s^2}}{\mathrm{d}t} = \overline{u'^2}\int_0^t R(\tau)\,\mathrm{d}\tau \tag{5.179}$$

The dynamic behavior of discrete particles in turbulence can be described by the BBO equation [Basset, 1888; Boussinesq, 1903; and Oseen, 1927] for the slow motion of a spherical particle in a fluid at rest. This equation was extended by Tchen (1947) to the case of a fluid moving with time-dependent velocity as

$$\frac{\pi}{6}d_\mathrm{p}^3\rho_\mathrm{p}\frac{\mathrm{d}u_\mathrm{p}}{\mathrm{d}t} = 3\pi\mu d_\mathrm{p}(u-u_\mathrm{p}) - \frac{\pi}{6}d_\mathrm{p}^3\frac{\partial p}{\partial x} + \frac{\pi}{12}d_\mathrm{p}^3\rho\left(\frac{\mathrm{d}u}{\mathrm{d}t} - \frac{\mathrm{d}u_\mathrm{p}}{\mathrm{d}t}\right)$$
$$+\frac{3}{2}d_\mathrm{p}^2\sqrt{\pi\rho\mu}\int_{t_0}^t \frac{\frac{\mathrm{d}u}{\mathrm{d}\tau} - \frac{\mathrm{d}u_\mathrm{p}}{\mathrm{d}\tau}}{\sqrt{t-\tau}}\,\mathrm{d}\tau \tag{5.180}$$

From the Navier–Stokes equation, we have

$$-\frac{\partial p}{\partial x} = \rho\left[\frac{\partial u}{\partial t} + u\frac{\partial u}{\partial x}\right] - \mu\frac{\partial^2 u}{\partial x^2} \tag{5.181}$$

The substantial derivative in a Lagrangian coordinate moving with the discrete particle can be expressed by

$$\frac{\mathrm{d}}{\mathrm{d}t} = \frac{\partial}{\partial t} + u_\mathrm{p}\frac{\partial}{\partial x} \tag{5.182}$$

Thus, the modified BBO equation becomes

$$\frac{du_p}{dt} = \frac{3\rho}{2\rho_p + \rho}\left[\frac{du}{dt} - \frac{2}{3}\frac{\mu}{\rho}\frac{\partial^2 u}{\partial x^2}\right] + \frac{36\mu}{(2\rho_p + \rho)d_p^2}(u - u_p)$$
$$+ \frac{2\rho}{2\rho_p + \rho}(u - u_p)\frac{\partial u}{\partial x} + \frac{18}{(2\rho_p + \rho)d_p}\sqrt{\frac{\rho\mu}{\pi}}\int_{t_0}^{t}\frac{\frac{du}{d\tau} - \frac{du_p}{d\tau}}{\sqrt{t-\tau}}\,d\tau \qquad (5.183)$$

Assuming the effect of Basset force is negligible and noting that in gas–solid flows $\rho_p/\rho \gg 1$, Eq. (5.183) reduces to

$$\frac{du_p}{dt} = \frac{u - u_p}{\tau_S} \qquad (5.184)$$

where τ_S is the Stokes relaxation time of the particle diffusion, defined by Eq. (3.39).

Both u and u_p may be represented by the Fourier integrals as

$$u = \int_0^\infty (\alpha\cos\omega t + \beta\sin\omega t)\,d\omega$$
$$u_p = \int_0^\infty (\gamma\cos\omega t + \delta\sin\omega t)\,d\omega \qquad (5.185)$$

where ω is the angular frequency and is equal to $2\pi f$. Substituting Eq. (5.185) into Eq. (5.184), we obtain the relations of α, β, γ, and δ as

$$\omega\tau_S\delta = \alpha - \gamma$$
$$\omega\tau_S\gamma = \delta - \beta \qquad (5.186)$$

Defining the one-sided autospectral density function $G(f)$ by

$$G(f) = 4\overline{u'^2}\int_0^\infty R(\tau)\cos 2\pi f\tau\,d\tau \qquad (5.187)$$

from the inverse Fourier transform, we have the corresponding expression of $R(\tau)$ as

$$R(\tau) = \frac{1}{\overline{u'^2}}\int_0^\infty G(f)\cos 2\pi f\tau\,df \qquad (5.188)$$

Note that $R(0) = 1$, which yields

$$\overline{u'^2} = \int_0^\infty G(f)\,df \qquad (5.189)$$

Thus, on the basis of Eq. (5.189), $G(f)\,df$ represents the contribution to the mean square of the turbulence fluctuation of velocity between frequencies f and $f + df$. When u is expressed by Eq. (5.185) $\overline{u}$ becomes zero, and thus u is equal to u'. It can be shown [Hinze, 1959] that

$$\overline{u'^2} = \frac{\pi}{2T}\int_0^\infty (\alpha^2 + \beta^2)\,d\omega \qquad (5.190)$$

where T is the integral time. Thus, $G(f)$ can be related to α and β by

$$G(f) = \frac{\pi^2}{T}(\alpha^2 + \beta^2) \qquad (5.191)$$

Hence, we have

$$\frac{G(\omega)}{G_{\mathrm{p}}(\omega)} = \frac{\alpha^2+\beta^2}{\gamma^2+\delta^2} \tag{5.192}$$

We assume that the Lagrangian correlation coefficient for fluid motion may be represented by an exponential function

$$R(t) = \exp\left(-\frac{t}{\tau_{\mathrm{f}}}\right) \tag{5.193}$$

where τ_{f} is the characteristic time of the fluid motion. The corresponding one-sided autospectral density function then becomes, from Eq. (5.187),

$$G(\omega) = \overline{u'^2}\frac{4\tau_{\mathrm{f}}}{1+\omega^2\tau_{\mathrm{f}}^2} \tag{5.194}$$

Thus, substituting Eq. (5.186) and Eq. (5.194) into Eq. (5.192), $G_{\mathrm{p}}(\omega)$ is obtained as

$$G_{\mathrm{p}}(\omega) = \overline{u'^2}\frac{4\tau_{\mathrm{f}}}{\left(1+\omega^2\tau_{\mathrm{f}}^2\right)\left(1+\omega^2\tau_{\mathrm{S}}^2\right)} \tag{5.195}$$

Consequently, we have

$$\overline{u_{\mathrm{p}}'^2} = \frac{1}{2\pi}\int_0^\infty G_{\mathrm{p}}(\omega)\,\mathrm{d}\omega = \frac{\tau_{\mathrm{f}}}{\tau_{\mathrm{f}}+\tau_{\mathrm{S}}}\overline{u'^2} \tag{5.196}$$

and

$$R_{\mathrm{p}}(t) = \frac{1}{2\pi\overline{u_{\mathrm{p}}'^2}}\int_0^\infty G_{\mathrm{p}}(\omega)\cos\omega t\,\mathrm{d}\omega = \frac{1}{\tau_{\mathrm{f}}-\tau_{\mathrm{S}}}\left(\tau_{\mathrm{f}}\exp\left(-\frac{t}{\tau_{\mathrm{f}}}\right)-\tau_{\mathrm{S}}\exp\left(-\frac{t}{\tau_{\mathrm{S}}}\right)\right) \tag{5.197}$$

from which we may conclude that the Lagrangian correlation coefficient for discrete particle motion is no longer a single exponential function. In terms of the definitions of the transport coefficients for turbulent diffusion given in Eq. (5.179), it can be shown that

$$\frac{D_{\gamma\mathrm{p}}}{D_\gamma} = 1+\left[\frac{1}{\left(\frac{\tau_{\mathrm{f}}}{\tau_{\mathrm{S}}}\right)^2-1}\right]\left[\frac{\exp\left(-\frac{t}{\tau_{\mathrm{S}}}\right)-\exp\left(-\frac{t}{\tau_{\mathrm{f}}}\right)}{1-\exp\left(-\frac{t}{\tau_{\mathrm{f}}}\right)}\right] \tag{5.198}$$

For a short diffusion time ($t \ll \tau_{\mathrm{f}}$ and $t \ll \tau_{\mathrm{S}}$), we have

$$\frac{D_{\gamma\mathrm{p}}}{D_\gamma} = \left(1+\frac{\tau_{\mathrm{S}}}{\tau_{\mathrm{f}}}\right)^{-1} \tag{5.199}$$

Equation (5.199) is known as the Hinze–Tchen equation.

Using the analogy of the kinetic theory of gases, from Eqs. (5.43), (5.46), and (5.49), it yields

$$\frac{\nu_{\mathrm{p}}}{\nu} = \frac{D_{\gamma\mathrm{p}}}{D_\gamma} = \left(1+\frac{\tau_{\mathrm{S}}}{\tau_{\mathrm{f}}}\right)^{-1} \tag{5.200}$$

where ν is the kinematic viscosity. The corresponding thermal conductivity can be given by

$$\frac{K_p}{K} = \frac{C_p}{C}\frac{D_{\gamma p}}{D_\gamma} = \frac{C_p}{C}\left(1 + \frac{\tau_S}{\tau_f}\right)^{-1} \tag{5.201}$$

where C is the specific thermal capacity of the gas and C_p is the specific thermal capacity of the particle.

5.3.4.2 *k–ε–k*$_p$ *Turbulence Model*

The gas–particle interaction may strongly enhance or attenuate the gas phase turbulence. Thus, a turbulence model for gas–solid flows needs to include the effect of gas–particle interactions. Here, we consider only isotropic turbulence and dilute gas–solid flows without phase changes. The flow is assumed to be steady and isothermal.

Following the approaches used in the single-phase k–ϵ model, the k and ϵ equations for dilute gas–solid flows can be expressed as [Huang and Zhou, 1991; Zhou, 1993]

$$\frac{\partial}{\partial x_j}(\rho(1-\alpha_p)U_j k) = \frac{\partial}{\partial x_j}\left(\frac{\mu_{\text{eff}}}{\sigma_k}\frac{\partial k}{\partial x_j}\right) + \mu_T \frac{\partial U_i}{\partial x_j}\left(\frac{\partial U_i}{\partial x_j} + \frac{\partial U_j}{\partial x_i}\right) + G_k - \rho(1-\alpha_p)\epsilon \tag{5.202}$$

and

$$\frac{\partial}{\partial x_j}(\rho(1-\alpha_p)U_j\epsilon) = \frac{\partial}{\partial x_j}\left(\frac{\mu_{\text{eff}}}{\sigma_\epsilon}\frac{\partial \epsilon}{\partial x_j}\right) + \frac{\epsilon}{k}\left[C_1\left\{\mu_T\frac{\partial U_i}{\partial x_j}\left(\frac{\partial U_i}{\partial x_j} + \frac{\partial U_j}{\partial x_i}\right) + G_k\right\} - C_2\rho(1-\alpha_p)\epsilon\right] \tag{5.203}$$

where G_k is the generation of k due to the gas–solid interaction. G_k can be expressed by

$$G_k = \frac{1}{\tau_p}\left[\rho_p\alpha_p(C_k\sqrt{kk_p} - 2k) + \frac{(U_i - U_{pi})}{\alpha_p}\frac{\mu_p}{\sigma_p}\frac{\partial \alpha_p}{\partial x_i}\right] \tag{5.204}$$

where C_k is an empirical parameter; k_p is the particle turbulent kinetic energy defined as

$$k_p = \frac{1}{2}\overline{u'_{pi}u'_{pi}} \tag{5.205}$$

and μ_p is the particle turbulent viscosity due to gas–particle interaction. μ_p is given by

$$\mu_p = C_{\mu p}\rho_p\alpha_p\frac{k_p^2}{|\epsilon_p|} \tag{5.206}$$

The particle turbulent kinetic energy is governed by its own transport equation. Similarly to the derivation of the k-equation, the k_p-equation is given by

$$\frac{\partial}{\partial x_j}(\rho_p\alpha_p U_{pj}k_p) = \frac{\partial}{\partial x_j}\left(\frac{\mu_p}{\sigma_p}\frac{\partial k_p}{\partial x_j}\right) + \mu_p\frac{\partial U_{pi}}{\partial x_j}\left(\frac{\partial U_{pi}}{\partial x_j} + \frac{\partial U_{pj}}{\partial x_i}\right) + \rho_p\epsilon_p \tag{5.207}$$

where ϵ_p is the rate of generation or destruction of k_p due to the gas–solid interaction. ϵ_p can be expressed by

$$\epsilon_p = \frac{1}{\tau_p}\left[\alpha_p(C_k\sqrt{kk_p} - 2k_p) - \frac{(U_i - U_{pi})}{\alpha_p\rho_p}\frac{\mu_p}{\sigma_p}\frac{\partial \alpha_p}{\partial x_i}\right] \tag{5.208}$$

The model given above is called the k–ϵ–k_p model, which can be used for dilute, nonswirling, nonbuoyant gas–solid flows. For strongly anisotropic gas–solid flows, the unified second-order moment closure model, which is an extension of the second-order moment closure model for single-phase flows [Zhou, 1993], may be used.

5.3.4.3 *Particle Viscosity Due to Interparticle Collisions*

When the particle concentration is high, the shear motion of particles leads to interparticle collisions. The transfer of momentum between particles can be described in terms of a pseudoshear stress and the viscosity of particle–particle interactions. Let us first examine the transfer of momentum in an elastic collision between two particles, as shown in Fig. 5.8(a). Particle 1 is fixed in space while particle 2 collides with particle 1 with an initial momentum in the x-direction. Assume that the contact surface is frictionless so that the rebound of the particle is in a form of specular reflection in the r–x-plane. The rate of change of the x-component of the momentum between the two particles is given by

$$m_2 U_2 - [m_2 U_2 \cos(\pi - 2\theta)] = 2 m_2 U_2 \cos^2\theta \tag{5.209}$$

where θ is the contact angle in the r–x-plane.

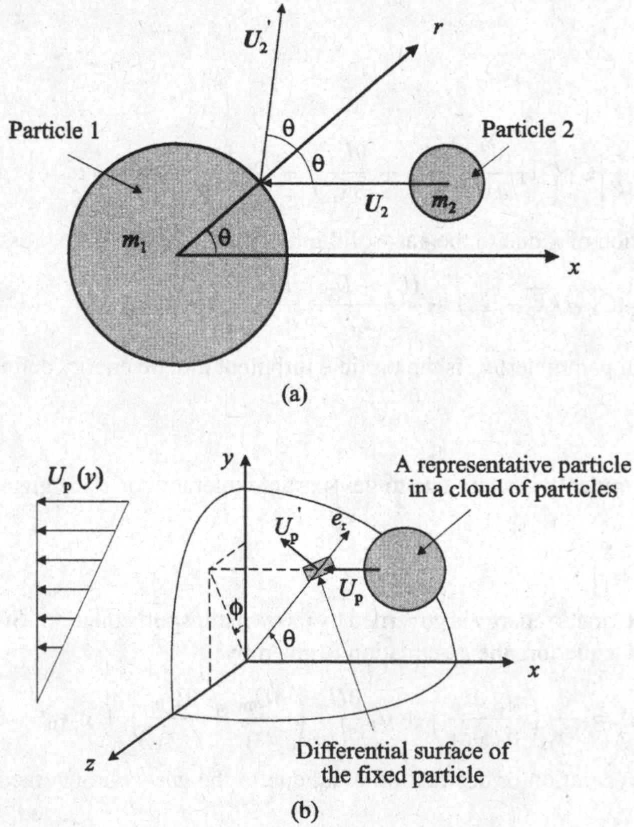

Figure 5.8. Impaction due to shear motion: (a) Elastic collision between two particles; (b) Elastic collision of a single sphere with a cloud of particles.

Consider a collision between two particles 1 and 2 with masses m_1 and m_2 and velocities U_1 and U_2, respectively. From Newton's law, we have

$$m_1 \frac{\mathrm{d}U_1}{\mathrm{d}t} = f_{12}; \qquad m_2 \frac{\mathrm{d}U_2}{\mathrm{d}t} = -f_{12} \tag{5.210}$$

where f_{12} is the collisional force. Equation (5.210) can be reduced to the form

$$m^* \frac{\mathrm{d}\Delta U}{\mathrm{d}t} = f_{12} \tag{5.211}$$

where $\Delta U = U_1 - U_2$, the relative velocity between the two particles; and m^* is the relative mass defined by Eq. (2.128). Thus, the collision between two moving particles is equivalent to the case where a particle collides with a fixed particle with the same relative mass and relative velocity.

Now consider a case of collision of a fixed single sphere with a cloud of particles as shown in Fig. 5.8(b). The sphere of radius r is subjected to a shear flow of a particle cloud with a velocity gradient of $\partial U_{\mathrm{p}}/\partial y$. The particle number density is denoted as n, and the mass of a particle is m. Select the velocity U_{p} as zero in the center plane of the sphere. The relative velocity is estimated by

$$\Delta U_{\mathrm{p}} = \frac{\partial U_{\mathrm{p}}}{\partial y} y = \frac{\partial U_{\mathrm{p}}}{\partial y} r \sin\theta \cos\phi \tag{5.212}$$

where ϕ is the azimuthal angle in the y–z plane. The differential area of impact on the sphere is $r^2 \sin\theta \,\mathrm{d}\theta \,\mathrm{d}\phi$. The projected area of impact normal to the incoming velocity U_{p} is $r^2 \sin\theta \cos\theta \,\mathrm{d}\theta \,\mathrm{d}\phi$ and the rate of change of the x-component of the momentum for a particle is $2m\Delta U_{\mathrm{p}} \cos^2\theta$ from Eqs. (5.209) and (5.212), assuming single scattering and specular reflection. The force acting on the semisphere is given by

$$F = 4 \int_0^{\frac{\pi}{2}} \int_0^{\frac{\pi}{2}} n(\Delta U_{\mathrm{p}})^2 m r^2 \sin\theta \cos^3\theta \,\mathrm{d}\theta \,\mathrm{d}\phi = \frac{\pi}{12} nm \left(\frac{\partial U_{\mathrm{p}}}{\partial y}\right)^2 r^4 \tag{5.213}$$

Consider a mixture of two groups of particles with radii a_1 and a_2, and corresponding number concentrations of n_1 and n_2. Assume that the group of particle 1 is subjected to a simple shear flow of particle 2. The x-component of the collision force between particle 1 and particle 2 can be obtained, from Eq. (5.213), as

$$F_{12} = \frac{\pi^2}{18} n_1 n_2 m^* \left(\frac{\partial U_{\mathrm{p}}}{\partial y}\right)^2 (a_1 + a_2)^7 \tag{5.214}$$

where $(a_1 + a_2)$ is the radius of impact.

By introducing an equivalent force, *i.e.*, equivalent to the x-component of the collision force, acting on a projected area normal to the y-axis, a shear stress of particle phase can be defined as

$$(\tau_{\mathrm{xy}})_{12} = \frac{2F_{12}}{\pi (a_1 + a_2)^2} = \frac{\pi}{9} n_1 n_2 m^* \left(\frac{\partial U_{\mathrm{p}}}{\partial y}\right)^2 (a_1 + a_2)^5 \tag{5.215}$$

Equation (5.215) can be extended to a generalized form as [Soo, 1989]

$$\tau_{12} = -\frac{\pi}{9} \frac{n_1 n_2 m_1 m_2}{(m_1 + m_2)} (a_1 + a_2)^5 (\Delta_{\mathrm{p2}} : \Delta_{\mathrm{p2}})^{\frac{1}{2}} \Delta_{\mathrm{p2}} = -\mu_{\mathrm{p12}} \Delta_{\mathrm{p2}} \tag{5.216}$$

where Δ_p has the Cartesian components

$$\Delta_{pij} = \frac{\partial U_{pi}}{\partial x_j} + \frac{\partial U_{pj}}{\partial x_i} \tag{5.217}$$

and

$$\Delta_p : \Delta_p = \Delta_{pij}\Delta_{pji} \tag{5.218}$$

The viscosity due to the collision between the group of particle 1 and the group of particle 2 is defined from Eq. (5.216) as

$$\mu_{p12} = \frac{\pi}{9}\frac{n_1 n_2 m_1 m_2}{(m_1 + m_2)}(a_1 + a_2)^5(\Delta_{p2} : \Delta_{p2})^{\frac{1}{2}} \tag{5.219}$$

For the special case of identical particles, Eq. (5.216) yields

$$\mu_p = \frac{1}{3}\alpha_p^2 \rho_p d_p^2 (\Delta_p : \Delta_p)^{\frac{1}{2}} \tag{5.220}$$

The thermal conductivity of the particle collisions, K_p, can be obtained on the basis of the kinetic theory. From Eqs. (5.43) and (5.49), for the identical particles, we have

$$K_p = \frac{c}{m}\mu_p = \frac{2c\alpha_p^2}{\pi d_p}(\Delta_p : \Delta_p)^{\frac{1}{2}} \tag{5.221}$$

So far, in this model, we assume that all the particles are elastic and the collision is of specular reflection on a frictionless smooth surface. For inelastic particles, we may introduce the restitution coefficient e, which is defined as the ratio of the rebound speed to the incoming speed in a normal collision. Therefore, for a collision of an inelastic particle with a frictionless surface as shown in Fig. 5.9, we have

$$\tan\theta' = \frac{U'_t}{U'_r} = \frac{\tan\theta}{e} \tag{5.222}$$

so the reflection angle θ' is larger than the incident angle θ; *i.e.*, the reflection is not specular. In Eq. (5.222) U'_t and U'_r are, respectively, the tangential and normal reflecting velocity after collision. The basic procedure in deriving the pseudostress and the resulting viscosity for inelastic particles is similar to that for elastic particles.

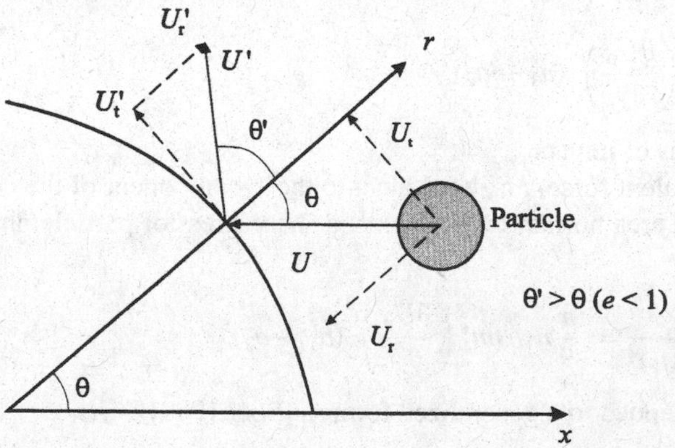

Figure 5.9. Inelastic impaction on a frictionless surface.

5.3.5 *Boundary Conditions of Particle Phase*

The boundary conditions for the gas phase are assumed to be of the same form as that for the single gas phase discussed in §5.2.5. For the particle phase, the boundary conditions are given as follows.

For the normal components of particle velocities, the boundary conditions may be expressed by using the coefficient of restitution as

$$U'_{\mathrm{pn}} = -eU_{\mathrm{pn}} \tag{5.223}$$

where U'_{pn} is the normal component of the rebound velocity of the particle. Hence, for elastic particles, $U'_{\mathrm{pn}} = -U_{\mathrm{pn}}$. For the tangential velocities of particles at a solid wall, slip conditions should be imposed; they may be expressed by

$$U'_{\mathrm{pt}} = fU_{\mathrm{pt}} \tag{5.224}$$

where f is a frictional factor related to the material properties of both wall and particles and the incident angle of impact. For frictionless surfaces, $f = 1$, which leads to $U'_{\mathrm{pt}} = U_{\mathrm{pt}}$.

The boundary conditions of particle temperature may be obtained from the heat transfer due to the collision of two bodies. Sun and Chen (1988) formulated the heat transfer per impact of two elastic particles by considering the collision of two elastic particles with different temperatures and assuming the heat conduction occurs only in the normal direction.

We may consider the solid wall as a very large particle with an infinite thermal capacity. Thus, for the collision between a particle and a solid wall, we have

$$T'_{\mathrm{p}} - T_{\mathrm{p}} = \frac{0.87 A_{\mathrm{c}} \sqrt{t_{\mathrm{c}}}}{\sqrt{\rho_{\mathrm{p}} C_{\mathrm{p}} K_{\mathrm{p}}}} (T_{\mathrm{w}} - T_{\mathrm{p}}) \tag{5.225}$$

where A_{c} is the maximum contact area given by Eq. (2.133) and t_{c} is the contact time duration given by Eq. (2.136).

For the boundary condition of particle concentration at the wall, a zero normal gradient condition is frequently adopted; that is

$$\left(\frac{\partial \alpha_{\mathrm{p}}}{\partial n} \right)_{\mathrm{w}} = 0 \tag{5.226}$$

Furthermore, it is assumed that there is no particle fluctuation on the wall so that the boundary condition of particle turbulent kinetic energy at the wall is given by

$$(k_{\mathrm{p}})_{\mathrm{w}} = 0 \tag{5.227}$$

5.4 Trajectory Modeling of Multiphase Flows

Trajectory models quantify the dynamic characteristics of particles in Lagrangian coordinates. The trajectory model is useful when the particle phase is so dilute that the description of particle behavior by continuum models may not be suitable.

Two basic trajectory models, *i.e.*, the deterministic trajectory model [Crowe *et al.*, 1977] and the stochastic trajectory model [Crowe, 1991], are introduced in this section. The deterministic approach, which neglects the turbulent fluctuation of particles, specifically, the turbulent diffusion of the mass, momentum, and energy of particles, is considered the most classical and simplest approach that uses the Lagrangian concept. However, the oversimplified assumptions for the transport behavior of the particles, especially on the energy transport processes, may yield inaccurate predictions of the particle number density distribution and the temperature distribution of the particles. The stochastic approach, with the aid of the Monte Carlo method, can directly simulate the instantaneous dynamic behavior of the particles on the basis of instantaneous momentum equations of the particles in Lagrangian coordinates. Depending on the choice of turbulence models and phase interaction mechanisms, this approach, in principle, can simulate the turbulence effect on the interactions between the gas and the particles. However, to obtain the statistical averages with reasonable accuracy, calculation of tens of thousands of trajectories, which sometimes requires enormous computing power, may be needed. It should be noted that in the modeling of gas–solid flows, the trajectory model is applied to account for the particle phase only, while the gas phase is usually described by the Eulerian model. Using the Lagrangian approach, the governing equations of the particle phase are expressed in the form of ordinary differential equations instead of partial differential equations.

To simplify the following analysis, we assume that (1) the particles are spherical and of identical size; (2) for the momentum interaction between the gas and solid phases, only the drag force in a locally uniform flow field is considered, *i.e.*, all other forces such as Magnus force, Saffman force, Basset force, and electrostatic force are negligible; and (3) the solids concentration is low so that particle–particle interactions are excluded.

5.4.1 *Deterministic Trajectory Models*

Consider that the turbulent diffusion of the particle mass, momentum, and energy can be neglected; *i.e.*, the particle viscosity μ_{p}, particle diffusivity D_{p}, and thermal conductivity of particle phase K_{p} are negligibly small. The governing equations of mass, momentum, and energy of the particle phase in the Lagrangian form can be obtained from the corresponding volume–time-averaged equations in the Eulerian form derived in §5.3.3, by setting all the correlation terms on the right-hand side of the particle-phase equations to zero. In this case, the time-averaged particle-phase equations in Lagrangian coordinates are formally identical to the instantaneous equations in Lagrangian coordinates.

5.4.1.1 *Mass Conservation Equation*

From Eq. (5.168), we have

$$\frac{\partial}{\partial t}(\alpha_{\mathrm{p}}\rho_{\mathrm{p}}) + \nabla\cdot(\alpha_{\mathrm{p}}\rho_{\mathrm{p}}\boldsymbol{U}_{\mathrm{p}}) = \Gamma_{\mathrm{p}} \tag{5.228}$$

If we express the density of the particle phase in terms of the product of the particle number density n and the mass of a particle m, *i.e.*, $\alpha_{\mathrm{p}}\rho_{\mathrm{p}} = nm$, and noting that $\Gamma_{\mathrm{p}} = n(\mathrm{d}m/\mathrm{d}t)$, we have

$$\frac{\partial n}{\partial t} + \nabla\cdot(n\boldsymbol{U}_{\mathrm{p}}) = 0 \tag{5.229}$$

For a steady flow, Eq. (5.229) can be reduced to

$$\nabla\cdot(n\boldsymbol{U}_{\mathrm{p}}) = 0 \tag{5.230}$$

Thus, integrating the preceding equation over any cross section of a stream tube of the particle phase and using the Gauss theorem, we can obtain

$$\int_A n\boldsymbol{U}_\mathrm{p} \cdot \mathrm{d}\boldsymbol{A} = \text{const} = N_\mathrm{p} \tag{5.231}$$

where $\boldsymbol{A}$ is the vector cross sectional area of the stream tube and N_p is the flow rate of the number of particles passing through A. Equation (5.231) reflects that the total particle number flux is conserved along the stream tube. Thus, it is concluded that the total particle number flux is conserved along the particle trajectory.

5.4.1.2 *Momentum Equation*

From Eq. (5.170), we have

$$\frac{\partial}{\partial t}(\alpha_\mathrm{p}\rho_\mathrm{p}\boldsymbol{U}_\mathrm{p}) + \nabla\cdot(\alpha_\mathrm{p}\rho_\mathrm{p}\boldsymbol{U}_\mathrm{p}\boldsymbol{U}_\mathrm{p}) = \alpha_\mathrm{p}\rho_\mathrm{p}\boldsymbol{f} + \boldsymbol{F}_\mathrm{Ap} + \boldsymbol{F}_{\Gamma\mathrm{p}} \tag{5.232}$$

where $\boldsymbol{F}_{\Gamma\mathrm{p}}$ is the momentum transfer due to the change of mass and is given by Eq. (5.125) as

$$\boldsymbol{F}_{\Gamma\mathrm{p}} = -\frac{1}{V}\int_{A_\mathrm{p}} \alpha_\mathrm{p}\rho_\mathrm{p}\boldsymbol{U}_\mathrm{p}(\boldsymbol{U}_\mathrm{p} - \boldsymbol{U}_\mathrm{s})\cdot\boldsymbol{n}_\mathrm{p}\,\mathrm{d}A \approx \boldsymbol{U}_\mathrm{p}\Gamma_\mathrm{p} \tag{5.233}$$

$\boldsymbol{F}_\mathrm{Ap}$ is the drag force between the gas and the solids, which can be expressed as

$$\boldsymbol{F}_\mathrm{Ap} = \frac{\alpha_\mathrm{p}\rho_\mathrm{p}}{\tau_\mathrm{rp}}(\boldsymbol{U} - \boldsymbol{U}_\mathrm{p}) \tag{5.234}$$

where τ_rp is the relaxation time of the particle. For a low relative motion, τ_rp may be obtained from Eq. (3.39). In Eq. (5.234), the effect of interparticle collisions is neglected since most applications of the trajectory model deal with very dilute multiphase flows.

Using Eq. (5.228), Eq. (5.232) becomes

$$\frac{\mathrm{d}\boldsymbol{U}_\mathrm{p}}{\mathrm{d}t_\mathrm{p}} = \boldsymbol{f} + \frac{1}{\tau_\mathrm{rp}}(\boldsymbol{U} - \boldsymbol{U}_\mathrm{p}) \tag{5.235}$$

where $\mathrm{d}/\mathrm{d}t_\mathrm{p}$ is defined as

$$\frac{\mathrm{d}}{\mathrm{d}t_\mathrm{p}} = \frac{\partial}{\partial t} + \boldsymbol{U}_\mathrm{p}\cdot\nabla \tag{5.236}$$

Moreover, the particle trajectory s can be obtained from

$$\boldsymbol{s} - \boldsymbol{s}_0 = \int_{t_0}^{t} \boldsymbol{U}_\mathrm{p}\,\mathrm{d}t \tag{5.237}$$

5.4.1.3 *Energy Equation*

From Eq. (5.171), we have

$$\frac{\partial}{\partial t}(\alpha_\mathrm{p}\rho_\mathrm{p}C_\mathrm{p}T_\mathrm{p}) + \nabla\cdot(\alpha_\mathrm{p}\rho_\mathrm{p}\boldsymbol{U}_\mathrm{p}C_\mathrm{p}T_\mathrm{p}) = J_\mathrm{ep} + Q_\mathrm{Ap} + Q_{\Gamma\mathrm{p}} \tag{5.238}$$

Using Eq. (5.228) and assuming that $Q_{\Gamma\mathrm{p}}$ can be approximated by

$$Q_{\Gamma\mathrm{p}} = -\frac{1}{V}\int_{A_\mathrm{p}} \rho_\mathrm{p}e_\mathrm{p}(\boldsymbol{U}_\mathrm{p} - \boldsymbol{U}_\mathrm{s})\cdot\mathrm{d}\boldsymbol{A} \approx C_\mathrm{p}T_\mathrm{p}\Gamma_\mathrm{p} \tag{5.239}$$

Eq. (5.238) is simplified to

$$\frac{\mathrm{d}T_{\mathrm{p}}}{\mathrm{d}t_{\mathrm{p}}} = \frac{1}{\alpha_{\mathrm{p}}\rho_{\mathrm{p}}C_{\mathrm{p}}}(J_{\mathrm{Ep}} + Q_{\mathrm{Ap}}) \tag{5.240}$$

Note that J_{ep} in Eq. (5.238) is replaced with J_{Ep} for Eq. (5.240), where J_{Ep} is the heat generated by thermal radiation per unit volume and Q_{Ap} is the heat transferred through the interface between gas and particles. Thus, once the gas velocity field is solved, the particle velocity, particle trajectory, particle concentration, and particle temperature can all be obtained directly by integrating Eqs. (5.235), (5.237), (5.231), and (5.240), respectively. Since the equations for the gas phase are coupled with those for the solid phase, final solutions of the governing equations may have to be obtained through iterations between those for the gas and solid phases.

5.4.2 *Stochastic Trajectory Models*

In trajectory models, the particle turbulent diffusion can be considered by calculating the instantaneous motion of particles in the turbulent flow field. In order to simulate the stochastic characteristics of the instantaneous gas velocity in a turbulent flow, it is required to generate random numbers in the calculation process.

The instantaneous motion of a particle is governed by

$$\frac{\mathrm{d}\boldsymbol{u}_{\mathrm{p}}}{\mathrm{d}t_{\mathrm{p}}} = \frac{1}{\tau_{\mathrm{rp}}}(\boldsymbol{u} - \boldsymbol{u}_{\mathrm{p}}) + \boldsymbol{f} \tag{5.241}$$

where $\boldsymbol{u}$ and $\boldsymbol{u}_{\mathrm{p}}$ are the instantaneous velocities of the gas and the particle, respectively. $\boldsymbol{f}$ is the body force acting on the unit mass of the particle. If we write Eq. (5.241) in a scalar form in cylindrical coordinates, the axial component of the equation becomes

$$\frac{\mathrm{d}u_{\mathrm{p}}}{\mathrm{d}t_{\mathrm{p}}} = \frac{1}{\tau_{\mathrm{rp}}}(U + u' - u_{\mathrm{p}}) + g_{\mathrm{z}} \tag{5.242}$$

while the radial component is

$$\frac{\mathrm{d}v_{\mathrm{p}}}{\mathrm{d}t_{\mathrm{p}}} = \frac{1}{\tau_{\mathrm{rp}}}(V + v' - v_{\mathrm{p}}) + \frac{w_{\mathrm{p}}^2}{r} + g_{\mathrm{r}} \tag{5.243}$$

and the tangential component is

$$\frac{\mathrm{d}w_{\mathrm{p}}}{\mathrm{d}t_{\mathrm{p}}} = \frac{1}{\tau_{\mathrm{rp}}}(W + w' - w_{\mathrm{p}}) - \frac{v_{\mathrm{p}}w_{\mathrm{p}}}{r} + g_{\phi} \tag{5.244}$$

where U, V, and W are the axial, radial, and tangential components, respectively, of the time-averaged gas velocity; u', v', and w' are the axial, radial, and tangential components, respectively, of the fluctuating gas velocity; and g_{z}, g_{r}, and g_{ϕ} are gravitational acceleration in the axial, radial, and azimuthal directions, respectively.

Assuming that the turbulence is isotropic, we have

$$\overline{u'^2} = \overline{v'^2} = \overline{w'^2} = \frac{2}{3}k \tag{5.245}$$

where k is the turbulent kinetic energy, which may be determined from the k–ϵ model. Furthermore, it is assumed that the local velocity fluctuations of the gas phase obey a

Gaussian probability density distribution. Thus, when a particle passes through a turbulent eddy, we have

$$u' = \zeta\sqrt{\frac{2}{3}k}, \qquad v' = \zeta\sqrt{\frac{2}{3}k}, \qquad w' = \zeta\sqrt{\frac{2}{3}k} \tag{5.246}$$

where ζ is a random number.

Substituting Eq. (5.246) into Eqs. (5.242), (5.243), and (5.244), we have the integrable instantaneous equations for the particle velocities

$$\frac{\mathrm{d}u_\mathrm{p}}{\mathrm{d}t_\mathrm{p}} = \frac{1}{\tau_\mathrm{rp}}\left(U + \zeta\sqrt{\frac{2}{3}k} - u_\mathrm{p}\right) + g_\mathrm{z} \tag{5.247}$$

$$\frac{\mathrm{d}v_\mathrm{p}}{\mathrm{d}t_\mathrm{p}} = \frac{1}{\tau_\mathrm{rp}}\left(V + \zeta\sqrt{\frac{2}{3}k} - v_\mathrm{p}\right) + \frac{w_\mathrm{p}^2}{r} + g_\mathrm{r} \tag{5.248}$$

and

$$\frac{\mathrm{d}w_\mathrm{p}}{\mathrm{d}t_\mathrm{p}} = \frac{1}{\tau_\mathrm{rp}}\left(W + \zeta\sqrt{\frac{2}{3}k} - w_\mathrm{p}\right) - \frac{v_\mathrm{p}w_\mathrm{p}}{r} + g_\phi \tag{5.249}$$

Thus, by coupling Eqs. (5.247), (5.248), and (5.249) with the governing equations for the gas phase, the instantaneous particle velocities can be obtained. Moreover, the stochastic trajectories of the particles can also be obtained by

$$z = z_\mathrm{p0} + \int_0^t u_\mathrm{p}\,\mathrm{d}t_\mathrm{p}, \qquad r = r_\mathrm{p0} + \int_0^t v_\mathrm{p}\,\mathrm{d}t_\mathrm{p}, \qquad \theta_\mathrm{p} = \theta_\mathrm{p0} + \int_0^t \frac{w_\mathrm{p}}{r}\,\mathrm{d}t_\mathrm{p} \tag{5.250}$$

In the numerical integration of a particle trajectory, the selection of the integral time step is important. A typical way of choosing the integral time step is based on the interacting duration between the turbulent eddy and the particle. This interacting duration τ_i may be determined by

$$\tau_\mathrm{i} = \min[\tau_\mathrm{e}, \tau_\mathrm{R}] \tag{5.251}$$

where τ_e is the eddy existence time, while τ_R is the flying time for the particle passing through the eddy. τ_e may be estimated from the k–ϵ model as

$$\tau_\mathrm{e} = \frac{l_\mathrm{e}}{\sqrt{\overline{u'^2} + \overline{v'^2} + \overline{w'^2}}} = \frac{l_\mathrm{e}}{\sqrt{2k}} \tag{5.252}$$

where l_e is the characteristic length of the turbulent eddy. Thus,

$$\tau_\mathrm{e} = \sqrt{\frac{3}{2}}C_\mu^{\frac{3}{4}}\frac{k}{\epsilon} \tag{5.253}$$

τ_R is determined by

$$\tau_\mathrm{R} = -\tau_\mathrm{rp}\ln\left(1 - \frac{l_\mathrm{e}}{\tau_\mathrm{rp}|\boldsymbol{U} - \boldsymbol{U}_\mathrm{p}|}\right) \tag{5.254}$$

and τ_i is taken to be τ_e when $l_\mathrm{e} > \tau_\mathrm{rp}|\boldsymbol{U} - \boldsymbol{U}_\mathrm{p}|$.

In computation using the stochastic trajectory model, the Monte Carlo approach is commonly employed. It is necessary to calculate several thousands, or even tens of thousands, of trajectories to simulate the particle flow field. The central issue in developing the stochastic trajectory model is how to model the instantaneous turbulent gas flow field. The method given uses the Gaussian distribution of velocity fluctuations. Others use the Langevin model, which is a statistical model [Elghobashi, 1994]. Recognizing the large computational requirement for the stochastic trajectory model, the alternative method is to calculate the most probable trajectories combined with the presumed probability density distribution functions of the particle velocity, concentration, and other quantities [Baxter, 1989].

5.5 Kinetic Theory Modeling for Collision-Dominated Dense Suspensions

When the particle concentration is high, the transport mechanism is significantly influenced or even dominated by interparticle collisions. For gravity-free suspensions of solid particles in a uniform shear flow, three distinct regimes were proposed by Bagnold (1954) on the basis of their rheological properties. These regimes are the macroviscous, the transitional, and the grain inertia regime, which are characterized by the Bagnold number, Ba, as

$$\mathrm{Ba} = \lambda^{\frac{1}{2}} \frac{d_{\mathrm{p}}^2 \rho_{\mathrm{p}}}{\mu} \left(\frac{\mathrm{d}U}{\mathrm{d}y} \right) = \frac{\text{inertial force}}{\text{viscous force}} \tag{5.255}$$

where λ is a parameter related to the solids fraction. The macroviscous regime is demarcated by $\mathrm{Ba} < 40$, where the fluid viscosity dominates the motion of solids, and the shear stress is linearly related to the shear rate. The particle motion in this regime can be described by using either the continuum modeling or the trajectory modeling. When $\mathrm{Ba} > 450$, where moving granular flows occur, the particle motion is largely governed by particle–particle interactions, and the shear stress depends on the square of the shear rate. This regime is thus defined as the grain inertia regime. The transition regime falls between the macroviscous and the grain inertia regimes [Hunt, 1989]. The preceding regime classifications have been validated both for shear flows [Savage, 1979; Savage and Sayed, 1984] and for hopper and chute flows [Zeininger and Brennen, 1985]. The study of grain inertia flows may be conducted in light of the kinetic theory modeling, where the interstitial gas plays little role in the momentum transport of solids.

Culick (1964) was among the earliest researchers who suggested an analogy between particle collisions in suspensions and molecular collisions in the kinetic theory of gases. However, the approach which rigorously followed the kinetic theories of gases for solid particles came to a halt as a result of the complexity surrounding the direct application of the Boltzmann equation in accounting for interparticle collisions. An alternative approach using simplified kinetic theories of gases based on mechanistically derived or intuitive relationships in place of the Boltzmann equation has been viewed as viable. For example, Savage and Jeffrey (1981), Jenkins and Savage (1983), and Lun *et al.* (1984) introduced elaborate collision models to determine the solid stress and other behavior of solids due to the interparticle collisions for collision-dominated dense suspensions. This approach has been applied to many gas–solid flow systems including fluidization and pneumatic transport [Gidaspow, 1993].

The physical condition of the kinetic theory of gases can be described by elastic collisions of monodispersed spheres with the Maxwellian velocity distribution in an infinite vacuum space. Therefore, for an analogy between particle–particle interactions and molecular interactions to be directly applicable, the following phenomena in gas–solid flows should not be regarded as significant in comparison to particle–particle interactions: the gas–particle interactions involving velocity-dependent forces and pressure-dependent forces, energy dissipation due to inelasticity and friction in particle collisions, and particle–wall interactions. In this section, the discussion is focused on the dense gas–solid suspensions, where collisions among particles are the only mechanism for the transport of mass, momentum, and energy.

5.5.1 *Dense-Phase Transport Theorem*

For a dense system of hard, smooth, and elastic spherical particles, a transport theorem based on the analogy of the kinetic theory of dense gases [Reif, 1965] may be derived. Define an ensemble average of any property ψ of a particle as

$$\langle\psi\rangle = \frac{1}{n}\int \psi f^{(1)}(\boldsymbol{r}, \boldsymbol{v}, t)\,\mathrm{d}\boldsymbol{v} \tag{5.256}$$

where n is the particle number density and $f^{(1)}(\boldsymbol{r}, \boldsymbol{v}, t)$ represents the single-particle velocity distribution function. Let us focus on the change of ψ in a fixed volume element $\mathrm{d}\boldsymbol{r}$ which is located between $\boldsymbol{r}$ and $\boldsymbol{r}+\mathrm{d}\boldsymbol{r}$ and contains $n\mathrm{d}\boldsymbol{r}$ particles. The rate of change of the ensemble average $\langle n\psi\,\mathrm{d}\boldsymbol{r}\rangle$ of the quantity ψ for all particles in $\mathrm{d}\boldsymbol{r}$ is affected by three factors: (1) the velocity $\boldsymbol{v}$ of each particle varies with time; (2) particles bearing ψ enter and leave the volume element $\mathrm{d}\boldsymbol{r}$; and (3) particles collide with each other in $\mathrm{d}\boldsymbol{r}$ [Jenkins and Savage, 1983]. Accordingly, the rate of increase of the ensemble average $\langle n\psi\,\mathrm{d}\boldsymbol{r}\rangle$ can be expressed by

$$\frac{\partial}{\partial t}\langle n\psi\rangle\,\mathrm{d}\boldsymbol{r} = A_{\mathrm{i}} + A_{\mathrm{f}} + A_{\mathrm{c}} \tag{5.257}$$

where A_{i} represents the rate of variation in the ensemble average of ψ due to the change of ψ for each particle inside $\mathrm{d}\boldsymbol{r}$; A_{f} accounts for the change of ψ due to the net flux of particles entering $\mathrm{d}\boldsymbol{r}$; and A_{c} describes the change of ψ due to particle collisions in $\mathrm{d}\boldsymbol{r}$.

In a time duration $\mathrm{d}t$, each particle changes its velocity by $\mathrm{d}\boldsymbol{v} = (\boldsymbol{F}/m)\,\mathrm{d}t$, where $\boldsymbol{F}$ is the total external force acting on an individual particle of mass m. The change in ψ can be given by

$$\frac{\partial\psi}{\partial v_{\mathrm{i}}}\frac{F_{\mathrm{i}}}{m}\,\mathrm{d}t = \frac{\boldsymbol{F}}{m}\cdot\frac{\partial\psi}{\partial\boldsymbol{v}}\,\mathrm{d}t \tag{5.258}$$

Hence, the rate of variation in the ensemble average of ψ in $\mathrm{d}\boldsymbol{r}$ is equal to

$$A_{\mathrm{i}} = \langle n\,\mathrm{d}\boldsymbol{r}\,D\psi\rangle = n\,\mathrm{d}\boldsymbol{r}\langle D\psi\rangle \tag{5.259}$$

where

$$D\psi = \frac{\boldsymbol{F}}{m}\cdot\frac{\partial\psi}{\partial\boldsymbol{v}} \tag{5.260}$$

A change of ψ in $\mathrm{d}\boldsymbol{r}$ is also possible through a net flux of particles entering the volume element. This increase rate is simply given by

$$A_{\mathrm{f}} = -\nabla\cdot\langle n\boldsymbol{v}\psi\,\mathrm{d}\boldsymbol{r}\rangle \tag{5.261}$$

The change of ψ due to collisions may be related to the changes of velocities of colliding particles in the form

$$\Delta\psi = \psi_1(\boldsymbol{r}, \boldsymbol{v}_1', t) + \psi_2(\boldsymbol{r}, \boldsymbol{v}_2', t) - \psi_1(\boldsymbol{r}, \boldsymbol{v}_1, t) - \psi_2(\boldsymbol{r}, \boldsymbol{v}_2, t) \tag{5.262}$$

The number of collisions is determined on the basis of three factors: (1) the scattering probability $\sigma(\boldsymbol{v}_1, \boldsymbol{v}_2 \rightarrow \boldsymbol{v}'_1, \boldsymbol{v}'_2)\, \mathrm{d}\boldsymbol{v}'_1\, \mathrm{d}\boldsymbol{v}'_2$; (2) the relative flux $|\boldsymbol{v}_1 - \boldsymbol{v}_2| f(\boldsymbol{r}, \boldsymbol{v}_1, t)\, \mathrm{d}\boldsymbol{v}_1$ of group 1 particles incident upon group 2 particles; and (3) the number of group 2 particles $f(\boldsymbol{r}, \boldsymbol{v}_2, t)\, \mathrm{d}\boldsymbol{r}\, \mathrm{d}\boldsymbol{v}_2$ which can do such scattering. Thus, A_c can be expressed as

$$A_\mathrm{c} = \frac{1}{2}\mathrm{d}\boldsymbol{r} \int \Delta\psi\, v_{12} f_1 f_2 \sigma\, \mathrm{d}\boldsymbol{v}_1\, \mathrm{d}\boldsymbol{v}_2\, \mathrm{d}\boldsymbol{v}'_1\, \mathrm{d}\boldsymbol{v}'_2 \tag{5.263}$$

where $v_{12} = |\boldsymbol{v}_1 - \boldsymbol{v}_2|$; $f_1 = f(\boldsymbol{r}, \boldsymbol{v}_1, t)$; $f_2 = f(\boldsymbol{r}, \boldsymbol{v}_2, t)$; and the factor of 1/2 accounts for the double counting of the colliding pairs in the preceding integration. The transport theorem, also known as Enskog's equation of change, can be obtained by substituting Eqs. (5.259), (5.261), and (5.263) into Eq. (5.257) as

$$\frac{\partial}{\partial t}\langle n\psi\rangle = n\langle D\psi\rangle - \nabla \cdot \langle n\boldsymbol{v}\psi\rangle + C(\psi) \tag{5.264}$$

where $C(\psi)$ is the collisional rate of increase of the mean of ψ per unit volume. From Eq. (5.263), $C(\psi)$ can be written as

$$C(\psi) = \frac{A_\mathrm{c}}{\mathrm{d}\boldsymbol{r}} = \frac{1}{2}\int \Delta\psi\, v_{12} f_1 f_2 \sigma\, \mathrm{d}\boldsymbol{v}_1\, \mathrm{d}\boldsymbol{v}_2\, \mathrm{d}\boldsymbol{v}'_1\, \mathrm{d}\boldsymbol{v}'_2 \tag{5.265}$$

For a system of hard, smooth, but inelastic particles of uniform diameter d_p, the collision term can be more explicitly expressed in terms of a collision transfer contribution and a source term [Lun *et al.*, 1984]. Consider a binary collision between the particles labeled 1 and 2 in Fig. 5.10. In time δt prior to the collision, particle 1 moves through a distance $\boldsymbol{v}_{12}\delta t$ relative to particle 2, where $\boldsymbol{v}_{12} = \boldsymbol{v}_1 - \boldsymbol{v}_2$. Thus, for a collision to occur within δt, the center of particle 1 must lie within the volume $d_\mathrm{p}^2 \delta\boldsymbol{k}(\boldsymbol{v}_{12}\cdot\boldsymbol{k})\delta t$. The probable number of collisions such that the center of particle 2 lies within the volume element $\delta\boldsymbol{r}$ and $\boldsymbol{v}_1$, $\boldsymbol{v}_2$, and $\boldsymbol{k}$ lie within the ranges $\delta\boldsymbol{v}_1$, $\delta\boldsymbol{v}_2$, and $\delta\boldsymbol{k}$ is

$$d_\mathrm{p}^2(\boldsymbol{v}_{12}\cdot\boldsymbol{k}) f^{(2)}(\boldsymbol{r} - d\boldsymbol{k}, \boldsymbol{v}_1; \boldsymbol{r}, \boldsymbol{v}_2; t)\delta\boldsymbol{k}\delta\boldsymbol{v}_1\delta\boldsymbol{v}_2\delta\boldsymbol{r}\delta t \tag{5.266}$$

where $f^{(2)}(\boldsymbol{v}_1, \boldsymbol{r}_1; \boldsymbol{v}_2, \boldsymbol{r}_2; t)$ is termed the complete pair distribution function and is defined such that $f^{(2)}(\boldsymbol{v}_1, \boldsymbol{r}_1; \boldsymbol{v}_2, \boldsymbol{r}_2; t)\delta\boldsymbol{v}_1\delta\boldsymbol{v}_2\delta\boldsymbol{r}_1\delta\boldsymbol{r}_2$ is the probability of finding a pair of particles in the volume element $\delta\boldsymbol{r}_1$, $\delta\boldsymbol{r}_2$ centered on the points $\boldsymbol{r}_1$, $\boldsymbol{r}_2$ and having velocities within the ranges $\boldsymbol{v}_1$ and $\boldsymbol{v}_1 + \delta\boldsymbol{v}_1$, and $\boldsymbol{v}_2$ and $\boldsymbol{v}_2 + \delta\boldsymbol{v}_2$, respectively.

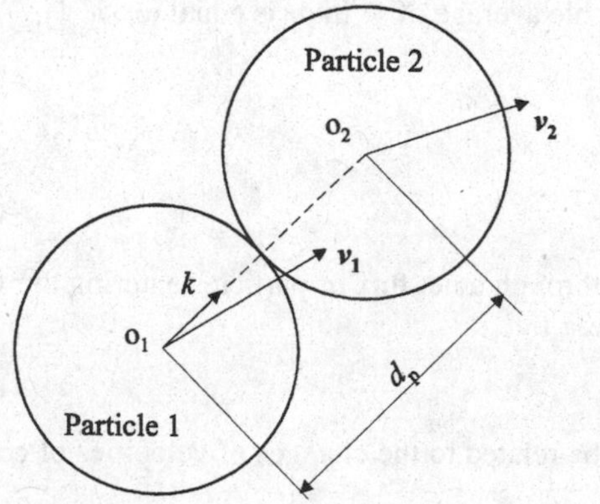

Figure 5.10. Binary collision of particles of diameter d_p.

During the collision, particle 2 gains a quantity $(\psi_2' - \psi_2)$ of ψ, where primed and unprimed quantities refer to the values after and before the collision, respectively. Considering only particles about to collide (*i.e.*, taking $\boldsymbol{v}_{12} \cdot \boldsymbol{k} > 0$), we can express the collisional rate of increase for the mean of ψ per unit volume as

$$C(\psi) = d_p^2 \int_{\boldsymbol{v}_{12}\cdot\boldsymbol{k}>0} (\psi_2' - \psi_2)(\boldsymbol{v}_{12} \cdot \boldsymbol{k}) f^{(2)}(\boldsymbol{r} - d\boldsymbol{k}, \boldsymbol{v}_1; \boldsymbol{r}, \boldsymbol{v}_2; t)\, d\boldsymbol{k}\, d\boldsymbol{v}_1\, d\boldsymbol{v}_2 \tag{5.267}$$

Similarly, by interchanging the subscripts 1 and 2 and replacing $\boldsymbol{k}$ by $-\boldsymbol{k}$, we can also write

$$C(\psi) = d_p^2 \int_{\boldsymbol{v}_{12}\cdot\boldsymbol{k}>0} (\psi_1' - \psi_1)(\boldsymbol{v}_{12} \cdot \boldsymbol{k}) f^{(2)}(\boldsymbol{r}, \boldsymbol{v}_1; \boldsymbol{r} + d\boldsymbol{k}, \boldsymbol{v}_2; t)\, d\boldsymbol{k}\, d\boldsymbol{v}_1\, d\boldsymbol{v}_2 \tag{5.268}$$

Expanding the pair-distribution function in a Taylor series, we have

$$\begin{aligned} f^{(2)}(\boldsymbol{r}, \boldsymbol{v}_1; \boldsymbol{r} + d\boldsymbol{k}, \boldsymbol{v}_2; t) &= f^{(2)}(\boldsymbol{r} - d\boldsymbol{k}, \boldsymbol{v}_1; \boldsymbol{r}, \boldsymbol{v}_2; t) \\ &\quad + \left(d\boldsymbol{k} \cdot \nabla - \frac{1}{2!}(d\boldsymbol{k} \cdot \nabla)^2 + \frac{1}{3!}(d\boldsymbol{k} \cdot \nabla)^3 + \cdots \right) \\ &\quad \times f^{(2)}(\boldsymbol{r}, \boldsymbol{v}_1; \boldsymbol{r} + d\boldsymbol{k}, \boldsymbol{v}_2; t) \end{aligned} \tag{5.269}$$

Substituting Eq. (5.269) into Eq. (5.268) and adding this to Eq. (5.267), we obtain an expression for $C(\psi)$ as

$$C(\psi) = -\nabla \cdot \boldsymbol{\theta} + \chi \tag{5.270}$$

where $\boldsymbol{\theta}$ is the collisional transfer contribution

$$\begin{aligned} \boldsymbol{\theta} = -\frac{d_p^3}{2} \int_{\boldsymbol{v}_{12}\cdot\boldsymbol{k}>0} & (\psi_1' - \psi_1)(\boldsymbol{v}_{12} \cdot \boldsymbol{k})\boldsymbol{k} \left[1 - \frac{1}{2}(d\boldsymbol{k} \cdot \nabla) + \frac{1}{3!}(d\boldsymbol{k} \cdot \nabla)^2 + \cdots \right] \\ & \times f^{(2)}(\boldsymbol{r}, \boldsymbol{v}_1; \boldsymbol{r} + d\boldsymbol{k}, \boldsymbol{v}_2; t)\, d\boldsymbol{k}\, d\boldsymbol{v}_1\, d\boldsymbol{v}_2 \end{aligned} \tag{5.271}$$

and χ is the source term

$$\chi = \frac{d_p^2}{2} \int_{\boldsymbol{v}_{12}\cdot\boldsymbol{k}>0} (\psi_2' + \psi_1' - \psi_1 - \psi_2)(\boldsymbol{v}_{12} \cdot \boldsymbol{k}) f^{(2)}(\boldsymbol{r} - d\boldsymbol{k}, \boldsymbol{v}_1; \boldsymbol{r}, \boldsymbol{v}_2; t)\, d\boldsymbol{k}\, d\boldsymbol{v}_1\, d\boldsymbol{v}_2 \tag{5.272}$$

Finally, for collisional dense phase suspensions of hard, smooth, and inelastic spherical particles, the transport theorem is expressed as

$$\frac{\partial}{\partial t}\langle n\psi \rangle = n\langle D\psi \rangle - \nabla \cdot \langle n\boldsymbol{v}\psi \rangle - \nabla \cdot \boldsymbol{\theta} + \chi \tag{5.273}$$

where $D\psi$, $\boldsymbol{\theta}$, and χ are given by Eqs. (5.260), (5.271), and (5.272), respectively.

5.5.2 Hydrodynamic Equations

The fundamental quantities which are conserved in a collision are the mass of a particle and each component of the total momentum of the colliding particles. Thus, conservation equations can be obtained by the use of the transport theorem. Although the energy equation can also be derived from the transport theorem, the total kinetic energy of the colliding particles is not conserved because of energy dissipation in the form of heat loss.

5.5.2.1 *Conservation of Mass*

Setting $\psi = m$, Eq. (5.273) leads to

$$\frac{\partial}{\partial t}(\alpha_p \rho_p) + \nabla \cdot (\alpha_p \rho_p \boldsymbol{U}_p) = 0 \tag{5.274}$$

where $\alpha_p \rho_p = nm$ and $\boldsymbol{U}_p = \langle \boldsymbol{v} \rangle$.

5.5.2.2 *Conservation of Momentum*

By taking $\psi = m\boldsymbol{v}$, we can obtain the local form of the balance equation of the linear momentum as

$$\alpha_p \rho_p \left(\frac{\partial \boldsymbol{U}_p}{\partial t} + \boldsymbol{U}_p \cdot \nabla \boldsymbol{U}_p \right) = -\nabla \cdot \boldsymbol{P}_p + \alpha_p \rho_p \frac{\boldsymbol{F}}{m} \tag{5.275}$$

where the total stress tensor $\boldsymbol{P}_p$ is the sum of the kinetic part $\boldsymbol{P}_k$ and the collisional part $\boldsymbol{P}_c$. $\boldsymbol{P}_k$ is given by

$$\boldsymbol{P}_k = \langle \alpha_p \rho_p \boldsymbol{u}\boldsymbol{u} \rangle \tag{5.276}$$

where $\boldsymbol{u} = \boldsymbol{v} - \langle \boldsymbol{v} \rangle = \boldsymbol{v} - \boldsymbol{U}_p$. $\boldsymbol{P}_c$ is given by

$$\boldsymbol{P}_c = \theta(m\boldsymbol{v}) = -\frac{1}{2} m d_p^3 \int_{\boldsymbol{v}_{12}\cdot\boldsymbol{k}>0} \left(\boldsymbol{v}_1' - \boldsymbol{v}_1\right)(\boldsymbol{v}_{12} \cdot \boldsymbol{k})\boldsymbol{k} \left[1 - \frac{1}{2}(\mathrm{d}\boldsymbol{k} \cdot \nabla) + \frac{1}{3!}(\mathrm{d}\boldsymbol{k} \cdot \nabla)^2 + \cdots \right]$$
$$\times f^{(2)}(\boldsymbol{r}_1, \boldsymbol{v}_1; \boldsymbol{r}_2, \boldsymbol{v}_2; t)\, \mathrm{d}\boldsymbol{k}\, \mathrm{d}\boldsymbol{v}_1\, \mathrm{d}\boldsymbol{v}_2 \tag{5.277}$$

5.5.2.3 *Equation of Energy*

Taking ψ to be $1/2mv^2$, it is noted that

$$\theta\left(\frac{1}{2} m v^2\right) = \boldsymbol{U}_p \cdot \boldsymbol{P}_p + \boldsymbol{q}_c \tag{5.278}$$

where $\boldsymbol{q}_c$ is the collisional contribution of the heat flux of fluctuation energy $\boldsymbol{q}_p$ and is expressed in terms of $1/2mu^2$ as given in the following:

$$\boldsymbol{q}_c = \theta\left(\frac{1}{2} m u^2\right)$$
$$= -\frac{1}{4} m d_p^3 \int_{\boldsymbol{v}_{12}\cdot\boldsymbol{k}>0} (u_1'^2 - u_1^2)(\boldsymbol{v}_{12} \cdot \boldsymbol{k})\boldsymbol{k} \left[1 - \frac{1}{2}(\mathrm{d}\boldsymbol{k} \cdot \nabla) + \frac{1}{3!}(\mathrm{d}\boldsymbol{k} \cdot \nabla)^2 + \cdots \right]$$
$$\times f^{(2)}(\boldsymbol{r}_1, \boldsymbol{v}_1; \boldsymbol{r}_2, \boldsymbol{v}_2; t)\, \mathrm{d}\boldsymbol{k}\, \mathrm{d}\boldsymbol{v}_1\, \mathrm{d}\boldsymbol{v}_2 \tag{5.279}$$

If we define T_c as the granular temperature to represent the specific kinetic energy of the velocity fluctuations or the translational fluctuation energy in such a way that

$$\frac{3}{2} T_c = \frac{1}{2} \langle u^2 \rangle \tag{5.280}$$

we obtain the energy equation in the form

$$\frac{3}{2}\alpha_{\mathrm{p}}\rho_{\mathrm{p}}\left(\frac{\partial T_{\mathrm{c}}}{\partial t}+\boldsymbol{U}_{\mathrm{p}}\cdot\nabla T_{\mathrm{c}}\right)=-\boldsymbol{P}_{\mathrm{p}}:\nabla\boldsymbol{U}_{\mathrm{p}}-\nabla\cdot\boldsymbol{q}_{\mathrm{p}}+\gamma \tag{5.281}$$

where γ is the collisional rate of dissipation per unit volume and is given by

$$\gamma=\chi\left(\frac{1}{2}mv^{2}\right)$$
$$=\frac{1}{2}md_{\mathrm{p}}^{2}\int_{\boldsymbol{v}_{12}\cdot\boldsymbol{k}>0}\left(v_{1}^{\prime 2}+v_{2}^{\prime 2}-v_{1}^{2}-v_{2}^{2}\right)(\boldsymbol{v}_{12}\cdot\boldsymbol{k})f^{(2)}(\boldsymbol{r}_{1},\boldsymbol{v}_{1};\boldsymbol{r},\boldsymbol{v}_{2};t)\,\mathrm{d}\boldsymbol{k}\,\mathrm{d}\boldsymbol{v}_{1}\,\mathrm{d}\boldsymbol{v}_{2} \tag{5.282}$$

and $\boldsymbol{q}_{\mathrm{p}}$ is the heat flux of fluctuation energy, which consists of the kinetic contribution $\boldsymbol{q}_{\mathrm{k}}$ and the collisional part $\boldsymbol{q}_{\mathrm{c}}$. The kinetic part $\boldsymbol{q}_{\mathrm{k}}$ is given by

$$\boldsymbol{q}_{\mathrm{k}}=\frac{1}{2}\langle\alpha_{\mathrm{p}}\rho_{\mathrm{p}}u^{2}\boldsymbol{u}\rangle \tag{5.283}$$

Consider the collision of two spherical particles that are identical, smooth, but inelastic. The loss of kinetic energy in the collision is given by Eq. (2.6)

$$\frac{1}{2}m\left(v_{1}^{\prime 2}+v_{2}^{\prime 2}-v_{1}^{2}-v_{2}^{2}\right)=\frac{1}{4}m(e^{2}-1)(\boldsymbol{k}\cdot\boldsymbol{v}_{12})^{2} \tag{5.284}$$

Hence, the collisional rate of dissipation and the collisional stress tensor are related to the coefficient of restitution by Eq. (5.284) and Eq. (5.277).

5.5.3 *Collisional Pair Distribution Function*

In order to evaluate the collisional integrals $\boldsymbol{P}_{\mathrm{c}}$, $\boldsymbol{q}_{\mathrm{c}}$, and γ, explicitly, it is important to know the specific form of the pair distribution function $f^{(2)}(\boldsymbol{v}_{1},\boldsymbol{r}_{1};\boldsymbol{v}_{2},\boldsymbol{r}_{2};t)$. The pair distribution function $f^{(2)}$ may be related to the single-particle velocity distribution function $f^{(1)}$ by introducing a configurational pair-correlation function $g(\boldsymbol{r}_{1},\boldsymbol{r}_{2})$. In the following, we first introduce the distribution functions and then derive the expression of $f^{(2)}$ in terms of $f^{(1)}$ by assuming $f^{(1)}$ is Maxwellian and particles are nearly elastic (*i.e.*, $1-e\ll 1$).

Define an L-particle configurational distribution function $n^{(\mathrm{L})}(\boldsymbol{r}_{1},\boldsymbol{r}_{2},\ldots,\boldsymbol{r}_{\mathrm{L}})$ such that $n^{(\mathrm{L})}(\boldsymbol{r}_{1},\boldsymbol{r}_{2},\ldots,\boldsymbol{r}_{\mathrm{L}})\delta\boldsymbol{r}_{1}\cdots\delta\boldsymbol{r}_{\mathrm{L}}$ is the probability of finding a particle in each of the volume elements $\delta\boldsymbol{r}_{1},\delta\boldsymbol{r}_{2},\ldots,\delta\boldsymbol{r}_{\mathrm{L}}$ centered on $\boldsymbol{r}_{1},\boldsymbol{r}_{2},\ldots,\boldsymbol{r}_{\mathrm{L}}$. The one-particle distribution function $n^{(1)}(\boldsymbol{r})$ is just the number density n of particles at $\boldsymbol{r}$. For a homogeneous bulk phase, we have

$$n^{(1)}(\boldsymbol{r})=\frac{N_{\mathrm{p}}}{V}=n \tag{5.285}$$

where N_{p} is the averaged number of particles in the volume V.

For an amorphous mass of particles there is no correlation between particles that are far apart. The joint probability of finding particles at $\boldsymbol{r}_{1}$ and $\boldsymbol{r}_{2}$ is simply the product of the individual probabilities. Let us define a configurational pair-correlation function $g(\boldsymbol{r}_{1},\boldsymbol{r}_{2})$ as

$$g(\boldsymbol{r}_{1},\boldsymbol{r}_{2})=\frac{n^{(2)}(\boldsymbol{r}_{1},\boldsymbol{r}_{2})}{n^{2}} \tag{5.286}$$

so that

$$g(\boldsymbol{r}_{1},\boldsymbol{r}_{2})=\frac{n^{(2)}(\boldsymbol{r}_{1},\boldsymbol{r}_{2})}{n^{2}}\approx 1\quad\text{for }|\boldsymbol{r}_{1}-\boldsymbol{r}_{2}|\gg d_{\mathrm{p}} \tag{5.287}$$

For a gas at equilibrium, *i.e.*, no mean deformation, there is a spatial homogeneity and thus $g(\boldsymbol{r}_1, \boldsymbol{r}_2)$ depends only on the separation distance $r = |\boldsymbol{r}_1 - \boldsymbol{r}_2|$. Then $g = g_0(r)$ is termed the radial distribution function, which may be interpreted as the ratio of the local number density at a distance r from the central particle to the bulk number density. For a system of identical spheres, the radial distribution function $g_0(r)$ at contact (*i.e.*, $r = d_\mathrm{p}$) can be expressed in terms of the volume fraction of solids α_p as

$$g_0(d_\mathrm{p}, \alpha_\mathrm{p}) = \frac{1}{(1-\alpha_\mathrm{p})} + \frac{3\alpha_\mathrm{p}}{2(1-\alpha_\mathrm{p})^2} + \frac{\alpha_\mathrm{p}^2}{2(1-\alpha_\mathrm{p})^3} \tag{5.288}$$

which is based on a semiempirical equation of state [Carnahan and Starling, 1969].

Assume that a complete pair distribution function can be expressed as the product of the spatial pair distribution function and the two single-particle velocity distribution functions. Thus, we have

$$f^{(2)}(\boldsymbol{v}_1, \boldsymbol{r}_1; \boldsymbol{v}_2, \boldsymbol{r}_2, t) = g(\boldsymbol{r}_1, \boldsymbol{r}_2, t) f^{(1)}(\boldsymbol{v}_1, \boldsymbol{r}_1, t) f^{(1)}(\boldsymbol{v}_2, \boldsymbol{r}_2, t) \tag{5.289}$$

The anisotropy in $g(\boldsymbol{r}_1, \boldsymbol{r}_2)$ may be determined by the use of a kinematic argument. Consider a bulk of particle subjected to a mean shear flow. The radial distribution function, which is spherical in equilibrium, becomes distorted into an ellipsoidal distribution as a result of the presence of the mean shear. Hence, in order for $g(\boldsymbol{r}_1, \boldsymbol{r}_2)$ to exhibit an anisotropy, $g(\boldsymbol{r}_1, \boldsymbol{r}_2)$ should depend not only on α_p, $\boldsymbol{r}_1$, and $\boldsymbol{r}_2$ but also on T_c, $\boldsymbol{v}_1$, and $\boldsymbol{v}_2$. For dimensional homogeneity, g can only be a function of α_p, $\boldsymbol{k} \cdot \boldsymbol{U}_{21}/T_\mathrm{c}^{1/2}$, and U_{21}^2/T_c. For a small deformation rate (or when the magnitude of $\boldsymbol{U}_{21}$ is small relative to $T_\mathrm{c}^{1/2}$), it is assumed that $g(\boldsymbol{r}_1, \boldsymbol{r}_2)$ takes the form [Jenkins and Savage, 1983]

$$g(\boldsymbol{r}_1, \boldsymbol{r}_2) = g_0\left[1 - \frac{\boldsymbol{k} \cdot \boldsymbol{U}_{21}}{\sqrt{\pi T_\mathrm{c}}}\right] \tag{5.290}$$

Assuming that the single-particle velocity distribution is Maxwellian and takes the form

$$f^{(1)}(\boldsymbol{v}, \boldsymbol{r}) = n\left(\frac{1}{2\pi T_\mathrm{c}}\right)^{3/2} \exp\left[-\frac{(\boldsymbol{v} - \boldsymbol{U}_\mathrm{p})^2}{2T_\mathrm{c}}\right] \tag{5.291}$$

we can express the complete pair distribution function at collision as

$$f^{(2)}(\boldsymbol{v}_1, \boldsymbol{r}_1; \boldsymbol{v}_2, \boldsymbol{r}_2) = g_0 n_1 n_2 \left(\frac{1}{4\pi^2 T_{\mathrm{c}1} T_{\mathrm{c}2}}\right)^{3/2} \left[1 - \frac{\boldsymbol{k} \cdot \boldsymbol{U}_{21}}{\sqrt{\pi T_\mathrm{c}}}\right]$$
$$\times \exp\left(-\left[\frac{(\boldsymbol{v}_1 - \boldsymbol{U}_{\mathrm{p}1})^2}{2T_{\mathrm{c}1}} + \frac{(\boldsymbol{v}_2 - \boldsymbol{U}_{\mathrm{p}2})^2}{2T_{\mathrm{c}2}}\right]\right) \tag{5.292}$$

Consider two spheres in contact at $\boldsymbol{r}$, as shown in Fig. 5.11, such that $\boldsymbol{r}_1 = \boldsymbol{r} - (d_\mathrm{p}/2)\boldsymbol{k}$ and $\boldsymbol{r}_2 = \boldsymbol{r} + (d_\mathrm{p}/2)\boldsymbol{k}$. We may expand $f^{(2)}$ in terms of the Taylor series about the point of contact $\boldsymbol{r}$. This allows the evaluation of the mean fields and the corresponding derivatives at the contact point. Discarding the terms with spatial derivatives higher than the first order, we can obtain the approximation of this expansion [Chapman and Cowling, 1970] as

$$f^{(2)}(\boldsymbol{v}_1, \boldsymbol{r}_1; \boldsymbol{v}_2, \boldsymbol{r}_2) = g_0 f_{10}^{(1)} f_{20}^{(1)} \left(1 - \frac{\mathrm{d}\boldsymbol{k}\boldsymbol{k} : \nabla \boldsymbol{U}_\mathrm{p}}{\sqrt{\pi T_\mathrm{c}}} + \frac{d_\mathrm{p}}{2}\boldsymbol{k} \cdot \nabla\left[\ln\frac{f_{10}^{(1)}}{f_{20}^{(1)}}\right]\right) \tag{5.293}$$

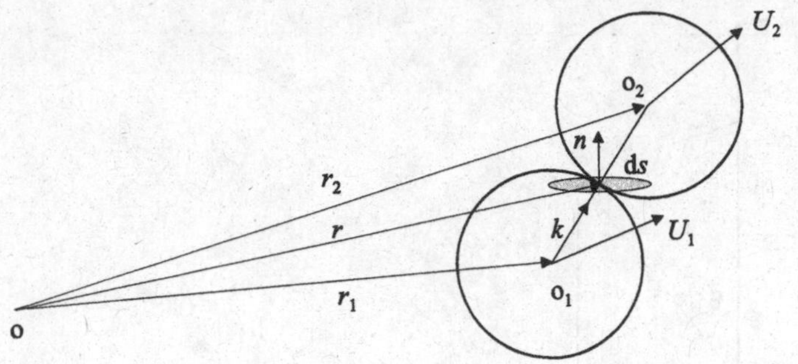

Figure 5.11. Schematic diagram of two spheres in contact at r.

where $f_{i0}^{(1)}$ (i = 1, 2) is

$$f_{i0}^{(1)} = n\left(\frac{1}{2\pi T_c}\right)^{3/2} \exp\left[-\frac{(v_i - U_p)^2}{2T_c}\right] \tag{5.294}$$

5.5.4 *Constitutive Relations*

Now, we proceed to the evaluation of the collisional integrals P_c, q_c, and γ by using Eq. (5.293) as the form of a collisional pair distribution function. Use the coordinates in Fig. 5.12, in which e_z is chosen to be parallel to the relative velocity v_{12}. θ and ϕ are the polar angles of k with respect to e_z and the plane of e_z and e_x, respectively. e_x, e_y, and e_z are the three mutually perpendicular unit vectors corresponding to each coordinate in Fig. 5.12. k, as mentioned in the previous section, is the unit normal on the collision point directed from the center of particle 1 to the center of particle 2. Thus, we have

$$k = e_x \sin\theta\cos\phi + e_y \sin\theta\sin\phi + e_z\cos\theta \tag{5.295}$$

The element of solid angle dk is given by

$$dk = \sin\theta \, d\theta \, d\phi \tag{5.296}$$

Note that the integrals are to be taken over all values of k for which $v_{12} \cdot k$ is positive. Since $v_{12} \cdot k = v_{12}\cos\theta$, the limits of integration are, for θ, 0, and $\pi/2$, and for ϕ, 0, and 2π. All terms of the integrand containing odd powers of $\sin\phi$ or $\cos\phi$ become zero.

It is noted that the bouncing velocities of the colliding particles may be expressed by

$$\begin{aligned} v_1' &= v_1 - \frac{1}{2}(1+e)(k \cdot v_{12})k \\ v_2' &= v_2 + \frac{1}{2}(1+e)(k \cdot v_{12})k \end{aligned} \tag{5.297}$$

The first-order approximation of the collisional stress tensor P_c may be simplified from Eq. (5.277) and Eq. (5.297) as

$$P_c \approx \frac{1}{4} m d_p^3 (1+e) \int_{v_{12}\cdot k>0} k(v_{12} \cdot k)^3 f^{(2)}(r_1, v_1; r_2, v_2; t)\, dk\, dv_1\, dv_2 \tag{5.298}$$

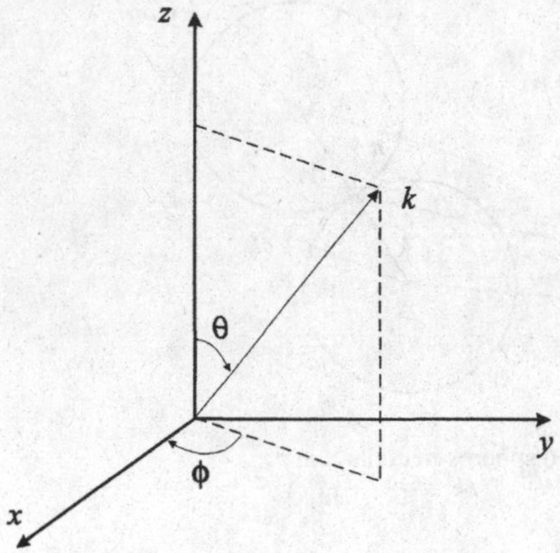

Figure 5.12. The coordinate system for the evaluation of collisional pair function.

Similarly, from Eq. (5.284), $\boldsymbol{q}_c$ can be expressed as

$$\boldsymbol{q}_c \approx -\frac{md_p^3}{4}\int_{\boldsymbol{v}_{12}\cdot\boldsymbol{k}>0}\left(u_1'^2 - u_1^2\right)\boldsymbol{k}\boldsymbol{v}_{12}\cdot\boldsymbol{k}f^{(2)}(\boldsymbol{r}_1,\boldsymbol{v}_1;\boldsymbol{r}_2,\boldsymbol{v}_2;t)\,\mathrm{d}\boldsymbol{k}\,\mathrm{d}\boldsymbol{v}_1\,\mathrm{d}\boldsymbol{v}_2 \tag{5.299}$$

where $u_1'^2 - u_1^2$ can be further expressed by Eq. (5.297) as

$$u_1'^2 - u_1^2 = -(1+e)\left[\left(\frac{\boldsymbol{v}_1+\boldsymbol{v}_2}{2} - \boldsymbol{U}_p\right)\cdot\boldsymbol{k}\right]\boldsymbol{k}\cdot\boldsymbol{v}_{12} - \frac{1-e^2}{4}(\boldsymbol{k}\cdot\boldsymbol{v}_{12})^2 \tag{5.300}$$

From Eq. (5.282) and Eq. (5.284), γ becomes

$$\gamma \approx -\frac{md_p^2}{4}(1-e^2)\int_{\boldsymbol{v}_{12}\cdot\boldsymbol{k}>0}(\boldsymbol{v}_{12}\cdot\boldsymbol{k})^2 f^{(2)}(\boldsymbol{r}_1,\boldsymbol{v}_1;\boldsymbol{r},\boldsymbol{v}_2;t)\,\mathrm{d}\boldsymbol{k}\,\mathrm{d}\boldsymbol{v}_1\,\mathrm{d}\boldsymbol{v}_2 \tag{5.301}$$

It is convenient to derive a set of basic integrals to aid in the integration of Eqs. (5.298), (5.299), and (5.301). From Eq. (5.295) and Eq. (5.296), we have [Chapman and Cowling, 1970; Jenkins and Savage, 1983]

$$\int \boldsymbol{k}(\boldsymbol{v}_{12}\cdot\boldsymbol{k})\,\mathrm{d}\boldsymbol{k} = \frac{2\pi}{3}\boldsymbol{v}_{12} \tag{5.302}$$

$$\int \boldsymbol{k}\boldsymbol{k}(\boldsymbol{v}_{12}\cdot\boldsymbol{k})^2\,\mathrm{d}\boldsymbol{k} = \frac{2\pi}{15}\left(2\boldsymbol{v}_{12}\boldsymbol{v}_{12} + v_{12}^2\boldsymbol{I}\right) \tag{5.303}$$

where $\boldsymbol{I}$ is the unit tensor.

If $\boldsymbol{a}$ is any vector independent of θ and ϕ, it can be obtained that

$$\int (\boldsymbol{a}\cdot\boldsymbol{k})(\boldsymbol{v}_{12}\cdot\boldsymbol{k})^3\,\mathrm{d}\boldsymbol{k} = \frac{2\pi}{5}v_{12}^2(\boldsymbol{a}\cdot\boldsymbol{v}_{12}) \tag{5.304}$$

and

$$\int (\boldsymbol{a}\cdot\boldsymbol{k})^2(\boldsymbol{v}_{12}\cdot\boldsymbol{k})^3\,\mathrm{d}\boldsymbol{k} = \frac{\pi}{12}v_{12}\left[3(\boldsymbol{a}\cdot\boldsymbol{v}_{12})^2 + v_{12}^2a^2\right] \tag{5.305}$$

Note that in Eq. (5.293)

$$f_{10}^{(1)} f_{20}^{(1)} = \frac{n^2}{2\pi T_c^3} \exp\left[-\frac{1}{2T_c}\left[(v_1 - U_p)^2 + (v_2 - U_p)^2\right]\right] \tag{5.306}$$

and

$$\ln\left(\frac{f_{20}^{(1)}}{f_{10}^{(1)}}\right) = \frac{1}{2T_c}\left[(v_1 - U_p)^2 - (v_2 - U_p)^2\right] \tag{5.307}$$

Hence, after some manipulation, the constitutive relations can be obtained [Jenkins and Savage, 1983; Lun *et al.*, 1984] as follows:
(1) Collisional flux of fluctuation energy $\boldsymbol{q}_c$

$$\boldsymbol{q}_c = -k\nabla T_c \tag{5.308}$$

where

$$k = 2\rho_p\alpha_p^2 g_0 d_p(1+e)\sqrt{\frac{T_c}{\pi}} \tag{5.309}$$

(2) Collisional stress tensor $\boldsymbol{P}_c$

$$\boldsymbol{P}_c = \left(\frac{k}{d_p}\sqrt{\pi T_c} - \frac{3k}{5}\operatorname{tr}\boldsymbol{D}\right)\boldsymbol{I} - \frac{6k}{5}\boldsymbol{D} \tag{5.310}$$

where $\boldsymbol{D}$ is the tensor defined as

$$\boldsymbol{D} = \frac{1}{2}\left[\nabla \boldsymbol{U}_p + (\nabla \boldsymbol{U}_p)^T\right] \tag{5.311}$$

where the superscript T denotes the transpose, and "tr" denotes the trace. Thus,

$$\operatorname{tr}\boldsymbol{D} = D_{ii} = D_{11} + D_{22} + D_{33} \tag{5.312}$$

where D_{ii} represents the Einstein summation.
(3) Rate of energy dissipation γ

$$\gamma = \frac{6(1-e)k}{d_p^2}\left[T_c - \left(\frac{\pi}{4} + \frac{1}{3}\right)d_p\sqrt{\frac{T_c}{\pi}}\operatorname{tr}\boldsymbol{D}\right] \tag{5.313}$$

So far, we are able to construct the constitutive equations for q_c, P_c, and γ. For moderate solids concentrations, we can neglect the kinetic contributions in comparison to the collisional ones. Thus, we can assume $\boldsymbol{P}_k \ll \boldsymbol{P}_c$ and $\boldsymbol{q}_k \ll \boldsymbol{q}_c$. Substituting the constitutive relations into Eqs. (5.274), (5.275), and (5.281), after neglecting the kinetic contributions, yields five equations for the five unknowns α_p, $\boldsymbol{U}_p$, and T_c (or $\langle u^2\rangle$). Hence, the closure problem is resolved.

Example 5.2 On the basis of the kinetic theory, which is used to model collision-dominated gas–solid flows, derive a general expression of solid stresses of elastic spheres in a simple shear flow.

Solution For elastic particles, $e = 1$. For a simple shear flow, $\boldsymbol{D}$ can be expressed as

$$\boldsymbol{D} = \frac{1}{2}\frac{\mathrm{d}U_\mathrm{p}}{\mathrm{d}y}\begin{pmatrix} 0 & 1 & 0 \\ 1 & 0 & 0 \\ 0 & 0 & 0 \end{pmatrix} \tag{E5.2}$$

Therefore, $\mathrm{tr}\,\boldsymbol{D} = 0$; and k is given by

$$k = 4\rho_\mathrm{p}\alpha_\mathrm{p}^2 g_0 d_\mathrm{p}\sqrt{\frac{T_\mathrm{c}}{\pi}} \tag{E5.3}$$

The collisional stress tensor $\boldsymbol{P}_\mathrm{c}$ can be obtained from Eq. (5.310) as

$$\boldsymbol{P}_\mathrm{c} = 4\alpha_\mathrm{p}^2 g_0\rho_\mathrm{p}T_\mathrm{c}\begin{pmatrix} 1 & 0 & 0 \\ 0 & 1 & 0 \\ 0 & 0 & 1 \end{pmatrix} - \frac{12}{5}\sqrt{\frac{T_\mathrm{c}}{\pi}}\alpha_\mathrm{p}^2 g_0\rho_\mathrm{p}d_\mathrm{p}\frac{\mathrm{d}U_\mathrm{p}}{\mathrm{d}y}\begin{pmatrix} 0 & 1 & 0 \\ 1 & 0 & 0 \\ 0 & 0 & 0 \end{pmatrix} \tag{E5.4}$$

The first term gives rise to the particle phase pressure, which, at low shear rates, is isotropic. The second term is the shear stresses, which indicate the effective viscosity in the form of

$$\mu_\mathrm{p} = \frac{12}{5}\sqrt{\frac{T_\mathrm{c}}{\pi}}\alpha_\mathrm{p}^2 g_0\rho_\mathrm{p}d_\mathrm{p} \tag{E5.5}$$

Thus, in this case, the elastic granular flow is Newtonian.

Example 5.3 Consider the two-dimensional gravity flow of granular materials down an inclined plane, as shown in Fig. E5.1 [Savage, 1983]. Assume that the flow is fully developed and collision-dominated with a free surface.

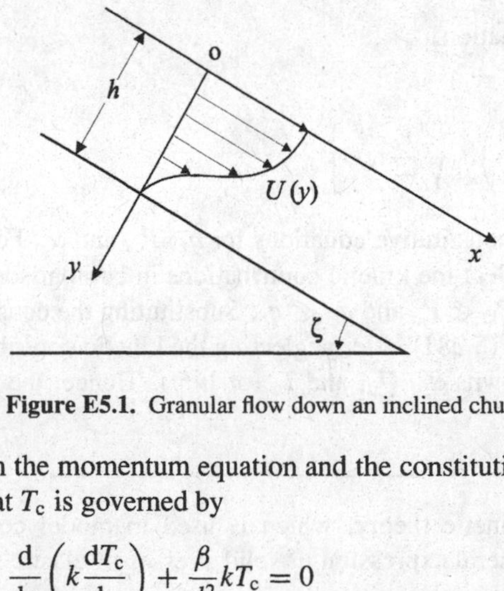

Figure E5.1. Granular flow down an inclined chute.

(1) From the momentum equation and the constitutive relation of collisional stress tensor, show that T_c is governed by

$$\frac{\mathrm{d}}{\mathrm{d}y}\left(k\frac{\mathrm{d}T_\mathrm{c}}{\mathrm{d}y}\right) + \frac{\beta}{d_\mathrm{p}^2}kT_\mathrm{c} = 0 \tag{E5.6}$$

where β is

$$\beta = \frac{5\pi}{3}\tan^2\zeta - 6(1-e) \tag{E5.7}$$

(2) Assume that there are no slip of solids and no flux of fluctuation energy at the wall. Determine the distributions of T_c, U_p, and α_p.

Solution (1) From the momentum equation Eq. (5.275) and the constitutive relation of collisional stress tensor Eq. (5.310), we have

$$P_{c_{yy}} = \frac{k}{d_p}\sqrt{\pi T_c} = g\cos\zeta\int_0^y \alpha_p\rho_p\,dy \tag{E5.8}$$

and

$$P_{c_{yx}} = -\frac{3k}{5}\frac{dU_p}{dy} = g\sin\zeta\int_0^y \alpha_p\rho_p\,dy \tag{E5.9}$$

Equations (E5.8) and (E5.9) lead to

$$\frac{dU_p}{dy} = -\frac{5\sqrt{\pi T_c}}{3d_p}\tan\zeta \tag{E5.10}$$

The fluctuation energy equation Eq. (5.281) can be expressed by

$$\frac{3}{2}\rho_p\frac{dT_c}{dt} = 0 = -P_{c_{xy}}\frac{dU_p}{dy} + \frac{d}{dy}\left(k\frac{dT_c}{dy}\right) - \gamma \tag{E5.11}$$

where the rate of dissipation is given by Eq. (5.313)

$$\gamma = \frac{6(1-e)}{d_p^2}kT_c \tag{E5.12}$$

Equation (E5.11) gives the relationship among the shear work, the energy flux gradient, and the rate of dissipation. Substituting Eq. (E5.9) and Eq. (E5.10) into Eq. (E5.11) yields Eq. (E5.6).

(2) From Eq. (E5.8), the following approximation holds

$$\frac{k}{d_p}\sqrt{\pi T_c} \approx g\cos\zeta\,\overline{\alpha}_p\rho_p y \tag{E5.13}$$

where the averaged value of α_p is defined by $\overline{\alpha}_p$ as given in the following:

$$\overline{\alpha}_p = \frac{1}{h}\int_0^h \alpha_p\,dy \tag{E5.14}$$

Substituting Eq. (E5.13) into Eq. (E5.6) and defining f as $T_c^{1/2}$, we have

$$\frac{d}{dy}\left(y\frac{df}{dy}\right) + \frac{\beta y}{2d_p^2}f = 0 \tag{E5.15}$$

which is the Bessel equation of zero order. At the free surface, there is no flux of fluctuation energy. Thus, it gives

$$\left(\frac{\mathrm{d}f}{\mathrm{d}y}\right)_{y=0} = 0 \tag{E5.16}$$

Note that for a fully developed flow, an energy flux from the granular materials to the wall corresponds to $\beta > 0$, whereas an energy flux from the wall to the granular materials corresponds to $\beta < 0$. For zero energy flux into the wall, $\beta = 0$, *i.e.*,

$$\tan\zeta = \tan\zeta_0 = \sqrt{\frac{18(1-e)}{5\pi}} \tag{E5.17}$$

where ζ_0 is the natural angle of repose.

Applying Eq. (E5.13) at the wall and using Eq. (5.309) yield the fluctuation-specific kinetic energy or the "granular temperature" T_c as

$$T_c = \frac{g\cos\zeta\,\overline{\alpha}_p h}{2\alpha_{pw}^2 g_0(\alpha_{pw})(1+e)} \tag{E5.18}$$

where α_{pw} is the volume fraction of the particles at the wall. Thus, from the nonslip condition at the wall, we can obtain the velocity profile from Eq. (E5.10) as

$$U_p = \frac{5}{3}\frac{\sqrt{\pi T_c}}{d_p}\tan\zeta_0(h-y) \tag{E5.19}$$

The distribution of volume fraction of the particles can also be obtained from Eq. (E5.13) and Eq. (E5.9) as

$$\frac{\alpha_p^2 g_0(\alpha_p)}{\alpha_{pw}^2 g_0(\alpha_{pw})} = \frac{y}{h} \tag{E5.20}$$

Two interesting points can be noted from this example. First, for elastic materials, $\zeta_0 = 0$ and it is impossible to have an energy flux from the wall to the granular materials. Second, for inelastic materials, if fluctuation energy is put into the materials by using a vibrating device, the slopes required for the chute flow can be considerably less than their natural angle of repose.

5.6 Equations for Flows Through Packed Beds

The physical understanding of gas flows through a packed bed is important for the study of the gas–solid transport system as it represents the limiting case of the gas–solid transport. A typical example is the minimum fluidization condition of gas–solid fluidization. In a gas–solid flow system, the presence of unsuspended particles is also common, and, thus, understanding the gas flow in a packed bed is essential.

Experiments were conducted by Darcy (1856) to study viscous flows through homogeneous porous media, which resulted in the well-known Darcy law. There are various models from which Darcy's law may be derived to describe gas flow through a packed bed

[Scheidegger, 1960; Bear, 1972]. Among them, the capillary tube model is considered to be the simplest. In this model, the porous medium is characterized by a bundle of straight circular capillary tubes with permeability of the tube bundle equal to that of the actual medium. Thus, the pressure drop through a single, straight circular tube can be described by the Hagen–Poiseuille equation derived for the steady tubular flow.

The Darcy law does not consider the effects of inertia. For a more general description of flows in a packed bed, particularly when the effects of inertia are important, Ergun (1952) presented a semiempirical equation covering a wide range of flow conditions, which is known as the Ergun equation [Ergun and Orning, 1949; Ergun, 1952].

5.6.1 *Darcy's Law*

To study the general behavior of viscous flow through homogeneous porous media, experiments were designed and performed by Darcy (1856). A schematic drawing of Darcy's experiment is shown in Fig. 5.13. A homogeneous filter bed, of length L in the test section and uniform cross-sectional area A, is filled with an incompressible liquid. The pressure difference for a given flow rate is measured by the difference of the manometer fluid level, Δh. By varying the experimental variables, the original form of Darcy's law can be deduced as

$$Q = -\frac{KA\Delta h}{L} \tag{5.314}$$

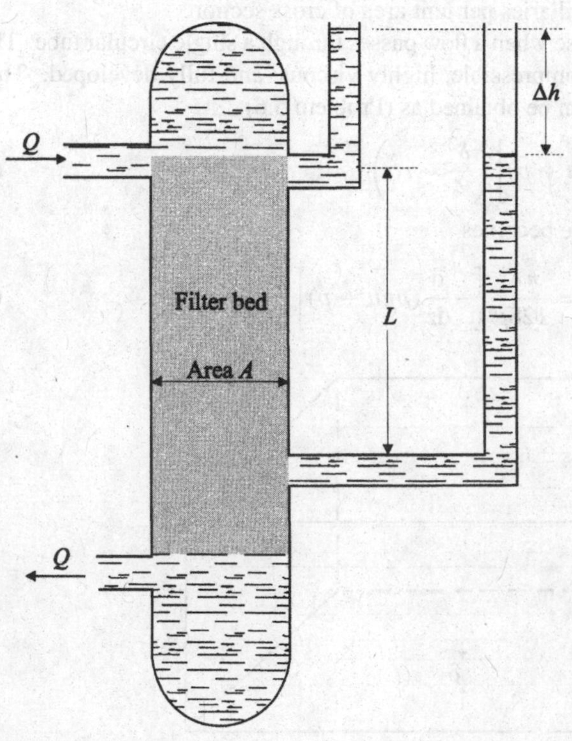

Figure 5.13. Schematic diagram of Darcy's filtration experiment (from Darcy, 1856).

where Q is the volumetric flow rate and K is a constant depending on the properties of the fluid and of the porous medium. Δh can be expressed in terms of the pressure difference ΔP. Furthermore, to separate the influence of the porous medium from that of the fluid, the constant K can be replaced by the ratio of the specific permeability of the porous medium k to the viscosity of the fluid μ. Hence, Eq. (5.314) becomes

$$Q = -\frac{k}{\mu} A \frac{\Delta P}{L} \tag{5.315}$$

or in a differential form

$$U = -\frac{k}{\mu} \frac{\mathrm{d}P}{\mathrm{d}z} \tag{5.316}$$

where U is the superficial velocity of the fluid.

Equations (5.315) and (5.316) are the commonly used forms of Darcy's law. It is noted that the gravitational effect is considered negligible in the Darcy's law analysis.

5.6.2 *Straight Capillaric Model*

The simplest capillaric model is the one representing a porous medium by a bundle of straight parallel capillaries of average diameter δ, as shown in Fig. 5.14. The equivalent voidage α can be related to the averaged diameter by

$$\alpha = \frac{\pi}{4} n \delta^2 \tag{5.317}$$

where n is the number of capillaries per unit area of cross section.

Let us first consider the case when a flow passes through a single circular tube. The flow is assumed to be steady, incompressible, highly viscous, and fully developed. Thus, the radial velocity distribution can be obtained as (Problem 5.6)

$$W = \frac{1}{4\mu}\left[-\frac{\mathrm{d}}{\mathrm{d}z}(\rho g h + p)\right]\left(\frac{\delta^2}{4} - r^2\right) \tag{5.318}$$

The flow rate through the tube becomes

$$Q = \int_0^{\frac{\delta}{2}} W 2\pi r \mathrm{d}r = \frac{\pi \delta^4}{128\mu}\left[-\frac{\mathrm{d}}{\mathrm{d}z}(\rho g h + p)\right] \tag{5.319}$$

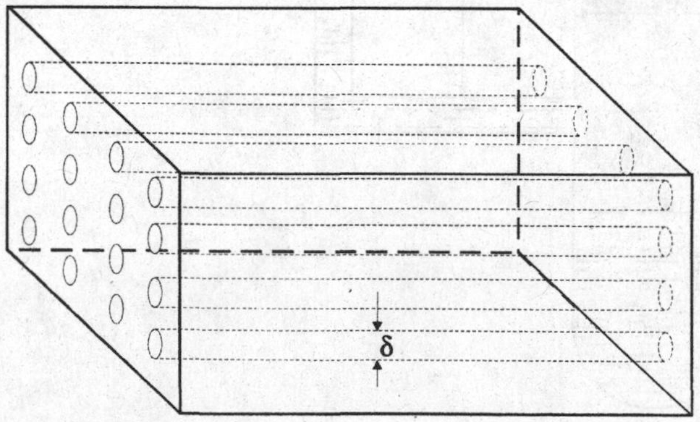

Figure 5.14. Straight capillaric model.

which is the Hagen–Poiseuille equation. Equation (5.319) indicates that the pressure drop is proportional to the flow rate (or the cross-sectional averaged velocity) for highly viscous pipe flows. The flow per unit area, or the superficial velocity in a porous medium, becomes

$$U = -\frac{n\pi\delta^4}{128\mu}\frac{\mathrm{d}}{\mathrm{d}z}(\rho gh + p) \tag{5.320}$$

By comparing the preceding equation with Darcy's law, the specific permeability k can be given as

$$k = \frac{\alpha\delta^2}{32} \tag{5.321}$$

It is noted that straight capillary tubes may not portray the complex structure of the porous medium. Thus, in practice, the factor 32 in Eq. (5.321) is commonly replaced by an empirical parameter known as tortuosity. The tortuosity accounts for the tortuous paths of the porous medium.

5.6.3 *Ergun's Equation*

The generalized relation for the pressure drop for flows through a packed bed was formulated by Ergun (1952). The pressure loss was considered to be caused by simultaneous kinetic and viscous energy losses. In Ergun's formulation, four factors contribute to the pressure drop. They are (1) fluid flow rate, (2) properties of the fluid (such as viscosity and density), (3) closeness (such as porosity) and orientation of packing, and (4) size, shape, and surface of the solid particles.

5.6.3.1 *Effect of Flow Rate*

The effect of the flow rate on the pressure drop for a flow through a packed bed can be analyzed using two different approaches. The first approach treats the flow as an assembly of many internal channel flows between the particles so that the pressure drop is due to the wall friction of the channel flows. In the second approach, the pressure drop of the flow through the packed bed is considered to be caused by the drag forces of the particles. The drag force of a single particle is proportional to the fluid velocity in the creeping flow regime and is proportional to the square of the velocity in the inertia regime. Thus, the pressure drop through a fixed bed is proportional to the fluid velocity at low flow rates, and approximately to the square of the velocity at high flow rates. As suggested by Reynolds (1900), for a general flow through a packed bed, the pressure drop can be expressed by the sum of two terms, *i.e.*, one proportional to the first power of the fluid velocity and the other proportional to the product of the density of the fluid and the second power of the fluid velocity

$$\frac{\Delta p}{L} = aU + b\rho U^2 \tag{5.322}$$

where a and b are factors to be determined.

5.6.3.2 *Effect of Fluid Properties*

For a limiting case where the velocity U approaches zero, Eq. (5.322) yields

$$\lim_{U\to 0}\frac{1}{U}\frac{\Delta p}{L} = a \tag{5.323}$$

With the aid of the straight capillaric model and according to the Hagen–Poiseuille equation and Darcy's law, a is proportional to the viscosity of the fluid. Thus, Eq. (5.322) becomes

$$\frac{\Delta p}{L} = a'\mu U + b\rho U^2 \tag{5.324}$$

5.6.3.3 *Effect of Voidage of Packing*

The voidage α may affect the pressure drop in a column containing solid particles through the parameter $(1-\alpha)^{\mathrm{i}}/\alpha^{\mathrm{j}}$, where i and j are exponents to be determined. As suggested by Ergun and Orning (1949), the flow rate of the fluid through the porous medium varies with the voidage of the medium. The values of a' and b in Eq. (5.324) vary with the flow regimes. The dependency of the viscous energy loss and the kinetic energy loss on the voidage can be illustrated by the Kozeny theory [Scheidegger, 1960] and the Burke and Plummer theory [Burke and Plummer, 1928], respectively, as discussed in the following.

A. KOZENY THEORY

Kozeny (1927) proposed equations accounting for the relationship between the pressure drop and the voidage of the porous medium under low flow rate conditions [Ergun and Orning, 1949]. A detailed discussion of the Kozeny theory is given in Scheidegger (1960). It is assumed that the porous medium can be represented by an ensemble of channels of various cross sections, but of definite length. The channel flow is horizontal, laminar, and fully developed. The inertial effect is negligibly small compared to the viscous effect.

Denote A_α as the pore cross-sectional area and A as the total cross-sectional area of the column. The voidage is

$$\alpha = \frac{A_\alpha}{A} \tag{5.325}$$

Now, consider the flow through a single straight channel of an irregular cross section. If a cross section of the tube is chosen in the x–y plane and the fluid flows in the z-direction, the equation of motion can be expressed as

$$\frac{\partial^2 W}{\partial x^2} + \frac{\partial^2 W}{\partial y^2} = \frac{1}{\mu}\frac{\mathrm{d}p}{\mathrm{d}z} \tag{5.326}$$

Introducing new variables

$$\xi = \frac{x}{\sqrt{A_\alpha}}, \qquad \eta = \frac{y}{\sqrt{A_\alpha}} \tag{5.327}$$

Eq. (5.326) becomes

$$\frac{\partial^2 W}{\partial \xi^2} + \frac{\partial^2 W}{\partial \eta^2} = \frac{A_\alpha}{\mu}\frac{\mathrm{d}p}{\mathrm{d}z} \tag{5.328}$$

which has the solution

$$W = -\frac{A_\alpha}{\mu}\frac{\mathrm{d}p}{\mathrm{d}z}\psi(\xi,\eta) \tag{5.329}$$

where the function ψ is given by

$$\psi(\xi, \eta) = -\frac{x^2 + y^2}{4A_\alpha} + \sum a_n \Phi_n(x + iy) \tag{5.330}$$

where Φ denotes a harmonic function. Moreover, assuming a no-slip condition at the wall for fluid flow, we have

$$\frac{x_0^2 + y_0^2}{4A_\alpha} = \sum a_n \Phi_n(x_0 + iy_0) \tag{5.331}$$

where (x_0, y_0) is a point on the boundary.

Let C be the circumference of the cross section of the pores and R be the equivalent radius of a circle having the same circumference, *i.e.*,

$$R = \frac{C}{2\pi} \tag{5.332}$$

Then, we may express Eq. (5.330) in polar coordinates r and ϕ as

$$\psi = \left(\frac{r\cos\phi}{\sqrt{A_\alpha}}, \frac{r\sin\phi}{\sqrt{A_\alpha}}\right) = \frac{C^2}{16\pi^2 A_\alpha} B \tag{5.333}$$

where B is defined as

$$B = \sum a_n \Phi_n(re^{i\phi}) \frac{4A_\alpha}{R^2} - \frac{r^2}{R^2} \tag{5.334}$$

The averaged interstitial velocity u can be obtained by

$$u = \frac{1}{A}\int W \mathrm{d}A = -\frac{C^2}{16\pi^2\mu}\frac{\mathrm{d}p}{\mathrm{d}z} B_\mathrm{m} \tag{5.335}$$

where

$$B_\mathrm{m} = \frac{1}{A}\int\int B r \,\mathrm{d}r\,\mathrm{d}\phi \tag{5.336}$$

If the expansion of the circumference is represented by $\beta = C^2/A_\alpha$, it is clear that B and u become smaller if β becomes larger. When β approaches infinity, u approaches 0. In this case, from the analysis based on Eq. (5.335), B_m must decrease faster than the expansion of $16\pi^2 c/\beta$. This requirement can be satisfied if B_m is expressed by

$$B_\mathrm{m} = \frac{16\pi^2 c}{\beta^\zeta} \quad \text{where } \zeta > 1 \tag{5.337}$$

c in Eq. (5.337) is known as the Kozeny constant, which depends only on the shape of the cross section.

Consider a packed bed whose fluid cross section consists of n single-pore cross sections with each having the cross-sectional area $A_{s\alpha}$ and circumference C_s, *i.e.*,

$$A_\alpha = nA_{s\alpha}, \qquad C = nC_s \tag{5.338}$$

Substituting Eq. (5.337) and Eq. (5.338) into Eq. (5.335) yields

$$u = -\frac{c}{\mu}\frac{\mathrm{d}p}{\mathrm{d}z}\frac{A_\alpha}{\beta^{\zeta-1}} = -\frac{c}{\mu}\frac{\mathrm{d}p}{\mathrm{d}z}\frac{A_{s\alpha}^\zeta}{C_s^{2\zeta-2}}\frac{1}{n^{\zeta-2}} \tag{5.339}$$

Since the interstitial velocity u should not depend on the number of pores, ζ must be equal to 2. Equation (5.339) then reduces to

$$u = -\frac{c}{\mu}\frac{\mathrm{d}p}{\mathrm{d}z}\frac{A_\alpha^2}{C^2} = -\frac{c}{\mu}\frac{\mathrm{d}p}{\mathrm{d}z}\frac{\alpha^2}{S^2} \tag{5.340}$$

where S is the specific surface area of the pore and is defined as

$$S = \frac{C}{A} \tag{5.341}$$

The superficial velocity U can be extended from Eq. (5.340) as

$$U = -\frac{c}{\mu}\frac{\mathrm{d}p}{\mathrm{d}z}\frac{\alpha^3}{S^2} \tag{5.342}$$

In order to generalize the expression of S for various packing materials, S is redefined as the total surface area per unit packed volume and is expressed by

$$S = S_0(1-\alpha) \tag{5.343}$$

where S_0 is the surface area exposed to the fluid per unit volume of solid materials. Equation (5.342) thus becomes

$$U = -\frac{c}{\mu S_0^2}\frac{\mathrm{d}p}{\mathrm{d}z}\frac{\alpha^3}{(1-\alpha)^2} \tag{5.344}$$

Therefore, for viscous energy loss at low flow rates, we have

$$\frac{\Delta p}{L} \propto \frac{(1-\alpha)^2}{\alpha^3} \tag{5.345}$$

B. BURKE AND PLUMMER THEORY

For kinetic energy loss at high flow rates, the theory of Burke and Plummer [Ergun and Orning, 1949] assumes that the total resistance of the packed bed can be treated as the sum of the resistances of the individual particles. For the fully developed turbulent flow, the drag force acting on an isolated spherical particle is

$$F_\mathrm{D} = f\rho d_\mathrm{p}^2 u^2 \tag{5.346}$$

where f is the friction factor and is a constant.

Consider a packed column filled with spheres of equal diameter d_p. The number of particles per unit of packed volume is given by

$$n = \frac{6}{\pi d_\mathrm{p}^3}(1-\alpha) \tag{5.347}$$

Assume the drag force, in the form of Eq. (5.346), is acting on each particle. The rate of work done by the drag force, or the power, is equal to the product of the interstitial velocity and the force acting on the particles. Hence, we have

$$P = nf\rho d_\mathrm{p}^2 u^2 u = 6\frac{f}{\pi}\frac{\rho}{d_\mathrm{p}}u^3(1-\alpha) \tag{5.348}$$

where P is the power per unit volume of packed bed. On the other hand, it is clear that P can also be related to the pressure gradient and superficial velocity by

$$P = \frac{\Delta p}{L} U \tag{5.349}$$

Combining Eq. (5.348) and Eq. (5.349) yields

$$\frac{\Delta p}{L} = 6 \frac{f}{\pi} \frac{\rho U^2}{d_\mathrm{p}} \frac{(1-\alpha)}{\alpha^3} \tag{5.350}$$

For spherical particles, the specific surface area, S_0, can be expressed in terms of the particle diameter, d_p, as

$$S_0 = \frac{6}{d_\mathrm{p}} \tag{5.351}$$

From Eqs. (5.350) and (5.351), it gives

$$\frac{\Delta p}{L} = \frac{f}{\pi} S_0 \rho U^2 \frac{(1-\alpha)}{\alpha^3} \tag{5.352}$$

which indicates

$$\frac{\Delta p}{L} \propto \frac{(1-\alpha)}{\alpha^3} \tag{5.353}$$

Therefore, by substituting Eq. (5.345) and Eq. (5.353) into Eq. (5.324), we obtain

$$\frac{\Delta p}{L} = a'' \mu U \frac{(1-\alpha)^2}{\alpha^3} + b'' \rho U^2 \frac{(1-\alpha)}{\alpha^3} \tag{5.354}$$

5.6.3.4 *Effects of Size, Shape, and Surface Area of Particles*

For viscous energy loss, from Kozeny's equation, the pressure drop is proportional to the square of the specific surface area of solids S_0. For kinetic energy loss, from Burke and Plummer's relation, the pressure drop is proportional to S_0. S_0 is related to the particle diameter by Eq. (5.351) for spherical particles; for nonspherical particles, the dynamic diameter (see §1.2) may be used for the particle diameter. The general form of the pressure drop can be expressed as

$$\frac{\Delta p}{L} = k_1 \frac{(1-\alpha)^2}{\alpha^3} \frac{\mu U}{d_\mathrm{p}^2} + k_2 \frac{(1-\alpha)}{\alpha^3} \frac{\rho U^2}{d_\mathrm{p}} \tag{5.355}$$

where k_1 and k_2 are universal constants which are determined experimentally. Equation (5.355) can be rearranged into a linear form

$$y = k_1 + k_2 x \tag{5.356}$$

where

$$y = \frac{\Delta p}{L} \frac{\alpha^3 d_\mathrm{p}^2}{(1-\alpha)^2 \mu U}, \qquad x = \frac{\rho U d_\mathrm{p}}{\mu (1-\alpha)} \tag{5.357}$$

By plotting (y/x) versus x in a logarithmic scale based on experimental data, the values of k_1 and k_2 given in Eq. (5.356) can be determined. The results shown in Fig. 5.15 indicate

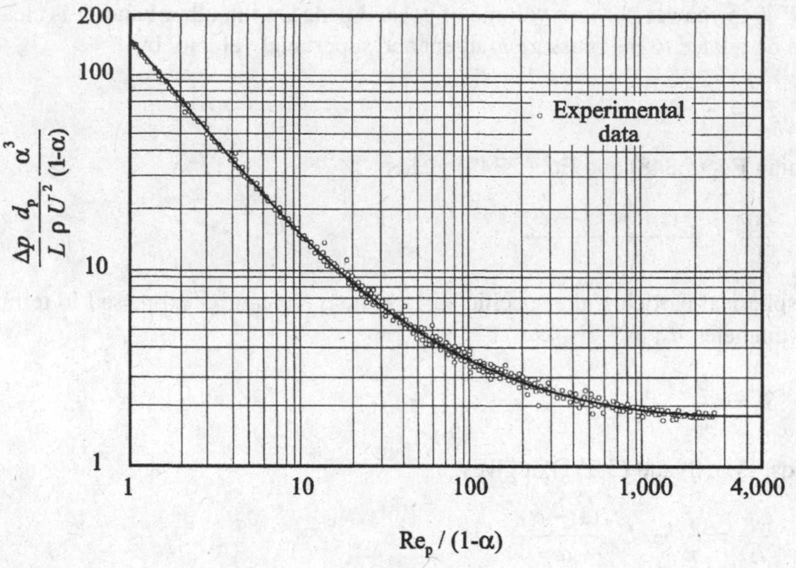

Figure 5.15. Pressure drop in fixed beds (from Ergun, 1952).

that k_1 and k_2 are 150 and 1.75, respectively. The Ergun equation is thus expressed by [Ergun, 1952]

$$\frac{\Delta p}{L} = 150\frac{(1-\alpha)^2}{\alpha^3}\frac{\mu U}{d_{\mathrm{p}}^2} + 1.75\frac{(1-\alpha)}{\alpha^3}\frac{\rho U^2}{d_{\mathrm{p}}} \tag{5.358}$$

5.7 Dimensional Analysis and Similarity

Dimensional analysis is an analytical method wherein a number of experimental variables that govern a given physical phenomenon reduce to form a smaller number of dimensionless variables. A phenomenon which varies as a function of k independent parameters can be reduced to a relationship between $(k - m)$ dimensionless parameters, where m is the number of dimensionally independent parameters.

The resulting dimensionless groups from the dimensional analysis can be used as the basis to account for the similarity or scaling relationships or laws for different design and operating situations. There are two common approaches in dimensional analysis to obtain the dimensionless variables: application of the Buckingham pi theorem and nondimensionalization of the governing equations and relevant boundary conditions. From these dimensionless variables, completely independent dimensionless parameters can be selected as the basis for the scaling law application. The validation of these scaling relationships depends on the appropriateness of the governing equations used and their related boundary conditions.

5.7.1 *Scaling Relationships for Pneumatic Transport of Dilute Suspensions*

Scaling for a dilute gas–solid suspension may represent a simple case of using the nondimensionalization method. The following example is a simplified version of the work of Chao (1982).

For pneumatic transport of solids in a dilute suspension, the effects of apparent mass, Basset force, diffusion, and electric charge of the particles may be ignored. Thus, the dynamic equation of a small particle in a gas medium is given by

$$\frac{\mathrm{d}\boldsymbol{U}_\mathrm{p}}{\mathrm{d}t} = \frac{1}{\tau_\mathrm{rp}}(\boldsymbol{U} - \boldsymbol{U}_\mathrm{p}) + \boldsymbol{g} \tag{5.359}$$

If we denote the displacement vector of the particle by $\boldsymbol{r}_\mathrm{p}$, we have

$$\boldsymbol{U}_\mathrm{p} = \frac{\mathrm{d}\boldsymbol{r}_\mathrm{p}}{\mathrm{d}t} \tag{5.360}$$

By introducing a reference length l and a reference velocity V, we may define the nondimensional quantities

$$\boldsymbol{r}_\mathrm{p}^* = \frac{\boldsymbol{r}_\mathrm{p}}{l}, \quad t^* = \frac{Vt}{l}, \quad \boldsymbol{U}^* = \frac{\boldsymbol{U}}{V}, \quad \boldsymbol{U}_\mathrm{p}^* = \frac{\boldsymbol{U}_\mathrm{p}}{V} \tag{5.361}$$

Thus, Eq. (5.359) becomes

$$\frac{\mathrm{d}\boldsymbol{U}_\mathrm{p}^*}{\mathrm{d}t^*} = \frac{1}{\mathrm{St}}(\boldsymbol{U}^* - \boldsymbol{U}_\mathrm{p}^*) + \frac{1}{\mathrm{Fr}}\boldsymbol{e}_\mathrm{g} \tag{5.362}$$

where $\boldsymbol{e}_\mathrm{g}$ is a unit vector in the direction of the gravitational acceleration; St is the Stokes number; and Fr is the Froude number. St and Fr are defined as

$$\mathrm{St} = \frac{V\tau_\mathrm{rp}}{l}, \qquad \mathrm{Fr} = \frac{V^2}{lg} \tag{5.363}$$

The dimensionless variable St is not always useful in the analysis because the relaxation time of the particle τ_rp in the expression depends on the particle Reynolds number Re_p in a general form

$$\frac{1}{\tau_\mathrm{rp}} = \frac{18\mu}{d_\mathrm{p}^2\rho_\mathrm{p}}C \tag{5.364}$$

where C is the non-Stokesian correction factor, which can be correlated as a function of Re_p. In the Stokes region, $C = 1$. The particle Reynolds number Re_p can be expressed as

$$\mathrm{Re}_\mathrm{p} = \frac{\rho d_\mathrm{p}}{\mu}|\boldsymbol{U} - \boldsymbol{U}_\mathrm{p}| = \mathrm{Re}_\mathrm{p}^*|\boldsymbol{U}^* - \boldsymbol{U}_\mathrm{p}^*| \tag{5.365}$$

where Re_p^* is the pseudoparticle Reynolds number, which is given by

$$\mathrm{Re}_\mathrm{p}^* = \frac{\rho V d_\mathrm{p}}{\mu} \tag{5.366}$$

In view of the preceding discussion concerning Eqs. (5.363) and (5.364), a similarity of St requires that, in addition to Re_p^*, another dimensionless parameter Ψ, given in the following, be similar

$$\Psi = \frac{\rho_\mathrm{p} V d_\mathrm{p}^2}{18\mu l} \tag{5.367}$$

where Ψ is termed the inertia parameter. Thus, for geometrically similar systems with kinematically similar boundary conditions, based on the model, the dynamic similarity can be obtained if Re_p^*, Ψ, and Fr are the same.

5.7.2 *Scaling Relationships for Fluidized Beds*

The derivation of scaling relationships for a dense gas–solid suspension in fluidized beds has been attempted on the basis of scale effects on the fundamental variables governing the dispersed phase and the interstitial gas dynamics [Horio *et al.*, 1986]. However, the scaling relationships for fluidized beds may be more generally established by using the nondimensionalization method accounting for the system behavior, as discussed later [Fitzgerald and Crane, 1980; Glicksman, 1984; 1988]. In the governing equations of this method, it is assumed that the interparticle forces, such as the collisional forces and the electrostatic forces, are negligible. In addition, the gas–solid suspension in a fluidized bed is treated as if in a steady-state laminar flow condition without phase change. The following derivation follows the approach used by Glicksman (1984). A limiting case for low Reynolds numbers is also considered.

The equations of conservation of mass for gas and particle phases are given, respectively, by

$$\nabla \cdot (\alpha \boldsymbol{U}) = 0 \tag{5.368}$$

$$\nabla \cdot [(1-\alpha)\boldsymbol{U}_{\mathrm{p}}] = 0 \tag{5.369}$$

The equation of motion of the gas can be written as

$$\boldsymbol{U} \cdot \nabla \boldsymbol{U} + \boldsymbol{g} + \frac{\nabla p}{\alpha \rho} + \frac{\beta}{\alpha \rho}(\boldsymbol{U} - \boldsymbol{U}_{\mathrm{p}}) = 0 \tag{5.370}$$

where $\beta(\boldsymbol{U} - \boldsymbol{U}_{\mathrm{p}})$ represents the drag force between the gas and the particles.

The equation of motion of the particles can be expressed, in a similar way, as

$$\boldsymbol{U}_{\mathrm{p}} \cdot \nabla \boldsymbol{U}_{\mathrm{p}} + \boldsymbol{g} - \frac{\beta}{(1-\alpha)\rho_{\mathrm{p}}}(\boldsymbol{U} - \boldsymbol{U}_{\mathrm{p}}) = 0 \tag{5.371}$$

The relevant boundary conditions for this system are given as follows:

(1) For particle velocities at solid walls such as the side wall of the bed in a batch-solids situation

$$\boldsymbol{U}_{\mathrm{pN}} = 0 \tag{5.372}$$

where $\boldsymbol{U}_{\mathrm{pN}}$ is the component of the particle velocity normal to the surface.

(2) For gas velocities at the base of the bed as well as at a height above the expanded bed surface when a screen or a fixed bed is used as the distributor

$$\boldsymbol{U} = U \boldsymbol{e}_{\mathrm{x}} \tag{5.373}$$

where U is the superficial gas velocity and $\boldsymbol{e}_{\mathrm{x}}$ is the unit vector along the axial direction. Along the bed walls

$$\boldsymbol{U} = 0 \tag{5.374}$$

(3) For the pressure at the side walls

$$\frac{\partial p}{\partial n} = 0 \tag{5.375}$$

Along the bottom surface, for a uniform distributor

$$p = p_{\mathrm{f}} + \frac{\beta U_{\mathrm{mf}} L_{\mathrm{mf}}}{\alpha_{\mathrm{mf}}} \tag{5.376}$$

where p_{f} represents the pressure in the freeboard region; U_{mf} is the minimum fluidization velocity; L_{mf} is the characteristic length at minimum fluidization; and α_{mf} is the bed voidage at minimum fluidization. Along the top surface, above the expanded bed surface

$$p = p_{\mathrm{f}} \tag{5.377}$$

By introducing the nondimensional quantities given next

$$\nabla^* = l\nabla, \qquad U^* = \frac{U}{U}, \qquad U_{\mathrm{p}}^* = \frac{U_{\mathrm{p}}}{U}, \qquad D^* = \frac{D}{l}, \qquad p^* = \frac{p}{\rho U^2} \tag{5.378}$$

the nondimensional equations, corresponding to Eqs. (5.368) through (5.371), become

$$\nabla^* \cdot (\alpha U^*) = 0 \tag{5.379}$$

$$\nabla^* \cdot \left[(1-\alpha)U_{\mathrm{p}}^*\right] = 0 \tag{5.380}$$

$$U^* \cdot \nabla^* U^* + \frac{gl}{U^2} + \frac{1}{\alpha}\nabla^*(p^*) + \frac{1}{\alpha}\frac{\beta l}{\rho U}(U^* - U_{\mathrm{p}}^*) = 0 \tag{5.381}$$

$$U_{\mathrm{p}}^* \cdot \nabla^* U_{\mathrm{p}}^* + \frac{gl}{U^2} - \frac{1}{(1-\alpha)}\frac{\rho}{\rho_{\mathrm{p}}}\frac{\beta l}{\rho U}(U^* - U_{\mathrm{p}}^*) = 0 \tag{5.382}$$

The dimensionless boundary conditions corresponding to Eqs. (5.372) through (5.377) can then be expressed as

$$U_{\mathrm{pN}}^* = 0 \tag{5.383}$$

$$U^* = 1 \tag{5.384}$$

$$U^* = 0 \tag{5.385}$$

$$\frac{\partial p^*}{\partial n^*} = 0 \tag{5.386}$$

The boundary condition for the pressure given in Eq. (5.376) can be rearranged as

$$(p - p_{\mathrm{f}}) = \beta U_{\mathrm{mf}} L_{\mathrm{mf}}/\alpha_{\mathrm{mf}} = \rho_{\mathrm{p}} g \alpha_{\mathrm{p}} l \tag{5.387}$$

Equations (5.377) and (5.387) can be nondimensionalized by the term ρU^2 as

$$\frac{(p - p_{\mathrm{f}})}{\rho U^2} = \frac{\beta U_{\mathrm{mf}} L_{\mathrm{mf}}}{\rho U^2 \alpha_{\mathrm{mf}}} = \frac{\rho_{\mathrm{p}}}{\rho}\left(\frac{gl}{U^2}\right)\alpha_{\mathrm{p}} \tag{5.388}$$

$$p^* = \frac{p_{\mathrm{f}}}{\rho U^2} \tag{5.389}$$

From Eqs. (5.379) through (5.389), the governing nondimensional parameters can be identified as

$$\frac{\beta l}{\rho U}, \qquad \frac{gl}{U^2}, \qquad \frac{D}{l}, \qquad \frac{\rho}{\rho_{\mathrm{p}}}, \qquad \frac{p_{\mathrm{f}}}{\rho U^2} \tag{5.390}$$

The nondimensionalized pressure term, $p_{\mathrm{f}}/\rho U^2$, can be ignored when the gas velocity is low compared to the speed of sound or when the effects of pressure on the thermodynamic properties are not significant. Note that U^*, α, and α_p are nondimensional dependent variables which are identical for two similar fluidized beds.

It should be noted that the coefficient β is not an independent parameter; it depends on the bed properties. In a fluidized bed, particles are closely spaced; it is assumed that the Ergun equation can be applied to account for the pressure drop in the bed. Thus, from Eq. (5.358), we have

$$\frac{\Delta p}{l} = \beta|\boldsymbol{U} - \boldsymbol{U}_p| = 150\frac{(1-\alpha)^2}{\alpha^2}\frac{\mu|\boldsymbol{U} - \boldsymbol{U}_p|}{d_p^2} + 1.75\frac{(1-\alpha)}{\alpha}\frac{\rho(\boldsymbol{U} - \boldsymbol{U}_p)^2}{d_p} \tag{5.391}$$

Rearranging this equation yields

$$\frac{\beta l}{\rho U} = 150\frac{(1-\alpha)^2}{\alpha^2}\frac{\mu l}{\rho U d_p^2} + 1.75\frac{(1-\alpha)}{\alpha}\frac{|\boldsymbol{U} - \boldsymbol{U}_p|}{d_p U}l \tag{5.392}$$

The second term in the right-hand side of Eq. (5.392) represents the fluid inertia, whereas the first term represents the viscous contribution to the pressure drop. At low Reynolds numbers, the Ergun equation can be simplified by neglecting the inertial term. Under this condition, Eq. (5.392) can be expressed as

$$\frac{\beta l}{\rho U} = 150\frac{(1-\alpha)^2}{\alpha^2}\left(\frac{\mu}{\rho U d_p}\right)\left(\frac{l}{d_p}\right) \tag{5.393}$$

Equation (5.393) shows that $(\beta l/\rho U)$ is only a function of $(\rho U d_p/\mu)$ and (d_p/l) for low Reynolds number cases. In this limit, the independent nondimensional scale-up parameters can be identified as

$$\frac{\rho U d_p}{\mu}, \qquad \frac{gl}{U^2}, \qquad \frac{d_p}{l}, \qquad \frac{D}{l}, \qquad \frac{\rho}{\rho_p} \tag{5.394}$$

Further discussion on scaling relationships for various fluid–particle systems is given in Glicksman *et al.* (1994).

Example 5.4 Design a geometrically similar laboratory-scale cold model fluidized bed to simulate the hydrodynamics of a large-scale fluidized bed combustor. Also specify the operating conditions for the cold model. The combustor is a square cross section column with a width of 1.0 m and a height of 6 m. The fluidized bed combustor is operated at a temperature of 1,150 K, a superficial gas velocity of 1.01 m/s, and a bed height of 1.06 m. Particles with a density of 2,630 kg/m^3 and a diameter of 677 μm are used for the combustor. The cold model is operated at a temperature of 300 K. Air is used for both the cold model and hot model fluidized beds. The physical properties of air are

$\rho_c = 1.16$ kg/m^3	$\rho_h = 0.3075$ kg/m^3
$\mu_c = 1.846 \times 10^{-5}$ kg/m · s	$\mu_h = 4.565 \times 10^{-5}$ kg/m · s
at 300 K and 1 atm	at 1,150 K and 1 atm

Solution The similarity analysis of Eqs. (5.378) through (5.394) indicates that for low Reynolds number (in this case, Reynolds number = 4.61), the dimensionless groups which must be matched between the model and the prototype are

$$\frac{\rho U d_p}{\mu}, \quad \frac{gl}{U^2}, \quad \frac{d_p}{l}, \quad \frac{D}{l}, \quad \frac{\rho}{\rho_p} \tag{E5.21}$$

Since the temperature influences only the physical properties of air, the design of the hot model can begin with the matching of the Reynolds number. The similarity condition

$$\left.\frac{U d_p \rho}{\mu}\right|_c = \left.\frac{U d_p \rho}{\mu}\right|_h \tag{E5.22}$$

requires that

$$\frac{U_c d_{pc}}{U_h d_{ph}} = \frac{\rho_h \mu_c}{\rho_c \mu_h} = 0.1072 \tag{E5.23}$$

The subscripts c and h refer to the cold and hot model, respectively. With the equality of the Froud number, U^2/gl, the scale factor for the superficial gas velocity can be expressed as

$$\frac{l_c}{l_h} = \left(\frac{U_c}{U_h}\right)^2 \tag{E5.24}$$

The condition of geometrical similarity requires

$$\frac{d_{pc}}{l_c} = \frac{d_{ph}}{l_h} \tag{E5.25}$$

Solving Eqs. (E5.23) through (E5.25) simultaneously yields that $U_c/U_h = 0.4750$ and $d_{pc}/d_{ph} = l_c/l_h = 0.2257$.

The similarity of D/l requires that

$$\frac{D_c}{D_h} = \frac{l_c}{l_h} = 0.2257 \tag{E5.26}$$

Similarly, we have

$$\frac{\rho_{pc}}{\rho_{ph}} = \frac{\rho_c}{\rho_h} = 3.77 \tag{E5.27}$$

In terms of the preceding scale factors, the operating conditions for both the hot and cold models can be obtained as given in Table E5.2.

Table E5.2. *The Operating Conditions for the Hot Model and Cold Model Fluidized Beds*

Operating conditions	Cold model	Hot model
Temperature (K)	300	1,150
Superficial gas velocity (m/s)	0.48	1.01
Bed height (m)	0.24	1.06
Particle density (kg/m^3)	9,921	2,630
Particle size (mm)	0.153	0.677
Bed width (or length) (m)	0.2257	1.0

Nomenclature

A	Cross-sectional area
A_c	Maximum contact area
A_k	Interfacial area of phase k
a	Particle radius
Ba	Bagnold number
C	Collisional rate of increase
C	Circumference
C	Specific thermal capacity of gas
C_1	Empirical constant in the k–ϵ model
C_2	Empirical constant in the k–ϵ model
C_{ek}	Empirical constant, defined by Eq. (5.162)
C_p	Specific thermal capacity of particle
C_μ	Empirical constant in the k–ϵ model
c	Specific heat of particles
c	Kozeny constant
D	Diffusivity
D	Column diameter or pipe diameter
D	Operator in the dense phase transport theorem
D_k	Eddy diffusivity of phase k
D_{Tk}	Eddy thermal diffusivity of phase k
$\boldsymbol{D}$	Collisional shear stress tensor
d_p	Particle diameter
E	Total energy per unit mass
E	Empirical constant, defined by Eq. (5.87)
e	Internal energy per unit mass
e	Coefficient of restitution
$\boldsymbol{e}$	Directive unit vector
$\boldsymbol{F}_{Ak}$	Momentum transfer vector across the interface per unit volume due to the pressure and viscous stresses of phase k
$\boldsymbol{F}_{\Gamma k}$	Momentum transfer vector across the interface per unit volume due to the mass generation of phase k
Fo	Fourier number
Fr	Froude number
f	Pair distribution function
f	Frictional factor
$f^{(1)}$	Single-particle velocity distribution function
$f^{(2)}$	Complete pair distribution function
$\boldsymbol{f}$	Body force vector acting on the unit mass of the particle
G	One-sided autospectral density function
G_k	Generation of phase k
g	Configurational pair-correlation function
g	Degeneracy of energy level
g	Gravitational acceleration
g_0	Radial distribution function
g_i	Degeneracy of energy level ϵ_i
g_r	Gravitational acceleration in the radial direction
g_z	Gravitational acceleration in the axial direction
g_ϕ	Gravitational acceleration in the azimuthal direction
h	Planck's constant
h	Vertical coordinate
$\boldsymbol{I}$	Unit tensor
J	Mass flux
J	Jacobian determinant
J_E	Rate of heat generation per unit volume by thermal radiation
J_e	Rate of heat generation per unit volume
$\boldsymbol{J}$	Flux vector of a physical quantity
$\boldsymbol{J}_q$	Heat flux vector
K	Thermal conductivity

K_k^T	Turbulent thermal conductivity of phase k	$\boldsymbol{P}_k$	Kinetic stress tensor of particles
k	Boltzmann constant	$\boldsymbol{P}_p$	Total stress tensor of particles
k	Collisional thermal conductivity	p	Pressure
k	Quantum number	p_f	Pressure in freeboard region
k	Specific permeability	p_m	Averaged pressure
k	Turbulent kinetic energy of the fluid phase	Q	Volumetric flow rate
k_p	Turbulent kinetic energy of particles	Q_{Ak}	Heat transfer across the interface
$\boldsymbol{k}$	Collisional vector	$Q_{\Gamma k}$	Internal energy transferred by the mass generation due to phase change
L	Bed length	q	Energy flux of particles
L	Geometric dimension	$\boldsymbol{q}_c$	Collisional heat flux vector
L_{mf}	Characteristic length at minimum fluidization	$\boldsymbol{q}_k$	Kinetic heat flux vector
l	Interparticle distance	$\boldsymbol{q}_p$	Total heat flux vector of fluctuation energy
l	Reference length	R	Lagrangian correlation coefficient
l_e	Characteristic length of turbulent eddy	R	Radius of averaging volume
l_m	Mixing length	R_{min}	Minimum radius of averaging volume
m	Mass of a particle	Re	Reynolds number
m^*	Relative mass, defined by Eq. (2.128)	Re_p	Particle Reynolds number
N	Number of particles	r	Radius
N_p	Number flow rate of the particle	r	Coordinate in r-space
n	Number of capillaries per unit area of cross section	S	Specific surface area
n	Total number of particles in the system	S_0	Surface area exposed to fluid per unit volume of solids
n	Configurational distribution function	St	Stokes number
n	Particle number density	s	Arbitrary curve path
n	Coordinate normal to a boundary	s	Distance
n_i	Number of particles having energy ϵ_i	T	Absolute temperature
P	Probability	T	Averaging time duration
P	Power per unit volume of the packed bed	T_c	Granular temperature
$\boldsymbol{P}$	Total pressure tensor	$\boldsymbol{T}$	Total stress tensor
$\boldsymbol{P}$	Position vector	t	Time
$\boldsymbol{P}_c$	Collisional stress tensor of particles	t_c	Contact time duration
		U	Superficial gas velocity
		U	Gas velocity component
		U_1	Velocity of particle 1
		U_2	Velocity of particle 2
		U_i	Gas velocity component in direction i

U_j Gas velocity component in direction j
U_{mf} Minimum fluidization velocity
U_n Gas velocity component normal to the wall
U_p Particle velocity component
U_{pi} Particle velocity component in direction i
U_{pj} Particle velocity component in direction j
U_r Normal impact velocity
U'_r Normal reflecting velocity after collision
U_s Speed of displacement of the interface
U_t Tangential impact velocity
U'_t Tangential reflecting velocity after collision
$\boldsymbol{U}$ Velocity vector of fluid
$\boldsymbol{U}^*$ Dimensionless gas velocity vector
$\boldsymbol{U}_k$ Velocity vector of phase k
$\boldsymbol{U}'_k$ Fluctuating velocity vector of phase k
$\boldsymbol{U}_p$ Velocity vector of particles
$\boldsymbol{U}^*_p$ Dimensionless particle velocity vector
u Interstitial gas velocity
u Instantaneous axial velocity component of gas
u' Fluctuating axial velocity component of gas
u'_i Gas fluctuating velocity component in direction i
u'_j Gas fluctuating velocity component in direction j
u_p Instantaneous axial velocity component of particle
u'_p Fluctuating axial velocity component of particle
u'_{pi} Particle fluctuating velocity component in direction i
$\boldsymbol{u}$ Relative velocity vector of a particle
V Gas velocity component
V Particle volume
V Reference velocity, defined by Eq. (5.361)
v Instantaneous velocity of a particle
v' Fluctuating radial velocity component of gas
v_p Instantaneous radial velocity component of particle
$\boldsymbol{v}$ Instantaneous velocity vector of a particle
W Gas velocity component
W Number of possible arrangements
W_{BE} Total number of possible arrangements for a certain set of n_i in Bose–Einstein statistics
$W_{i,BE}$ Number of possible arrangements for the *ith* level of energy for any n_i in Bose–Einstein statistics
W_{MB} Total number of possible arrangements for a certain set of n_i in the corrected Maxwell–Boltzmann statistics
w' Fluctuating tangential velocity component of gas
w_p Instantaneous tangential velocity component of particle
x Cartesian coordinate
Y Yield strength
y Cartesian coordinate
y Normal distance from wall
z Cartesian coordinate
z Cylindrical coordinate
z Partition function

Greek Symbols

α Volume fraction of the gas phase
α_p Volume fraction of the particle phase

Γ_k	Mass generation rate of phase k per unit volume
γ	Collisional rate of dissipation per unit volume
γ	Shear strain
γ'	Scalar quantity, defined by Eq. (5.176)
δ	Diameter of a capillary tube
δ	Relative deviation of calculated α_p
δ_{ij}	Kronecker delta
δ_w	Distance from wall
ϵ	Normal strain
ϵ	Dissipation rate of turbulent kinetic energy
ϵ	Quantum energy
ϵ	Wall emissivity
ϵ	Averaged energy of a particle
ϵ_i	Energy at quantum level i
ϵ_p	Generation or destruction rate of particle turbulent kinetic energy
ζ	Random number
θ	Cylindrical coordinate
θ	Spherical coordinate
θ	Contact angle
$\boldsymbol{\theta}$	Collisional transfer contribution vector
κ	Von Karman's constant
λ	Mean free path
λ	Parameter, defined by Eq. (5.255)
μ	Dynamic viscosity
μ	Transport coefficient of viscosity, defined by Eq. (5.43)
μ_{eff}	Effective viscosity
μ_T	Turbulent viscosity
μ'	Fluctuation of fluid viscosity
μ'	Deformation viscosity of fluid
μ_{ek}	Effective viscosity of phase k
ν	Kinematic viscosity
ρ	Density of fluid
ρ_p	Density of particles
$\boldsymbol{\sigma}$	Total stress tensor of the fluid phase
σ_{ek}	Empirical constant associated with the energy transport by fluctuations in density and temperature
σ_k	Empirical constant in the k–ϵ model
σ_ϵ	Empirical constant in the k–ϵ model
$\boldsymbol{\tau}$	Shear stress tensor
τ	Shear stress
τ_e	Eddy existence duration
τ_f	Characteristic time of fluid motion
τ_i	Interacting duration
τ_R	Flying time for a particle passing through an eddy
τ_{rp}	Relaxation time of particle motion
τ_S	Stokes relaxation time
Φ	Rate of generation of f per unit volume
$\bar{\Phi}$	Averaged variable
ϕ	Dissipation function
ϕ	Instantaneous variable
ϕ	Azimuthal angle
χ	Collisional source term
Ψ	Wave function
$\boldsymbol{\psi}$	Flux vector of f
ω	Angular frequency

References

Anderson, T. B. and Jackson, J. (1967). A Fluid Mechanical Description of Fluidized Beds. *I & EC Fund.*, **6**, 527.

Bagnold, R. A. (1954). Experiments on a Gravity-Free Dispersion of Large Solid Spheres in a Newtonian Fluid under Shear. *Proc. R. Soc. London*, **A225**, 49.

Basset, A. B. (1888). *Hydrodynamics*. Cambridge: Deighton, Bell; (1961) New York: Dover.

Baxter, L. L. (1989). *A Statistical-Trajectory Particle Dispersion Model*. Ph.D. Dissertation. Brigham Young University.

Bear, J. (1972). *Dynamics of Fluids in Porous Media*. New York: American Elsevier.

Boussinesq, J. (1877). Theory de L'ecoulement Tourbillant. *Mem. Pres. Par Div. Savants a L'acad. Sci., Paris*. **23**, 46.

Boussinesq, J. (1903). *Theorie Analytique de la Chaleur*, **2**. Paris: Gauthier-Villars.

Burke, S. P. and Plummer, W. B. (1928). Suspension of Macroscopic Particles in a Turbulent Gas Stream. *I & EC*., **20**, 1200.

Carnahan, N. F. and Starling, K. E. (1969). Equations of State for Nonattracting Rigid Spheres. *J. Chem. Phys*., **51**, 635.

Celmiņš, A. (1988). Representation of Two-Phase Flows by Volume Averaging. *Int. J. Multiphase Flow*, **14**, 81.

Chao, B. T. (1982). Scaling and Modeling. In *Handbook of Multiphase Systems*. Ed. G. Hetsroni. New York: Hemisphere; McGraw-Hill.

Chapman, S. and Cowling, T. G. (1970). *The Mathematical Theory of Nonuniform Gases*, 3rd ed. Cambridge: Cambridge University Press.

Crowe, C. T. (1991). The State-of-the-Art in the Development of Numerical Models for Dispersed Two-Phase Flows. *Proceedings of the First International Conference on Multiphase Flows*, Tsukuba. **3**, 49.

Crowe, C. T., Sharma, M. P. and Stock, D. E. (1977). Particle–Source–In Cell (PSI CELL) Model for Gas-Droplet Flows. *Trans. ASME, J. Fluids Eng*., **99**, 325.

Culick, F. E. C. (1964). Boltzmann Equation Applied to a Problem of Two-Phase Flow. *Phys. Fluids*, **7**, 1898.

Darcy, H. (1856). *Les Fontaines Publiques de la Ville de Dijon*. Paris: Victor Dalmon.

Delhaye, J. M. and Archard, J. L. (1976). *Proceedings of CSNI Specialists' Meeting*, Toronto. Ed. Bannerjee and Weaver. *AECL*, **1**, 5.

Elghobashi, S. E. (1994). Numerical Models for Gas–Particle Flows. *Seventh Workshop on Two-Phase Flow Predictions*, Erlangen.

Ergun, S. (1952). Fluid Flow Through Packed Columns. *Chem. Eng. Prog*., **48**, 89.

Ergun, S. and Orning, A. A. (1949). Fluid Flow Through Randomly Packed Columns and Fluidized Beds. *I & EC*., **41**, 1179.

Fitzgerald, T. J. and Crane S. D. (1980). Cold Fluidized Bed Modeling. *Proceedings of the Sixth International Conference on Fluidized Bed Combustion*, Atlanta, Georgia. **3**, 815. U.S. DOE.

Gidaspow, D. (1993). Hydrodynamic Modeling of Circulating and Bubbling Fluidized Beds. In *Particulate Two-Phase Flow*. Ed. M. C. Roco. Boston: Butterworth–Heinemann.

Glicksman, L. R. (1984). Scaling Relationships for Fluidized Beds. *Chem. Eng. Sci*., **39**, 1373.

Glicksman, L. R. (1988). Scaling Relationships for Fluidized Beds. *Chem. Eng. Sci*., **43**, 1419.

Glicksman, L. R., Hyre, M. R. and Farrell, P. A. (1994). Dynamic Similarity in Fluidization. *Int. J. Multiphase Flow*, **20/S**, 331.

Hinze, J. O. (1959). *Turbulence*. New York: McGraw-Hill.

Horio, M., Nonaka, A., Sawa, Y. and Muchi, I. (1986). A New Similarity Rule for Fluidized Bed Scale-Up. *AIChE J*., **32**, 1466.

Huang, X. Q. and Zhou, L. X. (1991). Simulation of 3-D Turbulent Recirculating Gas–Particle Flows by an Energy-Equation Model of Particle Turbulence. *FED*, **121**, *Gas–Solid Flows, ASME*, 261.

Hunt, M. L. (1989). Comparison of Convective Heat Transfer in Packed Beds and Granular Flows. In *Annual Review of Heat Transfer*. Ed. C. L. Tien. Washington: Hemisphere.

Ishii, M. (1975). *Thermo-Fluid Dynamic Theory of Two-Phase Flow*. Paris: Eyrolles.

Jenkins, J. T. and Savage, S. B. (1983). A Theory for the Rapid Flow of Identical, Smooth, Nearly Elastic, Spherical Particles. *J. Fluid Mech.*, **130**, 187.

Kozeny, J. (1927). Über kapillare Leitung des Wassers im Boden (Aufstieg, Versickerung und Anwendung auf die Bewässerung). *Ber. Wien. Akad.* **136a**, 271.

Launder, B. E. and Spalding, D. B. (1972). *Mathematical Models of Turbulence*. London: Academic Press.

Launder, B. E. and Spalding, D. B. (1974). The Numerical Computation of Turbulent Flows. *Computer Methods in Applied Mechanics and Engineering*, **3**, 269.

Lun, C. K. K., Savage, S. B. and Jeffery, D. J. (1984). Kinetic Theories for Granular Flow: Inelastic Particles in Couette Flow and Slightly Inelastic Particles in a General Flow Field. *J. Fluid Mech.*, **140**, 223.

Oseen, C. W. (1927). *Hydrodynamik*. Leipzig: Akademische Verlagsgescellschafe.

Patankar, S. V. (1980). *Numerical Heat Transfer and Fluid Flow*. Washington: Hemisphere.

Prandtl, L. (1925). Bericht uber Untersuchung zur ausgebildeten Turbulenz. *ZAMM*, **5**, 136.

Reif, F. (1965). *Fundamentals of Statistical and Thermal Physics*. New York: McGraw-Hill.

Reynolds, O. (1900). *Papers on Mechanical and Physical Subjects*. Cambridge: Cambridge University Press.

Savage, S. B. (1979). Gravity Flow of Cohesionless Granular Materials in Chutes and Channels. *J. Fluid Mech.*, **92**, 53.

Savage, S. B. (1983). Granular Flows at High Shear Rates. In *Theory of Dispersed Multiphase Flow*. Ed. R. E. Meyer. New York: Academic Press.

Savage, S. B. and Jeffery, D. J. (1981). The Stress Tensor in a Granular Flow at High Shear Rates. *J. Fluid Mech.*, **110**, 255.

Savage, S. B. and Sayed, M. (1984). Stresses Developed by Dry Cohesionless Granular Materials Sheared in an Annular Shear Cell. *J. Fluid Mech.*, **142**, 391.

Scheidegger, A. E. (1960). *The Physics of Flow Through Porous Media*. Toronto: University of Toronto Press.

Slattery, J. C. (1967a). General Balance Equation for a Phase Interface. *I & EC Fund.*, **6**, 108.

Slattery, J. C. (1967b). Flow of Viscoelastic Fluid Through Porous Media. *AIChE J.*, **13**, 1066.

Soo, S. L. (1965). Dynamics of Multiphase Flow Systems. *I & EC Fund.*, **4**, 426.

Soo, S. L. (1989). *Particulates and Continuum: Multiphase Fluid Dynamics*. New York: Hemisphere.

Sun, J. and Chen, M. M. (1988). A Theoretical Analysis of Heat Transfer due to Particle Impact. *Int. J. Heat & Mass Transfer*, **31**, 969.

Taylor, G. I. (1921). Diffusion by Continuous Movements. *Proc. London Math. Soc.*, **20**, 196.

Tchen, C. M. (1947). *Mean Value and Correlation Problems Connected with the Motion of Small Particles in a Turbulent Field*. Ph.D. Thesis. Delft University, Netherlands.

Whitaker, S. (1969). Advances in Theory of Fluid Motion in Porous Media. *I & EC.*, **61**, 4.

Zeininger, G. and Brennen, C. E. (1985). Interstitial Fluid Effects in Hopper Flows of Granular Materials. *ASME Cavitation and Multiphase Flow Forum*, Albuquerque, N. Mex.

Zhang, J., Nieh, S. and Zhou, L. (1992). A New Version of Algebraic Stress Model for Simulating Strongly Swirling Turbulent Flows. *Numerical Heat Transfer, Part B: Fundamentals*, **22**, 49.

Zhou, L. (1993). *Theory and Numerical Modeling of Turbulent Gas–Particle Flows and Combustion*. Boca Raton, Fla.: CRC Press.

Problems

5.1 Show that $\mathrm{d}J/\mathrm{d}t = (\nabla \cdot \boldsymbol{U})J$, where J is the Jacobian determinant.

5.2 Prove that the Maxwell–Boltzmann distribution can be expressed as Eq. (5.24). Several steps are suggested:

(1) Use Stirling's approximation formula ($\ln x! = x \ln x - x$) to show that

$$\ln W_{\mathrm{MB}} = n + \sum (n_{\mathrm{i}} \ln g_{\mathrm{i}} - n_{\mathrm{i}} \ln n_{\mathrm{i}})$$

(2) On the basis of the Lagrangian method (Lagrangian multipliers) and the conservation of total number of particles and total energy of the system, show that the Maxwell–Boltzmann distribution can take the form

$$n_{\mathrm{i}} = \frac{g_{\mathrm{i}}}{Ae^{B\epsilon_{\mathrm{i}}}}$$

where A and B are constants to be determined.

(3) From the conservation of total number of particles, show that

$$A = \frac{1}{n} \sum g_{\mathrm{i}} e^{-B\epsilon_{\mathrm{i}}}$$

(4) Note that the averaged translational energy of a single particle is

$$\overline{\epsilon} = \frac{3}{2} kT$$

Show that $B = 1/kT$.

5.3 Derive Eq. (5.38) on the basis of the Maxwell–Boltzmann velocity distribution.

5.4 Show that for a steady, incompressible, isothermal flow, the k-equation can be expressed by Eq. (5.75). The derivation procedure is suggested as follows:

(1) Subtract the time-averaged momentum equation from the instantaneous momentum equations.

(2) Multiply the equation by u'_{i}.

(3) Take the time average over the resulting equation.

5.5 Show that for a steady, incompressible, isothermal flow, the ϵ-equation can be expressed by Eq. (5.80). The derivation procedure is suggested as follows:

(1) Differentiate the instantaneous momentum equations with respect to x_{l}.

(2) Multiply the equation by $2\upsilon \partial U_{\mathrm{i}}/\partial x_{\mathrm{l}}$.

(3) Take the time average over the resulting equation.

(4) Differentiate the time-averaged momentum equations with respect to x_{l}.

(5) Multiply the equation by $2\upsilon \partial \overline{U}_{\mathrm{i}}/\partial x_{\mathrm{l}}$.

(6) Subtract (5) from (3).

5.6 Show that, for flow through a circular tube, the velocity distribution can be expressed by Eq. (5.318).

5.7 Design a scale prototype for a dilute solid fuel combustor with the scale of 1:10. The prototype (P) operates at the temperature $T_{\mathrm{p}} = 1{,}150$ K and the pressure $p_{\mathrm{p}} = 10$ atm. The averaged air velocity in the prototype $V_{\mathrm{P}} = 6$ m/s. The operating temperature of the model (M) is $T_{\mathrm{M}} = 300$ K. The physical properties of the combustion products are assumed to be the same as that of air. It is further assumed that the air obeys the equation of state of an ideal gas. Thus, $\mu_{\mathrm{P}} = 4.56 \times 10^{-5}$ kg/m·s and $\mu_{\mathrm{M}} = 1.85 \times 10^{-5}$ kg/m·s. Determine the particle size and operational pressure and velocity of the air in the model.

5.8 Applying the volume-averaging theorems to Eqs. (5.14) and (5.18), verify that the volume-averaged momentum and energy equations for phase k in a multiphase flow can be given by Eqs. (5.123) and (5.126), respectively.

5.9 Consider an isothermal, steady gas–solid laminar pipe flow under the microgravity condition. The flow is regarded as axisymmetric. Both the gas and particle phases can be treated as a pseudocontinuum. Derive the governing equations, in cylindrical coordinates, which account for the motion of the gas and particle phases.

5.10 A golf ball with diameter d_p travels in the air at uniform velocity $\boldsymbol{U} = u_0\mathbf{e}_x$. Consider the initial velocity of the ball $\boldsymbol{U}_{p0} = u_{p0}\mathbf{e}_x + v_{p0}\mathbf{e}_y$ with initial angular velocity $\boldsymbol{\Omega} = \Omega_0\mathbf{e}_z$. The gravity acts in the y-direction. The angular velocity is assumed to be constant during the motion of the ball. Derive the governing equations which describe the motion of the ball.

CHAPTER 6

Intrinsic Phenomena in a Gas–Solid Flow

6.1 Introduction

There are inherent phenomena associated with gas–solid flows. Such phenomena are of considerable interest to a wide spectrum of process applications involving gas–solid flows and can be exemplified by erosion, attrition, pressure-wave propagation, flow instability, and gas–solid turbulence modulation.

Mechanical erosion may cause severe damage to a pipe wall or any moving part in a gas–solid system. Particle attrition yields fine particles, which may alter the flow conditions of the system and may become a source of particulate emission. Understanding the basic modes of mechanical wear and the mechanism of attrition is essential to the control of their behavior. Pressure waves through a gas–solid suspension are directly related to nozzle flows (such as in jet combustion) and measurement techniques associated with acoustic waves and shock waves. Understanding the propagation of the speed of sound and normal shock waves in a dilute gas–solid suspension is, therefore, important. Flow instability such as wavelike motion represents an intrinsic feature of a gas–solid flow. An instability analysis can be made using the pseudocontinuum approach. For most gas–solid flows, particles move in a strong turbulent gas stream. The gas turbulence can be significantly modulated by the solid particles. Thus, understanding the particle-turbulence interactive phenomenon is necessary.

This chapter addresses the various phenomena indicated. In addition, the thermodynamic laws governing physical properties of the gas–solid mixture such as density, pressure, internal energy, and specific heat are introduced. The thermodynamic analysis of gas–solid systems requires revisions or modifications of the thermodynamic laws for a pure gas system. In this chapter, the equation of state of the gas–solid mixture is derived and an isentropic change of state is discussed.

6.2 Erosion and Attrition

The mechanical erosion of a solid surface such as a pipe wall in a gas–solid flow is characterized by the loss of solid material from the solid surface due to particle impacts. The collisions of the particles either with other particles or with a solid wall may lead to particle breakup, known as particle attrition. Pipe erosion and particle attrition are major concerns in the design of a gas–solid system and during the operation of such a system. The wear of turbine blades or pipe elbows due to the directional impact of dust or granular materials, the wear of mechanical sieves by the random impact of solids, and the wear of immersed pipes in a fluidized bed by both directional and random impacts are examples of the erosion phenomenon in industrial systems. The surface wear associated with the erosion phenomenon of a gas–solid flow has been exploited to provide beneficial industrial applications such as abrasive guns, as well.

The most commonly used pipe materials may be classified into four categories based on the mechanical erosion modes: metals such as copper, aluminum, and steel; ceramics and glasses; plastics such as polyvinyl chloride (PVC) and acrylic (Plexiglas); and rubbers.

6.2.1 *Ductile Erosion and Brittle Erosion*

When a particle strikes a solid surface, the extent and nature of the damage on the surface depend on the normal compressive force F_n, tangential cutting force F_t, area of contact A_c, duration of contact t_c, angle of incidence α_i, shape of the particle, and the materials of the particle and the solid surface. The mechanisms of mechanical erosion may be explained in terms of two basic modes, *i.e.*, the ductile mode and the brittle mode.

Ductile erosion is the loss of material from a solid surface due to the tangential cutting force of the particle impact. When a solid surface is struck by a particle, elastic and plastic deformations occur. The yield stress for ductile removal Y_D must be greater than the onset yield stress or one-half of the yield strength in simple tension [Timoshenko and Goodier, 1951]. Y_D is typically greater than or nearly equal to the pressure for plastic deformation, 0.35 percent of the modulus of elasticity [Goldsmith and Lyman, 1960].

A ductile rupture may be characterized by a maximum erosion occurring at some small angle of incidence, usually between 20° and 30°. Typical ductile damages due to single-particle impacts are illustrated in Fig. 6.1(a) [Hutchings, 1987]. In this figure, it can be seen that rounded particles striking at shallow impact angles lead to ploughed craters, while impacts from sharp-edged particles cause indentations with a pronounced chip of material still attached. In both cases, the chips can be detached by only a small number of subsequent impacts. In a surface damaged by multiple impacts of hard particles, flake- or platelet-shaped debris may be formed as a result of collisions with round particles, whereas bulky-shaped debris is likely to be caused by the impacts of angular particles.

Brittle erosion is the loss of material from a solid surface due to fatigue cracking and brittle cracking caused by the normal collisional force F_n. Materials with very limited capacity for elastic and plastic deformation, such as ceramics and glass, respond to particle impacts by fracturing. The yield stress for brittle failure Y_B for normal impacts is about

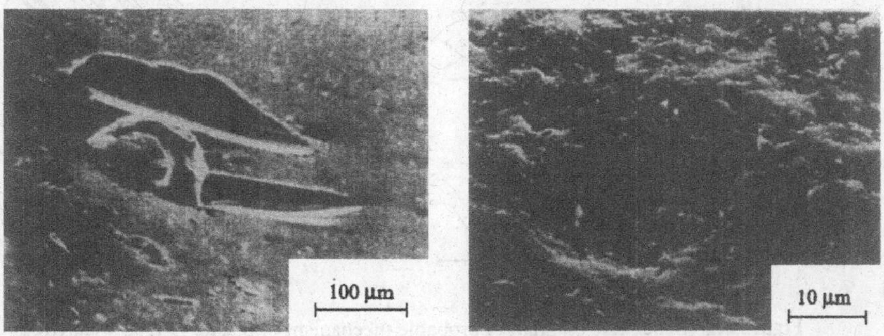

(a) (b)

Figure 6.1. Examples of damage sites due to single-particle impacts: (a) On mild steel, a ductile metal; (b) On polycrystalline alumina, a brittle ceramic (from Hutchings, 1987; reproduced with permission).

three times the yield stress [Tabor, 1951], or stress to cause fracture at static conditions, or 0.43 percent of the shear modulus. Y_B is also equal to $[(1 - 2\nu)/1.86]$ times the yield strength in simple tension, where ν is the Poisson ratio. A typical example of brittle erosion is shown in Fig. 6.1(b), where the surface is characterized by the formation of isolated pits from which some grains appear to have been removed by intergranular fracture [Hutchings, 1987].

The fractures on a plane surface, created by the collisions of hard spherical particles at low-impact velocities, may form a conical crack according to the Hertzian quasi-static stress theory. In a multiple-impact situation, the conical cracks meet those extending from neighboring impact sites, and then the brittle material becomes detached. Once appreciable damage is done, the cracking mechanism may be altered because the particles no longer strike on a plane surface; nevertheless the brittle removal continues by the successive formation and intersection of cracks.

A probable mechanism of erosion for plastics is illustrated in Fig. 6.2 [Briscoe and Evans, 1987]. Initially, a series of plastically deformed grooves can be formed by the abrasion of particle flows. The subsequent directional or random impacts of particles may push the deformed grooves from side to side. The fatigue limits of the plastics would eventually make the ridges between the grooves detach to form ribbonlike debris. Brittle cracks also occur when the wear tracks interact.

The essential material property of rubbers is their low elastic modulus, which ensures that the contact deformation remains elastic over a very wide range of contact conditions. The abrasive wear of rubbers is due to either fatigue of the material or tearing by a cutting force from impacts with sharp-edged particles.

In general, erosion via tangential cutting tends to be more prominent for ductile materials such as metals and plastics than for brittle materials such as ceramics and glasses. In practice, erosion is a result of the combined effect of both the ductile and brittle modes. The extent

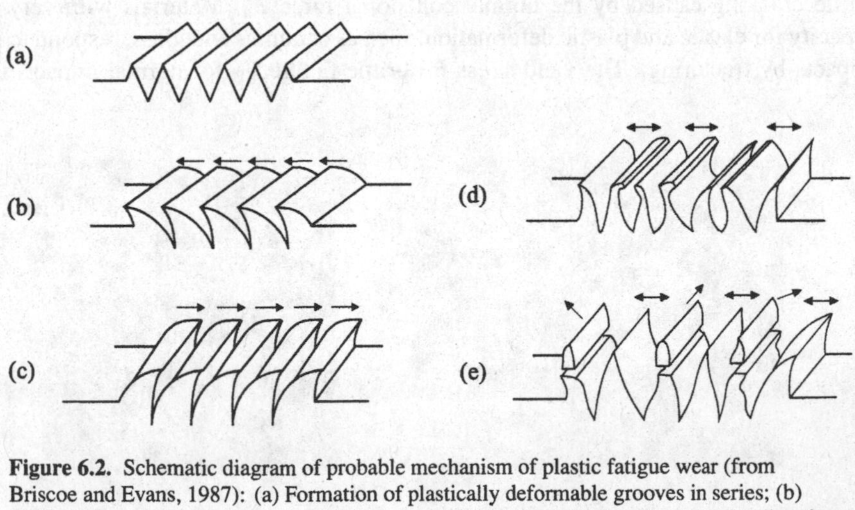

Figure 6.2. Schematic diagram of probable mechanism of plastic fatigue wear (from Briscoe and Evans, 1987): (a) Formation of plastically deformable grooves in series; (b) Deformation of the grooves pushed in one direction; (c) Sway back to the opposite direction; (d) Deterioration of ridges after repeated fluttering; (e) Detachment of ridges in the form of band-shaped debris.

of erosion for a small deviation from the elastic states can be accounted for by a linear combination of that due to the ductile and brittle modes.

6.2.2 *Locations of Erosive Wear*

Abrasive erosion in a straight pipe caused by particle impacts is usually insignificant compared to the wear at bends such as elbow connectors for pneumatic conveying pipelines and the outer wall near the entrance of a cyclone separator. The angle at which maximum erosion occurs primarily depends on the material properties. For ductile materials, the impact angles are typically in the range of 20° to 45°, with the smaller angles being related to more angular eroding particles. On the other hand, for brittle materials, maximum erosion is caused by brittle fractures under conditions of normal impingement where the impact angle approaches 90°. An example of both cases can be seen in Fig. 6.3, in which the eroding particles are 100 μm sharp-edged silicon carbide particles with a velocity of 150 m/s, the ductile material is aluminum, and the brittle material is alumina [Finnie *et al.*, 1967]. In this example, the erosion is defined as the mass of material removed per unit mass of colliding particles.

The location of the most severe erosion on the extrados of a pipe bend also depends on the particle flow pattern inside the bend section. Particle trajectories in a right angle connector are shown in Fig. 6.4. For large or heavy particles, the inertia of the particles plays a dominant role in determining the trajectories of these solids and the fluid drag forces only slightly affect their motion. Thus, the paths of large particles in a bend are nearly straight, leading to a steep angle of impact on the pipe wall (see Fig. 6.4, path a). For smaller or lighter particles, both the drag force and inertia are important in the particle motion, and thus the paths of the particles may deflect, with the particle–wall impact occurring at a

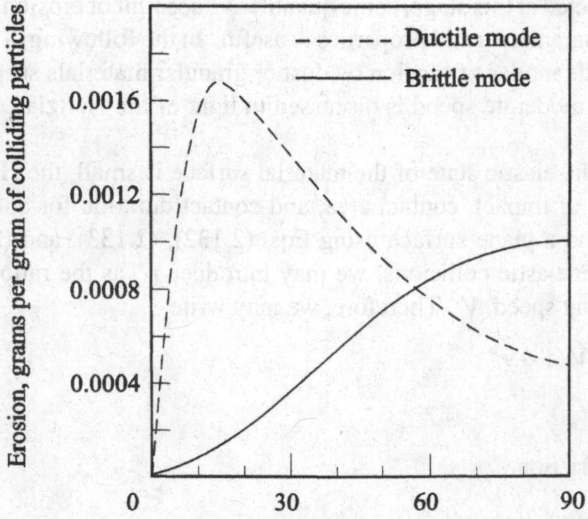

Figure 6.3. Dependence of erosion on impact angle (from Finnie *et al.*, 1967).

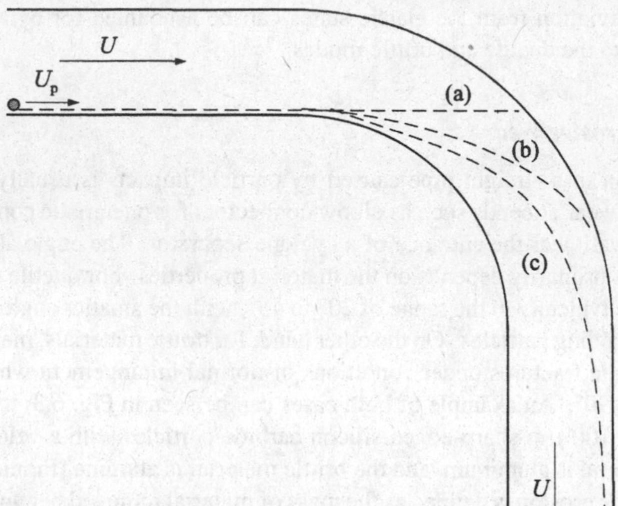

Figure 6.4. Particle trajectories in a right-angled pipe bend: (a) Heavy and/or large particle; (b) Medium-sized particle; (c) Light and/or small particle.

shallower angle, closer to the exit of the bend connector (see Fig. 6.4, path b). With further reduction of particle size or weight, the particles may closely follow the fluid motion and may not contact the wall at all (see Fig. 6.4, path c).

As mentioned, the erosion of a solid surface depends on the collisional force, angle of incidence, and material properties of both surface and particles. Although abrasive erosion rates cannot be precisely predicted at this stage, some quantitative account of erosion modes which relates various impact parameters and properties is useful. In the following, a simple model for the ductile and brittle modes of erosion by dust or granular materials suspended in a gas medium moving at a moderate speed is discussed in light of the Hertzian contact theory [Soo, 1977].

When the deviation from the elastic state of the material surface is small, the Hertzian theory can estimate the force of impact, contact area, and contact duration for collisions between spherical particles and a plane surface using Eqs. (2.132), (2.133), and (2.136), respectively. To account for inelastic collisions, we may introduce r^* as the ratio of the reflection speed to the incoming speed, V. Therefore, we may write

$$F_{\mathrm{n}}t_{\mathrm{c}} = 2.94(5/16)mV(1+r^*) \tag{6.1}$$

$$F_{\mathrm{t}} = fF_{\mathrm{n}} \tag{6.2}$$

where f is the coefficient of friction.

$$A_{\mathrm{c}}t_{\mathrm{c}} = A_0t_0(\sin\alpha_{\mathrm{i}})^{-\frac{1}{5}} \tag{6.3}$$

where α_{i} is the impact angle and

$$A_0t_0 = 2.94(15\pi^2/8)2^{-\frac{2}{5}}mV(k_1k_2)^{\frac{1}{2}}(1+r^*)^{\frac{4}{5}}N_{\mathrm{Im}}^{-\frac{1}{5}} \tag{6.4}$$

where $k = (1-\nu)/\pi E$; subscripts "1" and "2" denote each of the colliding pair; and N_{Im} is the impact number, defined by

$$N_{\mathrm{Im}} = \frac{5\pi^2}{2}\rho_{\mathrm{p}} V^2 \sqrt{k_1 k_2}\left(\sqrt{\frac{k_1}{k_2}} + \sqrt{\frac{k_2}{k_1}}\right)^4 \tag{6.5}$$

which correlates the dynamic and material properties of the impacting system.

In the following, the effects of gliding, scattering, and lifting of the particles by the gas stream are neglected and the mechanical efficiency of impact is assumed to be 100 percent. Thus, for ductile erosions, the wear in volume loss per impact equals the work of ductile wear by the tangential force divided by the energy required to remove a unit volume of material in ductile erosion mode. Therefore, we have

$$w_{\mathrm{D}} = \frac{1}{\epsilon_{\mathrm{D}}}\left(\frac{F_{\mathrm{t}}}{A_{\mathrm{c}}} - Y_{\mathrm{D}}\right) V \cos\alpha_{\mathrm{i}} A_{\mathrm{c}} t_{\mathrm{c}} = \frac{F_{\mathrm{t}} t_{\mathrm{c}} V}{\epsilon_{\mathrm{D}}} \cos\alpha_{\mathrm{i}} \left(1 - \frac{Y_{\mathrm{D}} A_0 t_0}{F_{\mathrm{t}} t_{\mathrm{c}}} (\sin\alpha_{\mathrm{i}})^{-\frac{1}{5}}\right) \tag{6.6}$$

where w_{D} represents the wear in volume loss per impact and ϵ_{D} denotes the energy required to remove a unit volume of material in ductile mode. In a dimensionless form, Eq. (6.6) can be expressed as

$$E_{\mathrm{D}} = \cos\alpha_{\mathrm{i}}\left[1 - K_{\mathrm{D}}(\sin\alpha_{\mathrm{i}})^{-\frac{1}{5}}\right] = E_{\mathrm{D}}(\alpha_{\mathrm{i}}, K_{\mathrm{D}}) \tag{6.7}$$

where E_{D} is a dimensionless ductile erosion parameter, defined by

$$E_{\mathrm{D}} = \frac{w_{\mathrm{D}}\epsilon_{\mathrm{D}}}{f F_{\mathrm{n}} t_{\mathrm{c}} V} \tag{6.8}$$

and K_{D} is the ductile resistance parameter, defined by

$$K_{\mathrm{D}} = \frac{Y_{\mathrm{D}}}{f}\frac{A_0 t_0}{F_{\mathrm{n}} t_{\mathrm{c}}} \tag{6.9}$$

Substituting Eqs. (6.1), (6.3), and (6.4) into Eqs. (6.8) and (6.9) yields

$$E_{\mathrm{D}} = \frac{w_{\mathrm{D}}\epsilon_{\mathrm{D}}}{2.94(5/16)(1+r^*) f m V^2} \tag{6.10}$$

and

$$K_{\mathrm{D}} = 3\pi\left(2^{\frac{3}{5}}\right)\frac{Y_{\mathrm{D}}}{f}\sqrt{k_1 k_2}(1+r^*)^{-\frac{1}{5}} N_{\mathrm{Im}}^{-\frac{1}{5}} \tag{6.11}$$

Equation (6.11) indicates that, since $(1+r^*)^{-1/5}$ is nearly unity, K_{D} is only related to the material properties and the collisional speed. Thus, K_{D} can be treated as a parametric constant and the function E_{D} characterizes the ductile wear at various angles of impact via Eq. (6.7). Hence, the maximum value of E_{D} occurs at an angle of impact α_{m} given by

$$K_{\mathrm{D}} = \frac{5(\sin\alpha_{\mathrm{m}})^{\frac{11}{5}}}{4\sin^2\alpha_{\mathrm{m}} + 1} \tag{6.12}$$

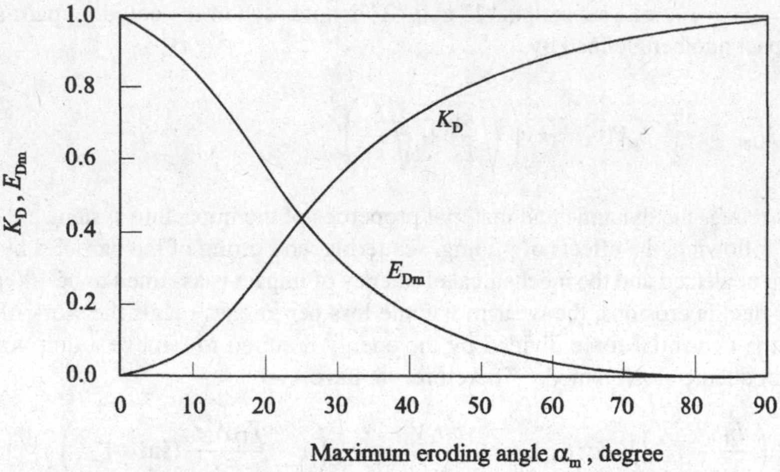

Figure 6.5. Relationship between the ductile resistance parameter and the angle of impact for maximum erosion (from Soo, 1977).

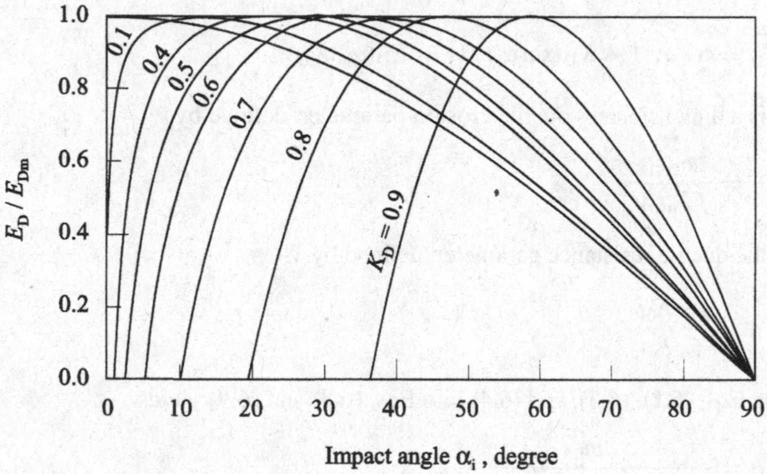

Figure 6.6. Normalized ductile erosion energy function varying with impact angle and ductile resistance parameter (from Soo, 1977).

In terms of Eqs. (6.12) and (6.7), the relations K_D and E_{Dm} versus α_m are given in Fig. 6.5, where $E_{Dm} = E_D(\alpha_m, K_D)$. The ratio of E_D to E_{Dm} at various angles of impact α_i and K_D is plotted in Fig. 6.6, which indicates the maximum eroding angle for a given K_D or the locations of the most severe ductile erosions for a given particle flow pattern.

Similarly, for brittle erosions, the wear in volume loss per impact w_B is expressed by

$$w_B = \frac{1}{\epsilon_B}\left(\frac{F_n}{A_c} - Y_B\right) V \sin\alpha_i A_c t_c = \frac{F_n t_c V}{\epsilon_B} \sin\alpha_i \left(1 - \frac{Y_B A_0 t_0}{F_n t_c}(\sin\alpha_i)^{-\frac{1}{5}}\right) \tag{6.13}$$

where ϵ_B is the energy required to remove a unit volume of material in brittle erosion mode. By putting Eq. (6.13) into a dimensionless form, we may define a dimensionless erosion parameter E_B for brittle erosion as

$$E_B = \frac{w_B \epsilon_B}{F_n t_c V} = \frac{w_B \epsilon_B}{2.94(5/16)(1+r^*)mV^2} \tag{6.14}$$

and a brittle resistance parameter K_B as

$$K_B = Y_B \frac{A_0 t_0}{F_n t_c} = 3\pi \left(2^{\frac{3}{5}}\right) Y_B \sqrt{k_1 k_2} (1+r^*)^{-\frac{1}{5}} N_{Im}^{-\frac{1}{5}} \tag{6.15}$$

so that

$$E_B = \sin\alpha_i \left[1 - K_B (\sin\alpha_i)^{-\frac{1}{5}}\right] = E_B(\alpha_i, K_B) \tag{6.16}$$

Equation (6.16) indicates that the maximum brittle wear occurs at $\alpha_i = 90°$, *i.e.*, normal collision. The variation of the ratio of E_B to E_{Bm} $(=E_B(\alpha_m, k_B))$ with α_i as a function of K_B is plotted on the basis of Eq. (6.16) as given in Fig. 6.7. The figure reflects that for a given K_B, the degree of abrasive damage due to brittle erosion may be estimated from the given particle flow pattern.

The preceding delineation of ductile and brittle modes by directional impacts represents the idealized components of surface wear by dust particles. In reality, most cases for surface wear result from the effects due to a combination of these two modes. Effects of gliding, scattering, and lifting of the particles by the gas stream may also significantly influence the impacts of particles. For example, the Saffman force due to the boundary layer motion of the gas may produce a gliding (barely touching) or lifting (nontouching) motion of particles, whereas the deflective motion of a gas flow may produce a centrifugal force on the particles, thereby increasing the compressive stresses for the particle impacts.

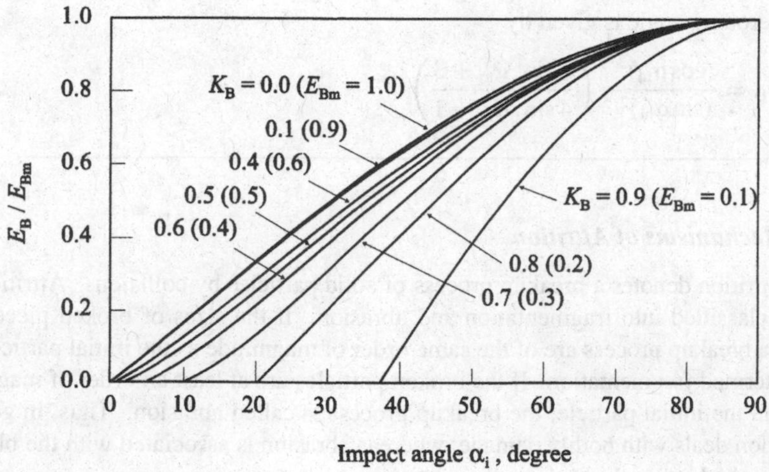

Figure 6.7. Brittle erosion energy function varied with impact angles and brittle resistance parameters (from Soo, 1977).

Example 6.1 In a survey of turbine bucket erosion, the friction coefficient can be expressed by [Soo, 1977]

$$f = f_0 \left(\frac{\cos \alpha_i}{\sin \alpha_i} \right)^{\frac{3}{5}} \tag{E6.1}$$

where

$$f_0 = \left(\frac{1 - r^*}{1 + r^*} \right)^{\frac{3}{5}} \tag{E6.2}$$

Assuming that the erosion is purely ductile, derive an expression for the impact angle yielding maximum wear.

Solution From Eq. (6.6), we have

$$w_D = \frac{f_0 F_n t_c V}{\epsilon_D} \cos \alpha_i \left(\frac{f}{f_0} - \frac{Y_D A_0 t_0}{f_0 F_n t_c} (\sin \alpha_i)^{-\frac{1}{5}} \right) \tag{E6.3}$$

Now we redefine E_D and K_D by

$$E_D = \frac{w_D \epsilon_D}{f_0 F_n t_c V} \tag{E6.4}$$

and

$$K_D = \frac{Y_D}{f_0} \frac{A_0 t_0}{F_n t_c} \tag{E6.5}$$

Substituting Eqs. (E6.1), (E6.4), and (E6.5) into Eq. (E6.3) yields

$$E_D = \cos \alpha_i \left\{ \left(\frac{\cos \alpha_i}{\sin \alpha_i} \right)^{\frac{3}{5}} - K_D (\sin \alpha_i)^{-\frac{1}{5}} \right\} \tag{E6.6}$$

Note that, at the maximum eroding angle α_m for a given K_D, $\mathrm{d}E_D/\mathrm{d}\alpha_i = 0$. Thus, the maximum eroding angle is given by

$$K_D = \frac{(\cos \alpha_m)^{3/5}}{(\sin \alpha_m)^{2/5}} \left(\frac{5 \sin^2 \alpha_m + 3}{4 \sin^2 \alpha_m + 1} \right) \tag{E6.7}$$

6.2.3 *Mechanisms of Attrition*

Attrition denotes a breakup process of solid particles by collisions. Attrition can be further classified into fragmentation and abrasion. If the sizes of broken pieces of a particle in a breakup process are of the same order of magnitude as the initial particle, this process is termed fragmentation. If the broken particles are at least an order of magnitude smaller than the initial particle, the breakup process is called abrasion. Thus, in general, fragmentation deals with bodily damage, whereas abrasion is associated with the blunting of corners and edges on the surface of a particle.

Particle attrition can be caused by the relative motion of mechanical parts such as a blade in the flow of bulk particles; by the impact of particles with a solid wall or with other particles; and by the volumetric change due to chemical reactions or phase change such as devolatilization of coal particles in a coal combustion process. The degradation in particle size by attrition would not only change the particle size distribution but also alter the flow pattern of solids, or even produce fine or ultrafine particles which usually present difficulties in transport, separation, and other handling processes.

For particle surface abrasions, a satisfactory empirical formulation was suggested by Gwyn (1969) as

$$w = \alpha \epsilon^{\beta} \tag{6.17}$$

where w is the attrited mass fraction of the sample particles; ϵ is the shear strain on the particle surface; and α and β are empirical constants. A theoretical basis for Eq. (6.17) was explored through a surface abrasion model [Paramanathan and Bridgwater, 1983], which indicates that the parameter β is related to the material properties. In the following the surface abrasion model is introduced.

Assume that the rate of abrasion is proportional to the reduction in radius of a particle raised to an arbitrary power, *i.e.*,

$$\frac{da}{d\epsilon} = -A(a_{\mathrm{i}} - a)^{-B} \tag{6.18}$$

where a_{i} is the initial radius of a particle and a is the radius of the particle at strain ϵ; A and B are constants depending on the attrition mode and material properties; and $B \neq -1$. Thus, by integrating Eq. (6.18), we have

$$\frac{(a_{\mathrm{i}} - a)^{B+1}}{B+1} = A\epsilon \tag{6.19}$$

Consider particles with an initial size distribution in the sieve range from a_1 to a_2 and define $a_{\mathrm{i}1}$ as the initial radius of a particle, which shrinks to a value a_1 at strain ϵ. Hence, we have

$$\frac{(a_{\mathrm{i}1} - a_1)^{B+1}}{B+1} = A\epsilon \tag{6.20}$$

As a result, the mass fraction attrited from other sizes to a_1, w, can be expressed by

$$w = \int_{a_1}^{a_{\mathrm{i}1}} f_{\mathrm{M}}(a_{\mathrm{i}})\, da_{\mathrm{i}} \tag{6.21}$$

where $f_{\mathrm{M}}(a_{\mathrm{i}})$ is the mass density function of initial particle sizes. Furthermore, if the material held on the top sieve produces negligible amounts of fines, the normalization of $f_{\mathrm{M}}(a_{\mathrm{i}})$ yields

$$\int_{a_1}^{a_2} f_{\mathrm{M}}(a_{\mathrm{i}})\, da_{\mathrm{i}} = 1 \tag{6.22}$$

Assuming that the initial size distribution is uniform, Eq. (6.22) gives

$$f_{\mathrm{M}}(a_{\mathrm{i}}) = \frac{1}{a_2 - a_1} \tag{6.23}$$

Therefore, we have

$$w = \frac{a_{\mathrm{i}1} - a_1}{a_2 - a_1} \tag{6.24}$$

Substituting Eq. (6.24) into Eq. (6.20) yields

$$\ln w = \frac{\ln A + \ln(B+1)}{B+1} - \ln(a_2 - a_1) + \frac{1}{B+1}\ln \epsilon \tag{6.25}$$

Now, defining α and β as

$$\ln \alpha = \frac{\ln A + \ln(B+1)}{B+1} - \ln(a_2 - a_1); \qquad \beta = \frac{1}{B+1} \tag{6.26}$$

Eq. (6.25) is reduced to Eq. (6.17).

6.3 Thermodynamic Properties of a Gas–Solid Mixture

The analysis of a multiphase flow system is complex, in part because of the difficulties in assessing the dynamic responses of each phase and the interactions between the phases. In some special cases, the gas–solid mixture can be treated as a single pseudohomogeneous phase in which general thermodynamic properties of a gas–solid mixture can be defined. This treatment provides an estimate for the "bulk" behavior of the gas–solid flow. The following treatment is based on the work of Rudinger (1980).

6.3.1 *Density, Pressure, and Equation of State*

Assume that the particles can be regarded as molecules of a second gaseous species. Hence, multiple or interactive collisions among particles, and forces between particles and gas molecules can be neglected. However, the finite volume of particles may be taken into account. It is shown later that with this type of treatment the mixture behaves as a van der Waals gas without correction for the interaction force between the molecules.

Denote α_p as the volume fraction of particles in a gas–solid mixture. The volume-averaged densities of both gas and particle phases are then given by

$$\langle\rho\rangle = (1-\alpha_p)\rho; \qquad \langle\rho_p\rangle = \alpha_p\rho_p \tag{6.27}$$

where ρ and ρ_p are the material densities of gas and particles, respectively. The bulk density of the mixture is defined by

$$\rho_m = \langle\rho\rangle + \langle\rho_p\rangle = (1-\alpha_p)\rho + \alpha_p\rho_p \tag{6.28}$$

It is convenient to express the particle loading by the particle mass fraction, which depends on the material densities of the phases as well as the volume fractions of the phases and is defined by

$$\phi = \frac{\langle\rho_p\rangle}{\rho_m} = \frac{\zeta\alpha_p}{1-\alpha_p+\zeta\alpha_p} \tag{6.29}$$

where ζ is the ratio of material density of particle to gas. Hence, an alternative form of Eq. (6.28) is

$$\frac{1}{\rho_m} = \frac{(1-\phi)}{\rho} + \frac{\phi}{\rho_p} \tag{6.30}$$

Since the particles are treated as a second "gas" phase, it is important to consider their contribution to the pressure of the mixture. The molecular weight of this "gas" is

$$w_\mathrm{p} = \left(\frac{\pi}{6} d_\mathrm{p}^3 \rho_\mathrm{p}\right) \frac{1}{m_\mathrm{H}} \tag{6.31}$$

where $m_\mathrm{H} = 1.66 \times 10^{-27}$ kg is the mass of a hydrogen atom. If the molecular weight of the gas is denoted by w_g, the molecular weight of the mixture is given by

$$\frac{1}{w_\mathrm{M}} = \frac{\phi}{w_\mathrm{p}} + \frac{(1-\phi)}{w_\mathrm{g}} \tag{6.32}$$

Thus, the ratio of the pressure of the gas p to the pressure of the mixture p_M becomes

$$\frac{p}{p_\mathrm{M}} = \frac{(1-\phi)/w_\mathrm{g}}{\phi/w_\mathrm{p} + (1-\phi)/w_\mathrm{g}} = \left(1 + \frac{\phi}{(1-\phi)} \frac{6 w_\mathrm{g} m_\mathrm{H}}{\pi d_\mathrm{p}^3 \rho_\mathrm{p}}\right)^{-1} \tag{6.33}$$

Note that the material density of a gas phase at a temperature of 300 K and pressure of 1 atm is given by

$$\rho_0 = m_\mathrm{H} w_\mathrm{g} L \tag{6.34}$$

where $L = 2.69 \times 10^{25}\ \mathrm{m}^{-3}$ is the Loschmidt number. Substituting Eq. (6.34) into Eq. (6.33) yields

$$\frac{p}{p_\mathrm{M}} = \left(1 + \frac{\phi}{(1-\phi)} \frac{6}{\pi d_\mathrm{p}^3 L} \frac{\rho_0}{\rho_\mathrm{p}}\right)^{-1} \tag{6.35}$$

If the contribution of the particles is less than 1 percent of the total pressure (*i.e.*, $p/p_\mathrm{M} > 0.99$), the particle diameter (measured in micrometers) must satisfy the condition

$$d_\mathrm{m} = 0.0192 \left(\frac{\rho_\mathrm{p}}{\rho_0} \frac{1-\phi}{\phi}\right)^{-1/3} \quad (\mu\mathrm{m}) \tag{6.36}$$

where d_m is the particle diameter below which the particle contribution to the pressure of the mixture exceeds 1 percent at the atmospheric pressure. The relationship between d_m and ϕ is shown in Fig. 6.8 for various values of ρ_p/ρ_0. It is noted that the contribution of the particles to the gas–solid mixture pressure can be neglected even at very high density and mass fraction so long as the particle is larger than a few hundredths of a micrometer. Thus, for practical purposes, the gas–solid mixture pressure in a gas–solid flow can be given by the gas-pressure alone.

Assume that the gas obeys the ideal gas law so that

$$p = \rho R_\mathrm{M} T \tag{6.37}$$

where R_M is the gas constant. For a gas–solid mixture in thermal equilibrium (*i.e.*, $T_\mathrm{p} = T$), the equation of state for the mixture may be obtained from Eq. (6.37) and Eq. (6.30) as

$$p = \frac{\rho_\mathrm{m}(1-\phi) R_\mathrm{M} T}{1 - \phi(\rho_\mathrm{m}/\rho_\mathrm{p})} \tag{6.38}$$

where ϕ, which is a constant in a closed system, is a parameter rather than a variable in the equation of state for the mixture. Hence, the equation of state for a gas–solid mixture does not have the same form as that for an ideal gas.

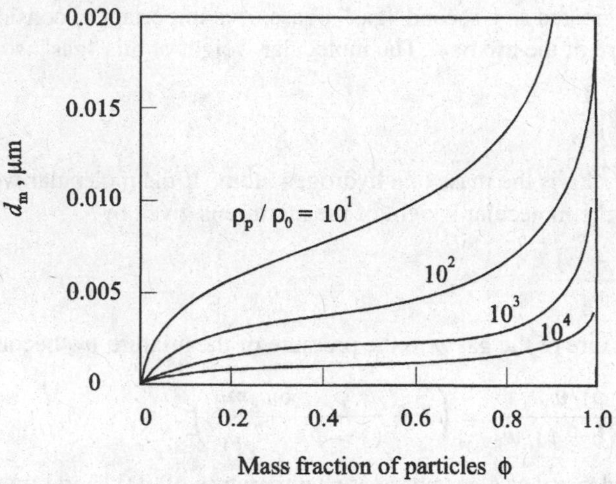

Figure 6.8. Relationship between d_m and ϕ (from Rudinger, 1980).

It should, however, be noted that the preceding results are valid only when the collisional interacting forces between the particles can be neglected; it is clear, however, that for high particle concentration conditions such as those in fluidized beds these forces cannot be neglected. A simple way to verify the significance of the interacting forces between the particles is to examine the average spacing between them in a given system. Assume that, on average, the particles are arranged in a regular cubic array of side l. The volume fraction occupied by a particle is determined by

$$\alpha_p = \frac{\pi}{6} \frac{d_p^3}{l^3} \tag{6.39}$$

As an example, for mass fraction $\phi = 0.1$ and $\zeta = 1{,}000$, the volume fraction α_p is about 10^{-4}. The average spacing between particles becomes about 17 particle diameters. Thus, for this example we can consider the direct interactions among particles to be insignificant.

Example 6.2 In a gas–solid mixture, the particle density is 2,400 kg/m^3 and the gas density is 1.2 kg/m^3. For a particle mass fraction of 99 percent, what is the corresponding particle volume fraction? What is the corresponding particle volume fraction, if the particle mass fraction is 50 percent?

Solution From Eq. (6.29), the particle volume fraction can be expressed in terms of material density ratio of particle to gas and particle mass fraction as

$$\alpha_p = \frac{\phi}{(1-\phi)\zeta + \phi} \tag{E6.8}$$

The material density ratio of particle to gas from the given conditions is 2,000. Thus, for a particle mass fraction of 99 percent, we have

$$\alpha_p = \frac{\phi}{(1-\phi)\zeta + \phi} = \frac{0.99}{0.01 \times 2{,}000 + 0.99} = 0.047 \tag{E6.9}$$

and for a particle mass fraction of 50 percent, the volume fraction of particles becomes

$$\alpha_p = \frac{\phi}{(1-\phi)\zeta + \phi} = \frac{0.5}{0.5 \times 2{,}000 + 0.5} = 5 \times 10^{-4} \tag{E6.10}$$

This example reflects that even for a very dilute suspension of gas–solid flow where the particle volume fraction is negligibly small, the particle mass fraction may still be significant, irrespective of the size of the particles.

6.3.2 *Internal Energy and Specific Heats*

The internal energy per unit volume of a mixture e_M is given by

$$e_M = (1-\phi)e + \phi e_p = (1-\phi)c_V T + \phi c T_p \tag{6.40}$$

where c_V is the specific heat of gas at constant volume and c is the specific heat of the particles. Hence, the specific heat of the mixture at a constant volume can be defined as

$$c_{VM} = (1-\phi)c_V + \phi c \tag{6.41}$$

Similarly, the specific heat of the mixture at constant pressure can be defined as

$$c_{pM} = (1-\phi)c_p + \phi c \tag{6.42}$$

From Eqs. (6.41) and (6.42), the ratio of the specific heats can be expressed as

$$\gamma_M = \frac{c_{pM}}{c_{VM}} = \gamma\left(\frac{1-\phi+\delta\phi}{1-\phi+\gamma\delta\phi}\right) \tag{6.43}$$

where δ denotes the ratio of the specific heats of particles to gas at constant pressure c/c_p.

On the basis of Eq. (6.43), the relationship between γ_M and ϕ is plotted for various values of δ, as given in Fig. 6.9. The figure shows that a large value of ϕ and/or δ leads to

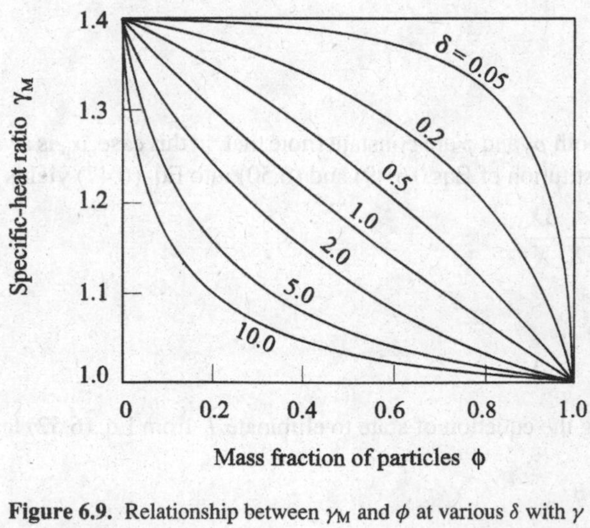

Figure 6.9. Relationship between γ_M and ϕ at various δ with $\gamma = 1.4$ (from Rudinger, 1980).

γ_M of unity, which implies an isothermal flow. Thus, the flow with a high particle loading and/or a high particle thermal capacity behaves as an isothermal flow because the large heat capacity of the particles significantly offsets the temperature variations in the gas induced by expansion or compression.

6.3.3 *Isentropic Change of State*

The equation of state in an isentropic process of a gas–solid mixture can be obtained in terms of an energy conservation relationship. When applying the first law of thermodynamics to a gas–solid mixture, we have

$$dq = de_M - \frac{p}{\rho_m^2} d\rho_m \tag{6.44}$$

where q is heat absorbed per unit mass of the mixture. Thus, for an isentropic process, we have

$$de_M = \frac{p}{\rho_m^2} d\rho_m \tag{6.45}$$

Assuming $T = T_p$, Eq. (6.40) yields

$$de_M = [(1-\phi)c_V + \phi c]\,dT \tag{6.46}$$

Substitution of Eqs. (6.46) and (6.38) into Eq. (6.45) gives

$$\frac{(1-\phi)c_V + \phi c}{(1-\phi)R_M} \frac{dT}{T} = \frac{1}{1-\phi\rho_m/\rho_p} \frac{d\rho_m}{\rho_m} \tag{6.47}$$

The relationship for the ideal gas gives

$$R_M = c_p - c_V \tag{6.48}$$

Therefore, we have

$$\frac{(1-\phi)c_V + \phi c}{(1-\phi)R_M} = \frac{c_{VM}}{c_{pM} - c_{VM}} = \frac{1}{\gamma_M - 1} \tag{6.49}$$

For convenience, let

$$X = \phi\frac{\rho_m}{\rho_p} \tag{6.50}$$

where, for a closed system, both ρ_p and ϕ are constant (note that, in this case, α_p is a variable rather than a constant). Substitution of Eqs. (6.49) and (6.50) into Eq. (6.47) yields

$$\frac{1}{\gamma_M - 1}\frac{dT}{T} = \frac{1}{1-X}\frac{dX}{X} \tag{6.51}$$

which gives

$$T\left(\frac{X}{1-X}\right)^{-(\gamma_M-1)} = A \tag{6.52}$$

where A is a constant. Using the equation of state to eliminate T from Eq. (6.52) leads to

$$p\left(\frac{X}{1-X}\right)^{-\gamma_M} = B \tag{6.53}$$

where B is a constant. In another form, Eq. (6.53) becomes

$$p\left(\frac{\rho_\mathrm{m}}{1-\phi\rho_\mathrm{m}/\rho_\mathrm{p}}\right)^{-\gamma_\mathrm{M}} = C \tag{6.54}$$

where C is a constant.

Equation (6.54) accounts for the change of state in the isentropic processes of a gas–solid mixture in which the effect of a finite particle volume is considered. An example using this equation to obtain the speed of sound in a gas–solid mixture is introduced in the next section.

6.4 Pressure Waves Through a Gas–Solid Suspension

The propagation of pressure waves such as acoustic wave, shock wave, and Prandtl–Meyer expansion through a gas–solid suspension is a phenomenon associated primarily with the transfer of momentum although certain processes of energy transfer such as kinetic energy dissipation and heat transfer between gas and solids almost always occur. Typical applications of the pressure wave propagation include the measurements of the solids concentration and flow rate by use of acoustic devices as well as detonation combustion such as in a rocket propellant combustor or in the barrel of a gun.

When a pressure wave travels through a gas–solid mixture, the interaction of the pressure wave with the suspension can be strong. In general, the traveling speed depends not only on the local concentration of solids and the properties of the gas and solid phases but also on the wave frequency. The propagation of different types of pressure waves through the same mixture behaves differently because of the difference in the dynamic response of the mixture. The speed of sound represents the propagation of periodic pressure waves of small amplitude. The speed of a shock wave, having a front of discontinued pressure rise followed by an extended relaxation zone, represents the propagation of a sharp pressure jump which produces a transition from one flow condition to another. For simplicity we consider only the cases of plane waves through a gas–solid suspension, in the following discussion.

6.4.1 *Acoustic Wave*

In order to obtain governing equations for the motion of particles and the speed of sound when an acoustic wave passes through a gas–solid suspension, it is necessary to determine the interacting force between the solid and gas phases which accounts for the relative oscillating motion of the particles. Under the condition of creeping flow, an expression for the interacting force between an oscillating solid sphere and an infinitely large incompressible fluid at rest was obtained by Lamb (1932). Thus, by substituting the resultant interacting force into the BBO equation (see §3.4), the oscillating motion of a spherical particle may be described by the modified equation. When this equation is coupled with the equation of a plane wave in a mixture, the speed of sound in a gas–solid suspension can be obtained.

First, consider the case of a ball pendulum oscillating in an infinite mass of fluid. The fluid is incompressible and originally at rest. It is convenient to set the origin at the mean position of the center, and the x-axis in the direction of the oscillation. For small motion of the fluid where the inertia can be neglected and in the absence of external forces, we have

$$\nabla \cdot \boldsymbol{U} = 0 \tag{6.55}$$

and

$$\frac{\partial \boldsymbol{U}}{\partial t} = -\frac{1}{\rho}\nabla p + \nu \nabla^2 \boldsymbol{U} \tag{6.56}$$

where $\boldsymbol{U}$ denotes the fluid velocity. The boundary condition on the surface of the sphere is given by

$$\begin{aligned} U &= U_{\mathrm{p}} = U_0 e^{i(\omega t + \phi)} \\ V &= V_{\mathrm{p}} = 0 \\ W &= W_{\mathrm{p}} = 0 \end{aligned} \tag{6.57}$$

where $\boldsymbol{U}_{\mathrm{p}}$ is the velocity of the sphere.

With a lengthy derivation, the analytic solution for the preceding problem was provided by Lamb (1932). The resultant force on the sphere due to oscillating motion is obtained as

$$\boldsymbol{F} = -\frac{1}{6}\pi \rho d_{\mathrm{p}}^3 \left(\frac{1}{2} + \frac{9}{4\sqrt{\mathrm{N}_\omega}}\right)\frac{\mathrm{d}\boldsymbol{U}}{\mathrm{d}t} - \frac{3}{8}\pi d_{\mathrm{p}}^3 \rho \frac{\omega}{\sqrt{\mathrm{N}_\omega}}\left(1 + \frac{1}{\sqrt{\mathrm{N}_\omega}}\right)\boldsymbol{U} \tag{6.58}$$

where N_ω is a dimensionless parameter expressed as

$$\mathrm{N}_\omega = \frac{\omega d_{\mathrm{p}}^2}{8\nu} \tag{6.59}$$

The first term on the right-hand side of Eq. (6.58) gives the correction for the inertia of the sphere, commonly referred to as the virtual mass term. It should be noted that, as a result of viscous and oscillating effects, the amount of replaced fluid mass increases. The restriction here is that $\mathrm{d}U_{\mathrm{p}}/\mathrm{d}t = \mathrm{i}\omega U_{\mathrm{p}}$, *i.e.*, pure oscillating motion of the particles. The second term is the drag force. When the oscillation period is made infinitely long (*i.e.*, $\omega \to 0$), Eq. (6.58) leads to the Stokes drag force.

Consider a gas–solid suspension which is in a state of steady dilute flow with no interparticle collision or contact. In this situation, the linear particle velocity is practically identical to the superficial particle velocity. The motion of a spherical particle in an oscillating flow field can thus be given by

$$\begin{aligned} \frac{\rho_{\mathrm{p}}}{\rho}\frac{\mathrm{d}\boldsymbol{U}_{\mathrm{p}}}{\mathrm{d}t} = {} & \left(\frac{1}{2} + \frac{9}{4\sqrt{\mathrm{N}_\omega}}\right)\frac{\mathrm{d}(\boldsymbol{U} - \boldsymbol{U}_{\mathrm{p}})}{\mathrm{d}t} + \frac{9}{4}\frac{\omega}{\sqrt{\mathrm{N}_\omega}}\left(1 + \frac{1}{\sqrt{\mathrm{N}_\omega}}\right)(\boldsymbol{U} - \boldsymbol{U}_{\mathrm{p}}) \\ & - \frac{1}{\rho}\nabla p + \frac{9}{d_{\mathrm{p}}}\sqrt{\frac{\nu}{\pi}}\int_{t_0}^{t}\frac{\frac{\mathrm{d}}{\mathrm{d}\tau}(\boldsymbol{U} - \boldsymbol{U}_{\mathrm{p}})}{\sqrt{t-\tau}}\,\mathrm{d}\tau \end{aligned} \tag{6.60}$$

which is a modified BBO equation derived by replacing the Stokes drag and carried mass terms in Eq. (3.101) by Eq. (6.58). If we further approximate the pressure term by

$$-\nabla p \approx \rho \frac{\mathrm{d}\boldsymbol{U}}{\mathrm{d}t} \tag{6.61}$$

Equation (6.60) becomes

$$\frac{\rho_\mathrm{p}}{\rho}\frac{\mathrm{d}\boldsymbol{U}_\mathrm{p}}{\mathrm{d}t} = \left(\frac{1}{2} + \frac{9}{4\sqrt{\mathrm{N}_\omega}}\right)\frac{\mathrm{d}(\boldsymbol{U} - \boldsymbol{U}_\mathrm{p})}{\mathrm{d}t} + \frac{9}{4}\frac{\omega}{\sqrt{\mathrm{N}_\omega}}\left(1 + \frac{1}{\sqrt{\mathrm{N}_\omega}}\right)(\boldsymbol{U} - \boldsymbol{U}_\mathrm{p})$$

$$+ \frac{\mathrm{d}\boldsymbol{U}}{\mathrm{d}t} + \frac{9}{d_\mathrm{p}}\sqrt{\frac{\nu}{\pi}}\int_{t_0}^{t}\frac{\frac{\mathrm{d}}{\mathrm{d}\tau}(\boldsymbol{U} - \boldsymbol{U}_\mathrm{p})}{\sqrt{t-\tau}}\,\mathrm{d}\tau \tag{6.62}$$

Now we consider the case of one-dimensional oscillation. If we neglect the effects due to the volume occupied by the solid particles and scattering, for a gas–solid suspension with solid-to-gas mass ratio of m_p, the equation of a plane wave in the mixture may be expressed by [Soo, 1990]

$$\left(\frac{\partial^2 U}{\partial t^2}\right) = \frac{\gamma p}{\rho}\left(\frac{\partial^2 U}{\partial z^2}\right) - \frac{6 m_\mathrm{p}}{\pi d_\mathrm{p}^3 \rho_\mathrm{p}}\left(\frac{\partial F}{\partial t}\right) \tag{6.63}$$

where F is the total particle–fluid interacting force (on the particle) including the carried mass force, Basset force, and drag force. F may be expressed by Eq. (6.58) when $\mathrm{Re_p}$ is in the Stokes regime.

For a harmonic oscillation, we may express U and U_p by

$$\begin{aligned} U &= U_0 e^{i(\omega t - kz)} e^{-\beta z} \\ U_\mathrm{p} &= U_{\mathrm{p}0} e^{i(\omega t + \phi(z))} \end{aligned} \tag{6.64}$$

where β is the attenuation coefficient. Thus, Eq. (6.62) becomes

$$U_\mathrm{p} = B(U - U_\mathrm{p}) - iA(U - U_\mathrm{p}) + \frac{\rho}{\rho_\mathrm{p}}U + C(1-i)(U - U_\mathrm{p}) \tag{6.65}$$

where

$$A = \frac{9}{4}\frac{\rho}{\rho_\mathrm{p}}\frac{1}{\sqrt{\mathrm{N}_\omega}}\left(1 + \frac{1}{\sqrt{\mathrm{N}_\omega}}\right), \qquad B = \frac{\rho}{\rho_\mathrm{p}}\left(\frac{1}{2} + \frac{9}{4}\frac{1}{\sqrt{\mathrm{N}_\omega}}\right), \qquad C = \frac{9}{4}\frac{\rho}{\rho_\mathrm{p}}\frac{1}{\sqrt{\mathrm{N}_\omega}} \tag{6.66}$$

A is the friction force coefficient; B is the inertia correction coefficient; and C is the resistance coefficient due to the Basset force. Equation (6.65) gives

$$\frac{U_\mathrm{p}}{U} = \frac{\eta^2 + (1+\xi)\left[\xi + \frac{\rho}{\rho_\mathrm{p}}\right] - i\left[1 - \frac{\rho}{\rho_\mathrm{p}}\right]\eta}{(1+\xi)^2 + \eta^2} \tag{6.67}$$

where $\xi = B + C$, and $\eta = A + C$. The speed of sound in the mixture can be obtained from Eq. (6.63) by noting that

$$\frac{\partial F}{\partial t} = \frac{\pi}{6} d_\mathrm{p}^3 \rho_\mathrm{p}\left(\frac{\partial^2 U_\mathrm{p}}{\partial t^2} - \frac{\rho}{\rho_\mathrm{p}}\frac{\partial^2 U}{\partial t^2}\right) \tag{6.68}$$

Substitution of Eq. (6.68) into Eq. (6.63) gives

$$\left(\frac{\partial^2 U}{\partial t^2}\right) = \frac{\gamma p}{\rho}\left(\frac{\partial^2 U}{\partial z^2}\right) - m_\mathrm{p}\left(\frac{\partial^2 U_\mathrm{p}}{\partial t^2} - \frac{\rho}{\rho_\mathrm{p}}\frac{\partial^2 U}{\partial t^2}\right) \tag{6.69}$$

Substituting the real part of Eq. (6.67) into Eq. (6.69) gives

$$\frac{\partial^2 U_p}{\partial t^2}\left[1 + m_p D - \frac{\rho}{\rho_p} m_p\right] = \frac{\gamma p}{\rho}\left(\frac{\partial^2 U_p}{\partial z^2}\right) \tag{6.70}$$

where D is the real part of Eq. (6.67) and is equal to

$$D = \frac{\eta^2 + (1+\xi)\left(\xi + \dfrac{\rho}{\rho_p}\right)}{(1+\xi)^2 + \eta^2} \tag{6.71}$$

Hence, the speed of sound in the mixture is given by

$$\frac{a_m^2}{a_g^2} = \left[1 + m_p\left(1 - \frac{\rho}{\rho_p}\right)\frac{(1+\xi)\xi + \eta^2}{(1+\xi)^2 + \eta^2}\right]^{-1} \tag{6.72}$$

where a_m is the speed of sound in the mixture and a_g is the speed of sound in the pure gas, which is given by

$$a_g^2 = \frac{\gamma p}{\rho} \tag{6.73}$$

Moreover, substituting Eq. (6.67) and the expression of U in Eq. (6.64) into Eq. (6.69), for $\beta a_g/\omega < 1$, an expression for $\beta a_g/\omega$ can be obtained as

$$\frac{\beta a_g}{\omega} = \frac{m_p}{2}\frac{a_m}{a_g}\left(1 - \frac{\rho}{\rho_p}\right)\frac{\eta}{(1+\xi)^2 + \eta^2} \tag{6.74}$$

An example of dispersion of sound in an air–magnesia mixture with solid-to-air mass ratio, m_p, of 0.3 for ρ_p/ρ $(=\zeta)$ of 100 and 1,000 is illustrated in Fig. 6.10 [Soo, 1960]. The figure gives the dispersion of sound for various values of N_ω when the heat transfer is neglected. Under the same experimental condition, the relationship between $\beta a_g/\omega$ and N_ω is shown in Fig. 6.11 [Soo, 1960]. It is noted that the preceding model is valid only

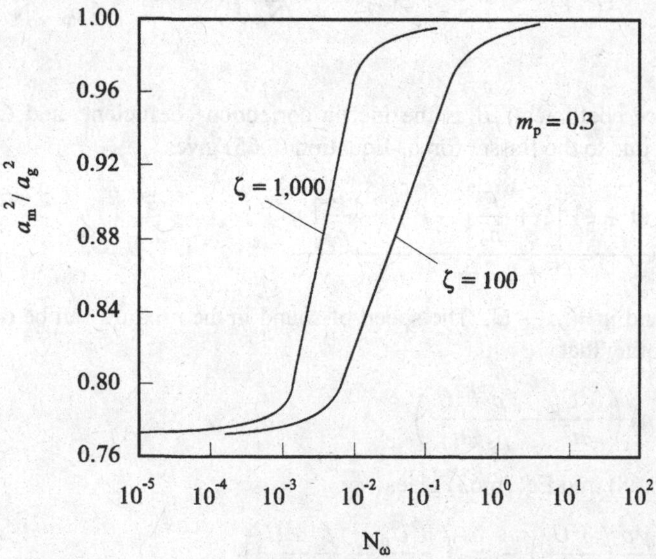

Figure 6.10. Dispersion of acoustic waves in the air–magnesia mixture (from Soo, 1960).

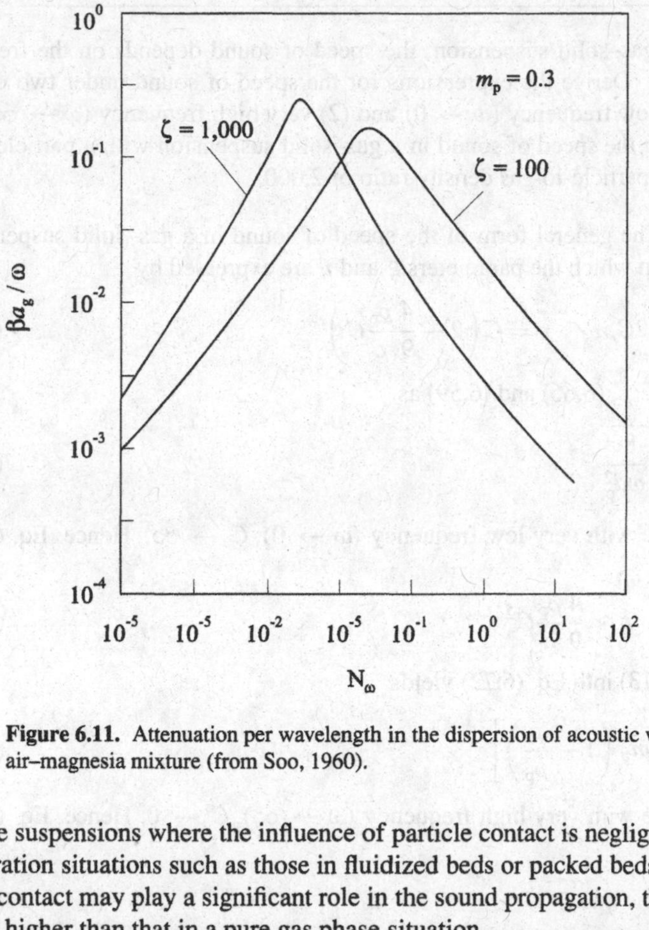

Figure 6.11. Attenuation per wavelength in the dispersion of acoustic waves in the air–magnesia mixture (from Soo, 1960).

for dilute suspensions where the influence of particle contact is negligible. For high solids concentration situations such as those in fluidized beds or packed beds where the effect of particle contact may play a significant role in the sound propagation, the speed of sound is possibly higher than that in a pure gas phase situation.

Another approach for estimating a_m is based on the pseudothermodynamic properties of the mixture, as suggested by Rudinger (1980). The equation for the isentropic changes of state of a gas–solid mixture is given by Eq. (6.53). Note that for a closed system the material density of particles and the mass fraction of particles can be treated as constant. Hence, in terms of the case for a single-phase fluid, the speed of sound in a gas–solid mixture can be expressed as

$$a_m^2 = \left(\frac{\partial p}{\partial \rho_m}\right)_S \tag{6.75}$$

which, with the substitution of Eq. (6.53), gives

$$a_m^2 = \frac{\phi}{\rho_p}\left(\frac{\partial p}{\partial X}\right)_S = \gamma_M p \frac{\phi}{\rho_p} \frac{1}{X(1-X)} \tag{6.76}$$

The subscript "s" denotes entropy. From Eqs. (6.28) and (6.50), the speed of sound in a mixture is obtained as

$$\frac{a_m^2}{a_g^2} = \frac{(1-\phi)}{(1-\alpha_p)^2}\left(\frac{1-\phi+\delta\phi}{1-\phi+\gamma\delta\phi}\right) \tag{6.77}$$

It is noted from Eq. (6.72) and Eq. (6.77) that $a_m < a_g$ under dilute suspension situations.

Example 6.3 In a gas–solid suspension, the speed of sound depends on the frequency of the acoustic wave. Derive the expressions for the speed of sound under two extreme conditions: (1) very low frequency ($\omega \to 0$) and (2) very high frequency ($\omega \to \infty$). Use the results to estimate the speed of sound in a gas–solid suspension with a particle-to-gas mass ratio of 1 and a particle-to-gas density ratio of 2,000.

Solution The general form of the speed of sound in a gas–solid suspension is given by Eq. (6.72), in which the parameters ξ and η are expressed by

$$\xi = \frac{1}{2}\frac{\rho}{\rho_{\mathrm{p}}} + 2C, \qquad \eta = C\left(2 + \frac{4}{9}\frac{\rho_{\mathrm{p}}}{\rho}C\right) \tag{E6.11}$$

where C is given by Eqs. (6.66) and (6.59) as

$$C = \frac{9}{4}\frac{\rho}{\rho_{\mathrm{p}}}\sqrt{\frac{8\nu}{\omega d_{\mathrm{p}}^2}} \tag{E6.12}$$

For the acoustic wave with very low frequency ($\omega \to 0$), $C \to \infty$. Hence, Eq. (E6.11) becomes

$$\xi \approx 2C, \qquad \eta \approx \frac{4}{9}\frac{\rho_{\mathrm{p}}}{\rho}C^2 \tag{E6.13}$$

Substituting Eq. (E6.13) into Eq. (6.72) yields

$$\frac{a_{\mathrm{m}}^2}{a_{\mathrm{g}}^2} = \left[1 + m_{\mathrm{p}}\left(1 - \frac{\rho}{\rho_{\mathrm{p}}}\right)\right]^{-1} \tag{E6.14}$$

For the acoustic wave with very high frequency ($\omega \to \infty$), $C \to 0$. Hence, Eq. (E6.11) becomes

$$\xi = \frac{1}{2}\frac{\rho}{\rho_{\mathrm{p}}}, \qquad \eta = 0 \tag{E6.15}$$

Substituting Eq. (E6.15) into Eq. (6.72) yields

$$\frac{a_{\mathrm{m}}^2}{a_{\mathrm{g}}^2} = \left[1 + \frac{m_{\mathrm{p}}}{2\rho_{\mathrm{p}}/\rho + 1}\left(1 - \frac{\rho}{\rho_{\mathrm{p}}}\right)\right]^{-1} \tag{E6.16}$$

For the gas–solid suspension with $m_{\mathrm{p}} = 1$ and $\rho_{\mathrm{p}}/\rho = 2{,}000$, from Eqs. (E6.14) and (E6.16), we have

$$\frac{a_{\mathrm{m}}}{a_{\mathrm{g}}} = \left[1 + m_{\mathrm{p}}\left(1 - \frac{\rho}{\rho_{\mathrm{p}}}\right)\right]^{-\frac{1}{2}} = \left[1 + 1 \times \left(1 - \frac{1}{2{,}000}\right)\right]^{-\frac{1}{2}} = 0.71 \tag{E6.17}$$

for very-low-frequency acoustic waves, and

$$\frac{a_{\mathrm{m}}}{a_{\mathrm{g}}} = \left[1 + \frac{m_{\mathrm{p}}}{2\rho_{\mathrm{p}}/\rho + 1}\left(1 - \frac{\rho}{\rho_{\mathrm{p}}}\right)\right]^{-\frac{1}{2}}$$

$$= \left[1 + \frac{1}{2 \times 2{,}000 + 1}\left(1 - \frac{1}{2{,}000}\right)\right]^{-\frac{1}{2}} = 1.00 \tag{E6.18}$$

for very-high-frequency acoustic waves.

This example shows that in a dilute gas–solid suspension, for a very-low-frequency perturbation, the particles may follow the changes in motion of the surrounding gas. The speed of sound in this case is known as the equilibrium speed of sound; it can be estimated from Eq. (E6.14). On the other hand, for a very-high-frequency perturbation in a dilute gas–solid suspension, the particles do not follow the changes in motion of the surrounding gas. In this case, the speed of sound is known as the frozen speed of sound and has almost the same value as that for the gas alone.

6.4.2 *Normal Shock Wave*

Consider the propagation of a one-dimensional normal shock wave in a gas medium heavily laden with particles. Select Cartesian coordinates attached to the shock front so that the shock front becomes stationary. The changes of velocities, temperatures, and pressures of gas and particle phases across the normal shock wave are schematically illustrated in Fig. 6.12, where the subscripts "1," "2," and "∞" represent the conditions in front of, immediately behind, and far away behind the shock wave front, respectively. As shown in Fig. 6.12, a nonequilibrium condition between particles and the gas exists immediately behind the shock front. Apparently, because of the finite rate of momentum transfer and heat transfer between the gas and the particles, a relaxation distance is required for the particles to gain a new equilibrium with the gas.

The elapsed time for particles to pass through the shock front may be approximated by d_p/U. Since U is of the same order of magnitude as the speed of sound in the gas, the ratio of the flying time to the Stokes relaxation time of a particle can be expressed by

$$\frac{\tau_p}{\tau_S} \approx \frac{d_p}{U\tau_S} \approx \frac{d_p}{a_g\tau_S} = \frac{18\mu}{d_p\rho_p\sqrt{\gamma R_M T}} \tag{6.78}$$

where τ_S is the Stokes relaxation time. Calculation using Eq. (6.78) based on some representative values reveals that, even for particles as small as 0.1 μm, the elapsed time for a particle to pass through the shock front is about three orders of magnitude smaller than the Stokes relaxation time [Rudinger, 1980]. Thus, as illustrated in Fig. 6.12, a particle passes through the shock front without yielding a significant change in its velocity. This is also the case for the change in the particle temperature when the particle passes through the shock front.

To analyze the shock wave behavior in a gas–solid flow, several assumptions are made:

(1) The gas follows the equation of state of an ideal gas.
(2) The particles are uniformly distributed in the gas upstream of the shock wave.
(3) The Brownian motion of the particles does not contribute to the pressure of the system so that the pressure of the mixture can be given by the gas pressure alone.
(4) The thermal diffusivity of the particles is high so that the internal temperature of a particle is uniform.
(5) The effect of thermal radiation is negligible. (It should be noted, however, that a cloud of particles is usually a better emitter and absorber of radiation than a pure gas. Thus, the hot particles downstream of the shock may preheat the cold particles upstream of the shock by radiation. This effect becomes significant as the surface area of the particles increases.)
(6) The particle volume fraction is much less than unity.

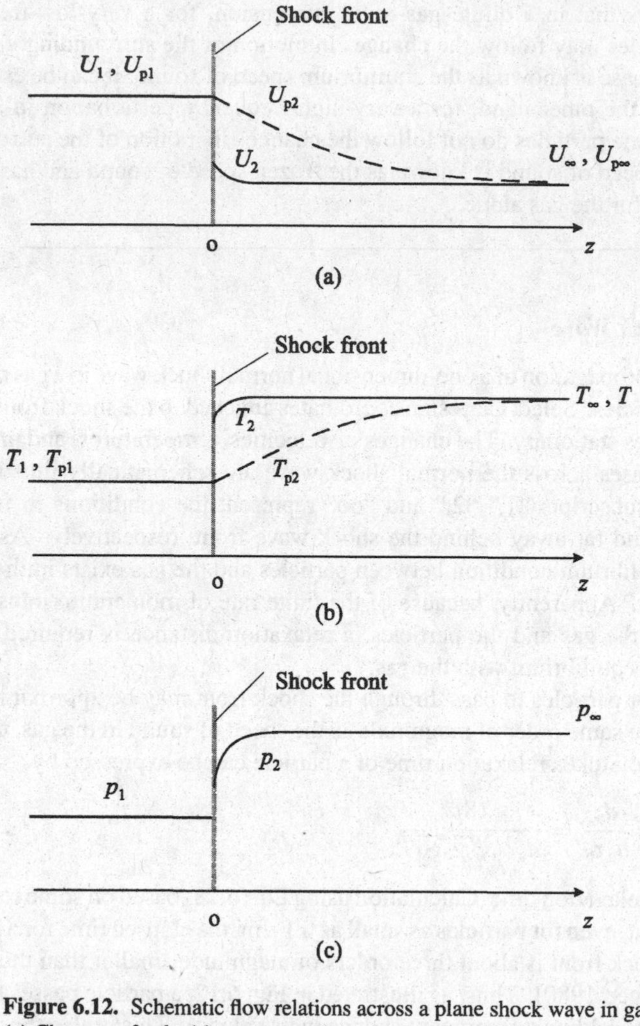

Figure 6.12. Schematic flow relations across a plane shock wave in gas–solid suspensions: (a) Changes of velocities across a plane shock wave; (b) Changes of temperatures across a plane shock wave; (c) Change of pressure across a plane shock wave.

The following analysis is based on the method suggested by Carrier (1958). The law of conservation of mass requires

$$J = (1 - \alpha_p)\rho U \approx \rho U = \text{const} \tag{6.79}$$

and

$$J_p = \alpha_p \rho_p U_p = \text{const} \tag{6.80}$$

where J and J_p are the mass fluxes of the gas and particles. The conservation of momentum of the gas–solid mixture yields

$$(\rho U)U + (\alpha_p \rho_p U_p)U_p + p = \text{const} \tag{6.81}$$

and the conservation of energy gives

$$\frac{1}{2}(\rho U)U^2 + \frac{1}{2}(\alpha_p\rho_p U_p)U_p^2 + (\rho U)c_p T + (\alpha_p\rho_p U_p)cT_p = \text{const} \tag{6.82}$$

Note that the equilibrium between the gas and solid phases indicates the conditions $U = U_p$ and $T = T_p$. Substituting these conditions as well as Eqs. (6.29), (6.37), and (6.79) into Eq. (6.81) and Eq. (6.82) gives

$$\frac{1}{1-\phi}U + \frac{R_M T}{U} = \text{const} \tag{6.83}$$

and

$$\frac{1}{1-\phi}U^2 + 2\left(c_p + \frac{\phi}{1-\phi}c\right)T = \text{const} \tag{6.84}$$

By use of Eq. (6.43), Eq. (6.84) becomes

$$\frac{1}{1-\phi}U^2 + \frac{2\gamma_M}{\gamma_M - 1}R_M T = \text{const} \tag{6.85}$$

Now, consider the velocities and temperatures in two equilibrium regions, one in front of the shock and the other far behind the shock. Applying Eq. (6.83) and Eq. (6.85) to these two regions yields

$$\frac{1}{1-\phi}U_1 + \frac{R_M T_1}{U_1} = \frac{1}{1-\phi}U_\infty + \frac{R_M T_\infty}{U_\infty} \tag{6.86}$$

and

$$\frac{1}{1-\phi}U_1^2 + \frac{2\gamma_M}{\gamma_M - 1}R_M T_1 = \frac{1}{1-\phi}U_\infty^2 + \frac{2\gamma_M}{\gamma_M - 1}R_M T_\infty \tag{6.87}$$

Combining Eq. (6.86) and Eq. (6.87) gives

$$U_\infty U_1 = \frac{2\gamma_M(1-\phi)}{\gamma_M + 1}R_M T_1 + \frac{\gamma_M - 1}{\gamma_M + 1}U_1^2 \tag{6.88}$$

Thus, U_∞ can be evaluated from Eq. (6.88). The gas speed just after the shock can also be obtained from Eq. (6.88), by setting $\phi = 0$, as

$$U_2 U_1 = \frac{2\gamma}{\gamma + 1}R_M T_1 + \frac{\gamma - 1}{\gamma + 1}U_1^2 \tag{6.89}$$

A comparison of Eq. (6.88) and Eq. (6.89) indicates that the states of the region far behind the shock and that just after the shock differ and vary with the particle mass fraction and/or particle thermal capacity. In the transition region in which the velocity and temperature of the particles are changing from an equilibrium before the shock to a new equilibrium after the shock, the changing processes can be described on the basis of the equation of motion and the equation of energy for the particle phase in the gas–solid mixture. Assuming that $U_2 = U_\infty$, $T_2 = T_\infty$, and only the drag force and the convective heat transfer are important, the equation of motion for a particle can be given by

$$\frac{\pi}{6}d_p^3\rho_p U_p \frac{dU_p}{dz} = -\frac{\pi}{4}d_p^2 \frac{C_D}{2}\rho(U_\infty - U_p)^2 \tag{6.90}$$

and the equation of energy for a particle can be given by

$$\frac{\pi}{6} d_p^3 \rho_p c U_p \frac{dT_p}{dz} = \pi d_p K \, Nu_p (T_\infty - T_p) \tag{6.91}$$

where Nu_p is the particle Nusselt number and K is the thermal conductivity of gas. The boundary conditions are given by

$$\begin{aligned} U_p|_{z=0} &= U_1; \qquad & T_p|_{z=0} &= T_1 \\ U_p|_{z\to\infty} (= U_{p\infty}) &= U_\infty; \qquad & T_p|_{z\to\infty} (= T_{p\infty}) &= T_\infty \end{aligned} \tag{6.92}$$

The rates of change in the particle velocity and temperature are therefore dependent on the values of the drag coefficient C_D and the heat transfer coefficient, respectively. In a gas–solid flow, both the drag coefficient and the heat transfer coefficient are functions of the particle Reynolds number and the volume fraction of solids. These relations are primarily represented by empirical or semiempirical relations (see §1.2 and §4.3). In a very dilute suspension, C_D and Nu_p for a single particle can be applied to Eq. (6.90) and Eq. (6.91). As an example, for a mixture of air and 10 μm glass beads with a particle to gas mass flow ratio of 0.2 and a Mach number of 1.5, on the basis of the equations given, different relations for drag and heat transfer coefficients are shown to significantly affect the pressures, velocities, and temperatures, as shown in Fig. 6.13 [Rudinger, 1969]. The

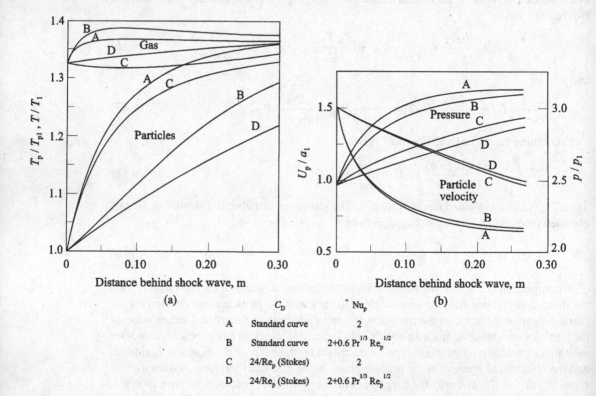

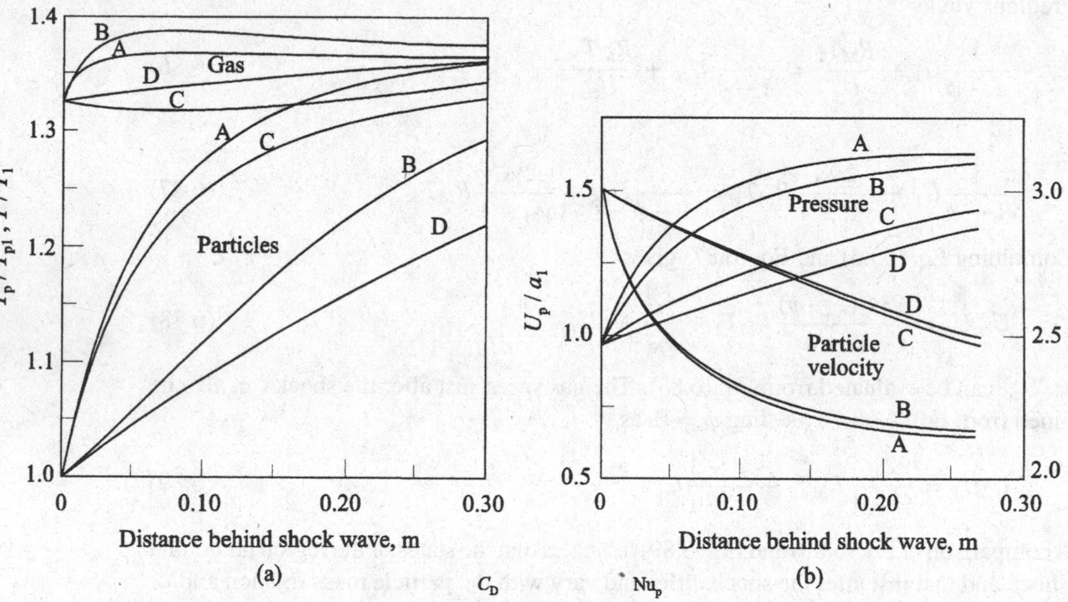

	C_D	Nu_p
A	Standard curve	2
B	Standard curve	$2+0.6\,Pr^{1/3}\,Re_p^{1/2}$
C	$24/Re_p$ (Stokes)	2
D	$24/Re_p$ (Stokes)	$2+0.6\,Pr^{1/3}\,Re_p^{1/2}$

Figure 6.13. Effect of C_D and Nu_p on the dynamic properties of phases over a plane shock in a gas–solid suspension with a Mach number of 1.5, 10 μm glass beads, and $m_p = 0.2$ (from Rudinger, 1969): (a) Temperature distributions; (b) Pressure and particle velocity distributions.

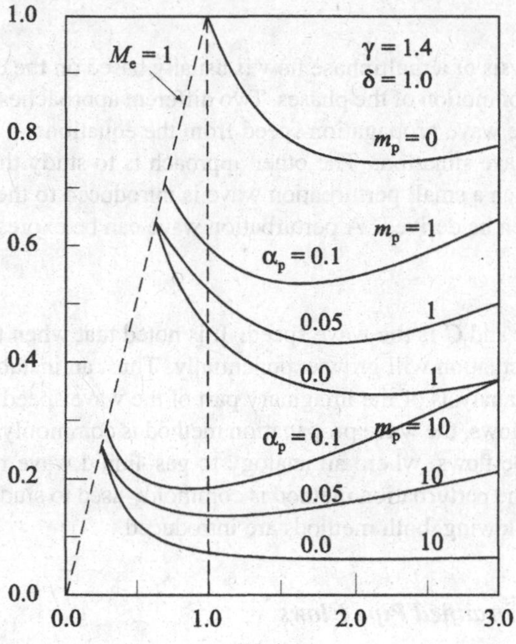

Figure 6.14. Effect of particle volume fraction on the equilibrium velocity behind a normal shock in a gas–solid suspension (from Rudinger, 1965).

effects of the particle volume fraction and particle to gas mass flow ratio on the equilibrium velocity, U_∞, behind a normal shock wave as a function of the velocity in front of the shock, U_1, are shown in Fig. 6.14 [Rudinger, 1965]. It is seen that the equilibrium velocity decreases with increasing mass ratio of particle to gas and increases with increasing particle volume fraction. It is noted that with an increase in the particle volume fraction, particle–particle collisions and other types of particle interactions become increasingly important and thus the assumption of negligible particle interactions used in the calculation is no longer feasible.

We introduce only the normal shock waves in an infinitely large, dilute, and uniformly distributed gas–solid suspension, which represents the simplest case for the propagation of a normal shock. For a normal shock passing through a one-dimensional duct with varied cross-sectional area, a set of equations for both gas and particle phases that may be solved by using the method characteristics can be derived [Soo, 1990]. Examples using the method of characteristics were reported for the centered expansion wave and shock wave generated by an impulsively driven piston in a suspension of 10 μm glass beads in air [Rudinger and Chang, 1964]. The normal shock wave analysis can be extended to oblique shocks. For oblique shocks, the shock front is not perpendicular to the direction of the fluid motion. However, for large particles, the velocity vectors of the gas and particles are no longer parallel after the oblique shock because of the inertia of the particles. Therefore, the flow after the shock becomes two-dimensional, thus requiring more complicated mathematical treatments [Morgenthaler, 1962; Peddieson, 1975; Chang, 1975].

6.5 Instability

An instability analysis of a multiphase flow is usually based on the continuum assumption and the equations of motion of the phases. Two different approaches are common. One approach is to study the wave propagation speed from the equations of motion using an analogy to the surface wave situation. The other approach is to study the growth rate of the perturbation wave when a small perturbation wave is introduced to the system. The criterion for stability can then be derived. A perturbation wave can be expressed by

$$y = y_0 e^{i(kr-Ct)} \tag{6.93}$$

where k is the wave number and C is the wave speed. It is noted that when the imaginary part of C is positive, the fluctuation will grow exponentially. Thus, an instability criterion can be determined from the analysis of the imaginary part of the wave speed.

In gas–solid multiphase flows, the wave propagation method is commonly used to study the stability of stratified pipe flows, where an analogy to gas–liquid wave motion with a free surface is prominent. The perturbation method is commonly used to study the stability of a fluidized bed. In the following, both methods are introduced.

6.5.1 *Wave Motions in Stratified Pipe Flows*

A fundamental account of unsteady motion in a dense suspension requires knowledge of phase interactions and origins of wavy stratified motion [Zhu, 1991; Zhu *et al.*, 1996]. Flow stratification involves a dense phase of solids flowing along the bottom of a horizontal pipe with wave motions, or a layer of the dense phase of solids flowing near the wall of a vertical pipe with wave motions. The former was analyzed, in part, on the basis of the theoretical work of de Crecy (1986) on stratified liquid–vapor flows.

6.5.1.1 *General Considerations for Stratified Pipe Flows*

To simplify the analysis of wave motions in stratified pipe flows, several assumptions can be made:

(1) A stratified pipe flow is composed of a dilute flow and a dense flow with an infinitesimal thickness of interface.
(2) Each phase (gas or solids) is continuous in any cross section.
(3) Regionally averaged flows are unsteady and one-dimensional.
(4) The area and shape of the cross section of the pipe are constant, and the pipe wall is impermeable.

An instantaneous regional average of phase j (j = gas or solids) in the region k is defined as

$$\langle x_{\mathrm{jk}} \rangle = \frac{1}{A_{\mathrm{k}}} \int_{A_{\mathrm{k}}(z,t)} x_{\mathrm{jk}} \, \mathrm{d}A \tag{6.94}$$

From Delhaye (1981), the instantaneous regional-averaged continuity equation projected along the tube axis can be expressed as

$$\frac{\partial}{\partial t} A_{\mathrm{k}} \langle \rho_{\mathrm{jk}} \rangle + \frac{\partial}{\partial z} A_{\mathrm{k}} \langle \rho_{\mathrm{jk}} W_{\mathrm{jk}} \rangle = - \int_{C(z,t)} \frac{\mathrm{d}m_{\mathrm{jk}}}{\mathrm{d}t} \frac{\mathrm{d}C}{\boldsymbol{n}_{\mathrm{k}} \cdot \boldsymbol{n}_{\mathrm{kc}}} \tag{6.95}$$

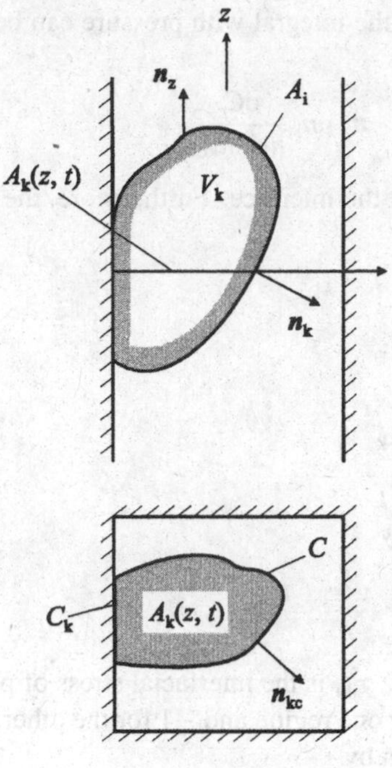

Figure 6.15. General configuration for regional averages.

and the momentum equation projected along the tube axis can be given by

$$\frac{\partial}{\partial t}A_k\langle\rho_{jk}W_{jk}\rangle + \frac{\partial}{\partial z}A_k\langle\rho_{jk}W_{jk}^2\rangle - A_k\langle\rho_{jk}f_{zj}\rangle + \frac{\partial}{\partial z}A_k\langle p_{jk}\rangle - \frac{\partial}{\partial z}A_k\langle \boldsymbol{n}_z\cdot\boldsymbol{\tau}_{jk}\cdot\boldsymbol{n}_z\rangle$$
$$= -\int_{C(z,t)}\frac{dm_{jk}}{dt}W_{jk}\frac{dC}{\boldsymbol{n}_k\cdot\boldsymbol{n}_{kc}} - \int_{C\cup C_k(z,t)}\boldsymbol{n}_z\cdot(\boldsymbol{n}_k\cdot(p_{jk}\boldsymbol{I}-\boldsymbol{\tau}_{jk}))\frac{dC}{\boldsymbol{n}_k\cdot\boldsymbol{n}_{kc}} \quad (6.96)$$

where A_k, A_i, C_k, C, $\boldsymbol{n}_z$, $\boldsymbol{n}_{kc}$, and $\boldsymbol{n}_k$ are defined in Fig. 6.15.

The integrals in the preceding equations can be simplified. It is noted that (1) $C(z,t)$ represents the common boundary of interface A_i with the cross section plane; (2) $C_k(z,t)+C(z,t)$ is the enclosed curve in the cross section and C_k is on the wall of the pipe; and (3) $\boldsymbol{n}_{kc}$, located in the cross section plane and directed away from region k, is the unit vector normal to C. The mass transfer rate through the interface is expressed by

$$\int_{C(z,t)}\frac{dm_{jk}}{dt}\frac{dC}{\boldsymbol{n}_k\cdot\boldsymbol{n}_{kc}} = \sum_{l=1}^{N-1}(\dot m_{jkl}-\dot m_{jlk})C \quad (6.97)$$

where $\dot m_{jkl}$ is the mass transfer rate from phase j to phase l in region k, and $\dot m_{jlk}$ is the mass transfer rate from phase l to phase j in region k. The momentum transfer rate due to mass transfer through the interface is given by

$$\int_{C(z,t)}\frac{dm_{jk}}{dt}W_{jk}\frac{dC}{\boldsymbol{n}_k\cdot\boldsymbol{n}_{kc}} = \sum_{l=1}^{N-1}(\dot m_{jkl}W_{jk}-\dot m_{jlk}W_{lk})C \quad (6.98)$$

From assumption (4), $\boldsymbol{n}_{\rm k} \cdot \boldsymbol{n}_{\rm z} = 0$ on $C_{\rm k}$. Hence, the integral with pressure can be written as

$$\int_{C \cup C_{\rm k}(z,t)} (\boldsymbol{n}_{\rm z} \cdot \boldsymbol{n}_{\rm k}) p_{\rm jk} \frac{{\rm d}C}{\boldsymbol{n}_{\rm k} \cdot \boldsymbol{n}_{\rm kc}} = P_{\rm jki} \int_{C(z,t)} \boldsymbol{n}_{\rm z} \cdot \boldsymbol{n}_{\rm k} \frac{{\rm d}C}{\boldsymbol{n}_{\rm k} \cdot \boldsymbol{n}_{\rm kc}} \tag{6.99}$$

where $P_{\rm jki}$ is the pressure of phase j in region k at the interface. Furthermore, the limiting form of the Gauss theorem gives [Delhaye, 1981]

$$\int_{C(z,t)} \boldsymbol{n}_{\rm z} \cdot \boldsymbol{n}_{\rm k} \frac{{\rm d}C}{\boldsymbol{n}_{\rm k} \cdot \boldsymbol{n}_{\rm kc}} = -\frac{\partial A_{\rm k}}{\partial z} \tag{6.100}$$

Equation (6.99) is then reduced to

$$\int_{C \cup C_{\rm k}(z,t)} (\boldsymbol{n}_{\rm z} \cdot \boldsymbol{n}_{\rm k}) p_{\rm jk} \frac{{\rm d}C}{\boldsymbol{n}_{\rm k} \cdot \boldsymbol{n}_{\rm kc}} = -P_{\rm jki} \frac{\partial A_{\rm k}}{\partial z} \tag{6.101}$$

Moreover, we may express the interfacial stress by

$$\int_{C(z,t)} \boldsymbol{n}_{\rm z} \cdot (\boldsymbol{n}_{\rm k} \cdot \boldsymbol{\tau}_{\rm jk}) \frac{{\rm d}C}{\boldsymbol{n}_{\rm k} \cdot \boldsymbol{n}_{\rm kc}} = -\epsilon_{\rm k} \tau_{\rm jki} C \tag{6.102}$$

where $\tau_{\rm jk}$ is the stress tensor of phase j in region k; $\tau_{\rm jki}$ is the interfacial stress of phase j in region k projected on the pipe axis; and $\epsilon_{\rm k} = 1$ for one region and -1 for the other; and the wall shear stress or frictional force at wall is given by

$$\int_{C_{\rm k}(z,t)} \boldsymbol{n}_{\rm z} \cdot (\boldsymbol{n}_{\rm k} \cdot \boldsymbol{\tau}_{\rm jk}) \frac{dC}{\boldsymbol{n}_{\rm k} \cdot \boldsymbol{n}_{\rm kc}} = -C_{\rm k} \chi_{\rm jf} \tau_{\rm jkw} \tag{6.103}$$

where $\chi_{\rm jf}$ is the fraction of the frictional perimeter of phase j, and $\tau_{\rm jkw}$ is the wall shear stress of phase j in region k projected on the pipe axis.

The volume fraction (or area fraction in this case) of phase j in the region k is

$$\alpha_{\rm jk} = \frac{A_{\rm jk}}{A_{\rm k}} \tag{6.104}$$

with

$$\sum \alpha_{\rm jk} = 1 \tag{6.105}$$

Now, define the intrinsic average ${}^i\langle x_{\rm jk}\rangle$ of phase j in the region k as

$${}^i\langle x_{\rm jk}\rangle = \frac{1}{A_{\rm jk}} \int_{A_{\rm jk}(z,t)} x_{\rm jk}\, {\rm d}A \tag{6.106}$$

The relationship between regional average and intrinsic average is

$$\langle x_{\rm jk}\rangle = \alpha_{\rm jk}{}^i\langle x_{\rm jk}\rangle \tag{6.107}$$

To simplify the problem further, we adopt a concept similar to that for the intrinsic averaged velocity and intrinsic velocity tensor as in Eqs. (5.129) and (5.130). Thus, Eqs. (6.95) and (6.96), respectively, become

$$\frac{\partial}{\partial t} A_{\rm k}\alpha_{\rm jk}{}^i\langle\rho_{\rm jk}\rangle + \frac{\partial}{\partial z} A_{\rm k}\alpha_{\rm jk}{}^i\langle\rho_{\rm jk}\rangle{}^i\langle W_{\rm jk}\rangle = \sum_{l=1}^{N-1} (\dot{m}_{\rm jlk} - \dot{m}_{\rm jkl}) C \tag{6.108}$$

and

$$\frac{\partial}{\partial t} A_k \alpha_{jk}{}^i \langle \rho_{jk} \rangle^i \langle W_{jk} \rangle + \frac{\partial}{\partial z} A_k \alpha_{jk}{}^i \langle \rho_{jk} \rangle^i \langle W_{jk} W_{jk} \rangle - A_k \alpha_{jk}{}^i \langle \rho_{jk} \rangle \langle f_z \rangle$$
$$+ \frac{\partial}{\partial z} A_k \alpha_{jk}{}^i \langle p_k \rangle - \frac{\partial}{\partial z} A_k \alpha_{jk}{}^i \langle \mathbf{n}_z \cdot \boldsymbol{\tau}_{jk} \cdot \mathbf{n}_z \rangle$$
$$= \sum_{l=1}^{N-1} (\dot{m}_{jlk} W_{jl} - \dot{m}_{jkl} W_{jk}) C - P_{jki} \frac{\partial A_k}{\partial z} - \epsilon_k \tau_{jki} C - C_k \chi_{jf} \tau_{jkw} \tag{6.109}$$

where the velocity tensor may be approximated as ${}^i\langle W_{jk} W_{jk} \rangle \approx {}^i\langle W_{jk} \rangle^i \langle W_{jk} \rangle$.

The expansion rate projected on the pipe axis for the gas phase (j = gas) may be expressed by

$$\frac{\partial}{\partial z} A_k \alpha_k{}^i \langle \mathbf{n}_z \cdot \boldsymbol{\tau}_k \cdot \boldsymbol{n}_z \rangle = F_k({}^i\langle W_{pk} \rangle - {}^i\langle W_k \rangle) \tag{6.110}$$

where τ_k is the stress tensor of the gas phase in region k. It is noted that the combination of the last two terms on the left-hand side of Eq. (6.109) represents the effective total stress projected along the pipe axis, $\boldsymbol{T}_{pk}$. Hence, for the solid phase in a dense suspension region, we have

$$\frac{\partial}{\partial z} A_k \alpha_{pk}{}^i \langle \boldsymbol{n}_z \cdot \boldsymbol{T}_{pk} \cdot \boldsymbol{n}_z \rangle = -\frac{\partial}{\partial z} \left(A_k \alpha_{pk} C_{pp} \frac{\partial^i \langle W_{pk} \rangle}{\partial z} \right) + F_{pk}({}^i\langle W_k \rangle - {}^i\langle W_{pk} \rangle) \tag{6.111}$$

where C_{pp} is the momentum transfer coefficient due to phase expansion.

For subsequent discussions, the symbol ${}^i\langle\ \rangle$ is removed with an understanding that the velocities and densities involved are intrinsically averaged.

6.5.1.2 *Stratified Horizontal Pipe Flows*

A stratified horizontal pipe flow is schematically illustrated in Fig. 6.16. In addition to the assumptions given in §6.5.1.1, other assumptions are introduced in the analysis of the stratified horizontal pipe flow:

(1) The pressure distribution over the cross section is induced by gravity.
(2) The mass transfer between phases can be neglected.
(3) The presence of particles in the dilute region is negligible, whereas particles in the dense region can be regarded as in a moving bed state.

In this section, we only discuss the case of stratified horizontal pipe flow in a rectangular duct. The expansion rate projected on the pipe axis for the gas phase (j = gas) is small (as

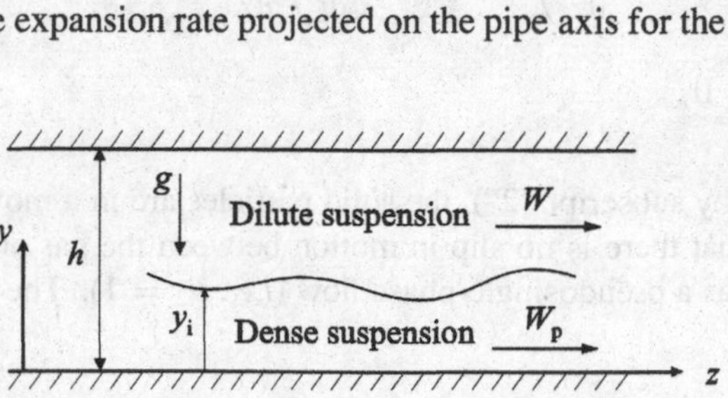

Figure 6.16. Phase interaction in stratified horizontal pipe flow.

is true for low-velocity channel flows), *i.e.*,

$$\frac{\partial}{\partial z} A_{\mathrm{k}} \alpha_{\mathrm{k}}{}^{i} \langle \boldsymbol{n}_{\mathrm{z}} \cdot \boldsymbol{\tau}_{\mathrm{k}} \cdot \boldsymbol{n}_{\mathrm{z}} \rangle \approx 0 \tag{6.112}$$

In the dilute region (denoted by subscript "1"), $\alpha_1 \approx 1$; the regional cross-sectional area is given by

$$\frac{A_1}{A} = 1 - \frac{y_{\mathrm{i}}}{h} \tag{6.113}$$

where A is the cross-sectional area of the duct; y_{i} is the depth of interface; and h is the height of the duct. From Assumption (1), the local pressure is given by

$$p_1 = P_{\mathrm{i}} - g \int_{y_{\mathrm{i}}}^{y} \rho \, \mathrm{d}y \tag{6.114}$$

where P_{i} is the pressure at the interface. The regional averaged pressure is thus obtained as

$$\langle p_1 \rangle = P_{\mathrm{i}} + \frac{1}{h - y_{\mathrm{i}}} \int_{y_{\mathrm{i}}}^{h} \left(g \int_{y}^{y_{\mathrm{i}}} \rho \, \mathrm{d}y \right) \mathrm{d}y = P_{\mathrm{i}} - \frac{\rho g (h - y_{\mathrm{i}})}{2} \tag{6.115}$$

Thus, Eqs. (6.108) and (6.109) yield

$$\frac{\partial A_1}{\partial t} + \frac{\partial}{\partial z}(A_1 W) = 0 \tag{6.116}$$

and

$$\begin{aligned} &\frac{\partial}{\partial t}(A_1 W) + \frac{\partial}{\partial z}(A_1 W^2) - \frac{g}{2}\frac{\partial}{\partial z}(A_1 (h - y_{\mathrm{i}})) + \frac{1}{\rho}\frac{\partial}{\partial z}(A_1 P_{\mathrm{i}}) \\ &\quad = \frac{P_{\mathrm{i}}}{\rho}\frac{\partial A_1}{\partial z} + \frac{1}{\rho}(\tau_{\mathrm{i}} C + C_1 \tau_{\mathrm{gw}}) \end{aligned} \tag{6.117}$$

where τ_{gw} is the wall shear stress of the gas phase in the dilute region projected on the pipe axis. Using Eqs. (6.113) and (6.116), we have

$$\frac{\partial}{\partial z}(A_1 W^2) = -2W \frac{\partial A_1}{\partial t} - W^2 \frac{\partial A_1}{\partial z} = \frac{A}{h}\left(2W \frac{\partial y_{\mathrm{i}}}{\partial t} + W^2 \frac{\partial y_{\mathrm{i}}}{\partial z} \right) \tag{6.118}$$

Equation (6.117) then becomes

$$\begin{aligned} &\frac{\partial}{\partial t}(A_1 W) + \frac{A}{h}\left(2W \frac{\partial y_{\mathrm{i}}}{\partial t} + W^2 \frac{\partial y_{\mathrm{i}}}{\partial z} \right) + Ag\left(1 - \frac{y_{\mathrm{i}}}{h} \right)\frac{\partial y_{\mathrm{i}}}{\partial z} + \frac{A_1}{\rho}\frac{\partial P_{\mathrm{i}}}{\partial z} \\ &\quad + \frac{1}{\rho}(\tau_{\mathrm{i}} C + C_1 \tau_{\mathrm{gw}}) = 0 \end{aligned} \tag{6.119}$$

In the dense region (denoted by subscript "2"), the solid particles are in a moving bed state. We may further assume that there is no slip in motion between the gas and solids so that we can treat the mixture as a pseudosingle-phase flow (*i.e.*, $\alpha_2 = 1$). The regional cross-sectional area is given by

$$\frac{A_2}{A} = \frac{y_{\mathrm{i}}}{h} \tag{6.120}$$

The regional averaged pressure of the mixture is expressed as

$$\langle p_2\rangle = P_\mathrm{i} + \frac{1}{y_\mathrm{i}}\int_0^{y_\mathrm{i}}\left(g\int_y^{y_\mathrm{i}}\rho_\mathrm{m}\,\mathrm{d}y\right)dy = P_\mathrm{i} + \frac{\rho_\mathrm{m} g y_\mathrm{i}}{2} \tag{6.121}$$

Thus, the continuity and momentum equations of the mixture can be obtained from Eqs. (6.108) and (6.109) as

$$\frac{\partial A_2}{\partial t} + \frac{\partial}{\partial z}(A_2 W_\mathrm{p}) = 0 \tag{6.122}$$

and

$$\frac{\partial}{\partial t}(A_2 W_\mathrm{p}) - \frac{A}{h}\left(2W_\mathrm{p}\frac{\partial y_\mathrm{i}}{\partial t} + W_\mathrm{p}^2\frac{\partial y_\mathrm{i}}{\partial z}\right) + gA\frac{y_\mathrm{i}}{h}\frac{\partial y_\mathrm{i}}{\partial z} + \frac{A_2}{\rho_\mathrm{m}}\frac{\partial P_\mathrm{i}}{\partial z}$$

$$+\frac{1}{\rho_\mathrm{m}}(-\tau_\mathrm{i} C + C_\mathrm{m}\tau_\mathrm{mw}) = 0 \tag{6.123}$$

where τ_mw is the wall shear stress of the mixture in the dense region projected on the pipe axis.

For the internal unsteady wave motion, it may be reasonable to assume that the total mass flow is constant, *i.e.*,

$$\rho A_1 W + \rho_\mathrm{m} A_2 W_\mathrm{p} = \text{const} \tag{6.124}$$

Combining Eq. (6.119) with Eq. (6.123) and using Eq. (6.124) yield a wave equation at the interface as

$$C_\mathrm{t}\frac{\partial y_\mathrm{i}}{\partial t} + C_\mathrm{z}\frac{\partial y_\mathrm{i}}{\partial z} = C_\mathrm{c} \tag{6.125}$$

where

$$C_\mathrm{t} = \frac{2(\rho W - \rho_\mathrm{m} W_\mathrm{p})}{h}$$

$$C_\mathrm{z} = \frac{(\rho W^2 - \rho_\mathrm{m} W_\mathrm{p}^2)}{h} + \rho g + (\rho_\mathrm{m} - \rho)\frac{y_\mathrm{i}}{h} \tag{6.126}$$

$$C_\mathrm{c} = -\frac{\partial P_\mathrm{i}}{\partial z} - \frac{(C_1\tau_\mathrm{gw} + C_\mathrm{m}\tau_\mathrm{mw})}{A}$$

According to the preceding model, there is always an unsteady flow since $C_\mathrm{t} \neq 0$. The wave speed of the interface is obtained as

$$U_\mathrm{i} = \frac{C_\mathrm{z}}{C_\mathrm{t}} = \frac{\rho W^2 - \rho_\mathrm{m} W_\mathrm{p}^2 + \rho g h + (\rho_\mathrm{m} - \rho) y_\mathrm{i}}{2(\rho W - \rho_\mathrm{m} W_\mathrm{p})} \tag{6.127}$$

In addition, a steady wave motion, $C_\mathrm{c} = 0$, leads to a balance between pressure drop and wall friction as

$$-\frac{\partial P_\mathrm{i}}{\partial z} = \frac{C_1\tau_\mathrm{gw} + C_\mathrm{m}\tau_\mathrm{mw}}{A} \tag{6.128}$$

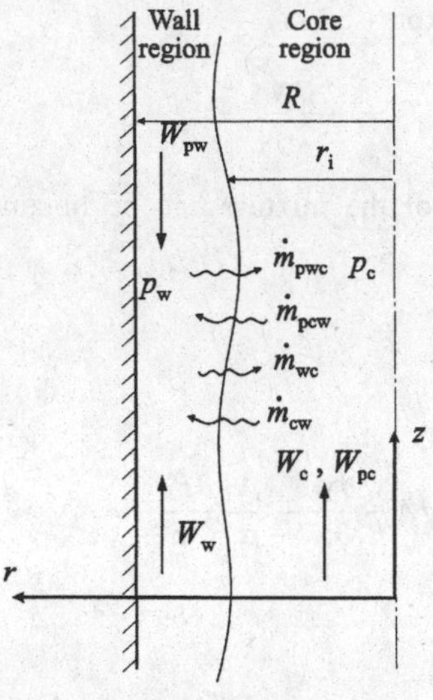

Figure 6.17. Phase interaction in stratified vertical pipe flow.

6.5.1.3 *Stratified Vertical Pipe Flows*

A stratified vertical pipe flow is schematically illustrated in Fig. 6.17. In addition to the assumptions given in §6.5.1.1, other assumptions are introduced in the analysis of the stratified vertical pipe flow:

(1) The flow is axially symmetric.
(2) The flow comprises a dilute suspension in the core region and a dense flow in the wall region.
(3) The solids concentration in each region is uniformly distributed.
(4) There is no phase change between the gas and solids.

In terms of Eqs. (6.108) and (6.109), the continuity and momentum equations of each phase are given in the following for each region.

A. GAS PHASE IN THE CORE REGION

$$\frac{\partial}{\partial t}A_c\alpha_c + \frac{\partial}{\partial z}A_c\alpha_c W_c = (\dot{m}_{wc} - \dot{m}_{cw})\frac{C}{\rho} \tag{6.129}$$

and

$$\frac{\partial}{\partial t}A_c\alpha_c W_c + \frac{\partial}{\partial z}A_c\alpha_c W_c^2 + A_c\alpha_c g + \frac{1}{\rho}\frac{\partial}{\partial z}A_c\alpha_c p_c$$
$$= \frac{F_{gpc}}{\rho}(W_{pc} - W_c) + (\dot{m}_{wc}W_w - \dot{m}_{cw}W_c)\frac{C}{\rho} + \frac{\alpha_c P_i}{\rho}\frac{\partial A_c}{\partial z} - \frac{\tau_{ci}C}{\rho} \tag{6.130}$$

where the first term on the right-hand side of Eq. (6.130) represents the drag force between the gas phase and the particle phase, and F_{gpc} is the coefficient of momentum transfer from particles to gas due to drag in the core region. τ_{ci} is the interfacial stress of the gas phase in the core region projected on the pipe axis.

B. PARTICLE PHASE IN THE CORE REGION

$$\frac{\partial}{\partial t}A_c\alpha_{pc} + \frac{\partial}{\partial z}A_c\alpha_{pc}W_{pc} = (\dot{m}_{pwc} - \dot{m}_{pcw})\frac{C}{\rho_p} \tag{6.131}$$

and

$$\frac{\partial}{\partial t}A_c\alpha_{pc}W_{pc} + \frac{\partial}{\partial z}A_c\alpha_{pc}W_{pc}^2 + A_c\alpha_{pc}g$$
$$= \frac{F_{pgc}}{\rho_p}(W_c - W_{pc}) + (\dot{m}_{pwc}W_{pw} - \dot{m}_{pcw}W_{pc})\frac{C}{\rho_p} - \frac{\tau_{pci}C}{\rho_p} \tag{6.132}$$

where τ_{pci} is the interfacial stress of the particle phase in the core region projected on the pipe axis and F_{pgc} is the coefficient of momentum transfer from gas to particles due to drag in the core region.

C. GAS PHASE IN THE WALL REGION

$$\frac{\partial}{\partial t}A_w\alpha_w + \frac{\partial}{\partial z}A_w\alpha_w W_w = (\dot{m}_{cw} - \dot{m}_{wc})\frac{C}{\rho} \tag{6.133}$$

and

$$\frac{\partial}{\partial t}A_w\alpha_w W_w + \frac{\partial}{\partial z}A_w\alpha_w W_w^2 + A_w\alpha_w g + \frac{1}{\rho}\frac{\partial}{\partial z}A_w\alpha_w p_w$$
$$= \frac{F_{gpw}}{\rho}(W_{pw} - W_w) + (\dot{m}_{cw}W_c - \dot{m}_{wc}W_w)\frac{C}{\rho} + \frac{\alpha_w P_i}{\rho}\frac{\partial A_w}{\partial z} + \frac{\tau_{wi}C}{\rho} - \frac{C_w\chi_{wg}\tau_w}{\rho} \tag{6.134}$$

where F_{gpw} is the coefficient of momentum transfer from particles to gas in the wall region; τ_{wi} is the interfacial stress of the gas phase in the wall region projected on the pipe axis; and τ_w is the wall shear stress of the gas phase projected on the pipe axis.

D. PARTICLE PHASE IN THE WALL REGION

$$\frac{\partial}{\partial t}A_w\alpha_{pw} + \frac{\partial}{\partial z}A_w\alpha_{pw}W_{pw} = (\dot{m}_{pcw} - \dot{m}_{pwc})\frac{C}{\rho_p} \tag{6.135}$$

and

$$\frac{\partial}{\partial t}A_w\alpha_{pw}W_{pw} + \frac{\partial}{\partial z}A_w\alpha_{pw}W_{pw}^2 + A_w\alpha_{pw}g + \frac{1}{\rho_p}\frac{\partial}{\partial z}\left(A_w\alpha_{pw}C_{pp}\frac{\partial W_{pw}}{\partial z}\right)$$
$$= \frac{F_{pgw}}{\rho_p}(W_w - W_{pw}) + (\dot{m}_{pcw}W_{pc} - \dot{m}_{pwc}W_{pw})\frac{C}{\rho_p} + \frac{\tau_{pwi}C}{\rho_p} - \frac{C_w\chi_{wp}\tau_{pw}}{\rho_p} \tag{6.136}$$

where F_{pgw} is the coefficient of momentum transfer from gas to particles in the wall region; τ_{pwi} is the interfacial stress of the particle phase in the wall region projected on the pipe axis; and τ_{pw} is the wall shear stress of the particle phase projected on the pipe axis. The last term on the left-hand side of Eq. (6.136) represents the expansion rate of the particle phase due to interparticle collisions.

Using the preceding continuity and momentum equations for each phase in each region, the equations for the interface can be obtained. For the gas phase in the core region, from

Eqs. (6.129) and (6.130), we have

$$\frac{\partial A_c}{\partial t} + \left(W_c + \frac{P_i - p_c}{W_c \rho}\right)\frac{\partial A_c}{\partial z} - \left(\frac{1}{W_c}\frac{\partial W_c}{\partial t} + \frac{g}{W_c} + \frac{1}{W_c \rho}\frac{\partial p_c}{\partial z}\right) A_c = 2\frac{\dot{Q}}{\alpha_c} - \frac{M_c}{\alpha_c W_c} \tag{6.137}$$

where

$$\dot{Q} = \frac{(\dot{m}_{wc} - \dot{m}_{cw})C}{\rho} \tag{6.138}$$

and

$$M_c = \frac{F_{gpc}}{\rho}(W_{pc} - W_c) + (\dot{m}_{wc} W_w - \dot{m}_{cw} W_c)\frac{C}{\rho} - \frac{\tau_{ci} C}{\rho} \tag{6.139}$$

For the particle phase in the core region, from Eqs. (6.131) and (6.132), we have

$$\frac{\partial A_c}{\partial t} + W_{pc}\frac{\partial A_c}{\partial z} - \left(\frac{1}{W_{pc}}\frac{\partial W_{pc}}{\partial t} + \frac{g}{W_{pc}}\right) A_c = 2\frac{\dot{Q}_p}{\alpha_{pc}} - \frac{M_{pc}}{\alpha_{pc} W_{pc}} \tag{6.140}$$

where

$$\dot{Q}_p = \frac{(\dot{m}_{pwc} - \dot{m}_{pcw})C}{\rho_p} \tag{6.141}$$

and

$$M_{pc} = \frac{F_{pgc}}{\rho_p}(W_c - W_{pc}) + (\dot{m}_{pwc} W_{pw} - \dot{m}_{pcw} W_{pc})\frac{C}{\rho_p} - \frac{\tau_{pci} C}{\rho_p} \tag{6.142}$$

In the wall region, note that $A_w = A - A_c$. Therefore, for the gas phase in the wall region, from Eqs. (6.133) and (6.134), we have

$$\begin{aligned} &\frac{\partial A_c}{\partial t} + \left(W_w + \frac{P_i - p_w}{W_w \rho}\right)\frac{\partial A_c}{\partial z} - \left(\frac{1}{W_w}\frac{\partial W_w}{\partial t} + \frac{g}{W_w} + \frac{1}{W_w \rho}\frac{\partial p_w}{\partial z}\right) A_c \\ &\quad = 2\frac{Q}{\alpha_w} + \frac{M_w}{\alpha_w W_w} - \left(\frac{1}{W_w}\frac{\partial W_w}{\partial t} + \frac{g}{W_w} + \frac{1}{W_w \rho}\frac{\partial p_w}{\partial z}\right) A \end{aligned} \tag{6.143}$$

where

$$M_w = \frac{F_{gpw}}{\rho}(W_{pw} - W_w) + (\dot{m}_{cw} W_c - \dot{m}_{wc} W_w)\frac{C}{\rho} + \frac{\tau_{wi} C}{\rho} - \frac{C_w \chi_{wg} \tau_w}{\rho} \tag{6.144}$$

For the particle phase in the wall region, from Eqs. (6.135) and (6.136), we have

$$\begin{aligned} &\frac{\partial A_c}{\partial t} + \left(W_{pw} - \frac{C_{pp}}{\rho_p W_{pw}}\frac{\partial W_{pw}}{\partial z}\right)\frac{\partial A_c}{\partial z} - \left(\frac{1}{W_{pw}}\frac{\partial W_{pw}}{\partial t} + \frac{g}{W_{pw}} + \frac{C_{pp}}{\rho_p W_{pw}}\frac{\partial^2 W_{pw}}{\partial z^2}\right) A_c \\ &\quad = 2\frac{\dot{Q}_p}{\alpha_{pw}} + \frac{M_{pw}}{\alpha_{pw} W_{pw}} - \left(\frac{1}{W_{pw}}\frac{\partial W_{pw}}{\partial t} + \frac{g}{W_{pw}} + \frac{C_{pp}}{\rho_p W_{pw}}\frac{\partial^2 W_{pw}}{\partial z^2}\right) A \end{aligned} \tag{6.145}$$

where

$$M_{pw} = \frac{F_{pgw}}{\rho_p}(W_w - W_{pw}) + (\dot{m}_{pcw} W_{pc} - \dot{m}_{pwc} W_{pw})\frac{C}{\rho_p} + \frac{\tau_{pwi} C}{\rho_p} - \frac{C_w \chi_{wp} \tau_{pw}}{\rho_p} \tag{6.146}$$

It is noted that Eqs. (6.137), (6.140), (6.143), and (6.145) refer to the identical interfacial area, which has the following common form of expression

$$\frac{\partial A_c}{\partial t} + F_1 \frac{\partial A_c}{\partial z} + F_2 A_c = F_3 \tag{6.147}$$

Therefore, the equation of the interface is a wave equation, reflecting that the flow is in an unsteady wave motion. Furthermore, because the corresponding coefficient of each term in these four equations should be equal, several relations among phase quantities result:

1. Relation for the slip velocity between phases in the core region

$$W_c - W_{pc} = \frac{p_c - P_i}{W_c \rho} \tag{6.148}$$

2. Relation for the slip velocity between phases in the wall region

$$W_w - W_{pw} = -\frac{C_{pp}}{\rho_p W_{pw}} \frac{\partial W_{pw}}{\partial z} + \frac{p_w - P_{wi}}{\rho W_w} \tag{6.149}$$

3. Relation for W_{pw} and W_{pc}

$$\frac{C_{pp}}{\rho_p W_{pw}} \frac{\partial W_{pw}}{\partial z} = W_{pw} - W_{pc} \tag{6.150}$$

4. Relation for W_c and W_{pc}

$$\left(\frac{1}{W_{pc}} \frac{\partial W_{pc}}{\partial t} - \frac{1}{W_c} \frac{\partial W_c}{\partial t} \right) + \left(\frac{1}{W_{pc}} - \frac{1}{W_c} \right) g = \frac{1}{\rho W_c} \frac{\partial p_c}{\partial z} \tag{6.151}$$

5. Relation for W_w and W_{pw}

$$\left(\frac{1}{W_{pw}} \frac{\partial W_{pw}}{\partial t} - \frac{1}{W_w} \frac{\partial W_w}{\partial t} \right) + \left(\frac{1}{W_{pw}} - \frac{1}{W_w} \right) g = \frac{1}{\rho W_w} \frac{\partial p_w}{\partial z} - \frac{C_{pp}}{\rho_p W_{pw}} \frac{\partial^2 W_{pw}}{\partial z^2} \tag{6.152}$$

6. Relation for W_{pw} and W_{pc}

$$\left(\frac{1}{W_{pw}} \frac{\partial W_{pw}}{\partial t} - \frac{1}{W_{pc}} \frac{\partial W_{pc}}{\partial t} \right) + \left(\frac{1}{W_{pw}} - \frac{1}{W_{pc}} \right) g + \frac{C_{pp}}{\rho_p W_{pw}} \frac{\partial^2 W_{pw}}{\partial z^2} = 0 \tag{6.153}$$

7. Relation between momentum transfer and mass transfer in the core region

$$2\frac{\dot{Q}}{\alpha_c} - \frac{M_c}{\alpha_c W_c} = 2\frac{\dot{Q}_p}{\alpha_{pc}} - \frac{M_{pc}}{\alpha_{pc} W_{pc}} \tag{6.154}$$

8. Relation between momentum transfer and mass transfer in the wall region

$$2\frac{\dot{Q}}{\alpha_w} + \frac{M_w}{\alpha_w W_w} = 2\frac{\dot{Q}_p}{\alpha_{pw}} + \frac{M_{pw}}{\alpha_{pw} W_{pw}} \tag{6.155}$$

9. Relation between momentum transfer and mass transfer of particle phase between two regions

$$2\dot{Q}_{\mathrm{p}}\left(1-\frac{\alpha_{\mathrm{pc}}}{\alpha_{\mathrm{pw}}}\right)W_{\mathrm{pc}} = M_{\mathrm{pc}} + M_{\mathrm{pw}}\frac{\alpha_{\mathrm{pc}}}{\alpha_{\mathrm{pw}}}\frac{W_{\mathrm{pc}}}{W_{\mathrm{pw}}} - \alpha_{\mathrm{pc}}\frac{W_{\mathrm{pc}}}{W_{\mathrm{pw}}}\left(\frac{\partial W_{\mathrm{pw}}}{\partial t} + g + \frac{C_{\mathrm{pp}}}{\rho_{\mathrm{p}}}\frac{\partial^2 W_{\mathrm{pw}}}{\partial z^2}\right)A \tag{6.156}$$

It is important to note that these 9 relationships are essential to the closure of the problem. There are a total of 14 unknowns, namely, W_{w}, W_{c}, W_{pc}, W_{pw}, $\dot{m}_{\mathrm{cw}}$, $\dot{m}_{\mathrm{wc}}$, $\dot{m}_{\mathrm{pcw}}$, $\dot{m}_{\mathrm{pwc}}$, p_{c}, p_{w}, P_{i}, A_{c}, α_{pc}, and α_{pw}. Eight independent equations representing four pairs of continuity and momentum equations (Eqs. (6.129) through (6.136)) are initially developed. Converting the momentum equations into wave equations and equating the coefficients of these wave equations yield a total of 14 independent equations, namely, 4 continuity equations, 1 wave equation, and 9 relations, and hence the closure of the problem.

6.5.2 *Continuity Wave and Dynamic Wave*

Waves in gas–solid flow generated by a perturbation to the flow can propagate in the gas–solid medium as indicated in §6.4. The propagation may yield continuous changes in local volume fractions of phases. It may also yield a step (sharp) change in local gas void fraction or solids holdup with a finite discontinuity, known as a shock (or shock wave). "Waves" or "shocks" exist in many different forms depending on the origin of the disturbance [Wallis, 1969]. Basically, two classes of waves can be distinguished, *i.e.*, continuity waves and dynamic waves. Although wave phenomena in gas–solid flow are two-dimensional or three-dimensional in nature, for simplicity only the one-dimensional wave is considered in the following discussion.

6.5.2.1 *Continuity Waves*

A continuity wave is generated when a perturbation in a gas–solid flow yields a local change in the solids holdup or gas voidage which propagates through the vessel at a certain velocity. In other words, the steady-state gas voidage or solids holdup simply propagates into another one, as illustrated in Fig. 6.18, and the propagation process does not introduce any dynamic effect of inertia or momentum [Wallis, 1969].

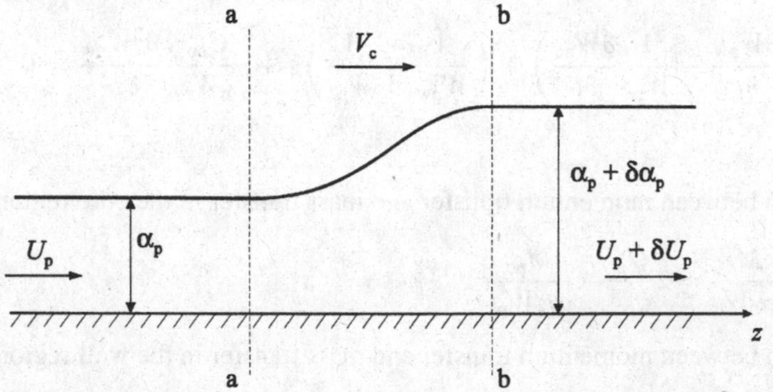

Figure 6.18. Continuity wave propagation due to the variation of volume fraction (from Wallis, 1969).

Consider an incompressible gas–solid flow in a tube of constant cross-sectional area. Denoting U, U_p, and U_m as the superficial velocities for gas, solids, and a mixture of gas and solids, respectively, U_m is related to U and U_p by

$$U_m = U + U_p \tag{6.157}$$

The continuity wave velocity can be derived by considering the solids holdup variation across a control volume enclosed by the aa and bb planes given in Fig. 6.18. The figure shows the propagation of a wave with velocity V_c from the aa plane where the solids holdup is α_p to the bb plane where the solids holdup is $\alpha_p + \delta\alpha_p$. A solids flow approaching the moving wave front would be equal to that leaving the moving front, *i.e.*,

$$U_p - \alpha_p V_c = U_p + \delta U_p - V_c(\alpha_p + \delta\alpha_p) \tag{6.158}$$

which yields

$$V_c = \frac{\partial U_p}{\partial \alpha_p} \tag{6.159}$$

Note that U_p can also be expressed by

$$U_p = u_p \alpha_p \tag{6.160}$$

where u_p is the particle velocity. Equations (6.159) and (6.160) yield

$$V_c = u_p + \alpha_p \frac{\partial u_p}{\partial \alpha_p} \tag{6.161}$$

An unsteady-state mass balance for solids in a differential element with a small perturbation in a system otherwise in equilibrium yields

$$\frac{\partial \alpha_p}{\partial t} + \frac{\partial(\alpha_p u_p)}{\partial z} = 0 \tag{6.162}$$

Equation (6.162) leads to

$$\frac{\left(\dfrac{\partial \alpha}{\partial t}\right)}{\left(\dfrac{\partial \alpha}{\partial z}\right)} + \alpha_p \frac{\left(\dfrac{\partial u_p}{\partial z}\right)}{\left(\dfrac{\partial \alpha_p}{\partial z}\right)} + u_p = 0 \tag{6.163}$$

Comparing Eqs. (6.161) and (6.163) yields

$$V_c = -\frac{\left(\dfrac{\partial \alpha}{\partial t}\right)_z}{\left(\dfrac{\partial \alpha}{\partial z}\right)_t} \tag{6.164}$$

Equation (6.164) expresses the continuity wave velocity as the ratio of the change of the voidage with time to the change in voidage with distance, which provides an alternative definition for V_c [Rietema, 1991].

Define the slip velocity V_s as $u_p - u$ where $u = U/\alpha$. In equilibrium conditions, U_m is constant throughout the system, indicating that $(\alpha u + \alpha_p u_p)$ or $(u_p - \alpha V_s)$ is constant.

Thus, Eq. (6.161) can be expressed by

$$V_c = u_p + \alpha_p \frac{\partial(\alpha V_s)}{\partial \alpha_p} \tag{6.165}$$

For batch-solid operation, $u_p = 0$. Thus, Eq. (6.165) can be reduced to

$$V_c = (1 - \alpha)\frac{\partial U}{\partial \alpha} \tag{6.166}$$

Consider a particulate fluidization system (see §9.3.2) where the Carman correlation given for $\alpha < 0.7$ [Carman, 1937] is applicable

$$\frac{U}{U_{pt}} = \frac{1}{10}\frac{\alpha^3}{(1-\alpha)} \tag{6.167}$$

Thus, for this system, V_c can be expressed from Eqs. (6.166) and (6.167) as

$$V_c = U\frac{(3 - 2\alpha)}{\alpha} \tag{6.168}$$

6.5.2.2 *Dynamic Waves*

Wallis (1969) defined the dynamic wave in one-phase flow as being that which occurs whenever there is a net force on the flowing medium produced by a concentration gradient. For a two-phase flow, *i.e.*, gas–solid flow, the flow medium refers to the gas phase and the concentration refers to the solids holdup. Thus, to analyze dynamic waves, one can examine the wave equation obtained from the perturbation of the momentum and mass balance equations for the gas and solid phases. The analyses given later for both §6.5.2.2 and §6.5.2.3 follow those of Rietema (1991).

The mass balance equations for gas and solid phases yield

$$\frac{\partial \alpha}{\partial t} + \frac{\partial(\alpha u)}{\partial z} = 0 \tag{6.169}$$

and

$$\frac{\partial(1-\alpha)}{\partial t} + \frac{\partial}{\partial z}[(1-\alpha)u_p] = 0 \tag{6.170}$$

In terms of the simplified two-fluid model, the momentum balance of the gas phase can be given as

$$\alpha\frac{\partial p}{\partial z} + \alpha\rho g + F_i = 0 \tag{6.171}$$

where p represents the pressure of the mixture of gas and solids, F_i is the interacting force between gas and solids, and the inertial terms in the gas phase are neglected. The pseudosteady-state condition is applied to Eq. (6.171) considering the assumption that the dynamic acceleration is negligible because of the low gas density and small variation of the gas velocity. The momentum balance of the solid phase can be expressed by

$$(1-\alpha)\rho_p\frac{\partial u_p}{\partial t} + (1-\alpha)\rho_p u_p\frac{\partial u_p}{\partial z} + (1-\alpha)\frac{\partial p}{\partial z} + \frac{\partial \sigma_z}{\partial z} + (1-\alpha)\rho_p g - F_i = 0 \tag{6.172}$$

where σ_z is the particle normal stress due to particle–particle interactions and the variation of particle shear stresses τ_{xz} and τ_{yz} is assumed to be negligible compared with the variation of σ_z in the z-direction. Furthermore, we may define the elasticity modulus of the particle phase as

$$E = -\frac{\dfrac{\partial \sigma_z}{\partial z}}{\dfrac{\partial \alpha}{\partial z}} \tag{6.173}$$

Combining Eqs. (6.171), (6.172), and (6.173) yields

$$(1-\alpha)\rho_p\left[\frac{\partial u_p}{\partial t} + u_p\frac{\partial u_p}{\partial z}\right] - E\frac{\partial \alpha}{\partial z} + (1-\alpha)(\rho_p - \rho)g - \frac{F_i}{\alpha} = 0 \tag{6.174}$$

Therefore, we obtain three independent equations (Eqs. (6.169), (6.170), and (6.174)) with three dependent variables, *i.e.*, α, u, and u_p.

In order to obtain an expression for the dynamic wave speed, consider a perturbation imposed on a steady-state flow yielding α, u, and u_p, as

$$\alpha = \alpha^0 + \alpha' \tag{6.175}$$

$$u = u^0 + u' \tag{6.176}$$

$$u_p = u_p^0 + u_p' \tag{6.177}$$

The superscript "0" indicates the steady-state value and the prime indicates the perturbed value. The perturbation of F_i/α yields

$$\frac{F_i}{\alpha} = \left(\frac{F_i}{\alpha}\right)^0 + \frac{\partial\left(\dfrac{F_i}{\alpha}\right)}{\partial \alpha}\alpha' + \frac{\partial\left(\dfrac{F_i}{\alpha}\right)}{\partial u}u' + \frac{\partial\left(\dfrac{F_i}{\alpha}\right)}{\partial u_p}u_p' \tag{6.178}$$

Consider a particulate fluidization system where $u_p^0 = 0$ and that the Carman correlation for F_i given in the following under the condition of $\alpha < 0.7$ [Carman, 1937] is applicable

$$F_i = \frac{180\mu(1-\alpha)^2}{d_p^2\alpha}(u - u_p) \tag{6.179}$$

Thus, Eqs. (6.169), (6.170), and (6.174) through (6.178) lead to

$$\frac{\partial^2\alpha'}{\partial t^2} - \frac{E}{\rho_p}\frac{\partial^2\alpha'}{\partial z^2} + \frac{180\mu}{d_p^2\rho_p}\frac{(1-\alpha^0)}{(\alpha^0)^3}\left[(3-2\alpha^0)u^0\frac{\partial\alpha'}{\partial z} + \frac{\partial\alpha'}{\partial t}\right] = 0 \tag{6.180}$$

The first two terms establish the basic wave equation describing the dynamic wave propagation with velocity V_d expressed by

$$V_d = \sqrt{\frac{E}{\rho_p}} \tag{6.181}$$

The coefficient in the third term, $(3 - 2\alpha^0)u^0$, is the continuity wave velocity as given by Eq. (6.168). Thus, Eq. (6.180) can be given by

$$\frac{\partial^2\alpha'}{\partial t^2} - V_d^2\frac{\partial^2\alpha'}{\partial z^2} + C\left(V_c\frac{\partial\alpha'}{\partial z} + \frac{\partial\alpha'}{\partial t}\right) = 0 \tag{6.182}$$

where

$$C = \frac{180\mu(1-\alpha^0)}{d_p^2\rho_p(\alpha^0)^3} \tag{6.183}$$

6.5.2.3 *Instability Analysis of a Batch-Solid Fluidization System*

The instability analysis of Eq. (6.182) for α' can yield a stability criterion accounting for the conditions under which bubbles will form in a fluidized bed. In this analysis, when bubbles form, the bed is regarded as unstable. Thus, the instability analysis gives rise to the condition for minimum bubbling of the bed (also see §9.4).

First, let us express the voidage perturbation α' in the form of a Fourier series as

$$\alpha' = \exp[at + i\omega_h(t - z/U_h)] \tag{6.184}$$

where e^{at} is the amplitude of perturbation; $a > 0$ indicates the perturbation will yield an unstable bed, while for $a < 0$ the perturbation will be damped out yielding a stable bed. U_h is the propagation velocity of the perturbation and ω_h is its frequency. Substituting Eq. (6.184) into Eq. (6.182) yields

$$(a + i\omega_h)^2 - V_d^2\left(\frac{i\omega_h}{U_h}\right)^2 + C\left(-i\omega_h\frac{V_c}{U_h} + a + i\omega_h\right) = 0 \tag{6.185}$$

which gives rise to

$$\omega_h^2 = \frac{C^2}{4}\left(\frac{V_c^2 - U_h^2}{U_h^2 - V_d^2}\right) \tag{6.186}$$

for the real part and

$$a = \frac{C}{2}\left(\frac{V_c}{U_h} - 1\right) \tag{6.187}$$

for the imaginary part. For the perturbation to grow, a has to be positive and, hence,

$$V_c > U_h \tag{6.188}$$

From Eq. (6.186), $\omega_h^2 > 0$ and thus we have

$$V_c^2 > U_h^2 > V_d^2 \tag{6.189a}$$

or

$$V_c^2 < U_h^2 < V_d^2 \tag{6.189b}$$

Equations (6.188) and (6.189) yield the criterion given for a bed to be unstable, *i.e.*, a bed with a bubble formation:

$$V_c > V_d \tag{6.190}$$

Likewise, for perturbation to be damped out, $a < 0$, and hence

$$V_c < U_h \tag{6.191}$$

Equations (6.189) and (6.191) yield the criterion given for a bed to be stable, *i.e.*, a bed without a bubble formation:

$$V_c < V_d \tag{6.192}$$

Thus, the condition for the minimum velocity for bubble formation or the minimum bubbling point can be obtained when

$$V_c = V_d \tag{6.193}$$

6.6 Particle–Turbulence Interaction

When particles are introduced into a turbulent flow, they are dispersed through turbulent fluctuations of the gas. The addition of particles can enhance or reduce the gas turbulence, affecting the behavior of the mean flow field (*e.g.*, drag reduction in pipe flows). In very dilute gas–solid flows (*e.g.*, $\alpha_p \ll 10^{-5}$), the particle dispersion by turbulence can be studied by assuming that the turbulent flow is not influenced by the presence of the particles (*i.e.*, one-way coupling). When the particle concentration is high (*e.g.*, $\alpha_p > 10^{-5}$) such that the particle momentum loss or gain transforming to the turbulence can no longer be neglected (*i.e.*, two-way coupling), the modulation of the turbulence by the particles should be considered.

One of the earliest models for turbulence modulation in homogeneous dilute particle-laden flows is Hinze's model (1972), in which the assumption of "vortex trapping" of particles is employed. On the basis of this model, the particle turbulent kinetic energy, k_p, is determined by the local gas turbulent kinetic energy k as

$$\frac{k}{k_p} = 1 + \frac{\tau_S}{\tau_e} \tag{6.194}$$

where τ_S is the Stokes relaxation time defined by Eq. (3.39) and τ_e is the eddy existence time defined by Eq. (5.253). On the other hand, the effect of particles on gas turbulence can be given by

$$\frac{k}{k_0} = \left(1 + \frac{\alpha_p \rho_p}{\rho}\right)^{-1} \tag{6.195}$$

where k_0 is the local gas turbulent kinetic energy in the absence of the particles. Equation (6.194) indicates that k_p is always smaller than k, and in the same turbulent flow, the larger the particle size the smaller the value of k_p. Moreover, according to Eq. (6.195), particles should always reduce gas turbulence. However, many experimental results, especially for large particles, do not support this indication [Tsuji and Morikawa, 1982; Lee and Durst, 1982; Tsuji *et al.*, 1984; Zhou and Huang, 1990]. In actual applications, particles are not always trapped by eddies, and turbulence modulation is a result of multiple particle–turbulence interaction mechanisms. Therefore, Hinze's model seems to be oversimplified to describe the general particle–turbulence interactions.

On the bais of an extensive review of experimental results of turbulence modulation in dilute suspension pipe flows and jet flows, Gore and Crowe (1989) proposed a critical ratio of particle diameter to a characteristic integral length scale of turbulence by the following relation

$$\left(\frac{d_p}{l_e}\right)_{cr} \approx 0.1 \tag{6.196}$$

where l_e is the Eulerian integral length scale. The turbulence intensity is enhanced when the ratio given in Eq. (6.196) is greater than the critical value and is suppressed when the ratio is less than the critical value. This criterion indicates that the turbulence is attenuated by small particles while it is amplified by large particles.

One predominant mechanism for the enhancement of turbulence in the presence of large particles is due to wake shedding, *i.e.*, vortex shedding from the particle wake. With $\mathrm{Re_p} < 110$, there is no wake shedding; with $\mathrm{Re_p} > 400$, wake shedding occurs. In the range $400 < \mathrm{Re_p} < 1{,}000$, the wake shedding frequency is given by [Achenbach, 1974]

$$\mathrm{St} = \frac{f d_\mathrm{p}}{|U - U_\mathrm{p}|} = 0.2 \tag{6.197}$$

where St is the Strouhal number and f is the wake shedding frequency. For a typical range of large particles (*e.g.*, $500\ \mu\mathrm{m} < d_\mathrm{p} < 5{,}000\ \mu\mathrm{m}$), the corresponding wake shedding frequencies are in the range where a considerable particle effect on the turbulence intensity is observed. On the basis of this observation, Hetsroni (1989) suggested that large particles, with $\mathrm{Re_p} > 400$, exhibit wake shedding, leading to an enhancement of the turbulence; *i.e.*, the wake shedding by large particles causes energy to be transferred from the average velocity to the fluctuating velocity.

In general, at least six mechanisms, which are not independent of each other, contribute to turbulence modulation in gas–solid suspension flows:

(1) Dissipation of turbulent kinetic energy by the particles.
(2) Increase in the apparent viscosity due to the presence of particles.
(3) Vortex shedding or the presence of the wake behind the particles.
(4) Gas moving with the particles as carried mass of the particles.
(5) Enhancement of the velocity gradients between particles.
(6) Preferential concentration of particles by turbulence, *i.e.*, the particles selectively concentrated in particular structures by turbulence may cause rapid attenuation of that structure or trigger a new instability [Squires and Eaton, 1990].

At this stage, complete modeling including all of the preceding mechanisms is not possible because of the complex coupling relationships of the turbulent interactions and the lack of knowledge of the sources of turbulence generation. Nevertheless, simple mechanistic modeling accounting for a few of the predominant mechanisms for the modulation of turbulence is possible. In the following, a simple mechanistic model developed by Yuan and Michaelides (1992) is introduced. This model includes only two predominant mechanisms of turbulence modulation in the dilute gas–solid flows, namely, (1) turbulence reduction due to the kinetic energy dissipation from an eddy, for the acceleration of a particle, and (2) turbulence enhancement due to the wake of the particle or the vortex shedding.

Consider a particle with velocity u_p entering an eddy of velocity u and interacting with it for a period of τ_i which can be estimated from Eq. (5.251). It is assumed that the velocity is scalar in the modeling and the only interacting force between the particle and gas is the drag force F_D. The rate of work done by the gas is equal to $F_\mathrm{D} u$ and the change of kinetic energy of the particle is $F_\mathrm{D} u_\mathrm{p}$. The rate of energy dissipation ϵ is thus expressed by

$$\epsilon = F_\mathrm{D}(u - u_\mathrm{p}) = \frac{\pi}{8} C_\mathrm{D} d_\mathrm{p}^2 \rho (u - u_\mathrm{p})^2 |u - u_\mathrm{p}| \tag{6.198}$$

To simplify the following calculation, the drag coefficient is expressed by

$$C_D = \frac{24C}{\mathrm{Re_p}} \tag{6.199}$$

where C, a function of $\mathrm{Re_p}$ in principle, is regarded as constant during the interaction.

During the particle–eddy interaction, the equation of motion for the particles can be expressed by

$$\frac{\pi}{6} d_p^3 \rho_p \frac{\mathrm{d}u_p}{\mathrm{d}t} = C_D \frac{\pi}{8} d_p^2 \rho (u - u_p)|u - u_p| \tag{6.200}$$

Combined with Eq. (6.199), Eq. (6.200) gives the approximation of the particle velocity during the time interval τ_i as

$$u_p = u_{p0} + (u - u_{p0})\left[1 - \exp\left(-\frac{Ct}{\tau_S}\right)\right], \quad t \le \tau_i \tag{6.201}$$

The total energy dissipation from the eddy, which equals the total work performed by the eddy on the particle during the particle–eddy interaction, is thus obtained by integrating Eq. (6.198) with respect to t

$$\Delta E_d = \frac{\pi}{12} d_p^3 \rho_p (u - u_{p0})^2 \left[1 - \exp\left(-\frac{2C\tau_i}{\tau_S}\right)\right] \approx \frac{\pi}{12} d_p^3 \rho_p V_s^2 \left[1 - \exp\left(-\frac{2C\tau_i}{\tau_S}\right)\right] \tag{6.202}$$

In this model, the velocity disturbance by the particle is from both the wake behind the particle ($\mathrm{Re_p} > 20$) and the vortex shedding ($\mathrm{Re_p} > 400$). Hence, the changes in the kinetic energy associated with the turbulence production are proportional to the difference between the squares of the two velocities and to the volume where the velocity disturbance originates. It is further assumed that the wake is half of a complete ellipsoid, with base diameter of d_p (same as the particle diameter) and wake length of l_w. Thus, the total energy production of the gas by the particle wake or vortex shedding is

$$\Delta E_p = \frac{\pi}{12} d_p^2 l_w \rho (u^2 - u_{p0}^2) \approx \frac{\pi}{12} d_p^2 l_w \rho V_s (2u - V_s) \tag{6.203}$$

Combined with Eq. (6.202), Eq. (6.203) yields the total turbulence modulation as

$$\Delta E_t = \Delta E_p - \Delta E_d = \frac{\pi}{12} d_p^2 l_w \rho V_s (2u - V_s) - \frac{\pi}{12} d_p^3 \rho_p V_s^2 \left[1 - \exp\left(-\frac{2C\tau_i}{\tau_S}\right)\right] \tag{6.204}$$

For the case of very small particles where $\tau_i \gg \tau_S$, the particle velocity approaches the gas velocity and the wake disappears. The asymptotic expansion of Eq. (6.204) for fine particles is thus given by

$$\Delta E_t = -\frac{\pi}{12} d_p^3 \rho_p V_s^2 \tag{6.205}$$

In the case of large particles where $\tau_i \ll \tau_S$, the particle velocity does not change appreciably during the particle–eddy interaction and the production term predominates. Consequently, the asymptotic value of ΔE_t for large particles equals ΔE_p. The asymptotic expansion of Eq. (6.202) shows that fine particles will cause turbulence reduction, which is proportional to the cube of the particle diameter; whereas large particles will predominantly cause an increase in turbulence, which is proportional to the square of the particle diameter. This indication is in agreement with the data compiled by Gore and Crowe (1989).

Despite the importance of particle–turbulence interaction in the dispersed multiphase flows, understanding of the phenomena is still lacking, in part because of the difficulties in obtaining comprehensive experimental data to substantiate the sources of turbulence generation. New avenues of gaining understanding may come from computational fluid mechanics using numerical simulations such as direct numerical simulation (DNS) [McLaughlin, 1994], large-eddy simulation (LES) and discrete vortex simulation (DVS). These numerical simulations may provide some insights into the physical mechanisms of the particle–turbulence interaction in dispersed multiphase flows.

Nomenclature

A	Constant, defined by Eq. (6.52)
A	Friction force coefficient
A_c	Contact area
A_c	Cross-sectional area in the core region
A_i	Interfacial area
A_k	Cross-sectional area of phase k
A_w	Cross-sectional area in the wall region
a	Particle radius
a_1	Speed of sound in a pure gas in front of the shock wave
a_g	Speed of sound in a pure gas
a_m	Speed of sound in a gas–solid mixture
B	Constant, defined by Eq. (6.53)
B	Inertia correction coefficient
C	Common boundary of interface with cross section plane
C	Coefficient of wave equation
C	Constant, defined by Eq. (6.54) and Eq. (6.183)
C	Resistance coefficient due to Basset force
C_D	Drag coefficient
C_k	Common boundary of pipe wall in region k with cross section plane
C_{pp}	Momentum transfer coefficient due to phase expansion
c	Specific heat of particles
c_p	Specific heat of gas at constant pressure
c_{pM}	Specific heat of mixture at constant pressure
c_V	Specific heat of gas at constant volume
c_{VM}	Specific heat of mixture at constant volume
D	Parameter, defined by Eq. (6.71)
d_m	Particle diameter below which the particle contribution to the pressure of mixture exceeds 1 percent at atmospheric pressure
d_p	Particle diameter
E	Modulus of elasticity of the particle phase
E_B	Brittle erosion parameter, defined by Eq. (6.14)
E_{Bm}	Brittle erosion parameter with $\alpha_i = \alpha_m$
E_D	Ductile erosion parameter, defined by Eq. (6.8)
E_{Dm}	Ductile erosion parameter with $\alpha_i = \alpha_m$
e_M	Internal energy per unit volume of mixture
F	Force
F_D	Drag force
F_{gpc}	Coefficient of momentum transfer from particles to gas in the core region

F_{gpw}	Coefficient of momentum transfer from particles to gas in the wall region
F_{i}	Interacting force between gas and particle
F_{k}	Coefficient of momentum transfer from particle to gas
F_{n}	Normal collisional force
F_{pgc}	Coefficient of momentum transfer from gas to particle in the core region
F_{pgw}	Coefficient of momentum transfer from gas to particle in the wall region
F_{pk}	Coefficient of momentum transfer from gas to particle
F_{t}	Tangential cutting force
f	Coefficient of friction
f	Force per unit mass
f	Wake shedding frequency
f_{M}	Mass density function
h	Height of the duct
$\boldsymbol{I}$	Unit tensor
J	Mass flux of gas
J_{p}	Mass flux of particles
K	Thermal conductivity
K_{B}	Brittle resistance parameter, defined by Eq. (6.15)
K_{D}	Ductile resistance parameter, defined by Eq. (6.9)
k	Parameter, defined as $(1-\nu)/\pi E$
k	Wave number
k	Turbulent kinetic energy of gas
k_0	Turbulent kinetic energy of gas in absence of the particles
k_{p}	Turbulent kinetic energy of particles
L	Loschmidt number
l	Interparticle distance
l_{e}	Eulerian integral length scale
l_{w}	Wake length
M_{c}	Parameter, defined by Eq. (6.139)
M_{e}	Equilibrium shock Mach number
M_{pc}	Parameter, defined by Eq. (6.142)
M_{pw}	Parameter, defined by Eq. (6.146)
M_{w}	Parameter, defined by Eq. (6.144)
$\dot{m}_{\mathrm{cw}}$	Gas mass transfer rate from the core region to the wall region
m_{H}	Mass of a hydrogen atom
m_{jk}	Mass of phase j in region k
$\dot{m}_{\mathrm{jkl}}$	Mass transfer rate from phase j to phase l in region k
$\dot{m}_{\mathrm{jlk}}$	Mass transfer rate from phase l to phase j in region k
m_{p}	Mass ratio of particle to gas
$\dot{m}_{\mathrm{pcw}}$	Particle mass transfer rate from the core region to the wall region
$\dot{m}_{\mathrm{pwc}}$	Particle mass transfer rate from the wall region to the core region
$\dot{m}_{\mathrm{wc}}$	Gas mass transfer rate from the wall region to the core region
$\mathrm{N_{Im}}$	Impact number, defined by Eq. (6.5)
N_{ω}	Parameter, defined by Eq. (6.59)
$\mathrm{Nu_p}$	Particle Nusselt number
P_{i}	Pressure at interface
P_{jki}	Pressure of phase j in region k at the interface
p	Pressure of gas
p_1	Pressure in front of the shock wave
p_1	Regional averaged pressure in dilute region
p_2	Regional averaged pressure in dense region
p_{c}	Pressure of gas in the core region
p_{jk}	Pressure of phase j in region k
p_{M}	Pressure of the mixture
p_{w}	Pressure of gas in the wall region
$\dot{Q}$	Parameter, defined by Eq. (6.138)
q	Heat absorbed by unit mass of gas-particle mixture
R_{M}	Gas constant

$\mathrm{Re_p}$ Particle Reynolds number, based on particle diameter and relative velocity
r Radial coordinate
r^* Ratio of reflection speed to incoming speed
St Strouhal number
T Absolute temperature of gas
T_1 Fluid temperature in front of the shock wave
T_p Absolute temperature of particle
T_p1 Particle temperature in front of the shock wave
t Time
t_c Contact time duration
U Superficial velocity of gas
U Gas velocity component
U_1 Gas velocity component in front of the shock wave
U_2 Gas velocity component immediately behind the shock wave
U_∞ Gas velocity component far behind the shock wave
$U_{\mathrm{p}\infty}$ Particle velocity component far behind the shock wave
$\boldsymbol{U}$ Velocity vector of fluid
U_h Velocity of perturbation
U_m Superficial velocity of the gas–solid mixture
U_p Superficial velocity of solids
U_p Particle velocity component
$\boldsymbol{U}_\mathrm{p}$ Velocity vector of particles
U_pt Particle terminal velocity
u Linear fluid velocity
u_1 Fluid velocity in front of the shock wave
u_p Particle velocity
V Incoming speed of particle before impact
V Gas velocity component
V_c Continuity wave velocity
V_d Dynamic wave velocity
V_k Volume occupied by phase k
V_p Particle velocity component
V_s Slip velocity
W Gas velocity component
W_c Axial gas velocity in the core region
W_jk Axial velocity of phase j in region k
W_k Axial velocity of gas phase in region k
W_lk Axial velocity of phase l in region k
W_p Axial particle velocity
W_pc Axial particle velocity in the core region
W_pk Axial particle velocity in region k
W_pw Axial particle velocity in the wall region
W_w Axial gas velocity in the wall region
w Attrited mass fraction of the sample particles
w_B Brittle wear in volume loss per impact
w_D Ductile wear in volume loss per impact
w_g Molecular weight of gas
w_M Molecular weight of a gas—particle mixture
w_p Molecular weight of "particulate gas"
X Parameter, defined by Eq. (6.50)
Y_B Yield stress for brittle removal
Y_D Yield stress for ductile removal
y_i Depth of interface
z Axial coordinate

Greek Symbols

α Volume fraction of gas phase
α Empirical constant, defined by Eq. (6.17)

α'	Voidage perturbation
α_c	Volume fraction of gas phase in the core region
α_i	Impact angle
α_{jk}	Volume fraction of phase j in region k
α_m	Maximum eroding impact angle
α_p	Volume fraction of the particle phase
α_{pc}	Volume fraction of particle phase in the core region
α_{pw}	Volume fraction of particle phase in the wall region
α_w	Volume fraction of gas phase in the wall region
β	Attenuation coefficient
γ	Ratio of specific heats of gas
γ_M	Ratio of specific heats of gas–particle mixture
ΔE_d	Total energy dissipation from the eddy during the particle–eddy interaction
ΔE_p	Total energy production of gas by the particle wake or vortex shedding
ΔE_t	Total turbulence modulation during the particle–eddy interaction
δ	Ratio of specific heats of particle to gas at constant pressure
ϵ	Shear strain on particle surface
ϵ	Rate of energy dissipation
ϵ_B	Energy required to remove a unit volume of material in brittle mode
ϵ_D	Energy required to remove a unit volume of material in ductile mode
ϵ_k	Sign parameter in a two-region model for region k
ζ	Ratio of density of particle to gas
η	Parameter, defined by Eq. (6.67)
ξ	Parameter, defined by Eq. (6.67)
μ	Dynamic viscosity of gas
ν	Kinematic viscosity of gas
ν	Poisson's ratio
ρ	Density of fluid
ρ_{jk}	Density of phase j in region k
ρ_m	Density of the mixture
ρ_p	Density of particles
σ	Normal stress
σ_z	Particle normal stress
τ_{ci}	Interfacial stress of the gas phase in the core region projected on the pipe axis
τ_e	Eddy existence time
τ_{gw}	Wall shear stress of the gas phase in the dilute region projected on the pipe axis
τ_i	Particle–eddy interaction time
τ_{jki}	Interfacial stress of phase j in region k projected on the pipe axis
τ_{jkw}	Wall shear stress of phase j in region k projected on the pipe axis
τ_{mw}	Wall shear stress of the mixture in the dense region projected on the pipe axis
τ_p	Flying time
τ_{pci}	Interfacial stress of the particle phase in the core region projected on the pipe axis
τ_{pw}	Wall shear stress of the particle phase projected on the pipe axis
τ_{pwi}	Interfacial stress of the particle phase in the wall region projected on the pipe axis
τ_S	Stokes relaxation time
τ_w	Wall shear stress of the gas phase projected on the pipe axis
τ_{wi}	Interfacial stress of the gas phase in the wall region projected on the pipe axis
$\boldsymbol{\tau}_{jk}$	Stress tensor of phase j in region k
$\boldsymbol{\tau}_k$	Stress tensor of the gas phase in region k
ϕ	Mass fraction of particles

χ_{jf}	Fraction of frictional perimeter of phase j	ω	Angular frequency
		ω_h	Frequency of perturbation

References

Achenbach, E. (1974). Vortex Shedding from Spheres. *J. Fluid Mech.*, **62**, 209.

Briscoe, B. J. and Evans, P. D. (1987). The Wear of Polymers by Particle Flows. In *Tribology in Particulate Technology*. Ed. Briscoe and Adams. Philadelphia: Adam Hilger.

Carman, P. C. (1937). Fluid Flow through Granular Beds. *Trans. Instn. Chem.* Engrs., **15**, 150.

Carrier, G. F. (1958). Shock Waves in a Dusty Gas. *J. Fluid Mech.*, **4**, 376.

Chang, S. S.-H. (1975). Nonequilibrium Phenomena in Dusty Supersonic Flow Past Blunt Bodies of Revolution. *Phys. Fluids*, **18**, 446.

de Crecy, F. (1986). Modeling of Stratified Two-Phase Flow in Pipes, Pumps, and Other Devices. *Int. J. Multiphase Flow*, **12**, 307.

Delhaye, J. M. (1981). Basic Equations for Two-Phase Flow Modeling. In *Two-Phase Flow and Heat Transfer in the Power and Process Industries*. Ed. Bergles, Collier, Delhaye, Hewitt and Mayinger. Washington: Hemisphere.

Finnie, I., Wolak, J. and Kabil, Y. (1967). Erosion of Metals by Solid Particles. *J. Mater. Sci.*, **2**, 682.

Goldsmith, W. and Lyman, P. T. (1960). The Penetration of Hard Steel Sphere into Plane Metal Surface. *Trans. ASME, J. Appl. Mech.*, **27**, 717.

Gore, J. P. and Crowe, C. T. (1989). Effect of Particle Size on Modulating Turbulent Intensity. *Int. J. Multiphase Flow*, **15**, 279.

Gwyn, J. E. (1969). On the Particle Size Distribution Function and the Attrition of Cracking Catalysts. *AIChE J.*, **15**, 35.

Hetsroni, G. (1989). Particle–Turbulence Interaction. *Int. J. Multiphase Flow*, **15**, 735.

Hinze, J. O. (1972). Turbulent Fluid and Particle Interaction. *Prog. Heat Mass Transfer*, **6**, 433.

Hutchings, I. M. (1987). Surface Impact Damage. In *Tribology in Particulate Technology.* Ed. Briscoe and Adams. Philadelphia: Adam Hilger.

Lamb, H. (1932). *Hydrodynamics*, 6th ed. Cambridge: Cambridge University Press.

Lee, S. L. and Durst, F. (1982). On the Motion of Particles in Turbulent Duct Flows. Int. *J. Multiphase Flow*, **8**, 125.

McLaughlin, J. B. (1994). Numerical Computation of Particle–Turbulence Interaction. *Int. J. Multiphase Flow*, **20/S**, 211.

Morgenthaler, J. H. (1962). Analysis of Two-Phase Flow in Supersonic Exhausts. In *Detonation and Two-Phase Flow*. Ed. Penner and Williams. New York: Academic Press.

Paramanathan, B. K. and Bridgwater, J. (1983). Attrition of Solids–II. *Chem. Eng. Sci.*, **38**, 207.

Peddieson, J. (1975). Gas–Particle Flow Past Bodies with Attached Shock Waves. *AIAA J.*, **13**, 939.

Rietema, K. (1991). *The Dynamics of Fine Powders*. London: Elsevier Applied Science.

Rudinger, G. (1965). Some Effects of Finite Particle Volume on Dynamics of Gas–Particle Mixture. *AIAA J.*, **3**, 1217.

Rudinger, G. (1969). Relaxation in Gas–Particle Flow. In *Nonequilibrium Flows*. Ed. P. P. Wegener. New York: Marcel Dekker.

Rudinger, G. (1980). *Fundamentals of Gas–Particle Flow.* New York: Elsevier Scientific.

Rudinger, G. and Chang, A. (1964). Analysis of Nonsteady Two-Phase Flow. *Phys. Fluids*, **7**, 1747.

Soo, S. L. (1960). Effect of Transport Process on Attenuation and Dispersion in Aerosols. *J. Acoustical Society of America*, **32**, 943.

Soo, S. L. (1977). A Note on Erosion by Moving Dust Particles. *Powder Tech.*, **17**, 259.
Soo, S. L. (1990). *Multiphase Fluid Dynamics*. Beijing: Science Press; Brookfield, USA: Gower Technical.
Squires, K. D. and Eaton, J. K. (1990). Particle Response and Turbulence Modification in Isotropic Turbulence. *Phys. Fluids*, **A2**, 1191.
Tabor, D. (1951). *The Hardness of Metals*. Oxford: Oxford University Press.
Timoshenko, S. and Goodier, J. E. (1951). *The Theory of Elasticity*, 2nd ed. New York: McGraw-Hill.
Tsuji, Y. and Morikawa, Y. (1982). LDV Measurements of an Air–Solid Two-Phase Flow in a Horizontal Pipe. *J. Fluid Mech.*, **120**, 385.
Tsuji, Y., Morikawa, Y. and Shiomi, H. (1984). LDV Measurements of an Air–Solid Two-Phase Flow in a Vertical Pipe. *J. Fluid Mech.*, **139**, 385.
Wallis, G. B. (1969). *One-Dimensional Two-Phase Flow*. New York: McGraw-Hill.
Yuan, Z. and Michaelides, E. E. (1992). Turbulence Modulation in Particulate Flows – a Theoretical Approach. *Int. J. Multiphase Flow*, **18**, 779.
Zhou, L. X. and Huang, X. Q. (1990). Predictions of Confined Turbulent Gas–Particle Jets by an Energy Equation Model of Particle Turbulence. *Science in China*, **33**, 52.
Zhu, C. (1991). *Dynamic Behavior of Unsteady Turbulent Motion in Pipe Flows of Dense Gas–Solid Suspensions*. Ph.D. Dissertation. University of Illinois at Urbana-Champaign.
Zhu, C., Soo, S. L. and Fan, L.-S. (1996). Wave Motions in Stratified Gas–Solid Pipe Flows. *Proceedings of ASME Fluids Engineering Division Summer Meeting*, San Diego, California. **1**, 565. ASME.

Problems

6.1 Consider the surface wear by random motion of particles in a fluidized bed. The random motion is assumed to have the intensity of $\langle u'^2 \rangle$. Determine the relationship between E_D and K_D for overall ductile wear and the relationship between E_B and K_B for overall brittle wear.

6.2 For a general surface erosion, the wear is composed of both ductile and brittle modes. Thus, the combined directional wear can be expressed as the weighted summation of both modes. Consider a case where the weighted factor for ductile wear is $\xi(\alpha_i) = A \sin \alpha_i$ and the weighted factor for brittle wear is $\eta(\alpha_i) = B \cos \alpha_i$. Both A and B are constant. What is the extent of the total erosion? Derive an expression for the maximum eroding angle in terms of K_D, K_B, A, and B.

6.3 Consider a closed system of a dust-laden air with an initial pressure of 1 atm. The dust particles are 10 μm in diameter and the particle density is 2,000 kg/m^3. The relative ratio of the specific heat of particle to gas is 50. The mass loading ratio of particle to gas is 10. Assuming the system undergoes an isentropic compression to a final state with a pressure of 20 atm, calculate the volume fraction of solids and estimate the averaged distance between particles in the final state.

6.4 Show that, from Eq. (6.74), the asymptotic value of the attenuation parameter for the acoustic wave propagation through a dilute gas–solid suspension can be expressed as

$$\frac{\beta a_g}{\omega} = \frac{2}{9} \frac{m_p}{\sqrt{1+m_p}} \frac{\rho_p}{\rho} N_\omega \qquad \text{(P6.1)}$$

for small N_ω, and

$$\frac{\beta a_g}{\omega} = \frac{9}{4} m_p \frac{\rho}{\rho_p} \frac{1}{\sqrt{N_\omega}} \tag{P6.2}$$

for large N_ω. Assume that $\rho_p/\rho \gg 1$ and $m_p(\rho/\rho_p) \ll 1$.

6.5 Consider an acoustic wave with a frequency of 1,000 Hz passing through dust-laden air at ambient conditions. Assume that $m_p = 0.5$, $d_p = 5\ \mu$m, $\rho_p/\rho = 1{,}500$, $\delta = 10$, $\nu = 1.5 \times 10^{-5}$ m²/s, and $\gamma = 1.4$. (a) Estimate the speed of sound using Eq. (6.72). Is the result close to the "frozen speed of sound" or the "equilibrium speed of sound?" (b) Estimate the speed of sound using Eq. (6.77). Compare the results between (a) and (b).

6.6 Considering a batch–solid fluidized bed, show that Eq. (6.180) can be derived from Eqs. (6.169), (6.170), and (6.174) through (6.179).

6.7 Assume that the bed expansion characteristics in gas–solid particulate fluidization can be described by an equation of the Richardson–Zaki form as given by

$$\frac{U}{U_{pt}} = \alpha^n \tag{P6.3}$$

where n is the modified Richardson–Zaki index for gas–solid systems. Express the elastic modulus E in terms of the gas and particle properties and the bed voidage for the minimum bubbling point.

6.8 For plastic beads with $\rho_p = 500$ kg/m³ falling in air at ambient conditions, estimate the range of variation of the shedding frequency of the particle wake when the particle Reynolds number, based on the particle terminal velocity, varies from 500 to 1,000. For this particle Reynolds number range, what is the corresponding range of variation for the particle sizes?

PART II

System Characteristics

CHAPTER 7

Gas–Solid Separation

7.1 Introduction

Separation processes are central to gas–solid flow systems concerning dust removal, particulate collection, sampling, particle recirculation, and other operations. Gas–solid separation can be achieved by application of the principles involving centrifugation, electrostatic effects, filtration, gravitational settling, and wet scrubbing. Gas–solid separators applying these principles employ rotary flow dust separators, electrostatic precipitators, filters, settling chambers, and scrubbers. In order to yield highly efficient solids collection or removal processes, multistage gas–solid separators based on a combination of several of these components are also commonly employed.

In this chapter, the collection mechanisms, types, and collection efficiencies of gas–solid separators are discussed. Specifically, separation by rotating flow, exemplified by a cyclone separator, is presented. Tangential flow cyclones, which are the most commonly used cyclones, are described. The principles of electrostatic precipitation are illustrated. Comments are presented with respect to the difficulties involved in obtaining an accurate estimation of the collection efficiency of an electrostatic precipitator. Factors contributing to these difficulties include the geometric complexity of the system, significant flow disturbance due to electric wind, and poor predictability of particle charges. Concepts on filtration are also introduced. A filtration process can collect particles of almost all sizes; however, the pressure drop across the filter may vary significantly with the internal structure of the filter, particle deposition mode (cake or in-depth deposition), and quantity of accumulated particles. Wet scrubbing, which produces wet sludge from the impaction of solids with liquid droplets, is also delineated in this chapter.

7.2 Separation by Rotating Flow

Separation based on rotating flow principles is one of the most common operations involved in gas–solid flows. This section describes the fundamental rotating flow principles and their applications to cyclone operation. The efficiency of dust collection in cyclones is also described.

7.2.1 *Mechanism and Type of Rotary Flow Dust Separators*

In a rotary flow dust separator, the rotating flow of the gas–solid suspension can be initiated as a result of the design of the separator, such as the tangential inlet of the separator, the guide vanes at the inlet, and relative rotating cylinders. As a result of the rotating flow, the particles are subjected to a centrifugal force which is usually at least two orders of magnitude greater than the gravitational force. Thus, even light particles can be easily directed toward the wall by the centrifugal force and collected in the dust bunker while the clean gas escapes through the gas outlet.

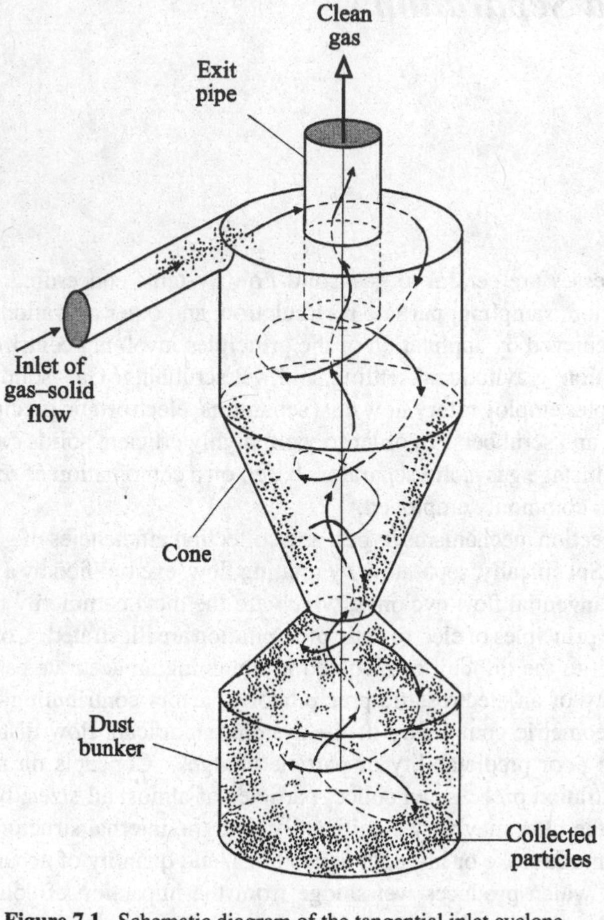

Figure 7.1. Schematic diagram of the tangential inlet cyclone.

Separating devices using this mechanism consist of reverse-flow cyclones and uniflow cyclones. The reverse-flow cyclones can be represented by the tangential inlet cyclone and the tornado dust collector, as shown in Figs. 7.1 and 7.2, respectively. The suspension flow is fed into the cyclone tangentially so that the flow begins rotating by following the inner contour of the cyclone. The particles accumulated on the wall slide down along the cone into the dust bunker. Typical uniflow cyclones include the axial flow cyclones and the rotating cylinder separators. In the axial flow cyclone, as shown in Fig. 7.3, a rotating flow is created by the guide vanes. The solid particles are forced toward the outer wall by the centrifugal force and are separated from the gas in the coaxial outlets. In the rotating cylinder separators represented by the rotex centrifugal separators, as shown in Fig. 7.4, the dust-laden gas flows through the rotating cylinders and starts rotating by the effect of wall friction at the cylinders. The separation of particles is achieved by the relative inertia between gas and particles and by the effect of gravity.

The cyclone dust collector is one of the simplest dust collectors: no moving parts and ease in maintenance. In spite of the simple structure, the centrifugal force on the solids in the cyclone dust collector can easily reach 300 to 2,000 times the gravitational force and a

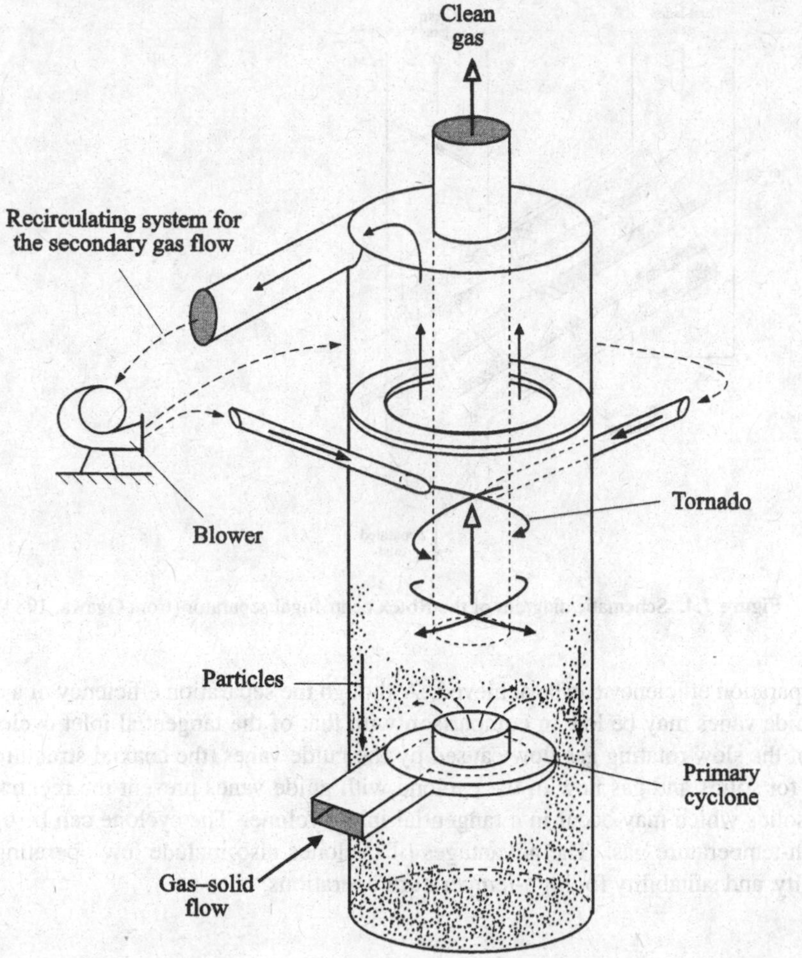

Figure 7.2. Schematic diagram of the tornado dust collector (after Ogawa, 1984).

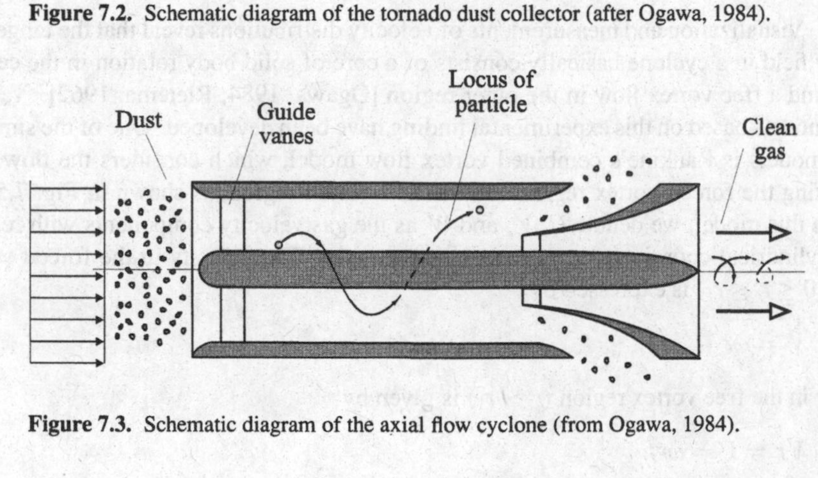

Figure 7.3. Schematic diagram of the axial flow cyclone (from Ogawa, 1984).

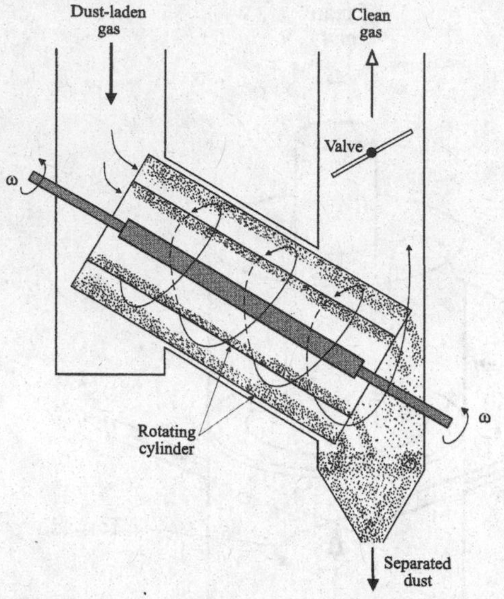

Figure 7.4. Schematic diagram of the Rotex centrifugal separator (from Ogawa, 1984).

high separation efficiency can be achieved. Although the separation efficiency of a cyclone with guide vanes may be low in comparison with that of the tangential inlet cyclone as a result of the slow rotating gas flow caused by the guide vanes, the coaxial structure of the outlets for solids and gas flow in the cyclone with guide vanes prevent the reentrainment of the solids which may occur in a tangential inlet cyclone. The cyclone can be operated for high-temperature gas. The advantages of cyclones also include low operating costs, reliability, and suitability for high-temperature operations.

7.2.2 *Flow Field in a Cyclone*

Visualization and measurements of velocity distributions reveal that the tangential velocity field in a cyclone basically consists of a core of solid body rotation in the central region and a free vortex flow in the outer region [Ogawa, 1984; Rietema, 1962]. Various vortex models based on this experimental finding have been developed. One of the simplest vortex models is Rankine's combined vortex flow model, which considers the flow field comprising the forced vortex region and the free vortex region, as shown in Fig. 7.5. To describe this model, we denote U, V, and W as the gas velocity components with respect to the cylindrical coordinates r, θ, and z, respectively. The velocity in the forced vortex region ($0 \leq r \leq r_f$) is expressed by

$$V = \omega r \tag{7.1}$$

and that in the free vortex region ($r > r_f$) is given by

$$Vr = \Gamma = \omega r_f^2 \tag{7.2}$$

where r_f is the radius of the free vortex, ω is the angular velocity, and Γ represents the

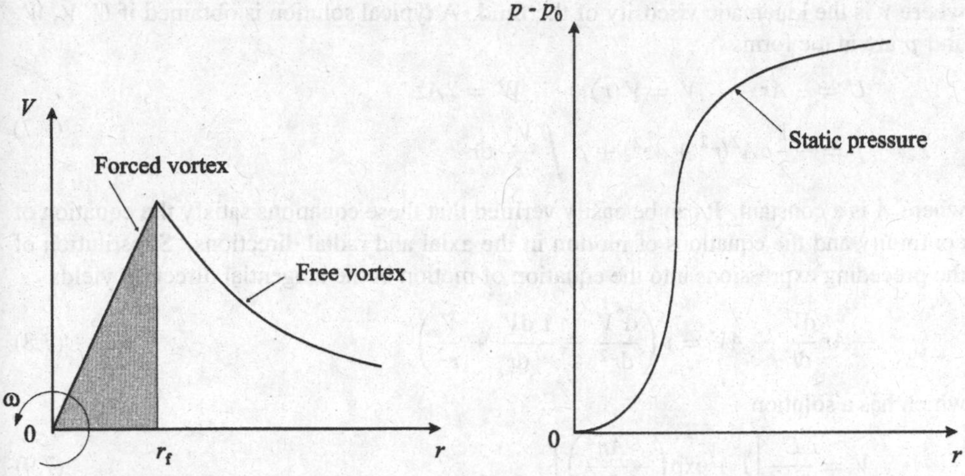

Figure 7.5. Tangential velocity and pressure distributions based on Rankine's combined vortex model.

circulation. The corresponding static pressure distribution can also be obtained as

$$p = p_0 + \frac{\rho\Gamma^2}{2r_{\mathrm{f}}^4}r^2 \tag{7.3}$$

for the forced vortex region and

$$p = p_0 + \frac{\rho\Gamma^2}{2}\left(\frac{2}{r_{\mathrm{f}}^2} - \frac{1}{r^2}\right) \tag{7.4}$$

for the free vortex region, where p_0 is the static pressure at the center of the cyclone and ρ is the gas density.

Another commonly used model is based on the general differential balance of mass and momentum [Burgers, 1948]. Consider a steady, incompressible, and axially symmetric flow in which the body forces are negligible. In cylindrical coordinates, the equation of continuity of the fluid can be given as

$$\frac{\partial U}{\partial r} + \frac{\partial W}{\partial z} + \frac{U}{r} = 0 \tag{7.5}$$

and the Navier–Stokes equations can be expressed by

$$\begin{aligned}
U\frac{\partial U}{\partial r} + W\frac{\partial U}{\partial z} - \frac{V^2}{r} &= -\frac{1}{\rho}\frac{\partial p}{\partial r} + \nu\left(\frac{\partial^2 U}{\partial r^2} + \frac{\partial^2 U}{\partial z^2} + \frac{1}{r}\frac{\partial U}{\partial r} - \frac{U}{r^2}\right) \\
U\frac{\partial V}{\partial r} + W\frac{\partial V}{\partial z} + \frac{UV}{r} &= \nu\left(\frac{\partial^2 V}{\partial r^2} + \frac{\partial^2 V}{\partial z^2} + \frac{1}{r}\frac{\partial V}{\partial r} - \frac{V}{r^2}\right) \\
U\frac{\partial W}{\partial r} + W\frac{\partial W}{\partial z} &= -\frac{1}{\rho}\frac{\partial p}{\partial z} + \nu\left(\frac{\partial^2 W}{\partial r^2} + \frac{\partial^2 W}{\partial z^2} + \frac{1}{r}\frac{\partial W}{\partial r}\right)
\end{aligned} \tag{7.6}$$

where ν is the kinematic viscosity of the fluid. A typical solution is obtained if U, V, W, and p are in the forms

$$U = -Ar, \qquad V = V(r), \qquad W = 2Az$$
$$p = -\frac{1}{2}\rho A^2(r^2 + 4z^2) + \rho \int \frac{V^2}{r}\,\mathrm{d}r \tag{7.7}$$

where A is a constant. It can be easily verified that these equations satisfy the equation of continuity and the equations of motion in the axial and radial directions. Substitution of the preceding expressions into the equation of motion in the tangential direction yields

$$-Ar\frac{\mathrm{d}V}{\mathrm{d}r} - AV = \nu\left(\frac{\mathrm{d}^2V}{\mathrm{d}r^2} + \frac{1}{r}\frac{\mathrm{d}V}{\mathrm{d}r} - \frac{V}{r^2}\right) \tag{7.8}$$

which has a solution

$$V = \frac{C}{2\pi r}\left[1 - \exp\left(-\frac{Ar^2}{2\nu}\right)\right] \tag{7.9}$$

where C is a constant representing the maximum circulation in the system. The circulation is thus given by

$$\Gamma = \frac{\mathrm{d}V}{\mathrm{d}r} + \frac{V}{r} = \frac{AC}{2\pi\nu}\exp\left(-\frac{Ar^2}{2\nu}\right) \tag{7.10}$$

Now, define the dimensionless velocities and coordinate as

$$U^* = \frac{U}{U_\mathrm{w}}, \qquad V^* = \frac{V}{V_\mathrm{w}}, \qquad r^* = \frac{r}{R} \tag{7.11}$$

where U_w and V_w are the radial and tangential gas velocities near the cyclone wall, respectively (just outside the wall surface boundary layer), and R is the radius of the cross section of the cyclone. Note that both V_w and R are functions of the axial coordinate z only. Neglecting the thickness of the wall boundary layer, the dimensionless velocity profiles are obtained from Eqs. (7.7) and (7.9) as

$$U^* = r^*, \qquad V^* = \frac{1}{r^*}\left(\frac{1 - \exp(-\mathrm{Re}_\mathrm{w} r^{*2}/2)}{1 - \exp(-\mathrm{Re}_\mathrm{w}/2)}\right) \tag{7.12}$$

where Re_w is the Reynolds number, defined as

$$\mathrm{Re}_\mathrm{w} = -\frac{RU_\mathrm{w}}{\nu} \tag{7.13}$$

In terms of Eq. (7.12), the variations of the radial profiles of the dimensionless tangential velocity V^* with Re_w are plotted as shown in Fig. 7.6. It is seen that the dimensionless tangential velocities reach a maximum at $r^* \leq 0.5$ for values of Re_w varying from 10 to 30.

In an actual operation, the gas–solid flow in a cyclone separator is turbulent. The turbulence effects on the distributions of the gas velocity and static pressure in the preceding two models are, however, excluded. The strongly swirling turbulent gas–solid flows in tangential inlet cyclones were accounted for numerically by Zhou and Soo (1990) using the k–ϵ turbulence model. Their typical results along with laser Doppler velocimetry (LDV) measurements for the radial variations of the axial velocity, tangential velocity, and pressure in the cyclone are shown in Fig. 7.7. It is seen that there are an upward flow region

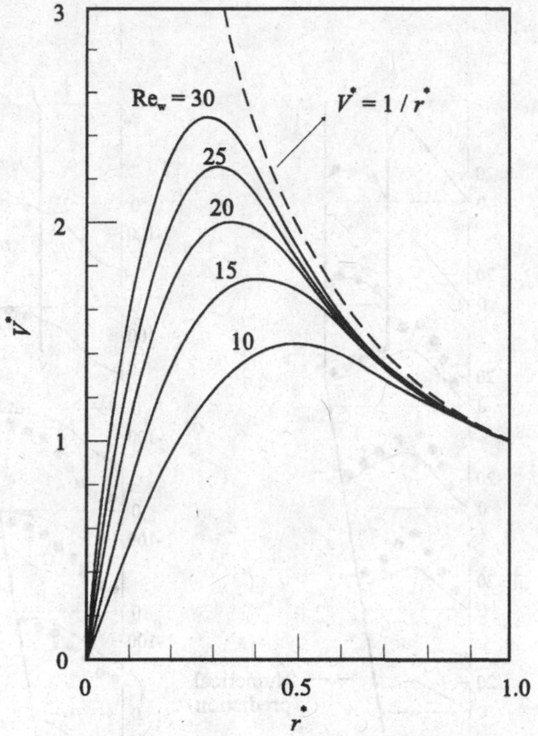

Figure 7.6. Vortex flow distributions (from Burgers, 1948).

in the near-axial core region and a downward flow region in the wall region. Also, the pressure increases radially in the cyclone. The k–ϵ model is seen to predict the velocity and pressure in the wall region well, but not so precisely in the near-axial core region because of the inability of the model to simulate the nonisotropic turbulence of strongly swirling flow.

For the engineering design of tangential inlet cyclones, the variation of the tangential velocity in the radial direction can be correlated in a form

$$Vr^{\beta} = \text{const} \tag{7.14}$$

where the vortex exponent β is related to the absolute gas temperature T and the cyclone radius R by [Alexander, 1949]

$$\beta = 1 - (1 - 0.74R^{0.14})\left(\frac{T}{283}\right)^{0.3} \tag{7.15}$$

For most tangential inlet cyclones, β varies from 0.5 to 0.7.

7.2.3 *Collection Efficiency of Cyclones*

The collection efficiency usually depends on the shape of the cyclone, size and density of the particles, and incoming tangential velocity. The collection efficiency of a

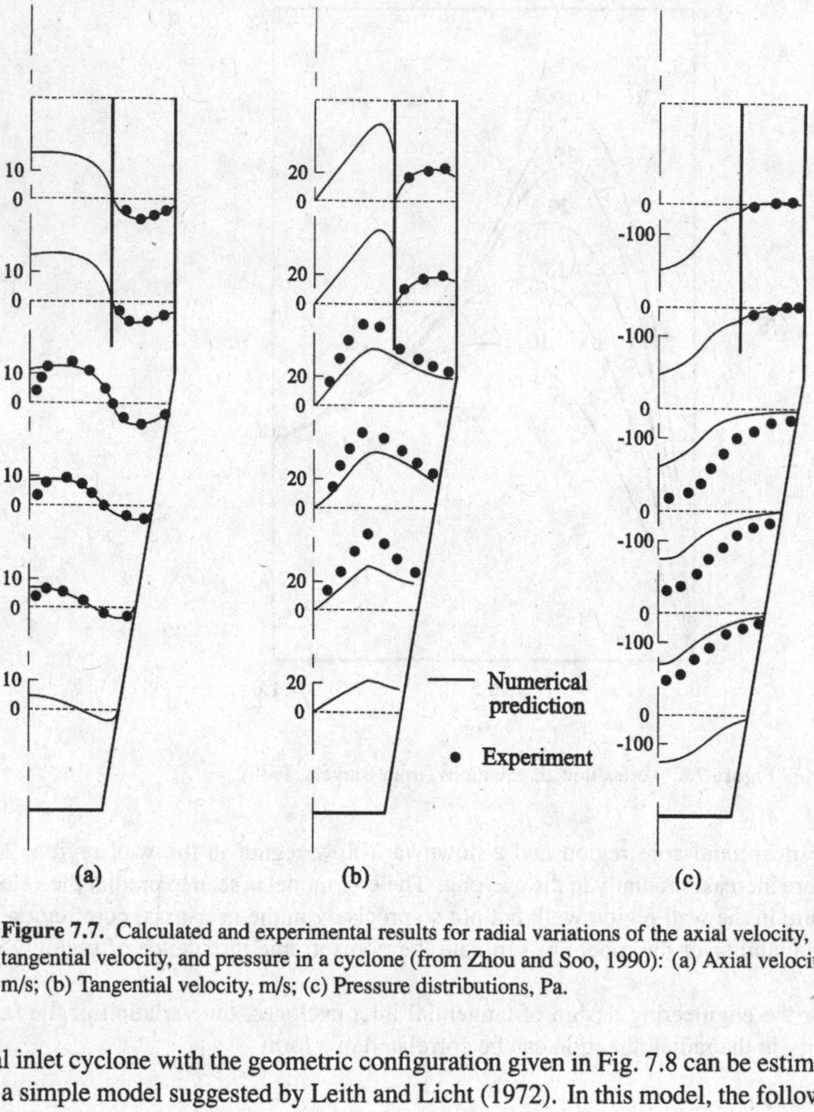

Figure 7.7. Calculated and experimental results for radial variations of the axial velocity, tangential velocity, and pressure in a cyclone (from Zhou and Soo, 1990): (a) Axial velocity, m/s; (b) Tangential velocity, m/s; (c) Pressure distributions, Pa.

tangential inlet cyclone with the geometric configuration given in Fig. 7.8 can be estimated by using a simple model suggested by Leith and Licht (1972). In this model, the following characteristics are assumed:

(1) Particles are spherical.
(2) The radial velocity of the gas is zero.
(3) The radial velocity of the particles is nearly constant.
(4) The drag force follows Stokes's law.
(5) There is no slip in the tangential direction between the particles and the gas.
(6) The tangential velocity can be expressed by Eq. (7.14).
(7) The effect of the electrostatic charge is negligible.
(8) The concentration of uncollected particles is uniformly distributed as a result of turbulent mixing.

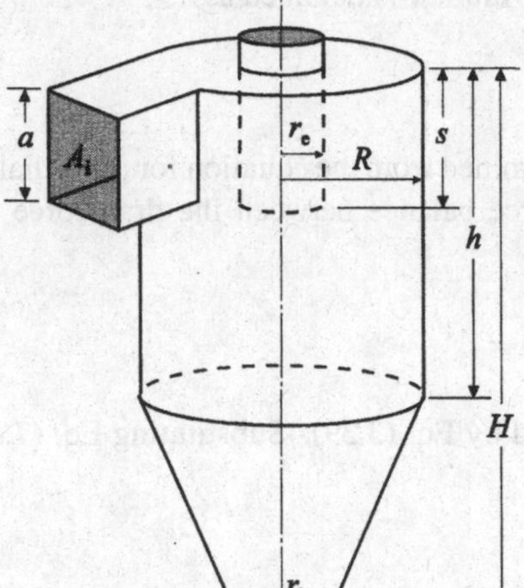

Figure 7.8. Geometric configuration of the tangential inlet cyclone.

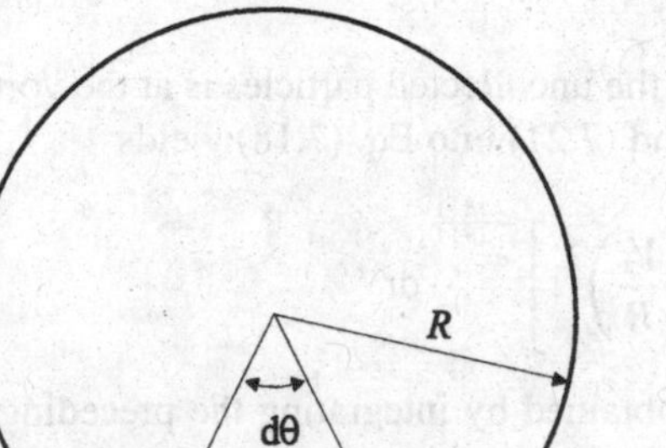

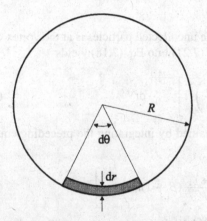

Figure 7.9. Horizontal cross section of a cyclone.

Consider a horizontal cross section of a cyclone, as shown in Fig. 7.9. During a time interval $\mathrm{d}t$, the number of particles collected, $\mathrm{d}N$, within a differential volume $\mathrm{d}r \cdot (r\,\mathrm{d}\theta) \cdot \mathrm{d}z$ is expressed by

$$-\mathrm{d}N = \frac{\mathrm{d}\theta}{2}[R^2 - (R - \mathrm{d}r)^2]n\,\mathrm{d}z \tag{7.16}$$

where n is the particle number density. The total number of particles in the sector from which particles are removed is

$$N = \frac{\mathrm{d}\theta}{2}R^2 n\,\mathrm{d}z \tag{7.17}$$

Therefore, the fraction of particles collected within dt is obtained as

$$-\frac{\mathrm{d}N}{N} = \frac{2R\,\mathrm{d}r - (\mathrm{d}r)^2}{R^2} \approx 2\frac{\mathrm{d}r}{R} \tag{7.18}$$

The relationship between dr and dt is determined from the equation for the radial motion of the particle, which is obtained by the force balance between the drag force and the centrifugal force, *i.e.*,

$$\frac{1}{\tau_{\mathrm{S}}}\frac{\mathrm{d}r}{\mathrm{d}t} = \frac{V^2}{r} \tag{7.19}$$

where τ_{S} is the Stokes relaxation time defined by Eq. (3.39). Substituting Eq. (7.14) into Eq. (7.19) yields

$$\frac{1}{\tau_{\mathrm{S}}}\frac{\mathrm{d}r}{\mathrm{d}t} = \frac{V_{\mathrm{i}}^2 R^{2\beta}}{r^{2\beta+1}} \tag{7.20}$$

which gives the traveling time t for particles moving from r_0 at $t = 0$ to r as

$$t = \frac{1}{2\tau_{\mathrm{S}}(\beta+1)}\left(\frac{R}{V_{\mathrm{i}}}\right)^2\left[\left(\frac{r}{R}\right)^{2\beta+2} - \left(\frac{r_0}{R}\right)^{2\beta+2}\right] \tag{7.21}$$

where V_{i} is the inlet velocity.

Assume that the initial radial position of the uncollected particles is at the vortex center where $r_0 = 0$. Substitution of Eqs. (7.20) and (7.21) into Eq. (7.18) yields

$$\frac{\mathrm{d}N}{N} = -2\tau_{\mathrm{S}}\left(\frac{V_{\mathrm{i}}}{R}\right)^2\left[2\tau_{\mathrm{S}}(\beta+1)\left(\frac{V_{\mathrm{i}}}{R}\right)^2 t\right]^{-\frac{2\beta+1}{2\beta+2}}\mathrm{d}t \tag{7.22}$$

Thus, the collection efficiency, η_{c}, can be obtained by integrating the preceding equation from $t = 0$ to the mean residence time, t_{m}, as

$$\eta_{\mathrm{c}} = \frac{N_0 - N_{\mathrm{m}}}{N_0} = 1 - \exp\left[-2\left(\frac{2\tau_{\mathrm{S}}V_{\mathrm{i}}^2}{R^2}(\beta+1)t_{\mathrm{m}}\right)^{\frac{1}{2\beta+2}}\right] \tag{7.23}$$

where N_0 is the number of particles at $t = 0$, and N_{m} is the remaining uncollected particles at time t_{m}.

The mean residence time t_{m} of gas can be related to the geometric configuration of the cyclone and the inlet flow rate. It is reasonable to take t_{m} as the average of the minimum residence time t_{m1} and the maximum residence time t_{m2}, where t_{m1} is the average time required for the gas to descend from the midlevel of the entrance to the exit at the bottom of the cyclone. Thus, we have

$$t_{\mathrm{m1}} = \frac{\pi(s - a/2)\left(R^2 - r_{\mathrm{e}}^2\right)}{V_{\mathrm{i}}A_{\mathrm{i}}} \tag{7.24}$$

where A_{i} is the cross-sectional area of inlet and s, a, R, and r_{e} are dimensions of the cyclone, defined in Fig. 7.8.

In order to estimate t_{m2}, we need to identify the lowest point to which the gas descends. This point does not necessarily correspond to the bottom of the cyclone as characterized by the depth H. The deepest descending length l is given by an empirical equation [Alexander, 1949] as

$$l = 7.3 r_e \left(\frac{R^2}{A_i} \right)^{\frac{1}{3}} \tag{7.25}$$

which is independent of the inlet flow rate. Hence, we have

$$t_{m2} = t_{m1} + \frac{1}{V_i A_i} \left[\pi R^2 (h - s) + \frac{\pi}{3} R^2 (l + s - h)(1 + \xi + \xi^2) - \pi r_e^2 l \right] \tag{7.26}$$

where

$$\xi = 1 - \left(1 - \frac{r_b}{R} \right) \left[\frac{s + l - h}{H - h} \right] \tag{7.27}$$

and h, H, and r_b are dimensions of the cyclone defined in Fig. 7.8.

A cyclone should be designed in such a way that its physical length $(H - s)$ is near l. If this physical length is longer than l, the space below the vortex turning point is wasted. If the physical length is shorter than l, the full separation potential of the cyclone is not realized. For a properly sized cyclone, t_{m2} is expressed by

$$t_{m2} = t_{m1} + \frac{1}{V_i A_i} \left[\pi R^2 (h - s) + \frac{\pi}{3} R^2 (H - h) \left(1 + \frac{r_b}{R} + \frac{r_b^2}{R^2} \right) - \pi r_e^2 (H - s) \right] \tag{7.28}$$

Therefore, t_m is given as

$$t_m = \frac{1}{2} (t_{m1} + t_{m2}) \tag{7.29}$$

where t_{m1} is given by Eq. (7.24) and t_{m2} is given by either Eq. (7.26) or Eq. (7.28). From Eqs. (7.29) and (7.23), the collection efficiency of a cyclone can be calculated.

It is noted that the particle–wall interaction in the boundary layer and the electrostatic effect due to the electrostatic charge carried by the particles may strongly affect particle collection and reentrainment in the cyclone and, consequently, affect the collection efficiency. In the presence of electrostatic charges of particles and an external electric field, the collection efficiency of a tangential inlet cyclone with a steep cone is given by [Soo, 1989]

$$\eta_c = 1 - \exp\left\{ -\frac{2\pi \sigma \tau_s H}{Q} \left(\frac{\Gamma_m^2}{R^2} + \frac{4\alpha_p \rho_p R^2}{\epsilon_0} \left(\frac{q}{m} \right)^2 + E \left(\frac{q}{m} \right) R \right) \right\} \tag{7.30}$$

where ρ_p is the density of particles; σ is the sticking probability of the particles to the wall or the layer of collected particles; Γ_m is the maximum circulation in the system; Q is the volumetric flow rate of gas; ϵ_0 is the permittivity of the vacuum; q/m is the charge-to-mass ratio of particles; and E is the electric field strength. The first term in the bracket on the right-hand side of Eq. (7.30) represents the collection by centrifugal force exerted by the vortex; the second term is the collection by space charge of the particles; and the last term accounts for the collection by the applied electric field. It can be shown that the sticking probability σ depends not only on the contact surfaces and deposit forces but also on the particle turbulent diffusivity and turbulent viscosity [Zhou and Soo, 1991].

The cut-off size of a cyclone, which is an important performance index of the cyclone, is defined as the size of particles below which the collection efficiency is less than 50 percent.

In terms of Eq. (7.23), the cut-off size of a cyclone, d_{pc}, in the absence of electrostatic charges in a gas–solid flow can be estimated by

$$d_{pc} = \frac{3R}{V_i}\sqrt{\frac{\mu}{\rho_p(\beta+1)t_m}\left(\frac{\ln 2}{2}\right)^{\beta+1}} \tag{7.31}$$

In the presence of electrostatic charges of particles and an external electric field, the cut-off size of a tangential inlet cyclone with a steep cone is given from Eq. (7.30) as

$$d_{pc}^2 = \frac{9(\ln 2)\mu Q}{\pi\rho_p\sigma H}\left(\frac{\Gamma_m^2}{R^2} + \frac{4\alpha_p\rho_p R^2}{\epsilon_0}\left(\frac{q}{m}\right)^2 + E\left(\frac{q}{m}\right)R\right)^{-1} \tag{7.32}$$

where μ is the viscosity of the gas.

Example 7.1 In a test of the performance of a cyclone, a dust-laden gas with the particle mass flow rate of 600 kg/h is used. Given that the dust escape rate is 30 kg/h, what is the overall collection efficiency? Plot the fractional efficiency curve to identify the cut-off size of this cyclone. The results of particle size analysis (by weight percentage [wt%]) at the inlet and outlet of the cyclone are given in Table E7.1.

Table E7.1. *Results of the Particle Size Analysis at the Inlet and Outlet of the Cyclone*

Size range	Particle size analysis (wt%)	
(μm)	Inlet	Outlet
0–5	1.5	28.5
5–10	1.7	20.0
10–15	2.2	13.0
15–20	3.4	10.0
20–30	9.5	14.7
30–40	15.0	8.5
40–50	19.0	3.3
50–60	22.0	1.3
>60	25.7	0.7

Solution The overall collection efficiency is

$$\eta_c = \frac{\text{mass rate of collected particles}}{\text{mass rate of inlet particles}} = \frac{600-30}{600} = 95 \text{ percent} \tag{E7.1}$$

The fractional collection efficiency at each size interval, η_{ci}, is calculated by using a similar formula, and the results are given in Table E7.2.

In terms of the averaged particle diameters and the corresponding fractional collection efficiency, the fractional efficiency curve is obtained as shown in Fig. E7.1, from which the cut-off size of this cyclone is estimated as 9 μm.

Table E7.2. *Fractional Collection Efficiency for Dust Collection in Each Particle Size Range*

d_p range (μm)	ave. of d_p (μm)	Inlet (wt%)	Inlet mass (kg/h)	Outlet (wt%)	Outlet mass (kg/h)	η_{ci} (%)
0–5	2.5	1.5	9.0	28.5	8.55	5.0
5–10	7.5	1.7	10.2	20.0	6.0	41.2
10–15	12.5	2.2	13.2	13.0	3.9	70.5
15–20	17.5	3.4	20.4	10.0	3.0	85.3
20–30	25	9.5	57.0	14.7	4.4	92.3
30–40	35	15.0	90.0	8.5	2.6	97.1
40–50	45	19.0	114.0	3.3	1.0	99.1
50–60	55	22.0	132.0	1.3	0.4	99.7
>60	60	25.7	154.2	0.7	0.2	99.9

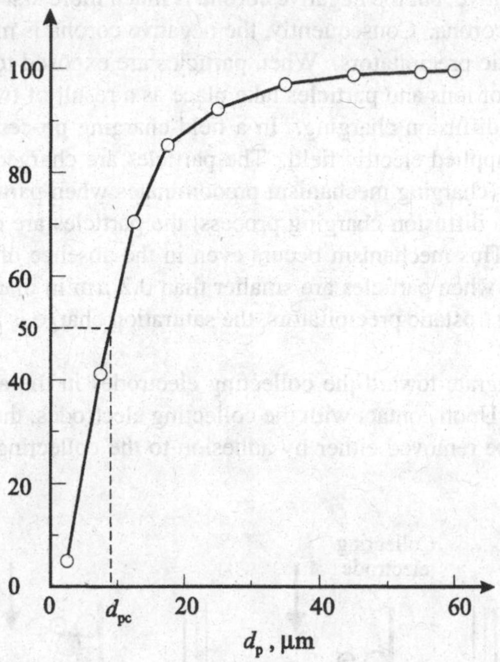

Figure E7.1. Fractional collection efficiency for a cyclone.

7.3 Electrostatic Precipitation

An electrostatic precipitator is a gas–solid separator in which particles suspended in a gas stream are charged and removed by electrostatic force. Since the separation force is directly applied to the particles without the need of accelerating the gas phase, an electrostatic precipitator usually requires much less hydraulic power than other separation systems. Hence, electrostatic precipitation is widely recognized as an important technique of gas–solid separation. This technique is characterized by low pressure drop, relative high collection efficiency for fine particles, and high adaptability to a wide variety of effluents, either wet, dry, or corrosive.

7.3.1 *Mechanism of an Electrostatic Precipitator*

The complete process of particle removal by electrostatic precipitation commonly includes three steps: (1) particle charging; (2) particle migration in an external electric field; and (3) particle discharging and collection. This removal process is illustrated in Fig. 7.10.

The electrostatic charges of particles in a gas stream can be acquired naturally through triboelectric charging via mutual impacts or by collisions with other solid materials such as pipe walls and valves (see §3.5.2). However, to reach a high collection efficiency in an electrostatic precipitator, the particles are typically artificially charged to a saturated degree by passing through a corona discharge area where the gas is highly ionized. The corona can be either positive or negative, but the negative corona is much more stable and is easier to control than the positive corona. Consequently, the negative corona is more frequently used in industrial electrostatic precipitators. When particles are exposed to unipolar ions in an electric field, impacts of ions and particles take place as a result of two basic mechanisms: field charging and diffusion charging. In a field-charging process, the motions of ions are induced by the applied electric field. The particles are charged as a result of the ion bombardments. This charging mechanism predominates when particles are larger than 1 μm in diameter. In a diffusion-charging process, the particles are charged by the Brownian motion of ions. This mechanism occurs even in the absence of external electric fields and predominates when particles are smaller than 0.2 μm in diameter. In most cases for particles inside electrostatic precipitators, the saturation charge is attained within a second.

The charged particles migrate toward the collecting electrodes in the applied electric field by the Coulomb force. Upon contact with the collecting electrodes, the particles lose their charges and thus can be removed either by adhesion to the collecting surface or by

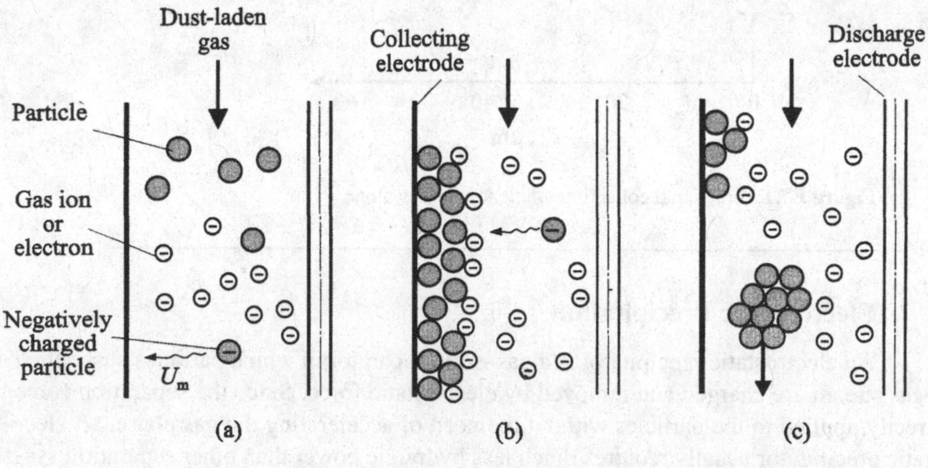

Figure 7.10. Mechanism of fine solids collection in electrostatic precipitation (after Ogawa, 1984): (a) Particle charging; (b) Particle migration; (c) Particle discharging and collection.

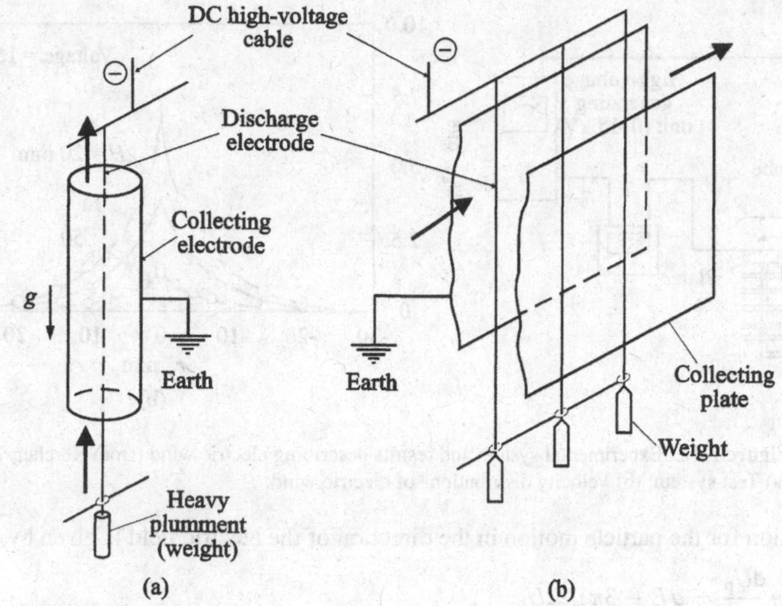

Figure 7.11. Basic types of electrodes (after Ogawa, 1984): (a) Cylinder-type precipitator; (b) Plate-type precipitator.

sliding into the discharge hoppers. Thus, the separation of particles from their carrying gas is achieved. Typical corona dischargers in electrostatic precipitators are either thin metal wires or metal needles, and typical collecting electrodes are plates or cylinders. Electrostatic precipitators may thus be divided into two types, cylinder-type and plate-type, as shown in Fig. 7.11. The cylinder-type electrostatic precipitators, sometimes used as wet precipitators, may achieve a collection efficiency as high as 99.9 percent. The plate-type precipitators, commonly used for dry particle-laden gas with large flow rates, can reach a collection efficiency of 95 percent [Ogawa, 1984].

Similarly to other separation methods, the electrostatic precipitation method has its own shortcomings. For certain types of particles, electrostatic charging is difficult without additional chemical conditioning. To handle a high flow rate of particulate-laden gas, the size of the precipitator, *i.e.*, total area of collecting electrodes, has to be large enough to ensure a high collection efficiency. The method requires a regulated high-voltage power supply, which is usually above 10 kV, and cleaning of the collecting plates is generally difficult. Consequently, the equipment and operating costs of this separator are relatively high.

7.3.2 *Migration Velocity and Electric Wind*

To estimate the particle migration velocity, it is assumed that (1) particles are spherical and have the same size; (2) all particles are charged to the same extent; (3) the particle motion is governed by the Coulomb force and the Stokes drag only; and (4) the direction of the applied electric field is perpendicular to the direction of the suspension flow.

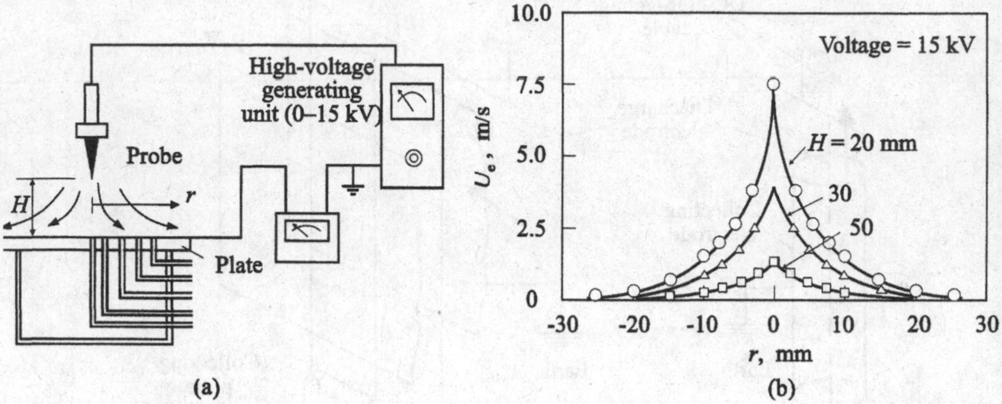

Figure 7.12. Experimental system and results describing electric wind (from Kercher, 1969): (a) Test system; (b) Velocity distributions of electric wind.

The equation for the particle motion in the direction of the electric field is given by

$$m\frac{dU_p}{dt} = qE - 3\pi\mu d_p U_p \tag{7.33}$$

where m is the particle mass and q is the electric charge carried by the particle. The particle velocity, U_p, is obtained by integrating Eq. (7.33) as

$$U_p = \frac{qE}{3\pi\mu d_p}\left[1 - \exp\left(-\frac{t}{\tau_S}\right)\right] \tag{7.34}$$

Define the migration velocity U_m as the terminal velocity of particles in the electric field. From Eq. (7.34), we have

$$U_m = \frac{qE}{3\pi\mu d_p} \tag{7.35}$$

The value of U_m measured in a practical situation is higher than that calculated from the preceding equation because of the existence of the electric wind from the discharge electrode to the collecting electrode. The electric wind is caused by the motion of ions induced by the strong corona discharge field. The intensity of the electric wind is related to both the properties of the gas and the intensity of the electric field. As illustrated in Fig. 7.12, the fluid velocity of electric wind can be on the order of 10 m/s [Kercher, 1969]. Hence, the motion of particle migration in the electrostatic precipitator can be strongly affected by the electric wind.

It can be shown that for fine particles ($d_p < 10\ \mu$m) in an electrostatic precipitator the migration velocity of the particle is usually much higher than the corresponding particle terminal velocity in the gravitational field. For this reason, fine particles are commonly separated by electrostatic precipitation.

7.3.3 *Collection Efficiency of Electrostatic Precipitators*

The collection efficiency of the wire-plate-type of electrostatic precipitator can be estimated by the widely used Deutsch theory [Deutsch, 1922]. To obtain the Deutsch equation, the following assumptions are made: (1) because of the turbulent mixing effect, particle concentration at any cross section of the precipitator is uniformly distributed;

(2) gas velocity is constant everywhere except in the boundary layer near the walls; (3) U_m is constant and is relatively small compared to the gas velocity; (4) no particle reentrainment occurs inside the precipitator; and (5) no disturbing electrical effect on the gas flow occurs.

Consider a one-dimensional flow of charged particles passing through the gap between two parallel rectangular plates acting as collecting electrodes. The plates are L in length, H in height, with a distance between the plates of l_w. The mass of particles collected, dM, over a distance dx is expressed by

$$dM = 2H\,dx\,\alpha_p \rho_p U_m\,dt = 2H\alpha_p\rho_p \frac{U_m}{U}\,dx^2 \tag{7.36}$$

where α_p is the particle volume fraction. On the other hand, from the mass balance of particles, we have

$$dM = -d\alpha_p\,\rho_p l_w H\,dx \tag{7.37}$$

Combining Eq. (7.36) with Eq. (7.37) yields

$$\frac{d\alpha_p}{\alpha_p} = -\frac{2U_m}{U}\frac{dx}{l_w} \tag{7.38}$$

which leads to

$$\frac{\alpha_{po}}{\alpha_{pi}} = \exp\left(-\frac{2U_m}{U}\frac{L}{l_w}\right) \tag{7.39}$$

where α_{pi} and α_{po} are the particle volume fractions at the inlet and the outlet, respectively. Therefore, the collection efficiency of an electrostatic precipitator, η_e, is obtained as

$$\eta_e = \frac{\alpha_{pi} - \alpha_{po}}{\alpha_{pi}} = 1 - \exp\left(-\frac{U_m A}{Q}\right) \tag{7.40}$$

where A is the area of the electrode. Equation (7.40) is called the Deutsch equation. It indicates that the collection efficiency of an electrostatic precipitator increases with increasing area of the electrodes and the migration velocity, and decreasing flow rate of gas. The actual collection efficiency of an electrostatic precipitator is also greatly affected by the adhesion of particles on the collecting surface. This adhesion may dramatically change the original electric field by sharply increasing the resistivity of the electrode; severe "coating" and high resistivity of particles may even reverse the ionization or reverse the corona discharge [White, 1963; Svarovsky, 1981; Ogawa, 1984].

Example 7.2 Consider a dust-laden gas passing through an electrostatic precipitator. Assume that all particles have the same charge-to-mass ratio, q/m, and the electric field is uniform. Use the Deutsch equation to estimate the overall collection efficiency if the particle size distribution based on weight percentage (wt%) can be expressed by the mass density function, f_M, as

$$f_M = \begin{cases} 0 & 0 < d_p < d_{pm} - \delta \\ \dfrac{3}{4\delta}\left(1 - \dfrac{(d_p - d_{pm})^2}{\delta^2}\right) & d_{pm} - \delta \le d_p \le d_{pm} + \delta \\ 0 & d_{pm} + \delta < d_p < \infty \end{cases} \tag{E7.2}$$

where d_{pm} and δ are the density function parameters.

Solution From Eq. (7.35), the migration velocity can be expressed as

$$U_{\mathrm{m}} = \frac{\rho_{\mathrm{p}} E}{18\mu}\left(\frac{q}{m}\right) d_{\mathrm{p}}^2 \tag{E7.3}$$

Therefore, the fractional collection efficiency is given by Eq. (7.40) as

$$\eta_{\mathrm{ei}} = 1 - \exp\left(-\frac{\rho_{\mathrm{p}} E}{18\mu}\left(\frac{q}{m}\right)\frac{A}{Q} d_{\mathrm{p}}^2\right) \tag{E7.4}$$

Given the mass density function of particles, the overall collection efficiency of an electrostatic precipitator can be expressed as

$$\eta_{\mathrm{e}} = \int_0^\infty f_{\mathrm{M}} \eta_{\mathrm{ei}} \, \mathrm{d}d_{\mathrm{p}} \tag{E7.5}$$

To simplify the derivation, let

$$x = \frac{d_{\mathrm{p}}}{\delta}; \qquad \beta = \frac{d_{\mathrm{pm}}}{\delta}; \qquad \xi^2 = \frac{\rho_{\mathrm{p}} \delta^2 E}{18\mu}\left(\frac{q}{m}\right)\frac{A}{Q} \tag{E7.6}$$

The overall collection efficiency of this electrostatic precipitator is

$$\begin{aligned}
\eta_{\mathrm{e}} &= \int_{\beta-1}^{\beta+1} \frac{3}{4}[1 - (x-\beta)^2][1 - \exp(-\xi^2 x^2)]\,\mathrm{d}x \\
&= 1 - \frac{3}{4}\int_{\beta-1}^{\beta+1} [(1-\beta^2) + 2\beta x - x^2]\exp(-\xi^2 x^2)\,\mathrm{d}x \\
&= 1 - \frac{3\sqrt{\pi}}{8\xi}\left(1 - \beta^2 - \frac{1}{2\xi^2}\right)\{\mathrm{erf}(\xi(\beta+1)) - \mathrm{erf}(\xi(\beta-1))\} \\
&\quad + \frac{3(\beta+1)}{8\xi^2}\exp(-\xi^2(\beta-1)^2) - \frac{3(\beta-1)}{8\xi^2}\exp(-\xi^2(\beta+1)^2)
\end{aligned} \tag{E7.7}$$

7.4 Filtration

In filtration, solid particles are separated from the gas stream when the gas–solid suspension passes through a single-layer screen, a multiple-layer screen, or a porous, permeable medium which retains the solid particles. This separation method is simple and has a high efficiency in collecting particles of all sizes. However, in this type of separation method, a high pressure drop may develop as a result of the accumulation of collected particles in the filter, which may yield high energy consumption and require high material strength of filters for the operation.

7.4.1 *Mechanisms of Filtration and Types of Filters*

On the basis of particle collection characteristics, filtration can be classified into cake filtration and depth filtration. In cake filtration, particles are deposited on the front surface of the collecting filter, as shown in Fig. 7.13(a). Filtration of this type is achieved mainly as an effect of screening. This is by far the most common type employed in the chemical and process industries. In depth filtration, particles flow through the filter and are collected

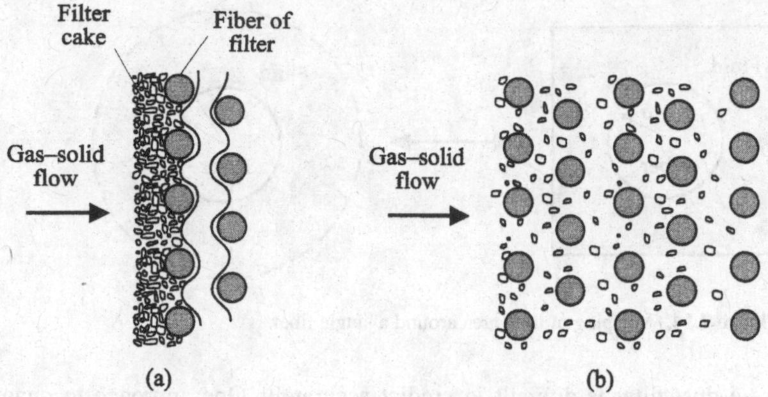

Figure 7.13. Basic types of filtration: (a) Cake filtration; (b) Depth filtration.

during their passage, as illustrated in Fig. 7.13(b). Examples of filters in cake filtration include fabric filters, sieves, nuclepore filters, and membrane filters. Filters for depth filtration include packed beds, meshes, fluidized beds, and glass-fiber filters. In both cake filtration and depth filtration, continuous particle collection increases the flow resistance through the filter, and, thus, the collection efficiency decreases. Efficient operation of the filter requires periodic removal of the particles deposited on the filter or replacement of the filter.

Filtration is a physical separation whereby particles are removed from the fluid and retained by the filters. Three basic collection mechanisms involving fibers are inertial impaction, interception, and diffusion. In collection by inertial impaction, the particles with large inertia deviate from the gas streamlines around the fiber collector and collide with the fiber collector. In collection by interception, the particles with small inertia nearly follow the streamline around the fiber collector and are partially or completely immersed in the boundary layer region. Subsequently, the particle velocity decreases and the particles graze the barrier and stop on the surface of the collector. Collection by diffusion is very important for fine particles. In this collection mechanism, particles with a zig-zag Brownian motion in the immediate vicinity of the collector are collected on the surface of the collector. The efficiency of collection by diffusion increases with decreasing size of particles and suspension flow rate. There are also several other collection mechanisms such as gravitational sedimentation, induced electrostatic precipitation, and van der Waals deposition; their contributions in filtration may also be important in some processes.

Filters can be grouped on the basis of different characteristics. For example, they can be grouped on the basis of the design capacity of the filters according to the particle concentration or loading [Svarovsky, 1981]; or they can be grouped on the basis of the materials of the filters, *e.g.*, fabric or nonfabric [Cooper and Freeman, 1982]. Most filters in use are bag filters, which are fabric. The common fabric materials include cotton, polyester, wool, asbestos, glass, acrylic, polytetrafluoroethylene (Teflon), poly (m-phenylene isophthalate) (Nomex), polycaprolactam (Nylon), and polypropylene.

7.4.2 *Pressure Drop in a Filter*

The flow pattern in a filter is complex as a result of the intricate structure of the fiber, which is further complicated by the particle deposition. Consequently, the pressure

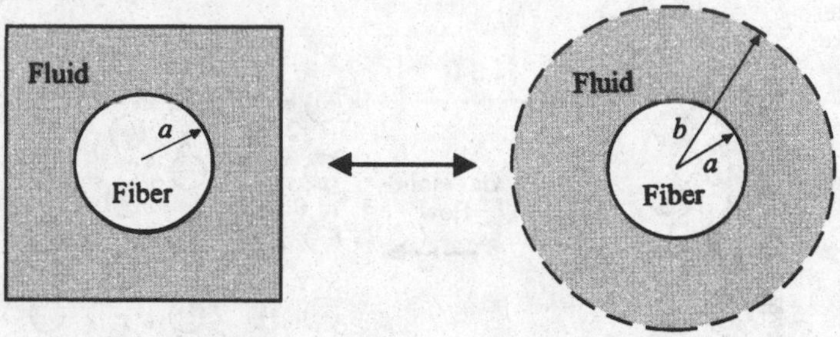

Figure 7.14. Mapping of flow area around a single fiber.

drop across a dust filter is difficult to predict accurately. One approach to quantify the pressure drop is to treat the flow as comprising several simple flows around the fibers.

Assume that the fibers in a filter are cylindrical; they are parallel to each other and are uniformly assembled. Consider cake filtration in which particles are collected with the deposited particles forming a layer of porous structure as shown in Fig. 7.13(a). Thus, to account for the total pressure drop, three basic flow modes are pertinent: (1) flow is parallel to the axis of fibers; (2) flow is perpendicular to the axis of the cylinder; and (3) flow passes through a layer of a homogeneous porous medium. In the analysis of the first two modes given later, Happel's model [Happel, 1959] is used, while for the third mode, Ergun's approach [Ergun, 1952] is used.

First, consider the case where the flow is parallel to the cylinders. It is assumed that the fluid is moving through the annular space between the cylinder of radius a and the fluid envelope of equivalent radius b, as shown in Fig. 7.14. Assume that the fluid motion is in the creeping flow regime so that inertia terms can be omitted from the Navier–Stokes equations. Thus, in cylindrical coordinates, we have

$$\frac{1}{r}\frac{\mathrm{d}}{\mathrm{d}r}\left(r\frac{\mathrm{d}W}{\mathrm{d}r}\right) = \frac{1}{\mu}\frac{\mathrm{d}p}{\mathrm{d}z} \tag{7.41}$$

with the boundary conditions

$$W|_{r=a} = 0; \qquad \left(\frac{\mathrm{d}W}{\mathrm{d}r}\right)_{r=b} = 0 \tag{7.42}$$

The axial fluid velocity inside the annular space is obtained as

$$W = -\frac{1}{4\mu}\frac{\mathrm{d}p}{\mathrm{d}z}\left(a^2 - r^2 + 2b^2\ln\frac{r}{a}\right) \tag{7.43}$$

The fluid flow rate through the entire annulus is given by

$$Q = -\frac{\pi}{8\mu}\frac{\mathrm{d}p}{\mathrm{d}z}\left(4a^2b^2 - a^4 - 3b^4 + 4b^4\ln\frac{b}{a}\right) \tag{7.44}$$

Denote α_f as the volume fraction of fibers and A as the total cross-sectional area. In this case, $\alpha_\mathrm{f} = a^2/b^2$ and $A = \pi b^2$. From Eq. (7.44), the pressure drop over the filter thickness L is obtained as

$$\frac{\Delta p}{L} = \frac{Q\mu}{Aa^2}\frac{4\alpha_\mathrm{f}}{(2\alpha_\mathrm{f} - \alpha_\mathrm{f}^2/2 - 3/2 - \ln\alpha_\mathrm{f})} \tag{7.45}$$

Now, consider the case where the flow is perpendicular to the cylinders. The cylinder may be treated as being stationary as a result of the force balance between the pressure gradient and the drag force. Thus, the pressure drop may be obtained from the drag force on a single cylinder. Alternatively, we can consider the case where a cylinder of radius a moves perpendicular to its axis in a fluid cell of radius b. It is assumed that no shear stress exists on the outside cylindrical fluid surface. The drag force in this case is noted to be the same as that for a fluid moving perpendicular to a stationary cylinder of an identical radius. In cylindrical coordinates, the continuity equation can be given as

$$\frac{\partial U}{\partial r} + \frac{U}{r} + \frac{1}{r}\frac{\partial V}{\partial \theta} = 0 \tag{7.46}$$

where U and V are the radial and tangential components of the fluid velocity, respectively. The momentum equation is given by

$$\begin{aligned} \frac{\partial p}{\partial r} &= \mu\left(\nabla^2 U - \frac{U}{r^2} - \frac{2}{r^2}\frac{\partial V}{\partial \theta}\right) \\ \frac{1}{r}\frac{\partial p}{\partial \theta} &= \mu\left(\nabla^2 V - \frac{V}{r^2} + \frac{2}{r^2}\frac{\partial U}{\partial \theta}\right) \end{aligned} \tag{7.47}$$

It is convenient to introduce the stream function, ψ, defined by

$$U = \frac{1}{r}\frac{\partial \psi}{\partial \theta}; \qquad V = -\frac{\partial \psi}{\partial r} \tag{7.48}$$

Equation (7.47) reduces to

$$\nabla^4 \psi = 0 \tag{7.49}$$

which has a general solution of the form

$$\psi = \sin\theta\left(\frac{1}{8}C_1 r^3 + \frac{1}{2}C_2 r\left(\ln r - \frac{1}{2}\right) + C_3 r + \frac{C_4}{r}\right) \tag{7.50}$$

where the C's are constants to be determined by the boundary conditions. The boundary conditions in this case are

$$\begin{aligned} &U|_{r=a} = U_0\cos\theta; \qquad V|_{r=a} = -U_0\sin\theta \\ &U|_{r=b} = 0; \qquad \left(\frac{\partial V}{\partial r} + \frac{1}{r}\frac{\partial U}{\partial \theta} - \frac{V}{r}\right)_{r=b} = 0 \end{aligned} \tag{7.51}$$

where U_0 is the velocity of the moving cylinder. Therefore, the C's are determined as

$$C_1 = \frac{4a^2 U_0}{(a^4+b^4)}\left[\ln\frac{b}{a} + \frac{a^4}{a^4+b^4} - \frac{1}{2}\right]^{-1} \tag{7.52}$$

$$C_2 = -2U_0\left[\ln\frac{b}{a} + \frac{a^4}{a^4+b^4} - \frac{1}{2}\right]^{-1} \tag{7.53}$$

$$C_3 = U_0\left(\ln b - \frac{1}{2}\right)\left[\ln\frac{b}{a} + \frac{a^4}{a^4+b^4} - \frac{1}{2}\right]^{-1} \tag{7.54}$$

$$C_4 = \frac{-a^2 b^4 U_0}{2(a^4+b^4)}\left[\ln\frac{b}{a} + \frac{a^4}{a^4+b^4} - \frac{1}{2}\right]^{-1} \tag{7.55}$$

Therefore, the drag force on a single cylinder, F_D, is found to be

$$F_D = -4\pi\mu U_0 \left[\ln\frac{b}{a} + \frac{a^4}{a^4+b^4} - \frac{1}{2}\right]^{-1} \tag{7.56}$$

Furthermore, the averaged pressure gradient over a single cylinder may be estimated from the force balance on the cylinder as

$$\frac{dp}{dz} = \frac{F_D}{\pi b^2} = -\frac{4\mu U_0}{b^2}\left[\ln\frac{b}{a} + \frac{a^4}{a^4+b^4} - \frac{1}{2}\right]^{-1} \tag{7.57}$$

Note that the volume fraction of the filter is $\alpha_f = a^2/b^2$ and the superficial velocity is $U = Q/A$. Thus, for a flow which is transverse to the cylindrical fibers, the pressure drop over a filter of thickness L is given by

$$\frac{\Delta p}{L} = -\frac{8Q\mu\alpha_f}{a^2 A}\left(\ln\alpha_f + \frac{1-\alpha_f^2}{1+\alpha_f^2}\right)^{-1} \tag{7.58}$$

For flow through a layer of a porous medium, the pressure drop can be estimated from Eq. (5.358), as discussed in §5.6. In the case of a random assemblage of fibers with a layer of dust, the total pressure drop for a fabric filter in cake filtration may be expressed in the form

$$\Delta p_t = \Delta p_3 + \gamma_1\Delta p_1 + \gamma_2\Delta p_2 \tag{7.59}$$

where Δp_1, Δp_2, and Δp_3 are the pressure drops contributed by each of the three different basic flow modes given. γ_1 and γ_2 are the weighting factors. In random assemblage flows, it is necessary to give twice the weight to the correlation for the flow perpendicular to cylinders as to that for the flow parallel to cylinders since the model for the flow perpendicular to cylinders does not distinguish between crossed and parallel arrangements of the cylinders [Happel, 1959].

Example 7.3 Consider a clean house bag filter filled with randomly assembled fibers. The porosity of the filter is 0.97, and the fibers have an effective diameter of 10 μm. The length of the filter is 0.1 m. If the flow velocity of the dust-laden air is 0.2 m/s, estimate the pressure drop across the filter. The viscosity of air is 1.8×10^{-5} kg/m · s.

Solution The overall pressure drop is a combined result of flow parallel to the fibers and flow perpendicular to the fibers. The pressure drop of flow parallel to the fibers is estimated from Eq. (7.45) as

$$\begin{aligned}\Delta p_\parallel &= \frac{16LU\mu\alpha_f}{d_f^2(2\alpha_f - \alpha_f^2/2 - 3/2 - \ln\alpha_f)} \\ &= \frac{16\times 0.1\times 0.2\times 1.8\times 10^{-5}\times 0.03}{10^{-10}(2\times 0.03 - (0.03)^2/2 - 1.5 - \ln 0.03)} = 836\ \text{Pa}\end{aligned} \tag{E7.8}$$

and the pressure drop of flow perpendicular to the fibers is estimated from Eq. (7.58) as

$$\Delta p_\perp = -\frac{32LU\mu\alpha_f}{d_f^2}\left(\ln\alpha_f + \frac{1-\alpha_f^2}{1+\alpha_f^2}\right)^{-1}$$
$$= -\frac{32\times 0.1\times 0.2\times 1.8\times 10^{-5}\times 0.03}{10^{-10}}\left(\ln 0.03 + \frac{1-0.03^2}{1+0.03^2}\right)^{-1}$$
$$= 1{,}378 \text{ Pa} \tag{E7.9}$$

For a flow through a random assemblage of fibers, it is necessary to give twice the weight to the pressure drop from the flow perpendicular to fibers as that from the flow parallel to fibers; therefore, the overall pressure drop is obtained as

$$\Delta p_t = \Delta p_\parallel + 2\Delta p_\perp = 836 + 2\times 1{,}378 = 3{,}592 \text{ Pa} \tag{E7.10}$$

7.4.3 *Collection Efficiency of Fabric Filters*

The collection efficiency of a fiber is defined as the ratio of the number of particles striking the fiber to the number which would strike it if the streamlines were not diverted [Dorman, 1966]. For particles with $d_p > 1\ \mu$m, the particles are collected primarily by inertial impaction and interception. Collection by diffusion is only important for submicron particles. In the following, we only discuss the methodology of collection efficiency by interception. Details about the filtration of fine and ultrafine powders or granular filtration in general can be found in Davies (1973), Fuchs (1964), Dickey (1961), Matteson and Orr (1987), and Tien (1989).

Consider the case of flow transverse to a cylindrical fiber, as shown in Fig. 7.15. The

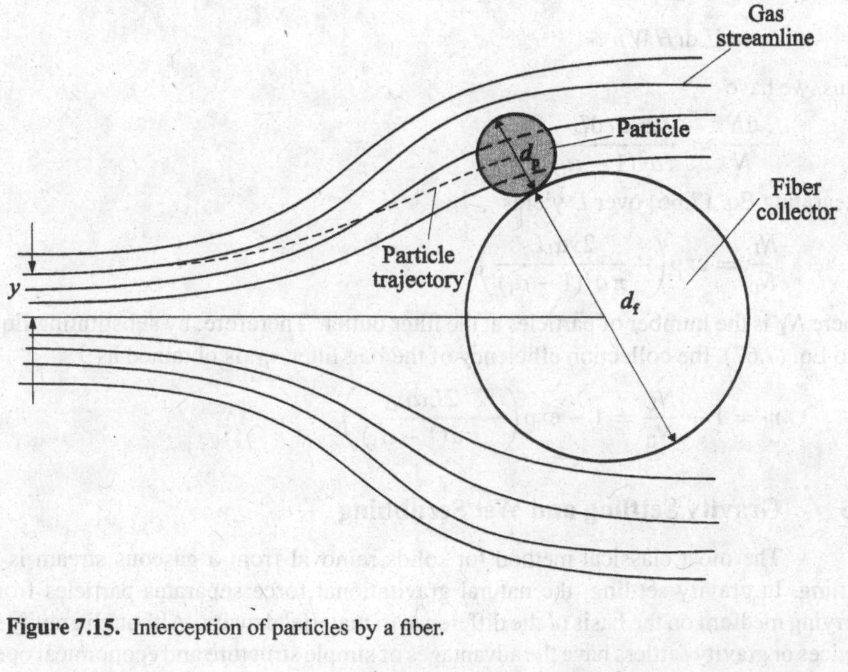

Figure 7.15. Interception of particles by a fiber.

collection efficiency of a fiber, η_i, can be directly given by its definition

$$\eta_i = \frac{y}{a} \tag{7.60}$$

where y is the limiting width of the particle stream to be collected. y can be calculated from the particle trajectories in the flow around the fiber. In general, the equation of motion for the particle is coupled with the momentum equation for the gas phase. The complexity of the coupled equations requires, in most cases, that the particle trajectory be solved numerically.

To determine the collection efficiency for a bag filter of thickness L, height H, and width W, it is assumed that the fibers are loosely packed and, thus, the interaction between the fibers can be neglected. Consider the flow moving transverse to the cylinders. The interstitial velocity u is given by

$$u = \frac{U}{1-\alpha_f} \tag{7.61}$$

where U is the superficial velocity of the dust-laden gas. In an element of thickness dL, the number of fibers N_f is

$$N_f = \frac{\mathrm{d}LH\alpha_f}{\pi a^2} \tag{7.62}$$

Within a time interval dt, the number of particles collected by a single fiber is

$$-\mathrm{d}N_1 = 2yu\,\mathrm{d}tWn \tag{7.63}$$

Thus, within a time interval dt the total number of particles collected by all fibers in the element becomes

$$-\mathrm{d}N = 2yu\,\mathrm{d}tWnN_f \tag{7.64}$$

The total number of particles entering within time dt is

$$N = U\,\mathrm{d}tHWn \tag{7.65}$$

Thus, we have

$$-\frac{\mathrm{d}N}{N} = \frac{2y\alpha_f\,\mathrm{d}L}{\pi a^2(1-\alpha_f)} \tag{7.66}$$

Integrating Eq. (7.66) over L yields

$$\frac{N_L}{N_0} = \exp\left(-\frac{2y\alpha_f L}{\pi a^2(1-\alpha_f)}\right) \tag{7.67}$$

where N_L is the number of particles at the filter outlet. Therefore, by substituting Eq. (7.60) into Eq. (7.67), the collection efficiency of the bag filter, η_f, is obtained as

$$\eta_f = 1 - \frac{N_L}{N_0} = 1 - \exp\left(-\frac{2L\alpha_f\eta_i}{\pi a(1-\alpha_f)}\right) \tag{7.68}$$

7.5 Gravity Settling and Wet Scrubbing

The most classical method for solids removal from a gaseous stream is gravity settling. In gravity settling, the natural gravitational force separates particles from their carrying medium on the basis of the difference in material density. Although gravity settling devices or gravity settlers have the advantages of simple structure and economical operation, their development has been nearly halted by the low collection efficiency (typically 20–60 percent for $d_p > 40\ \mu m$) and the large collection chamber requirement compared to that of cyclones, fabric filters, wet scrubbers, and electrostatic precipitators. Nevertheless, gravity settlers are still used in industry as precleaners to remove large particles.

Wet scrubbing is another technique to remove solids from a gas stream. In wet scrubbing, solid particulates collide with liquid droplets generated from the nozzles in the scrubbers. The collected particles and liquid droplets then form a slurry which flows to the slurry collection chamber. Water is the most commonly used working liquid in wet scrubbing. The scrubbers can be installed inside existing separators such as cyclones, gravity settlers, and bed filters, or as an independent unit. The advantages of wet scrubbing include low pressure drop, relatively high efficiency for fine particle collection, no particle reentrainment, and possible removal of some gaseous pollutants simultaneously by using chemical solutions as the collecting liquid. The disadvantage of this separation method is the necessity for the treatment of the wet sludge generated from the wet scrubbing process.

7.5.1 *Gravity Settling Chambers*

In gravity settling separation, the particle-laden gas is fed horizontally into a large expansion chamber. The enlargement of the cross section of the flow stream significantly reduces the gas velocity. The particles settle downward on the collection surface or hopper collectors at the bottom of the chamber. The decrease in the gas stream velocity reduces the reentrainment of collected particles and increases the residence time of the particles so that they may have sufficient time to settle by gravity.

There are two basic types of gravity settlers: the simple expansion chamber or horizontal flow settling chamber (Fig. 7.16(a)) and the multiple-tray settling chamber or Howard settling chamber (Fig. 7.16(b)). In the multiple-tray settling chamber, the presence of thin collection plates yields a slightly larger linear gas velocity and significantly shorter vertical settling distance compared to those in the simple expansion chamber. Thus, the overall collection efficiency of the multiple-tray settling chamber is higher than that of the simple expansion chamber. For a given collection efficiency, the size of the multiple-tray settler is smaller than that of the simple expansion settler. A slight variation of the gravity settling chamber, known as an elutriator, is also frequently used in industry. A system of three elutriators in series is schematically illustrated in Fig. 7.17.

For a given type of gravity settler, the pick-up velocity and the minimum diameter of particles which can be collected must be defined. The pick-up velocity is the onset velocity for particle reentrainment and depends on the particle diameter. The gas flow rate is to be controlled in such a way that the average gas velocity in the settling chamber is much smaller than the pick-up velocity. The minimum diameter of collected particles can be evaluated by the following analysis.

Consider a small spherical particle settling in a two-dimensional (2-D) rectangular-shaped simple expansion chamber. The settling chamber is of length L and height H. Assume that (1) the gas velocity U is uniformly distributed throughout the chamber; (2) there is no slip between the particle and gas in the horizontal direction; (3) the particle settles down at its terminal velocity; and (4) the particle is subjected to the Stokes drag. Thus, the residence time of the small particle, t_r, is given by

$$t_r = \frac{L}{U} \tag{7.69}$$

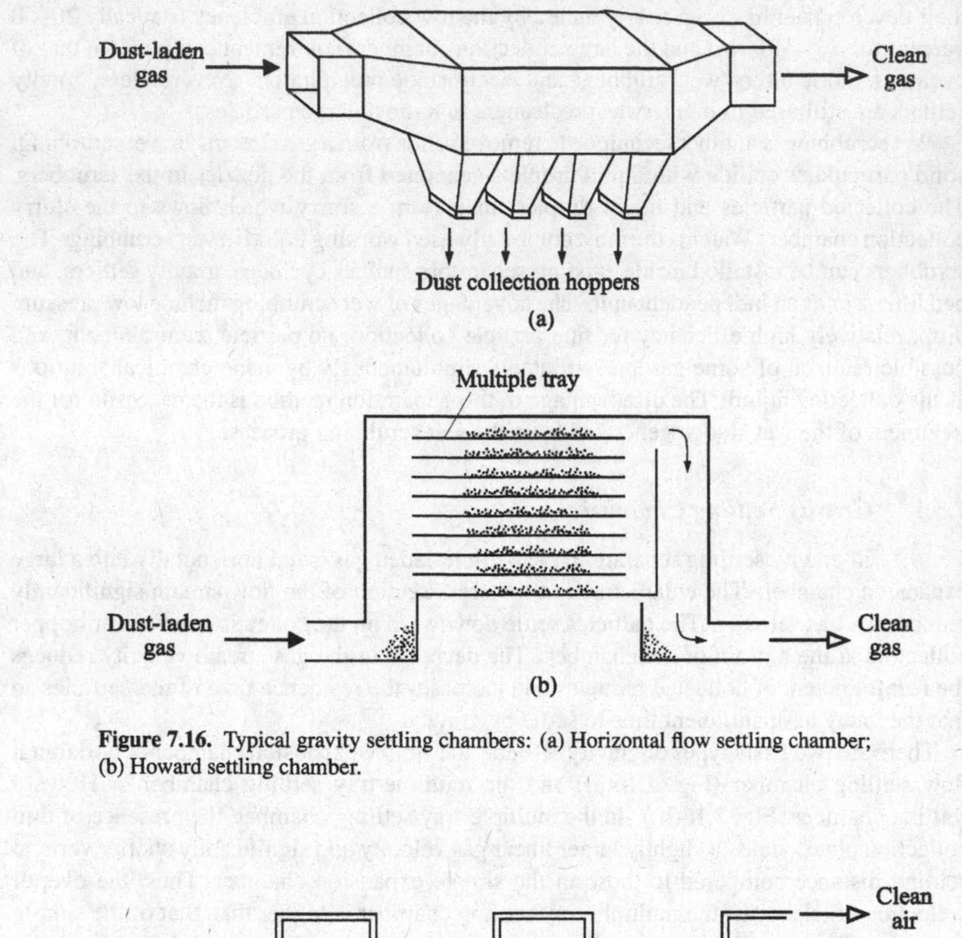

Figure 7.16. Typical gravity settling chambers: (a) Horizontal flow settling chamber; (b) Howard settling chamber.

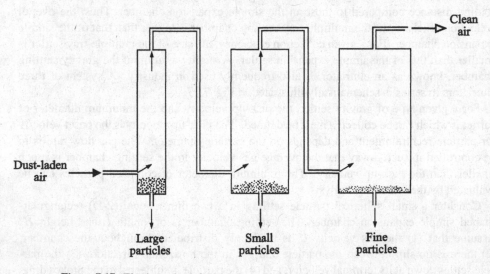

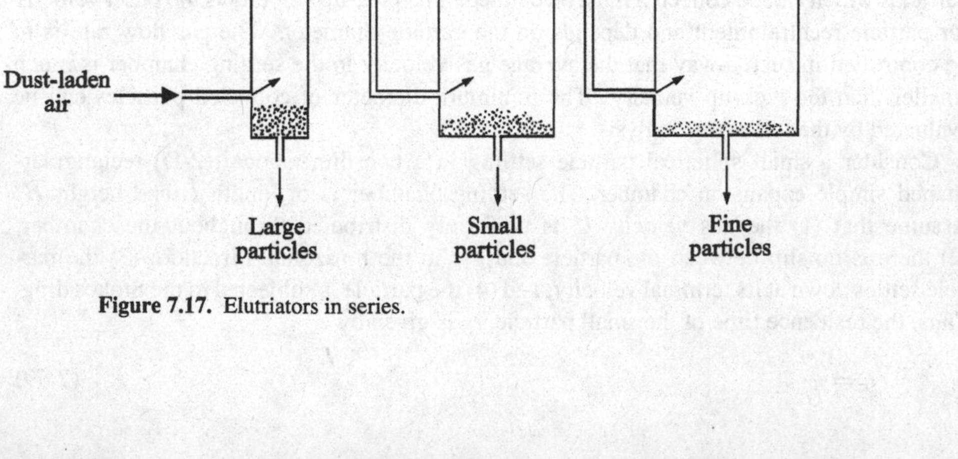

Figure 7.17. Elutriators in series.

and the settling time, t_s, can be estimated by

$$t_s = \frac{y}{U_{pt}} \tag{7.70}$$

where y is the initial vertical distance between the particle and the collection plate and U_{pt} is the particle terminal velocity given by Eq. (1.7). To ensure that the particles are collected, the maximum settling time of the particle, t_{sm}, should be less than the particle residence time in the chamber, *i.e.*,

$$t_{sm} = \frac{H}{U_{pt}} \le t_r \tag{7.71}$$

Substituting Eq. (1.7) and Eq. (7.69) into Eq. (7.71) leads to the following expression for the diameter of the particle that can be collected as

$$d_p^2 \ge \frac{18\mu U H}{L(\rho_p - \rho)g} \tag{7.72}$$

When the gas velocity in the settler is sufficiently high, the particles deposited on the collection plates can be picked up by the gas stream. This reentrainment can substantially reduce the collection efficiency. By neglecting interparticle friction, the pick-up velocity, U_{pp}, may be estimated as [Zenz and Othmer, 1960]

$$U_{pp} = \sqrt{\frac{4gd_p}{3}\left(\frac{\rho_p}{\rho} - 1\right)} \tag{7.73}$$

In actual applications, the gas flow in a gravity settler is often nonuniform and turbulent; the particles are polydispersed; and the flow is beyond the Stokes regime. In this case, the particle settling behavior and hence the collection efficiency can be described by using the basic equations introduced in Chapter 5, which need to be solved numerically. One common approach is to use the Eulerian method to represent the gas flow and the Lagrangian method to characterize the particle trajectories. The random variations in the gas velocity due to turbulent fluctuations and the initial entering locations and sizes of the particles can be accounted for by using the Monte Carlo simulation. Examples of this approach were provided by Theodore and Buonicore (1976).

7.5.2 *Mechanisms of Scrubbing and Types of Scrubbers*

Wet scrubbing uses liquid droplets to remove fine dust in a gas stream. In all types of wet scrubbing, the basic cleaning mechanism involves the attachment of particles to the droplets. The function of the droplets in scrubbers is similar to that of spherical fibers in filtration. Likewise, the primary collection mechanisms in scrubbing are similar to those in filtration, *i.e.*, inertial impaction, interception, and diffusion [*e.g.*, Fan, 1989]. Secondary collection mechanisms include thermophoresis due to temperature gradients, coagulation of particles due to particle electrification, and particle growth due to liquid condensation.

Almost all particle-separating devices can be converted into wet scrubbers by adding liquid spraying systems. Three types of commonly used scrubbers are the spray chamber, cyclonic scrubber, and venturi scrubber. Figure 7.18 shows a simple spray chamber in which water is sprayed through a series of nozzles into a settling chamber. The dust-laden gas is fed from the bottom of the chamber and exits from the upper portion of the chamber.

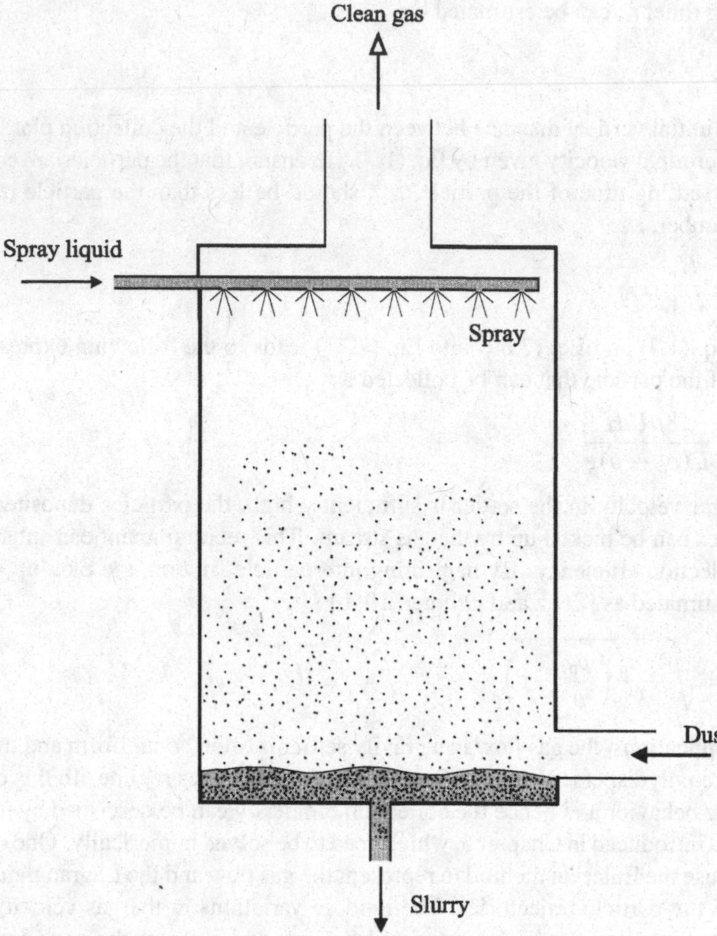

Figure 7.18. Spray chamber.

The collection of particles is achieved in a countercurrent flow between the water droplets and the particulates. In a cyclonic scrubber, water is injected into the cyclone chamber from sprayers located along the central axis, as shown in Fig. 7.19. The water droplets capture particles mainly in the cross-flow motion and are thrown to the wall by centrifugal force, forming a layer of slurry flow moving downward to the outlet at the bottom of the cyclone. Another type of scrubber employs a venturi, as shown in Fig. 7.20. The velocity of the gas–solid suspension flow is accelerated to a maximum value at the venturi throat. The inlet of the water spray is located just before the venturi throat so that the maximum difference in velocity between droplets and particles is obtained to achieve higher collection efficiency by inertial impaction. A venturi scrubber is usually operated with a particle collector such as a settling chamber or cyclone for slurry collection.

7.5.3 *Modeling for Scrubbing and Collection Efficiency*

In wet scrubbing, water droplets collide with particles in three flow modes of operation: (1) cross-flow; (2) cocurrent flow; and (3) countercurrent flow. Here the mathematical

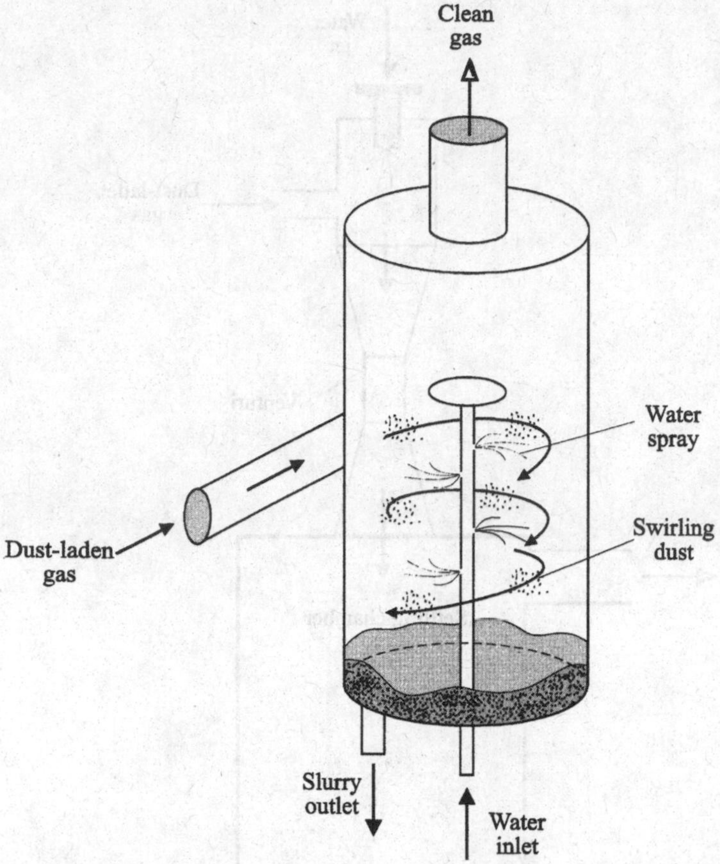

Figure 7.19. Cyclone scrubber.

model proposed by Fonda and Herne [Ogawa, 1984] is introduced to describe each of these flow modes.

A model for wet scrubbing in a cross-flow is illustrated in Fig. 7.21. Consider a rectangular scrubbing domain of length L, height H, and width of unity in Cartesian coordinates. Assume that the gas–solid suspension flow is moving horizontally, and that the solid particles are spherical and of uniform size. The particle concentration across any plane perpendicular to the flow is assumed to be uniform. The water droplets fall vertically and are uniformly distributed in the flow system.

The mass balance of particles in the volume element in Fig. 7.21 can be given as

$$U_p H \rho_p n - U_p H \rho_p \left(n + \frac{dn}{dx} dx \right) - U_w H \eta_w \alpha_w \left(\frac{3}{2d_w} \right) \rho_p n \, dx = 0 \qquad (7.74)$$

where d_w is the averaged diameter of the water droplets; α_w is the volume fraction of droplets; $3/2d_w$ is the ratio of droplet cross-sectional area to droplet volume so that $(3/2d_w)H\,dx\,\alpha_w$ represents the total cross-sectional area of the water droplets in that volume element; η_w is the collection efficiency of a single water droplet; U_p and U_w are velocities of particles and

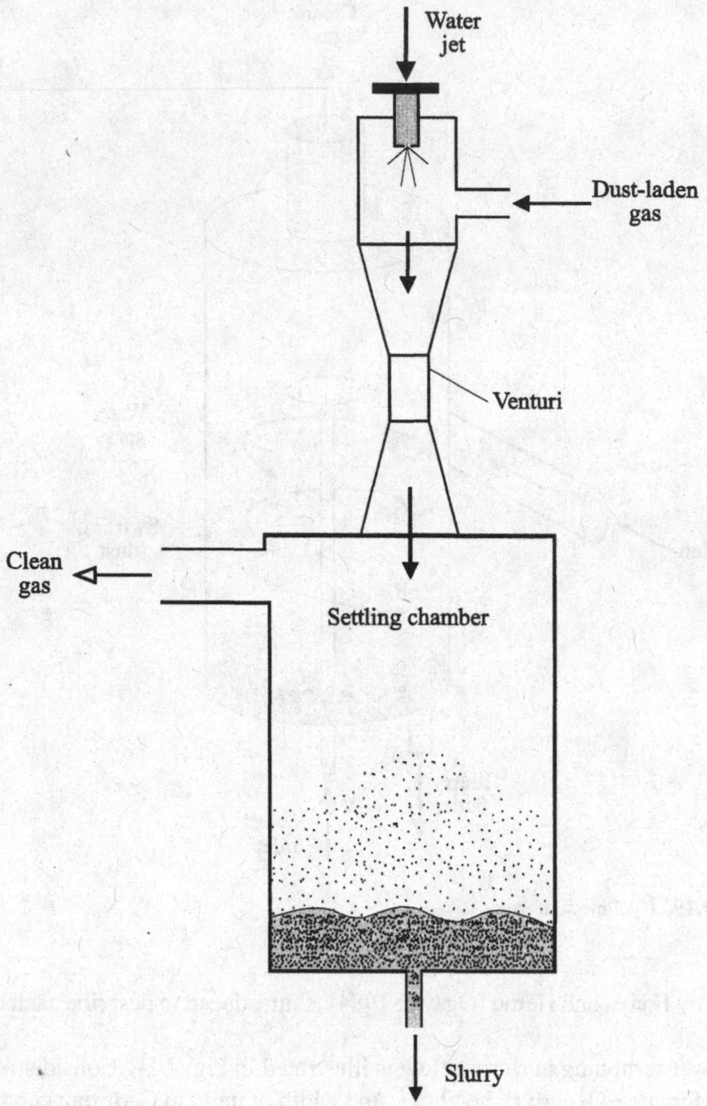

Figure 7.20. Venturi scrubber.

droplets, respectively; and ρ_p is the particle density. Thus, we have

$$\frac{dn}{n} = -\frac{3}{2d_w}\frac{U_w}{U_p}\eta_w\alpha_w\, dx = -\frac{3}{2d_w}\frac{Q_w}{Q_p}\frac{H}{L}\eta_w\alpha_{pi}\, dx \tag{7.75}$$

where Q_p and Q_w are the flow rates of the particles and the droplets, respectively; and α_{pi} is the particle volume fraction at inlet. The total collection efficiency, η_s, is thus obtained from the integration of Eq. (7.75) as

$$\eta_s = 1 - \frac{n_o}{n_i} = 1 - \exp\left(-\frac{3}{2}\frac{Q_w}{Q_p}\frac{H}{d_w}\eta_w\alpha_{pi}\right) \tag{7.76}$$

where n_i and n_o denote the particle number densities at inlet and outlet, respectively.

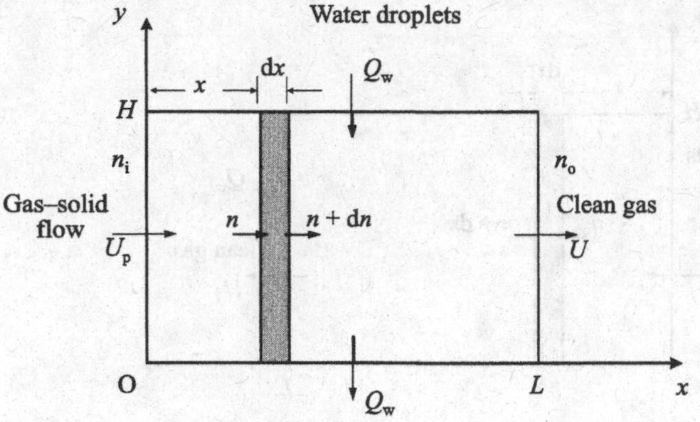

Figure 7.21. Cross-flow mode.

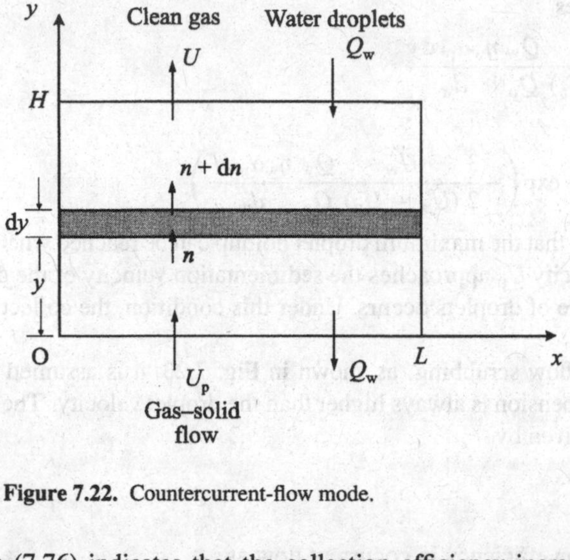

Figure 7.22. Countercurrent-flow mode.

Equation (7.76) indicates that the collection efficiency increases with the height of the scrubbing chamber and with increasing ratio of flow rate of droplets to flow rate of particles.

For the vertical countercurrent scrubbing system shown in Fig. 7.22, the gas–solid suspension flows upward from the bottom of the chamber and the water droplets fall downward. It is assumed that both the gas–solid suspension velocity and droplet velocity are unchanged in the process of scrubbing. It is also assumed that the no-slip condition exists between the particles and gas. The particle concentration across any horizontal plane is assumed to be uniform.

Similarly to the case of cross-flow, the mass balance of particles in the volume element shown in Fig. 7.22 is expressed by

$$U_p L \rho_p n - U_p L \rho_p \left(n + \frac{dn}{dy} \, dy \right) - U_w L \eta_w \alpha_w \left(\frac{3}{2d_w} \right) \rho_p n \, dy = 0 \tag{7.77}$$

where α_w is

$$\alpha_w = \frac{Q_w}{(U_w - U_p)L} \tag{7.78}$$

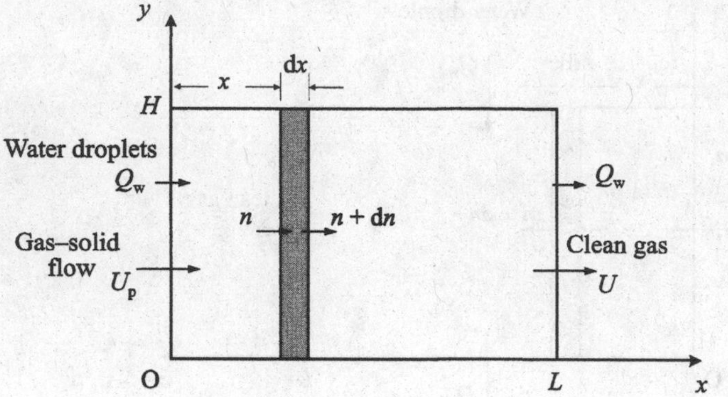

Figure 7.23. Cocurrent-flow mode.

Equation (7.77) thus becomes

$$\frac{\mathrm{d}n}{n} = -\frac{3}{2}\frac{U_w}{(U_w - U_p)}\frac{Q_w}{Q_p}\frac{\eta_w \alpha_{pi}\,\mathrm{d}y}{d_w} \tag{7.79}$$

which leads to

$$\eta_s = 1 - \frac{n_o}{n_i} = 1 - \exp\left(-\frac{3}{2}\frac{U_w}{(U_w - U_p)}\frac{Q_w}{Q_p}\frac{\eta_w \alpha_{pi} H}{d_w}\right) \tag{7.80}$$

It is noted from Eq. (7.78) that the maximum droplet holdup can be reached when the upward gas–solid mixture velocity U_p approaches the sedimentation velocity of the droplets, assuming that no coalescence of droplets occurs. Under this condition, the collection efficiency is maximum.

In the case of cocurrent flow scrubbing, as shown in Fig. 7.23, it is assumed that the velocity of the gas–solid suspension is always higher than the droplet velocity. The volume fraction of droplets is then given by

$$\alpha_w = \frac{Q_w}{(U_p - U_w)L} \tag{7.81}$$

which results in the collection efficiency in cocurrent flow as

$$\eta_s = 1 - \exp\left(-\frac{3}{2}\frac{U_w}{(U_p - U_w)}\frac{Q_w}{Q_p}\frac{\eta_w \alpha_{pi} H}{d_w}\right) \tag{7.82}$$

It is interesting to note that the volume fraction and collection efficiency in cocurrent scrubbing are in the same form as those in countercurrent scrubbing even though the flow directions of the dust-laden gas in these two cases are opposite to each other.

Nomenclature

A	Constant, defined by Eq. (7.7)	a	Radius of fiber
A	Area of electrode	a	Dimension, defined by Fig. 7.8
A_i	Area, dimension, defined by Fig. 7.8	b	Equivalent radius of flow cell around a single fiber

C	Constant, defined by Eq. (7.9)
d_f	Diameter of fiber
d_p	Particle diameter
d_{pc}	Cut-off size of a cyclone
d_{pm}	Parameter, defined by Eq. (E7.2)
d_w	Diameter of water droplet
E	Electric field strength
F_D	Drag force
f_M	Mass density function
g	Gravitational acceleration
H	Height of a rectangular collecting plate
H	Cyclone dimension, defined by Fig. 7.8
H	Height of a scrubber
H	Height of the settling chamber
H	Height of the bag filter
h	Cyclone dimension, defined by Fig. 7.8
L	Length of a rectangular collecting plate
L	Length of a scrubber
L	Thickness of the filter
L	Length of the settling chamber
l	Deepest descending length
l_w	Distance between two collecting plates
m	Mass of a particle
N	Number of particles
N_0	Number of particles at $t = 0$
N_1	Number of particles collected by a single fiber
N_f	Number of fibers
N_L	Number of particles at outlet
N_m	Number of particles at $t = t_m$
n	Particle number density
n_i	Particle number density at the separator inlet
n_o	Particle number density at the separator outlet
p	Pressure
p_0	Static pressure at the center of the cyclone
Q	Gas flow rate
Q_p	Particle flow rate
Q_w	Droplet flow rate
q	Charge carried by a particle
R	Radius of cross section of the cyclone
R	Maximum radius of the cyclone
Re_w	Reynolds number based on U_w
r	Radial coordinate
r^*	Dimensionless radial coordinate
r_0	Radial location of particles at $t = 0$
r_b	Cyclone dimension, defined by Fig. 7.8
r_e	Cyclone dimension, defined by Fig. 7.8
r_f	Radius of free vortex
s	Cyclone dimension, defined by Fig. 7.8
T	Absolute gas temperature
t	Time
t_m	Mean residence time
t_{m1}	Minimum residence time
t_{m2}	Maximum residence time
t_r	Residence time of a particle
t_s	Settling time of a particle
t_{sm}	Maximum settling time of a particle
U	Superficial gas velocity
U	Gas velocity component
U^*	Dimensionless radial gas velocity
U_0	Velocity of the moving cylinder
U_e	Velocity of electric wind
U_m	Particle migration velocity in the electric field
U_p	Particle velocity
U_{pp}	Pick-up velocity
U_{pt}	Particle terminal velocity
U_w	Radial gas velocity near the cyclone wall
U_w	Velocity of droplets

u Interstitial gas velocity
V Gas velocity component
V^* Dimensionless gas velocity component
V_i Tangential gas velocity at the cyclone inlet
V_w Tangential gas velocity near the cyclone wall
W Gas velocity component
W Width of the bag filter
y Limiting width of the particle stream
y Initial vertical distance between the particle and collection plate
z Axial coordinate

Greek Symbols

α_f Fiber volume fraction
α_p Particle volume fraction
α_{pi} Particle volume fraction at inlet
α_{po} Particle volume fraction at outlet
α_w Volume fraction of droplets
β Vortex exponent
Γ Circulation
Γ_m Maximum circulation
γ Weighting factors, defined by Eq. (7.59)
Δp Pressure drop
δ Parameter, defined by Eq. (E7.2)
ϵ_0 Permittivity of a vacuum
η_c Collection efficiency of a cyclone
η_{ci} Fractional collection efficiency at size intervals
η_e Collection efficiency of an electrostatic precipitator
η_f Collection efficiency of a filter
η_i Collection efficiency of a fiber
η_s Collection efficiency of a scrubber
η_w Collection efficiency of a droplet
θ Tangential coordinate
μ Dynamic viscosity of gas
ν Kinematic viscosity of gas
ξ Parameter, defined by Eq. (7.27)
ξ Parameter, defined by Eq. (E7.6)
ρ Density of gas
ρ_p Density of particles
τ_S Stokes relaxation time
σ Sticking probability
ψ Stream function
ω Angular velocity

References

Alexander, R. M. (1949). Fundamentals of Cyclone Design and Operation. *Proc. Australas. Inst. Min. Met.*, **152**, 203.

Burgers, J. M. (1948). A Mathematical Model Illustrating the Theory of Turbulence. In *Advances in Applied Mechanics*, 1. Ed. von Mises and von Karman. New York: Academic Press.

Cooper, D. W. and Freeman, M. P. (1982). Separation. In *Handbook of Multiphase Systems*. Ed. G. Hetsroni. New York: McGraw-Hill.

Davies, C. N. (1973). *Air Filtration*. London: Academic Press.

Deutsch, W. (1922). Bewegung und Ladung der Electrizitätsträger im Zylinderkondensator. *Ann. Phys.*, **68**, 335.

Dickey, G. D. (1961). *Filtration*. New York: Reinhold.

Dorman, R. G. (1966). Filtration. In *Aerosol Science*. Ed. C. N. Davies. New York: Academic Press.

Ergun, S. (1952). Fluid Flow Through Packed Columns. *Chem. Eng. Prog.*, **48**, 89.

Fan, L.-S. (1989). *Gas–Liquid–Solid Fluidization Engineering*. Stoneham, Mass.: Butterworths.

Fuchs, N. A. (1964). *The Mechanics of Aerosol*. Oxford: Pergamon Press.

Happel, J. (1959). Viscous Flow Related to Arrays of Cylinders. *AIChE J.*, **5**, 174.

Kercher, H. (1969). Elektischer Wind: Rücksprüken und Staubwiderstand als Einfluβgröβen im Elektrofilter. *Staub*, **29**, 314.

Leith, D. and Licht, W. (1972). The Collection Efficiency of Cyclone Type Particle Collectors: A New Theoretical Approach. *AIChE. Symp. Ser.*, **68**(126), 196.
Matteson, M. J. and Orr, C. (1987). *Filtration: Principles and Practices*, 2nd ed. New York: Marcel Dekker.
Ogawa, A. (1984). *Separation of Particles from Air and Gases*, **II**. Boca Raton, Fla.: CRC Press.
Rietema, K. (1962). *Proceedings of Symposium on Interaction Between Fluid and Particles.* London: Institute of Chemical Engineers.
Soo, S. L. (1989). *Particulates and Continuum: Multiphase Fluid Dynamics*. New York: Hemisphere.
Svarovsky, L. (1981). *Solid–Gas Separation*. New York: Elsevier Scientific.
Theodore, L. and Buonicore, A. J. (1976). *Industrial Air Pollution Control Equipment for Particulates*. Cleveland, Ohio: CRC Press.
Tien, C. (1989). *Granular Filtration of Aerosols and Hydrosols*. Stoneham, Mass.: Butterworths.
White, H. J. (1963). *Industrial Electrostatic Precipitation*. Reading, Mass.: Addison-Wesley.
Zenz, F. A. and Othmer, D. F. (1960). *Fluidization and Fluid–Particle Systems*. New York: Reinhold.
Zhou, L. X. and Soo, S. L. (1990). Gas–Solid Flow and Collection of Solids in a Cyclone Separator. *Powder Tech.*, **63**, 45.
Zhou, L. X. and Soo, S. L. (1991). On Boundary Conditions of Particle Phase and Collection Efficiency in Cyclones. *Powder Tech.*, **64**, 213.

Problems

7.1 Show that Eq. (7.7) represents a set of nontrivial solutions of Eq. (7.6).

7.2 A tangential inlet cyclone with steep cone is operated under the following conditions: $R = 0.12$ m, $Q = 0.20$ m³/s, $\Gamma_m = 4.2$ m²/s, $H = 0.87$ m, $d_p = 10\ \mu$m, $\rho_p = 2{,}000$ kg/m³, $q/m = 10^{-4}$ C/kg, $\alpha_p \rho_p / \rho = 4$, and $\sigma = 0.05$. The carrier gas is air at ambient conditions. Calculate the collection efficiency. Assuming all conditions except particle diameter remain unchanged, estimate the cut-off size of this cyclone.

7.3 A cyclone has an overall collection efficiency of 90 percent for dust-laden air at 1 atm and 25°C. The mass loading of dust to air is 5. The results of the particle size analysis are given in Table P7.1.

Table P7.1. *Data on Size Distribution of Particles at Inlet and Outlet of the Cyclone*

Size range (μm)	Particle size analysis (wt%) Inlet	Outlet
0–5	8.0	76.0
5–10	1.4	12.9
10–15	1.9	4.5
15–20	2.1	2.1
20–25	2.1	1.5
25–30	2.0	0.7
30–35	2.0	0.5
35–40	2.0	0.4
40–45	2.0	0.3
>45	76.5	1.1

Plot the fractional efficiency curve and estimate the cut-off diameter of this cyclone.

7.4 Consider a parallel-plate electrostatic precipitator across which a 10 kV potential is applied. The interplate spacing is 1 cm. The particle density is 2,000 kg/m^3. The total surface area is 10 m^2, and the flow rate is 0.5 m^3/s. Use the test results provided in Table P7.2 to estimate

Table P7.2. *Data on Properties of Particles Collected in the Electrostatic Precipitator*

d_p (μm)	U_m (cm/s)	wt%($< d_p$)
5	2.2	—
10	5.9	10
20	8.3	50
40	11.2	90
80	13.8	99.6
>80	14.0	—

(a) Charge-to-mass distribution versus particle size
(b) Overall collection efficiency
(c) Size distribution (wt%) at the outlet

7.5 Consider a clean house bag filter filled with randomly assembled fibers. The porosity of the filter is 0.97, and the fibers have an effective diameter of 10 μm. The length of the filter is 0.1 m. The viscosity of air is 1.8×10^{-5} kg/m · s. For a velocity of the dust-laden air of 0.2 m/s, and a collection efficiency of a single fiber of 1.50 percent, estimate the overall efficiency of the filter. To reach a collection efficiency of 99.99 percent, what is the minimum length of this type of filter? At this minimum length, what is the overall pressure drop?

7.6 On the basis of the Navier–Stokes equation given in Eq. (7.6) for the gas flow in a cyclone separator, derive the static pressure distribution equations, *i.e.*, Eqs. (7.3) and (7.4), for this separator. It is assumed that the radial component of the gas velocity is negligible.

CHAPTER 8

Hopper and Standpipe Flows

8.1 Introduction

Granular materials in the food, pharmaceutical, chemical, and fuel industries are routinely processed through hoppers and standpipes. When the solids flow in a vertical or inclined pipe is dominated by gravity and the solids are in a nearly packed condition, the flow is known as a standpipe flow or downcomer flow. In a standpipe flow, the cross-sectional area of the standpipe is unchanged and the gas pressure at the bottom is higher than that at the top. The downward transport may also be achieved with higher pressure at the top. In this case, the solids transport is referred to as a discharged flow. The discussion of discharged flow is, however, beyond the scope of this chapter; readers interested in discharged flow may refer to the work of Berg (1954), and Knowlton *et al.* (1986, 1989).

Hoppers can be classified into two categories, namely, mass flow hoppers and core (or funnel) flow hoppers. A mass flow hopper is characterized by the simultaneous movement of every particle of the material during the discharge process, as shown in Fig. 8.1(a). On the other hand, in a core flow hopper, there exist stagnation zones in the material discharge process, as shown in Fig. 8.1(b). Advantages of mass flow hoppers are overwhelming compared to those of core flow hoppers. Problems like flooding, segregation, inconsistent flow rate, arching, and piping are less likely to occur in mass flow hoppers. There are two main flow patterns of hopper flows. The first type is illustrated by conical or axisymmetrical hoppers where stresses, stress distribution, and flow pattern of solids are symmetric about the central axis of the hopper. The second type is the two-dimensional or plane-strain hoppers, in which stresses and solids flow are independent of the third transverse dimension.

In this chapter, the mechanics of hopper flow and standpipe flow along with their operational characteristics are described. Problems such as segregation, inconsistent flow rate, arching, and piping that disrupt and obstruct the flow of bulk solids in hoppers are discussed. Remedies with respect to the use of flow-promoting devices such as vibrators and aerating jets to reinitiate the flow are presented. The importance of the flowability of the solids to be handled in relation to the hopper design is emphasized.

8.2 Powder Mechanics in Hopper Flows

The onset of powder motion in a hopper is due to stress failure in powders. Hence, the study of a hopper flow is closely related to the understanding of stress distribution in a hopper. The cross-sectional averaged stress distribution of solids in a cylindrical column was first studied by Janssen (1895). Walker (1966) and Walters (1973) extended Janssen's analysis to conical hoppers. The local distributions of static stresses of powders can only be obtained by solving the equations of equilibrium. From stress analyses and suitable failure criteria, the rupture locations in granular materials can be predicted. As a result, the flowability of granular materials in a hopper depends on the internal stress distributions determined by the geometry of the hopper and the material properties of the solids.

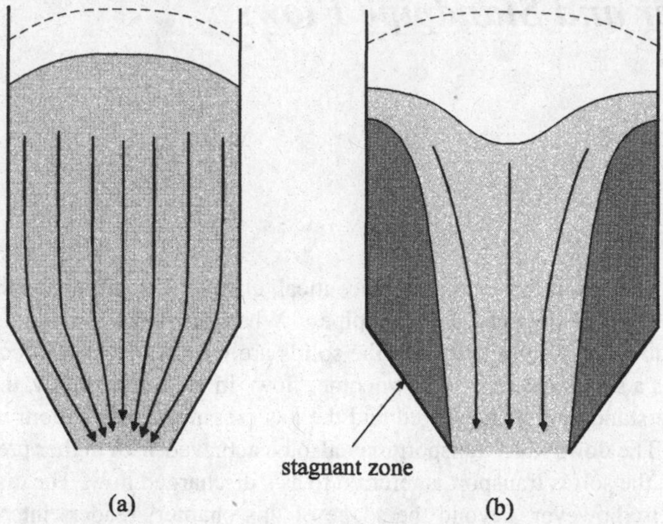

Figure 8.1. Hopper types based on flow patterns: (a) Mass flow hopper; (b) Core (funnel) flow hopper.

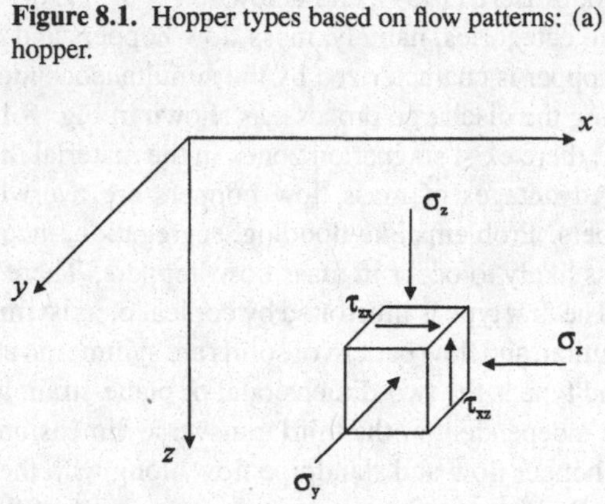

Figure 8.2. Stress components in a plane–strain problem.

8.2.1 *Mohr Circle for Plane Stresses*

Consider a plane-strain problem in the x–z plane, as shown in Fig. 8.2. The stress tensor, expressed in Cartesian coordinates, takes the form

$$\boldsymbol{T} = \begin{bmatrix} \sigma_x & 0 & \tau_{xz} \\ 0 & \sigma_y & 0 \\ \tau_{zx} & 0 & \sigma_z \end{bmatrix} \tag{8.1}$$

where the σ's are normal stresses (compressive stresses are considered as positive) and the τ's are shear stresses. From Hooke's law (Eq. (1.50)) and assuming no displacement in the y-direction, we have

$$\sigma_y = \nu(\sigma_x + \sigma_z) \tag{8.2}$$

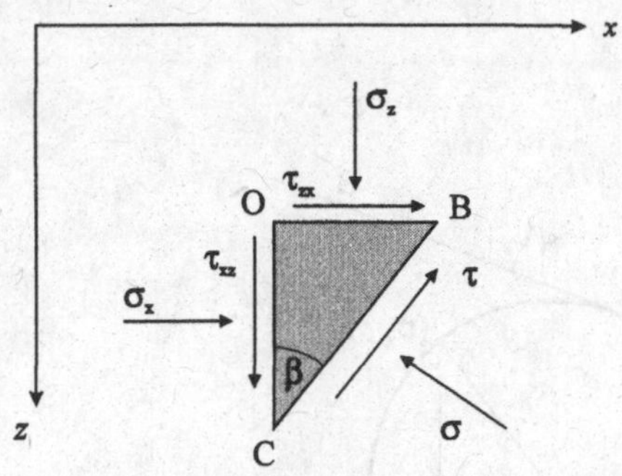

Figure 8.3. Equilibrium of stress components on a differential element.

where ν is Poisson's ratio. In addition, from the conservation of angular momentum, $\boldsymbol{T}$ is found to be a symmetric tensor so that $\tau_{xz} = \tau_{zx}$. Thus, the plane stress tensor in Eq. (8.1) depends solely on σ_x, σ_z, and τ_{xz}.

Since a tensor is invariant with respect to the selection of the reference coordinates, the stresses on any plane can be expressed in terms of σ_x, σ_z, and τ_{xz}. As shown in Fig. 8.3, the force balance of the differential element results in the stress relationships on the BC plane, as given by

$$\begin{aligned} \sigma &= \sigma_x \cos^2\beta + \sigma_z \sin^2\beta + 2\tau_{xz} \sin\beta\cos\beta \\ \tau &= \tau_{xz}(\cos^2\beta - \sin^2\beta) + (\sigma_z - \sigma_x)\sin\beta\cos\beta \end{aligned} \tag{8.3}$$

where β is the angle between the normal of the BC plane and the x-axis. From Eq. (8.3), two perpendicular planes can be found on which shear stresses vanish (*i.e.*, $\tau = 0$). The directions of these planes are known as the principal directions and the corresponding normal stresses as the principal stresses. The angle for principal directions, β_{pr}, is determined from Eq. (8.3) as

$$\beta_{pr} = \frac{1}{2}\tan^{-1}\left(\frac{2\tau_{xz}}{\sigma_x - \sigma_z}\right) \tag{8.4}$$

which yields the principal stresses as

$$\sigma_{1,3} = \frac{\sigma_x + \sigma_z}{2} \pm \sqrt{\left(\frac{\sigma_x - \sigma_z}{2}\right)^2 + \tau_{xz}^2} \tag{8.5}$$

If the principal directions are taken as the x- and z-axes, Eq. (8.3) is reduced to

$$\begin{aligned} \sigma &= \sigma_x \cos^2\beta + \sigma_z \sin^2\beta \\ \tau &= (\sigma_z - \sigma_x)\sin\beta\cos\beta \end{aligned} \tag{8.6}$$

which is an equation of a circle in the σ–τ coordinates, as shown in Fig. 8.4. This circle is known as the Mohr circle. The direction and magnitude of stresses on any plane can also be determined graphically from the Mohr circle. As shown in Fig. 8.4, the normal stresses on the principal planes are of maximum and minimum values.

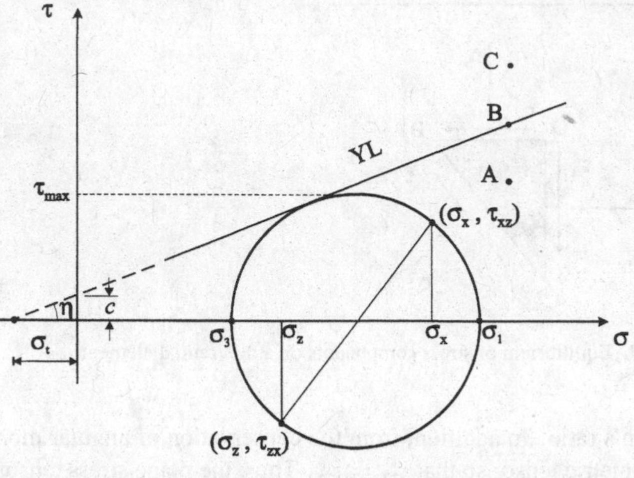

Figure 8.4. Mohr circle and Mohr–Coulomb failure envelope.

8.2.2 *Mohr–Coulomb Failure Criterion and Coulomb Powders*

The most common failure criterion for granular materials is the Mohr–Coulomb failure criterion. Mohr introduced his theory for rupture in materials in 1910. According to his theory, the material fails along a plane only when a critical combination of normal and shear stresses exists on the failure plane. This critical combination, known as the Mohr–Coulomb failure criterion, is given by

$$\tau = c + \sigma \tan \eta \tag{8.7}$$

where c is the cohesion defined as the resistance of the material to shear under zero normal load and is a result of the intermolecular cohesive forces, frictional forces, and other forces acting on the material. η is the angle of internal friction of the material, which corresponds to the maximum static friction condition as the bulk solids start to slide on themselves at the state of incipient failure.

The Mohr–Coulomb failure criterion can be recognized as an upper bound for the stress combination on any plane in the material. Consider points A, B, and C in Fig. 8.4. Point A represents a state of stresses on a plane along which failure will not occur. On the other hand, failure will occur along a plane if the state of stresses on that plane plots a point on the failure envelope, like point B. The state of stresses represented by point C cannot exist since it lies above the failure envelope. Since the Mohr–Coulomb failure envelope characterizes the state of stresses under which the material starts to slide, it is usually referred to as the yield locus, YL.

A rigid-plastic powder which has a linear yield locus is called a Coulomb powder. Most powders have linear yield loci, although, in some cases, nonlinearity appears at low compressive stresses. A relation between the principal stresses in a Coulomb powder at failure can be found from the Mohr circle in Fig. 8.4 as

$$\frac{(\sigma_3 + \sigma_t)}{(\sigma_1 + \sigma_t)} = \frac{1 - \sin \eta}{1 + \sin \eta} \tag{8.8}$$

where σ_t is the tensile strength of the powder. For cohesionless powders ($\sigma_t = c = 0$), Eq. (8.8) reduces to

$$\frac{\sigma_3}{\sigma_1} = \frac{1 - \sin \eta}{1 + \sin \eta} \tag{8.9}$$

An important application of this equation is to distinguish between two extreme failure conditions, known as the active and passive failures. First, the active and passive states of stress may be explained as follows: Consider a cohesionless Coulomb powder. If the powder is assembled in a large container in successive horizontal layers without disturbance, there will be no shear stresses along the horizontal and vertical planes inside the powder because of the symmetry of the problem. Thus, at any point, the horizontal and vertical normal stresses are the principal stresses at that point. In this case, if the major principal stress is the horizontal stress, σ_h, the powder is said to be in a passive state of stress. On the other hand, if the major principal stress is the vertical stress, σ_v, the powder is in an active state of stress. Thus, Eq. (8.9) can be written for each state as

$$\begin{aligned} \frac{\sigma_h}{\sigma_v} &= \frac{1 + \sin \eta}{1 - \sin \eta} = K_p \quad \text{for passive states where } \sigma_h > \sigma_v \\ \frac{\sigma_h}{\sigma_v} &= \frac{1 - \sin \eta}{1 + \sin \eta} = K_a \quad \text{for active states where } \sigma_v > \sigma_h \end{aligned} \tag{8.10}$$

where K_p and K_a are the ratios between the principal stresses in the passive and active states, respectively. When the Mohr circle for an active or passive state of stress touches the yield locus, a state of active or passive failure takes place along the plane represented in the Mohr circle by the radial line which passes the point of intersection between the circle and the yield locus, as shown in Fig. 8.5.

During the filling of a hopper an active state of stress exists since the powder tends to compress vertically. This case is also referred to as the "static state" of stress since the powder is at rest. On the other hand, during discharging from the hopper, the powder tends to expand in the vertical direction. In that case, the horizontal stress becomes the major principal stress, leading to a state of passive stress. The passive state of stress in this case may also be referred to as the "dynamic state" of stress because of the flow of powder.

8.2.3 *Static Stress Distributions in Standpipes and Hoppers*

So far, we understand that the flowability of powders depends on their "failure" stresses from the Mohr–Coulomb failure criterion. Therefore, analyses of powder flows

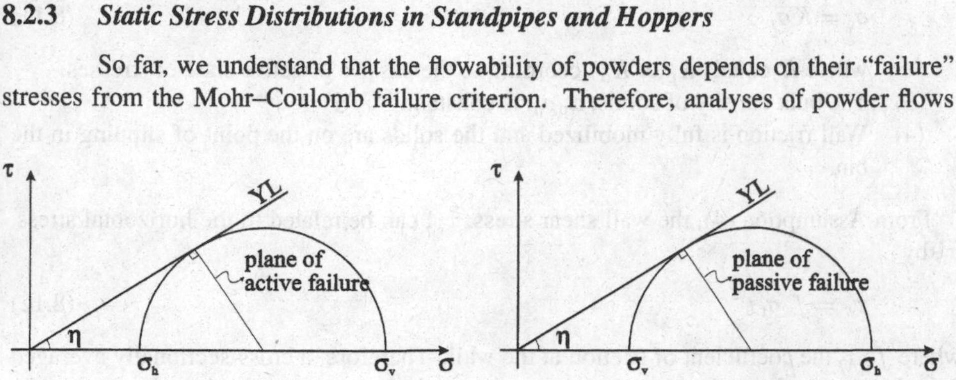

Figure 8.5. Mohr circles for two cases: (a) Active failure; (b) Passive failure.

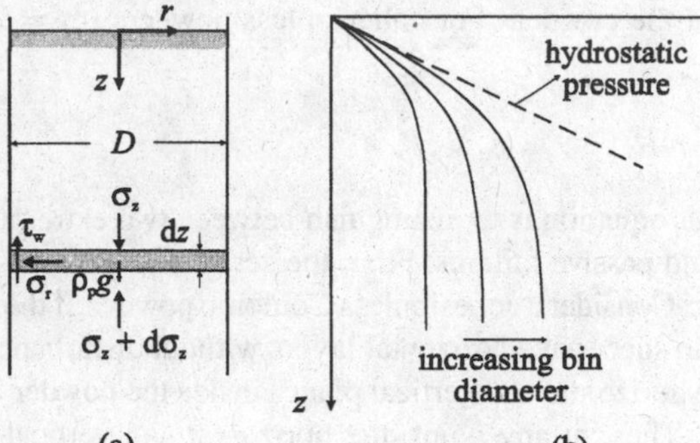

Figure 8.6. Janssen's model for stress distribution in cylindrical bins: (a) Schematic of the forces applied to a differential slice; (b) Wall pressure distribution; Eq. (8.16).

in bins and hoppers should be closely related to the stress analyses of the systems. By treating a powder as a pseudocontinuum solid, the static stress distribution inside the hopper or standpipe, in both active and passive states, may be obtained from the equations of equilibrium and the appropriate boundary conditions. In the following sections, simple analytical methods to evaluate cross-sectionally averaged stresses in a cylindrical vertical column and a conical hopper are presented. For more details on these methods and other more sophisticated ones, the reader may refer to specialized references on the subject [*e.g.*, Horne and Nedderman, 1978; Nedderman and Tüzün, 1979; Drescher, 1991; Nedderman, 1992].

8.2.3.1 *Averaged Stresses in a Cylindrical Column*

A simple model for the estimation of stress distribution in cohesionless bulk solids stored in a tall cylindrical bin was presented by Janssen in 1895. The system of consideration is shown in Fig. 8.6(a). In Janssen's model, it is assumed that

(1) Stresses in the bulk solids are functions of z only.
(2) σ_r and σ_z are principal stresses, which are related by

$$\sigma_r = K\sigma_z \tag{8.11}$$

where K equals K_p or K_a according to the passive or active state of stresses.
(3) The bulk density of solids, $\alpha_p\rho_p$, is constant.
(4) Wall friction is fully mobilized and the solids are on the point of slipping in the bin.

From Assumption (4), the wall shear stress, τ_w, can be related to the horizontal stress, σ_r, by

$$\tau_w = f_w\sigma_r \tag{8.12}$$

where f_w is the coefficient of friction at the wall. Therefore, a cross-sectionally averaged force balance in the z-direction leads to

$$\frac{d\sigma_z}{dz} = \alpha_p\rho_p g - \frac{4\tau_w}{D} = \alpha_p\rho_p g - \frac{4f_wK}{D}\sigma_z \tag{8.13}$$

where D is the diameter of the cylinder bin. The boundary condition is

$$\sigma_z = 0 \qquad \text{at } z = 0 \tag{8.14}$$

Thus, σ_z is obtained as

$$\sigma_z = \frac{\alpha_p \rho_p g D}{4 f_w K}\left[1 - \exp\left(\frac{-4 f_w K}{D} z\right)\right] \tag{8.15}$$

Substituting Eq. (8.15) into Assumption (2) yields

$$\sigma_r = \frac{\alpha_p \rho_p g D}{4 f_w}\left[1 - \exp\left(\frac{-4 f_w K}{D} z\right)\right] \tag{8.16}$$

Equations (8.15) and (8.16) show that, as z increases, σ_z and σ_r approach the maximum values of

$$\sigma_{z\max} = \frac{\alpha_p \rho_p g D}{4 f_w K}, \qquad \sigma_{r\max} = \frac{\alpha_p \rho_p g D}{4 f_w} \tag{8.17}$$

Moreover, the horizontal wall pressure, σ_r, may be estimated from Eq. (8.16). The results are illustrated in Fig. 8.6(b).

It should be noted that the stresses in Janssen's model are cross-sectionally averaged. Therefore, Janssen's results in principle are not applicable for local stress analysis. This argument can be clearly shown by the inconsistency of Eqs. (8.15) and (8.16) with the equations of equilibrium.

8.2.3.2 *Averaged Stresses in a Storage Hopper*

Janssen's analysis can be extended to the stress analysis of conical hoppers. Walker (1966) proposed a differential slice with vertical sides, as shown in Fig. 8.7(a), where the cross-sectionally averaged force balance in the vertical direction results in

$$N\frac{\sigma_z}{z} - \frac{d\sigma_z}{dz} = \alpha_p \rho_p g \tag{8.18}$$

where σ_z is the average vertical stress on the horizontal plane and N is a parameter related to the geometry of the hopper system. In Walker's model, N is given by

$$N = \frac{2B\Delta}{\tan \varphi_w} \tag{8.19}$$

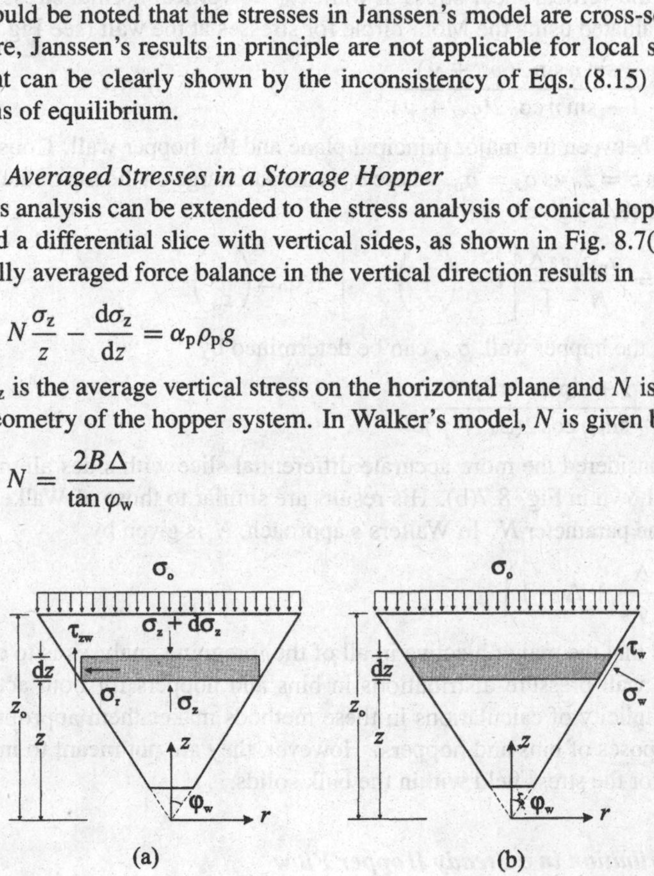

Figure 8.7. Types of differential slices used for hopper analysis: (a) Walker (1966); (b) Walters (1973).

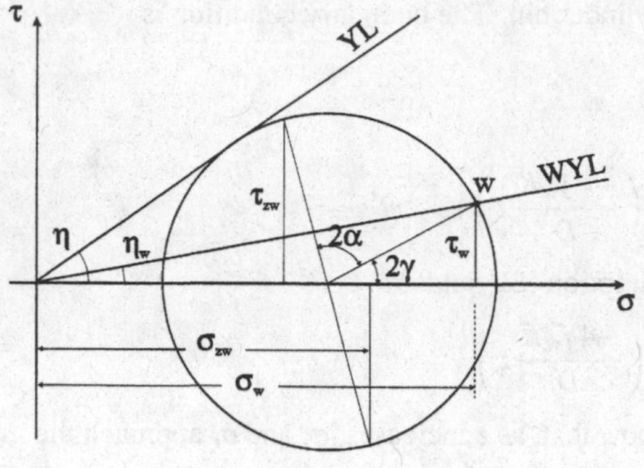

Figure 8.8. Mohr circle for the state of stresses near the hopper wall.

where φ_w is the hopper half-apex angle; Δ is a distribution factor, defined as the ratio of the vertical normal stress at wall, σ_{zw}, to cross-sectional averaged vertical normal stress, σ_z; and B is the ratio of the vertical shear stress at wall, τ_{zw}, to vertical normal stress at wall, σ_{zw}, which can be evaluated using the Mohr circle for stresses at the wall (see Fig. 8.8) as

$$B = \frac{\tau_{zw}}{\sigma_{zw}} = \frac{\sin\eta \sin 2(\varphi_w + \gamma)}{1 - \sin\eta \cos 2(\varphi_w + \gamma)} \tag{8.20}$$

where γ is the angle between the major principal plane and the hopper wall. Consider the boundary condition at $z = z_o$ as $\sigma_z = \sigma_o$, where σ_o is the surcharge on the top of the hopper. Then, integrating Eq. (8.18) gives

$$\sigma_{zw} = \Delta\sigma_z = \frac{\alpha_p \rho_p g z \Delta}{N - 1}\left[1 - \left(\frac{z}{z_o}\right)^{N-1}\right] + \sigma_o \Delta \left(\frac{z}{z_o}\right)^N \tag{8.21}$$

The normal stress on the hopper wall, σ_w, can be determined by

$$\sigma_w = \sigma_{zw} \frac{1 + \sin\eta \cos 2\gamma}{1 - \sin\eta \cos 2(\varphi_w + \gamma)} \tag{8.22}$$

Walters (1973) considered the more accurate differential slice with sides aligned with the hopper walls as shown in Fig. 8.7(b). His results are similar to those of Walker with a slight difference in the parameter N. In Walters's approach, N is given by

$$N = 2\left(\frac{B\Delta}{\tan\varphi_w} + \Delta - 1\right) \tag{8.23}$$

It should be noted that the main objective in all of the foregoing analyses is to estimate the upper bounds of wall pressure distributions in bins and hoppers for both active and passive stresses. Simplicity of calculations in these methods makes them appropriate for structural design purposes of bins and hoppers. However, they are not meant to introduce a complete solution for the stress field within the bulk solids.

8.2.4 *Stress Distribution in a Steady Hopper Flow*

For a slow and steady motion of powders, the local stresses inside the powder are approximated to be pseudostatic and, consequently, are governed by the equation of equilibrium as

$$\nabla \cdot \boldsymbol{T} + \boldsymbol{F} = 0 \tag{8.24}$$

where $\boldsymbol{F}$ represents the body force. Let us consider a gravity flow of a powder in a hopper. We may express the equation of equilibrium in cylindrical coordinates with the z-axis in the same direction as the gravity in the form

$$\frac{\partial \sigma_r}{\partial r} + \frac{\partial \tau_{rz}}{\partial z} + \frac{\sigma_r - \sigma_\theta}{r} = 0 \tag{8.25}$$

and

$$\frac{\partial \sigma_z}{\partial z} + \frac{\partial \tau_{rz}}{\partial r} + \frac{\tau_{rz}}{r} + \alpha_p \rho_p g = 0 \tag{8.26}$$

The axisymmetric nature of conical hoppers results in $\epsilon_\theta = 0$ and, according to Eq. (2.20), $\sigma_\theta = (\sigma_r + \sigma_z)\nu$. To close the problem, one more relation is required. In a continuum solid, this relation is known as the compatibility requirement, *i.e.*, the relationship of strains. This relation, with the aid of constitutive relations between stress and strain (*e.g.*, Hooke's law), provides an additional equation for stress so that the problem can be closed. However, the compatibility relation for a continuum solid may not be extendable to the cases of powders. Thus, additional assumptions or models are needed to yield another equation for stresses in powders.

It was recognized by Jenike (1961) that the powder undergoes a continuous shear deformation without change of local stresses in a steady hopper flow. This indicates that the bulk solid in a steady motion can be regarded as a plastic, as illustrated in Fig. 8.9. Mohr circles which represent the steady flow of any packed particulate materials under different consolidating conditions are tangential to a straight line (known as the effective yield locus) that passes through the origin as shown in Fig. 8.9. Thus,

$$\sin \eta_k = \frac{\sigma_1 - \sigma_3}{\sigma_1 + \sigma_3} \tag{8.27}$$

where η_k is the kinematic angle of internal friction. Jenike (1961) also introduced the theory of radial stress field in conical hoppers. In this theory, the stress is considered to be a function only of the radial distance from the virtual apex of the hopper. Hence, with appropriate boundary conditions, the local stress distributions in a steady hopper flow can be obtained by solving the system of equations which consists of the equation of equilibrium,

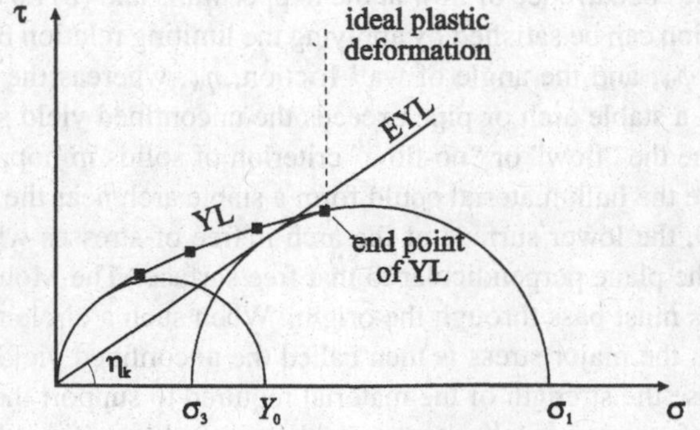

Figure 8.9. Effective yield locus and evaluation of Y_0 and σ_1 for steady hopper flows.

Eq. (8.24), and the modified Mohr–Coulomb yield (or failure) criterion, Eq. (8.27). It should be noted that other yield criteria, such as the von Mises criterion, are used to model the flow of bulk solids in hoppers, and more conditions may need to be imposed, such as the Lévy flow rule, in order to close the system of equations [Cleaver and Nedderman, 1993].

8.2.5 *Flowability of Powders in Hopper Design*

Powder flowability is a major concern in hopper design to prevent problems such as arching, insufficient flow rate and capacity, piping, segregation, and flooding from occurring inside hoppers. The initiation of powder flow is due to the incipient failure where the forces acting on the solid produce stresses in excess of the strength of the solid. The rupture of powders depends on the mechanical properties of bulk solids, including [Johanson, 1981]

(1) Angle of internal friction, η_s
(2) Kinematic angle of internal friction, η_k
(3) Kinematic angle of wall friction, η_w
(4) Bulk density, $\alpha_p \rho_p$
(5) Unconfined yield strength, Y_0
(6) Major principal stress, σ_1
(7) Cohesion, c

The angle of friction is related to the corresponding coefficient of friction by

$$\eta_i = \tan^{-1} f_i, \quad i = s, k, \text{ and } w \tag{8.28}$$

In addition to the foregoing properties, there are other important flow properties such as the angle of repose, angle of fall, compressibility, and dispersibility which provide additional measures for the flowability of granular materials when discharged out of the hopper [Carr, 1965]. In general, flow properties are influenced by the conditions under which they are handled, such as moisture content, storage time at rest, and consolidating pressure. All of these properties can be determined with simple bench-top equipment such as the Jenike translatory shear cell [Jenike, 1961], the rotational split-level shear cell [Peschl, 1989], and other shear testing devices [Kamath *et al.*, 1993].

A simple approach for the design of mass flow hoppers was suggested by Jenike (1964a); it is based on two conditions: (a) occurrence of flow at the hopper walls and (b) no arching and/or piping. The first condition can be satisfied by applying the limiting relation between the angle of the hopper wall, φ_w, and the angle of wall friction, η_w, whereas the second one can be met if the stress in a stable arch or pipe exceeds the unconfined yield strength of the solids. In order to define the "flow" or "no-flow" criterion of solids in hoppers, let us first examine the case where the bulk material could form a stable arch near the hopper outlet. As shown in Fig. 8.10, the lower surface of the arch is free of stresses while the major stress, σ_a, is acting on the plane perpendicular to that free surface. The Mohr circle representing this state of stress must pass through the origin. When such a circle touches the yield locus of the material, the major stress is then called the unconfined yield stress, Y_0, which represents in this case the strength of the material required to support the stable arch. Thus, the critical state of stress $\sigma_a = Y_0$ represents that needed to initiate the flow and make the arch collapse. Dependence of the minimum size of hopper outlet required to

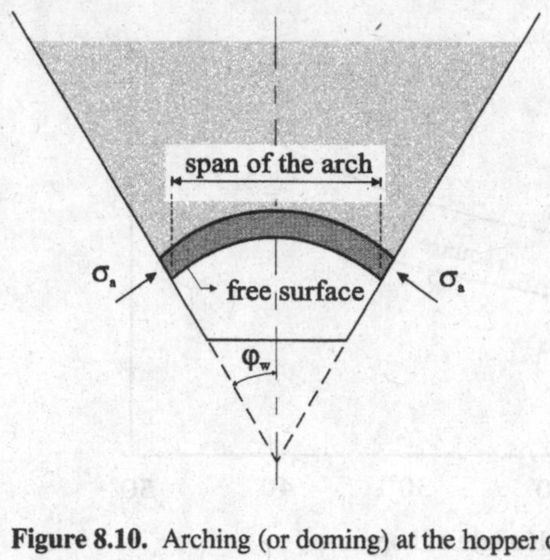

Figure 8.10. Arching (or doming) at the hopper outlet.

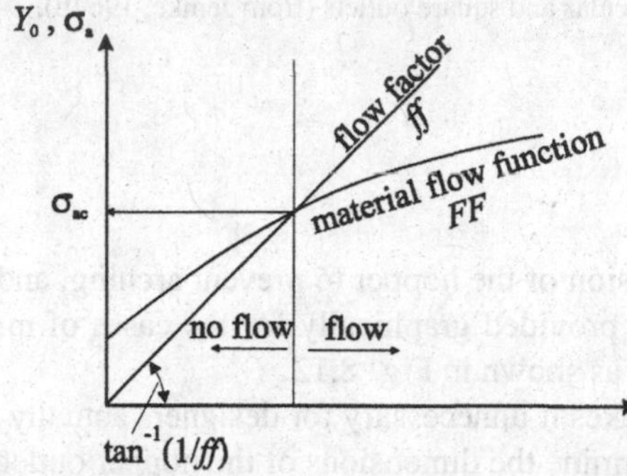

Figure 8.11. Flow factor *ff* and material flow function *FF*.

prevent arching on the unconfined yield stress of the material can be obtained by studying the equilibrium of forces in a stable arch (see Problem 8.3).

It is convenient to introduce the concepts of material flow function, *FF*, and flow factor, *ff*. The material flow function, *FF*, relates the unconfined yield stress, Y_0, to the corresponding major consolidating stress, σ_1, and is determined experimentally from the yield locus of the material, as shown in Fig. 8.9. The material flow function is presented as a plot of Y_0 versus σ_1, as shown in Fig. 8.11. The flow factor, *ff*, is defined by

$$ff = \frac{\sigma_1}{\sigma_a} \tag{8.29}$$

and it can be regarded as a measure of the stresses that might develop when flow is interrupted by the formation of a stable arch. Jenike (1964a) observed that the ratio σ_1/σ_a is constant for a given φ_w, η_k, and η_w. Thus, when plotted on the same coordinate with the flow function, the flow factor is a straight line passing through the origin with a slope of $1/ff$, as shown in Fig. 8.11. The intersection between *ff* and *FF* represents the critical condition marking flow or no-flow of solids. The flow criterion is satisfied for the conditions where *FF* lies below *ff*. The critical value of σ_a corresponding to the point of intersection can be used to determine the minimum width of the hopper outlet required to ensure the collapse of the

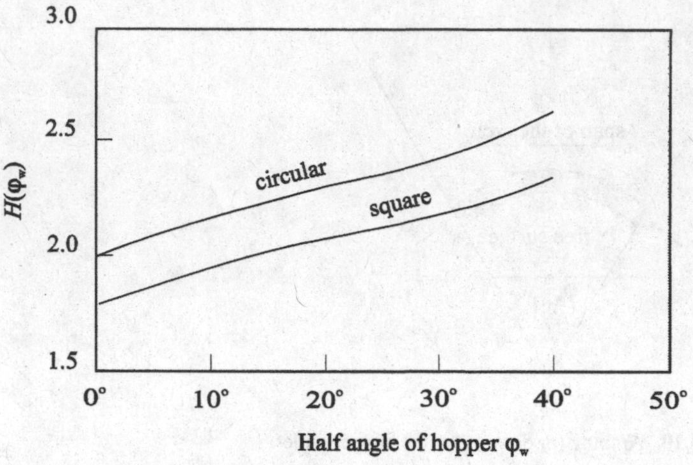

Figure 8.12. Function $H(\varphi_w)$ for circular and square outlets (from Jenike, 1964b).

arch by the relation [Jenike, 1964b]

$$d_{min} = \frac{\sigma_{ac} H(\varphi_w)}{\alpha_p \rho_p g} \tag{8.30}$$

where d_{min} is the minimum outlet dimension of the hopper to prevent arching, and $H(\varphi_w)$ is a function of the hopper angle and is provided graphically for the cases of mass flow hoppers with circular and square outlets, as shown in Fig. 8.12.

The introduction of the flow factor makes it unnecessary for designers actually to solve the stress field equations in order to determine the dimensions of the hopper outlet. Jenike (1964b) and several codes of practice for silo design [*e.g.*, BMHB, 1988] provide designers with graphs, obtained numerically, of *ff* as a function of φ_w, η_k, and η_w. The standard design procedure of hoppers using these design graphs is illustrated in the following example.

Example 8.1 Three sets of yield loci under different consolidation conditions are obtained for a sample of powder of bulk density 1,500 kg/m^3, as shown in Fig. E8.1. If a conical hopper is to be designed, determine the wall slope of the hopper and the opening size necessary to ensure a steady mass flow. The angle of wall friction is 15°. The design diagram for mass flow conical hoppers is given in Fig. E8.2 [BMHB, 1988].

Solution The kinematic angle of internal friction can be determined from the Mohr circle, which is tangential to the yield locus at the end point. This Mohr circle yields the major consolidating stress σ_1 and minor consolidating stress σ_3. Thus, η_k is found to be 30°, either from Eq. (8.27) or from a tangent of the Mohr circle which passes through the origin, as shown in Fig. E8.1.

In order to determine *FF*, two Mohr circles are drawn for each yield locus, as shown in Fig. E8.1. The circle passing through the origin and being tangential to the yield locus gives the unconfined yield strength, Y_0, whereas the circle which is tangential to the yield locus at the end point yields the corresponding σ_1. Thus, from the three sets of yield loci, three sets of Y_0 and σ_1 are obtained. Consequently, *FF* is determined as shown in Fig. E8.3.

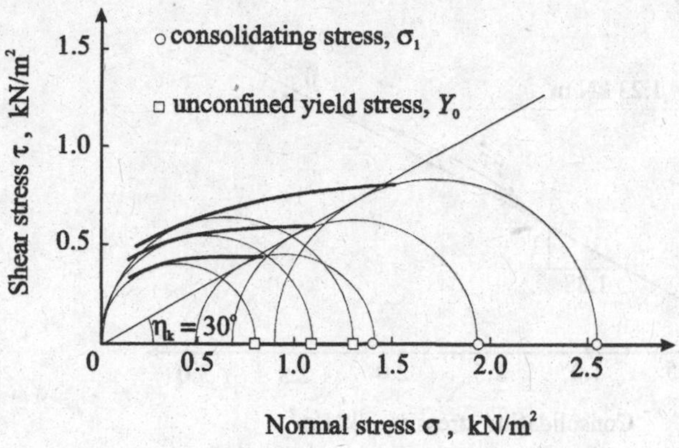

Figure E8.1. Shear test results and flow property measurements.

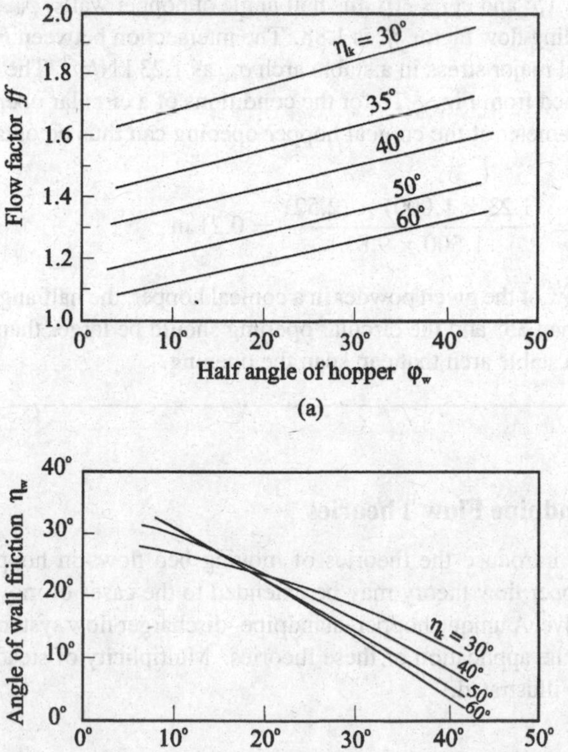

Figure E8.2. Design diagram for mass flow conical hopper (from BMHB, 1988): (a) Flow factor; (b) Angle of wall friction.

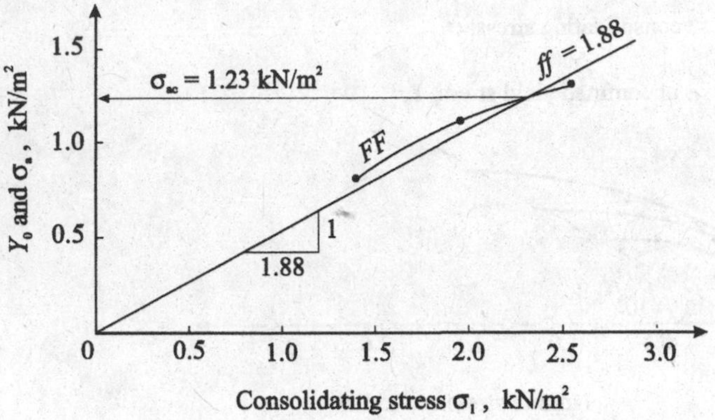

Figure E8.3. Flow factor *ff* and material flow function *FF*.

From Fig. E8.2, for $\eta_w = 15°$ and $\eta_k = 30°$, the half angle of hopper walls, φ_w, is found to be 35° and the corresponding flow factor, *ff*, is 1.88. The intersection between *FF* and *ff* in Fig. E8.3 gives the critical major stress in a stable arch σ_{ac} as 1.23 kN/m^2. The value of $H(\varphi_w) = 2.52$ can be obtained from Fig. 8.12 for the conditions of a circular opening and $\varphi_w = 35°$. The minimum diameter of the conical hopper opening can thus be obtained by using Eq. (8.30) as

$$d_{min} = \frac{\sigma_{ac} H(\varphi_w)}{\alpha_p \rho_p g} = \frac{(1.23 \times 1{,}000) \times (2.52)}{1{,}500 \times 9.81} = 0.21 \text{ m} \tag{E8.1}$$

Therefore, to ensure mass flow of the given powder in a conical hopper, the half angle of the hopper should be no more than 35° and the circular opening should be larger than 0.21 m to prevent the formation of a stable arch that can span the opening.

8.3 Hopper and Standpipe Flow Theories

In this section, we introduce the theories of moving bed flows in hoppers and general standpipes. The hopper flow theory may be extended to the cases of moving bed flows through a discharge valve. A unique hopper–standpipe–discharger flow system is then discussed as an example of the application of these theories. Multiplicity of steady flows in a standpipe system is also illustrated.

8.3.1 *Moving Bed Flows in a Feed Hopper*

Let us consider moving bed flows through a conical feed hopper where the gas moves upward relative to the moving bed of solids, although the gas motion relative to the hopper wall may still be downward. It is convenient to write equations of motion in spherical coordinates with the origin at the vertex and the polar axis along the axis of symmetry of the cone. The flow is assumed to be invariant with respect to the azimuthal angle, θ. Moreover, the volume fraction of gas in the moving bed, α_{mb}, is constant.

The continuity equations for the gas and the particles, respectively, are

$$\frac{\partial(\rho u_{\mathrm{R}})}{\partial R}+\frac{2\rho u_{\mathrm{R}}}{R}+\frac{1}{R}\frac{\partial(\rho u_{\varphi})}{\partial\varphi}+\frac{\rho u_{\varphi}}{R}\cot\varphi=0 \tag{8.31}$$

$$\frac{\partial u_{\mathrm{Rp}}}{\partial R}+\frac{2u_{\mathrm{Rp}}}{R}+\frac{1}{R}\frac{\partial u_{\varphi\mathrm{p}}}{\partial\varphi}+\frac{u_{\varphi\mathrm{p}}}{R}\cot\varphi=0 \tag{8.32}$$

For the momentum equation of gas in a moving bed flow, the inertial and gravitational terms can be neglected. The pressure drop is therefore balanced by the drag forces, which give

$$\frac{\partial p}{\partial R}=\beta(u_{\mathrm{Rp}}-u_{\mathrm{R}}) \tag{8.33}$$

$$\frac{1}{R}\frac{\partial p}{\partial\varphi}=\beta(u_{\varphi\mathrm{p}}-u_{\varphi}) \tag{8.34}$$

where β is a coefficient which depends on the voidage of the mixture. The momentum equations of solids are expressed by [Nguyen *et al.*, 1979; Chen *et al.*, 1984]

$$\begin{aligned}&\frac{\partial\sigma_{\mathrm{R}}}{\partial R}+\frac{1}{R}\frac{\partial\tau_{\mathrm{R}\varphi}}{\partial\varphi}+\frac{1}{R}[2\sigma_{\mathrm{R}}-\sigma_{\theta}-\sigma_{\varphi}+\tau_{\mathrm{R}\varphi}\cot\varphi]+(1-\alpha_{\mathrm{mb}})\rho_{\mathrm{p}}g\cos\varphi\\&\quad+(1-\alpha_{\mathrm{mb}})\rho_{\mathrm{p}}\left[u_{\mathrm{Rp}}\frac{\partial u_{\mathrm{Rp}}}{\partial R}+\frac{u_{\varphi\mathrm{p}}}{R}\frac{\partial u_{\mathrm{Rp}}}{\partial\varphi}-\frac{u_{\varphi\mathrm{p}}^{2}}{R}\right]+\beta(u_{\mathrm{Rp}}-u_{\mathrm{R}})=0\end{aligned} \tag{8.35}$$

$$\begin{aligned}&\frac{\partial\tau_{\mathrm{R}\varphi}}{\partial R}+\frac{1}{R}\frac{\partial\sigma_{\varphi}}{\partial\varphi}+\frac{1}{R}[(\sigma_{\varphi}-\sigma_{\theta})\cot\varphi+3\tau_{\mathrm{R}\varphi}]-(1-\alpha_{\mathrm{mb}})\rho_{\mathrm{p}}g\sin\varphi\\&\quad+(1-\alpha_{\mathrm{mb}})\rho_{\mathrm{p}}\left[u_{\mathrm{Rp}}\frac{\partial u_{\varphi\mathrm{p}}}{\partial R}+\frac{u_{\varphi\mathrm{p}}}{R}\frac{\partial u_{\varphi\mathrm{p}}}{\partial\varphi}+\frac{u_{\mathrm{Rp}}u_{\varphi\mathrm{p}}}{R}\right]+\beta(u_{\varphi\mathrm{p}}-u_{\varphi})=0\end{aligned} \tag{8.36}$$

To close the problem, constitutive relations of powders must be introduced for the internal connections of components of the stress tensor of solids and the linkage between the stresses and velocities of solids. It is assumed that the bulk solid material behaves as a Coulomb powder so that the isotropy condition and the Mohr–Coulomb yield condition may be used. In addition, σ_θ has to be formulated with respect to the other stress components.

The isotropy condition is that in the (R, φ) plane, the principal directions of stresses and the rate of deformation coincide [Jenike, 1961]. Thus, the isotropy condition is written as

$$\frac{\sigma_{\mathrm{R}}-\sigma_{\varphi}}{\tau_{\mathrm{R}\varphi}}=\frac{\dfrac{\partial u_{\mathrm{Rp}}}{\partial R}-\dfrac{1}{R}\dfrac{\partial u_{\varphi\mathrm{p}}}{\partial\varphi}-\dfrac{u_{\mathrm{Rp}}}{R}}{\dfrac{1}{2}\left(\dfrac{\partial u_{\varphi\mathrm{p}}}{\partial R}-\dfrac{u_{\varphi\mathrm{p}}}{R}+\dfrac{1}{R}\dfrac{\partial u_{\mathrm{Rp}}}{\partial\varphi}\right)} \tag{8.37}$$

The Mohr–Coulomb yield condition is given by

$$\left(\frac{\sigma_{\mathrm{R}}-\sigma_{\varphi}}{2}\right)^{2}+\tau_{\mathrm{R}\varphi}^{2}=\left(\frac{\sigma_{\mathrm{R}}+\sigma_{\varphi}}{2}\right)^{2}\sin^{2}\eta_{\mathrm{k}} \tag{8.38}$$

Here, the Haar–von Karman hypothesis, which requires σ_θ to be equal to one of the principal stresses in the (R, φ) plane, is adopted. Thus, it is reasonable to assume σ_θ to be equal to the major principal stress since the deformation of solids in a converging cone is compressive in the azimuthal direction when the solids move down the hopper. This gives rise to

$$\sigma_\theta = \frac{\sigma_R + \sigma_\varphi}{2}(1 + \sin\eta_k) \tag{8.39}$$

So far, we have nine equations (Eqs. (8.31)–(8.39)) for nine variables (u_R, u_φ, u_{Rp}, $u_{\varphi p}$, p, σ_R, σ_θ, σ_φ, and $\tau_{R\varphi}$). Therefore, the problem is closed and, in principle, can be solved numerically. However, analytical solutions are available with further simplifications of the model equations.

To obtain an analytical solution of the preceding problem, the stresses in terms of the mean stress σ_m and the stress angle ψ may be expressed by [Sokolovskii, 1965]

$$\begin{aligned} \sigma_R &= \sigma_m(1 - \sin\eta_k \cos 2\psi); \qquad \sigma_\varphi = \sigma_m(1 + \sin\eta_k \cos 2\psi) \\ \tau_{R\varphi} &= -\sigma_m \sin\eta_k \sin 2\psi; \qquad \sigma_\theta = \sigma_m(1 + \sin\eta_k) \end{aligned} \tag{8.40}$$

where $\sigma_m = (\sigma_R + \sigma_\varphi)/2$ and $\pi/2 - \psi$ is the acute angle between the radial direction and the major principal stress axis in the (R, φ) plane. Furthermore, variables may be approximated by using a perturbation technique with respect to the polar angle φ [Nguyen *et al.*, 1979]. The expansion of variables in powers of φ terminates at terms of $o(\varphi)$. Thus, we have

$$\begin{aligned} &\sigma_m \approx \sigma_m(R); \qquad \psi \approx 0; \qquad u_R \approx u_R(R); \qquad u_\varphi \approx 0 \\ &u_{Rp} \approx u_{Rp}(R); \qquad u_{\varphi p} \approx 0; \qquad p \approx p(R); \qquad \rho \approx \rho(R) \end{aligned} \tag{8.41}$$

Hence, Eq. (8.31) and Eq. (8.32) reduce to

$$\frac{d(\rho u_R)}{dR} + \frac{2\rho u_R}{R} = 0 \tag{8.42}$$

$$\frac{du_{Rp}}{dR} + \frac{2u_{Rp}}{R} = 0 \tag{8.43}$$

Integrating the preceding equations yields

$$\rho u_R = \frac{\rho_i U_{Ri} R_i^2}{\alpha_{mb} R^2} = \frac{\rho_o U_{Ro} R_o^2}{\alpha_{mb} R^2} \tag{8.44}$$

$$u_{Rp} = \frac{U_{Rpi} R_i^2}{(1 - \alpha_{mb}) R^2} = \frac{U_{Rpo} R_o^2}{(1 - \alpha_{mb}) R^2} \tag{8.45}$$

where the subscripts "i" and "o" denote the inlet and outlet locations, respectively, and the U's represent the superficial velocities.

Similarly, for an ideal gas in an isothermal process, Eq. (8.33) becomes

$$\frac{dp}{dR} = \frac{\beta R_i^2}{R^2}\left(\frac{U_{Rpi}}{1 - \alpha_{mb}} - \frac{U_{Ri}}{\alpha_{mb}}\frac{p_i}{p}\right) \tag{8.46}$$

which gives the pressure distribution in the mass hopper flow as

$$p_i - p + B \ln\left(\frac{p_i - B}{p - B}\right) = A\left(\frac{1}{R} - \frac{1}{R_i}\right) \tag{8.47}$$

where A and B are defined by

$$A = \beta R_i^2 \frac{U_{Rpi}}{1 - \alpha_{mb}}; \qquad B = \frac{U_{Ri}}{U_{Rpi}} \frac{(1 - \alpha_{mb})}{\alpha_{mb}} p_i \tag{8.48}$$

It is noted that the trigonometric functions in Eq. (8.35) can be expanded in powers of φ as

$$\cos 2\psi \approx 1; \qquad \sin 2\psi \approx 2\psi_{\mathrm{w}}\frac{\varphi}{\varphi_{\mathrm{w}}}; \qquad \cot\varphi \approx \frac{1}{\varphi} \tag{8.49}$$

where φ_{w} is the half angle of the feed hopper and ψ_{w} is the value of ψ at the wall of the feed hopper. Equation (8.35) is thus simplified to

$$(1-\sin\eta_{\mathrm{k}})\frac{\mathrm{d}\sigma_{\mathrm{m}}}{\mathrm{d}R} - \frac{4\sigma_{\mathrm{m}}}{R}\left(1+\frac{\psi_{\mathrm{w}}}{\varphi_{\mathrm{w}}}\right)\sin\eta_{\mathrm{k}} + (1-\alpha_{\mathrm{mb}})\rho_{\mathrm{p}}g + (1-\alpha_{\mathrm{mb}})\rho_{\mathrm{p}}u_{\mathrm{Rp}}\frac{\mathrm{d}u_{\mathrm{Rp}}}{\mathrm{d}R} + \beta(u_{\mathrm{Rp}}-u_{\mathrm{R}}) = 0 \tag{8.50}$$

Substituting Eqs. (8.44) and (8.45) into the preceding equation gives

$$(1-\sin\eta_{\mathrm{k}})\frac{\mathrm{d}\sigma_{\mathrm{m}}}{\mathrm{d}R} - \frac{4\sigma_{\mathrm{m}}}{R}\left(1+\frac{\psi_{\mathrm{w}}}{\varphi_{\mathrm{w}}}\right)\sin\eta_{\mathrm{k}} + (1-\alpha_{\mathrm{mb}})\rho_{\mathrm{p}}g - \frac{2\rho_{\mathrm{p}}U_{\mathrm{Rpi}}^2 R_{\mathrm{i}}^4}{(1-\alpha_{\mathrm{mb}})R^5} + \frac{A}{R^2}\left(1-\frac{B}{p}\right) = 0 \tag{8.51}$$

On the basis of Eq. (8.51), distributions of the mean stress σ_{m} can be determined and, thus, stress distributions inside the hopper can be evaluated from Eq. (8.40).

Example 8.2 Consider a conical feed hopper with half angle 10°. The inlet and outlet diameters of the hopper are 1.0 and 0.5 m, respectively. The particles, 200 μm beads with a density of 2,300 kg/m^3, are fed into the hopper at the mass flow rate of 100 kg/m$^2\cdot$s under an inlet pressure of 10^5 Pa. The particles are in a moving bed motion with a particle volume fraction of 0.5. The pressure at the hopper outlet is 1.2×10^5 Pa. Determine the air flow rate in this hopper flow. The coefficient β is given as 102,700 kg/m$^3\cdot$s.

Solution In spherical coordinates with origin at the vertex, the distances of the hopper inlet and outlet from the vertex are given by

$$R_{\mathrm{i}} = \frac{r_{\mathrm{i}}}{\tan\varphi_{\mathrm{w}}} = \frac{0.5}{\tan 10^\circ} = 2.84\ \mathrm{m}, \qquad R_{\mathrm{o}} = \frac{r_{\mathrm{o}}}{\tan\varphi_{\mathrm{w}}} = \frac{0.25}{\tan 10^\circ} = 1.42\ \mathrm{m} \tag{E8.2}$$

The superficial velocity of particles at the inlet is calculated by

$$U_{\mathrm{Rpi}} = \frac{(\alpha_{\mathrm{p}}\rho_{\mathrm{p}}u_{\mathrm{p}})}{\rho_{\mathrm{p}}} = \frac{-100}{2{,}300} = -0.043\ \mathrm{m/s} \tag{E8.3}$$

where the minus sign indicates that the particles flow in the opposite direction of the polar axis.

The relationship between the inlet and outlet pressures is given by Eq. (8.47). Substituting all the given conditions into Eq. (8.47) leads to

$$10^5 - 1.2\times10^5 + B\ln\left(\frac{10^5-B}{1.2\times10^5-B}\right) = 1.027\times10^5\times2.84^2\left(\frac{-0.043}{0.5}\right)\left(\frac{1}{1.42}-\frac{1}{2.84}\right) \tag{E8.4}$$

from which B is obtained as 0.23×10^5 Pa. Thus, from Eq. (8.48), the superficial velocity of gas at the inlet is solved as

$$U_{\mathrm{Ri}} = U_{\mathrm{Rpi}}\frac{\alpha_{\mathrm{mb}}}{(1-\alpha_{\mathrm{mb}})}\frac{B}{p_{\mathrm{i}}} = -0.043\times\frac{0.5}{1-0.5}\frac{0.23\times10^5}{10^5} = -0.01\ \mathrm{m/s} \tag{E8.5}$$

The overall gas flow rate in the hopper is thus obtained by

$$Q = |U_{\mathrm{Ri}}|\pi r_{\mathrm{i}}^2 = 0.01\times\pi\times0.5^2 = 0.0079\ \mathrm{m^3/s} \tag{E8.6}$$

8.3.2 *Standpipe Flows*

As defined in §8.1, standpipe flow refers to the downward flow of solids with the aid of gravitational force against a gas pressure gradient. Although the gas flow with respect to the downward-flowing solids is in the upward direction, the actual direction of flow of gas relative to the wall can be either upward or downward [Knowlton, 1986]. Solids fed into the standpipe are often from hoppers, cyclones, or fluidized beds. A standpipe can be either vertical or inclined and its outlet can be simply an orifice or be connected to a valve or a fluidized bed. There can be aeration along the side of the standpipe. Figure 8.13 illustrates a typical standpipe application in catalytic cracking and regeneration systems. In this application, solids are fed into the vertical standpipes from fluidized beds and through slide valves. It should be noted that side aeration is employed for this standpipe flow.

The model given here for flows in a vertical standpipe follows closely the approach of Ginestra *et al.* (1980), Chen *et al.* (1984), and Jackson (1993). In this model, it is assumed that

(1) Both solids and gas are regarded as a pseudocontinuum throughout the standpipe system.
(2) Motions of solids and gas are steady and one-dimensional (in the axial direction).
(3) Solids can flow in either moving bed mode or dilute suspension mode. Solids stresses among particles and between particle and pipe wall are considered in the moving bed flows but neglected in the dilute suspension flows.
(4) The gas can be regarded as an ideal gas, and the transport process is isothermal.

The cylindrical coordinate system is selected for the standpipe, as shown in Fig. 8.14. For a one-dimensional steady motion of solids, the momentum equation of the particle phase can be written as

$$\rho_{\mathrm{p}}(1-\alpha)u_{\mathrm{zp}}\frac{\mathrm{d}u_{\mathrm{zp}}}{\mathrm{d}z} = \rho_{\mathrm{p}}g(1-\alpha) - \frac{\mathrm{d}\sigma_{\mathrm{pz}}}{\mathrm{d}z} - \frac{2\tau_{\mathrm{pw}}}{R_{\mathrm{s}}} + F_{\mathrm{D}} \tag{8.52}$$

where σ_{pz} is the normal stress of solids; τ_{pw} is the shear stress of solids at the pipe wall; F_{D} is the drag force per unit volume; and R_{s} is the radius of the standpipe. Under low particle Reynolds number conditions where the drag is dominated by the viscous effects, F_{D} can be expressed in terms of the relative velocity of gas and solids as

$$F_{\mathrm{D}} = \beta(u_{\mathrm{z}} - u_{\mathrm{zp}}) \tag{8.53}$$

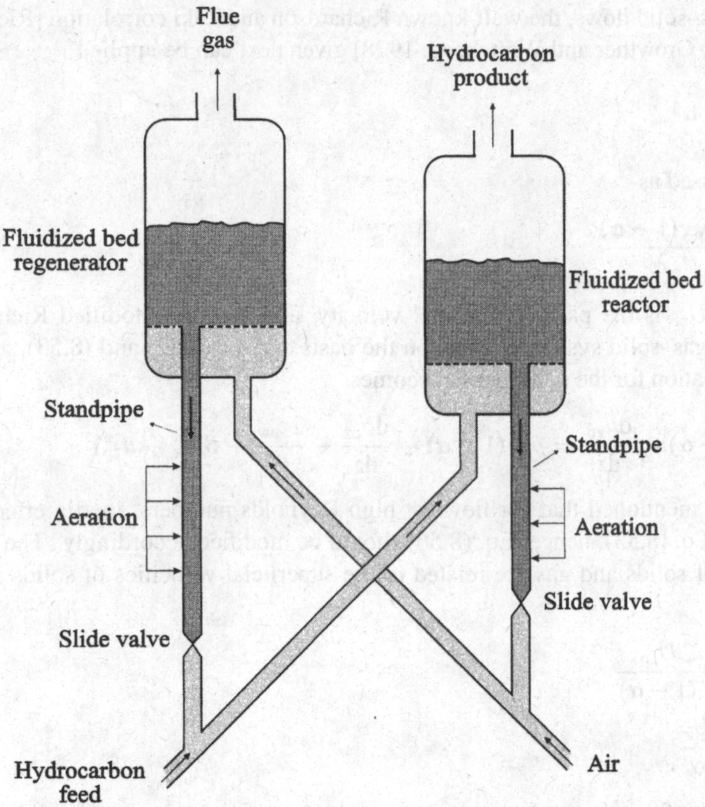

Figure 8.13. Typical solids flow loop in a catalytic cracker with underflow standpipes.

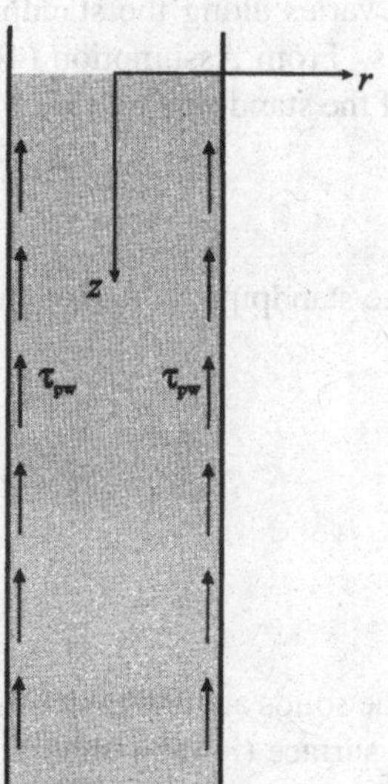

Figure 8.14. One-dimensional standpipe flow.

For uniform gas–solid flows, the well-known Richardson and Zaki correlation [Richardson and Zaki, 1954; Growther and Whitehead, 1978] given next can be applied

$$\frac{U}{U_{\mathrm{pt}}} = \alpha^n \tag{8.54}$$

β can be expressed as

$$\beta = \frac{\rho_{\mathrm{p}} g(1-\alpha)}{U_{\mathrm{pt}}\alpha^{n-1}} \tag{8.55}$$

In Eq. (8.54), U_{pt} is the particle terminal velocity, and n is the modified Richardson–Zaki index for gas–solid systems. Thus, on the basis of Eqs. (8.52) and (8.53), a general momentum equation for the solid phase becomes

$$\rho_{\mathrm{p}}(1-\alpha)u_{\mathrm{zp}}\frac{\mathrm{d}u_{\mathrm{zp}}}{\mathrm{d}z} = \rho_{\mathrm{p}} g(1-\alpha) - \frac{\mathrm{d}\sigma_{\mathrm{pz}}}{\mathrm{d}z} - \frac{2\tau_{\mathrm{pw}}}{R_{\mathrm{s}}} + \beta(u_{\mathrm{z}} - u_{\mathrm{zp}}) \tag{8.56}$$

It should be mentioned that for flows at high Reynolds numbers, inertia effects must be included in Eq. (8.53), hence, Eq. (8.56) should be modified accordingly. The interstitial velocities of solids and gas are related to the superficial velocities of solids and gas, respectively, by

$$U_{\mathrm{zp}} = \frac{U_{\mathrm{zp}}}{(1-\alpha)} \tag{8.57a}$$

$$u_{\mathrm{z}} = \frac{U_{\mathrm{z}}}{\alpha} \tag{8.57b}$$

It is noted that from the continuity equation, the superficial velocity of solids remains constant since the cross-sectional area of the standpipe is unchanged and the solid material is incompressible. However, the superficial gas velocity varies along the standpipe as a result of the pressure variation and compressibility of gas. From Assumption (4), U_{z}, ρ, and p at any location can be related to those at the inlet of the standpipe by

$$\frac{U_{\mathrm{z0}}}{U_{\mathrm{z}}} = \frac{\rho}{\rho_0} = \frac{p}{p_0} \tag{8.58}$$

where subscript "0" denotes the location at the inlet of the standpipe ($z = 0$). Therefore, Eqs. (8.57b) and (8.58) yield

$$u_{\mathrm{z}} = \frac{U_{\mathrm{z0}} p_0}{\alpha p} \tag{8.59}$$

Since the particles are sliding down at the wall, we have

$$\tau_{\mathrm{pw}} = f_{\mathrm{w}}\sigma_{\mathrm{prw}} \tag{8.60}$$

where f_{w} is the coefficient of kinematic friction between the solids and the pipe wall, while σ_{prw} represents the normal stress in the solids on the wall surface (in the radial direction). From Assumption (2), the normal stress in the radial direction is constant. Thus, σ_{prw} can be related to σ_{pz} by Eq. (8.11) as

$$\sigma_{\mathrm{prw}} = \sigma_{\mathrm{pr}} = K\sigma_{\mathrm{pz}} \tag{8.61}$$

where K is given by Eq. (8.10). Thus, substitution of Eqs. (8.57a), (8.59), (8.60), and (8.61) into Eq. (8.56) gives

$$\frac{\rho_p U_{zp}^2}{(1-\alpha)^2}\frac{d\alpha}{dz} = \rho_p g(1-\alpha) - \left[\frac{d\sigma_{pz}}{dz} + \frac{2 f_w K \sigma_{pz}}{R_s}\right] + \beta\left[\frac{U_{z0} p_0}{\alpha p} - \frac{U_{zp}}{1-\alpha}\right] \tag{8.62}$$

Equation (8.62) is the general momentum balance equation. For moving bed solids transport in the standpipe, α is constant with a value of α_{mb}. Thus, Eq. (8.62) can be simplified to

$$\frac{d\sigma_{pz}}{dz} + \frac{2 f_w K \sigma_{pz}}{R_s} = \rho_p g(1-\alpha_{mb}) + \beta\left[\frac{U_{z0} p_0}{\alpha_{mb} p} - \frac{U_{zp}}{1-\alpha_{mb}}\right] \tag{8.63}$$

It is noted that, in the preceding equation, the drag force represented by the second term on the right-hand side may be positive or negative depending on the compressibility of gas reflected in the value of p. Thus, σ_{pz} may increase or decrease along the pipe depending on the effect of the drag force. For a suspension flow mode of solids transport, σ_{pz} is negligibly small. Consequently, Eq. (8.62) reduces to

$$\frac{U_{zp}^2}{g(1-\alpha)^3}\frac{d\alpha}{dz} = 1 + \frac{\beta}{\rho_p g(1-\alpha)}\left[\frac{U_{z0} p_0}{\alpha p} - \frac{U_{zp}}{1-\alpha}\right] \tag{8.64}$$

By equating the pressure gradient and the gas–particle drag force, the general momentum balance for the gas phase can be expressed as

$$\frac{dp}{dz} = \beta(u_{zp} - u_z) \tag{8.65}$$

In an alternative form, Eq. (8.65) becomes

$$\frac{dp}{dz} = \beta\left[\frac{U_{zp}}{1-\alpha} - \frac{U_{z0} p_0}{\alpha p}\right] \tag{8.66}$$

Equation (8.66) is directly applicable to the suspension flow mode of solids transport. For a moving bed flow, Eq. (8.66) can be simplified to

$$\frac{dp}{dz} = \beta\left[\frac{U_{zp}}{1-\alpha_{mb}} - \frac{U_{z0} p_0}{\alpha_{mb} p}\right] \tag{8.67}$$

For suspension flows, solutions for α and p can be obtained by solving numerically the governing equations, *i.e.*, Eqs. (8.64) and (8.66). For moving bed flows, analytical expressions for the solutions of the governing equations, *i.e.*, Eqs. (8.63) and (8.67), can be obtained. Integrating Eq. (8.67), p can be expressed in a dimensionless form

$$p^* - 1 + a \ln\left[\frac{p^* - a}{1-a}\right] = bz^* \tag{8.68}$$

where

$$p^* = \frac{p}{p_0}; \quad z^* = \frac{z}{H}; \quad a = \frac{U_{z0}(1-\alpha_{mb})}{U_{zp}\alpha_{mb}}; \quad b = \frac{H\beta U_{zp}}{p_0(1-\alpha_{mb})} \tag{8.69}$$

where H is the height of the standpipe. Integrating Eq. (8.63) yields

$$\sigma^* = \sigma_0^* \exp(-cz^*) + \left\{ \int_0^{z*} \left[1 + d\left(\frac{a}{p^*} - 1 \right) \right] \exp(cx)\,\mathrm{d}x \right\} \exp(-cz^*) \tag{8.70}$$

where

$$\sigma^* = \frac{\sigma_{\mathrm{pz}}}{H\rho_{\mathrm{p}}g(1-\alpha_{\mathrm{mb}})}; \qquad c = \frac{2f_{\mathrm{w}}KH}{R_{\mathrm{s}}}; \qquad d = b\frac{p_0}{\rho_{\mathrm{p}}gH(1-\alpha_{\mathrm{mb}})} \tag{8.71}$$

The bulk flux of solids in a standpipe depends not only on the geometry of the standpipe but also on its feed devices and flow control valve, located at the top and bottom of the standpipe, respectively. The flow pattern of solids in the pipe may vary with the pipe distance, depending on the variation of the slip velocity in the pipe, $u_{\mathrm{zp}} - u_{\mathrm{z}}$. The effect of gas compression due to pressure increase may also be very significant. As a result, ρ varies along the standpipe, which in turn changes α and, consequently, the slip velocity. Thus, for a standpipe flow, it is possible that several flow patterns exist in various axial locations as solids travel through the standpipe.

8.3.3 *Hopper–Standpipe–Discharger Flow*

As mentioned, the flow rate in a standpipe depends on the solid feed device as well as the flow control valve. In this section, we discuss the gas–solid flows in a simple standpipe system where the feed device is a mass flow hopper and the solid flow regulator is a discharge orifice [Chen *et al.*, 1984]. As shown in Fig. 8.15, the entrance of the vertical standpipe is connected to a conical hopper feeder of half angle φ_{w}, while a discharge orifice is located at the bottom of the standpipe. Here, two representative solids flow patterns are considered. One is a dilute suspension flow, and the other is a solid moving bed. In this case, the following additional assumptions are needed:

(1) The angle of the feed hopper is small so that a continued flow of solids to the standpipes can be established without creating a stagnant region in the hopper.
(2) The solid flow through the discharge orifice is characterized by the flow through a conical converging channel due to the accumulation of solids on the orifice plate. The half angle of the channel is the angle of friction for the particles, as shown in Fig. 8.15.
(3) For a given type of solids, the control variables of the standpipe system are the orifice opening and the pressure drop from the top of the hopper to the bottom of the orifice.

The general system in Fig. 8.15 consists of six sections or regions: the hopper section, the transition region I (between hopper and standpipe), the dilute transport region in the standpipe, the moving bed region in the standpipe, the transition region II (between the standpipe and discharge orifice), and the discharge section.

In the hopper section (1–2), the pressure drop is given by

$$p_2 - p_1 + B\ln\left(\frac{p_2 - B}{p_1 - B}\right) = A\left(\frac{1}{R_1} - \frac{1}{R_2}\right) \tag{8.72}$$

where $R_2 = R_{\mathrm{s}}/\sin\varphi_{\mathrm{w}}$, and A and B are given by Eq. (8.48) with the subscript "i" representing "2." The velocities of gas and particles are given by Eqs. (8.44) and (8.45) with the subscripts "i" and "o" representing "2" and "1," respectively.

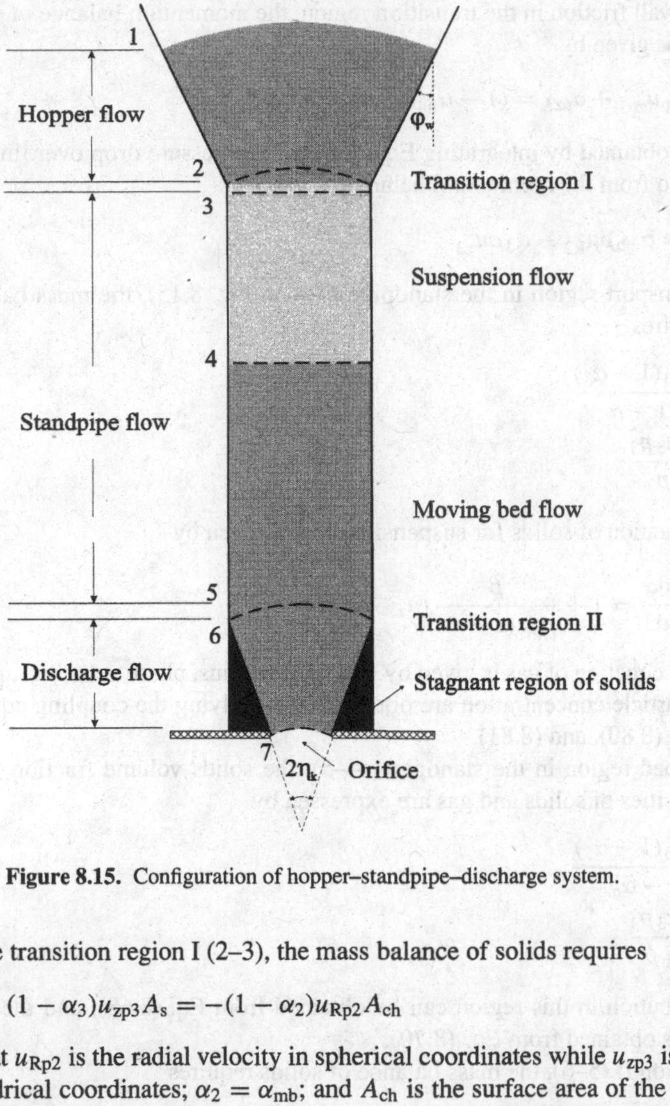

Figure 8.15. Configuration of hopper–standpipe–discharge system.

In the transition region I (2–3), the mass balance of solids requires

$$(1-\alpha_3)u_{zp3}A_s = -(1-\alpha_2)u_{Rp2}A_{ch} \tag{8.73}$$

Note that u_{Rp2} is the radial velocity in spherical coordinates while u_{zp3} is the axial velocity in cylindrical coordinates; $\alpha_2 = \alpha_{mb}$; and A_{ch} is the surface area of the spherical cap near the exit of the hopper, whereas A_s represents the cross-sectional area of the standpipe. A_{ch} is related to A_s by

$$\frac{A_{ch}}{A_s} = \frac{2(1-\cos\varphi_w)}{\sin^2\varphi_w} \tag{8.74}$$

Thus, Eq. (8.73) becomes

$$-\frac{u_{zp3}}{u_{Rp2}} = \left(\frac{1-\alpha_{mb}}{1-\alpha_3}\right)\frac{2(1-\cos\varphi_w)}{\sin^2\varphi_w} \tag{8.75}$$

Similarly, the mass balance of gas (assumed incompressible in this region) requires

$$-\frac{u_{z3}}{u_{R2}} = \frac{\alpha_{mb}}{\alpha_3}\frac{2(1-\cos\varphi_w)}{\sin^2\varphi_w} \tag{8.76}$$

By neglecting the wall friction in the transition region, the momentum balance of solids in the axial direction is given by

$$(1-\alpha_3)\rho_p u_{zp3}^2 + \sigma_{pz3} = (1-\alpha_{mb})\rho_p u_{Rp2}^2 + \sigma_{pR2} \tag{8.77}$$

where σ_{pR2} can be obtained by integrating Eq. (8.51). The pressure drop over this region can also be obtained from the momentum balance of gas as

$$p_3 - p_2 = \alpha_{mb}\rho u_{R2}^2 - \alpha_3 \rho u_{z3}^2 \tag{8.78}$$

In the dilute transport region in the standpipe (3–4 in Fig. 8.15), the mass balance of solids and gas requires

$$u_{zp} = \frac{u_{zp3}(1-\alpha_3)}{1-\alpha} \tag{8.79}$$

$$u_z = \frac{u_{z3}\alpha_3 p_3}{\alpha p} \tag{8.80}$$

The momentum equation of solids for suspension flow is given by

$$\frac{u_{zp}^2}{g(1-\alpha)}\frac{d\alpha}{dz} = 1 + \frac{\beta}{\rho_p g(1-\alpha)}(u_z - u_{zp}) \tag{8.81}$$

and the momentum equation of gas is given by Eq. (8.65). Thus, phase velocities, pressure distribution, and particle concentration are obtained from solving the coupling equations: Eqs. (8.65), (8.79), (8.80), and (8.81).

In the moving bed region in the standpipe (4–5), the solids volume fraction remains constant. The velocities of solids and gas are expressed by

$$u_{zp} = \frac{u_{zp3}(1-\alpha_3)}{1-\alpha_{mb}} \tag{8.82}$$

$$u_z = \frac{u_{z3}\alpha_3 p_3}{\alpha_{mb} p} \tag{8.83}$$

The pressure distribution in this region can be obtained from Eq. (8.68), and the normal stress distribution is obtained from Eq. (8.70).

In transition region II (5–6), the mass balance of solids requires

$$u_{zp5} A_s = -u_{Rp6} A_{co} \tag{8.84}$$

Note that u_{zp5} is the axial velocity in the cylindrical coordinate while u_{Rp6} is the radial velocity in the spherical coordinate; A_s represents the cross-sectional area of the standpipe whereas A_{co} is the surface area of the spherical cup near the inlet of the discharge orifice region. A_{co} is related to A_s by

$$\frac{A_{co}}{A_s} = \frac{2(1-\cos\eta_k)}{\sin^2\eta_k} \tag{8.85}$$

Thus,

$$-\frac{u_{zp5}}{u_{Rp6}} = -\frac{u_{z5}}{u_{R6}} = \frac{2(1-\cos\eta_k)}{\sin^2\eta_k} \tag{8.86}$$

By neglecting the wall friction in the transition region, the momentum balance of solids in the axial direction is given by

$$(1-\alpha_{\mathrm{mb}})\rho_{\mathrm{p}}u_{\mathrm{zp5}}^2+\sigma_{\mathrm{pz5}}=(1-\alpha_{\mathrm{mb}})\rho_{\mathrm{p}}u_{\mathrm{Rp6}}^2+\sigma_{\mathrm{pR6}} \tag{8.87}$$

The pressure drop over this region is obtained from the momentum balance of gas as

$$p_6-p_5=\alpha_{\mathrm{mb}}\rho u_{\mathrm{z5}}^2-\alpha_{\mathrm{mb}}\rho u_{\mathrm{R6}}^2 \tag{8.88}$$

In the discharge orifice region (6–7 in Fig. 8.15), the pressure drop is given by

$$p_7-p_6+B\ln\left(\frac{p_7-B}{p_6-B}\right)=A\left(\frac{1}{R_6}-\frac{1}{R_7}\right) \tag{8.89}$$

where $R_6=R_{\mathrm{s}}/\sin\eta_{\mathrm{k}}$ and $R_7=r_{\mathrm{d}}/\sin\eta_{\mathrm{k}}$. r_{d} is the radius of the discharge orifice. A and B are given by Eq. (8.48) with the subscript "i" representing "6." The velocities of gas and particles are given by Eqs. (8.44) and (8.45) with the subscripts "i" and "o" representing "6" and "7," respectively.

Thus, the total pressure drop over the system is

$$\Delta p=p_1-p_7=\sum_{\mathrm{i}=1}^{6}(p_{\mathrm{i}}-p_{\mathrm{i}+1}) \tag{8.90}$$

and the mass flow rates depend on the size of the discharge orifice.

8.3.4 *Multiplicity of Steady Standpipe Flows*

For a standpipe system, different flow patterns of steady-state flows may exist, depending upon the ranges of operational parameters of the system. This phenomenon is commonly known as the steady-state multiplicity. We introduce the general concepts of the steady-state multiplicity based on the standpipe system in §8.3.3, in which the solids in the feed hopper are kept in a steady moving bed motion.

There are four possible flow regimes for standpipe flows, as shown in Fig. 8.16. Regime 1 represents a moving bed flow mode throughout the standpipe. If a suspension flow occurs, the gas volume fraction α increases with z as the particle accelerates away from the hopper outlet until the gravity and drag forces are in balance. Then, α decreases as a result of gas compression until α_{mb} is reached and the moving bed flow begins. If the location where α reaches α_{mb} is in the region above the discharge cone, the flow is in Regime 2, while if it is inside the discharge cone region, then the flow is in Regime 3. If it never reaches α_{mb} within the length of the standpipe, the flow is in Regime 4. The flow behavior in each of the four regimes can thus be simulated using the model described in §8.3.3. Consequently, the multiplicity of steady standpipe flows for this standpipe system can be characterized by two parameters, namely, the ratio of discharge opening diameter to the standpipe diameter γ and the pressure drop over the system Δp.

For a system with a fixed γ, the steady-state multiplicity may be revealed by the curve of superficial velocity of solids U_{p} versus Δp, as illustrated in Fig. 8.17, where a hysteresis loop is formed between two branch (upper and lower) curves [Chen *et al.*, 1984]. The upper branch corresponds to Regimes 2, 3, or 4, whereas the lower one corresponds to Regime 1. At Δp less than $(\Delta p)_1$, the system is operated on the upper branch of the curves, while it is operated on the lower branch of the curves at Δp larger than $(\Delta p)_2$. Between these two

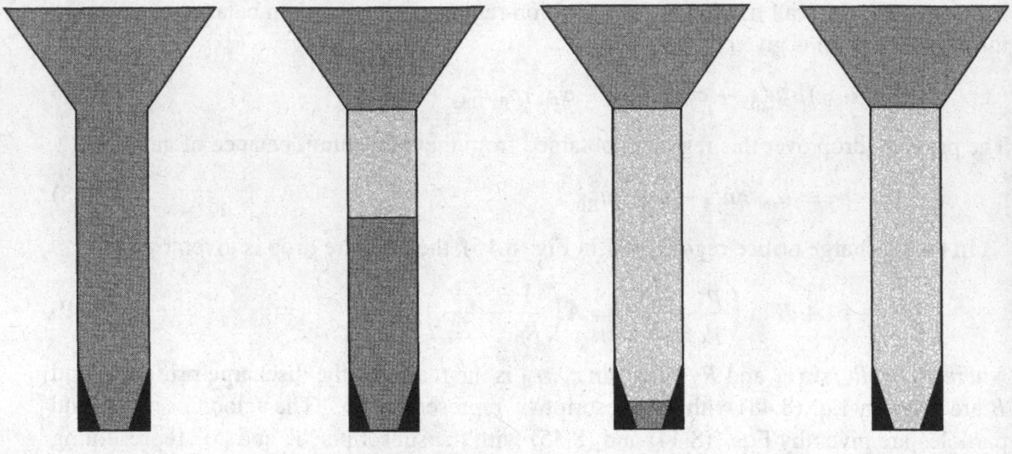

Figure 8.16. Flow regimes for standpipe without aeration (from Chen *et al.*, 1984).

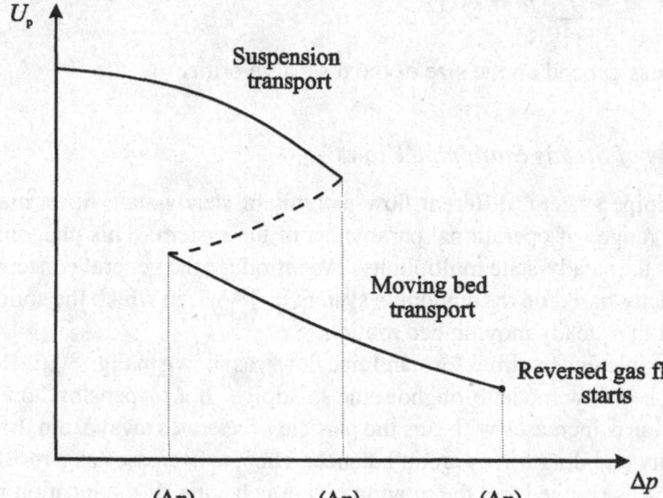

Figure 8.17. Schematic illustration of multiplicity of steady standpipe flows (after Chen *et al.*, 1984).

values of Δp, the system can be operated at one of the multiple superficial velocities of solids. Furthermore, the transition from Regime 1 to other regimes (or vice versa) may not be continuous. For a system initially operated in Regime 1, as Δp decreases, U_p follows the lower curve until the point of $(\Delta p)_1$ where a sudden rise of U_p to the upper curve takes place, indicating a sudden transition from Regime 1 to Regime 2 where a suspension section in the standpipe occurs. In contrast, if the system initially has a section of suspension flow, as Δp increases, U_p follows the upper curve until the point of $(\Delta p)_2$, where a sudden drop of U_p to the lower curve takes place. This indicates a transition from Regime 2 to Regime 1 where the solids suddenly fill the suspension section into a moving bed. In addition, it is noted that the standpipe system may not be operated at a very high Δp. The upper limit $(\Delta p)_3$ corresponds to the start of upward gas flow relative to the pipe wall.

In standpipe systems, aeration from the side is often implemented to provide a lubricating effect for the solids flow. Aeration is also used in process applications to provide a stripping gas to remove undesirable gas entrained by the solid particles in the standpipe flow. With side aeration, the multiplicity of flows in the standpipe becomes much more complicated. The flow patterns of solids and gas depend not only on Δp and γ but also on the aeration locations and aeration rates. For a single aeration point, the possible number of flow regimes increases to 12 [Mountziaris and Jackson, 1991].

8.3.5 *Leakage Flow of Gas in a Standpipe*

Standpipes are used in the gas–solid circulating flow systems in which a cyclone–standpipe–valve configuration for the separation and reintroduction of particles, as shown in Fig. 8.18, is constantly encountered. The satisfactory operation of the cyclone–standpipe system depends on a small leakage flow of gas up the standpipe. A simple method to estimate

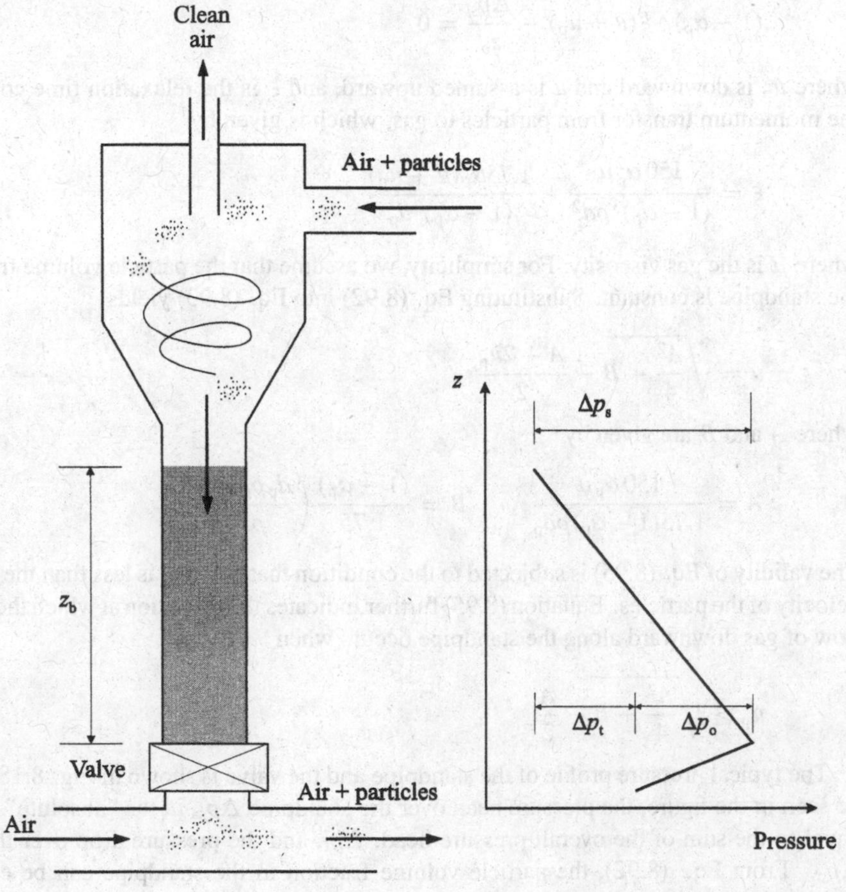

Figure 8.18. Schematic configuration of a cyclone–standpipe–valve system.

this leakage flow of gas in the standpipe is delineated by Soo and Zhu (1993) as given in the following.

Consider the flow of cohesionless material in the system. The particle mass flow through a valve can be modeled as the flow of solids through an orifice where the Bernoulli relation holds [Li *et al.*, 1982]. Thus the particle mass flow in the standpipe can be expressed as

$$\alpha_p \rho_p u_p = C_o \frac{A_o}{A_s} \sqrt{2\rho_p \alpha_{pmf} \Delta p_o} \tag{8.91}$$

where C_o is an orifice coefficient, A_o is the valve opening area, A_s is the cross-sectional area of the standpipe, and α_{pmf} is the volume fraction of particles at minimum fluidization. When the volume fraction of particles in the standpipe is higher than 0.2, the pressure head due to gravity in the standpipe can be estimated by [Chong *et al.*, 1987]

$$\Delta p_s = \alpha_p \rho_p g z_b \tag{8.92}$$

where z_b is the height of bulk solids in the standpipe.

The momentum equation of gas in the standpipe can be expressed as [Soo, 1989]

$$(1 - \alpha_p)\rho\, \xi (u + u_p) - \frac{\Delta p_s}{z_b} = 0 \tag{8.93}$$

where u_p is downward and u is assumed upward, and ξ is the relaxation time constant of the momentum transfer from particles to gas, which is given by

$$\xi = \frac{150\,\alpha_p^2 \mu}{(1 - \alpha_p)^3 \rho d_p^2} + \frac{1.75\alpha_p (u + u_p)}{(1 - \alpha_p)^2 d_p} \tag{8.94}$$

where μ is the gas viscosity. For simplicity, we assume that the particle volume fraction in the standpipe is constant. Substituting Eq. (8.92) into Eq. (8.93) yields

$$u = \sqrt{\frac{A^2}{4} + B} - \frac{A + 2u_p}{2} \tag{8.95}$$

where A and B are given by

$$A = \frac{150\,\alpha_p \mu}{1.75(1 - \alpha_p)\rho d_p}, \qquad B = \frac{(1 - \alpha_p)}{1.75} \left(\frac{d_p \rho_p g}{\rho} \right) \tag{8.96}$$

The validity of Eq. (8.95) is subjected to the condition that $|u + u_p|$ is less than the terminal velocity of the particles. Equation (8.95) further indicates the condition at which the leakage flow of gas downward along the standpipe occurs when

$$u_p > \sqrt{\frac{A^2}{4} + B} - \frac{A}{2} \tag{8.97}$$

The typical pressure profile of the standpipe and the valve is shown in Fig. 8.18. As can be seen in the figure, the pressure head over the standpipe Δp_s, in the "absolute" sense, is equal to the sum of the overall pressure head, Δp_t, and the pressure drop over the valve, Δp_o. From Eq. (8.92), the particle volume fraction in the standpipe can be estimated from the measured height of the bulk particles. Thus, given the particle mass flow rate, the leakage flow of gas in the standpipe can be estimated from Eq. (8.95).

Example 8.3 In an operation of gas–solid circulating flow in a cyclone–standpipe–valve system, the particles are 100 μm glass beads with a density of 2,500 kg/m^3. The particle volume fraction in the standpipe is 0.55. The gas is air with a viscosity of 1.8×10^{-5} kg/m·s and a density of 1.2 kg/m^3. The particle mass flow is 70 kg/m^2·s. The height of solids in the standpipe is 1.4 m. The total pressure head over the standpipe and valve is 4,500 Pa. Estimate the leakage flow of air in the standpipe. If the particle volume fraction at minimum fluidization is 0.5 and the area ratio of valve opening to pipe cross section is 0.6, what is the orifice coefficient of this valve?

Solution The interstitial velocity of particles in the standpipe is obtained by

$$u_{\mathrm{p}} = \frac{(\alpha_{\mathrm{p}}\rho_{\mathrm{p}}u_{\mathrm{p}})}{\rho_{\mathrm{p}}\alpha_{\mathrm{p}}} = \frac{70}{2{,}500 \times 0.55} = 0.051 \text{ m/s} \tag{E8.7}$$

From Eq. (8.96), we have

$$A = \frac{150\,\alpha_{\mathrm{p}}\mu}{1.75(1-\alpha_{\mathrm{p}})\rho d_{\mathrm{p}}} = \frac{150 \times 0.55 \times 1.8 \times 10^{-5}}{1.75 \times 0.45 \times 1.2 \times 10^{-4}} = 15.71 \text{ m/s}$$

$$B = \frac{(1-\alpha_{\mathrm{p}})}{1.75}\left(\frac{d_{\mathrm{p}}\rho_{\mathrm{p}}g}{\rho}\right) = \frac{0.45}{1.75}\left(\frac{10^{-4} \times 2{,}500 \times 9.8}{1.2}\right) = 0.525 \text{ m}^2/\text{s}^2 \tag{E8.8}$$

Thus, the interstitial velocity of leakage flow of air is estimated from Eq. (8.95) as

$$u = \sqrt{\frac{A^2}{4} + B} - \frac{A + 2u_{\mathrm{p}}}{2} = \sqrt{\frac{15.71^2}{4} + 0.525} - \frac{15.71 + 2 \times 0.051}{2} = -0.018 \text{ m/s} \tag{E8.9}$$

The minus sign means that the leakage flow of air is directed downward.

The pressure drop over the valve is given by

$$\Delta p_{\mathrm{o}} = \alpha_{\mathrm{p}}\rho_{\mathrm{p}}gz_{\mathrm{b}} - \Delta p_{\mathrm{t}} = 0.55 \times 2{,}500 \times 9.8 \times 1.4 - 4{,}500 = 14{,}365 \text{ Pa} \tag{E8.10}$$

Thus, on the basis of Eq. (8.91), the orifice coefficient of the valve can be calculated as

$$\begin{aligned} C_{\mathrm{o}} &= \alpha_{\mathrm{p}}\rho_{\mathrm{p}}u_{\mathrm{p}}\left(\frac{A_{\mathrm{o}}}{A}\sqrt{2\rho_{\mathrm{p}}\alpha_{\mathrm{pmf}}\Delta p_{\mathrm{o}}}\right)^{-1} \\ &= 70 \times (0.6\sqrt{2 \times 2{,}500 \times 0.5 \times 14{,}365})^{-1} = 0.019 \end{aligned} \tag{E8.11}$$

8.4 Types of Standpipe Systems

Standpipes are applied to fluidized bed systems in several ways. The following describes industrial applications of standpipes. An application of nonmechanical valves which control the solids flow rates is also given.

8.4.1 *Overflow and Underflow Standpipes*

Standpipes are commonly grouped into overflow and underflow types, as illustrated in Fig. 8.19. An overflow standpipe is structured with solids overflow from the top of a

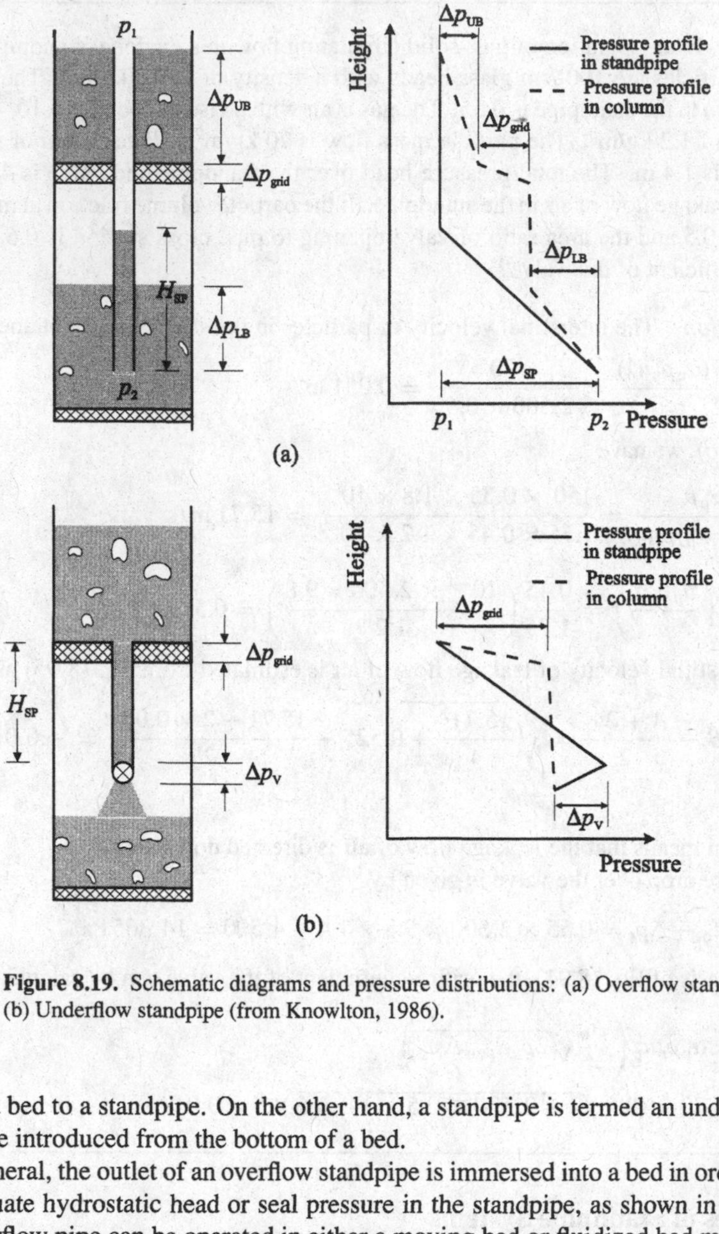

Figure 8.19. Schematic diagrams and pressure distributions: (a) Overflow standpipe; (b) Underflow standpipe (from Knowlton, 1986).

fluidized bed to a standpipe. On the other hand, a standpipe is termed an underflow type if solids are introduced from the bottom of a bed.

In general, the outlet of an overflow standpipe is immersed into a bed in order to provide an adequate hydrostatic head or seal pressure in the standpipe, as shown in Fig. 8.19(a). The overflow pipe can be operated in either a moving bed or fluidized bed mode. Usually, solids occupy only a fraction of the pipe near the standpipe outlet. The solids flow rate in the pipe depends on the overflow rate from the fluidized bed. Consequently, no valve is needed at the pipe outlet. In the overflow standpipe system where solids transfer from one fluidized bed to the other, the pressure drops across the lower fluidized bed, the grid (distributor), the upper fluidized bed, and the height of solids in the standpipe. For a given standpipe system and given particles, the pressure distribution profile depends on the gas velocity. The difference in the pressure distribution reflects the difference in solids concentration, bed height for the lower bed, solids concentration for the upper bed, and solids holdup height in the standpipe.

Solids in an underflow standpipe are often discharged to a freeboard region of a lower fluidized bed via a valve which regulates the solids flow rate, as illustrated in Fig. 8.19(b). Solids in the standpipe are usually in a moving bed mode with an adequate height to provide a seal pressure for the downward solids flow. If the solids size is not uniformly distributed in the fluidized bed, the underflow pipe will basically discharge small particles. The pressure distribution in Fig. 8.19(b) shows the pressure drops across the gas distributor in the upper fluidized bed, underflow standpipe, discharge valve, and freeboard region at the bottom of the standpipe. The gradients of the pressures across the distributor and standpipe are higher than those of other parts of pressure distributions shown in the figure.

For an underflow standpipe system, the pressure distribution varies with the gas velocity. By defining the slip velocity U_s as $U_p - U$, Kojabashian (1958) indicated that, because of gas compression, when the decrease of voidage is greater than the increase of gas density, U_p decreases and U increases, and hence U_s decreases. In a bubbling fluidized standpipe flow, when U_s decreases to a value less than the minimum fluidization velocity, defluidization in the standpipe occurs. In this case, if the gas velocity relative to the solid velocity is upward or U_s is positive and directed upward, the pressure in the defluidized region increases with decreasing pipe height, as illustrated in Fig. 8.20(a). On the other hand, if the gas

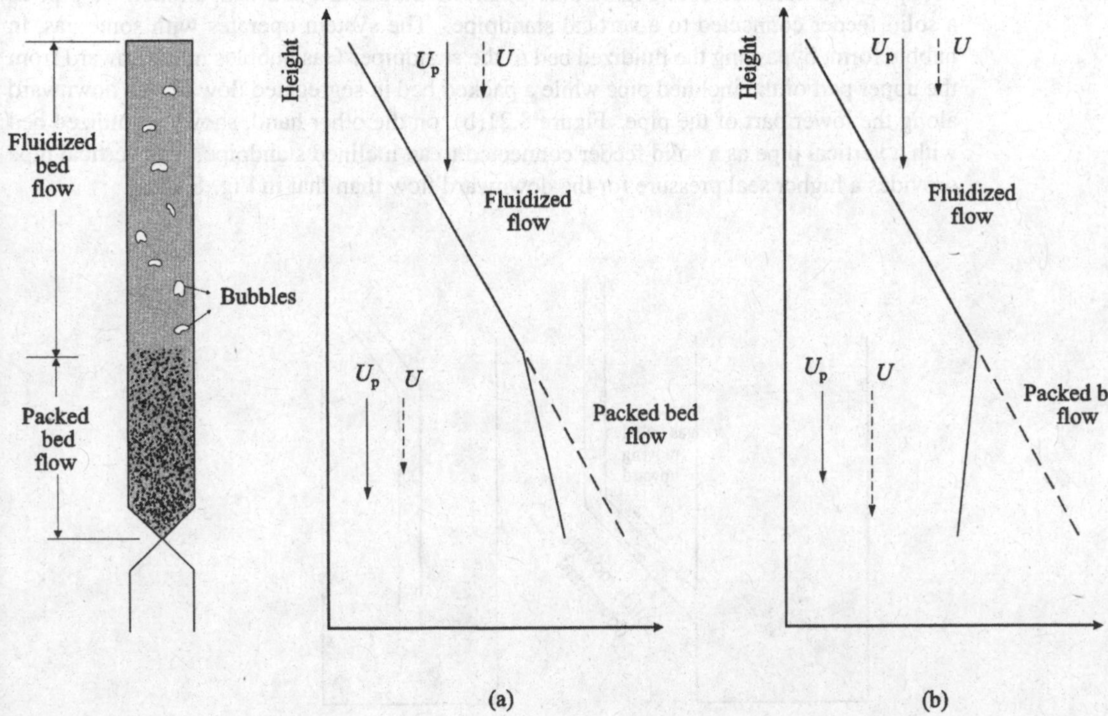

Figure 8.20. Pressure profiles in an underflow standpipe (from Knowlton, 1986): (a) Without pressure reversal; (b) With pressure reversal.

velocity relative to the solid velocity is downward or U_s is negative and directed downward, the pressure in the defluidized region decreases with decreasing pipe height (*i.e.*, pressure reversal occurs), as shown in Fig. 8.20(b).

In Fig. 8.20(b), there are two distinctive regions for pressure distributions, namely, a standpipe flow condition (*i.e.*, solids flow against the gas pressure gradient) and a discharge flow condition (*i.e.*, solids flow coinciding with the gas pressure gradient). However, considering the overall gas pressure gradient between the inlet and outlet, the overall pipe flow is a standpipe flow. It is noted that the solids flow needs to overcome the pressure drop further through the flow control valve. Hence, it is necessary to have the overall pressure buildup sufficiently high to yield a sealing effect which prohibits the gas from flowing back into the standpipe.

The pressure drop in the overflow and underflow standpipes can be predicted using Eqs. (8.64) through (8.67) considering either the moving bed flow or the suspension flow. The pressure drops across the fluidized bed and grid or distributor can be evaluated, respectively, by Eqs. (9.7) and (P9.7) given in Chapter 9.

8.4.2 *Inclined Standpipe and Nonmechanical Valves*

In practice, it is common to have standpipes with inclined configurations. Figure 8.21 illustrates two examples of applications using an inclined standpipe for solids transport to a fluidized bed. Figure 8.21(a) shows a fluidized bed with an inclined pipe as a solid feeder connected to a vertical standpipe. The system operates with some gas, in bubble form, bypassing the fluidized bed to the standpipe. Gas bubbles move upward from the upper part of the inclined pipe while a packed bed in segregated flow moves downward along the lower part of the pipe. Figure 8.21(b), on the other hand, shows a fluidized bed with a vertical pipe as a solid feeder connected to an inclined standpipe. The vertical pipe provides a higher seal pressure for the downward flow than that in Fig. 8.21(a).

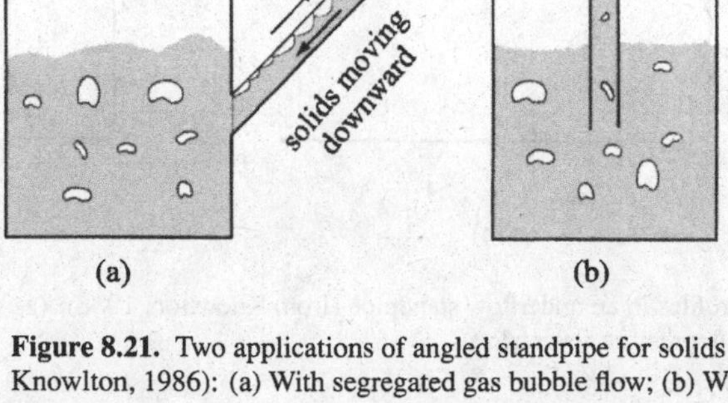

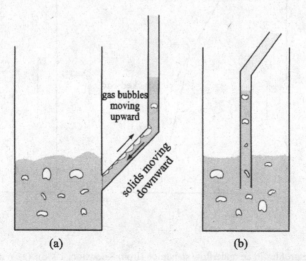

Figure 8.21. Two applications of angled standpipe for solids flow to a fluidized bed (from Knowlton, 1986): (a) With segregated gas bubble flow; (b) Without segregated gas bubble flow.

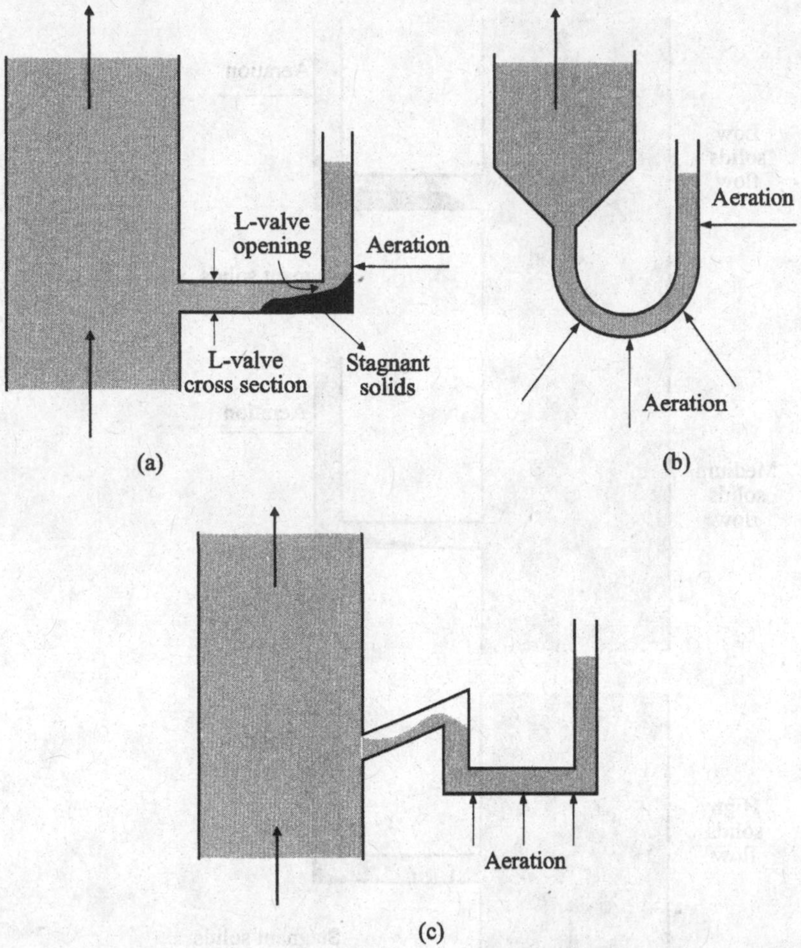

Figure 8.22. Nonmechanical valves and loopseal: (a) L-valve; (b) J-valve; (c) Loopseal.

In standpipe systems, the solids flow rate can be controlled by using nonmechanical valves (*e.g.*, L-valve and J-valve), as shown in Fig. 8.22(a) and (b). In such arrangements, only aeration gas is needed to control the solids flow. Arrangements of this nature are known as nonmechanical valves; as the name implies, they use nonmechanical means to control the solids flow rate. By adjusting the gas aeration rate, the solids flow into a dense or dilute environment can be controlled [Knowlton, 1986]. Using an L-valve for low solids flow rates, a stagnant solids region will form and the size of the region will reduce as the solids flow rate increases, as shown in Fig. 8.23. The loopseal shown in Fig. 8.22(c) provides a viable means for gas seal when solids are transferred to a vessel. This is achieved by moving up the solids in a short standpipe at the end of the L-valve and introducing the solids to the vessel by an inclined pipe. The solids flow rate can also be controlled by aeration along the loopseal.

The L-valve is the most commonly used nonmechanical valve. The pressure drop across the L-valve can be evaluated by Eqs. (10.9) and (10.10), given in Chapter 10.

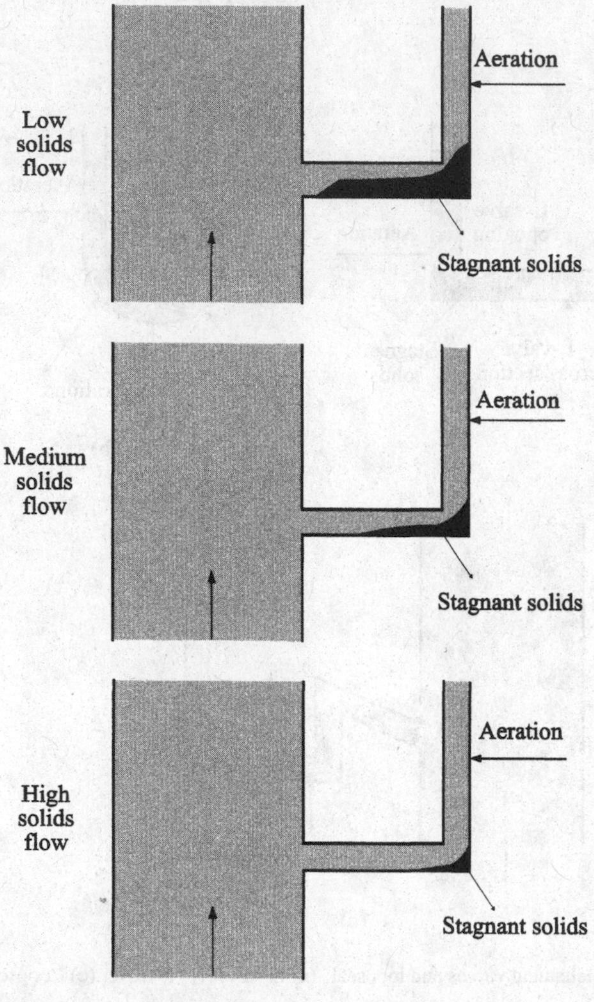

Figure 8.23. Stagnant regions in an L-valve for three solids flow rates (from Knowlton, 1986).

Nomenclature

A	Parameter, defined by Eq. (8.48)
A	Parameter, defined by Eq. (8.96)
A_o	Valve opening area
A_{ch}	Surface area of spherical cap at outlet from feed hopper
A_{co}	Surface area of spherical cap at inlet from discharge cone
A_s	Cross-sectional area of standpipe
B	Ratio of vertical shear stress at wall to vertical normal stress at wall
B	Parameter, defined by Eq. (8.48)
B	Parameter, defined by Eq. (8.96)
C_o	Orifice coefficient
c	Cohesion
D	Column diameter
d_{min}	Minimum outlet dimension of the hopper

d_p	Particle diameter
$\boldsymbol{F}$	Body force vector
F_D	Drag force
FF	Material flow function
ff	Flow factor
f_k	Kinematic coefficient of friction
f_s	Static coefficient of friction
f_w	Coefficient of friction at wall
g	Gravitational acceleration
H	Height of standpipe
$H(\varphi_w)$	Function, defined graphically in Fig. 8.12
K_a	Principal stress ratio in an active state
K_p	Principal stress ratio in a passive state
N	Parameter, defined by Eq. (8.19) or Eq. (8.23)
n	Modified Richardson–Zaki index for gas–solid systems
p	Pressure
p^*	Dimensionless pressure
Q	Gas flow rate
R_s	Radius of standpipe
r_d	Radius of discharge orifice
$\boldsymbol{T}$	Stress tensor
U	Superficial gas velocity
U_p	Superficial particle velocity
U_{pt}	Particle terminal velocity
U_{zp}	Superficial particle velocity in one-dimensional axial flow
U_R	Superficial gas velocity in one-dimensional radial flow
U_s	Slip velocity
U_z	Superficial gas velocity in one-dimensional axial flow
u	Interstitial gas velocity
u_p	Particle velocity
u_R	Gas radial velocity component
u_{Rp}	Particle radial velocity component
u_z	Gas velocity in the axial direction
u_{zp}	Particle velocity in the axial direction
u_φ	Gas circumferential velocity component
$u_{\varphi p}$	Particle circumferential velocity component
Y_o	Unconfined yield strength
z_b	Height of solids in standpipe

Greek Symbols

α	Volume fraction of the gas phase
α_{mb}	Volume fraction of gas in a moving bed
α_p	Volume fraction of the particle phase
α_{pmf}	Volume fraction of solids at minimum fluidization
β	Coefficient for the drag between the particulate phase and interstitial gas
β	Angle between x-axis and the normal to an inclined plane
β_{pr}	Angle for principal directions
γ	Angle between the major principal stress and hopper wall
γ	Ratio of discharge opening diameter to standpipe diameter
Δ	Ratio of vertical normal stress at wall to averaged vertical normal stress
Δp_o	Pressure drop over orifice
Δp_s	Pressure drop over standpipe
Δp_t	Overall pressure drop
η	Angle of internal friction of solids
η_k	Kinematic angle of internal friction of solids
η_s	Static angle of internal friction of solids
η_w	Kinematic angle of wall friction

ξ	Relaxation constant, defined by Eq. (8.94)	σ_h	Horizontal normal stress
ρ	Density of gas	σ_m	Mean stress
ρ_p	Density of particles	σ_o	Surcharge on top of the hopper
σ	Normal stress	σ_t	Tensile strength
σ_a	Major principal stress in a stable arc which spans the hopper outlet	σ_v	Vertical normal stress
σ_{ac}	Critical major principal stress in a stable arc which spans the hopper outlet	τ	Shear stress
σ_i	Principal stress (i = 1, 2, or 3)	μ	Dynamic viscosity of gas
		ν	Poisson's ratio
		φ_w	Hopper half-apex angle
		ψ	Angle between R-direction and the normal to the major principal stress axis

References

Berg, C. H. O. (assigned to Union Oil Company of California) (1954). U.S. Patent 2,684,868.

BMHB (British Material Handling Board) (1988). *Draft Design Code of Silos, Bins, Bunkers and Hoppers*. Berkshire: Milton Keynes.

Carr, R. L. (1965). Evaluating Flow Properties of Solids. *Chemical Engineering*, **72** (2), 163.

Chen, Y. M., Rangachari, S. and Jackson, R. (1984). Theoretical and Experimental Investigation of Fluid and Particle Flow in a Vertical Standpipe. *I&EC Fund.*, **23**, 354.

Chong, Y. O., Teo, C. S. and Leung, L. S. (1987). Recent Advances in Standpipe Flow. *J. Pipelines*, **6**, 121.

Cleaver, J. A. S. and Nedderman, R. M. (1993). Theoretical Prediction of Stress and Velocity Profiles in Conical Hoppers. *Chem. Eng. Sci.*, **48**, 3693.

Drescher, A. (1991). *Analytical Methods in Bin-Load Analysis*. Amsterdam: Elsevier.

Ginestra, J.C., Rangachari, S. and Jackson, R. (1980). A One-Dimensional Theory of Flow in a Vertical Standpipe. *Powder Tech.*, **27**, 69.

Growther, M. E. and Whitehead, J. C. (1978). Fluidization of Fine Particles at Elevated Pressure. In *Fluidization*. Ed. Davidson and Kearins. Cambridge: Cambridge University Press.

Horne, R. M. and Nedderman, R. M. (1978). Stress Distribution in Hoppers. *Powder Tech.*, **19**, 243.

Jackson, R. (1993). Gas–Solid Flow in Pipes. In *Particulate Two Phase Flow*. Ed. M. C. Roco. Boston: Butterworth-Heinemann.

Janssen, H. A. (1895). Versuche üder Getreidedruck in Silozellen. *Vercin Deutcher Ingenieure*, **39**, 1045.

Jenike, A. W. (1961). *Gravity Flow of Bulk Solids*. Bulletin No. 108, Utah Engineering Experiment Station.

Jenike, A. W. (1964a). Steady Gravity Flow of Frictional–Cohesive Solids in Converging Channels. *Trans. ASME, J. Appl. Mech.*, **31**, 5.

Jenike, A. W. (1964b). *Storage and Flow of Solids*. Bulletin No. 123, Utah Engineering Experiment Station.

Johanson, J. R. (1981). How to Predict and Use the Properties of Bulk Solids. In *Solids Handling*. Ed. McNaughton and the Staff of Chemical Engineering. New York: McGraw-Hill.

Kamath, S., Puri, V. M., Manbeck, H. B. and Hogg, R. (1993). Flow Properties of Powders Using Four Testers: Measurement, Comparison and Assessment. *Powder Tech.*, **76**, 277.

Knowlton, T. M. (1986). Solids Transfer in Fluidized Systems. In *Gas Fluidization Technology*. Ed. D. Geldart. New York: Wiley.

Knowlton, T. M., Findlay, J. G., Sishtla, C. and Chan, I. (1986). Solids Pressure Reduction Without Lockhoppers: The Restricted Pipe Discharge System. *1986 AIChE Annual Meeting*, Miami, Fla.: Nov. 2–7.

Knowlton, T. M., Findlay, J. G. and Chan, I. (1989). Continuous Depressurization of Solids Using a Restricted Pipe Discharge System. *1989 AIChE Annual Meeting*, San Francisco, Calif.: Nov. 5–10.

Kojabashian, C. (1958). *Properties of Dense-Phase Fluidized Solids in Vertical Downflow*. Ph.D. Dissertation. Massachusetts Institute of Technology.

Li, X. G., Liu, D. L. and Kwauk, M. (1982). Pneumatically Controlled Multistage Fluidized Beds. II. *Proceedings of Joint Meeting of Chemical Engineers, SIESC and AIChE*. Beijing: Chemical Industry Press.

Mountziaris, T. J. and Jackson, R. (1991). The Effects of Aeration on the Gravity Flow of Particles and Gas in Vertical Standpipes. *Chem. Eng. Sci.*, **46**, 381.

Nedderman, R. M. (1992). *Statics and Kinematics of Granular Materials*. Cambridge: Cambridge University Press.

Nedderman, R. M. and Tüzün, U. (1979). A Kinematic Model for the Flow of Granular Materials. *Powder Tech.*, **22**, 243.

Nguyen, T. V., Brennen, C. and Sabersky, R. H. (1979). Gravity Flow of Granular Materials in Conical Hoppers. *Trans. ASME, J. Appl. Mech.*, **46**, 529.

Peschl, I. A. S. Z. (1989). Measurement and Evaluation of Mechanical Properties of Powders. *Powder Handling & Processing*, **1**, 135.

Richardson, J. F. and Zaki, W. N. (1954). Sedimentation and Fluidization. *Trans. Instn. Chem. Engrs.*, **32**, 35.

Sokolovskii, V. V. (1965). *Statics of Granular Media*. New York: Pergamon Press.

Soo, S. L. (1989). *Particulates and Continuum: Multiphase Fluid Dynamics*. New York: Hemisphere.

Soo, S. L. and Zhu, C. (1993). Unsteady Motion of Dense Suspensions and Rheological Behavior. In *Particulate Two-Phase Flow*. Ed. M. C. Roco. Boston: Butterworth-Heinemann.

Walker, D. M. (1966). An Approximate Theory for Pressures and Arching in Hoppers. *Chem. Eng. Sci.*, **21**, 975.

Walters, J. K. (1973). A Theoretical Analysis of Stresses in Axially-Symmetric Hoppers and Bunkers. *Chem. Eng. Sci.*, **28**, 779.

Problems

8.1 A cohesionless material of weight density 12 kN/m^3 and an angle of internal friction of 30° is stored in a tall cylindrical bin of diameter 1.8 m and an angle of wall friction 22°. Use Janssen's formula to determine the depths at which the radial stress in the material reaches 70 percent of its maximum value for both active and passive cases.

8.2 A conical hopper of a half angle 20° and an angle of wall friction 25° is used to store a cohesionless material of bulk density 1,900 kg/m^3 and an angle of internal friction 45°. The top surface of the material lies at a level 3.0 m above the apex and is free of loads. Apply Walker's method to determine the normal and shear stresses on the wall at a height of 1.2 m above the apex if the angle between the major principal plane at that height and the hopper wall is 30°. Assume a distribution factor of 1.1.

8.3 Considering the equilibrium of a stable arch, show that the minimum diameter of the orifice in a conical hopper, d_{min}, which is defined to account for the arching, can be related to the unconfined yield stress of the bulk material by the relation

$$d_{min} = 2Y_0/\alpha_p \rho_p g \tag{P8.1}$$

Neglect the force exerted on the arch by the weight of the material above the arch in the equilibrium equation.

8.4 It is required to design a mass flow conical hopper with the volume capacity of 100 m^3 to store a cohesionless material of bulk density 1,700 kg/m^3 and an angle of internal friction 40°. Four sets of shear tests have been conducted on the material and results for the unconfined yield strength and the corresponding consolidating stress are as follows:

	(1)	(2)	(3)	(4)
Y_0 (kN/m^2)	0.90	1.35	1.75	1.95
σ_1 (kN/m^2)	1.10	1.75	2.75	3.25

Determine the half angle of the hopper and the minimum outlet diameter for the two cases where the hopper wall material has an angle of wall friction of (i) 10° and (ii) 30°.

8.5 Consider a conical hopper flow where there is no-slip between the gas and the particle. Determine the distribution of mean stress of particles in the hopper flow. The particles can be assumed to be in a moving bed condition.

8.6 For a fluidized standpipe, the drag force of particles balances the pressure head as a result of the weight of solids. If the Richardson and Zaki form of equation (Eq. (8.55)) is proposed for the drag force, derive an expression for the leakage flow of gas in this standpipe. Discuss the effect of particle size on the leakage flow assuming all other conditions are maintained constant.

CHAPTER 9

Dense-Phase Fluidized Beds

9.1 Introduction

Contact schemes of gas–solid systems can be classified on the basis of the state of solids motion. For a batch–solids system, the gas at a low velocity merely percolates through the voids between packed particles while the particles remain motionless. The solids in this case are in the fixed bed state. With an increase in the gas velocity, particles move apart and become suspended; the bed then enters the gas fluidization state. Under relatively low gas velocities, a dense gas–solid suspension characterizes the bed; on the bed surface, some solids entrainment occurs. Gas bubbles or voids of bubble shapes can usually be clearly distinguished in the dense suspension. Under relatively high gas velocities, significant solids entrainment occurs and bubbles or voids are much less distinguishable in the dense suspension. These states of the beds are referred to as dense-phase fluidization. Further increase in the gas velocity will intensify the solids entrainment and the dense bed surface gradually becomes undistinguishable. At high gas velocities, particles are fully entrained, and the whole bed transforms to a "lean" gas–solid suspension. This state of fluidization is maintained by means of external recirculation of the entrained particles. Such operation is termed lean-phase fluidization. This chapter focuses on dense-phase fluidization. Lean-phase fluidization or transport is discussed in Chapters 10 and 11. As indicated in the Preface, the correlation equations presented in the text, unless otherwise noted, are given in SI units.

Dense-phase fluidized beds with bubbles represent the majority of the operating interests although the beds may also be operated without bubbles. The bubbling dense-phase fluidized bed behavior is fluidlike. The analogy between the bubble behavior in gas–solid fluidized beds and that in gas–liquid bubble columns is often applied. Dense-phase fluidized beds generally possess the following characteristics, which promote their use in reactor applications:

(1) Capability of continuous operation and transport of solids in and out of the bed
(2) High heat transfer rates from bed to surface and from gas to particle leading to temperature uniformity in the bed
(3) High mass transfer rates from gas to particle
(4) Applicability over a wide range of particle properties and high solids mixing rates
(5) Simplicity in geometric configuration and suitability for large-scale operation

Examples of applications of gas–solid dense-phase fluidization are given in Table 9.1.

9.2 Particle and Regime Classifications and Fluidized Bed Components

The transport behavior of fluidized bed systems can be described in light of the properties of the fluidized particles and the flow regimes. In the following, particle and regime classifications along with general components in a fluidized bed are given.

Table 9.1. *Examples of Applications of Gas–Solid Dense-Phase Fluidization*

Types of applications			
Physical operations	Chemical syntheses	Metallurgical and mineral processes	Other applications
Heat exchange	Phthalic anhydride synthesis	Uranium processing	Coal combustion
Solids blending	Acrylonitrile synthesis	Reduction of iron oxide	Coal gasification
Particle coating	Maleic anhydride synthesis	Pyrolysis of oil shale	Fluidized catalytic cracking
Drying	Ethylene dichloride synthesis	Roasting of sulfide ores	Incineration of solid waste
Adsorption	Methanol to gasoline process	Crystalline silicon production	Cement clinker production
Solidification & granulation	Vinyl acetate synthesis	Titanium dioxide production	Microorganism cultivation
Particle growth	Olefin polymerization	Calcination	

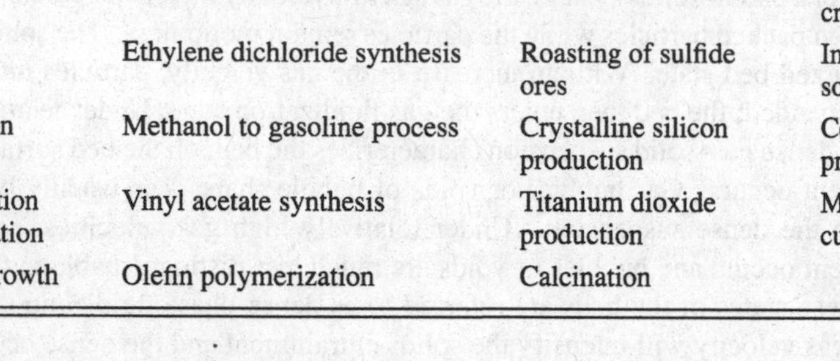

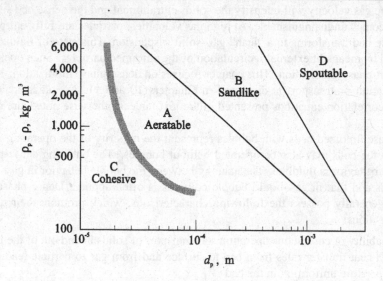

Figure 9.1. Geldart's classification of fluidized particles (from Geldart, 1973).

9.2.1 *Classification of Fluidized Particles*

Particles can be classified into four groups (*i.e.*, Groups A, B, C, and D) based on their fluidization behavior [Geldart, 1973]. This classification, known as Geldart's classification, is shown in Fig. 9.1, where particles are classified in terms of the density difference between the particles and the gas $(\rho_p - \rho)$, and the average particle diameter d_p.

Figure 9.1 was obtained empirically and has been widely adopted in the fundamental research and design of gas–solid fluidized beds. However, it should be noted that some fluidization variables other than $(\rho_p - \rho)$ and d_p may alter the demarcation of particle groups. For example, for fluidization using different gases, the cohesive nature of fluidized particles due to the surface or interparticle contact forces can vary significantly because of the difference in the type of gases adsorbed on the particles. Likewise, gas velocity, temperature, and pressure may also dramatically affect the fluidized particle characteristics, which in turn alter the particle group demarcation.

Group C in Fig. 9.1 comprises small particles ($d_p < 20\ \mu$m) which are cohesive. In fluidization with Group C particles, the interparticle contact forces such as the van der Waals force, capillary forces, and electrostatic forces dominate the hydrodynamic forces. The interparticle contact forces are affected by particle properties such as hardness, electrical conductivity, magnetic susceptibility, surface asperities, and moisture content, as well as by gas properties. Group C particles are difficult to fluidize and gas channeling is the most common characteristic when fluidizing these particles. When Group C particles are fluidized, the bed expansion can be very high.

Group A particles, with a typical size range of 30–100 μm, are readily fluidized. In fluidization with Group A particles, the hydrodynamic forces are important; the interparticle contact forces may also play a significant role. Beds with Group A particles can be operated in both the particulate fluidization regime where bubbles are not present, and the bubbling fluidization regime where bubbles are present (see §9.2.2 for regime classification). Thus, for Group A particles, the gas velocity for minimum fluidization, U_{mf}, is smaller than that for the minimum bubbling, U_{mb}. For bubbling fluidization with Group A particles, there exists a maximum stable bubble size.

For Group B particles, there is no particulate fluidization regime. In this case, U_{mf} equals U_{mb}. The bubble size increases with the bed height and bed expansion is moderate. For Group B particle fluidization, there exists no maximum stable bubble size.

Group D comprises coarse particles ($d_p > 1$ mm) which are commonly processed by spouting. When Group D particles are fluidized, the bed expansion is low and the particle mixing is not as good as that for Group A and B particles.

For demarcation between Group C and A particles, Molerus (1982) proposed a semiempirical criterion based on a balance between the van der Waals force and the hydrodynamic forces as given by

$$\left[g(\rho_p - \rho)d_p^3\right]_{CA} = 0.1 K F_H \tag{9.1}$$

where F_H is the cohesive force between particles with a value ranging from 8.8×10^{-8} N (for hard material) to 3.7×10^{-7} N (for soft material) and K is a constant ($K = 0.01$). The subscript CA refers to the C–A boundary.

For demarcation between Group A and B particles, an empirical equation with air as the fluidizing gas under ambient conditions was suggested by Geldart (1973):

$$[(\rho_p - \rho)^{1.17} d_p]_{AB} = 0.906 \tag{9.2}$$

Equation (9.2) can be modified to apply to elevated pressures and temperatures as well as to different gases as given by [Grace, 1986]

$$\mathrm{Ar}_{AB} = 1.03 \times 10^6 \left(\frac{\rho_p - \rho}{\rho}\right)^{-1.275} \tag{9.3}$$

where the subscript AB represents the A–B boundary; and Ar is the Archimedes number (see Eq. (9.12)).

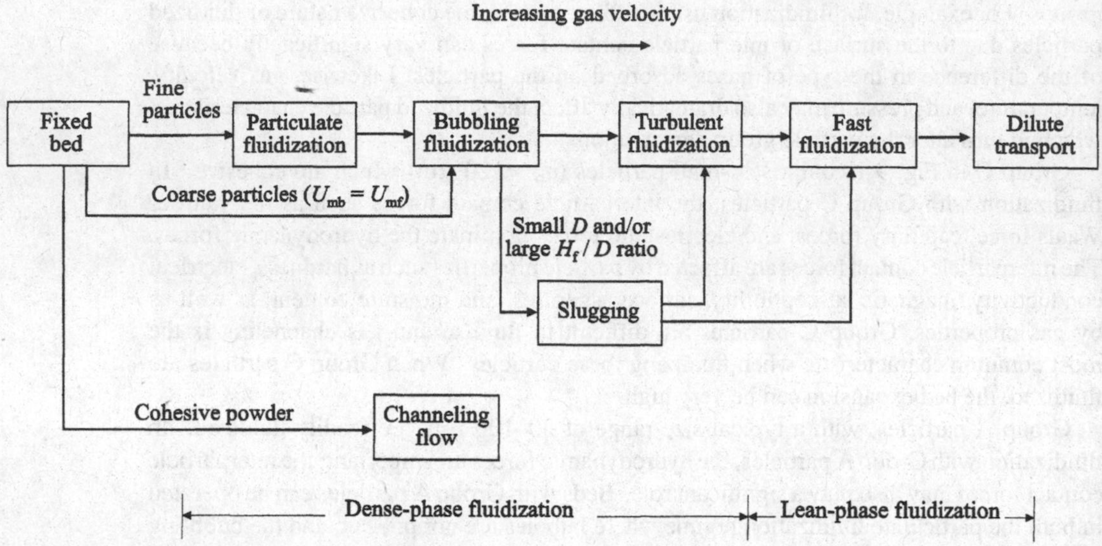

Figure 9.2. Interrelationship of various regimes including fixed bed, dense-phase fluidization, and lean-phase fluidization.

For demarcation between Group B and D particles, Grace (1986) considered the fast bubble characteristics for Group B particles in which the bubble rise velocity is higher than the interstitial gas velocity, and the slow bubble characteristics for Group D particles in which the bubble rise velocity is lower than the interstitial gas velocity (see §9.4.2). Thus, by setting the bubble rise velocity and the interstitial gas velocity equal, a criterion for demarcation between Group B and D particles can be obtained as [Grace, 1986]

$$\mathrm{Ar_{BD}} = 1.581 \times 10^7 \frac{\rho/(\rho_p - \rho)}{[1 - 219\rho/(\rho_p - \rho)]^2} \quad \text{for } \frac{\rho_p - \rho}{\rho} > 219 \tag{9.4}$$

9.2.2 *Fluidization Regimes*

Regime classification for dense-phase fluidization in general can be based on bubble behavior. Figure 9.2 shows the interrelationship of the regimes in dense-phase fluidization with those in lean-phase fluidization, which covers fast fluidization (Chapter 10) and dilute transport (Chapter 11). Dense-fluidization regimes include particulate fluidization, bubbling fluidization, and turbulent fluidization. In a broad sense, "dense-phase fluidization" also encompasses the slugging, spouting, and channeling regimes or conditions of operation. The fundamental distinction between these regimes or conditions of operation is briefly described later.

The particulate fluidization regime, as shown in Fig. 9.3(a), is bounded by the minimum fluidization velocity and the minimum bubbling velocity. In particulate fluidization, all the gas passes through the interstitial space between the fluidizing particles without forming bubbles. The bed appears homogeneous. This regime exists only in a bed with Group A particles under a narrow operating range of gas velocities. At high pressures or with gases of high density, the operating range of this regime expands. In a bed of coarse particles

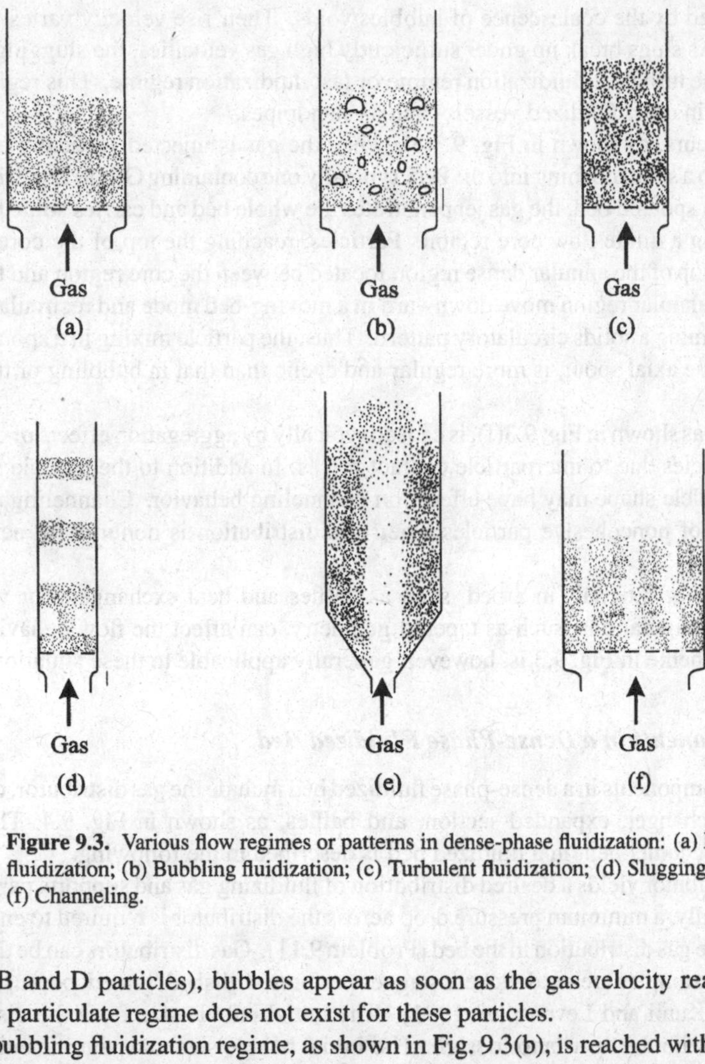

Figure 9.3. Various flow regimes or patterns in dense-phase fluidization: (a) Particulate fluidization; (b) Bubbling fluidization; (c) Turbulent fluidization; (d) Slugging; (e) Spouting; (f) Channeling.

(Group B and D particles), bubbles appear as soon as the gas velocity reaches U_{mf} and, thus, the particulate regime does not exist for these particles.

The bubbling fluidization regime, as shown in Fig. 9.3(b), is reached with an increase in the gas velocity beyond U_{mb}. Bubbles form and induce vigorous motion of the particles. In the bubbling fluidization regime, bubble coalescence and breakup take place. With increasing gas velocity, the tendency of bubble coalescence is enhanced. Two distinct phases, *i.e.*, the bubble phase and the emulsion phase, are present in this regime.

The turbulent fluidization regime, as shown in Fig. 9.3(c), is realized when the gas velocity increases beyond the bubbling fluidization regime. In the turbulent fluidization regime, the bubble and emulsion phases may become indistinguishable with increased uniformity of the suspension. In contrast to that in the bubbling regime, the tendency for bubble breakup is enhanced as the gas velocity increases in this regime, resulting in reduced presence of large bubbles. In addition, the surface of the bed becomes very diffused. Increased freeboard activities are evidenced in this regime.

Slugging, as shown in Fig. 9.3(d), refers to the fluidization regime which is characterized by the presence of bubbles or slugs with sizes comparable to the bed diameter. Slugging occurs only in a fluidized bed of a small bed diameter and/or large height/diameter ratio.

Slugs are formed by the coalescence of bubbles/voids. Their rise velocity varies with the bed diameter. As slugs break up under sufficiently high gas velocities, the slugging regime transforms to the turbulent fluidization regime or fast fluidization regime. This regime may also be present in other fluidized vessels such as standpipes.

Spouting occurs, as shown in Fig. 9.3(e), when the gas is injected vertically at a high velocity through a small opening into the bed, typically one containing Group D particles. At the center of the spouted bed, the gas jet penetrates the whole bed and carries some particles upward, forming a dilute flow core region. Particles, reaching the top of the core region, fall back to the top of the annular dense region located between the core region and the wall. Particles in the annular region move downward in a moving-bed mode and recirculate to the core region, forming a solids circulatory pattern. Thus, the particle mixing in a spouted bed, as induced by the axial spout, is more regular and cyclic than that in bubbling or turbulent fluidized beds.

Channeling, as shown in Fig. 9.3(f), is caused typically by aggregation effects of cohesive (Group C) particles due to interparticle contact forces. In addition to the particle size and density, the particle shape may have effects on channeling behavior. Channeling can also occur in a bed of noncohesive particles when gas distribution is nonuniform across the distributor.

The presence of internals in a bed, such as baffles and heat exchangers, or variation in geometric configurations, such as tapered geometry, can affect the flow behavior. The classification scheme in Fig. 9.3 is, however, generally applicable to these situations.

9.2.3 *Components in a Dense-Phase Fluidized Bed*

Bed components in a dense-phase fluidized bed include the gas distributor, cyclone, dipleg, heat exchanger, expanded section, and baffles, as shown in Fig. 9.4. The basic function of each component in a fluidized bed is described in the following.

The gas distributor yields a desired distribution of fluidizing gas and supports particles in the bed. Generally, a minimum pressure drop across the distributor is required to ensure the uniformity of the gas distribution in the bed [Problem 9.11]. Gas distributors can be designed in many ways; examples are sandwiched, stagger perforated, dished, grated, porous, tuyere, and cap types [Kunii and Levenspiel, 1991]. The initial bubble size from the distributor strongly varies with the distributor design. The initial bubble size may significantly affect the bubble size in the bed.

The cyclone separates solid particles from the outlet gas. Several cyclones may be combined to form a multistage cyclone system. More than one multistage cyclone system can be placed inside or outside the bed. Other types of gas–solid separators (*e.g.*, filters) are also used for gas–solid separation although the cyclone is most widely adopted in fluidized beds. The dipleg/standpipe returns the particles separated by the cyclone into the dense bed. The outlet of a dipleg may be located in the freeboard or immersed in the dense bed.

The heat exchanger removes generated heat from or adds required heat to the fluidized bed by a cooling fluid or heating fluid. Heat exchangers can either be immersed in the dense bed or be placed both in the dense bed and in the freeboard, which is the region above the dense bed. Possible erosion and corrosion of heat exchangers require special attention.

The expanded section reduces the linear gas velocity in the freeboard so that settling of the particles carried by the fluidizing gas can be efficiently achieved. The expanded section

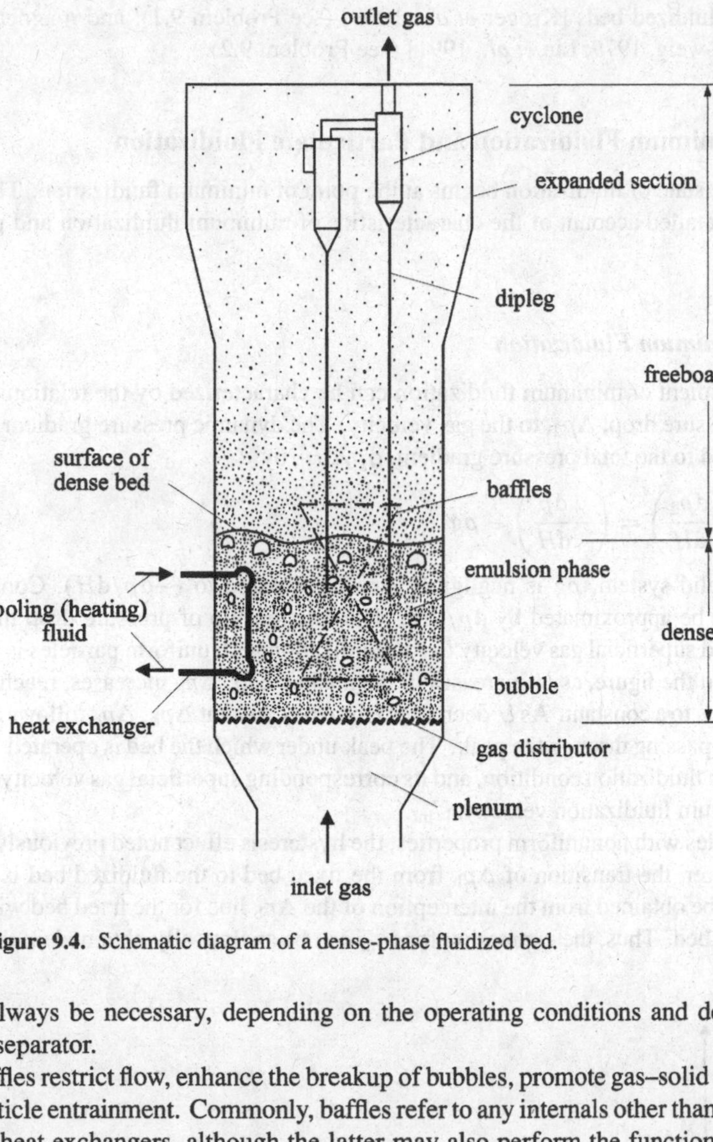

Figure 9.4. Schematic diagram of a dense-phase fluidized bed.

may not always be necessary, depending on the operating conditions and design of the gas–solid separator.

The baffles restrict flow, enhance the breakup of bubbles, promote gas–solid contact, and reduce particle entrainment. Commonly, baffles refer to any internals other than diplegs and immersed heat exchangers, although the latter may also perform the functions of baffles. Like heat exchangers, baffles can be installed in the dense bed and/or in the freeboard. Baffles can be designed with a number of variations, *e.g.*, horizontal and vertical grates, fins of different sizes and directions, solid or mesh, or even constructed with a shape like a pagoda [Jin *et al.*, 1980]. The benefits of baffles are more distinct for beds of coarse particles (Groups B and D) than for fine particle beds (Group A), because the bubbles in the latter case are smaller.

Dense-phase fluidization can also be conducted in the presence of force fields other than a gravitational field. Such force fields include vibrational, acoustic, centrifugal, and magnetic fields. Operations with applications of these fields are known, respectively, as vibrofluidized beds [Mori *et al.*, 1992], acoustic fluidized beds [Montz *et al.*, 1988; Chirone *et al.*, 1992], centrifugal fluidized beds [Kroger *et al.*, 1979] (see Problem 9.1), and magnetofluidized beds [Rosensweig, 1979; Liu *et al.*, 1991] (see Problem 9.2).

9.3 Minimum Fluidization and Particulate Fluidization

The state of fluidization begins at the point of minimum fluidization. This section provides a detailed account of the characteristics of minimum fluidization and particulate fluidization.

9.3.1 *Minimum Fluidization*

Incipient or minimum fluidization can be characterized by the relationship of the dynamic pressure drop, Δp_d, to the gas velocity. The dynamic pressure gradient, $\mathrm{d}p_d/\mathrm{d}H$, can be related to the total pressure gradient, $\mathrm{d}p/\mathrm{d}H$, by

$$\left(-\frac{\mathrm{d}p_d}{\mathrm{d}H}\right) = \left(-\frac{\mathrm{d}p}{\mathrm{d}H}\right) - \rho g \tag{9.5}$$

For a gas–solid system, ρg is negligibly small compared to $(-\mathrm{d}p/\mathrm{d}H)$. Consequently, $\mathrm{d}p_d/\mathrm{d}H$ can be approximated by $\mathrm{d}p/\mathrm{d}H$. The relationship of pressure drop through the bed, Δp_b, and superficial gas velocity U for fluidization with uniform particles is illustrated in Fig. 9.5. In the figure, as U increases in the packed bed, Δp_b increases, reaches a peak, and then drops to a constant. As U decreases from the constant Δp_b, Δp_b follows a different path without passing through the peak. The peak under which the bed is operated is denoted the minimum fluidization condition, and its corresponding superficial gas velocity is defined as the minimum fluidization velocity, U_{mf}.

For particles with nonuniform properties, the hysteresis effect noted previously may also occur; however, the transition of Δp_b from the fixed bed to the fluidized bed is smoother and U_{mf} can be obtained from the interception of the Δp_b line for the fixed bed with that for the fluidized bed. Thus, the expression for U_{mf} can be analytically obtained on the basis of

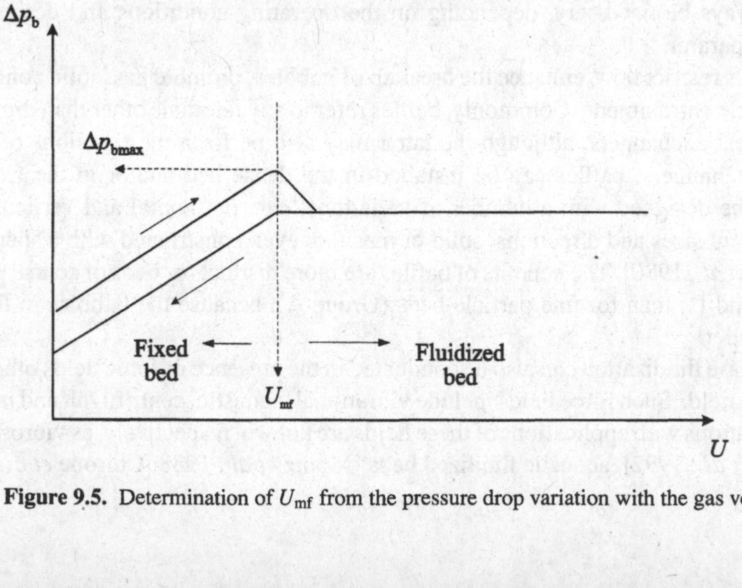

Figure 9.5. Determination of U_{mf} from the pressure drop variation with the gas velocity.

the equivalence of the pressure drop for a fixed bed and that for a fluidized bed under the minimum fluidization condition.

The pressure drop in the fixed bed can be described by the Ergun equation, as given by Eq. (5.377). Under the minimum fluidization condition, we have

$$\frac{\Delta p_\mathrm{b}}{H_\mathrm{mf}} = 150\frac{(1-\alpha_\mathrm{mf})^2}{\alpha_\mathrm{mf}^3}\frac{\mu U_\mathrm{mf}}{\varphi^2 d_\mathrm{p}^2} + 1.75\frac{(1-\alpha_\mathrm{mf})}{\alpha_\mathrm{mf}^3}\frac{\rho U_\mathrm{mf}^2}{\varphi d_\mathrm{p}} \tag{9.6}$$

where φ is the sphericity of the particles and H_mf is the bed height at minimum fluidization. It is noted that most fluidized particles are not spherical, and for most cases, the equivalent particle diameter refers to the volume diameter (Eq. (1.3)). For a bed with a wide range of particle sizes, DeBroucker's mean diameter (Eq. (1.40)) may be used for Eq. (9.6).

In a fully fluidized bed, the pressure drop (cross-sectionally averaged) counterbalances the weight of the pseudocontinuum of the gas–solid mixture, which yields

$$-\frac{\mathrm{d}p}{\mathrm{d}H} = [\rho_\mathrm{p}(1-\alpha) + \rho\alpha]g \tag{9.7}$$

Usually, the bed voidage α is considered independent of the bed height and is given by

$$\alpha = 1 - \frac{M_\mathrm{p}}{\rho_\mathrm{p} A H_\mathrm{f}} \tag{9.8}$$

where M_p is the total mass of particles in the bed; A is the cross-sectional area of the bed; and H_f is the expanded bed height.

From Eqs. (9.5) and (9.7), we have

$$-\frac{\mathrm{d}p_\mathrm{d}}{\mathrm{d}H} = (\rho_\mathrm{p} - \rho)(1-\alpha)g \tag{9.9}$$

Under the minimum fluidization condition, Eq. (9.9) gives

$$\frac{\Delta p_\mathrm{b}}{H_\mathrm{mf}} = \frac{\Delta p_\mathrm{d}}{H_\mathrm{mf}} = (\rho_\mathrm{p} - \rho)(1-\alpha_\mathrm{mf})g \tag{9.10}$$

Equating Eqs. (9.6) and (9.10) results in

$$\mathrm{Ar} = 150\frac{(1-\alpha_\mathrm{mf})}{\alpha_\mathrm{mf}^3\varphi^2}\mathrm{Re}_\mathrm{pmf} + \frac{1.75}{\alpha_\mathrm{mf}^3\varphi}\mathrm{Re}_\mathrm{pmf}^2 \tag{9.11}$$

where the Archimedes number (Ar) and the particle Reynolds number at minimum fluidization (Re_pmf) are defined as

$$\mathrm{Ar} = \frac{\rho(\rho_\mathrm{p} - \rho)g d_\mathrm{p}^3}{\mu^2} \tag{9.12}$$

and

$$\mathrm{Re}_\mathrm{pmf} = \frac{\rho U_\mathrm{mf} d_\mathrm{p}}{\mu} \tag{9.13}$$

Equation (9.11) can be written as $\mathrm{Ar} = k_1\mathrm{Re}_\mathrm{pmf} + k_2\mathrm{Re}_\mathrm{pmf}^2$. In terms of this form, a frequently used semiempirical correlation relating Re_pmf to Ar is given by [Wen and Yu, 1966]

$$\mathrm{Re}_\mathrm{pmf} = \sqrt{(33.7)^2 + 0.0408\mathrm{Ar}} - 33.7 \tag{9.14}$$

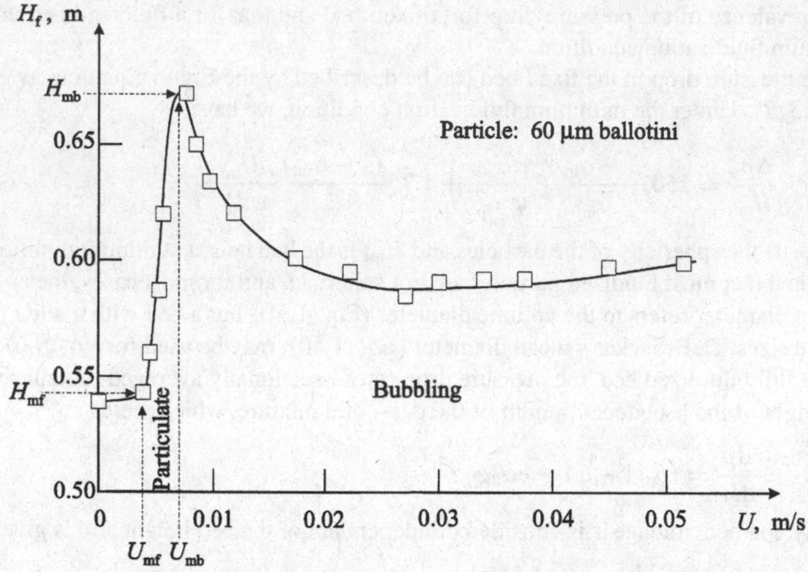

Figure 9.6. Bed expansion curve for typical Group A particles (from Geldart, 1986).

Following a similar approach, other correlations have been proposed [*e.g.*, Babu *et al.*, 1978; Thonglimp, 1981; Grace, 1982].

9.3.2 *Particulate Fluidization*

For a bed with Group A particles, bubbles do not form when the gas velocity reaches U_{mf}. The bed enters the particulate fluidization regime under this condition. The operation under the particulate fluidization regime is characterized by a smooth bed expansion with an apparent uniform bed structure for $U_{mf} < U < U_{mb}$, where U_{mb} is the superficial gas velocity at the minimum bubbling condition. The height of the bed expansion in terms of α can be estimated by [Abrahamsen and Geldart, 1980a]

$$\frac{\alpha^3}{1-\alpha}\frac{(\rho_p-\rho)gd_p^2}{\mu} = 210(U-U_{mf}) + \frac{\alpha_{mf}^3}{1-\alpha_{mf}} + \frac{(\rho_p-\rho)gd_p^2}{\mu} \tag{9.15}$$

The gas velocity for the upper bound of the particulate fluidization regime is U_{mb}. The operational range of the particulate fluidization regime is quite narrow, as shown in Fig. 9.6. It is seen that for $U > U_{mf}$, H_f monotonically increases with U until the minimum bubbling point is reached. Further increase in the gas velocity leads to a decrease in the bed height as bubbling fluidization begins.

The particulate fluidized regime will occur when the following criterion is met [Wilhelm and Kwauk, 1948]

$$\mathrm{Fr}_{mf} < 0.13 \tag{9.16}$$

where the Froude number at minimum fluidization Fr_{mf} is defined by

$$\mathrm{Fr}_{mf} = \frac{U_{mf}^2}{gd_p} \tag{9.17}$$

Considering more parameters, an improved criterion was suggested by Romero and Johanson (1962) as

$$\mathrm{Fr_{mf}}\,\mathrm{Re_{pmf}}\left(\frac{\rho_p-\rho}{\rho}\right)\frac{H_{mf}}{D}<100 \tag{9.18}$$

9.4 Bubbling Fluidization

In a bubbling fluidized bed, the behavior of bubbles significantly affects the flow or transport phenomena in the bed, including solids mixing, entrainment, and heat and mass transfer. Theories accounting for the multibubble behavior are primarily based on those for single-bubble behavior. Hence, in this section, fundamentals pertaining to single-bubble behavior in a fluidized bed such as the bubble wake, cloud, shape, and rise velocity are introduced. A theory describing the flow field surrounding a single rising bubble and characteristics of bubble formation and jetting is presented. In addition, the behavior of multibubbles in free bubbling conditions including bubble interaction, bubble size, rise velocity, and volume fraction in the bed is discussed in conjunction with the two-phase theory of fluidization. It should be noted that free bubbling refers to bubbling fluidization without the internals constraint.

9.4.1 *Onset of Bubbling*

Bubbles are formed as a result of the inherent instability of gas–solid systems. The instability of a gas–solid fluidized bed is characterized by fast growth in local voidage in response to a system perturbation. Because of the instability in the bed, the local voidage usually grows rapidly into a shape resembling a bubble. Although it is not always true, the initiation of the instability is usually perceived to be the onset of bubbling, which marks the transition from particulate fluidization to bubbling fluidization. The theoretical explanation of the physical origin behind and prediction of the onset of the instability of gas–solid fluidized beds has been attempted [*e.g.*, Anderson and Jackson, 1968; Verloop and Heertjes, 1970; Rietema and Piepers, 1990]. The efforts have been focused on the primary forces behind the stability among the interparticle contact forces, particle–fluid interaction forces, and particle–particle interaction via particle velocity fluctuation.

In theoretical analyses, Jackson (1963) predicted that a fluidized bed would always be unstable if the interaction forces among particles were not considered. Garg and Pritchett (1975) indicated that a fluidized bed may be stabilized with the inclusion of a particle concentration gradient driven force in the momentum equation of the particle phase. A simple criterion for bed instability, based on a stabilization mechanism of the particle–fluid interaction forces, was established by Foscolo and Gibilaro (1984). Batchelor (1988) established another simple criterion for bed stabilization due to hydrodynamic forces by considering the particle diffusion effect due to random fluctuation of the particle velocity. It is noted, however, that, for some Group A particles (*e.g.*, fluid catalytic cracking [FCC] and fine sand particles), the inherent hysteresis phenomenon in the variation of pressure drop with gas velocity in the particulate fluidization regime given in Fig. 9.5 has been strongly linked to the interparticle contact forces as the factor responsible for bed stability for fine particles [Tsinontides and Jackson, 1993].

While the issue concerning origin of the bed instability is yet to be clarified, a mechanistic interpretation of the minimum bubbling point based on the wave concept appears viable. In terms of Wallis's concepts of continuity wave and dynamic wave, the minimum bubbling point in a fluidized bed corresponds to a condition in which the continuity wave velocity and dynamic wave velocity are equal (see §6.5.2.3 and Problem 9.3). Accordingly, the minimum bubbling velocity U_{mb} can be expressed in terms of the elasticity modulus of the bed as [Rietema, 1991]

$$U_{mb} = \sqrt{\frac{E}{\rho_p}\left(\frac{\alpha_{mb}}{3 - 2\alpha_{mb}}\right)} \tag{9.19}$$

It should be noted that the elasticity modulus E is not merely a property of the solid material in the bed. In general, E is a complex function of the structure of packing, material properties of packing particles, particle size, and particle contact and cohesion forces between particles.

In practice, the minimum bubbling velocity can be estimated by [Abrahamsen and Geldart, 1980a]

$$U_{mb} = 2.07\exp(0.716\,\phi_f)\frac{d_p\rho^{0.06}}{\mu^{0.347}} \tag{9.20}$$

where ϕ_f is the mass fraction of the fine powder smaller than 45 μm. The preceding expression becomes invalid when U_{mb} calculated is less than U_{mf}.

9.4.2 *Single Bubble in a Fluidized Bed*

In the experimental study of single-bubble behavior, the gas–solid fluidized bed to which a single bubble is introduced is usually maintained at the minimum fluidization condition. Characteristics of a single gas bubble in a gas–solid fluidized bed are similar to those in a liquid medium [Clift and Grace, 1985; Fan and Tsuchiya, 1990; Krishna, 1993].

9.4.2.1 *Bubble Configurations*

Most bubbles in gas–solid fluidized beds are of spherical cap or ellipsoidal cap shape. Configurations of two basic types of bubbles, fast bubble (clouded bubble) and slow bubble (cloudless bubble), are schematically depicted in Fig. 9.7. The cloud is the region established

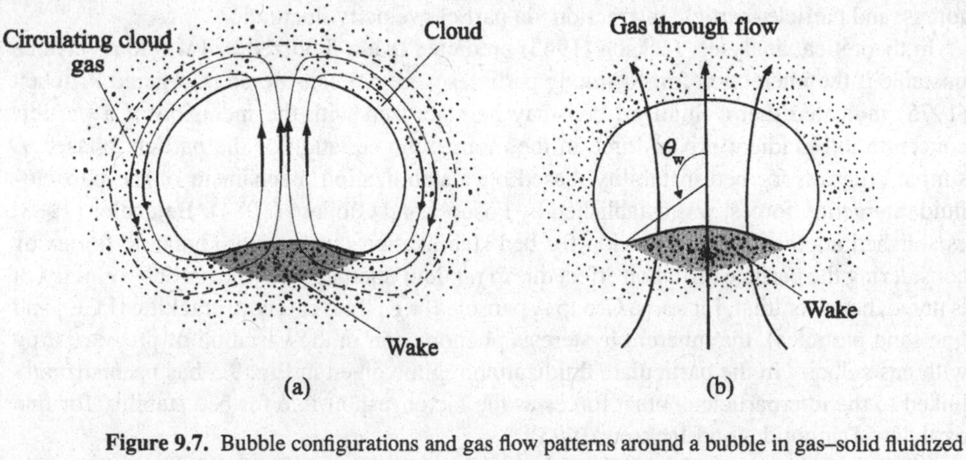

Figure 9.7. Bubble configurations and gas flow patterns around a bubble in gas–solid fluidized beds: (a) Fast bubble (clouded bubble) $U_b > U_{mf}/\alpha_{mf}$; (b) Slow bubble (cloudless bubble) $U_b < U_{mf}/\alpha_{mf}$.

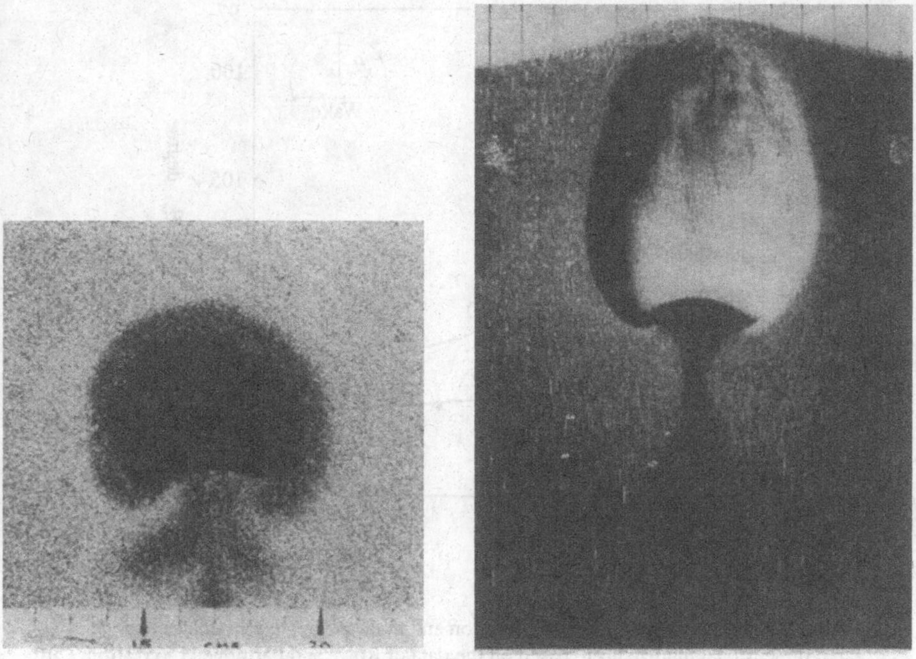

(a) (b)

Figure 9.8. (a) An NO_2 bubble rising in a two-dimensional ballotini bed showing the cloud region of the bubble; (b) A two-dimensional bubble rising through a layer of black particles showing the bubble wake in a gas–solid fluidized bed (from Rowe, 1971; reproduced with permission).

by the gas which circulates in a closed loop between the bubble and its surroundings. The cloud phase can be visualized with the aid of a color tracer gas bubble. For example, when a dark brown NO_2 bubble is injected into the bed (as shown in Fig. 9.8(a)), the light brown color surrounding the bubble represents the cloud region [Rowe, 1971]. When the bubble rise velocity is higher than the interstitial gas velocity, a "clouded" bubble, in which the circulatory flow of gas takes place between the bubble and the cloud, as shown in Fig. 9.7(a), forms. The cloud size decreases as the bubble rise velocity increases (Problem 9.4). As the bubble rise velocity becomes significantly higher than the interstitial gas velocity (*i.e.*, $U_b \gg U_{mf}/\alpha_{mf}$), the cloud size becomes so thin that most of the gas circulates inside the bubble. Bubbles in fluidized beds with Group A or B particles are typically characterized by clouded bubbles. When the interstitial gas velocity is greater than the bubble rise velocity, it yields a "cloudless" bubble in which the emulsion phase gas flows through the bubble phase, as shown in Fig. 9.7(b). This gas through flow in the bubble is also known as invisible bubble flow, which is distinguished from visible bubble flow. Bubbles in Group D particle fluidized beds are typically characterized by cloudless bubbles.

The wake angle θ_w defined in Fig. 9.7 is closely related to the shape of a bubble. A smaller θ_w represents a flatter bubble, whereas a large θ_w represents a more rounded one. A large wake angle θ_w yields a smaller wake fraction f_w which is defined as the volume ratio of wake to bubble. As shown in Fig. 9.9, θ_w is related to particle diameter d_p. An increase in

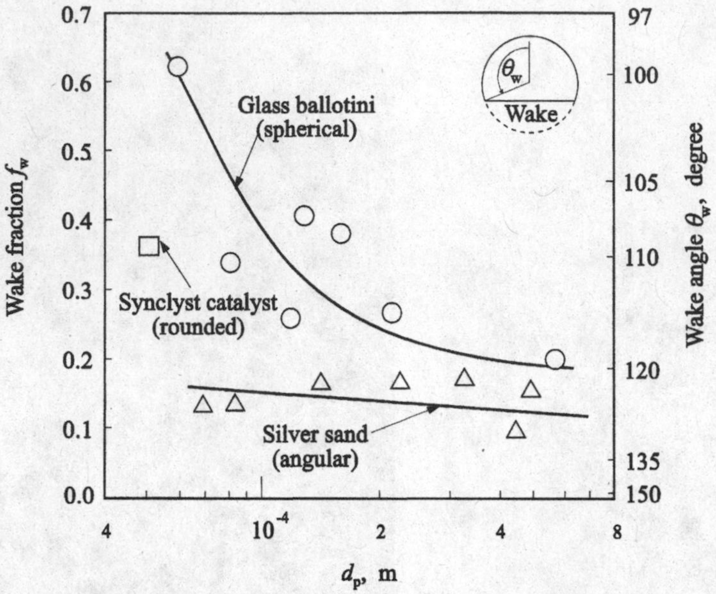

Figure 9.9. Relationship of the wake fraction and the wake angle for bubbles in a three-dimensional fluidized bed, based on the data of Rowe and Partridge (1965) (from Clift and Grace, 1985).

d_p results in an increase in θ_w [Grace, 1970; Clift *et al.*, 1978]. It is noted that the apparent viscosity of the bed measured experimentally increases with particle diameter for Group A and Group B particles [Geldart, 1986].

The bubble wake plays an important role in solids movement or solids mixing in the bed and the freeboard. A bubble wake in a single-phase fluid is defined as the streamline enclosed region beneath the bubble base. In a gas–solid fluidized bed, the emulsion phase can be treated as a pseudosingle-phase fluid. Hence, a bubble wake is defined as the region enclosed by streamline of the pseudofluid behind the bubble base. In the bed, the wake rises with the bubble and thereby provides an essential means for global solids circulation and induces axial solids mixing. In the freeboard, particles carried along in the bubble wake are the primary origin of the particles there (see §9.6). The bubble wake fractions in gas–solid fluidized beds are generally much smaller than those in low-viscosity liquid systems.

Figure 9.8(b) shows a photograph of a large two-dimensional bubble followed by a wake of black particles. It is seen that the bubble wake has a cusp-shaped end. Compared to the geometry of the bubble wake in gas–liquid or slurry bubble column systems [Fan and Tsuchiya, 1990], that of the bubble wake in a gas–solid fluidized bed is more regular and distinct. It is noted that the precise flow structure inside the bubble wake in a gas–solid fluidized bed is not fully clear. Studies of bubble wake lead to two different views. The first view is conceived on the basis of an analogy of a gas bubble in liquids with that in gas–solid fluidized beds [Clift, 1986]. Re_b in a gas–solid fluidized bed, defined in terms of the apparent density and apparent viscosity of the gas–solid suspension, is typically on the order of 100, on the basis of the apparent viscosity of gas–solid suspensions on the order of 1 kg/m·s [Grace, 1970; Clift and Grace, 1985]. With this order of magnitude for Re_b under the analogy, the rising bubble in a gas–solid fluidized bed would yield a vortical motion of gas and particles in its wake. The second view indicates that the wake flow is nonvortical, with gas and solids entering the wake from its outer boundary, flowing inward radially toward the central region, and then directed vertically downward out of the wake [Kozanoglu and Levy, 1991]. This view could be interpreted to indicate that for a bubble with large θ_w, which is common in gas–solid fluidized beds, the curvature around the bubble edge is not sufficiently large to generate a significant amount of vorticity [Leal, 1989; Fan and Tsuchiya, 1990] and, thus, the accumulation of vorticity in the wake does not occur. Since the flow pattern in the wake strongly depends on the curvature of the bubble edge, which varies with the shape of the bubble, a complete analogy of bubbles in liquids with those in a gas–solid fluidized bed requires a similarity in the Weber number in addition to similarity in Re_b. The Weber number varies with the surface tension around the bubble. However, surface tension does not exist for bubbles in a gas–solid fluidized bed, and, thus, the analogy discussed in the first view may not strictly hold. For the second view, computation of the vorticity based on a realistic bubble shape with a precise edge curvature is needed to quantify the vorticity generation. Thorough experiments are necessary to clarify this important, complex wake flow problem.

In gas–solid fluidized beds wake shedding has also been observed. The shed wake fragments are banana-shaped and the shedding may occur at fairly regular intervals [Rowe and Partridge, 1965; Rowe, 1971].

9.4.2.2 *Flow Around a Bubble in a Fluidized Bed*

To describe the particle and gas flows around the bubble in fluidized beds, the pioneering model of Davidson and Harrison (1963) is particularly noteworthy because of its fundamental importance and relative simplicity. On the basis of some salient features of this model, a number of other models were developed [*e.g.*, Collins, 1965; Stewart, 1968; Jackson, 1971]. The material introduced later follows Davidson and Harrison's approach.

Consider a bubble rising in a fluidized bed. It is assumed that the bubble is solids-free, is spherical, and has a constant internal pressure. Moreover, the emulsion phase is assumed to be a pseudocontinuum, incompressible, and inviscid single fluid with an apparent density of $\rho_p(1-\alpha_{mf})+\rho\alpha_{mf}$. It should be noted that the assumption of incompressibility of the mixture is not strictly valid as voidage in the vicinity of the bubble is higher than that in the emulsion phase [Jackson, 1963; Yates *et al.*, 1994]. With these assumptions, the velocity and pressure distributions of the "fluid" in a uniform potential flow field around a bubble, as portrayed by Fig. 9.10, can be given as [Davidson and Harrison, 1963]

$$\begin{aligned} U_{pf} &= -U_{b\infty}\left(1-\frac{R_b^3}{r^3}\right)\cos\theta \\ V_{pf} &= U_{b\infty}\left(1+\frac{R_b^3}{2r^3}\right)\sin\theta \end{aligned} \tag{9.21}$$

and

$$p_{pf} = p_{pf}|_{z=0} + \left(\frac{\partial p_{pf}}{\partial z}\right)_\infty \left(r-\frac{R_b^3}{r^2}\right)\cos\theta \tag{9.22}$$

where the subscripts "pf" and "∞" represent the pseudofluid and the undisturbed condition, respectively; $U_{b\infty}$ is the rise velocity of an isolated bubble; and R_b is the bubble radius. The pressure far away from the rising bubble in a fluidized bed at minimum fluidization can

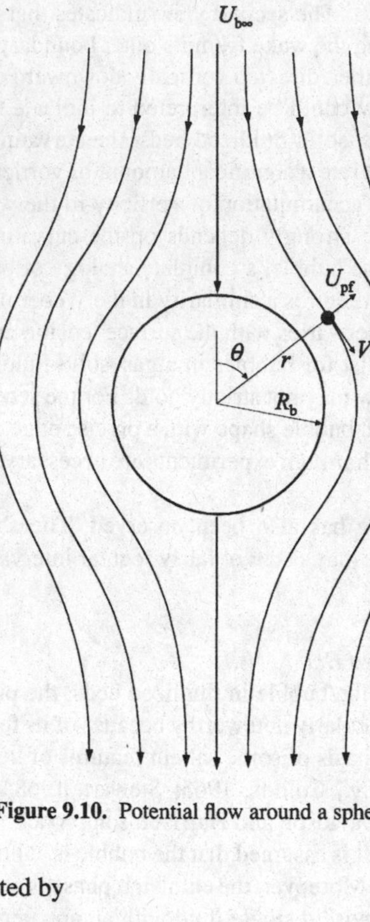

Figure 9.10. Potential flow around a spherical bubble in a two-dimensional projection.

be estimated by

$$\left(\frac{\partial p_{pf}}{\partial z}\right)_{\infty} \approx \frac{\Delta p_b}{H_f} = -(\rho_p - \rho)(1-\alpha_{mf})g \tag{9.23}$$

Thus, Eq. (9.22) becomes

$$p_{pf} = p_{pf}|_{z=0} - (\rho_p - \rho)(1-\alpha_{mf})g\left(r - \frac{R_b^3}{r^2}\right)\cos\theta \tag{9.24}$$

From the discussion in §6.3.1, it is reasonable to neglect the solids contributions to the total pressure. Hence, the pressure of gas $p \approx p_{pf}$. Furthermore, it is assumed that the particle velocity $\boldsymbol{U}_p \approx \boldsymbol{U}_{pf}$. In order to determine the local gas velocity, the pseudofluid medium, *i.e.*, the emulsion phase, is treated as a porous medium so that the Darcy law (*i.e.*, Eq. (5.316)) can be applied. Thus, the local gas velocity is obtained as

$$\begin{aligned} U &= U_p - \frac{k}{\mu}\frac{\partial p}{\partial r} \\ V &= V_p - \frac{k}{\mu}\frac{\partial p}{r\partial\theta} \end{aligned} \tag{9.25}$$

where k is the specific permeability of the porous medium. In the undisturbed region which

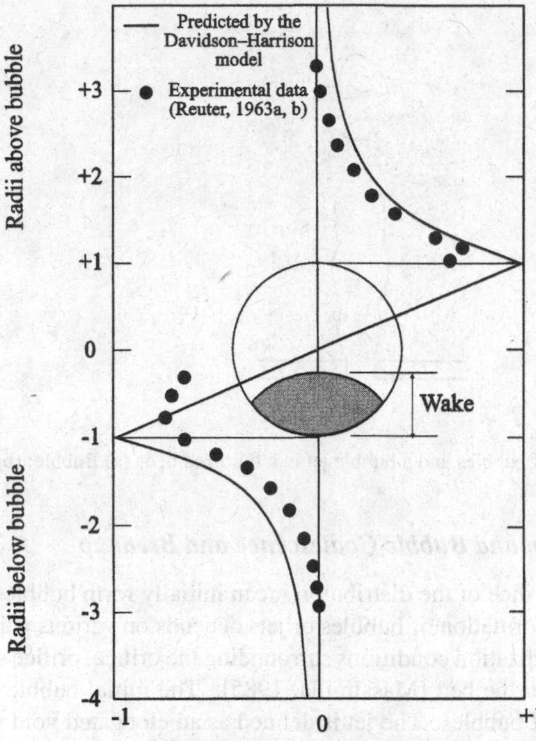

Figure 9.11. Pressure distribution in the vicinity of a rising three-dimensional bubble with a comparison of the experimental data by Reuter (1963a, b) and the Davidson–Harrison model prediction (from Stewart, 1968).

is far away from the bubble, it gives

$$\frac{U_{\mathrm{mf}}}{\alpha_{\mathrm{mf}}} = U_{\mathrm{b}\infty} - \frac{k}{\mu}\left(\frac{\partial p}{\partial z}\right)_{\infty} \tag{9.26}$$

Substituting Eqs. (9.23), (9.24), and (9.26) into Eq. (9.25) yields

$$\begin{aligned} U &= \left[\frac{R_{\mathrm{b}}^3}{r^3}\left(U_{\mathrm{b}\infty} + 2\frac{U_{\mathrm{mf}}}{\alpha_{\mathrm{mf}}}\right) - \left(U_{\mathrm{b}\infty} - \frac{U_{\mathrm{mf}}}{\alpha_{\mathrm{mf}}}\right)\right]\cos\theta \\ V &= \left[\frac{R_{\mathrm{b}}^3}{r^3}\left(\frac{U_{\mathrm{b}\infty}}{2} + \frac{U_{\mathrm{mf}}}{\alpha_{\mathrm{mf}}}\right) + \left(U_{\mathrm{b}\infty} - \frac{U_{\mathrm{mf}}}{\alpha_{\mathrm{mf}}}\right)\right]\sin\theta \end{aligned} \tag{9.27}$$

Figure 9.11 shows a comparison between the measured dynamic pressure and the calculated results based on Eq. (9.24). The profile indicates that there is a local high-pressure region near the bubble nose and a local low-pressure region around the bubble base, *i.e.*, the wake region. The low pressure in the wake region promotes pressure-induced bubble coalescence in the bed.

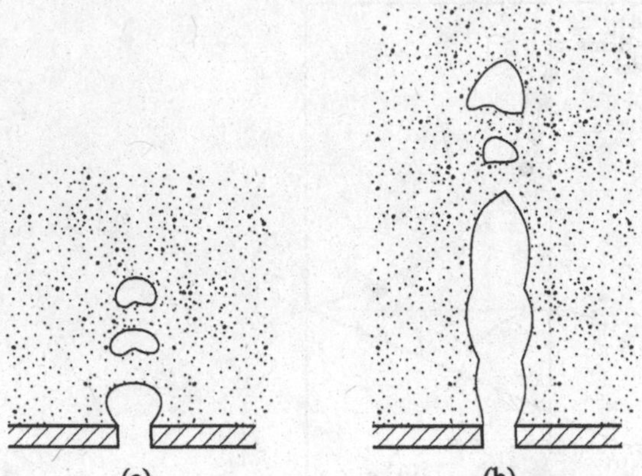

Figure 9.12. Formation of bubbles and a bubble jet in a fluidized bed: (a) Bubble; (b) Jet.

9.4.3 *Bubble/Jet Formation and Bubble Coalescence and Breakup*

When gas enters the orifice of the distributor, it can initially form bubbles or jets, as illustrated in Fig. 9.12. The formation of bubbles or jets depends on various parameters including types of particles, fluidization conditions surrounding the orifice, orifice size, and presence of internals or walls in the bed [Massimilla, 1985]. The initial bubble or jet is then transformed into a chain of bubbles. The jet is defined as an elongated void which is appreciably larger than a bubble; it extends permanently to some distance from the orifice into the inner bed. In general, bubbles tend to form in the presence of small particles, such as Group A particles [Rowe *et al.*, 1979]; jets tend to form in the presence of large particles, such as Group D particles, when the emulsion phase is not sufficiently fluidized or when internals are present and disrupt the solids flow to the orifice region [Massimilla, 1985].

The initial bubble can be formed either directly with little necking or with distinguishable necking; the neck is a small elongated void which provides a path for the gas flow for initial bubble formation. In the presence of neck, when the initial bubble is detached from the neck, the neck size may reduce and then grow to form the next bubble. A jet undergoes dynamic variations in location due to pressure fluctuations of the bed. It is noted that jet and spout (see §9.8) are different in that the spout penetrates the bed and is the main source for bed aeration, whereas the jet usually ends in the distributor region and is part of the bubble formation process in the bed and the bed (or emulsion phase) aeration occurs mainly through bubble action.

Bubbles may coalesce to form large bubbles or break up into smaller ones, as shown in Fig. 9.13. This bubble interaction leads to bubbling behavior in a multibubble system significantly different from that in a single-bubble system. The coalescence of bubbles in a gas–solid medium is similar to that in a liquid or liquid–solid medium [Fan and Tsuchiya, 1990]. The coalescence usually takes place with the trailing bubble overtaking the leading bubble through its wake region as a result of the regional minimum pressure (see §9.4.2.2). When two bubbles coalesce, the volume of the resulting bubble may be greater than the sum of the two coalescing bubbles [Clift and Grace, 1985]. It is noted that a high-voidage region which is different from the cloud region may exist around a bubble [Yates *et al.*, 1994]. Upon coalescence, the gas in this high-voidage region may be added to the volume

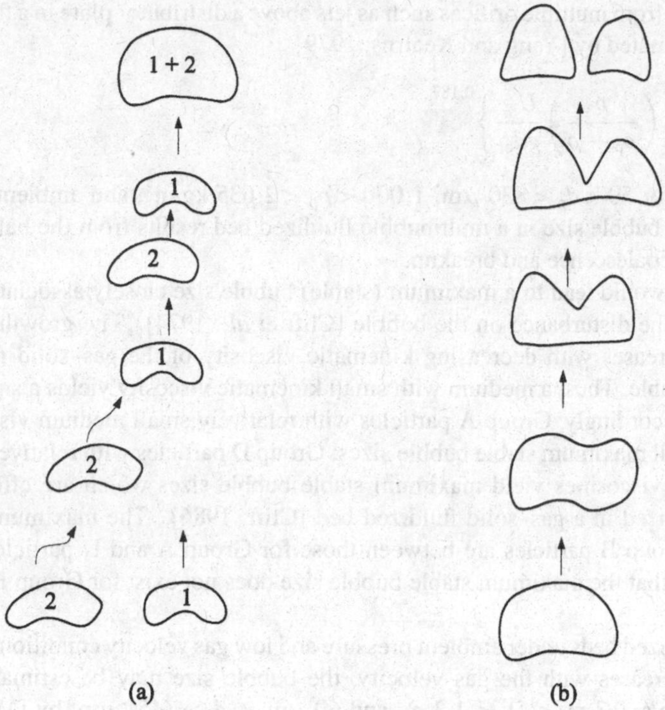

Figure 9.13. Bubble coalescence and breakup processes: (a) Bubble coalescence; (b) Bubble breakup.

of the resulting bubble. On the other hand, the breakup of a large bubble may start from an indentation on the upper boundary of the bubble resulting from the disturbance induced by relative motion with the particles, which may eventually break the bubble. Similarly to bubble coalescence, when a bubble splits, the total volume of the new bubbles can be less than the volume of the original parent bubble [Grace and Venta, 1973].

9.4.4 *Bubble/Jet Size and Bubble Rise Velocity*

The volume-equivalent diameter, or volume diameter (see Eq. (1.3)), of an initial bubble formed at a single upward orifice, d_{bi}, may be related to the volumetric gas flow rate through the orifice Q_{or} by [Davidson and Harrison, 1963; Miwa *et al.*, 1972]

$$d_{\mathrm{bi}} = 1.38 Q_{\mathrm{or}}^{0.4} g^{-0.2} \tag{9.28}$$

For a jet discharging from a single jet nozzle, the jet penetration length L_{j}, which is the averaged distance from the nozzle to the tip of the elongated void, may be estimated, for $50 < d_{\mathrm{p}} < 3{,}800$ μm, $208 < \rho_{\mathrm{p}} < 11{,}750$ kg/m^3, and ambient conditions, by [Yang and Keairns, 1978]

$$\frac{L_{\mathrm{j}}}{d_{\mathrm{or}}} = 6.5\sqrt{\left(\frac{\rho}{\rho_{\mathrm{p}} - \rho}\right)\frac{U_{\mathrm{or}}^2}{g d_{\mathrm{or}}}} \tag{9.29}$$

For jets discharging from multiple orifices such as jets above a distributor plate in a fluidized bed, L_j may be estimated by [Yang and Keairns, 1979]

$$\frac{L_j}{d_{or}} = 15\left\{\left(\frac{\rho}{\rho_p - \rho}\right)\frac{U_{or}^2}{gd_{or}}\right\}^{0.187} \tag{9.30}$$

Equation (9.30) is for $50 < d_p < 830$ μm, $1{,}000 < \rho_p < 2{,}635$ kg/m^3, and ambient conditions. The ultimate bubble size in a multibubble fluidized bed results from the balance of the rates of bubble coalescence and breakup.

Bubble stability would lead to a maximum (stable) bubble size closely associated with the growth rate of the disturbance on the bubble [Clift *et al.*, 1974]. The growth rate of the disturbance increases with decreasing kinematic viscosity of the gas–solid medium surrounding the bubble. Thus, a medium with small kinematic viscosity yields a small stable bubble size. Accordingly, Group A particles with relatively small medium viscosities yield relatively small maximum stable bubble sizes; Group D particles with relatively large medium kinematic viscosities yield maximum stable bubble sizes which are effectively too large to be realized in a gas–solid fluidized bed [Clift, 1986]. The maximum stable bubble sizes for Group B particles are between those for Group A and D particles. It is generally regarded that the maximum stable bubble size does not exist for Group B and D particles.

In bubbling fluidized beds under ambient pressure and low gas velocity conditions where the bubble size increases with the gas velocity, the bubble size may be estimated, for $U - U_{mf} < 0.48$ m/s, 0.3 m $< D < 1.3$ m, and 60 μm $< d_p < 450$ μm, by [Mori and Wen, 1975]

$$d_b = d_{bm} - (d_{bm} - d_{bi})\exp\left(-0.3\frac{H}{D}\right) \tag{9.31}$$

where H is the vertical distance from the gas distributor; d_b is the volume diameter of bubbles averaged over a cross-sectional area of the bed; d_{bi} is the initial bubble size at the distributor; and d_{bm} is the maximum bubble size. d_{bi} is given by

$$d_{bi} = \begin{cases} 1.38g^{-0.2}[A_d(U - U_{mf})]^{0.4} & \text{for perforated plates} \\ 0.376(U - U_{mf})^2 & \text{for porous plates} \end{cases} \tag{9.32}$$

where A_d is the distributor plate area per orifice evaluated by dividing the total distributor plate area by the total number of orifices on the distributor. d_{bm} can be expressed by

$$d_{bm} = 1.49[D^2(U - U_{mf})]^{0.4} \tag{9.33}$$

The bubble size may also be estimated on the basis of the correlation of Darton *et al.* (1977) under conditions of free bubbling beds without slugging and maximum stable bubble size as

$$d_b = 0.54(U - U_{mf})^{0.4}(H + 4\sqrt{A_d})^{0.8}g^{-0.2} \tag{9.34}$$

For dense-phase fluidization at high gas velocities and/or under elevated pressures, the bubble size may not always increase with the gas velocity. With 0.1 MPa $< p < 7.1$ MPa (in either the bubbling or the turbulent regime), the bubble size can be empirically correlated by [Cai *et al.*, 1994]

$$d_b = 0.38H^{0.8}\left(\frac{p}{p_a}\right)^{0.06}(U - U_{mf})^{0.42}$$
$$\times \exp\left[-1.4 \times 10^{-4}\left(\frac{p}{p_a}\right)^2 - 0.25(U - U_{mf})^2 - 0.1\frac{p}{p_a}(U - U_{mf})\right] \tag{9.35}$$

where p_a is the ambient pressure. Equation (9.35) covers the fluidization range for Group B particles and the lower part of Group D particles. This equation does not distinguish the distributor effects. The volume diameter of bubbles averaged over the entire bed, d_{bb}, can be given by

$$d_{bb} = \frac{1}{H_f}\int_0^{H_f} d_b\,dH \tag{9.36}$$

Example 9.1 In a fluidized bed of Group B particles with $H_f = 1$ m, calculate the average bubble sizes in the bed under the conditions of 0.1 MPa $< p <$ 2.1 MPa and 0.05 m/s $< U - U_{mf} <$ 1.8 m/s.

Solution Since the operating conditions cover both the low and high gas velocity ranges, Eq. (9.35) is used. Thus, d_{bb} can be obtained in terms of Eq. (9.36) as

$$d_{bb} = 0.21H_f^{0.8}\left(\frac{p}{p_a}\right)^{0.06}(U - U_{mf})^{0.42}$$
$$\times \exp\left[-1.4 \times 10^{-4}\left(\frac{p}{p_a}\right)^2 - 0.25(U - U_{mf})^2 - 0.1\frac{p}{p_a}(U - U_{mf})\right] \tag{E9.1}$$

The average bubble sizes calculated by Eq. (E9.1) under the given conditions are shown in a three-dimensional plot given in Fig. E9.1. It is seen that the bubble size variations are quite different in the bubbling regime and the turbulent regime, which are clearly demarcated by the onset velocity of the transition from the bubbling regime to the turbulent regime, *i.e.*, U_c (see §9.5). The applicable ranges of Eqs. (9.31) and (9.34) are also indicated in the figure.

The rise velocity of a single spherical cap bubble in an infinite liquid medium can be described by the Davies and Taylor equation [Davies and Taylor, 1950] (Problem 9.6). Experimental results indicate that the Davies and Taylor equation is valid for large bubbles ($d_{b\infty} > 0.02$ m, in general) with bubble Reynolds numbers greater than 40, while for bubbles in fluidized beds, the bubble Reynolds numbers are typically on the order of 10 or less [Clift, 1986]. By analogy, the rise velocity of an isolated single spherical cap bubble in an infinite gas–solid medium can be expressed in terms of the volume bubble diameter by [Davidson and Harrison, 1963]

$$U_{b\infty} = 0.71\sqrt{gd_{b\infty}} \tag{9.37}$$

In applying Eq (9.37), the infinite medium condition can be assumed when $d_b/D < 0.125$;

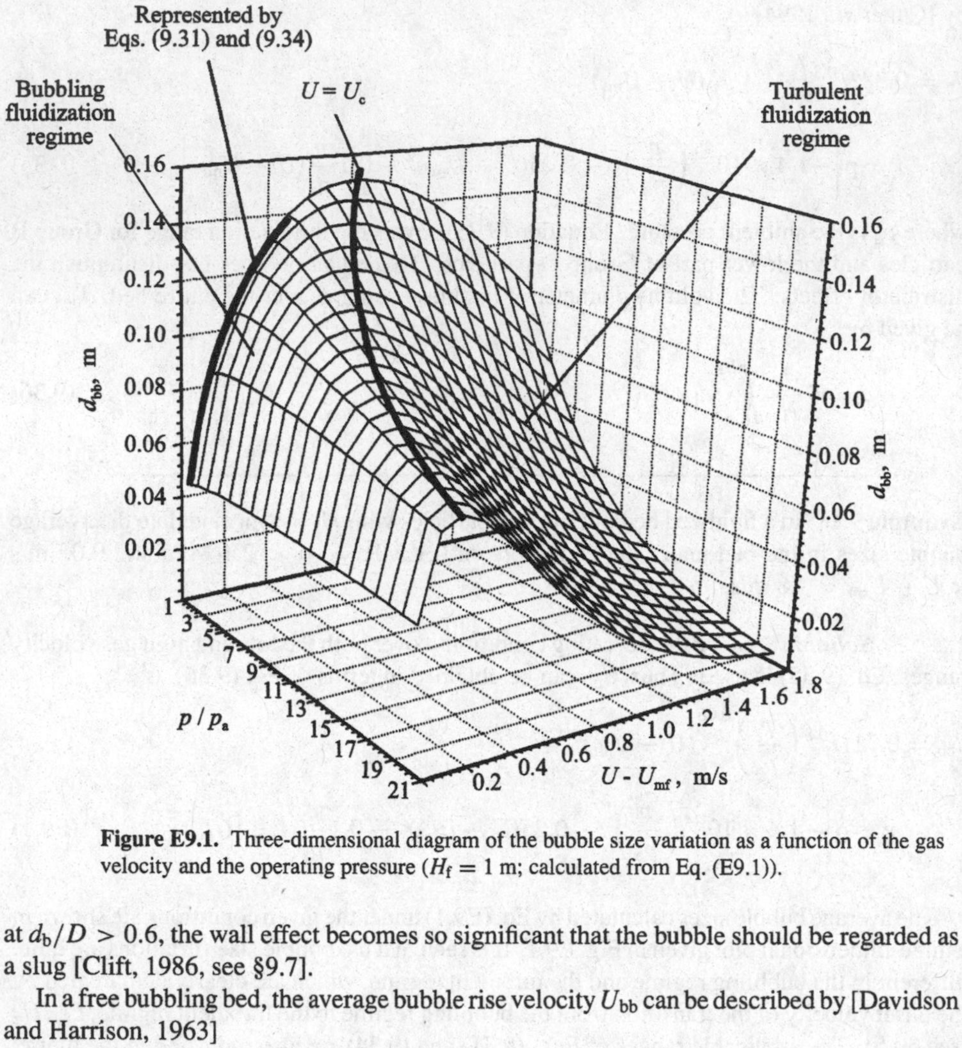

Figure E9.1. Three-dimensional diagram of the bubble size variation as a function of the gas velocity and the operating pressure ($H_f = 1$ m; calculated from Eq. (E9.1)).

at $d_b/D > 0.6$, the wall effect becomes so significant that the bubble should be regarded as a slug [Clift, 1986, see §9.7].

In a free bubbling bed, the average bubble rise velocity U_{bb} can be described by [Davidson and Harrison, 1963]

$$U_{bb} = U - U_{mf} + 0.71\sqrt{gd_{bb}} \tag{9.38}$$

where $(U - U_{mf})$ is added to the rise velocity of a single bubble to account for multibubble effects.

9.4.5 *Gas Flow Division and Bed Expansion*

The distribution of gas flow in the fluidized bed is important for the analysis of the fundamental characteristics of transport properties in the bed. One common method to estimate the gas flow division is based on the two-phase theory of fluidization, which divides the superficial gas flow in the bed into two subflows, *i.e.*, bubble phase flow and emulsion phase flow, as shown in Fig. 9.14. According to the theory, the flow velocity can be generally expressed as

$$U = \alpha_b U_{bb} + U_{em} \tag{9.39}$$

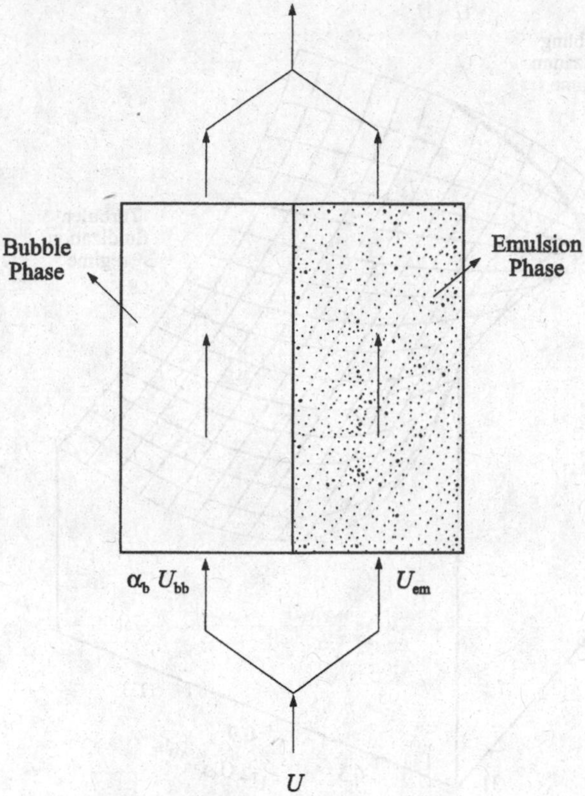

Figure 9.14. Gas flow distribution in the bed based on the two-phase model.

where U_{em} is the superficial gas velocity for the emulsion phase flow and $\alpha_b U_{bb}$ is that for the bubble phase flow. The variation of the relative magnitude of each flow depends on the gas velocity and gas and particle properties. The simple two-phase theory is to express U_{em} as U_{mf} and $\alpha_b U_{bb}$ as $(U - U_{mf})$ [Toomey and Johnstone, 1952]. Thus, the volumetric gas flow rate to the bubble phase, Q_b, becomes

$$Q_b = (U - U_{mf})A \tag{9.40}$$

Equation (9.40) may overestimate the actual bubble phase flow in a fluidized bed as a result of two prevailing effects of the gas flow pattern: (1) significant emulsion phase flow or invisible gas flow through the bubble and (2) larger interstitial gas flow in the emulsion phase than U_{mf} [Clift and Grace, 1985]. The theoretical analyses of invisible gas flow through the bubble indicate that the extent of the invisible gas flow through the 2-D and 3-D bubbles may vary from $0.5U_{mf}$ to $4U_{mf}$, depending on the specific stream function used in characterizing the flow field [Davidson and Harrison, 1963; Partridge and Rowe, 1966].

To account for the reduced bubble flow, Eq. (9.40) can be modified as [Peters *et al.*, 1982]

$$Q_b = Y(U - U_{mf})A \tag{9.41}$$

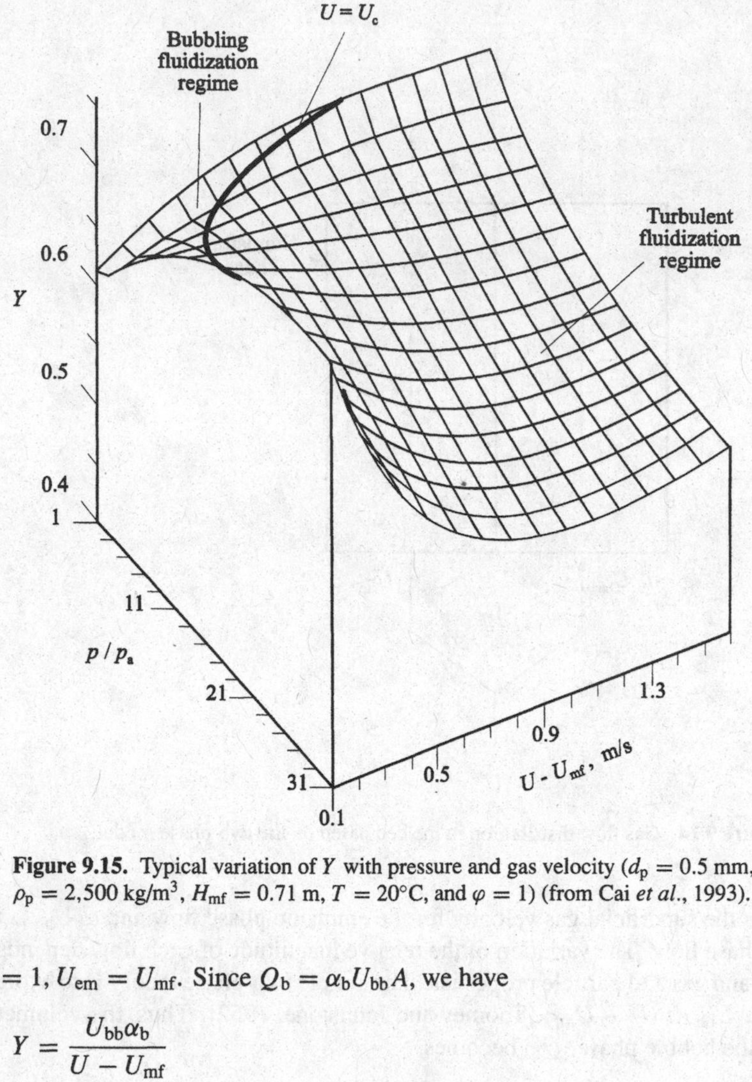

Figure 9.15. Typical variation of Y with pressure and gas velocity ($d_p = 0.5$ mm, $\rho_p = 2{,}500$ kg/m^3, $H_{mf} = 0.71$ m, $T = 20°$C, and $\varphi = 1$) (from Cai *et al.*, 1993).

With $Y = 1$, $U_{em} = U_{mf}$. Since $Q_b = \alpha_b U_{bb} A$, we have

$$Y = \frac{U_{bb}\alpha_b}{U - U_{mf}} \tag{9.42}$$

Figure 9.15 illustrates the variation of Y with the gas velocity and pressure, which shows that the deviation of the gas flow distribution from the simple two-phase theory (Eq. (9.40)) becomes significant with high gas velocities at high pressures. In other words, more gas flows to the emulsion phase under these conditions. Under very low gas velocities in the bubbling regime, the prediction of Q_b by Eq. (9.40) becomes more accurate for high pressures. Empirically, Y can also be estimated by $2.27\text{Ar}^{-0.21}$ [Geldart, 1986].

In dense-phase fluidization, the expansion of the bed is caused by two factors: emulsion phase expansion and presence of bubbles. Thus, the bed expansion height H_f can be expressed as a linear addition of the contributions of these factors

$$H_f = H_{em} + H_f \alpha_b \tag{9.43}$$

where H_{em} is the expansion height of the emulsion phase, which is larger than H_{mf} for fine particles (Group A). For Group B and D particles, $H_{em} = H_{mf}$. Under very high operating pressures, some Group B particles may yield $H_{em} > H_{mf}$. From Eqs. (9.39) and (9.43), it gives

$$H_f = \frac{H_{em}U_{bb}}{U_{bb} - (U - U_{em})} \tag{9.44}$$

For the emulsion phase expansion of Group A particles, H_{em} can be correlated by [Abrahamsen and Geldart, 1980b]

$$\frac{H_{em}}{H_{mf}} = \frac{2.54\rho^{0.016}\mu^{0.066}\exp(0.09\phi_f)}{d_p^{0.1}g^{0.118}(\rho_p - \rho)^{0.118}H_{mf}^{0.043}} \tag{9.45}$$

Equation (9.45) assumes that there is no velocity effect on H_{em}.

When U_{em} or U_{bb} in Eq. (9.44) is unknown, the determination of bed expansion height, H_f, can be estimated by empirical correlations. At ambient temperature, the correlation for H_f of Babu *et al.* (1978) covers a wide range of particle properties ($0.05 < d_p < 2.87$ mm; $257 < \rho_p < 3{,}928$ kg/m^3) and operating pressures ($0.1 < p < 7.0$ MPa)

$$\frac{H_f}{H_{mf}} = 1 + \frac{14.3(U - U_{mf})^{0.738}d_p^{1.006}\rho_p^{0.376}}{U_{mf}^{0.937}\rho^{0.126}} \tag{9.46}$$

When operated under high-temperature (up to 1,258 K) conditions, the bed expansion can be estimated by [Cai *et al.*, 1993]

$$\frac{H_f}{H_{mf}} = 1 + \frac{21.4\left(U - U_{mf}^*\right)^{0.738}d_p^{1.006}\rho_p^{0.376}}{\left(U_{mf}^*\right)^{0.937}\left(w_g\dfrac{p}{p_a}\right)^{0.126}} \tag{9.47}$$

where w_g is the molecular weight of gas and U_{mf}^* is the minimum fluidization velocity at ambient temperature. Equation (9.47) covers the same ranges of d_p, ρ_p, and p as Eq. (9.46) does.

Example 9.2 Determine the gas flow rate in the bubble phase and emulsion phase for a fluidized bed combustor under the following two operating conditions: (a) $T = 1{,}173$ K, $p = 1.1$ MPa, $D = 1.2$ m, $M_p = 1{,}600$ kg, $d_p = 0.51$ mm, $\varphi = 0.84$, $\rho_p = 2{,}422$ kg/m^3, and $U = 1.0$ m/s; (b) $T = 293$ K, $p = 0.1$ MPa, and other conditions are the same as (a).

Solution (a) Knowing d_p, ρ_p, ρ, and μ, the minimum fluidization velocity under both the given and ambient conditions can be obtained from Eq. (9.14) as $U_{mf} = 0.077$ m/s and $U_{mf}^* = 0.122$ m/s, respectively. Knowing d_p, ρ_p, ρ, μ, and φ, the voidage at the minimum fluidization condition can be calculated from Eq. (9.11) as $\alpha_{mf} = 0.422$. Thus, the bed height at minimum fluidization is given by

$$H_{mf} = \frac{M_p}{A_t\rho_p(1 - \alpha_{mf})} = \frac{1{,}600}{\dfrac{\pi}{4} \times 1.2^2 \times 2{,}422 \times (1 - 0.422)} = 1.01 \text{ m} \tag{E9.2}$$

The expansion height of the bed can be calculated by Eq. (9.47) as

$$H_f = 1.01\left\{1 + \frac{21.4 \times (1.0 - 0.122)^{0.738} \times (510 \times 10^{-6})^{1.006} \times 2{,}422^{0.376}}{0.122^{0.937} \times (28.9 \times 11)^{0.126}}\right\} = 1.63 \text{ m} \tag{E9.3}$$

The bubble fraction in the bed can be obtained from Eq. (9.43) using $H_{em} = H_{mf}$ for Group B particles as

$$\alpha_b = 1 - \frac{H_{mf}}{H_f} = 1 - \frac{1.01}{1.63} = 0.380 \tag{E9.4}$$

The average bubble size in the bed can be calculated by Eq. (E9.1) as

$$d_{bb} = 0.21 \times 1.63^{0.8} \times (1.0 - 0.077)^{0.06} \times \exp[-1.4 \times 10^{-4} \times 11^2 - 0.25 \times (1.0 - 0.077)^2 - 0.1 \times 11 \times (1.0 - 0.077)] = 0.100 \text{ m} \tag{E9.5}$$

The average bubble rise velocity in the bed can be obtained by Eq. (9.38) as

$$U_{bb} = 1.0 - 0.077 + 0.71 \times \sqrt{9.807 \times 0.100} = 1.63 \text{ m/s} \tag{E9.6}$$

Thus, Y can be evaluated from Eq. (9.42) as given by

$$Y = \frac{0.380 \times 1.63}{1.0 - 0.077} = 0.671 \tag{E9.7}$$

(b) Repeating the procedure for the conditions of 293 K and 0.1 MPa yields $Y = 0.754$. The gas flow distributions between the two phases for these two cases can be summarized as follows:

Conditions	Bubble phase, $\alpha_b U_{bb}$	Emulsion phase, $U - Y(U - U_{mf})$
1,173 K, 1.1 MPa	0.619 m/s	0.381 m/s
293 K, 0.1 MPa	0.662 m/s	0.338 m/s

9.5 Turbulent Fluidization

The turbulent regime is often regarded as the transition regime from bubbling fluidization to lean-phase fluidization. At relatively low gas velocities, bubbles are present in the turbulent regime, while at relatively high gas velocities in the turbulent regime, the clear boundary of bubbles disappears and the nonuniformity of solids concentration distribution yields distinct gas voids which become less distinguishable as the gas velocity further increases toward lean-phase fluidization.

9.5.1 *Regime Transition and Identification*

In bubbling fluidization, bubble motion becomes increasingly vigorous as the gas velocity increases. This behavior can be reflected in the increase of the amplitude of the pressure fluctuations in the bed. With further increase in the gas velocity, the fluctuation will reach a maximum, decrease, and then gradually level off, as shown in Fig. 9.16. This fluctuation variation marks the transition from the bubbling to the turbulent regime.

The onset velocity of the transition to the turbulent regime is commonly defined to be the gas velocity corresponding to the peak, U_c, whereas the leveling point, U_k, may be recognized as the onset of the turbulent regime proper [Yerushalmi and Cankurt, 1979].

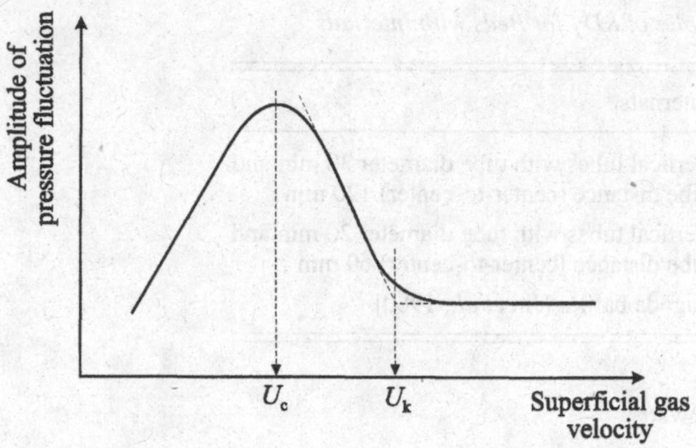

Figure 9.16. Variations of pressure fluctuation with the gas velocity for dense-phase fluidized beds with FCC particles (from Yerushalmi and Cankurt, 1979).

However, on the basis of direct observation of bed phenomena, appreciable variations in bubble behavior occur at gas velocities around U_c. Specifically, the bubble interaction is dominated by bubble coalescence at gas velocities less than U_c, while it is dominated by bubble breakup at gas velocities greater than U_c. Furthermore, there is no significant flow structure change observed at gas velocities from U_c to U_k and even higher [Brereton and Grace, 1992]. At U_k, the bed is usually already highly turbulent and may even be close to the transition to lean-phase fluidization [Bi and Fan, 1992]. U_k was reported to vary with axial location [Chehbouni *et al.*, 1994]. The plot in Fig. 9.16 for the amplitude of pressure fluctuation appears to be typical only of Group A particles. For Group B and D particles, the amplitude of pressure fluctuation at velocities beyond U_c may decrease without having a distinct leveling-off point for U_k [Satija and Fan, 1985]. Thus, for all practical purposes, U_c is defined as the onset velocity for the transition to the turbulent regime. The upper bound of the turbulent regime is discussed in Chapter 10.

Examining the variation of bubble behavior across U_c, a criterion for the bubbling–turbulent regime transition is proposed by Cai *et al.* (1992) as

$$\left(\frac{\partial^2 N_B}{\partial U^2}\right)_{U=U_c} = 0 \tag{9.48}$$

where N_B is the bubble number per unit bed volume. The conditions for the bubbling and the turbulent regimes can be expressed by

$$\begin{aligned} \frac{\partial^2 N_B}{\partial U^2} &< 0 \quad \text{for the bubbling regime} \\ \frac{\partial^2 N_B}{\partial U^2} &> 0 \quad \text{for the turbulent regime} \end{aligned} \tag{9.49}$$

Thus, to predict the onset velocity for the regime transition using Eq. (9.48), a mechanistic model accounting for N_B for a given fluidized bed operation is needed (Problem 9.9).

The specific flow and transport characteristics of turbulent fluidization, which are distinctly different from those of bubbling fluidization operated at relatively low gas velocities,

Table 9.2. *Examples of KD_f for Beds with Internals*

KD_f	Internals
2.32×10^{-3}	Vertical tubes with tube diameter 30 mm and tube distance (center-to-center) 120 mm
1.64×10^{-3}	Vertical tubes with tube diameter 20 mm and tube distance (center-to-center) 60 mm
3.42×10^{-4}	Pagoda baffles [Jin *et al.*, 1980]

can be described as follows:

(1) The upper surface of the turbulent bed exists but becomes more diffused with a large particle concentration in the freeboard.
(2) The bubble size is small. Bubbles split and redisperse frequently and often appear in more irregular shapes.
(3) Bubbles/voids move violently, rendering it difficult to distinguish the emulsion (continuous) phase from the bubble/void (discrete) phase in the bed.
(4) Bubble motion appears to be more random with enhanced interphase exchange and hence intimate gas–solid contact and high heat and mass transfer.

9.5.2 *Determination of Transition Velocity*

While the efforts to develop a mechanistic model for predicting U_c are limited, empirical correlations are commonly employed. For Group A and B particles within the range of $293 < T < 773$ K and $0.1 < p < 0.8$ MPa, the correlation of Cai *et al.* (1989) can be used:

$$\frac{U_c}{\sqrt{gd_p}} = \left(\frac{\mu_a}{\mu}\right)^{0.2}\left[\left(\frac{\rho_a}{\rho}\right)\left(\frac{\rho_p - \rho}{\rho}\right)\left(\frac{KD_f}{d_p}\right)\right]^{0.27} \tag{9.50}$$

where KD_f is the parameter accounting for the effects of the bed geometry and internals. For beds without internals, KD_f can be expressed by

$$KD_f = D\left(\frac{0.211}{D^{0.27}} + \frac{2.42 \times 10^{-3}}{D^{1.27}}\right)^{\frac{1}{0.27}} \tag{9.51}$$

Equation (9.51) is applicable for 57 mm $\leq D \leq$ 475 mm. For beds with internals, values for KD_f are given in Table 9.2, which shows that KD_f is highly dependent on internals geometry.

Equation (9.50) shows the functionality dependency of U_c on the operating variables. The equation reflects that the turbulent regime transition occurs at a lower gas velocity when the particle size is smaller or the operating pressure is higher. The influence of internals or baffles on U_c is also revealed in the equation as well as in Fig. 9.17 for pagoda baffles and vertical tubes. The significant reduction of U_c in a bed with internals or baffles is evident. This is due to the fact that the internals or baffles induce bubble breakup, leading to the dominance of bubble interactions via breakup over coalescence at a relatively lower gas velocity. The magnitude of the pressure fluctuation reduces in the presence of the baffles and hence smoother operation results in baffled beds than in beds without baffles.

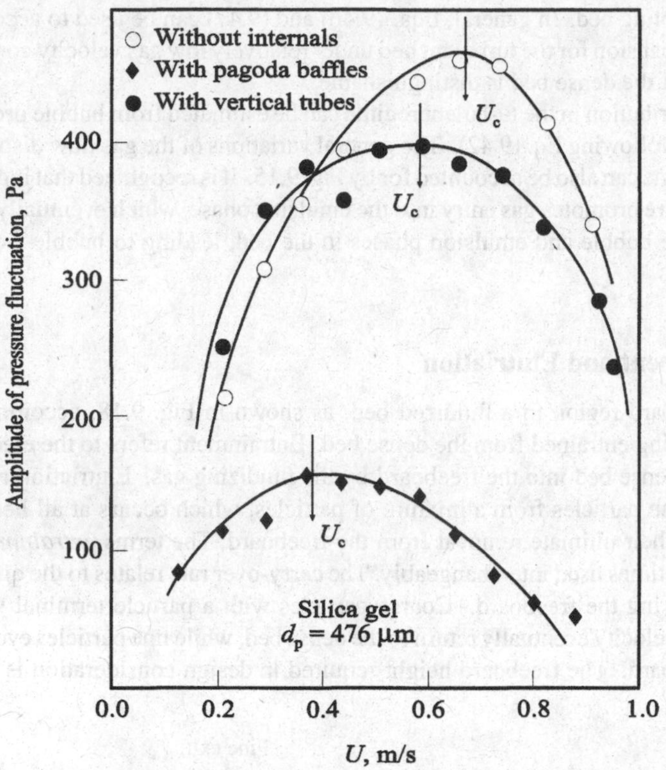

Figure 9.17. Effect of immersed internals on the flow regime transition (from Cai, 1989).

9.5.3 *Hydrodynamic Characteristics*

As a result of the existence of bubbles/voids in turbulent fluidized beds, the hydrodynamic behavior of turbulent fluidization under relatively low gas velocity conditions is similar, to a certain extent, to that of bubbling fluidization. However, distinct differences exist under relatively high gas velocity conditions, thus rendering many correlations previously developed for the bubbling regime invalid for the turbulent regime.

The bubble/void size in the turbulent regime tends to decrease with an increase in the gas velocity due to the predominance of bubble breakup over bubble coalescence, as shown in Fig. E9.1. This trend is opposite to that exhibited in the bubbling regime. However, similarly to the bubbling regime, an increase in the operating pressure at a constant gas velocity or constant excess gas velocity (*i.e.*, $U - U_{mf}$) decreases the bubble size. Thus, the bubble/void size in the turbulent bed can eventually be reduced to such a magnitude, with sufficiently high gas velocities and high pressures, that it marks a gradual transition to lean-phase bubble-free fluidization. For bubble/void size prediction, Eq. (9.35) can be applied to both the bubbling and turbulent fluidization regimes in beds without internals. For bubble rise velocity prediction, Eq. (9.38) can be employed provided that bubble/void sizes are known.

As more small bubbles/voids with low rise velocities and long residence times are present in the turbulent bed, they give rise to more significant dense bed expansion in the turbulent bed than in the bubbling bed. In general, Eqs. (9.46) and (9.47) can be used to account for the height of bed expansion for the turbulent bed under relatively low gas velocity conditions where the surface of the dense bed is distinguishable.

The gas flow distribution in the turbulent regime can be estimated from bubble properties and bed expansion, following Eq. (9.42). The general variations of the gas flow distribution in the turbulent regime can also be accounted for by Fig. 9.15. It is recognized that increasing the operating pressure promotes gas entry into the emulsion phase, which eventually results in undistinguishable bubble and emulsion phases in the bed, leading to bubble-free lean-phase fluidization.

9.6 Entrainment and Elutriation

The freeboard region in a fluidized bed, as shown in Fig. 9.18, accommodates particles that are being entrained from the dense bed. Entrainment refers to the ejection of particles from the dense bed into the freeboard by the fluidizing gas. Elutriation refers to the separation of fine particles from a mixture of particles, which occurs at all heights of the freeboard, and their ultimate removal from the freeboard. The terms *entrainment* and *elutriation* are sometimes used interchangeably. The carry-over rate relates to the quantities of the particles leaving the freeboard. Coarse particles with a particle terminal velocity higher than the gas velocity eventually return to the dense bed, while fine particles eventually exit from the freeboard. The freeboard height required in design consideration is usually

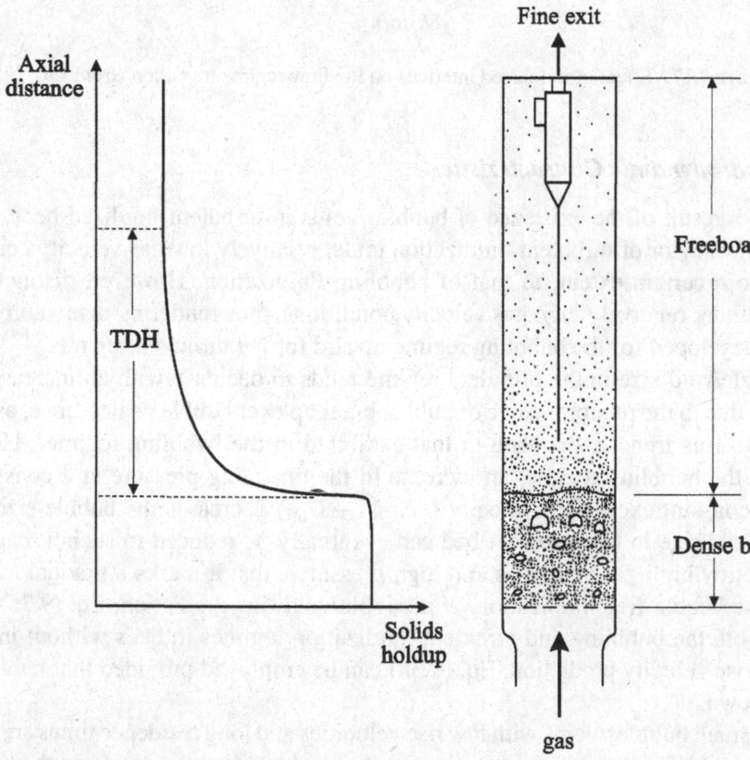

Figure 9.18. Solids holdup distribution in the freeboard region of a fluidized bed.

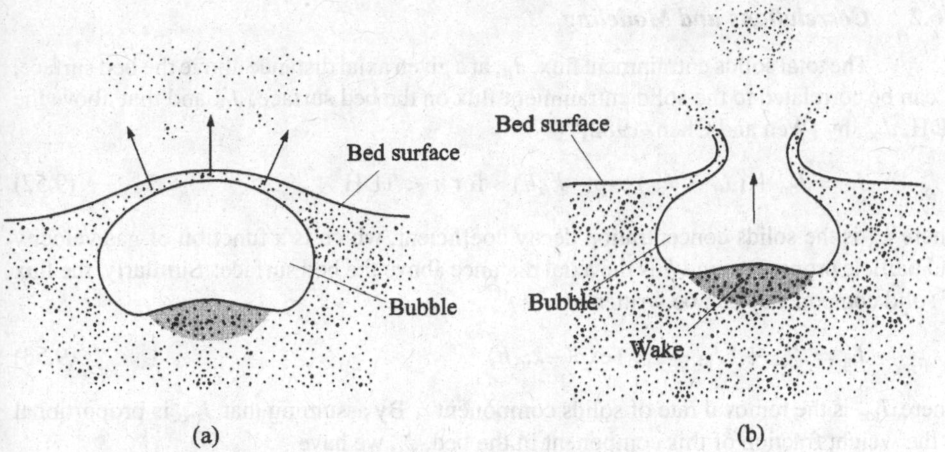

Figure 9.19. Two main mechanisms of solids injection into the freeboard (from Kunii and Levenspiel, 1991): (a) From the bubble roof; (b) From the bubble wake.

higher than the transport disengagement height (TDH), defined as a height beyond which the solids holdup, solids entrainment, or carry-over rate remains nearly constant. The nonuniformity of the solids holdup distribution in the freeboard region, as indicated in Fig. 9.18, is in part due to the nonuniformity of particle properties in the dense bed. As shown in the figure, coarse particles are present primarily in the region of the freeboard close to the dense bed surface, whereas the upper part of the freeboard consists mainly of fine particles.

9.6.1 *Mechanisms of Solids Ejection into the Freeboard*

Particles are ejected into the freeboard via two basic modes: (1) ejection of particles from the bubble roof and (2) ejection of particles from the bubble wake, as illustrated in Fig. 9.19. The roof ejection occurs when the bubble approaches the surface of the bed, and a dome forms on the surface. As the bubble further approaches the bed surface, particles between the bubble roof and surface of the dome thin out [Peters *et al.*, 1983]. At a certain dome thickness, eruption of bubbles with pressure higher than the surface pressure takes place, ejecting the particles present on top of the bubble roof to the freeboard. In wake ejection, as the bubble erupts on the surface, the inertia effect of the wake particles traveling at the same velocity as the bubble promptly ejects these particles to the freeboard. The gas leaving the bed surface then entrains these ejected particles to the freeboard.

In free bubbling fluidization, especially at a high velocity, bubble coalescence frequently takes place near the bed surface. The coalescence due to an accelerated traveling bubble yields a significant ejection of wake particles from the leading bubble. The wake ejection in this case becomes the dominant source of particles present in the freeboard in free bubbling fluidized beds [Pemberton, 1982]. The wake ejection in the turbulent regime is much more pronounced than that in the bubbling regime because of higher bubble rise velocities. The fluid–particle and particle–particle interaction and gravitational effects are key factors dictating the solids holdup profile in Fig. 9.18. Particles may also form agglomerates or clusters in the freeboard.

9.6.2 *Correlations and Modeling*

The total solids entrainment flux, J_h, at a given axial distance above the bed surface, h, can be correlated to the solid entrainment flux on the bed surface, J_0, and that above the TDH, J_∞, by [Wen and Chen, 1982]

$$J_h = J_\infty + (J_0 - J_\infty)\exp(-k_A h) \quad \text{for } h < \text{TDH} \tag{9.52}$$

where k_A is the solids concentration decay coefficient, which is a function of gas velocity and particle properties, and h is the axial distance above the bed surface. Similarly, the flux of solid component i can be written as

$$J_{ih} = J_{i\infty} + (J_{i0} - J_{i\infty})\exp(-k_{iA} h) \tag{9.53}$$

where $J_{i\infty}$ is the removal rate of solids component i. By assuming that $J_{i\infty}$ is proportional to the weight fraction of this component in the bed, ϕ_i, we have

$$J_{i\infty} = -\frac{1}{A}\frac{\mathrm{d}M_i}{\mathrm{d}t} = k_{iB}\phi_i \tag{9.54}$$

where M_i is the mass of the solid component i in the bed, k_{iB} is the elutriation rate constant of component i, and A is the cross-sectional area of the bed. Likewise, for total solid particles, the equation becomes

$$J_\infty = -\frac{1}{A}\frac{\mathrm{d}M_p}{\mathrm{d}t} = \sum k_{iB}\phi_i \tag{9.55}$$

where k_{iB} can be estimated by [Geldart *et al.*, 1979]

$$\frac{k_{iB}}{\rho U} = 23.7\exp\left(-5.4\frac{U_{pti}}{U}\right) \tag{9.56}$$

for small particles ($60 < d_p < 350$ μm) in the bubbling or turbulent fluidization regime ($0.6 < U < 3$ m/s); or by [Colakyan and Levenspiel, 1984]

$$k_{iB} = 0.011\rho_p\left(1 - \frac{U_{pti}}{U}\right)^2 \tag{9.57}$$

for elutriation of relatively coarse particles ($0.3 < d_p < 1$ mm) at high gas velocity ($0.9 < U < 3.7$ m/s).

The TDH for Group A particles can be determined by [Horio *et al.*, 1980]

$$\text{TDH} = 4.47\sqrt{d_{bs}} \tag{9.58}$$

where d_{bs} is the bubble size at the bed surface. Once given TDH at a gas velocity U_1, the TDH at a different gas velocity U_2 can be calculated by [Fournol *et al.*, 1973]

$$\frac{\text{TDH}_2}{\text{TDH}_1} = \frac{U_2^2}{U_1^2} \tag{9.59}$$

The variation of the TDH with respect to the bed diameter as well as the gas velocity is illustrated by Fig. 9.20. It shows that a higher TDH results from an increase in the gas velocity at a given D. In addition, the TDH may also increase under elevated pressures [Chan and Knowlton, 1984].

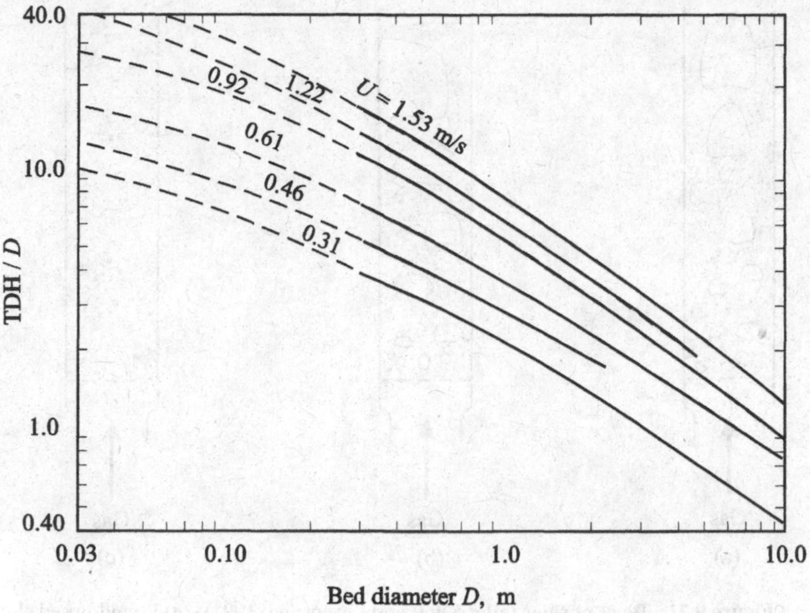

Figure 9.20. Variation of TDH with bed diameter for FCC particles (from Zenz and Weil, 1958).

9.7 Slugging

When bubbles grow to sizes comparable to the bed diameter, slugging occurs, as shown in Fig. 9.21. Slugging is most frequently encountered in small-diameter beds with large bed heights, especially when large/heavy particles are employed. There are requirements of a minimum slugging velocity and a minimum bed height for a slug flow to take place. The minimum gas velocity for slugging can be estimated by [Stewart and Davidson, 1967]

$$U_{\mathrm{msl}} = U_{\mathrm{mf}} + 0.07\sqrt{gD} \tag{9.60}$$

whereas the minimum bed height required for slugging is given by [Baeyens and Geldart, 1974]

$$H_{\mathrm{msl}} = 1.34D^{0.175} \tag{9.61}$$

for $0.05 < D < 0.3$ m, $850 < \rho_{\mathrm{p}} < 2{,}800$ kg/m^3, and $0.055 < d_{\mathrm{p}} < 3.38$ mm. Unlike a constant pressure drop pattern for the bubbling fluidized bed, the total pressure drop in a slugging bed increases with an increase in the gas velocity as a result of strong particle–wall friction and momentum dissipation from the gas to the solids.

9.7.1 *Shapes and Rise Velocities of Single Slugs*

The slugs may appear in different forms. As shown in Fig. 9.21, Type (a) is the round-nosed slug, which occurs in systems of fine particles; Type (b) is the wall slug (also known as the half slug), which takes place in beds with rough walls, large d_{p}/D ratios,

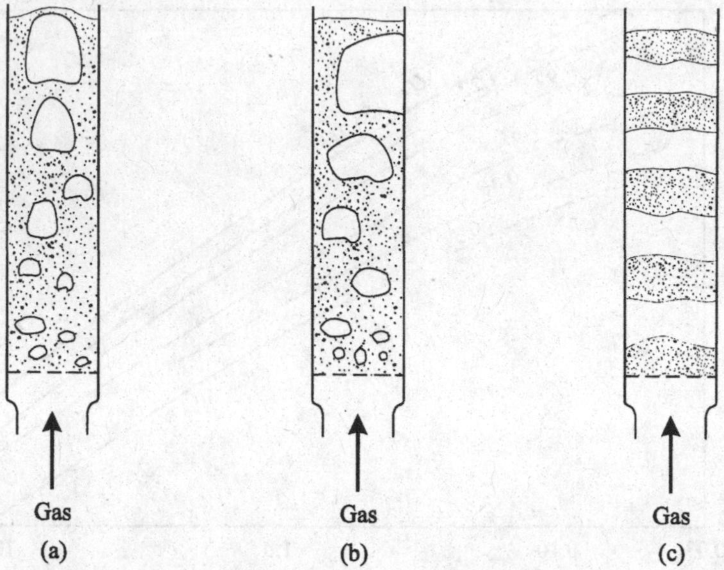

Figure 9.21. Types of slugs (after Kunii and Levenspiel, 1991): (a) Round-nosed slug; (b) Wall slug (half slug); (c) Square-nosed slug (or plug).

angular particle shapes, and relatively high gas velocities; and Type (c) is the square-nosed slug, which appears in coarse particle systems where the particle bridging effect is significant.

The rise velocity of a single isolated Type (a) slug can be given by [Hovmand and Davidson, 1971]

$$U_{\mathrm{sl}\infty} = 0.35\sqrt{gD} \tag{9.62}$$

The Type (b) slug can be treated as a full round-nosed slug in a column of diameter $2D$. Thus, the typical rising velocity is approximately $\sqrt{2}$ times the velocity given by Eq. (9.62). However, for Type (c) slug (plug), the situation becomes quite different. The particles will "rain" through the plug continuously, and the particles are in a state of bridging instead of truly fluidizing. The plug rising velocity is then a function of interparticle forces, internal friction of particles, and gas velocity. In general, the rise velocity of a plug is lower than that calculated from Eq. (9.62).

9.7.2 *Continuous Slugging*

In a continuous slugging flow, particularly of Types (a) and (b), two adjacent slugs can coalesce to form a larger slug in a way similar to but simpler than that of bubble coalescence. If the distance between the tail of the leading slug and the nose of the following slug is larger than $2D$, the leading slug has almost no effect on the trailing one [Hovmand and Davidson, 1971]. Coalescence between slugs is less frequent than that in the comparable bubbling bed. Furthermore, a smaller wake volume and more ordered wake motion or less wake dispersion for the slugging bed are observed than in the bubbling bed.

Applying the simple two-phase theory to the slugging bed, the average slug velocity in continuous slugging flows can be expressed by [Ormiston *et al.*, 1965]

$$U_{sl} = U_{sl\infty} + U - U_{mf} \tag{9.63}$$

The maximum height of expansion of the slugging bed, H_{max}, can also be obtained by

$$H_{max} = H_{mf} + H_{max}\alpha_{sl} \tag{9.64}$$

where α_{sl} is obtained by considering the overall gas balance as

$$U = U_{sl}\alpha_{sl} + U_{mf} \tag{9.65}$$

Substituting Eqs. (9.63) and (9.65) into Eq. (9.64) yields

$$H_{max} = \frac{H_{mf}U_{sl}}{U_{sl\infty}} \tag{9.66}$$

Example 9.3 Verify Stewart's criterion (Eq. (9.60)), which states that slug flow occurs if

$$U - U_{mf} \geq 0.07\sqrt{gD} \tag{E9.8}$$

with the following assumptions (see Fig. E9.2):
(1) The slug is approximated as a paraboloid revolution so that the slug volume is given by

$$V_{sl} = \frac{\pi}{8}D^3 \tag{E9.9}$$

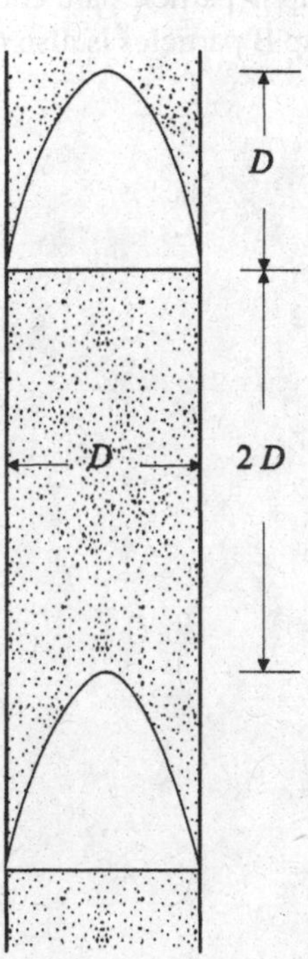

Figure E9.2. Schematic representation of the slug flow conditions given in Example 9.3.

(2) The distance between consecutive slugs is $2D$.

Solution Substituting Eqs. (9.62) and (9.63) into Eq. (9.65) gives

$$U - U_{\mathrm{mf}} = 0.35\left(\frac{\alpha_{\mathrm{sl}}}{1-\alpha_{\mathrm{sl}}}\right)\sqrt{gD} \tag{E9.10}$$

The slug voidage in a continuous stream with separation distance $2D$, as shown in Fig. E9.2, is given by

$$\alpha_{\mathrm{sl}} = \frac{\text{volume of slug}}{\text{total volume}} = \frac{(\pi/8)D^3}{(\pi/4)D^2(2D+D)} = \frac{1}{6} \tag{E9.11}$$

Substituting Eq. (E9.11) into Eq. (E9.10) yields the condition for slug flow given by Eq. (E9.8).

9.8 Spouted Beds

In a spouted bed, gas enters the bed through a jet nozzle of diameter D_{i}, forming a spout of diameter D_{s} in the center of the bed, as shown in Fig. 9.22. The surrounding annular region forms a downward-moving bed. Particles are entrained into the spout from the bottom and side wall of the spout. Part of the gas seeps into the annular region through the spout wall, whereas the other part leaves the bed from the top of the spout. The particles carried into the spout disengage from the gas in a solid disengagement fountain just above the bed and then return to the top of the annular region. Group D particles are commonly used for the spouted bed operation, although spouting of Group B particles is also possible

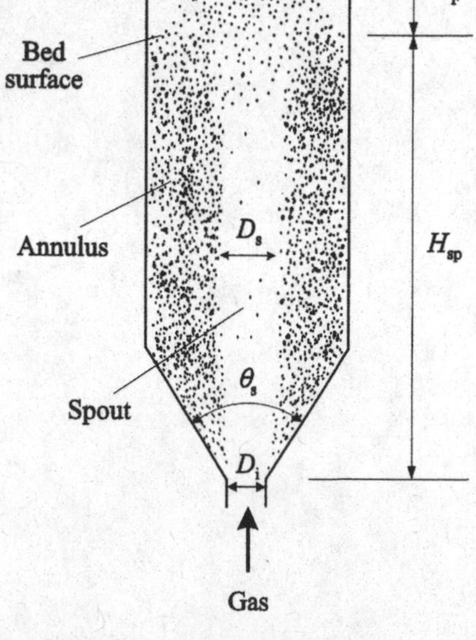

Figure 9.22. Schematic diagram of a spouted bed (after Mathur and Epstein, 1974).

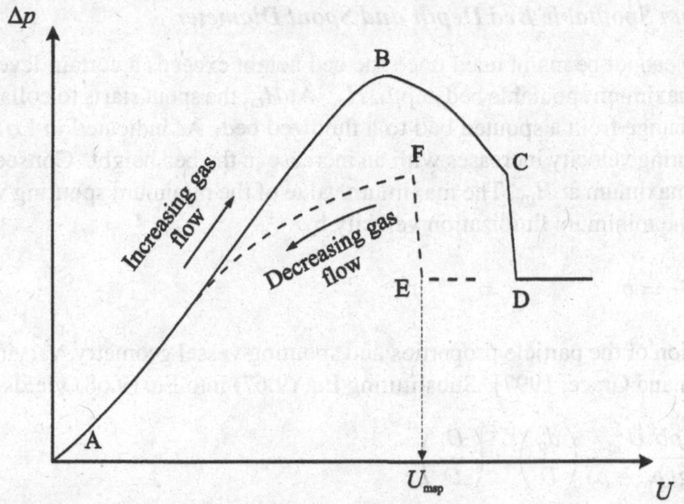

Figure 9.23. Variation of the pressure drop with the gas flow rate in a spouted bed (from Mathur and Epstein, 1974).

if the nozzle diameter does not exceed 25.4 times the particle diameter [Chandnani and Epstein, 1986]. In some cases, a fluid-spout or aerated-spout bed operation is conducted with secondary gas, which is introduced into the bed through the side wall of the conical base to increase the gas–particle contact in the annular region.

9.8.1 *Onset of Spouting*

The process of spouting onset can be schematically depicted as in Fig. 9.23, which shows the variation of the pressure drop with the gas flow rate in the process. With an increase in the gas velocity, a central dilute flow core region due to gas penetration is formed under the bed surface, and the pressure drop increases, following the line AB. As the gas velocity further increases, the pressure drop decreases until point C, when the core region penetrates the entire bed, forming a spout. A slight further increase in the gas velocity leads to a steep decrease in the pressure drop to point D, beyond which the pressure drop levels off. However, as the gas velocity decreases from point D, the pressure drop traces a different path (DEFA), marking a hysteresis loop. The minimum spouting velocity, U_{msp}, is defined as the gas velocity at point E.

Unlike the minimum fluidization velocity, the minimum spouting velocity U_{msp} depends not only on the particle and gas properties but also on the bed geometry and static bed height. By reducing the gas velocity or by increasing the bed height, spouting can be diminished. U_{msp} can be determined by [Mathur and Gishler, 1955]

$$U_{msp} = \left(\frac{d_p}{D}\right)\left(\frac{D_i}{D}\right)^{1/3}\left(\frac{2gH_{sp}(\rho_p - \rho)}{\rho}\right)^{1/2} \quad \text{for } D < 0.4\text{ m} \qquad (9.67)$$

where H_{sp} is the spouted bed height. For $D > 0.4$ m, U_{msp} can be estimated by multiplying the value of U_{msp}, calculated in terms of Eq. (9.67), by a factor of $2D$ [Fane and Mitchell, 1984].

9.8.2 *Maximum Spoutable Bed Depth and Spout Diameter*

Spouting cannot be maintained once the bed height exceeds a certain level, commonly known as maximum spoutable bed depth, H_m. At H_m, the spout starts to collapse and the bed starts to change from a spouted bed to a fluidized bed. As indicated in Eq. (9.67), the minimum spouting velocity increases with an increase in the bed height. Consequently, U_{msp} achieves its maximum at H_m. The maximum value of the minimum spouting velocity can be related to the minimum fluidization velocity by

$$\frac{(U_{msp})_{max}}{U_{mf}} = b \tag{9.68}$$

where b is a function of the particle properties and spouting vessel geometry, varying from 0.9 to 1.5 [Epstein and Grace, 1997]. Substituting Eq. (9.67) into Eq. (9.68) yields

$$H_m = \frac{\rho b^2 U_{mf}^2}{2g(\rho_p - \rho)} \left(\frac{d_p}{D}\right)^{-2} \left(\frac{D_i}{D}\right)^{-2/3} \tag{9.69}$$

The spout diameter varies along the bed height. The average spout diameter D_s can be estimated by [McNab, 1972]

$$D_s = 2.0 J^{0.49} D^{0.68} / \rho_b^{0.41} \tag{9.70}$$

where J is the mass flux of gas and ρ_b is the bulk density of particles. This equation is valid only for ambient temperature. For a spouted bed of $D < 0.16$ m at elevated temperatures, D_s is given by [Wu *et al.*, 1987]

$$D_s = 5.61 \left[\frac{J^{0.433} D^{0.583} \mu^{0.133}}{(\rho_b \rho g)^{0.283}}\right] \tag{9.71}$$

The voidage of the annular region is near the minimum spouting condition, which is very close to α_{mf}. Thus, the annular region can be treated as a loosely packed bed. The voidage in the spout decreases roughly linearly with the height [Epstein and Grace, 1997].

9.8.3 *Fountain Height*

The fountain height is the distance from the surface of the annular region to the fountain top. The shape of the fountain is approximately parabolic for most cases. To estimate the fountain height, H_F, Grace and Mathur (1978) proposed an expression based on a momentum integral over the fountain, *i.e.*,

$$H_F = \alpha_O^{1.46} \left[\frac{\rho_p V_{SHO}^2}{2g(\rho_p - \rho)}\right] \tag{9.72}$$

where V_{SHO} is the particle velocity along the spout axis at the bed surface and α_O is the spout voidage at the bed surface level.

9.8.4 *Gas Flow Distribution*

The axial gas flow distribution in the annulus in a spouted bed under the condition $H_{sp} = H_m$ can be derived by considering the boundary condition of $U_a = U_{mf}$ at $z_s = H_m$, as [Mamuro and Hattori, 1970; Problem 9.10]

$$\frac{U_a}{U_{mf}} = 1 - \left(1 - \frac{z_s}{H_m}\right)^3 \tag{9.73}$$

where U_a is the superficial gas velocity in the annulus, based on the annular cross-sectional area, in the spouted bed at z_s. Epstein *et al.* (1978) noted that though Eq. (9.73) was derived on the assumption that $H_{sp} = H_m$, it applies unchanged even when $H_{sp} < H_m$. That is, the flow in the annulus at any bed level is independent of H_{sp}. When $H_{sp} < H_m$, the superficial gas velocity in the annulus, based on the annulus cross-sectional area, in the spouted bed at $z_s = H_{sp}$, *i.e.*, U_{aH}, can be expressed from Eq. (9.73) by [Epstein *et al.*, 1978]

$$\frac{U_{aH}}{U_{mf}} = 1 - \left(1 - \frac{H_{sp}}{H_m}\right)^3 \tag{9.74}$$

Nomenclature

A	Cross-sectional area of the bed
A_0	Total cross-sectional area of the orifice in the distributor
Ar	Archimedes number
A_a	Cross-sectional area of annulus in a spouted bed
A_d	Distributor plate area per orifice
C_d	Drag coefficient
d_b	Volume diameter of bubble
d_{bb}	Averaged volume diameter of bubble in the bed
d_{bi}	Initial volume diameter of bubble formed at the distributor
d_{bm}	Maximum volume diameter of bubble
d_{bs}	Volume diameter of bubble at the bed surface
$d_{b\infty}$	Volume diameter of an isolated bubble
d_{or}	Diameter of the orifice
d_p	Particle diameter
D	Bed diameter
D_i	Diameter of the entering orifice of a spouted bed
D_s	Spout diameter
E	Young's modulus
F_H	Cohesive force between adjacent particles
Fr_{mf}	Froude number at minimum fluidization
f_w	Wake fraction, defined as the volume ratio of wake to bubble
g	Gravitational acceleration
h	Axial distance above the bed surface
H	Vertical distance from gas distributor
H_{em}	Emulsion phase expansion height
H_F	Fountain height of a spouted bed
H_f	Expanded bed height
H_m	Maximum spoutable bed depth
H_{max}	Maximum expanding height of a slugging bed
H_{mb}	Bed height at minimum bubbling
H_{mf}	Bed height at minimum fluidization
H_{msl}	Minimum bed height for slugging
H_{sp}	Spouted bed height
J	Mass flux of gas
J_h	Total solids entrainment flux
J_0	Solid entrainment flux on the bed surface
J_∞	Solid entrainment flux above the TDH
KD_f	Parameter, defined by Eq. (9.50)

k	Specific permeability of a porous medium
k_A	Solids concentration decay coefficient
k_{iA}	Solids concentration decay constant of component i
k_{iB}	Elutriation rate constant of component i
L_j	Penetration length of a jet
M_i	Total mass of the particle component i in the bed
M_p	Total mass of the particles in the bed
N_B	Bubble number per unit bed volume
p	Total pressure
p_a	Ambient pressure
p_d	Dynamic pressure
p_{pf}	Pressure of pseudofluid
Q_b	Volumetric gas flow rate to the bubble phase
Q_{or}	Volumetric gas flow rate through the orifice
R_b	Bubble radius
R_c	Radius of bubble cloud
Re_b	Bubble Reynolds number
Re_{pf}	Particle Reynolds number
Re_{pmf}	Particle Reynolds number at minimum fluidization velocity
r	Radial coordinate
r	Radial distance
r_c	Radius of frontal curvature of a spherical cap bubble
r_i	Inner radius of the centrifugal fluidized bed
r_o	Outer radius of the centrifugal fluidized bed
t	Time
TDH	Transport disengagement height
U	Gas velocity component in the r-direction
U	Superficial gas velocity
U_a	Superficial gas velocity in the annulus, based on the cross-sectional area of the annulus, in the spouted bed at z_s
U_{aH}	Superficial gas velocity in the annulus, based on the cross-sectional area of the annulus, in the spouted bed at H_{sp}
U_b	Local bubble rise velocity
$U_{b\infty}$	Rise velocity of an isolated bubble
U_{bb}	Average bubble rise velocity in the whole bed
U_c	Transition velocity between bubbling and turbulent fluidization
U_{em}	Superficial gas velocity in the emulsion phase
U_k	Gas velocity corresponding to the pressure fluctuation leveling point in the pressure fluctuation versus velocity plot
U_{mb}	Superficial gas velocity at minimum bubbling conditions
U_{mf}	Superficial gas velocity for minimum fluidization
U_{mf}^*	Superficial gas velocity for minimum fluidization at ambient conditions
U_{msl}	Minimum gas velocity for slugging
U_{msp}	Minimum spouting velocity
U_{or}	Gas velocity at the orifice
U_p	Particle radial velocity component in the r-direction
U_{pf}	Velocity component of pseudofluid in the r-direction
U_{pti}	Terminal velocity of particle component i
U_{sl}	Average slug rise velocity in the bed
$U_{sl\infty}$	Velocity of an isolated slug
V	Gas velocity component in the θ-direction

V_p Particle velocity component in the θ-direction

V_{pf} Velocity component of pseudofluid in the θ-direction

V_{SHO} Particle velocity along the spout axis at the bed surface

w_g Molecular weight of gas

Y Parameter, defined by Eq. (9.41)

z_s Axial distance from gas inlet of the spouted bed

Δp_b Pressure drop across the bed

θ Polar coordinate

θ_s Conical angle of the spouted bed

θ_w Wake angle

μ Dynamic viscosity

μ_a Viscosity at ambient conditions

μ_e Apparent viscosity of a gas–solid suspension

ρ Density of fluid

ρ_a Gas density at ambient conditions

ρ_b Bulk density of particles

ρ_e Apparent density of a gas–solid suspension

ρ_p Density of particles

σ Vertical stress

ϕ_f Mass fraction of particles less than 45 μm

ϕ_i Weight fraction of particle component i in the bed

φ Sphericity of particles

ω Angular velocity of the centrifugal fluidized bed

Greek Symbols

α Bed voidage

α_b Volume fraction of the bed occupied by bubbles

α_{mb} Bed voidage at minimum bubbling conditions

α_{mf} Bed voidage at minimum fluidization

α_O Spout voidage at the bed surface

α_{sl} Volume fraction of the bed occupied by slugs

References

Abrahamsen, A. R. and Geldart, D. (1980a). Behavior of Gas Fluidized Beds of Fine Powders. Part I. Homogeneous Expansion. *Powder Tech.*, **26**, 35.

Abrahamsen, A. R. and Geldart, D. (1980b). Behavior of Gas Fluidized Beds of Fine Powders. Part II. Voidage of the Dense Phase in Bubbling Beds. *Powder Tech.*, **26**, 47.

Anderson, T. B. and Jackson, R. (1968). A Fluid Mechanical Description of Fluidized Beds: Stability of the State of Uniform Fluidization. *I&EC Fund.*, **7**, 12.

Babu, S. P., Shah, B. and Talwalker, A. (1978). Fluidization Correlations for Coal Gasification Materials: Minimum Fluidization Velocity and Fluidized Bed Expansion Ratio. *AIChE Symp. Ser.*, **74**(176), 176.

Baeyens, J. and Geldart, D. (1974). An Investigation into Slugging Fluidized Beds. *Chem. Eng. Sci.*, **29**, 255.

Batchelor, G. K. (1988). A New Theory of the Instability of a Uniform Fluidized Bed. *J. Fluid Mech.*, **193**, 75.

Bi, H.-T. and Fan, L.-S. (1992). Existence of Turbulent Regime in Gas–Solid Fluidization. *AIChE J.*, **38**, 297.

Brereton, C. M. H. and Grace, J. R. (1992). The Transition to Turbulent Fluidization. *Trans. Instn. Chem. Engrs.*, **70**, Part A, 246.

Cai, P. (1989). *Flow Regime Transition in Dense-Phase Fluidized Beds*. Ph.D. Dissertation. Tsinghua University (a cooperative program with Ohio State University), People's Republic of China.

Cai, P., Chen, S. P., Jin, Y., Yu, Z. Q. and Wang, Z. W. (1989). Effect of Operating Temperature and Pressure on the Transition from Bubbling to Turbulent Fluidization. *AIChE Symp. Ser.*, **85** (270), 37.

Cai, P., DeMichele, G., Traniello Gradassi, A. and Miccio, M. (1993). A Generalized Method for Predicting Gas Flow Distribution Between the Phases in FBC. In *FBC's Role in the World Energy Mix*. Ed. L. Rubow. New York: ASME.

Cai, P., Jin, Y., Yu, Z. Q. and Fan, L.-S. (1992). Mechanistic Model for Onset Velocity Prediction for Regime Transition from Bubbling to Turbulent Fluidization. *I&EC Res.*, **31**, 632.

Cai, P., Jin, Y., Yu, Z. Q. and Wang, Z. W. (1990). Mechanism of Flow Regime Transition from Bubbling to Turbulent Fluidization. *AIChE J.*, **36**, 955.

Cai, P., Schiavetti, M., DeMichele, G., Grazzini, G. C. and Miccio, M. (1994). Quantitative Estimation of Bubble Size in PFBC. *Powder Tech.*, **80**, 99.

Chan, I. H. and Knowlton, T. M. (1984). The Effect of System Pressure on the Transport Disengaging Height (TDH) Above Bubbling Gas-Fluidized Beds. *AIChE Symp. Ser.*, **80** (241), 24.

Chandnani, P. P. and Epstein, N. (1986). Spoutability and Spouting Destabilization of Fine Particles with a Gas. In *Fluidization V*. Ed. Ostergaard and Sorensen. New York: Engineering Foundation.

Chehbouni, A., Chaouki, J., Guy, C. and Klvana, D. (1994). Characterization of the Flow Transition Between Bubbling and Turbulent Fluidization. *I&EC Res.*, **33**, 1889.

Chirone, R., Massimilla, L. and Russo, S. (1992). Bubbling Fluidization of a Cohesive Powder in an Acoustic Field. In *Fluidization VII*. Ed. Potter and Nicklin. New York: Engineering Foundation.

Clift, R. (1986). Hydrodynamics of Bubbling Fluidized Beds. In *Gas Fluidization Technology*. Ed. D. Geldart. New York: John Wiley & Sons.

Clift, R. and Grace, J. R. (1985). Continuous Bubbling and Slugging. In *Fluidization*, 2nd ed. Ed. Davidson, Clift and Harrison. London: Academic Press.

Clift, R., Grace, J. R. and Weber, M. E. (1974). Stability of Bubbles in Fluidized Beds. *I&EC Fund.*, **13**, 45.

Clift, R., Grace, J. R. and Weber, M. E. (1978). *Bubbles, Drops and Particles*. New York: Academic Press.

Colakyan, M. and Levenspiel, O. (1984). Elutriation from Fluidized Beds. *Powder Tech.*, **38**, 223.

Collins, R. (1965). The Rise Velocity of Davidson's Fluidization Bubble. *Chem. Eng. Sci.*, **20**, 788.

Darton, R. C., La Nauze, R. D., Davidson, J. F. and Harrison, D. (1977). Bubble Growth Due to Coalescence in Fluidized Beds. *Trans. Instn. Chem. Engrs.*, **55**, 274.

Davidson, J. F. and Harrison, D. (1963). *Fluidized Particles*. Cambridge: Cambridge University Press.

Davies, L. and Taylor, G. I. (1950). The Mechanics of Large Bubbles Rising Through Extended Liquids and Through Liquids in Tubes. *Proc. R. Soc. London*, **A200**, 375.

Epstein, N. and Grace, J. R. (1997). Spouting of Particulate Solids. In *Handbook of Powder Science and Technology*, 2nd ed. Ed. Fayed and Otten. New York: Chapman & Hall.

Epstein, N., Lim, C. J. and Mathur, K. B. (1978). Data and Models for Flow Distribution and Pressure Drop in Spouted Beds. *Can. J. Chem. Eng.*, **56**, 436.

Fan, L.-S. and Tsuchiya, K. (1990). *Bubble Wake Dynamics in Liquids and Liquid–Solid Suspensions*. Boston: Butterworths.

Fane, A. G. and Mitchell, R. A. (1984). Minimum Spouting Velocity of Scaled-up Beds. *Can. J. Chem. Eng.*, **62**, 437.

Foscolo, P. U. and Gibilaro, L. G. (1984). A Fully Predictive Criterion for the Transition Between Particulate and Aggregative Fluidization. *Chem. Eng. Sci.*, **39**, 1667.

Fournol, A. B., Bergougnou, M. A. and Baker, C. G. I. (1973). Solids Entrainment in a Large Gas Fluidized Bed. *Can. J. Chem. Eng.*, **51**, 401.

Gabavcic, Z. B., Vukovic, D. V., Zdanski, F. K. and Littman, H. (1976). Fluid Flow Pattern, Minimum Spouting Velocity and Pressure Drop in Spouted Beds. *Can. J. Chem. Eng.*, **54**, 33.

Garg, S. K. and Pritchett, J. W. (1975). Dynamics of Gas-Fluidized Beds. *J. Appl. Phys.*, **46**, 4493.

Geldart, D. (1973). Types of Gas Fluidization. *Powder Tech.*, **7**, 285.

Geldart, D. (1986). *Gas Fluidization Technology*. New York: John Wiley & Sons.

Geldart, D., Cullinan, J., Georghiades, S., Gilvray, D. and Pope, D. J. (1979). The Effect of Fines on Entrainment from Gas Fluidized Beds. *Trans. Instn. Chem. Engrs.*, **57**, 269.

Grace, J. R. (1970). The Viscosity of Fluidized Beds. *Can. J. Chem. Eng.*, **48**, 30.

Grace, J. R. (1982). Fluidized-Bed Hydrodynamics. In *Handbook of Multiphase Systems*. Ed. G. Hetsroni. Washington, D.C.: Hemisphere.

Grace, J. R. (1986). Contacting Modes and Behavior Classification of Gas–Solid and Other Two-Phase Suspensions. *Can. J. Chem. Eng.*, **64**, 353.

Grace, J. R. and Mathur, K. B. (1978). Height and Structure of the Fountain Region Above Spouted Beds. *Can. J. Chem. Eng.*, **56**, 533.

Grace, J. R. and Venta, J. (1973). Volume Changes Accompanying Bubble Splitting in Fluidized Beds. *Can. J. Chem. Eng.*, **51**, 110.

Horio, M., Taki, A., Hsieh, Y. S. and Muchi, I. (1980). Elutriation and Particle Transport Through the Freeboard of a Gas–Solid Fluidized Bed. In *Fluidization*. Ed. Grace and Matsen. New York: Plenum.

Hovmand, S. and Davidson, J. F. (1971). Pilot Plant and Laboratory Scale Fluidized Reactors at High Gas Velocities: The Relevance of Slug Flow. In *Fluidization*. Ed. Davidson and Harrison. London: Academic Press.

Jackson, R. (1963). The Mechanics of Fluidized Beds: The Motion of Fully Developed Bubbles. *Trans. Instn. Chem. Engrs.*, **41**, 22.

Jackson, R. (1971). Fluid Mechanical Theory. In *Fluidization*. Ed. Davidson and Harrison. New York: Academic Press.

Jin, Y., Yu, Z. Q., Shen, J. Z. and Zhang, L. (1980). Pagoda-Type Vertical Internal Baffles in Gas-Fluidized Beds. *Int. Chem. Eng.*, **20**, 191.

Kozanoglu, B. and Levy, E. K. (1991). Transient Mixing of Homogeneous Solids in a Bubbling Fluidized Bed. *AIChE Symp. Ser.*, **87** (281), 58.

Krishna, R. (1993). Analogies in Multiphase Reactor Hydrodynamics. In *Encyclopedia of Fluid Mechanics*. Supplement 2. *Advances in Multiphase Flow*. Ed. N. P. Cheremisinoff. Houston: Gulf Publishing.

Kroger, D. G., Levy, E. K. and Chen, J. C. (1979). Flow Characteristics in Packed and Fluidized Rotating Beds. *Power Tech.*, **24**, 9.

Kunii, D. and Levenspiel, O. (1991). *Fluidization Engineering*, 2nd ed. Boston: Butterworth-Heinemann.

Leal, L. G. (1989). Vorticity Transport and Wake Structure for Bluff Bodies at Finite Reynolds Number. *Phys. Fluids A, Fluid Dynamics*, **1**, 124.

Liu, Y. A., Hamby, R. K. and Colberg, R. D. (1991). Fundamental and Practical Developments of Magnetofluidized Beds: A Review. *Powder Tech.*, **64**, 3.

Mamuro, T. and Hattori, H. (1970). Flow Pattern of Fluid in Spouted Beds: Correction. *J. Chem. Eng. Japan*, **3**, 119.

Massimilla, L. (1985). Gas Jets in Fluidized Beds. In *Fluidization*, 2nd ed. Ed. Davidson, Clift and Harrison. London: Academic Press.

Mathur, K. B. and Epstein, N. (1974). *Spouted Beds*. New York: Academic Press.

Mathur, K. B. and Gishler, P. E. (1955). A Technique for Contacting Gases with Coarse Solid Particles. *AIChE J.*, **1**, 157.

Matsen, J. M., Rossetti, S. J. and Halow, J. S. (1986). Fluidized Beds and Gas Particle Systems. In *Encyclopedia of Chemical Processing and Design*. Ed. J. J. McKetta. New York: Marcel Dekker.

McNab, G. S. (1972). Prediction of Spouted Diameter. *Brit. Chem. Eng. Proc. Technol.*, **17**, 532.

Miwa, K., Mori, S., Kato, T. and Muchi, I. (1972). Behavior of Bubbles in Gaseous Fluidized Beds. *Int. Chem. Eng.*, **12**, 187.

Molerus, O. (1982). Interpretation of Geldart's Type A, B, C and D Powders by Taking into Account Interparticle Cohesion Forces. *Powder Tech.*, **33**, 81.

Montz, K. W., Beddow, J. K. and Butler, P. B. (1988). Adhesion and Removal of Particulate Contaminants in a High-Decibel Acoustic Field. *Powder Tech.*, **55**, 133.

Mori, S., Iwasaki, N., Mizutani, E. and Okada, T. (1992). Vibro-Fluidization of Two-Component Mixtures of Group C Particles. In *Fluidization VII*. Ed. Potter and Nicklin. New York: Engineering Foundation.

Mori, S. and Wen, C. Y. (1975). Estimation of Bubble Diameter in Gaseous Fluidized Beds. *AIChE J.*, **21**, 109.

Ormiston, R. M., Mitchell, F. R. G. and Davidson, J. F. (1965). The Velocities of Slugs in Fluidized Beds. *Trans. Instn. Chem. Engrs.*, **43**, T209.

Partridge, B. A. and Rowe, P. N. (1966). Analysis of Gas Flow in a Bubbling Fluidized Bed When Cloud Formation Occurs. *Trans. Instn. Chem. Engrs.*, **44**, 349.

Pemberton, S. T. (1982). *Entrainment from Fluidized Beds*. Ph.D. Dissertation. Cambridge University.

Peters, M. H., Fan, L.-S. and Sweeney, T. L. (1982). Reactant Dynamics in Catalytic Fluidized Bed Reactors with Flow Reversal of Gas in the Emulsion Phase. *Chem. Eng. Sci.*, **37**, 553.

Peters, M. H., Fan, L.-S. and Sweeney, T. L. (1983). Study of Particle Ejection in the Freeboard Region of a Fluidized Bed with an Image Carrying Probe. *Chem. Eng. Sci.*, **38**, 481.

Reuter, H. (1963a). Druckverteilung um Blasen im Gas-Feststoff-Flieβbett. *Chem.-Ing.-Tech.*, **35**, 98.

Reuter, H. (1963b). Mechanismus der Blasen im Gas-Feststoff-Flieβbett. *Chem.-Ing.-Tech.*, **35**, 219.

Rietema, K. (1991). *The Dynamics of Fine Powders*. London: Elsevier Applied Science.

Rietema, K. and Pipers, H. W. (1990). The Effect of Interparticle Forces on the Stability of Gas-Fluidized Beds. I. Experimental Evidence. *Chem. Eng. Sci.*, **45**, 1627.

Romero, J. B. and Johanson, L. N. (1962). Factors Affecting Fluidized Bed Quality. *Chem. Eng. Prog. Symp. Ser.*, **58**(38), 28.

Rosensweig, R. E. (1979). Fluidization: Hydrodynamic Stabilization with a Magnetic Field. *Science*, **204**, 57.

Rosensweig, R. E. (1985). *Ferro Hydrodynamics*. Cambridge: Cambridge University Press.
Rowe, P. N. (1971). Experimental Properties of Bubbles. In *Fluidization*. Ed. Davidson and Harrison. New York: Academic Press.
Rowe, P. N., Macgillivray, H. J. and Cheesman, D. J. (1979). Gas Discharge from an Orifice into a Gas Fluidised Bed. *Trans. Instn. Chem. Engrs.*, **57**, 194.
Rowe, P. N. and Partridge, B. A. (1965). An X-Ray Study of Bubbles in Fluidized Beds. *Trans. Instn. Chem. Engrs.*, **43**, 157.
Satija, S. and Fan, L.-S. (1985). Characteristics of the Slugging Regime and Transition to the Turbulent Regime for Fluidized Beds of Large Coarse Particles. *AIChE J.*, **31**, 1554.
Stewart, P. S. B. (1968). Isolated Bubbles in Fluidized Beds: Theory and Experiment. *Trans. Instn. Chem. Engrs.*, **46**, T60.
Stewart, P. S. B. and Davidson, J. F. (1967). Slug Flow in Fluidized Beds. *Powder Tech.*, **1**, 61.
Thonglimp, V. (1981). *Contribution à L'étude Hydrodynamique des Couches Fluidiseés pour un Gaz Vitesse Minimale de Fluidisation et Expansion*. Docteur-Ingenieur thesis. Institut National Polytechnique, Toulouse.
Toomey, R. D. and Johnstone, H. F. (1952). Gaseous Fluidization of Solid Particles. *Chem. Eng. Prog.*, **48**, 220.
Tsinontides, S. C. and Jackson R. (1993). The Mechanics of Gas Fluidized Beds with an Interval of Stable Fluidization. *J. Fluid Mech.*, **255**, 237.
Verloop, J. and Heertjes, P. M. (1970). Shock Waves as a Criterion for the Transition from Homogeneous to Heterogeneous Fluidization. *Chem. Eng. Sci.*, **25**, 825.
Wen, C. Y. and Chen, L. H. (1982). Fluidized Bed Freeboard Phenomena: Entrainment and Elutriation. *AIChE J.*, **28**, 117.
Wen, C. Y. and Yu, Y. H. (1966). Mechanics of Fluidization. *Chem. Eng. Prog. Symp. Ser.*, **62**(62), 100.
Wilhelm, R. H. and Kwauk, M. (1948). Fluidization of Solid Particles. *Chem. Eng. Prog.*, **44**, 201.
Wu, S. W. M., Lim, C. J. and Epstein, N. (1987). Hydrodynamics of Spouted Beds at Elevated Temperatures. *Chem. Eng. Commun.*, **62**, 251.
Yang, W. C. and Keairns, D. L. (1978). Momentum Dissipation of and Gas Entrainment into a Gas–Solid Two-Phase Jet in a Fluidized Bed. In *Fluidization*. Ed. Davidson and Keairns. Cambridge: Cambridge University Press.
Yang, W. C. and Keairns, D. L. (1979). Estimating the Jet Penetration Depth of Multiple Vertical Grid Jets. *I&EC Fund.*, **18**, 317.
Yates, J. G., Cheesman, D. J. and Sergeev, Y. A. (1994). Experimental Observations of Voidage Distribution Around Bubbles in a Fluidized Bed. *Chem. Eng. Sci.*, **49**, 1885.
Yerushalmi, J. and Cankurt, N. T. (1979). Further Studies of the Regimes of Fluidization. *Powder Tech.*, **24**, 187.
Zenz, F. A. and Weil, N. A. (1958). A Theoretical-Empirical Approach to Mechanism of Particle Entrainment from Fluidized Beds. *AIChE J.*, **4**, 472.

Problems

9.1 The dense-phase fluidized bed can be operated in the presence of an external force field other than the gravitational field, such as the centrifugal or magnetic field. Consider a centrifugal fluidized bed in which the dense bed is situated in a rotary cylinder (see Fig. P9.1) [*e.g.*, Kroger *et al.*, 1979]. Gas enters the bed through the cylindrical wall to fluidize the particles inside the cylinder. Generally, the centrifugal acceleration is significantly larger than the gravitational acceleration so that the dense bed expands toward the central axis of the cylinder. Particulate and bubbling fluidization in the radial direction can be achieved in such a cylinder. Derive the relationship between the pressure drop and the minimum fluidization velocity for a centrifugal fluidized bed shown in Fig. P9.1.

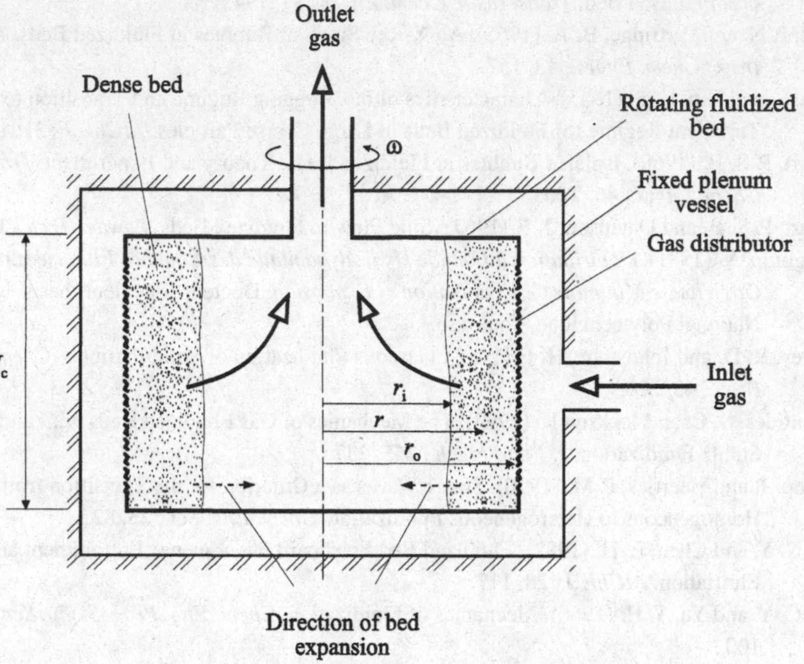

Figure P9.1. Cross-sectional view of the centrifugal fluidized bed.

9.2 Fluidization behavior of a dense bed with ferromagnetic particles can be significantly altered by exerting a magnetic field on the bed (see Fig. P9.2) [*e.g.*, Rosensweig, 1979, 1985]. When applying an alternating magnetic field to the fluidized bed, the gas bubble formation may be impeded and, thus, the particulate fluidization regime can be largely extended. Solids can flow through this magnetically "stabilized" bed with almost no backmixing. The magnetofluidized bed has been applied to aerosol filtration, petroleum refining, and fermentation. Under the conditions of no bubbles and of a uniform magnetic field, derive the continuity and momentum equations for the gas and the particles in a magnetofluidized bed and interpret the physical meaning of each term in the equations.

9.3 The minimum bubbling velocity, U_{mb}, given in Eq. (9.19) can be derived from the instability analysis based on the wave concept as described in §6.5.2. A wave in a gas–solid medium can be generated by perturbation. A continuity wave yields a local change in the solids holdup or gas voidage which propagates through the vessel at a certain velocity. The propagation process does not introduce any dynamic effect of inertia or momentum. The continuity wave velocity V_c for batch-solid operation can be expressed by

$$V_c = (1-\alpha)\frac{\partial U}{\partial \alpha} \tag{P9.1}$$

where U and α are related by

$$\frac{U}{U_{pt}} = \frac{1}{10}\frac{\alpha^3}{1-\alpha} \tag{P9.2}$$

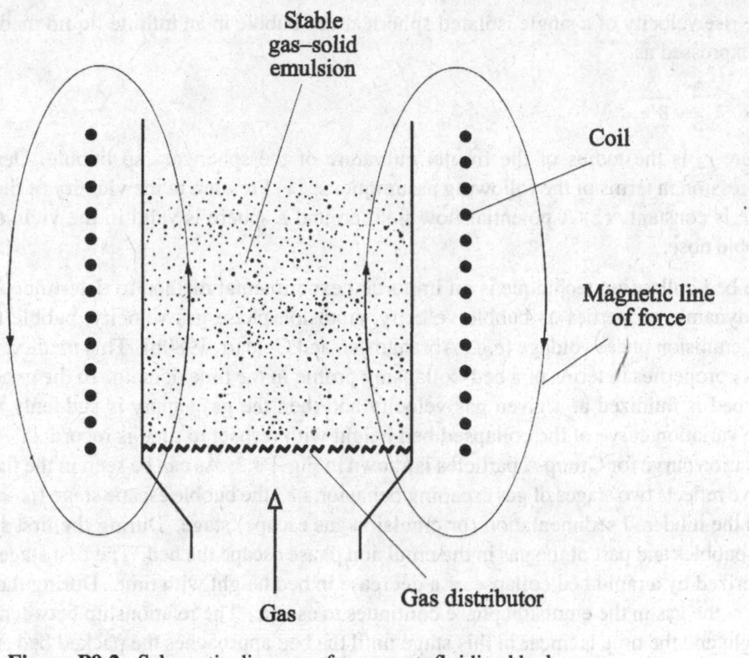

Figure P9.2. Schematic diagram of a magnetofluidized bed.

The propagation of a dynamic wave produces dramatic changes in local solids holdup and momentum in a gas–solid flow. Thus, to analyze the dynamic wave, one can examine the wave equation obtained for the perturbation of the momentum and mass balance equations for each of the gas and solid phases. Accordingly, the dynamic wave velocity, V_d, can be expressed by

$$V_d = \sqrt{\frac{E}{\rho_p}} \tag{P9.3}$$

The condition for the minimum velocity for bubble formation corresponds to $V_c = V_d$. Derive Eq. (9.19) on the basis of the preceding information.

9.4 Derive the "cloud" thickness and volume ratio of cloud to bubble for a single bubble in a fluidized bed in terms of the Davidson–Harrison (1963) model. Since the cloud is the region established by the gas circulating in a closed loop between the bubble and its surrounding, the following assumptions can be employed: (1) zero radial gas velocity outside the surface of the cloud sphere ($r = R_c$); (2) uniform gas velocity at far distance from the bubble, *i.e.*, $U_{b\infty} = U_{mf}/\alpha_{mf}$.

9.5 For a fluidized bed with a bed diameter of 0.15 m, a porous plate distributor, and Group B particles, calculate the bubble sizes at (1) $H = 0.25$ m and $U - U_{mf} = 0.028$ m/s and (2) $H = 0.4$ m, and $U - U_{mf} = 0.039$ m/s under the operating pressure of 0.1 to 6.1 MPa using Eqs. (9.31), (9.34), and (9.35). Compare the calculated results with the following experimental data [after Cai *et al.*,1994]:

p (MPa)	0.1	0.6	1.1	1.6	2.1	3.1	4.1	5.1	6.1
d_b (mm) at (1)	27	30	30	31	29	29	27	26	22
d_b (mm) at (2)	45	47	52	49	45	45	39	—	—

9.6 The rise velocity of a single isolated spherical cap bubble in an infinite liquid medium can be expressed as

$$U_{b\infty} = \frac{2}{3}\sqrt{gr_c} \tag{P9.4}$$

where r_c is the radius of the frontal curvature of the spherical cap bubble. Derive this expression in terms of the following assumptions: (1) Pressure in the vicinity of the bubble nose is constant. (2) A potential flow field around a sphere is valid in the vicinity of the bubble nose.

9.7 The bed-collapsing technique is an important experimental method to determine such hydrodynamic properties as bubble velocity, emulsion phase gas velocity, bubble fraction, and emulsion phase voidage [*e.g.*, Abrahamsen and Geldart, 1980b]. This method portrays these properties in terms of a bed-collapsing profile in the time domain. In the experiment, the bed is fluidized at a given gas velocity and then the gas supply is suddenly shut off. The variation curve of the collapsed bed height with respect to time is recorded. A typical variation curve for Group A particles is shown in Fig. P9.3. As can be seen in the figure, the curve reflects two stages of gas escaping behavior, *i.e.*, the bubble escape stage ($t_0 < t < t_1$) and the hindered sedimentation (or emulsion gas escape) stage. During the first stage, all the bubbles and part of the gas in the emulsion phase escape the bed. The first stage is characterized by a rapid bed collapse or a decrease in bed height with time. During the second stage, the gas in the emulsion phase continues to escape. The relationship between the bed height and the time is linear in this stage until the bed approaches the packed bed state. On the basis of the information provided, determine U_{bb}, α_{em}, α_b, and U_{em} from Fig. P9.3.

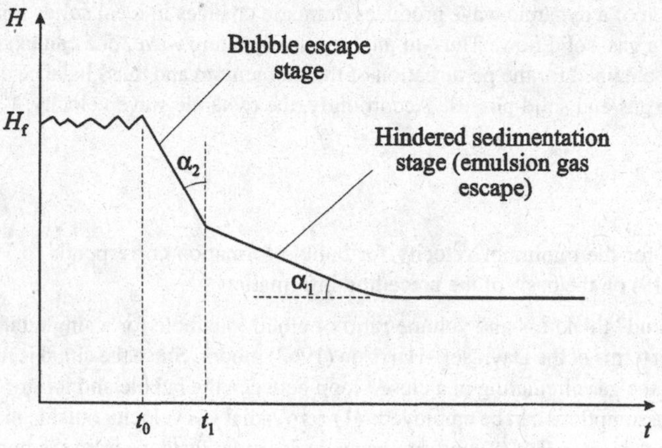

Figure P9.3. Typical results from bed-collapsing experiments.

9.8 The wall effect on the performance of fluidized beds can be reflected by the differences of the overall hydrodynamic behavior in beds with different diameters. Practically, in order to scale up a pilot fluidized bed to a full-size commercial unit, the size of the pilot unit should be large enough to exclude the wall effect. A critical scale-up diameter (D_{cs}) for the pilot fluidized bed is thus defined as the minimum bed diameter beyond which the bed behavior is almost independent of the bed size. Calculate the critical scale-up diameter for fluidized beds without internals, using the following criterion:

$$\frac{|(U_c)_{D=\infty} - (U_c)_{D=D_{cs}}|}{(U_c)_{D=\infty}} \leq 2 \text{ percent} \tag{P9.5}$$

9.9 It was observed by Cai *et al.* (1990) that as the predominant bubble interaction varies from coalescence to breakup, the prevailing flow regime undergoes a transition from bubbling to turbulent fluidization. On the basis of this observation, a criterion for the regime transition can be established according to the variation of the bubble number per unit bed volume, N_B, with the gas velocity, as given by Eq. (9.48). Express Eq. (9.48) in terms of the operating variables of the bed, using the simple two-phase theory (Eq. (9.40)). Assume that the relationship between the bed voidage, α, and the operating variables is known to be

$$\alpha = f(U, \rho, \mu, d_p, \rho_p) \tag{P9.6}$$

9.10 Consider the superficial gas velocity U_a in the annular region of a spouted bed (see Fig. P9.4). U_a is defined in terms of the cross-sectional area of the annulus. Assume that (1) the annular region is a uniformly packed bed; (2) the flow is only a function of height z_s; (3) Darcy's law holds for the pressure drop; (4) the Janssen model given in Eq. (8.11) can be applied in such a way that the vertical solids stress in the annulus at z_s is proportional to the square of the radial gas velocity on the inner wall of the annular region; and (5) the friction at the vertical boundaries of the annulus can be neglected. (a) Derive the governing equation for U_a. (b) Solve the equation from (a) for U_a (see Eq. (9.73)). The following boundary conditions are applicable for the condition of $H_{sp} = H_m$: (1) at $z_s = H_m$, $U_a = U_{mf}$; (2) at $z_s = H_m$, $d\sigma/dz_s = 0$; (3) at $z_s = 0$, $U_a = 0$.

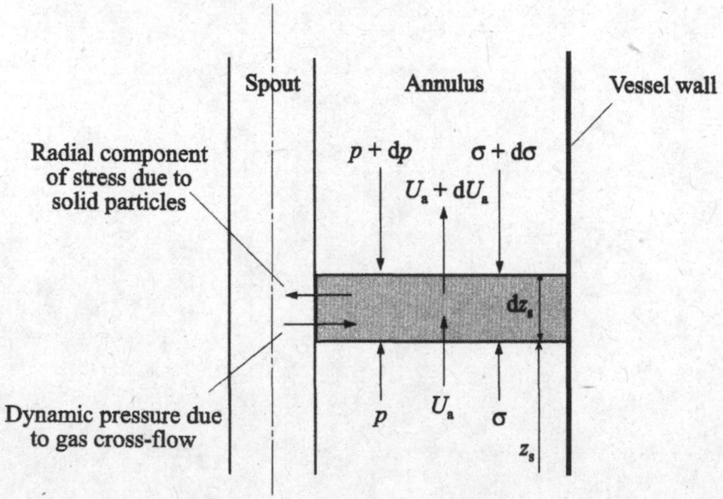

Figure P9.4. Schematic diagram illustrating the force balance in the annular region of a spouted bed (from Mamuro and Hattori, 1970).

9.11 In industrial fluidized bed reactors, the bed height is commonly fixed by an overflow weir. Thus, as the gas velocity, U, increases from U_{mf}, the apparent bed density decreases. An important design principle for the gas distributor is to ensure its sufficient pressure drop for a uniform gas distribution, *i.e.*, without gas channeling, and stable bed operation. Specifically, the total pressure drop across both the distributor and the bed should be in an increasing trend with an increase in the gas velocity. Suppose that the pressure drop across a perforated distributor, $\Delta p_{\text{distributor}}$, with a total orifice area of A_0 can be expressed by

$$\Delta p_{\text{distributor}} = \frac{\rho(UA/A_0)^2}{2C_d^2} \tag{P9.7}$$

where C_d is the drag coefficient. Verify that for a uniform gas distribution and stable bed operation, the ratio of the pressure drop across the distributor, $\Delta p_{\text{distributor}}$, to that across the bed, Δp_b, satisfies the inequality [Matsen *et al.*, 1986]

$$\frac{\Delta p_{\text{distributor}}}{\Delta p_b} \geq \frac{U}{2U_b} \tag{P9.8}$$

Assume that the simple two-phase theory can be applied and the following relationship holds

$$\frac{\partial U_{b\infty}}{\partial U} = 0 \tag{P9.9}$$

where $U_{b\infty}$ is the rise velocity of an isolated bubble.

CHAPTER 10

Circulating Fluidized Beds

10.1 Introduction

As discussed in Chapter 9, dense-phase fluidization other than particulate fluidization is characterized by the presence of an emulsion phase and a discrete gas bubble/void phase. At relatively low gas velocities in dense-phase fluidization, the upper surface of the bed is distinguishable. As the gas velocity increases, the bubble/void phase gradually becomes indistinguishable from the emulsion phase. The bubble/void phase eventually disappears and the gas evolves into the continuous phase with further increasing gas velocities. In a dense-phase fluidized bed, the particle entrainment rate is low and increases with increasing gas velocity. As the gas flow rate increases beyond the point corresponding to the disappearance of the bubble/void phase, a drastic increase in the entrainment rate of the particles occurs such that a continuous feeding of particles into the fluidized bed is required to maintain a steady solids flow. Fluidization at this state, in contrast to dense-phase fluidization, is denoted lean-phase fluidization.

Lean-phase fluidization encompasses two flow regimes, *i.e.*, the fast fluidization regime and the dilute transport regime. The fast fluidization regime is characterized by a heterogeneous flow structure, whereas the dilute transport regime is characterized by a homogeneous flow structure. The circulating fluidized bed (CFB) is a fluidized bed system in which solid particles circulate between the riser and the downcomer. The fast fluidization regime is the principal regime under which the circulating fluidized bed is operated. The operating variables for a circulating fluidized bed system include both the gas flow rate and the solids circulation rate, in contrast to the gas flow rate only in a dense-phase fluidized bed system. The solids circulation is established by a high velocity of gas flow. Note that the dilute transport regime is discussed in Chapter 11, in the context of pneumatic conveying.

The flow properties, such as solids holdup, can be established with various combinations of gas flow rate and solids circulation rate [Lim *et al.*, 1995]. High operating gas velocities for lean-phase fluidization yield a short contact time between the gas and solid phases, which is desirable for process applications involving chemical reactions. The system has been widely used in the petrochemical industry (*e.g.*, in fluid catalytic cracking (FCC) [King, 1992]) and the utility industry (*e.g.*, in coal combustion [Basu and Fraser, 1991]). Other commercial applications of the CFB systems include maleic anhydride production from n-butane [Contractor *et al.*, 1993] and Fischer–Tropsch synthesis of liquid fuels from carbon monoxide and hydrogen [Shingles and McDonald, 1988]. The operating conditions of the CFB systems can be significantly different, depending on process applications. For example, for coal combustion, the gas velocity and solids flow rate are typically 5–8 m/s and less than 40 kg/m^2 · s, respectively; for FCC, on the other hand, the gas velocity and solids flow rate are substantially higher, *e.g.*, 15–20 m/s (at the riser exit) and greater than 300 kg/m^2 · s, respectively [Werther, 1993]. The flow behavior of the CFB systems can also be significantly different, depending on design or operating variations such as insertion of internals [Wu *et al.*, 1990; Jiang *et al.*, 1991; Zheng *et al.*, 1991], superimposition of different fluidized particles [Johnson, 1982; Toda *et al.*, 1983; Fan *et al.*, 1985; Satija and Fan, 1985], and tangential gas injection [Ilias *et al.*, 1988].

This chapter discusses basic flow characteristics of a CFB operation. Material is introduced to illustrate the flow phenomena and operating constraints of the riser and the effects caused by the intricate interplay with the accompanying components in an integrated gas–solid flow loop. The phenomenological analyses of the system are a major part of this chapter. More rigorous analyses involving computational fluid dynamics are being developed. The flow regimes and their transitions, hydrodynamic behavior, solids flow structure, and flow models are specifically described. Again, as indicated in the Preface, the correlation equations presented in the text, unless otherwise noted, are given in SI units.

10.2 System Configuration

The riser, gas–solid separator, downcomer, and solids flow control device are the four integral parts of a CFB loop. The riser is the main component of the system. In the riser, gas and solids commonly flow cocurrently upward, although they can also flow cocurrently downward. This chapter covers only the cocurrent gas and solid upward operation. In this operation, as shown in Fig. 10.1, the fluidizing gas is introduced at the bottom of the riser, where solid particles from the downcomer are fed via a control device and carried upward in the riser. Particles exit at the top of the riser into gas–solid separators. Separated particles then flow to the downcomer and return to the riser.

The entrance and exit geometries have significant effects on the gas and solid flow behavior in the riser. The efficiency of the gas–solid separator can affect the particle size distribution and solids circulation rate in the system. In a CFB system, particle separation is typically achieved by cyclones (see Chapter 7). The downcomer provides hold volume

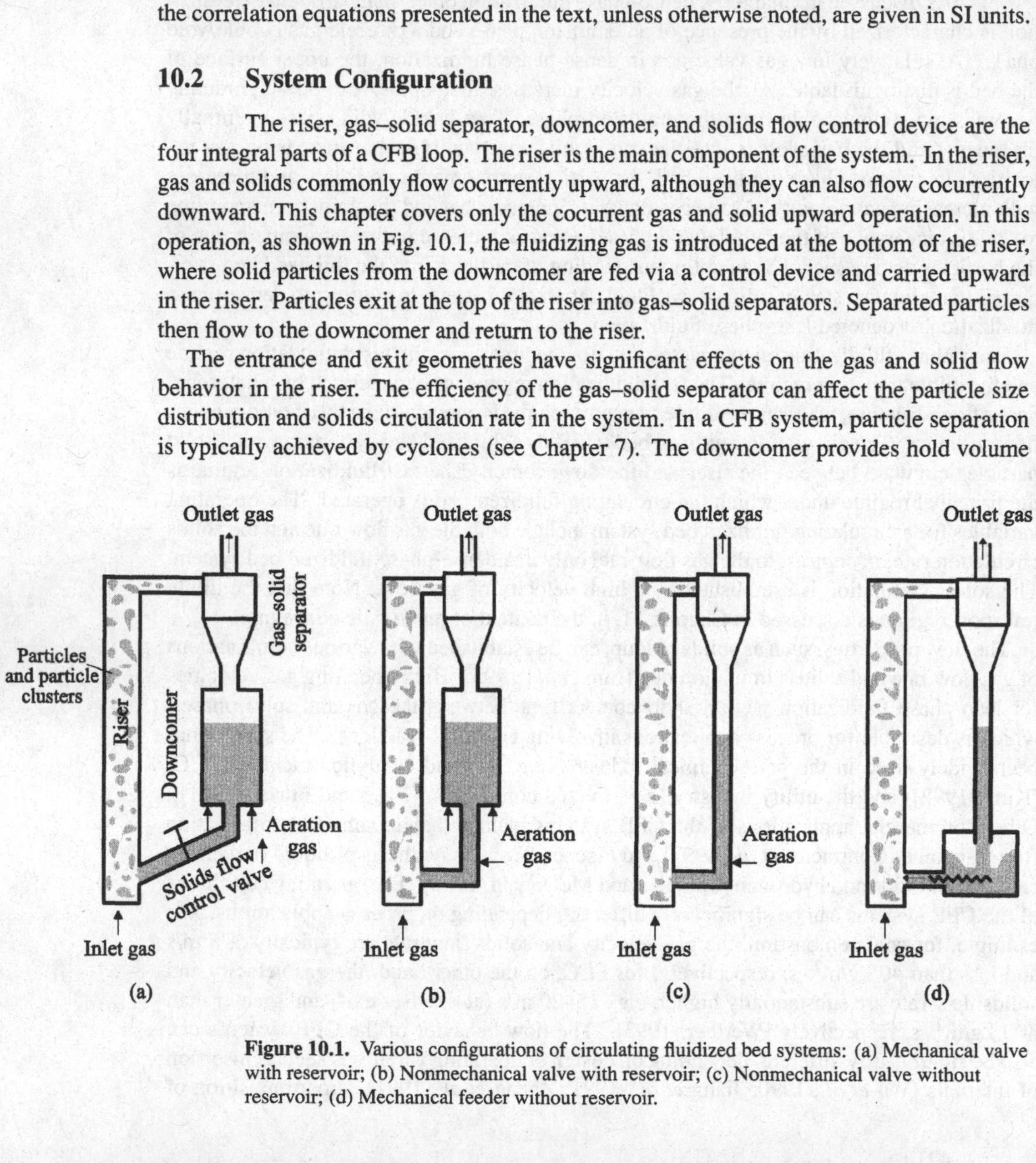

Figure 10.1. Various configurations of circulating fluidized bed systems: (a) Mechanical valve with reservoir; (b) Nonmechanical valve with reservoir; (c) Nonmechanical valve without reservoir; (d) Mechanical feeder without reservoir.

and a static pressure head for particle recycling to the riser. The downcomer can be a large reservoir which aids in regulating the solids circulation rate (see Figs. 10.1(a) and (b)). In applications involving chemical reactions, the downcomer may be used as a heat exchanger or as a spent solid regenerator. The downcomer can also simply be a standpipe (see Chapter 8) which provides direct solids passage to the riser (see Figs. 10.1(c) and (d)).

The key to smooth operation of a CFB system is the effective control of the solids recirculation rate to the riser. The solids flow control device serves two major functions, namely, sealing riser gas flow to the downcomer and controlling solids circulation rate. Both mechanical valves or feeders (see Figs. 10.1(a) and (d)) and nonmechanical valves (see Figs. 10.1(b) and (c)) are used to perform these functions. Typical mechanical valves are rotary, screw, butterfly, and sliding valves. Nonmechanical valves include L-valves, J-valves (see Chapter 8), V-valves, seal pots, and their variations. Blowers and compressors are commonly used as the gas suppliers. Operating characteristics of these gas suppliers which are directly associated with the dynamics and instability of the riser operation must be considered (see §10.3.3.2).

10.3 Flow Regimes and Transitions Between Regimes

The flow behavior in the riser varies with gas velocity, solids circulation rate, and system geometry. On the basis of the flow behavior, the fast fluidization regime can be distinguished from neighboring regimes.

10.3.1 *Flow Regimes and Regime Diagrams*

The fast fluidization regime is characterized by a dense region at the bottom of the riser and a dilute region above it [Li and Kwauk, 1980]. The interrelationship of the fast fluidization regime with other fluidization regimes in dense-phase fluidization and with the dilute transport regime is reflected in the variations of the pressure drop per unit length of the riser, $\Delta p/\Delta z$; gas velocity; and solids circulation rate [Bai *et al.*, 1993], as given in Fig. 10.2.

The upper diagram of Fig. 10.2 shows the typical variations of the pressure drop per unit length with the gas velocity for a constant solids circulation rate of J_{p1} at a lower part of the riser, $(\Delta p/\Delta z)_{\mathrm{L}}$ (curve a–b–c–d), and an upper part of the riser, $(\Delta p/\Delta z)_{\mathrm{U}}$ (curve a–b–c–d′). It is seen in the figure that at high gas velocities the wall friction is the dominant component for $(\Delta p/\Delta z)$; $(\Delta p/\Delta z)$ increases with an increase in gas velocity. At relatively low gas velocities, the solids holdup is the dominant component for $(\Delta p/\Delta z)$; $(\Delta p/\Delta z)$ decreases with an increase in gas velocity. Point b represents the minimum $\Delta p/\Delta z$ marking the onset of the dilute transport regime where the pressure drop is dominated by the wall friction. In a dilute flow (curve a–b), there is no axial variation in $(\Delta p/\Delta z)$; *i.e.*, $(\Delta p/\Delta z)_{\mathrm{L}}$ and $(\Delta p/\Delta z)_{\mathrm{U}}$ are the same for a given gas velocity, noting that the location for $(\Delta p/\Delta z)_{\mathrm{L}}$ is beyond the acceleration region. Curve b–c represents the transition from the dilute transport regime to the fast fluidization regime; in the transition, the annular-core flow pattern prevails with little axial variation in $(\Delta p/\Delta z)$. In the fast fluidization regime (curve c–d or c–d′), there is a variation in $(\Delta p/\Delta z)$ between the upper and the lower parts of the riser; *i.e.*, $(\Delta p/\Delta z)_{\mathrm{L}}$ and $(\Delta p/\Delta z)_{\mathrm{U}}$ are different. The variation of $(\Delta p/\Delta z)_{\mathrm{L}}$ with the gas velocity is much sharper than that of $(\Delta p/\Delta z)_{\mathrm{U}}$. At point d, the flow transition to dense-phase fluidization occurs. The gas velocities corresponding to points c and d or d′ are

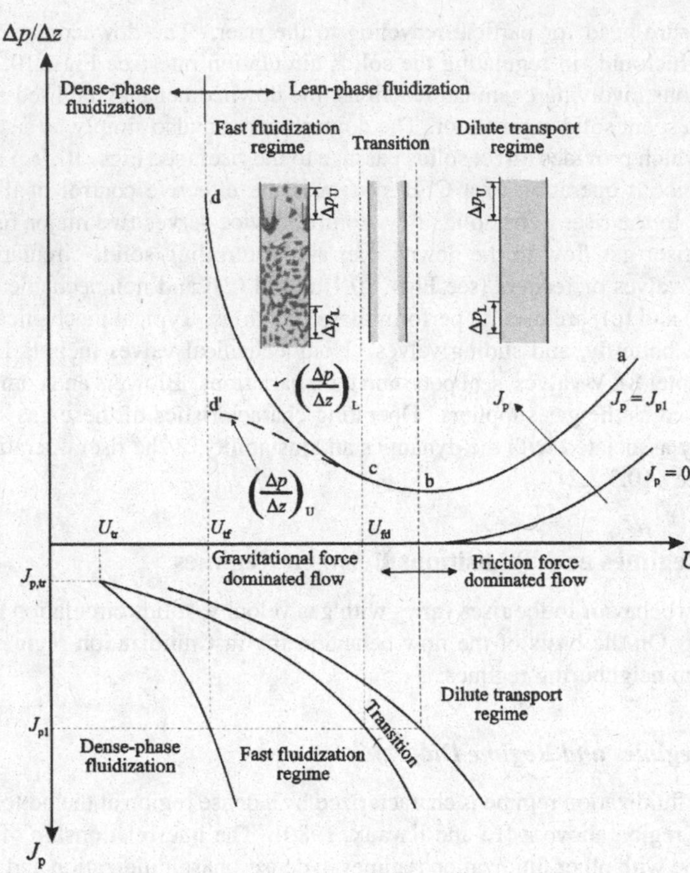

Figure 10.2. Variations of pressure drop per unit riser length with solids circulation rate and gas velocity for various fluidization regimes (after Bai *et al.*, 1993).

denoted U_{fd} and U_{tf}, respectively. For a given gas velocity, $\Delta p/\Delta z$ increases as the solids circulation rate (J_p) increases. The curve of $\Delta p/\Delta z$ for $J_p = 0$ in the diagram represents the wall friction due to gas flow alone.

The lower diagram of Fig. 10.2 reveals the effect of the gas velocity U, at the solids circulation rate of J_{p1}, on the flow regime map plotted in terms of J_p versus U. It is seen in the diagram that as J_p decreases, the applicable range of the gas velocity for the fast fluidization regime decreases; eventually, this regime diminishes when the solids circulation rate and the gas velocity reach $J_{p,tr}$ and U_{tr}, respectively. Therefore, for a system to be operated in the fast fluidization regime, the solids circulation rate and the gas velocity need to be maintained greater than $J_{p,tr}$ and U_{tr}, respectively. With decreasing gas velocity at J_{p1}, the transition from fast fluidization to dense-phase fluidization can be a fuzzy or a sharp process, depending on variables that include column size and particle properties. For a small column with large particles, the transition to dense-phase fluidization goes through choking (Problem 10.7) and is a relatively sharp process which is reflected by a steep slope of the $(\Delta p/\Delta z)_L$ curve at U_{tf}, as given in the figure. For a large column with small particles, the transition to dense-phase fluidization does not go through choking (also noted as the nonchoking transition; see §10.3.2.2) and is a fuzzy process; in this case, the slope of the $(\Delta p/\Delta z)_L$ curve at U_{tf} would be less steep. Choking occurs in a vertical gas–solid flow system when the solids circulation rate reaches the saturation solids carrying capacity at a given gas velocity, leading to a collapse of a relatively dilute suspension to a relatively dense suspension [Zenz and Othmer, 1960; Bi *et al.*, 1993; also see §11.2.4]. The dense suspension in the choking condition is usually characterized by the presence of solids slugs [Yang, 1975] yielding considerable pressure fluctuations.

10.3.2 *Regime Transition Determination*

The boundaries of the fast fluidization regime can be determined from three independent approaches: phenomenological, statistical, and structural. The phenomenological approach utilizes the relationship among the pressure drop or bed voidage, gas velocity, and solids circulation rate, based on the overall fluidization behavior. This approach expresses the regime boundaries in terms of the transport velocity, gas velocity, and solids circulation rates, as given in Fig. 10.2. The statistical approach considers that the property fluctuations are dictated by bubbles under low-velocity conditions and solids assemblage under high-velocity conditions. This approach utilizes the amplitude and standard deviation of the voidage, pressure, or differential pressure fluctuations as well as their correlation and power spectra as criteria to define the regime boundaries. The structural approach recognizes the differences in flow structure of the flow system in the mesoscale such as bubbles and clusters and defines the flow regimes on the basis of these differences. In the following, the quantitative accounts of the transport velocity and the regime boundaries are discussed.

10.3.2.1 *Transport Velocity*

As noted in §10.3.1, the transport velocity, U_{tr}, marks the lower limit of the gas velocity for fast fluidization operation. The characteristics and the prediction of U_{tr} are given in the following discussion.

A. CHARACTERISTICS OF THE TRANSPORT VELOCITY

The transport velocity can be characterized from the voidage and gas velocity relationship for a riser flow with Group A particles. In general, the voidage–gas velocity curve in a logarithmic plot spanning several flow regimes exhibits a straight line within each flow regime [Avidan and Yerushalmi, 1982]. Figure 10.3 shows such a relationship for FCC particles. It is seen that a sharp change in the slope occurs at a point noted as U_{tr} which marks the onset of the dependency of the solids circulation rate on the voidage–gas velocity relationship.

The transport velocity can also be evaluated from the variations of the local pressure drop per unit length $(\Delta p/\Delta z)$ with respect to the gas velocity and the solids circulation rate, J_p. An example of such a relationship is shown in Fig. 10.4. It is seen in the figure that, along the curve AB, the solids circulation rates are lower than the saturation carrying capacity of the flow. Particles with low particle terminal velocities are carried over from the riser, while others remain at the bottom of the riser. With increasing solids circulation rate, more particles accumulate at the bottom. At point B in the curve, the solids fed into the riser are balanced by the saturated carrying capacity. A slight increase in the solids circulation rate yields a sharp increase in the pressure drop (see curve BC in Fig. 10.4). This behavior reflects the collapse of the solid particles into a dense-phase fluidized bed. When the gas

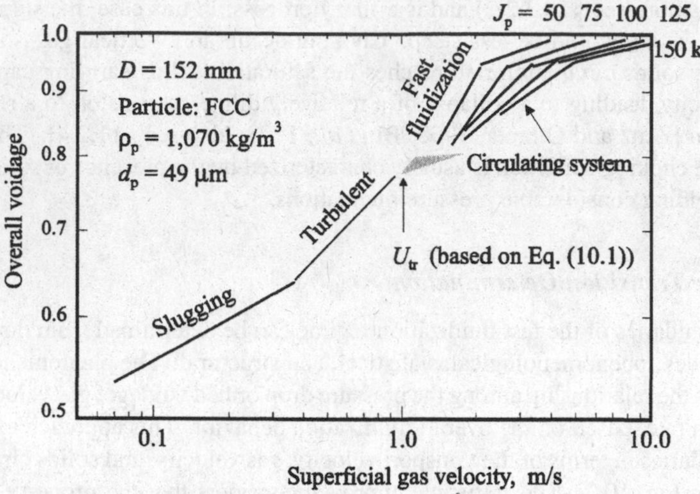

Figure 10.3. Characteristics of the overall voidage in various fluidization regimes for FCC particles (after Avidan and Yerushalmi, 1982).

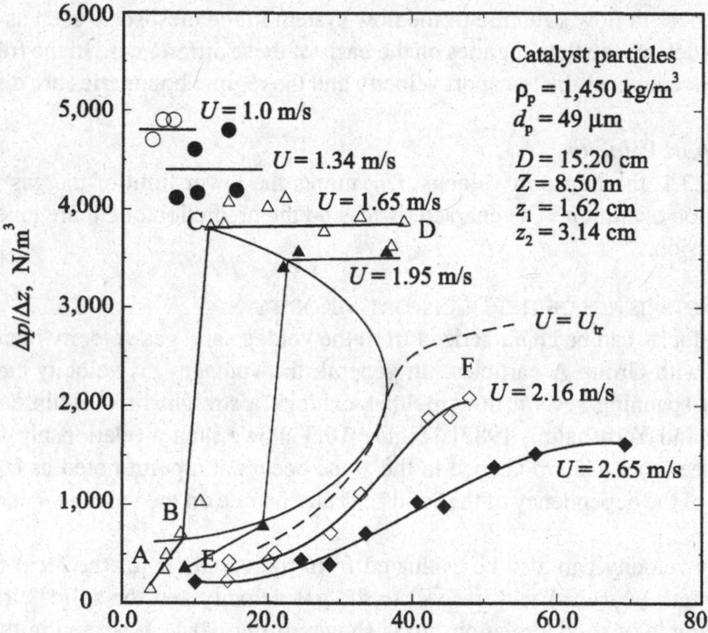

Figure 10.4. Local pressure drop as a function of solids circulation rate at various gas velocities (after Yerushalmi and Cankurt, 1979).

velocity is equal to or higher than the transport velocity, *e.g.*, curve EF in Fig. 10.4, there is no longer a sharp increase in the slope of the J_p versus $(\Delta p/\Delta z)$ relationship. Thus, U_{tr} is characterized as the lowest gas velocity at which the discontinuity in the curve of J_p versus $(\Delta p/\Delta z)$ disappears [Yerushalmi and Cankurt, 1979].

B. Prediction of the Transport Velocity

There has been no definitive theory proposed to predict the transport velocity. A simple empirical equation proposed by Bi and Fan (1992), given next, can be used for estimation of the transport velocity:

$$\mathrm{Re}_{tr} = 2.28\mathrm{Ar}^{0.419} \tag{10.1}$$

where $\mathrm{Re}_{tr} = \rho U_{tr} d_p/\mu$ and Ar is defined by Eq. (9.12). This correlation, covering Re_{tr} from 2.42 to 2,890, is applicable for particles of sizes from 23.6 to 5,000 μm and of densities from 660 to 4,510 kg/m^3, *i.e.*, Groups A, B, and D particles, in risers of diameters from 0.05 to 0.3 m. Equation (10.1) is graphically shown in Fig. 10.5. In the figure, Geldart's demarcations of Groups A, B, and D particles in terms of Ar which are applicable for particles with $(\rho_p - \rho)/\rho$ ranging from about 1,000 to 2,000 [Grace, 1986] are also shown along with the variation of the Reynolds number based on single particle terminal velocity, Re_t, with Ar. This figure indicates that for Ar less than 125 (Group A particles), for a given Ar, the Re_t is lower than Re_{tr}, and for $125 < \mathrm{Ar} < 1.45 \times 10^5$ (Group B particles), Re_t is virtually of the same order as Re_{tr}. However, at Ar larger than 1.45×10^5 (Group D particles), Re_t is greater than Re_{tr}. Equation (10.1) clearly indicates that U_{tr} varies solely with the gas and solid particle properties.

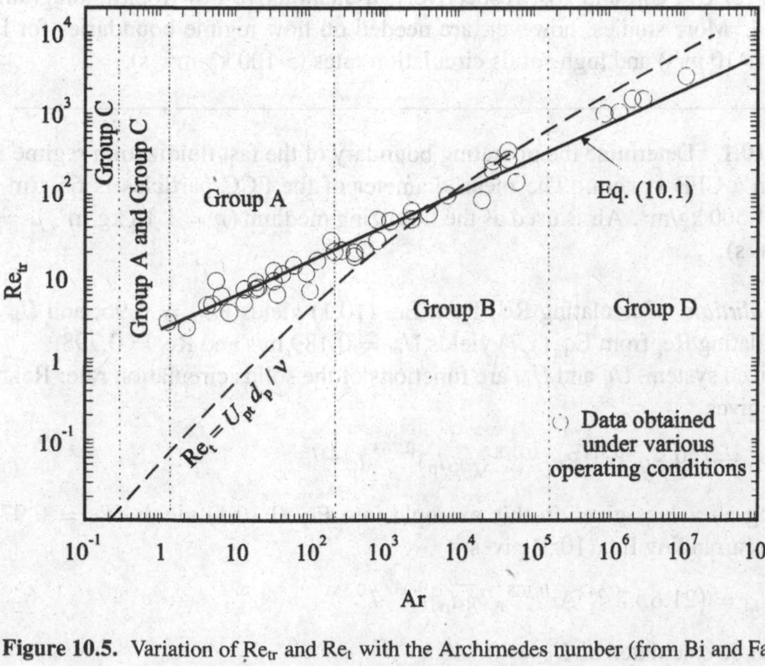

Figure 10.5. Variation of Re_{tr} and Re_t with the Archimedes number (from Bi and Fan, 1992).

10.3.2.2 *Minimum and Maximum Gas Velocities for Fast Fluidization*

As noted, the fast fluidization operation is confined by a certain range of gas velocities and solids circulation rates. For a given solids circulation rate in the fast fluidization regime, the gas velocity needs to be greater than U_{tr}. The analysis of the regime boundaries for fast fluidization relies primarily on empirical approaches. An empirical correlation which is applicable to both choking and nonchoking transitions from dense-phase fluidization to fast fluidization, *i.e.*, lower bound, was proposed by Bi and Fan (1991):

$$\frac{U_{tf}}{\sqrt{gd_p}} = 39.8\left(\frac{J_p}{\rho U_{tf}}\right)^{0.311} \mathrm{Re}_t^{-0.078} \tag{10.2}$$

This correlation can be applied to both Group A and Group B particles in small risers ($D < 0.3$ m). It is noted that operation with choking is unstable. Another type of unstable operation is caused by system design and operation (see §10.3.3.2). The lower bound of the gas velocity due to instability caused by system design and operation is greater than or equal to that due to choking.

The following empirical correlation can be used to estimate the upper bound of fast fluidization operation for Group A particles in risers of $D < 0.2$ m [Bi and Fan, 1991]:

$$\frac{U_{fd}}{\sqrt{gd_p}} = 21.6\left(\frac{J_p}{\rho U_{fd}}\right)^{0.542} \mathrm{Ar}^{0.105} \tag{10.3}$$

The fast fluidization regime is thus bounded by $U \geq U_{tr}$ and $J_{p,min} < J_p < J_{p,max}$ for a given gas velocity ($U > U_{tr}$), or $U_{tf} < U < U_{fd}$ for a given solids circulation rate. Note that replacing U_{tf} and U_{fd} with U for a given value of U, the values of J_p obtained from Eqs. (10.3) and (10.2) represent $J_{p,min}$ and $J_{p,max}$, respectively. The values for $J_{p,min}$ and $J_{p,max}$ would merge to a value $J_{p,tr}$ when U approaches U_{tr}. Thus, using Eqs. (10.1), (10.2), and (10.3) for U_{tr}, U_{tf}, and U_{fd}, respectively, a quantitative flow regime diagram can be constructed. More studies, however, are needed on flow regime boundaries for high gas velocities (>10 m/s) and high solids circulation rates (>100 kg/m$^2\cdot$s).

Example 10.1 Determine the operating boundary of the fast fluidization regime for FCC particles in a CFB system. The mean diameter of the FCC particles is 65 μm and the density is 1,500 kg/m^3. Air is used as the fluidizing medium ($\rho = 1.18$ kg/m^3, $\mu = 1.82 \times 10^{-5}$ kg/m$\cdot$s).

Solution Calculating Re_{tr} from Eq. (10.1) yields $\mathrm{Re}_{tr} = 6.96$, and $U_{tr} = 1.65$ m/s. Calculating Re_t from Eq. (1.7) yields $U_{pt} = 0.189$ m/s and $\mathrm{Re}_t = 0.798$.

For a given system, U_{tf} and U_{fd} are functions of the solids circulation rate. Rearranging Eq. (10.2) gives

$$U_{tf} = \left(39.8\rho^{-0.311}\mathrm{Re}_t^{-0.078}\sqrt{gd_p}\right)^{0.763} J_p^{0.237} \tag{E10.1}$$

Substituting all values given in this example into Eq. (E10.1) yields $U_{tf} = 0.978 J_p^{0.237}$. Likewise, rearranging Eq. (10.3) gives

$$U_{fd} = \left(21.6\rho^{-0.542}\mathrm{Ar}^{0.105}\sqrt{gd_p}\right)^{0.649} J_p^{0.351} \tag{E10.2}$$

and U_{fd} can be calculated as $U_{fd} = 0.763 J_p^{0.351}$

The calculated results of U_{tf} and U_{fd} as functions of J_p are plotted in Fig. E10.1 with U_{tf} replaced by U_{tr}, the value for J_p ($=J_{p,tr}$) can be calculated from Eq. (E10.1) as 9.09 kg/m$^2 \cdot$ s. The values for U_{tr} and $J_{p,tr}$ are also indicated in the figure. At gas velocities higher than U_{tr}, the lower bound of the fast fluidization regime marks the transition to dense-phase fluidization. The bed undergoes transition from the fast fluidization regime to the dilute phase transport as the gas velocity reaches U_{fd}. The solids flow pattern and axial voidage profiles pertaining to these regimes are also given in Fig. E10.1.

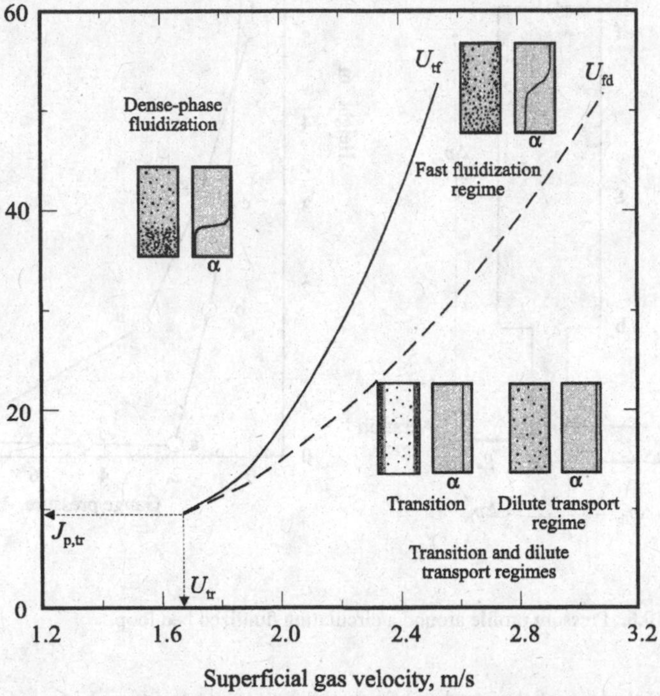

Figure E10.1. Fast fluidization regime boundaries for Example 10.1.

10.3.3 *Operable Fluidization Regimes*

The riser cannot be considered as an isolated entity in the CFB loop. When no particles are introduced into or discharged from the loop, the solids mass flow rate in the riser is equal to that in the downcomer. For a given quantity of solids in the loop, the presence of fewer solids in the riser implies the presence of more solids in the downcomer. Likewise, the pressure drop across the riser must be balanced by that imposed by the flow through its accompanying components such as the downcomer and the recirculation device. Furthermore, the flow characteristics of the riser can be significantly affected by the behavior of the accompanying components in the loop.

10.3.3.1 *Overall Pressure Balance Around a CFB Loop*

Typical pressure profiles in a CFB loop with a nonmechanical valve (see Chapter 8) are shown in Fig. 10.6. In this figure, line a–b–c–d represents the pressure drop across the riser,

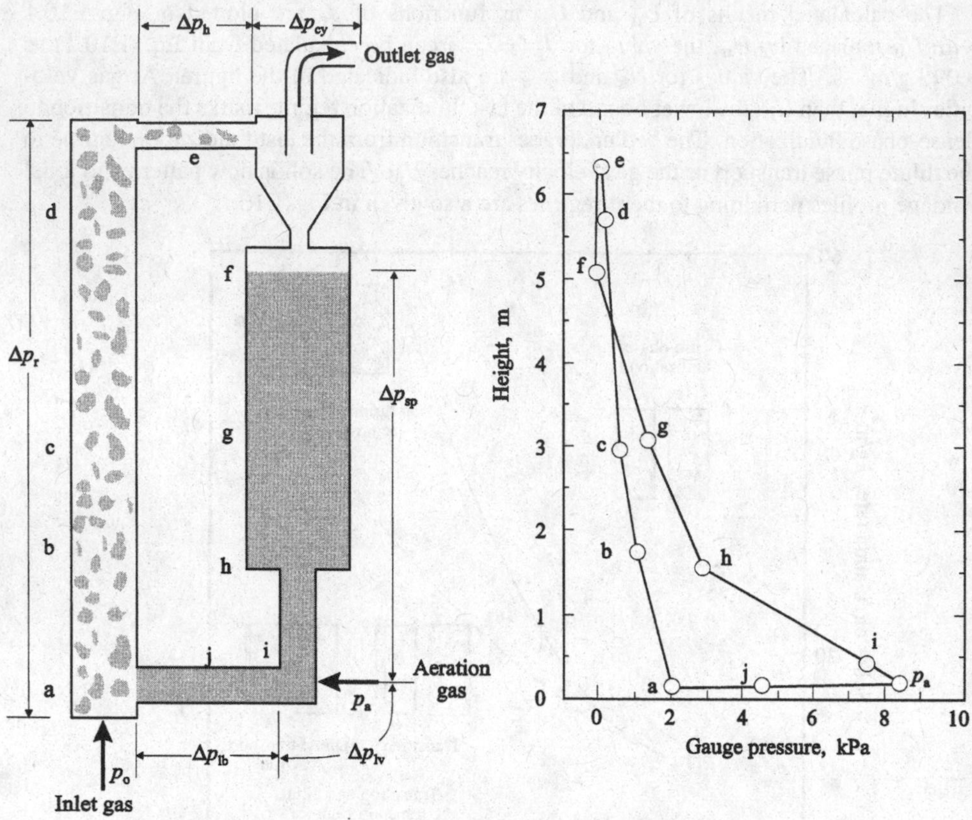

Figure 10.6. Pressure profile around a circulating fluidized bed loop.

indicating a smooth drop of the gradient from the bottom to the top. Line e–f illustrates a small pressure drop across the cyclone and connecting pipes, and line f–g–h–i–p_a depicts a pressure buildup by gravity of the particles in the downcomer and the hopper above it. Line p_a–j–a describes the pressure drop across the solids recirculation and control devices. The pressure drops across the downcomer, the solids circulation and control device, and the riser are major elements in the pressure balance around the CFB loop. The measurements of Rhodes and Laussmann (1992) showed that $\Delta p_{sp}/\Delta p_r = 2$–$5$ and $(\Delta p_{lv} + \Delta p_{lb})/\Delta p_r =$ 1–5 under the operating range of the gas velocity from 2 to 5 m/s, and the solids circulation rate from 5 to 50 kg/m$^2 \cdot$s for nonporous alumina particles with a density of 2,456 kg/m^3 and a mean diameter of 75 μm (see notations defined in the figure). The balance of the pressure around the CFB loop requires quantitative information for the pressure drops in each component. In the following, the pressure drop on each loop component is described.

A. Pressure Drop Across the Riser

For the gas flow in a riser, energy in the gas phase is partially transferred into the solids phase through gas–particle interactions and is partially dissipated as a result of friction. Under most operating conditions, gravitational effects dominate overall gas phase energy consumption. Thus, neglecting the particle acceleration effects, the pressure drop in the riser can be approximated by the equation

$$\Delta p_{\mathrm{r}} = \int_0^Z \rho_{\mathrm{p}}(1-\overline{\alpha})g\,\mathrm{d}z = \rho_{\mathrm{p}}(1-\alpha_{\mathrm{o}})gZ \tag{10.4}$$

where Z is the riser height.

B. Pressure Drop Through the Cyclone

The pressure drop through a cyclone has been analyzed in Chapter 7. Empirically, the pressure drop can be expressed by

$$\Delta p_{\mathrm{cy}} = k\rho U_{\mathrm{cy}}^2 \tag{10.5}$$

where U_{cy} is the inlet gas velocity. The value for the coefficient, k, may vary from 1 to 20, depending on the design [Perry *et al.*, 1984].

C. Pressure Drop Across the Downcomer

Maintaining the downcomer or a standpipe under a moving bed condition allows a large pressure buildup along the downcomer. For a given quantity of particles in the downcomer, the pressure at the bottom of the downcomer is closely associated with the relative velocity between the gas and particles. The pressure drop rises with the relative velocity as the particles are in a moving packed state. The maximum pressure drop in the downcomer is established when particles are fluidized, a state which can be expressed in terms of the pressure drop under the minimum fluidization condition as

$$\Delta p_{\mathrm{sp}}|_{\max} = L_{\mathrm{sp}}(1-\alpha_{\mathrm{mf}})\rho_{\mathrm{p}}g \tag{10.6}$$

where L_{sp} is the height of the solids in the downcomer. Details of the pressure drop analysis in a downcomer or standpipe are given in Chapter 8.

D. Pressure Drop Through the Solids Flow Control Devices

The solids flow in the downcomer can be either in a dense fluidized state or in a moving packed state. If the particles in the downcomer are fluidized, the pressure drop through the mechanical solids flow control devices can be expressed as [Jones and Davidson, 1965]

$$\Delta p_{\mathrm{cv}} = \frac{1}{2\rho_{\mathrm{p}}(1-\alpha_{\mathrm{mf}})}\left(\frac{W_{\mathrm{wp}}}{C_{\mathrm{o}}A_{\mathrm{o}}}\right)^2 \tag{10.7}$$

where W_{wp} is the solids feeding rate and A_{o} is the opening area for the mechanical valve. The coefficient, C_{o}, can be taken as 0.7~0.8 over a variety of systems and control device opening configurations [Rudolph *et al.*, 1991]. For moving packed flow through an orifice, the pressure drop across the orifice can be calculated from [Leung and Jones, 1978]

$$W_{\mathrm{wp}} = A_{\mathrm{o}}\rho_{\mathrm{p}}(1-\alpha_{\mathrm{mf}})\left(\frac{gD_{\mathrm{v}}}{\tan\delta}\right)^{0.5} + C_{\mathrm{o}}A_{\mathrm{o}}[2\rho_{\mathrm{p}}(1-\alpha_{\mathrm{mf}})\Delta p_{\mathrm{cv}}]^{0.5} \tag{10.8}$$

where δ is the angle of internal friction of particles.

The solids flow rate can be controlled by nonmechanical valves such as the L-valve, as noted in Chapter 8. The L-valve has a long horizontal leg. Thus, it is convenient to characterize the pressure drop across an L-valve by two terms. One term is the pressure drop through the elbow (Δp_{lv}). This term can be described by the equations developed for the mechanical valve because the solids flow patterns between the two are similar except that the opening area for the L-valve is actuated by the amount of external aeration and the amount of gas entrainment into the valve [Yang and Knowlton, 1993]. The other term is the pressure drop caused by the gas–solid flow in the horizontal leg (Δp_{lb}). Solids are transferred into the horizontal leg with very low initial velocity and are then accelerated. There is no reliable correlation for the pressure drop which occurs during the acceleration of the solids. Yang and Knowlton (1993) suggested the following equation to describe the pressure drop for an L-valve, including both the elbow and the horizontal leg contributions:

$$(\Delta p_{\text{lv}} + \Delta p_{\text{lb}}) = \frac{1}{2\rho_{\text{p}}(1-\alpha_{\text{mf}})}\left(\frac{W_{\text{wp}}}{C_{\text{o}}A_{\text{o}}}\right)^2 \tag{10.9}$$

$C_{\text{o}} = 0.5$ can be used here. The opening area for the L-valve (see Fig. 8.22(a)), A_{o}, is determined by the external aeration rate and can be evaluated by

$$A_{\text{o}} = \frac{Q_{\text{t}} - 0.177(\pi D_{\text{h}}^2/4)L_{\text{h}}}{0.710U_{\text{pt}}} \tag{10.10}$$

where D_{h} is the diameter of the horizontal section of the L-valve and L_{h} is the horizontal length of the L-valve. Equation (10.10) is applicable to the following conditions: $D_{\text{h}} \leq$ 152.4 mm; $W_{\text{wp}} < 2.5$ kg/s; 175 μm $\leq d_{\text{p}} \leq$ 509 μm; 1,230 kg/m^3 $\leq \rho_{\text{p}} \leq$ 4, 150 kg/m^3. The total gas flow rate, Q_{t}, consists of external aeration, Q_{ext}, and the gas entrained by the moving particles in the L-valve, Q_{ent}. The external aeration rate, Q_{ext}, is an operating variable. Q_{ent} can be approximated in most cases by the following equation, assuming no-slip between gas and moving particles [Yang and Knowlton, 1993]:

$$Q_{\text{ent}} = \frac{W_{\text{wp}}}{\rho_{\text{p}}(1-\alpha_{\text{mf}})}\alpha_{\text{mf}} \tag{10.11}$$

Any changes in the gas velocity and/or solids circulation rate in the riser will change the pressure drop in each component of the system. Figure 10.7 shows the effects of the solids circulation rate on the pressure drop in each component for a given solids inventory in a CFB system with an L-valve. Comparisons of the pressure drops for each component clearly indicate the important roles of ($\Delta p_{\text{lv}} + \Delta p_{\text{lb}}$) and Δp_{sp} in the loop pressure balance. Generally, the pressure drop across the downcomer, Δp_{sp}, is larger than that across the L-valve. Δp_{sp} shows a slight increase with the solids circulation rate. For a given solids circulation rate, Δp_{sp} remains essentially unchanged with an increase in the gas velocity. A decrease in the pressure drop across the L-valve with an increase in the solids circulation rate is also indicated in the figure. As expected, Δp_{r} increases with an increase in the solids circulation rate.

10.3.3.2 *Operating Constraints of Fast Fluidization*

In practice, unstable operation may occur at a higher gas velocity than that at choking or at nonchoking transition to dense-phase fluidization. Thus, the minimum operable gas velocity for a given solids circulation rate can be higher than U_{tf} for fast fluidization operation in some CFBs. The factors contributing to this unstable situation are

(1) Insufficient pressure head established at the bottom of the downcomer
(2) Limited solids circulation rate which can be delivered by the downcomer
(3) Insufficient pressure head provided by gas suppliers

These factors can be illustrated in two aspects as given in the following.

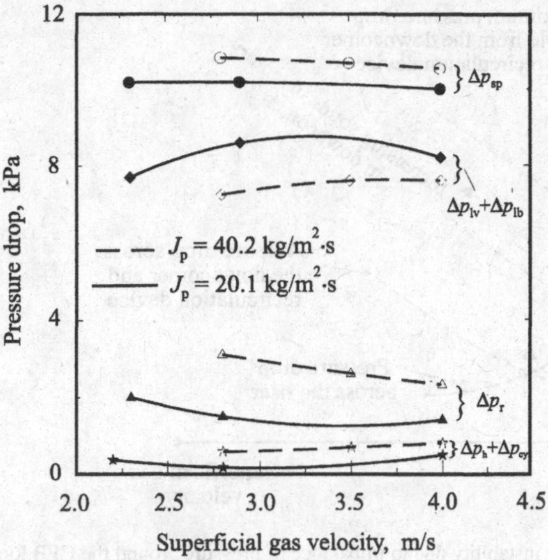

Figure 10.7. Effects of gas velocity on the pressure drop of various system components in a CFB loop (after Rhodes and Laussmann, 1992).

A. Unstable Operations Related to Downcomer or Solids Flow Control Devices

Operating constraints related to the downcomer or solids flow control devices in CFB systems include limitations concerning the maximum available pressure drop across the downcomer and the maximum solids circulation rate, which can be delivered by the downcomer and the solids circulation and control device.

In CFB systems, as noted, the pressure drop in the downcomer is affected by the flow in the other components. Under normal operating conditions, the particles in the downcomer are in a moving bed state; the pressure drops across the downcomer and solids control device adjust accordingly, on the basis of the resulting relative velocity between gas and particles. However, there is a maximum pressure drop that the downcomer can develop, which is equal to the pressure drop of a fluidized bed at its incipient fluidization state, $(\Delta p/L_{sp})_{mf}$. Under normal operating conditions, the pressure drop across the downcomer increases with the relative velocity. When the relative velocity is higher than the minimum fluidization velocity of the bed materials, bubbles (or slugs) will be generated. The bubble motion, in turn, may hinder the particle flow and reduce the solids flow rate, leading to unstable feeding in the system (see Chapter 8). In this case, the operation of the CFB loop is limited by the solids feeding rate from the downcomer if the feeding rate at the incipient fluidization condition is less than the solids circulation rate desired for a given gas velocity.

The constraint of the pressure drop across the downcomer is graphically illustrated in Fig. 10.8. For a given solids inventory in the downcomer and given gas and solids flow rates, the pressures at the bottom of the riser and the downcomer can be determined at steady state. Under normal operating conditions (point A in the figure), the pressure drop across the riser is balanced by the pressure drop across the recirculation loop. If a small reduction in gas velocity takes place, the flow in the riser responds by moving upward along the pressure drop curve of the riser to point B. On point B of line AB, the decrease in the gas velocity causes the pressure drop across the riser to rise by δP_r, which has to be balanced by the

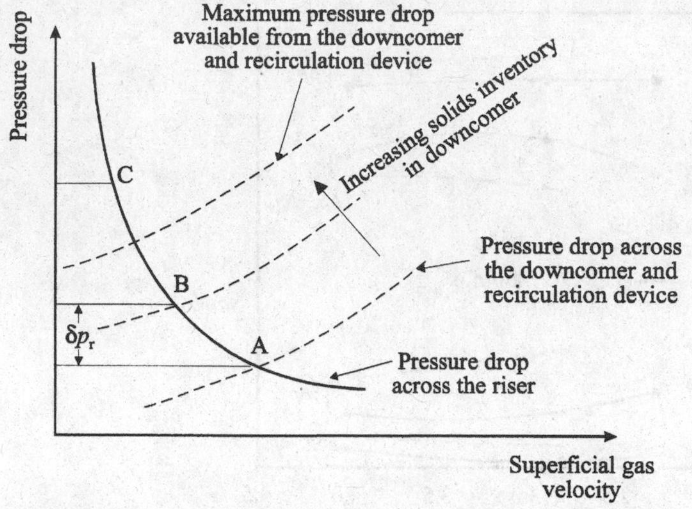

Figure 10.8. Operational instability due to imbalance of pressure around the CFB loop (after Bi and Zhu, 1993).

pressure drop through the circulating loop. Within a certain range of operating conditions, the pressure drop across the downcomer can be adjusted accordingly, and the operating state can be shifted to a new balance point, *i.e.*, point B. However, when the pressure drop through the riser is higher than the maximum pressure drop that is available from the downcomer and solids recirculation devices, *e.g.*, the solids flow control valve, the system starts to oscillate, causing unstable operation (point C) [Bi and Zhu, 1993]. It is noted that unstable operation related to downcomer limitations will not be present when a screw feeder is used. Using the screw feeder, the downcomer is decoupled hydrodynamically from the riser [Hirama *et al.*, 1992], and the control of the solids feeding rate is through mechanical means which do not depend on the downcomer pressure drop.

It appears that the fast fluidization regime may exist only within a narrow range in a system when the operating instability indicated is encountered. Improvements in operating strategy and unit design can expand the operating range of the gas velocity for the fast fluidization regime. These improvements include increased solids inventory level in the downcomer and reduced resistance through the solids recirculation device. For a given system, the operable limits can be predicted by considering the hydrodynamics of each component in the loop. This calculation procedure is illustrated in Example 10.2.

Example 10.2 A circulating fluidized bed is operated using FCC particles with a mean diameter of 65 μm and a density of 1,500 kg/m^3. The following geometric and operating conditions are given:

$\alpha_{\mathrm{mf}} = 0.48$, $\mu = 1.82 \times 10^{-5}$ kg/m · s, $\rho = 1.18$ kg/m^3, $U_{\mathrm{pt}} = 0.189$ m/s

Riser: $D = 100$ mm, $Z = 8$ m, $A_{\mathrm{r}} = 7.85 \times 10^{-3}$ m^2

Downcomer: $D_{\mathrm{sp}} = 200$ mm, $A_{\mathrm{sp}} = 3.14 \times 10^{-2}$ m^2

L-valve: $D_{\mathrm{h}} = 60$ mm, $L_{\mathrm{h}} = 300$ mm, $L_{\mathrm{z}} = 500$ mm, $A_{\mathrm{l}} = 2.83 \times 10^{-3}$ m^2

Solids inventory: $M = 20$, 30, and 40 kg

Also, Q_{ext} can be estimated by [Rhodes and Cheng, 1994]

$$Q_{\text{ext}} = \frac{\frac{\pi}{4}U_{\text{mf}}}{1{,}240}\left(J_{\text{P}}\frac{D^2}{D_{\text{h}}} + 1{,}107D_{\text{h}}^2\right) \tag{E10.3}$$

where A_{l} is the cross-sectional area of the L-valve; A_{r} is the cross-sectional area of the riser; A_{sp} is the cross-sectional area of the downcomer; L_{z} is the length of the vertical section of the L-valve; D is the riser diameter; and D_{sp} is the diameter of the downcomer. Express the pressure drops around the riser, cyclone, downcomer, and L-valve in the circulating fluidized bed loop. Also calculate the maximum operable solids circulation rate as a function of gas velocity.

Solution The maximum solids circulation rate through an L-valve is limited by the pressure drop across the downcomer under minimum fluidizing conditions. Thus, the pressure balance can be written in terms of voidage and solids inventory in the downcomer as (see Fig. 10.6)

$$\rho_{\text{p}}g(1-\alpha_{\text{mf}})L_{\text{sp}} = \Delta p_{\text{lv}} + \Delta p_{\text{lb}} + \rho_{\text{p}}g(1-\alpha_{\text{o}})Z + \Delta p_{\text{h}} + \Delta p_{\text{cy}} \tag{E10.4}$$

Rearranging this equation yields

$$1-\alpha_{\text{o}} = (1-\alpha_{\text{mf}})\frac{L_{\text{sp}}}{Z} - \frac{\Delta p_{\text{lv}} + \Delta p_{\text{lb}} + \Delta p_{\text{h}} + \Delta p_{\text{cy}}}{\rho_{\text{p}}gZ} \tag{E10.5}$$

Calculate Δp_{cy} from Eq. (10.5) by taking the coefficient k as 10 and inlet velocity U_{cy} as U, i.e.,

$$\Delta p_{\text{cy}} = k\rho U_{\text{cy}}^2 = 11.8U^2 \tag{E10.6}$$

$(\Delta p_{\text{lv}} + \Delta p_{\text{lb}})$ can be evaluated from Eq. (10.9) on the basis of the following information:
(a) Solids circulation rate through the L-valve is

$$W_{\text{wp}} = \frac{\pi}{4}D^2 J_{\text{p}} = 7.85 \times 10^{-3} J_{\text{p}} \tag{E10.7}$$

(b) With W_{wp} from Eq. (E10.7) and values provided by this example, Eq. (10.11) can be expressed by

$$Q_{\text{ent}} = \frac{W_{\text{wp}}}{\rho_{\text{p}}(1-\alpha_{\text{mf}})}\alpha_{\text{mf}} = 4.838 \times 10^{-6} J_{\text{p}} \tag{E10.8}$$

(c) From $Q_{\text{t}} = Q_{\text{ent}} + Q_{\text{ext}}$ and Eqs. (E10.3) and (E10.8), Eq. (10.10) becomes

$$A_{\text{o}} = \frac{Q_{\text{t}} - 0.177(\pi D_{\text{h}}^2/4)L_{\text{h}}}{0.710U_{\text{pt}}} = \frac{4.838 \times 10^{-6}J_{\text{p}} + Q_{\text{ext}} - 1.501 \times 10^{-4}}{0.134} \tag{E10.9}$$

Substituting the expressions obtained into Eq. (10.9), $(\Delta p_{\text{lv}} + \Delta p_{\text{lb}})$ can be given as

$$\Delta p_{\text{lv}} + \Delta p_{\text{lb}} = \frac{1}{2\rho_{\text{p}}(1-\alpha_{\text{mf}})}\left(\frac{W_{\text{wp}}}{C_{\text{o}}A_{\text{o}}}\right)^2 = 2.840 \times 10^3\left(\frac{J_{\text{p}}}{4.838J_{\text{p}} + 10^6 Q_{\text{ext}} - 150.1}\right)^2 \tag{E10.10}$$

The solids inventory level in the downcomer can be evaluated from a mass balance. Neglecting the solids inventory in the connecting tubes and the cyclone, L_{sp} can be obtained as

$$L_{sp} = \frac{M/\rho_p - L_h A_l(1-\alpha_{mf}) - L_z A_l(1-\alpha_{mf}) - Z A_r(1-\alpha_o)}{A_{sp}(1-\alpha_{mf})}$$
$$= 0.0408M - 3.846(1-\alpha_o) - 0.072 \tag{E10.11}$$

On the basis of the calculated values from the preceding expression, and assuming $\Delta p_h \approx 0$, Eq. (E10.5) gives rise to

$$1-\alpha_o = 2.65\times10^{-3}M - 0.25(1-\alpha_o) - 4.68\times10^{-3}$$
$$-0.024\left(\frac{J_p}{4.838J_p + 10^6 Q_{ext} - 150.1}\right)^2 - 1.007\times10^{-4}U^2 \tag{E10.12}$$

Rearranging Eq. (E10.12) yields

$$1-\alpha_o = 2.12\times10^{-3}M - 8.05\times10^{-5}U^2$$
$$-0.0193\left(\frac{J_p}{4.838J_p + 10^6 Q_{ext} - 150.1}\right)^2 - 3.744\times10^{-3} \tag{E10.13}$$

The sum on the right-hand side of Eq. (E10.13) is a function of operating variables. To determine the operating limit for the solids circulation rate, another equation linking the bed voidage, solids circulation rate, and gas velocity is necessary.

Consider a riser where the voidage in the bottom dense region is uniform and the top dilute region behaves as the freeboard of a dense-phase fluidized bed. The axial profile of the voidage in the top dilute region can be expressed by [Kunii and Levenspiel, 1990] (see §10.4.1)

$$\frac{\alpha^* - \bar{\alpha}}{\alpha^* - \alpha_a} = e^{-a(z-z_i)} \tag{E10.14}$$

where a is a decay constant and can be taken as 0.5 for this example. Integrating Eq. (E10.14) over the top dilute region and combining with the voidage in the bottom dense region give the overall voidage as

$$1-\alpha_o = \frac{\alpha_e - \alpha_a}{aZ} - \left(1 - \frac{z_i}{Z}\right)(\alpha^* - \alpha_a) + (1-\alpha_a) \tag{E10.15}$$

α_e in Eq. (E10.15) is the voidage at the riser exit, which can be evaluated by $1 - J_p/(U - U_{pt})\rho_p$. The height of the bottom dense region, z_i, can be determined from Eq. (E10.14) as

$$z_i = Z - \frac{1}{a}\ln\left(\frac{\alpha^* - \alpha_a}{\alpha^* - \alpha_e}\right) \tag{E10.16}$$

Substituting Eq. (E10.16) into Eq. (E10.15) yields

$$1-\alpha_o = (1-\alpha_a) + \frac{1}{aZ}\left[(\alpha_e - \alpha_a) - (\alpha^* - \alpha_a)\ln\left(\frac{\alpha^* - \alpha_a}{\alpha^* - \alpha_e}\right)\right] \tag{E10.17}$$

α_o in Eq. (E10.17) is also a function of the gas velocity and solids circulation rate.

Equations (E10.13) and (E10.17) can be solved simultaneously to obtain the relationship between the minimum operable gas velocity and solids circulation rate, as given in Fig. E10.2. To illustrate the dependence of operable velocities on system designs, the results for three solids inventory heights in the downcomer are given as shown in Fig. E10.2. It is seen in the figure that for a given solids circulation rate, a higher inventory level yields a lower minimum operable velocity.

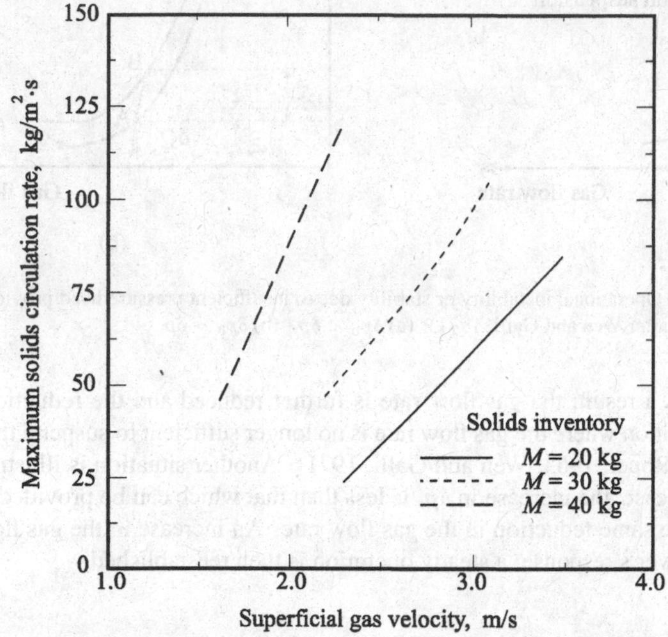

Figure E10.2. Maximum solids circulation rates as a function of gas velocity for Example 10.2.

B. UNSTABLE OPERATION RELATED TO THE BLOWER OR COMPRESSOR

Another operational limit in the CFB system involves gas suppliers. Three types of gas suppliers, *i.e.*, a reciprocating compressor, a blower with throttle valve, and a compressor, are commonly used in the CFB system. For blower operation, as the gas flow rate decreases, the pressure head of the blower increases. For compressor operation, the pressure head of the compressor can be maintained constant with variable gas flow rates. The interactive behavior between a CFB system and a blower can be illustrated in Fig. 10.9, where dashed curves refer to the blower characteristics and solid curves refer to the riser pressure drop. At point A, the pressure drop across the riser matches the pressure head provided by a blower; thus, a stable operation can be established. Since the pressure drop across the riser in fast fluidization increases with a decrease in the gas flow rate at a given solids circulation rate, a reduction in the gas flow rate causes the pressure drop to move upward on the curve in the figure to point B with an increase in the pressure drop of δp_r. In the case shown in Fig. 10.9(a), with the same reduction in the gas flow rate, *i.e.*, δQ, the increase in the pressure drop, δp_r, from point A to point B is greater than that which can be provided by

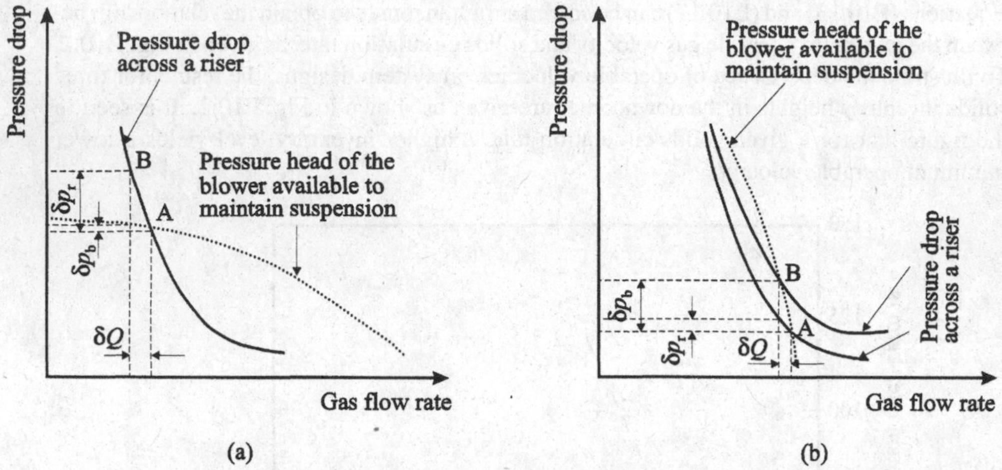

Figure 10.9. Operational instability or stability due to insufficient pressure head provided by gas blowers (after Wen and Galli, 1971): (a) $\delta p_b < \delta p_r$; (b) $\delta p_b > \delta p_r$.

the blower, δp_b. As a result, the gas flow rate is further reduced and the reduction may continue to the condition where the gas flow rate is no longer sufficient to suspend the solid particles [Doig and Roper, 1963; Wen and Galli, 1971]. Another situation is illustrated in Fig. 10.9(b). In this case, the increase in δp_r is less than that which can be provided by the blower, δp_b, with the same reduction in the gas flow rate. An increase in the gas flow rate results from the blower's response; a steady operation is then reestablished.

10.4 Hydrodynamic Behavior in a Macroscale

The hydrodynamics of a circulating fluidized bed can be analyzed from both the macroscopic and mesoscopic points of view. The nonuniformity of the solids concentration in the radial and axial directions represents macroscopic behavior. The existence of solid clusters characterizes mesoscopic behavior (see §10.5). The hydrodynamic behavior in a macroscale is discussed in the following.

10.4.1 *Axial Profiles of Cross-Sectional Averaged Voidage*

The axial profile of the cross-sectional averaged voidage in the riser is typically S-shaped [Li and Kwauk, 1980], as shown in Fig. 10.10 for Group A particles. This profile reflects an axial solids concentration distribution with a dense region at the bottom and a dilute region at the top of the riser. The boundary between the two regions is marked by the inflection point in the profile. An increase in the gas flow rate at a given solids circulation rate reduces the dense region [from (a) to (c) in Fig. 10.10], whereas an increase in the solids circulation rate at a given gas flow rate results in an expansion of the dense region [from (c) to (a) in Fig. 10.10]. When the solids circulation rate is very low and/or the gas velocity is very high, the dilute region covers the entire riser (see curve (d) in Fig. 10.10). For given gas and solids flow rates, particles of high density or large size yield a low voidage in the bottom of the riser.

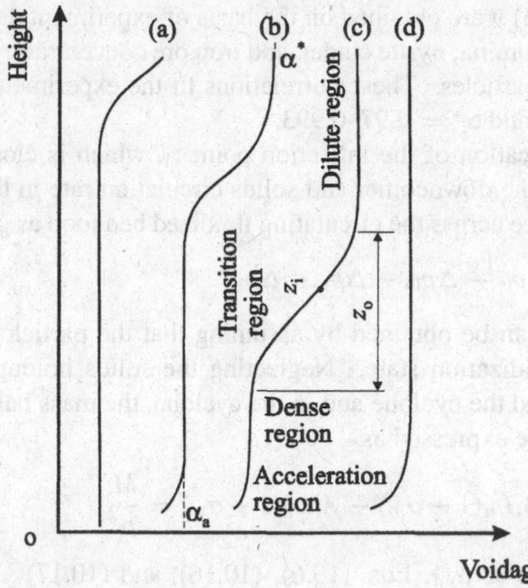

Figure 10.10. Typical axial voidage profiles for Group A particles (after Li and Kwauk, 1980; Yang, 1992).

The S-shaped profiles of the voidage can be represented by an equation of the following form [Li and Kwauk, 1980]:

$$\frac{\overline{\alpha} - \alpha_a}{\alpha^* - \overline{\alpha}} = \exp\left(\frac{z - z_i}{Z_0}\right) \tag{10.12}$$

Equation (10.12) contains four parameters: α_a, α^*, Z_0, and z_i. The value of Z_0 represents the length of the transition region between the dilute and dense regions. Z_0 approaching 0 indicates a clear interface between the dense and dilute region, while Z_0 approaching infinity implies a uniform axial profile of the voidage. Z_0 can be empirically correlated by [Kwauk, 1992]

$$Z_0 = 500\exp[-69(\alpha^* - \alpha_a)] \tag{10.13}$$

where α_a is the asymptotic voidage in the bottom dense region and α^* is the asymptotic voidage in the top dilute region. Their values can be obtained from the following empirical correlations [Kwauk, 1992]:

$$1 - \alpha_a = 0.2513\left(\frac{18\mathrm{Re}_a + 2.7\mathrm{Re}_a^{1.687}}{\mathrm{Ar}}\right)^{-0.4037}; \qquad \mathrm{Re}_a = \frac{d_p\rho}{\mu}\left(U - \frac{J_p}{\rho_p}\frac{\alpha_a}{1-\alpha_a}\right) \tag{10.14}$$

for α_a, and

$$1 - \alpha^* = 0.05547\left(\frac{18\mathrm{Re}^* + 2.7\mathrm{Re}^{*1.687}}{\mathrm{Ar}}\right)^{-0.6222}; \qquad \mathrm{Re}^* = \frac{d_p\rho}{\mu}\left(U - \frac{J_p}{\rho_p}\frac{\alpha^*}{1-\alpha^*}\right) \tag{10.15}$$

for α^*. Equations (10.14) and (10.15) were obtained on the basis of experimental data for FCC catalyst, fine alumina, coarse alumina, pyrite cinder, and iron ore concentrate, ranging from Group A to weakly Group B particles. These correlations fit the experimental data well in the range of $\alpha_a = 0.85$–0.93 and $\alpha^* = 0.97$–0.993.

Of primary significance is the location of the inflection point z_i, which is closely associated with the solids quantity in the downcomer and solids circulation rate in the CFB system. Consider the pressure balance across the circulating fluidized bed loop as given by

$$(1 - \alpha_o)\rho_p g Z = \Delta p_{sp} - \Delta p_{lv} - \Delta p_{lb} - \Delta p_h - \Delta p_{cy} \tag{10.16}$$

The pressure drops in Eq. (10.16) can be obtained by assuming that the particles in the downcomer are in the incipient fluidization state. Neglecting the solids holdup in the connecting tube between the riser and the cyclone and in the cyclone, the mass balance of solid particles in the CFB loop can be expressed as

$$A_r Z(1 - \alpha_o) + A_{sp} L_{sp}(1 - \alpha_{sp}) + A_l L_h(1 - \alpha_{lh}) + A_l L_z(1 - \alpha_{lz}) = \frac{M}{\rho_p} \tag{10.17}$$

Assuming the voidage in an L-valve as α_{mf}, Eqs. (10.6), (10.16), and (10.17) yield the following expression for $(1 - \alpha_o)$:

$$1 - \alpha_o = \frac{Mg - (\Delta p_{lv} + \Delta p_{lb} + \Delta p_h + \Delta p_{cy})A_{sp} - A_l(1 - \alpha_{mf})(L_h + L_z)\rho_p g}{\rho_p g(A_r + A_{sp})Z} \tag{10.18}$$

Thus, the inflection point can be obtained by substituting α_o from Eq. (10.18) into Eq. (10.21) given in §10.4.3.

Recent experimental observations have indicated that the voidage in the bottom dense region is very uniform along the bed height, and the voidage increases exponentially along the bed height in the top dilute region. Thus, alternatively, the top dilute region can be regarded to be the freeboard of the dense-phase fluidized bed. Consequently, the well-established entrainment model (see §9.6), which was originally developed for particle entrainment in a freeboard region of a bubbling bed, can be adopted to describe the top dilute region [Rhodes and Geldart, 1987; Kunii and Levenspiel, 1990].

The axial profile of the voidage or solids concentration given in Fig. 10.10 is influenced not only by the gas velocity, solids circulation rate, and particle properties, but also by the riser entrance and exit geometries. For smooth entrance and exit geometries, the end effects are minimized and axial voidage exhibits a typical S-shaped profile, shown in Fig. 10.10. The shape of the profile may vary, however, for nonsmooth entrance and exit geometries. Figure 10.11 shows the gas and solid flow rate effects on the axial voidage profiles for an abrupt exit. Considerable impaction of particles on the top end of the riser yields a sharp decrease in the voidage in the exit region as evident in the figure. It is seen in the figure that the axial voidage profile varies more significantly when the gas and solid flow rates are higher.

10.4.2 *Radial Profiles of Voidage and Solids Flux*

A comprehensive account of radial voidage distribution requires a recognition of the lateral movement of solids in addition to their axial movement. One of the most important and yet least understood aspects of riser hydrodynamics is the lateral solids distribution mechanism [Kwauk, 1992]. Typical experimental findings for the radial voidage

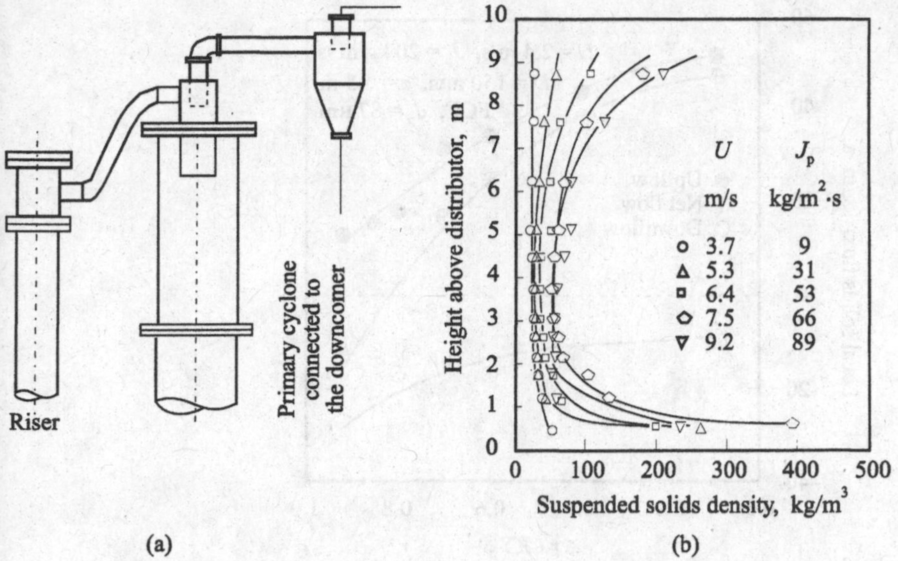

Figure 10.11. Effects of gas and solids flow rates on the axial profile of the cross-sectional averaged voidage in a riser with an abrupt exit (from Brereton and Grace, 1993b): (a) Abrupt exit geometry; (b) Voidage profiles.

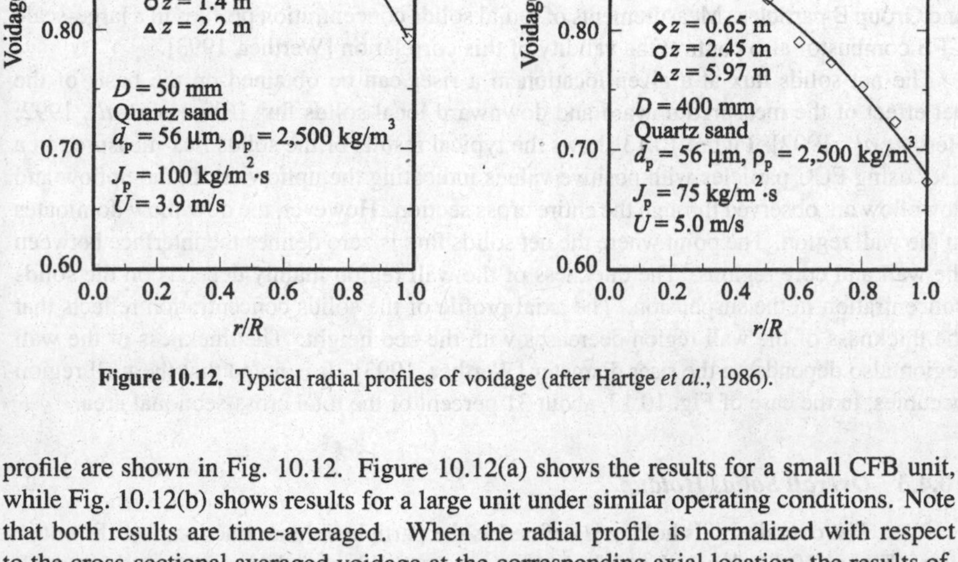

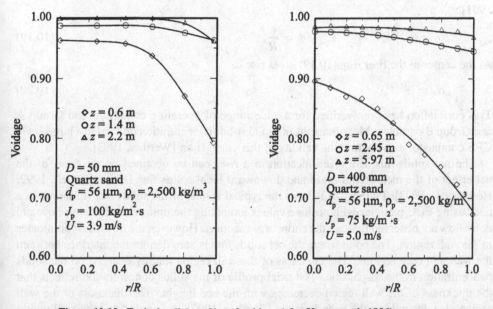

Figure 10.12. Typical radial profiles of voidage (after Hartge *et al.*, 1986).

profile are shown in Fig. 10.12. Figure 10.12(a) shows the results for a small CFB unit, while Fig. 10.12(b) shows results for a large unit under similar operating conditions. Note that both results are time-averaged. When the radial profile is normalized with respect to the cross-sectional averaged voidage at the corresponding axial location, the results of

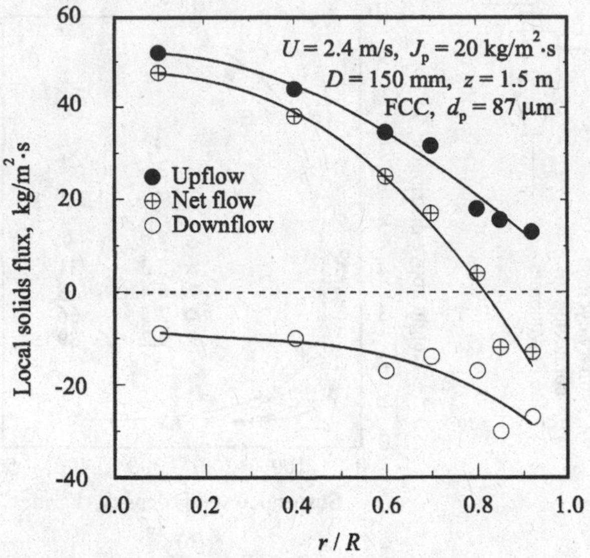

Figure 10.13. Radial profiles of local solids flux (after Herb *et al.*, 1992).

Figs. 10.12(a) and (b) along with other data can be empirically correlated by [Zhang *et al.*, 1991]

$$\alpha = \overline{\alpha}^{(0.191+\phi^{2.5}+3\phi^{11})}; \qquad \phi = \frac{r}{R} \tag{10.19}$$

At the center of the riser, Eq. (10.19) gives rise to

$$\alpha_c = \overline{\alpha}^{0.191} \tag{10.20}$$

This correlation has been verified for a wide range of operating conditions and Group A and Group B particles. Measurements of radial solids concentration profiles in a large-scale CFB combustor also confirm the validity of this correlation [Werther, 1993].

The net solids flux at a given location in a riser can be obtained on the basis of the net effect of the measured upward and downward local solids flux [Rhodes *et al.*, 1992; Herb *et al.*, 1992]. Figure 10.13 shows the typical results of the solids flux measured in a riser using FCC particles with positive values indicating the upflow. Both the upflow and downflow are observed through the entire cross section. However, the downflow dominates in the wall region. The point where the net solids flux is zero defines the interface between the wall and core regions. The thickness of the wall region mainly depends on the solids concentration in the suspension. The axial profile of the solids concentration reflects that the thickness of the wall region decreases with the bed height. The thickness of the wall region also depends on the riser diameter [Werther, 1993]. It is noted that the wall region occupies, in the case of Fig. 10.13, about 31 percent of the total cross-sectional area.

10.4.3 *Overall Solids Holdup*

The overall solids holdup is defined as the particle volume fraction over the entire riser. Figure 10.14 shows the typical effect of the solids circulation rate on the overall

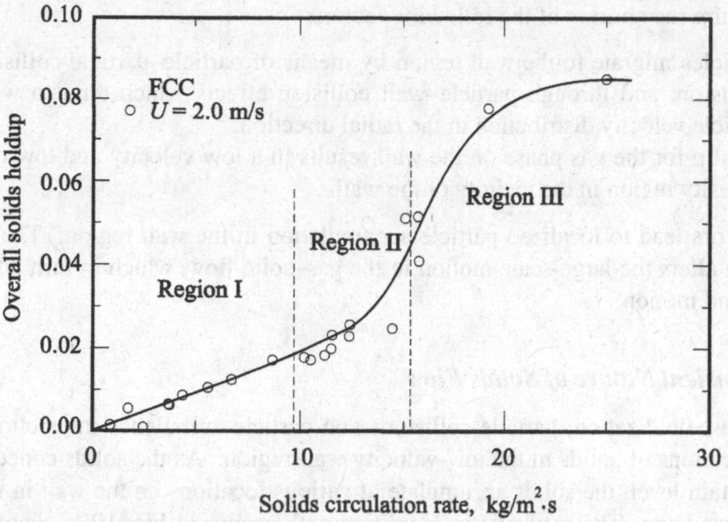

Figure 10.14. Overall solids holdup as a function of solids circulation rate for Group A particles (after Jiang *et al.*, 1993).

solids holdup for FCC particles. In the figure, three distinct regions can be identified on the basis of riser flow behavior. In Region I, there is a nearly linear relationship between the solids circulation rate and the overall solids holdup. The implication of this relationship is that the average particle velocity essentially remains constant with the change in the solids circulation rate for a given gas velocity. Region I represents the dilute transport regime. In Region II, the high solids circulation rate leads to the formation of a particle wall layer and particle refluxing, resulting in a steeper increase in the solids holdup as J_p increases. This region is narrow at low gas velocities and becomes relatively wider at higher gas velocities. The flow in this region is characterized by core-annular flow. Region III represents the fast fluidization regime. The flow at the bottom of the riser is in a turbulencelike state. This flow region is characterized by a decrease in the slope of the solids holdup versus J_p relationship.

The prediction of overall solids holdups can be obtained by integrating the axial profile of the solids concentration over the axial length. In terms of Eq. (10.12), the overall solids holdup can be expressed as

$$\frac{\alpha_o - \alpha_a}{\alpha^* - \alpha_a} = \frac{Z_0}{Z}\ln\left(\frac{1 + \exp[(Z - z_i)/Z_0]}{1 + \exp(-z_i/Z_0)}\right) \tag{10.21}$$

The equation is applicable for $\alpha_a = 0.85$–0.93. For comparison, the typical solids concentration ranges for the bubbling and turbulent regimes are, respectively, 0.4–0.55 and 0.22–0.4 [Kunii and Levenspiel, 1990].

10.5 Local Solids Flow Structure

The preceding sections are concerned with the time-averaged flow behavior in a macroscale. The time-variant flow behavior is complex. Analyses of the instantaneous flow structure require recognition of the following factors:

(1) Particles migrate to the wall region by means of particle–particle collisions and diffusion, and through particle–wall collision effects which tend to widen the particle velocity distribution in the radial direction.
(2) No-slip for the gas phase on the wall results in a low velocity and low turbulent intensity region in the vicinity of the wall.

These factors lead to localized particle accumulation in the wall region. The particle accumulation alters the large-scale motion in the gas–solid flow, which in turn affects the cluster size and motion.

10.5.1 *Transient Nature of Solids Flow*

In fast fluidization, particle collisions and particle–turbulence interactions yield high concentrations of solids in the low-velocity wall region. As the solids concentration reaches a certain level, the solids accumulate at various locations on the wall in the form of a thin, dense layer. The solids layer, identified with arrows in Fig. 10.15, is of a wavy shape. As the wavy solids layer builds up, more particles are moving into the wall region. The wavy solids layer moves downward as a result of the net effect of gravitational and drag forces imposed by the flow stream in the core region. When the layer becomes highly wavy and the frontal face (leading edge of the layer) becomes bluff, particles in the wavy layer are swept away from the wall region in much the same way as turbulent bursts occurring in the boundary layer [Praturi and Brodkey, 1978]. The bursting process is quite sudden compared to the growing process, and considerable amounts of solids are entrained from the wall region to the core region during this process. A complete cycle of the evolution

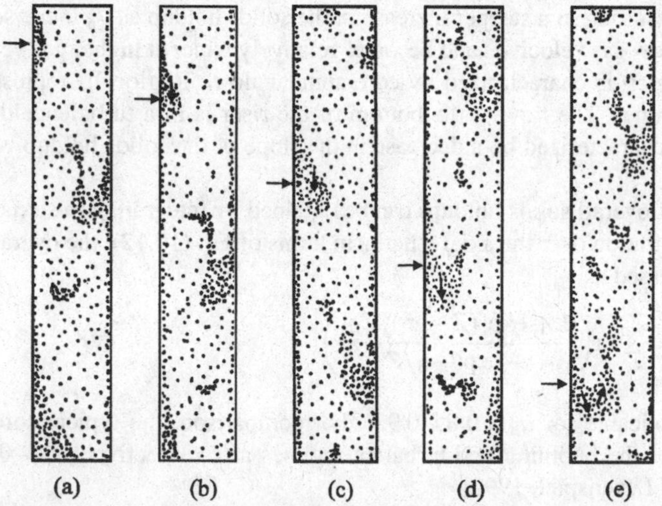

Figure 10.15. Evolution process of wavelike solids flow structure inferred from the solids concentration measurements and visual observations (from Jiang *et al.*, 1993): (a) Particle interaction yielding a thin, dense, wavy solids layer in the wall region; (b) Solids accumulation and descent of enlarged solids layers; (c) Shaping as a bluff solids layer; (d) Solids layer bursting; (e) Solids sweeping to the core region.

of such a wavy solids layer is illustrated in Fig. 10.15. The measurement of local voidage suggests that the solids wavy layer is fragmented into relatively small clusters during the ejecting process as a result of large eddies. The bursting causes a surge in the local solids concentration, which in turn enhances the local turbulence. Thus, the bursting phenomenon evolves into a large-scale structure characteristic of the core region, while the flow in the core region affects the frequency of bursting. The existence of the transient wavy solids layer and the event of bursting can be detected from the measurements of the instantaneous local solids concentration variation [Jiang *et al.*, 1993], visualization, or computation [Gidaspow *et al.*, 1989]. The preceding represents one cluster formation mechanism in the top dilute region of the riser. As research continues in this area, more will be unraveled on the general mechanisms of wavy flow and clustering phenomena.

It is noted that a cluster as referred to here is a lump of solid particles over which flow properties such as voidage do not vary substantially. It is formed mainly as a result of hydrodynamic effects. The mechanism of particle clustering is different from that of agglomeration, in which particles adhere to one another mainly by surface attraction (*e.g.*, van der Waals force and electrostatic forces), and mechanical or chemical interaction [Horio and Clift, 1992].

10.5.2 *Characterization of Intermittent Solids Flow*

As noted, a high degree of voidage fluctuation is associated with the bursting of wavy solids layer into clusters. The intermittency of the bursting and the cluster velocity are elaborated in the following.

10.5.2.1 *Intermittency Index*

The intermittent nature of the local solids flow behavior in a riser can be quantified on the basis of the intermittency index defined as [Brereton and Grace, 1993a]

$$\gamma = \frac{\sigma}{\sigma_s} \tag{10.22}$$

where σ is the standard deviation of the voidage fluctuations at a given point in the riser and σ_s is the standard deviation of the voidage fluctuations for an ideal cluster flow (see Fig. 10.16(a)) with an identical time-averaged voidage at the same location. The ideal cluster flow refers to a fluidization state in which the flow field comprises only clusters and voids. In the ideal cluster flow, the local voidages are either 1 or α_{mf}, yielding [Brereton and Grace, 1993a] the standard deviation of the voidage variation for this flow as

$$\sigma_s = \sqrt{(1-\alpha)(\alpha - \alpha_{mf})} \tag{10.23}$$

where α is the time-averaged local voidage. For fast fluidization in a riser, the values for γ vary between 0 and 1. For an ideal cluster flow, the value for γ is 1 (see Fig. 10.16(a)). Figure 10.16 shows two other ideal gas–solid flow structures, *i.e.*, core-annular flow (Fig. 10.16(b)) and uniform gas–solid flow (Fig. 10.16(c)). It is noted that at a given location there is no time variation in the voidage in either one of the flows and, thus, the value of the intermittency index for these ideal flows is zero.

The general trend of the intermittency index obtained experimentally for, as an example, $U = 6.5$ m/s and $J_p = 42$ kg/m$^2 \cdot$ s (case (a)) and 62 kg/m$^2 \cdot$ s (case (b)), is given in Fig. 10.17. The results indicate that the intermittency index varies between 0.1 to 0.7. These values imply that the gas–solid flow in the riser is never solely in the ideal

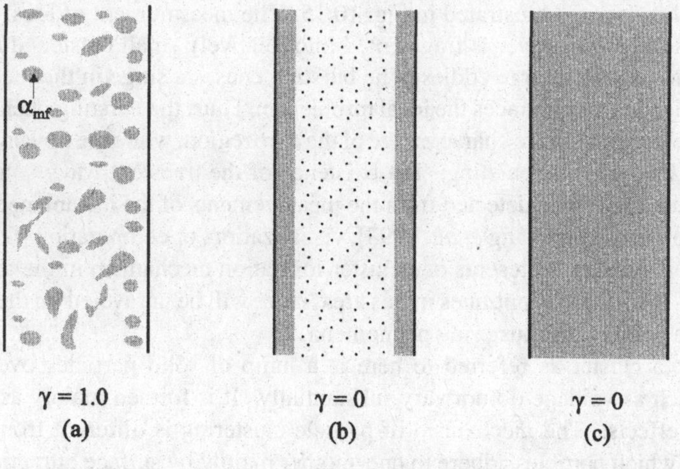

Figure 10.16. Illustration of the definition of intermittency index for three limiting cases (after Brereton and Grace, 1993a): (a) Ideal cluster flow; (b) Core-annular flow; (c) Uniform dispersed flow.

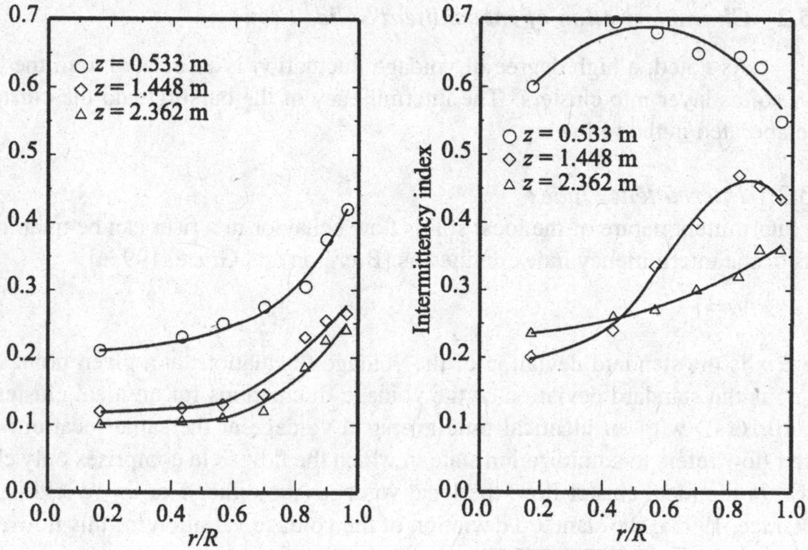

Figure 10.17. Intermittency index versus radial position at three axial positions (from Brereton and Grace, 1993): (a) $U = 6.5$ m/s, $J_p = 42$ kg/m$^2 \cdot$ s; (b) $U = 6.5$ m/s, $J_p = 62$ kg/m$^2 \cdot$ s.

core-annular nor in ideal clustering structure [Brereton and Grace, 1993a]. In the core-annular region of the real riser flow, the increase in the intermittency index along the radial direction reflects a variation in the solids flow pattern from a relatively uniform flow in the core region to a transient wavy flow in the annular region. When the dense region is well established at the bottom of the riser, the clusterlike flow structure dominates across the entire cross section, yielding a large value of the intermittency index with a small variation along the radial direction (see case (b) in the figure with $z = 0.533$ m). These results indicate that for high solids concentrations, the clusterlike flow dominates the core-annular flow. The core-annular flow pattern prevails for low solids concentrations, especially in the top dilute region of the riser.

10.5.2.2 *Descending Velocity and Characteristic Length of Clusters*

As a result of the turbulent flow nature, there are significant variations in size and shape of the wavy solids layers and of the ejecting locations of the wavy solids layer. Thus, the velocities measured for descending clusters vary appreciably for a given gas velocity and solids circulation rate. The experimental results obtained by Schnitzlein and Weinstein (1988) and Jiang *et al.* (1993) reveal a slight decrease of the descending velocity with the gas velocity. The descending velocity is about 1.8 m/s over a range of gas velocities from 1.5 to 5.0 m/s and is not sensitive to the solids circulation rate. Bader *et al.* (1988) and Horio *et al.* (1988) have also estimated the falling velocity to be in the range of 0.5 to 1.8 m/s for FCC particles. For sand particles, Wu *et al.* (1991), who recorded the solids flow by using high-speed cinematography at the wall, reported that there were widely distributed velocities with an average of 1.26 m/s and that the velocity was independent of the solids concentration.

The coverage time for each wavy solid layer multiplied by the descending velocity gives the characteristic length of the wavy solids layer. Jiang *et al.* (1993) reported that the most probable length observed is around 20 cm.

10.6 Mathematical Models of Fast Fluidization

On the basis of the observations in the macroscale, the flow of a fast fluidized bed can be represented by the core-annulus flow structure in the radial direction, and coexistence of a bottom dense region and a top dilute region in the axial direction. Particle clusters are an indication of the heterogeneity in the mesoscale. A complete characterization of the hydrodynamics of a CFB requires the determination of the voidage and velocity profiles. There are a number of mathematical models accounting for the macro- or mesoaspects of the flow pattern in a CFB that are available. In the following, basic features of several types of models are discussed.

10.6.1 *Models Based on the Concept of Cluster*

The cluster concept evolved as a result of the recognition of a large slip velocity between the gas and solid particles in the circulating fluidized bed [Yerushalmi *et al.*, 1978; Yerushalmi and Cankurt, 1979]. Li and Kwauk (1980) and Li *et al.* (1988) proposed that the flow field in the fast fluidized beds consists of a dilute suspension phase and a cluster phase. Considering the size of cluster in fast fluidization to be larger than the Kolmogorov dissipative scale, Li *et al.* (1988) assumed that the cluster diameter is inversely proportional to the total energy input into the system and developed a model to estimate the cluster diameter based on the assumption that the steady-state condition follows the energy minimization principle. Under typical operating conditions of a fast fluidized bed, the results calculated from these models showed that the cluster diameter is in the range of 20 to 80 mm. A wide range of cluster sizes (as large as 80 mm) was also experimentally observed by Horio and Kuroki (1994) using particle imaging techniques and Soong *et al.* (1995) using optical probe techniques.

Consideration of particle clustering effects is of relevance to the modeling of the hydrodynamic behavior in the riser. Thus, when the cluster properties are known for a given riser operation, models based on the concept of clusters can be applied.

10.6.2 *Models Based on the Core-Annular Flow Structure*

Various models based on the core-annular flow structure have been proposed. Bolton and Davidson (1988) extended Rhodes's entrainment model [Rhodes and Geldart, 1987] by taking into account the falling particle film adjacent to the wall. Horio *et al.* (1988) demonstrated that the annular flow structure cannot be rigorously explained without taking the cluster effects into consideration. Thus, they incorporated the cluster concept into the annular model originally developed by Nakamura and Capes (1973) for pneumatic transport systems. In the model of Horio *et al.* (1988), the equation of Richardson and Zaki (1954) was applied to estimate the effective cluster size in the wall or annular region and core region. To close the model, they used the minimum pressure drop principle. Senior and Brereton (1992) proposed a core-annular model based on the observed physical conditions for the annular region. The mechanisms and relationship for interchange of solid particles between the annular and the core regions were given.

The consideration of the core-annular structure is of significance to the modeling of the hydrodynamic behavior in the top dilute region of the riser. Thus, when the dilute region is predominantly present in the fast fluidized bed, models based on the core-annular structure can be applied. The core-annular model is a one- (axial)-dimensional model. The mass and momentum equations for the annular and the core regions can be obtained by using a radial-property-averaging procedure such as that given by Delhaye *et al.* (1981). The resulting equations are, however, complex. Simplified models have been proposed. The model by Bolton and Davidson (1988) is simple to apply and is feasible in its mechanistic account of the large gas–solid slip velocity in the system; this model is, therefore, introduced here.

Figure 10.18 shows the conceptual configuration of Bolton and Davidson's model. In the model, the flow structure is idealized as a dilute core region surrounded by a dense particle film which falls adjacent to the wall. The mathematical model for a circular riser can be based on the following assumptions.

(1) The flow is a one-dimensional, steady flow.
(2) The particle film region is of annular cross section with an outer diameter of D and an inner diameter of D_c, and the core region is of circular cross section with a diameter D_c.
(3) The core region consists of a dilute upward-flowing suspension of solids with a velocity of u_{pc} and a volume fraction of α_{pc}.
(4) The annular region consists of a dense downward-flowing suspension of solids with a velocity of u_{pw} and a volume fraction of α_{pw}.
(5) All gas passes through the core at a velocity of u_{fc}.
(6) There are no radial variations in terms of velocity and volume fraction across either the annular region or the core region.
(7) Particles are transported by turbulent diffusion from the core region to the surface of the falling film, where they are trapped and are carried downward. The implication of this assumption is that the net upward particle flow will decrease along the riser height in the core.

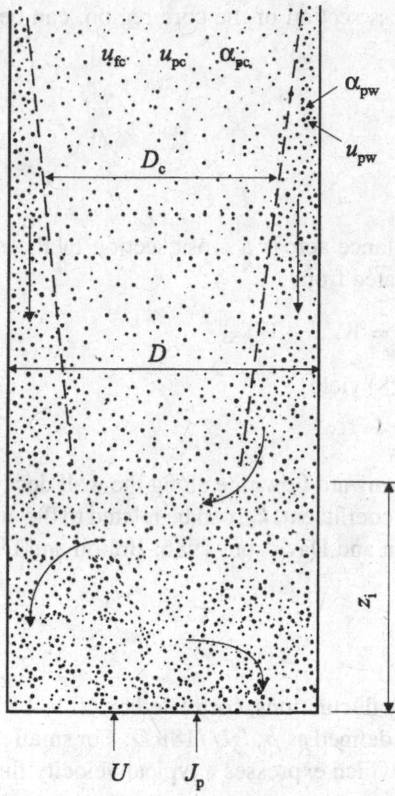

Figure 10.18. Conceptual configuration of the core-annular model.

Considering an element of dz in the riser, the mass balance of the solids in the core region gives

$$\frac{1}{4}Du_{pc}\frac{d\alpha_{pc}}{dz} = -k_d(\alpha_{pc} - \alpha_{pc\infty}) \tag{10.24}$$

where k_d is the deposition coefficient accounting for the turbulent diffusion of particles in the core region to the wall region; $\alpha_{pc\infty}$ is an equilibrium solids concentration in the core region where deposition and entrainment occur at the same rate. Note that Eq. (10.24) is developed by assuming that D_c can be approximated by D. The changes in the particle velocity in the core region are likely to be small. Therefore, it is reasonable to assume that u_{pc} is independent of the bed height. Also, assuming that particles do not interfere with each other significantly, the upward solids velocity in the core region can be taken as

$$u_{pc} = u_{fc} - U_{pt} \tag{10.25}$$

Equation (10.24) can be solved with respect to α_{pc} to yield

$$\alpha_{pc} = \alpha_{pc\infty} + (\alpha_{pc0} - \alpha_{pc\infty})e^{-K_d z} \tag{10.26}$$

where $K_d = 4k_d/Du_{pc}$. α_{pc0} is the solids concentration in the core region at $z = 0$, and $\alpha_{pc\infty}$ represents the asymptotic concentration at large z. The upward flow of solids, W_e, defined as the total flow rate through the cross section of the core region, can be expressed as

$$W_e = \frac{\pi}{4} D_c^2 u_{pc} \alpha_{pc} \tag{10.27}$$

From Eqs. (10.26) and (10.27), it yields

$$W_e = W_{e\infty} + (W_{e0} - W_{e\infty}) e^{-K_d z} \tag{10.28}$$

On the other hand, in terms of the mass balance across a cross section of the riser, the upward solids flow rate W_e can also be evaluated from

$$\frac{\pi}{4} D^2 J_p = W_e - W_w = W_{e0} - W_{w0} = W_{e\infty} - W_{w\infty} \tag{10.29}$$

Substituting W_w in Eq. (10.29) into Eq. (10.28) yields

$$W_w = W_{w\infty} + (W_{w0} - W_{w\infty}) e^{-K_d z} \tag{10.30}$$

Equation (10.30) implies that the particle downward flow rate along the wall decays exponentially with the bed height. The deposition coefficient, k_d, can be related to the amplitude of the gas velocity fluctuations as [Pemberton and Davidson, 1986; Bolton and Davidson, 1988] (Problem 10.8)

$$k_d = \frac{0.1\sqrt{\pi} u'}{1 + \mathrm{St}/12} \tag{10.31}$$

where u' is the amplitude of turbulent velocity fluctuations, correlated by $U(1\text{–}2.8\,\mathrm{Re}^{-1/8})$, and $\mathrm{Re} = UD\rho_p/\mu$. St is the Stokes number, defined as $\rho_p d_p^2 U/18\mu D$. For small particles, the Stokes number is small, and $u' = 0.1U$, which expresses a typical velocity fluctuation for a pipe flow. These simplifications give

$$k_d = 0.01\sqrt{\pi} U \tag{10.32}$$

Thus, K_d can be given as

$$K_d = \frac{0.04\sqrt{\pi}}{D} \frac{U}{u_{pc}} \tag{10.33}$$

Once W_e and W_w are known, the cross-sectional averaged solids holdup, α_p, can be calculated by using the following equation

$$\alpha_p = \frac{W_e}{\rho_p A u_{pc}} + \frac{W_w}{\rho_p A u_{pw}} \tag{10.34}$$

where u_{pw} is the falling velocity of the film and can be estimated on the basis of the information given in §10.5.2.2.

Example 10.3 A riser is of 0.15 m in diameter and 8 m in height. Particles with a mean diameter of 200 μm and a density of 384 kg/m^3 are used in the riser, which operates at U = 2.21 m/s and J_p = 3.45 kg/m$^2\cdot$s. The gas used is air. For this operating condition, Davidson (1991) reported a particle downward velocity, u_{pw}, of 0.5 m/s and a particle downward flow rate, W_w, of 0.2 kg/s in the annular region. Assume that the solids volume fraction in the central core region, α_{pc}, is 0.015. Calculate the cross-sectionally averaged solids holdup and the decay constant, K_d, defined in Eq. (10.33) in terms of the core-annular model.

Solution Assuming that the slip velocity between particles and air in the core region is equal to the particle terminal velocity and D_c can be approximated by D, the particle velocity in the core region can be determined from Eq. (10.25), *i.e.*,

$$u_{pc} = u_{fc} - U_{pt} = \frac{U}{1 - \alpha_{pc}} - U_{pt} = \frac{2.21}{1 - 0.015} - 0.331 = 1.91 \text{ m/s} \qquad (E10.18)$$

where U_{pt} is calculated from Eq. (1.7). W_e can be calculated from Eq. (10.29), which is rearranged to yield

$$W_e = \frac{\pi}{4} D^2 J_p + W_w = \frac{\pi}{4} 0.15^2 \times 3.45 + 0.2 = 0.261 \text{ kg/s} \qquad (E10.19)$$

With the expression of u_{pc} in Eq. (E10.18) and the value of W_e in Eq. (E10.19) for the operating condition given, Eq. (10.34) becomes

$$\alpha_p = \frac{1}{\rho_p A}\left(\frac{W_e}{u_{pc}} + \frac{W_w}{u_{pw}}\right) = \frac{1}{384 \times 0.0176}\left(\frac{0.261}{u_{pc}} + \frac{0.2}{0.5}\right)$$
$$= \frac{3.86 \times 10^{-2}}{u_{pc}} + 5.92 \times 10^{-2} = 0.079 \qquad (E10.20)$$

The decay constant, K_d, can be calculated from Eq. (10.33) as

$$K_d = \frac{0.04\sqrt{\pi}}{D}\frac{U}{u_{pc}} = \frac{0.04\sqrt{\pi}}{0.15}\frac{2.21}{1.91} = 0.55 \text{ m}^{-1} \qquad (E10.21)$$

10.6.3 *Models Based on the Axial Profiles of Solids Holdup*

These types of models extend the approach of the conventional fluidized bed with a freeboard of varying solids concentrations [Kunii and Levenspiel, 1990] to the circulating fluidized bed. The two-phase theory used for the bubbling bed is adopted to account for the bottom dense region, while the entrainment model proposed by Wen and Chen (1982) is extended to the top dilute region of the fast fluidized bed. Some descriptions of the model, considering the coexistence of the bottom dense region and top dilute region, are given in §10.4.1. In the absence of the radial variation information, this type of model can provide a reasonable approximation of the gross hydrodynamic behavior.

10.6.4 *Two-Phase Flow Models and Computational Fluid Dynamics*

One-dimensional flow models are adopted in the early stages of model development for predicting the solids holdup and pressure drop in the riser. These models consider the steady flow of a uniform suspension. Four differential equations, including the gas continuity equation, solids phase continuity equation, gas–solid mixture momentum equation, and solids phase momentum equation, are used to describe the flow dynamics. The formulation of the solids phase momentum equation varies with the models employed [*e.g.*, Arastoopour and Gidaspow, 1979; Gidaspow, 1994]. The one-dimensional model does not simulate the prevailing characteristics of radial nonhomogeneity in the riser. Thus, two- or three-dimensional models are required.

For the motion of a gas–solid suspension in the riser, both the gas and particle velocities have local averaged and random components. Thus, it is desirable to develop a mechanistic model which incorporates a variety of interactive effects due to both the gas and particle velocity components (see Chapter 5) as given in the following [Sinclair and Jackson, 1989]:

(1) Interactions between particles and gas resulting from the mean slip velocity that yield the drag force driving the nonrandom part of the particle motion
(2) Interactions of the particles with the fluctuating component of the gas velocity, which lead to particle turbulent diffusion and induce an exchange of the kinetic energy between the fluctuating components of the velocity of the two phases, which results in either damping the fluctuations of the gas velocity and enhancing fluctuations in particle velocity, or vice versa
(3) Interactions of the fluctuating part of the particle motion with the mean particle motion through interparticle collisions, which generate pressure and stresses in the particle assembly, consequently yielding apparent viscosity of the particle phase
(4) Interactions between the turbulent fluctuations of the gas velocity and the mean motion of the gas, which yield the gas Reynolds stresses

The two-fluid models considering these effects for the gas–solid flow in a riser can be described by Eqs. (5.168) through (5.170). In these models, the gas and solid are treated as two interpenetrating fluids (see §5.3).

For equation closure, constitutive equations for the solid phase stresses are needed. From the physical point of view, the solid stresses are a by-product of the microscale mechanisms of the momentum transfer between the particles. The random motion of particles generates an effective pressure in the solid phase, together with an effective viscosity which resists shearing of the particle assembly. The kinetic energy of this random motion can be analogous to that of the thermal motion of molecules in a gas, and correspondingly it can be characterized by a granular temperature defined as proportional to the mean square of the random component of the particle velocity. Both the effective pressure and the viscosity are functions of the granular temperature. Thus, it is required to have separate differential equations representing a balance for the pseudothermal energy; that is, the energy of random particle motion (see §5.5).

To simulate the granular flow while adopting the gas kinetic analogy, Sinclair and Jackson (1989) pointed out that lateral segregation is induced by stresses which are caused by interactions between the random motion and mean motion of the particle in the particle assembly. This consideration led to the development of a model for fully developed vertical flows by taking into account the particle phase interactions. Louge *et al.* (1991) considered the effect of gas phase turbulence in the modeling; however, it appears that interactions among particles are the key to lateral solids segregation. Pita and Sundaresan (1993) have proposed a two-fluid model for developing flow, and their computation revealed profound effects of the entry configuration on the internal recirculation and lateral segregation. Tsuo and Gidaspow (1990) have presented a two-dimensional unsteady-state model. Their simulation characterizes the transient cluster motion for the dilute flow and the annular flow structure for the dense flow and indicates that the transient integration does not converge toward a steady state over a long period, reflecting the dynamic behavior of the flow. Furthermore, their model indicates that the two-fluid model with the kinetic energy balance of the particle velocity fluctuations can be used to account for the general behavior of the riser flow. An improved model, however, is needed to describe the detailed flow structure in the core and annular regions and the dynamics of the cluster formation.

Nomenclature

A_o Opening area for the mechanical valve or L-valve

A_l Cross-sectional area of the L-valve

A_r Cross-sectional area of the riser

A_{sp} Cross-sectional area of the downcomer

Ar Archimedes number

a Decay constant, defined by Eq. (E10.14)

b Coefficient, defined by Eq. (P10.8)

C_o Valve aperture discharge coefficient

D Riser diameter

D_c Diameter of the central core region in the core-annular flow model

D_g Gas turbulent diffusion coefficient

D_h Diameter of the horizontal section of the L-valve

D_p Particle turbulent diffusion coefficient

D_{sp} Diameter of the downcomer

D_v Diameter of the solids flow control valve

d_p Particle diameter

F_f Friction force

g Gravitational acceleration

J_p Solids circulation rate or solids flux

$J_{p,\,min}$ Minimum solids circulating rate for transition from the fast fluidization regime to the dilute transport regime

$J_{p,\,max}$ Maximum solids circulating rate for transition from the turbulent regime or choking to the fast fluidization regime

J_{pr} Solids flux at radial position r

$J_{p,\,tr}$ Solids circulation rate corresponding to the condition of $U = U_{tr}$

k Coefficient, defined by Eq. (10.5)

k_d Deposition coefficient, defined by Eq. (10.24)

K_d Parameter, defined as $4k_d/Du_{pc}$

L_h Horizontal length of the L-valve

L_{sp} Solids inventory height in the downcomer or standpipe

L_z Length of the vertical section of the L-valve

M Solids inventory in the CFB loop

m Coefficient, defined by Eq. (P10.8)

p_a Pressure at the L-valve inlet

p_o Pressure at the riser inlet

Q_{ent} Gas entrainment rate in the solids flow control valve

Q_{ext} External gas aeration rate to the solids flow control valve

Q_t Total gas aeration rate to the solids flow control valve

R Radius of the riser

Re_t Particle Reynolds number, based on particle diameter and particle terminal velocity

Re_{tr} Particle Reynolds number, based on particle diameter and transport velocity

r Radial position

St Stokes number

U Superficial gas velocity
U_{cy} Gas velocity at the inlet of a cyclone
U_{fd} Upper bound of the gas velocity for the fast fluidization regime
U_{mf} Minimum fluidization velocity
U_{pt} Particle terminal velocity
U_{tf} Lower bound of the gas velocity for the fast fluidization regime
U_{tr} Transport velocity
u Linear gas velocity
u' Gas velocity fluctuation
u_{fc} Gas velocity in the core region of the core-annular model
u_p Particle velocity
u_{pc} Particle velocity in the core region of the core-annular model
u_{pw} Particle velocity in the wall region of the core-annular model
W_e Solids entrainment rate in the core region
W_{e0} Solids entrainment rate in the core region at $z = 0$
$W_{e\infty}$ Asymptotic limit of the solids entrainment rate in the core region
W_w Solids flow rate in the wall region
W_{w0} Solids flow rate in the wall region at $z = 0$
W_{wp} Solids feeding rate
$W_{w\infty}$ Asymptotic limit of the solids flow rate in the wall region
z Distance above the distributor
z_i Location of inflection point for fast fluidization
Z Riser height
Z_0 Characteristic length, defined by Eq. (10.12)

Greek Symbols

α Local voidage
$\overline{\alpha}$ Cross-sectional averaged voidage
α^* Asymptotic voidage in the upper dilute region
α_a Asymptotic voidage in the lower dense region
α_c Local voidage at the center of a riser
α_e Voidage at the riser exit
α_{mf} Bed voidage at minimum fluidization
α_{lh} Voidage in the horizontal section of the L-valve
α_{lz} Voidage in the vertical section of the L-valve
α_o Overall voidage
α_p Cross-sectional averaged solids holdup
α_{pc} Solids volume fraction in the central core region
α_{pc0} Asymptotic solids volume fraction in the central core region at $z = 0$
$\alpha_{pc\infty}$ Asymptotic solids volume fraction in the central core region, defined by Eq. (10.24)
α_{pw} Volume fraction of the solid phase in the wall region
γ Intermittency index, defined by Eq. (10.22)
$\Delta p/\Delta z$ Local pressure drop
Δp_{cy} Pressure drop across a cyclone
Δp_{cv}. Pressure drop across a solids flow control valve
Δp_h Pressure drop between the riser exit and the inlet of a cyclone
Δp_{lb} Pressure drop across the

	horizontal section of the L-valve	μ	Dynamic viscosity of gas
Δp_{lv}	Pressure drop across the elbow of the L-valve	ρ	Density of fluid
Δp_r	Pressure drop across a riser	ρ_p	Density of particles
Δp_{sp}	Pressure drop across a downcomer or standpipe	σ	Standard deviation of solids concentration fluctuations at a given point
δ	Angle of internal friction	σ_s	Standard deviation of solids concentration fluctuations for fully segregated two-phase flow with identical time-averaged solids concentration at the same point
δp_b	Pressure perturbation in the blower	ν	Kinematic viscosity of gas
δp_r	Pressure perturbation in the riser		
δQ	Perturbation in the gas flow rate		

References

Arastoopour, H. and Gidaspow, D. (1979). Vertical Pneumatic Conveying Using Four Hydrodynamic Models. *I & EC Fund.*, **18**, 123.

Avidan, A. A. and Yerushalmi, J. (1982). Bed Expansion in High Velocity Fluidization. *Powder Tech.*, **32**, 223.

Bader, R., Findlay, J. and Knowlton, T. M. (1988). Gas–Solids Flow Pattern in a 30.5cm Diameter Circulating Fluidized Bed. In *Circulating Fluidized Bed Technology II*. Ed. Basu and Large. Toronto: Pergamon Press.

Bai, D., Jin, Y. and Yu, Z. (1993). Flow Regimes in Circulating Fluidized Beds. *Chem. Eng. Technol.*, **16**, 307.

Basu, P. and Fraser, S. A. (1991). *Circulating Fluidized Bed Boilers: Design and Operations*. Boston: Butterworths.

Bi, H. T. and Fan, L.-S. (1991). Regime Transition in Gas–Solid Circulating Fluidized Beds. *1991 AIChE Annual Meeting*, Los Angeles, Calif., Nov. 17–22.

Bi, H. T. and Fan, L.-S. (1992). Existence of Turbulent Regime in Gas–Solid Fluidization. *AIChE J.*, **38**, 297.

Bi, H. T., Grace, J. R. and Zhu, J.-X. (1993). Types of Choking in Vertical Pneumatic Systems. *Int. J. Multiphase Flow*, **19**, 1077.

Bi, H. T. and Zhu, J. (1993). Static Instability Analysis of Circulating Fluidized Beds and Concept of High-Density Risers. *AIChE J.*, **39**, 1272.

Bolton, L. W. and Davidson, J. F. (1988). Recirculation of Particles in Fast Fluidized Risers. In *Circulating Fluidized Bed Technology II*. Ed. Basu and Large. Toronto: Pergamon Press.

Brereton, C. and Grace, J. R. (1993a). Microstructural Aspects of the Behavior of Circulating Fluidized Bed. *Chem. Eng. Sci.*, **48**, 2565.

Brereton, C. and Grace, J. R. (1993b). End Effects in Circulating Fluidized Bed Hydrodynamics. In *Circulating Fluidized Bed Technology IV*. Ed. A. A. Avidan. New York: AIChE Publications.

Contractor, R. M., Patience, G. S., Garnett, D. I., Horowitz, H. S., Sisler, G. M. and Bergna, H. E. (1993). A New Process for n-Butane Oxidation to Maleic Anhydride Using a Circulating Fluidized Bed Reactor. In *Circulating Fluidized Bed Technology IV*. Ed. A. A. Avidan. New York: AIChE Publications.

Davidson, J. F. (1991). The Two-Phase Theory of Fluidization: Successes and Opportunities. *AIChE Symp. Ser.*, **87** (281), 1.

Davidson, J. F. and Harrison, D. (1963). *Fluidized Particles*. Cambridge: Cambridge University Press.

Delhaye, J. M., Giot, M. and Riethmuller, M. L. (1981). *Thermohydraulics of Two-Phase System for Industrial Design and Nuclear Engineering*. Washington, D.C.: Hemisphere; New York: McGraw-Hill.

Doig, I. D. and Roper, G. H. (1963). The Minimum Gas Rate for Dilute-Phase Solids Transportation in Gas Stream. *Australian Chem. Eng.*, **1**, 9.

Fan, L.-S., Toda, M. and Satija, S. (1985). Apparent Drag Reduction Phenomenon in the Defluidized Packed Dense Bed of the Multisolid Pneumatic Transport Bed. *Chem. Eng. Sci.*, **40**, 809.

Gidaspow, D. (1994). *Multiphase Flow and Fluidization: Continuum and Kinetic Theory Descriptions*. San Diego, Calif.: Academic Press.

Gidaspow, D., Tsuo, Y. P. and Luo, K. M. (1989). Computed and Experimental Cluster Formation and Velocity Profiles in Circulating Fluidized Beds. In *Fluidization VI*. Ed. Grace, Shemilt and Bergougnou. New York: Engineering Foundation.

Grace, J. R. (1986). Contacting Modes and Behavior Classification of Gas–Solid and Other Two-Phase Suspensions. *Can. J. Chem. Eng.*, **64**, 353.

Hartge, E.-U., Li, Y. and Werther, J. (1986). Analysis of the Local Structure of the Two Phase Flow in a Fast Fluidized Bed. In *Circulating Fluidized Bed Technology*. Ed. P. Basu. Toronto: Pergamon Press.

Herb, B., Dou, S., Tuzla, K. and Chen, J. C. (1992). Solid Mass Fluxes in Circulating Fluidized Beds. *Powder Tech.*, **70**, 197.

Hirama, T., Takeuchi, H. and Chiba, T. (1992). On the Definition of Fast Fluidization in a Circulating Fluidized Bed Riser. In *Fluidization VII*. Ed. Potter and Nicklin. New York: Engineering Foundation.

Horio, M. and Clift, R. (1992). A Note on Terminology: Clusters and Agglomerates. *Powder Tech.*, **70**, 196.

Horio, M. and Kuroki, H. (1994). Three-Dimensional Flow Visualization of Dilutely Dispersed Solids in Bubbling and Circulating Fluidized Beds. *Chem. Eng. Sci.*, **49**, 2413.

Horio, M., Morishita, K., Tachibana, O. and Murata, N. (1988). Solid Distribution and Movement in Circulating Fluidized Beds. In *Circulating Fluidized Bed Technology II*. Ed. Basu and Large. Toronto: Pergamon Press.

Ilias, S., Ying, S., Mathur, G. D. and Govind, R. (1988). Studies on a Swirling Circulating Fluidized Bed. In *Circulating Fluidized Bed Technology II*. Ed. Basu and Large. Toronto: Pergamon Press.

Jiang, P. J., Bi, H., Jean, R.-H. and Fan, L.-S. (1991). Baffle Effects on Performance of Catalytic Circulating Fluidized Bed Reactor. *AIChE J.*, **37**, 1392.

Jiang, P. J., Cai, P. and Fan, L.-S. (1993). Transient Flow Behavior in Fast Fluidization. In *Circulating Fluidized Bed Technology IV*. Ed. A. A. Avidan. New York: AIChE Publications.

Johnson, E. P. (1982). Better Fluid-Bed Units Ready to Make Debuts. *Chemical Engineering*, **89** (7), 39.

Jones, D. R. M. and Davidson, J. F. (1965). The Flow of Particles from a Fluidized Bed Through an Orifice. *Rheologica Acta*, **4**, 180.

King, D. (1992). Fluidized Catalytic Crackers: An Engineering Review. In *Fluidization VII*. Ed. Potter and Nicklin. New York: Engineering Foundation.

Kunii, D. and Levenspiel, O. (1990). Entrainment of Solids from Fluidized Beds. I. Holdup of Solids in the Freeboard. II. Operation of Fast Fluidized Beds. *Powder Tech.*, **61**, 193.

Kwauk, M. (1992). *Fluidization: Idealized and Bubbleless, with Applications*. Beijing: Science Press.

Leung, L. S. and Jones, P. J. (1978). Flow of Gas–Solid Mixture in Standpipes: A Review. *Powder Tech.*, **20**, 145.

Li, Y. and Kwauk, M. (1980). The Dynamics of Fast Fluidization. In *Fluidization*. Ed. Grace and Matsen. New York: Plenum.

Li, J., Li, Y. and Kwauk, M. (1988). Energy Transport and Regime Transition in Particle–Fluid Two-Phase Flow. In *Circulating Fluidized Bed Technology II*. Ed. Basu and Large. Toronto: Pergamon Press.

Lim, K. S., Zhu, J. X. and Grace, J. R. (1995). Hydrodynamics of Gas–Solid Fluidization. *Int. J. Multiphase Flow*, **21/s**, 141.

Louge, M., Mastorakos, E. and Jenkins, J. T. (1991). The Role of Particle Collisions in Pneumatic Transport. *J. Fluid Mech.*, **231**, 345.

Nakamura, K. and Capes, C. E. (1973). Vertical Pneumatic Conveying: A Theoretical Study of Uniform and Annular Particle Flow Models. *Can. J. Chem. Eng.*, **51**, 39.

Pemberton, S. T. and Davidson, J. F. (1986). Elutriation from Fluidized Beds. II. Disengagement of Particles from Gas in the Freeboard. *Chem. Eng. Sci.*, **41**, 253.

Perry, R. H., Green, D. W. and Maloney, J. O. (1984). *Perry's Chemical Engineer's Handbook*, 6th ed. New York: McGraw-Hill.

Pita, J. A. and Sundaresan, S. (1993). Developing Flow of a Gas–Particle Mixture in a Vertical Riser. *AIChE J.*, **39**, 541.

Praturi, A. and Brodkey, R. S. (1978). A Stereoscopic Visual Study of Coherent Structures in Turbulent Shear Flow. *J. Fluid Mech.*, **89**, 251.

Rhodes, M. J. and Cheng, H. (1994). Operation of an L-Valve in a Circulating Fluidized Bed of Fine Solids. In *Circulating Fluidized Bed Technology IV*. Ed. A. A. Avidan. New York: AIChE Publications.

Rhodes, M. J. and Geldart, D. (1987). A Model for the Circulating Fluidized Bed. *Powder Tech.*, **53**, 155.

Rhodes, M. J. and Laussmann, P. (1992). A Study of the Pressure Balance Around the Loop of a Circulating Fluidized Bed. *Can. J. Chem. Eng.*, **70**, 625.

Rhodes, M. J., Wang, X. S., Cheng, H., Hirama, T. and Gibbs, B. M. (1992). Similar Profiles of Solids Flux in Circulating Fluidized Bed Risers. *Chem. Eng. Sci.*, **47**, 1635.

Richardson, J. F. and Zaki, W. N. (1954). Sedimentation and Fluidization, Part I. *Trans. Instn. Chem. Engrs.*, **43**, 35.

Rudolph, V., Chong, Y. O. and Nicklin, D. J. (1991). Standpipe Modelling for Circulating Fluidized Beds. In *Circulating Fluidized Bed Technology III*. Ed. Basu, Horio and Hasatani. Oxford: Pergamon Press.

Satija, S. and Fan, L.-S. (1985). Characteristics of Slugging Regime and Transition to Turbulent Regime for Fluidized Beds of Large Coarse Particles. *AIChE J.*, **31**, 1554.

Schnitzlein, M.G. and Weinstein, H. (1988). Flow Characterization in High-Velocity Fluidized Beds Using Pressure Fluctuations. *Chem. Eng. Sci.*, **43**, 2605.

Senior, R. C. and Brereton, C. M. M. (1992). Modelling of Circulating Fluidized-Bed Solids Flow and Distribution. *Chem. Eng. Sci.*, **47**, 281.

Shingles, T. and McDonald, A. F. (1988). Commercial Experience with Synthol CFB Reactors. In *Circulating Fluidized Bed Technology II*. Ed. Basu and Large. Toronto: Pergamon Press.

Sinclair, J. L. and Jackson, R. (1989). The Effect of Particle–Particle Interaction on the Flow of Gas and Particles in a Vertical Pipe. *AIChE J.*, **35**, 1473.

Soong, C. H., Tuzla, K. and Chen, J. C. (1995). Experimental Determination of Cluster Size and Velocity in Circulating Fluidized Bed. In *Fluidization VIII* (preprints). Ed. Large and Laguerie. Tours, France: Conference Publication.

Tsuo, Y. P. and Gidaspow, D. (1990). Computation of Flow Pattern in Circulating Fluidized Beds. *AIChE J.*, **36**, 885.

Toda, M., Satija, S. and Fan, L.-S. (1983). Fundamental Characteristics of a Multisolid Pneumatic Transport Bed: Minimum Fluidization Velocity of the Dense Bed. In *Fluidization IV*. Ed. Kunii and Toei. New York: Engineering Foundation.

Weinstock, J. (1978). Analytical Approximations in the Theory of Turbulent Diffusion. *Phys. Fluids*, **21**, 887.

Wen, C. Y. and Chen, L. H. (1982). Fluidized Bed Freeboard Phenomena: Entrainment and Elutriation. *AIChE J.*, **28**, 117.

Wen, C. Y. and Galli, A. F. (1971). Dilute Phase Systems. In *Fluidization*. Ed. Davidson and Harrison. London: Academic Press.

Werther, J. (1993). Fluid Mechanics of Large-Scale CFB Units. In *Circulating Fluidized Bed Technology IV*. Ed. A. A. Avidan. New York: AIChE Publications.

Wu, R. L., Lim, C. J., Grace, J. R. and Brereton, C. M. H. (1991). Instantaneous Local Heat Transfer and Hydrodynamics in a Circulating Fluidized Bed. *Int. J. Heat & Mass Transfer*, **34**, 2019.

Wu, R. L., Grace, J. R. and Lim, C. J. (1990). A Model for Heat Transfer in Circulating Fluidized Beds. *Chem. Eng. Sci.*, **45**, 3389.

Yang, W. C. (1975). A Mathematical Definition of Choking Phenomenon and a Mathematical Model for Predicting Choking Velocity and Choking Voidage. *AIChE J.*, 21, 1013.

Yang, W. C. (1976). A Criterion for Fast Fluidization. *Proceedings of the Third International Conference on the Pneumatic Transport of Solids in Pipes (Pneumotransport 3)*, Bath, England. E5 - 49.

Yang, W. C. (1992). The Hydrodynamics of Circulating Fluidized Beds. In *The Supplements of the Encyclopedia of Fluid Mechanics*. Ed. N. P. Cheremininoff. Houston: Gulf Publishing.

Yang, W. C. and Knowlton, T. M. (1993). L-Valve Equations. *Powder Tech.*, **77**, 49.

Yerushalmi, J. and Cankurt, N. T. (1979). Further Studies of the Regimes of Fluidization. *Powder Tech.*, **24**, 187.

Yerushalmi, J., Cankurt, N. T., Geldart, D. and Liss, B. (1978). Flow Regimes in Vertical Gas–Solids Contact System. *AIChE Symp. Ser.*, **74** (176), 1.

Yuu, S., Yasukouchi, N., Hirosawa, Y. and Jotaki, T. (1978). Particle Turbulent Diffusion in a Dust Laden Round Jet. *AIChE J.*, **24**, 509.

Zenz, F. A. and Othmer, D. F. (1960). *Fluidization and Fluid–Particle Systems*. New York: Reinhold.

Zhang, W., Tung, Y. and Johnsson, F. (1991). Radial Voidage Profiles in Fast Fluidized Beds of Different Diameters. *Chem. Eng. Sci.*, **46**, 3045.

Zheng, C. G., Tung, Y., Xia, Y. S., Bin, H. and Kwauk, M. (1991). Voidage Redistribution by Ring Internals in Fast Fluidization. In *Fluidization'91 Science and Technology*. Ed. Kwauk and Hasatani. Beijing: Science Press.

Problems

10.1 Calculate the transport velocity of particles with
(1) $d_p = 80\ \mu m$ and $\rho_p = 1{,}500\ kg/m^3$ (Group A)
(2) $d_p = 800\ \mu m$ and $\rho_p = 1{,}500\ kg/m^3$ (Group B)
(3) $d_p = 4$ mm and $\rho_p = 1{,}010\ kg/m^3$ (Group D)
for a fluidizing medium with $\rho = 1.18\ kg/m^3$ and $\mu = 1.82 \times 10^{-5}\ kg/m \cdot s$

10.2 Assuming the riser flow to be characterized by one-dimensional steady flow and applying the two-fluid model to the flow, one can obtain the momentum equation for the gas–solid mixture as

$$(1-\overline{\alpha})\rho_p u_p \frac{du_p}{dz} + \overline{\alpha}\rho u \frac{du}{dz} = -[\overline{\alpha}\rho + (1-\overline{\alpha})\rho_p]g - \frac{dp}{dz} - F_f \quad \text{(P10.1)}$$

where F_f is the friction force. Use assumptions that the variation of gas momentum and the friction can be neglected to prove the following equation and discuss the effects of the axial variations of voidage on the pressure drop:

$$\frac{dp}{dz} = \left(\frac{J_p}{1-\overline{\alpha}}\right)^2 \frac{1}{\rho_p}\frac{d(1-\overline{\alpha})}{dz} - [\overline{\alpha}\rho + (1-\overline{\alpha})\rho_p]g \quad \text{(P10.2)}$$

10.3 In the core-annular flow model, the mass balance of the solid phase can be expressed as

$$\frac{J_p}{\rho_p} = \eta^2 \alpha_{pc} u_{pc} + (1-\eta^2) u_{pw} \alpha_{pw}, \quad \eta = \frac{D_c}{D} \quad \text{(P10.3)}$$

and

$$(1-\overline{\alpha}) = \alpha_{pc}\eta^2 + \alpha_{pw}(1-\eta^2) \quad \text{(P10.4)}$$

Derive the following equation from Eqs. (P10.3) and (P10.4)

$$(1-\overline{\alpha})\rho_p = \frac{J_p}{u_{pc}} + (1-\eta^2)\left(1 - \frac{u_{pw}}{u_{pc}}\right)\alpha_{pw}\rho_p \quad \text{(P10.5)}$$

10.4 Calculate the axial profiles of the cross-sectional averaged voidage in a CFB system with FCC particles of $d_p = 200\ \mu m$ and $\rho_p = 1{,}500\ kg/m^3$ ($\alpha_{mf} = 0.48$; $U_{pt} = 0.88$ m/s) under the following conditions:

Riser diameter $D = 80$ mm, and height $Z = 8$ m
Downcomer diameter $D_{sp} = 250$ mm
L-valve diameter $D_h = 50$ mm, length $L_h = 300$ mm, and $L_z = 500$ mm
Cross-sectional area of the cyclone $A_c = 6{,}000\ mm^2$
Solids inventory $M = 60$ kg
Gas velocity $U = 3.5$ and 4.5 m/s
Solids circulation rate $J_p = 80\ kg/m^2 \cdot s$

The properties of the fluidizing medium are $\rho = 1.18\ kg/m^3$ and $\mu = 1.82 \times 10^{-5}\ kg/m \cdot s$. Also, discuss the solids inventory effects on the voidage profiles.

10.5 The mass balance equations for a gas–solid flow in a riser can be described as

$$\frac{\partial}{\partial t}(\rho\alpha) + \frac{1}{r}\frac{\partial}{\partial r}(r\rho\alpha u_r) + \frac{\partial}{\partial z}(\rho\alpha u_z) = 0 \quad \text{(P10.6)}$$

for the gas phase and

$$\frac{\partial}{\partial t}[\rho_p(1-\alpha)] + \frac{1}{r}\frac{\partial}{\partial r}[r\rho_p(1-\alpha)u_{pr}] + \frac{\partial}{\partial z}[\rho_p(1-\alpha)u_{pz}] = 0 \quad \text{(P10.7)}$$

for the solid phase. Here u_z and u_r are linear gas velocities in the z and r directions, respectively. u_{pz} and u_{pr} are particle velocities in the z and r directions, respectively. For the core-annular model, the one-dimensional continuity equations for the core and the annular regions can be obtained from volume averaging over Eqs. (P10.6) and (P10.7). With the assumptions given in §10.6.2, derive one-dimensional continuity equations for the core and annular regions based on Eqs. (P10.6) and (P10.7).

10.6 The reduced solids flux profile, J_{pr}/J_p, can be empirically correlated by [Rhodes *et al.*, 1992]

$$\frac{J_{pr}}{J_p} = 1 + b\left[1 - \left(\frac{r}{R}\right)^m\right] - \frac{m}{m+2}b \tag{P10.8}$$

where J_{pr} is the solids flux at radial position r. Assuming that all gas goes through the core region and the slip velocity between particle and gas in the core region is equal to the particle terminal velocity, determine the parameter b in the preceding equation. Also prove that the thickness of the annular region can be expressed by

$$D_c = D\left(\frac{1}{b} + \frac{2}{m+2}\right)^{1/m} \tag{P10.9}$$

10.7 In the Davidson and Harrison's (1963) maximum stable bubble size model, the bubble disintegration takes place when the relative velocity between the bubble and the particles exceeds the particle terminal velocity. Considering that, for a vertical gas–solid flow system, choking occurs when the maximum stable bubble size is equal to the column size, Yang (1976) obtained the following choking criterion for fine particles fluidization:

$$\frac{U_{pt}}{\sqrt{gD}} \le 0.35 \quad \text{without choking} \tag{P10.10}$$

$$\frac{U_{pt}}{\sqrt{gD}} > 0.35 \quad \text{with choking} \tag{P10.11}$$

Derive Eqs. (P10.10) and (P10.11) on the basis of the conditions stated. In deriving these equations, consider also the simple two-phase theory (see §9.4.5), which gives the particle velocity above the gas slug as $U - U_{mf}$.

10.8 The particle turbulent diffusion coefficient D_p can be related to the gas turbulent diffusion coefficient D_g by [Yuu *et al.*, 1978]

$$D_p/D_g = 1 + \mathrm{St}/12 \tag{P10.12}$$

D_g can be expressed empirically by [Weinstock, 1978]

$$D_g = 0.2\sqrt{\pi}u^*/k_0 \tag{P10.13}$$

In Eq. (P10.13), k_0 is the eddy number, which is assumed to be $D/4$, and $u^* = u'/2$. On the basis of the information given, verify that the deposition coefficient k_d can be expressed by Eq. (10.31) if the Sherwood number, defined as $k_d D/D_p$, for the turbulent flow is assumed to be 4.

CHAPTER 11

Pneumatic Conveying of Solids

11.1 Introduction

In solids processing systems, it is commonly required to transport solids from one location to another. Solids can be transported via various means, including (1) pneumatic conveying, in which solids are transported in a pipe or channel by gas flow through blowing or suction; (2) gravity chutes, where solids transport downward by gravitational force; (3) air slides, where solids, partially suspended in a channel by the upward flow of air through a porous wall, flow at a small angle to the horizontal; (4) belt conveyors, where solids are conveyed horizontally, or at small angles to the horizontal, on a continuous moving belt; (5) screw conveyors, in which solids are transported in a pipe or channel by a rotating helical impeller; (6) bucket elevators, in which solids are carried upward in buckets attached to a continuously moving vertical belt; and (7) vibratory conveyors, where solids flow is activated by jigging action provided by angled spring supports.

This chapter focuses on pneumatic conveying as the means for solids transport. Examples of solid materials that are commonly transported via pneumatic conveying are flour, granular chemicals, lime, soda ash, plastic chips, coal, gunpowder pellets, ores, and grains [Stoess, 1983; Williams, 1983; Konrad, 1986; Soo, 1990; Marcus *et al.*, 1990]. Materials to be conveyed are usually dry and readily free-flowing. Some sluggish and damp materials can also be free-flowing with proper aeration. The advantages of pneumatic conveying include extreme flexibility in routing and spacing, safe working conditions, and low maintenance cost. The disadvantages include relatively high power consumption compared to that of other bulk solids transport systems and significant wearing and abrasive effects due to high collision velocity of solids in the dilute-phase conveying systems. The transport phenomena associated with the pneumatic conveying of solids are complex, and, therefore, special requirements are needed in the design and operation of those systems. Variables governing the phenomena encompass the gas velocity, characteristics of the solids (size, density, distribution, and shape), solids loading, pipe size and configuration, solids feeding device, and transport direction [Yang, 1987].

11.2 Classifications of Pneumatic Conveying Systems

Pneumatic conveying systems can be classified on the basis of the angle of inclination of pipelines, operational modes (*i.e.*, negative- or positive-pressure operation), and flow characteristics (*i.e.*, dilute or dense phase transport; steady or unsteady transport). A practical pneumatic conveying system is often composed of several vertical, horizontal, and inclined pipelines. Multiple flow regimes may coexist in a given operational system.

11.2.1 *Horizontal and Vertical Transport*

In a horizontal conveying pipeline, motion of the particles is not always straight in the horizontal direction. Particles constantly fall down to the bottom of the pipe by gravity, slide along or collide with the bottom wall of the pipe, and reenter the gas stream. Consequently, the particle concentration is higher at the bottom than at the top of the pipe except for dilute-phase flows of fine particles at very high transport velocities. Thus, when the gravity effect is significant, the gas–solid horizontal pipe flows are essentially two-dimensional regardless of the shape of the pipe [Roco and Shook, 1984; Soo and Mei, 1987]. In a horizontal pneumatic transport, the gas flow must overcome the drag forces, the frictions, and the forces associated with particle reentrainment and acceleration. To establish a fully suspended particle flow, a high gas velocity is required to provide upward lift of the particles at the bottom of the pipe.

In vertical pneumatic transport, particles are always suspended in the gas stream mainly because the direction of the gravity is in line with that of the gas flow. Hence, to transport the same solids mass flow rate in a suspended flow, it takes lower gas velocity for upward vertical conveying than for horizontal conveying. For most cases in vertical pneumatic conveying, the radial particle concentration distribution is near-uniform and, thus, gas and solids can be reasonably treated as being in one-dimensional flow. The drag force, wall friction, and gravity combine to produce a pressure drop in upward vertical conveying that is higher than that in horizontal conveying for given gas and solids flow rates.

Inclined pipelines require the highest gas velocity to convey the same amount of solid materials when compared to both vertical and horizontal conveying. In this case, gas flow has to overcome the forces associated with horizontal conveying and the tendency of materials to slide back down the incline. In practice, a slope of less than 15° or more than 80° from the horizontal need not be considered an incline [Williams, 1983].

11.2.2 *Negative- and Positive-Pressure Pneumatic Conveyings*

There are two major operational modes of pneumatic conveyor systems, *i.e.*, negative-pressure or vacuum conveying and positive-pressure conveying, as shown in Fig. 11.1. The operating pressure for the former is lower than ambient pressure while it is higher for the latter. Vacuum conveyors are in general used for transport of materials from several feeding points to a common collection point. Because of the employment of exhaust systems in this type of operation, vacuum conveyors have a definite limitation of operating pressure (about 0.4 atm) which restricts the capacity and the length of a pipeline. Negative-pressure conveying is extremely useful in the transport of toxic and hazardous materials. This type of conveyance not only provides dust-free feeding but also prevents the escape of solids through leakage (if any) in the pipeline. Positive-pressure conveying is commonly used to handle larger solids loadings and longer conveying distances without theoretical limitations. This type of operation is ideally suited for applications with multiple discharges, where solids are picked up from one location and delivered to several receiving locations. As expected, positive-pressure systems are in general more complicated than negative-pressure systems. Both negative- and positive-pressure operational modes are used for horizontal as well as vertical pneumatic transport.

In a pneumatic conveying system, either positive or negative, materials are first fed into a conveying pipeline and then carried along the pipeline by the gas flow. After reaching the destination, particles are separated from the gas stream and collected in receiving bins. The

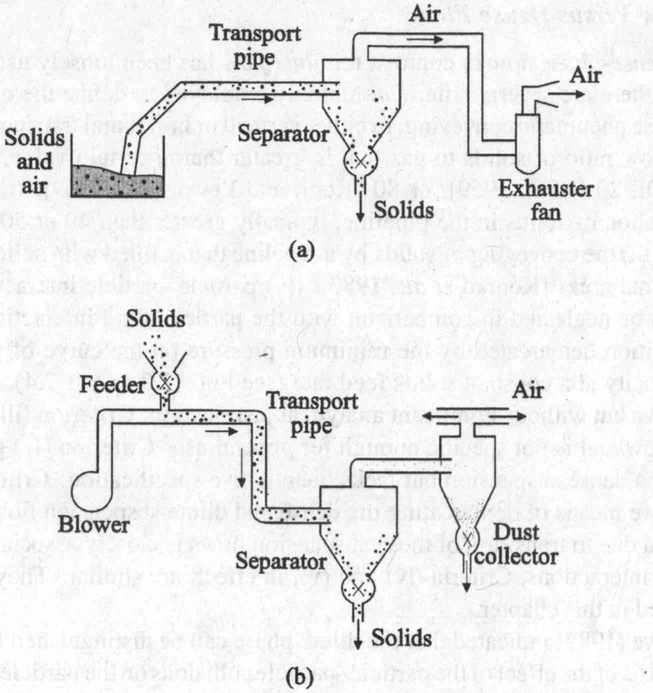

Figure 11.1. Pneumatic conveying systems: (a) Negative conveying system; (b) Positive conveying system.

major components in a pneumatic conveying system therefore include a gas mover, particle feeder, conveying conduit, gas–solid separator, and solids collector.

Blowers, fans, and vacuum pumps can be used as gas movers to provide necessary energy to transport the solids. In a negative-pressure conveying system, the vacuum pump is located downstream of the material collection device. In the positive-pressure conveying systems, the blowers or fans are installed upstream of the solids feeding equipment. The selection of a gas mover is primarily based on gas flow rate and pressure drop requirements for the successful transport of solids.

Uniform feeding of the solids and mixing of the solids and carrier gas are essential to the effective operation of pneumatic conveying. In a negative-pressure system, the material feeding and gas and solids mixing are relatively independent of the feeding arrangement; in a positive-pressure system, an air-lock type of solids feeder must be used to discharge the materials into the conveying lines. Rotary feeders are most frequently used in the vacuum- or low-positive-pressure systems. The particle size and abrasive properties have an important bearing on the design of the feeding vanes. A high solids feeding rate can be obtained by using a fluidized bed or a standpipe feeder.

For a positive-pressure system with dust-free solids, the gas–solid mixture can be directly charged into terminal receiving bins, and the exiting gas passing through the receiving bins can be discharged through the top vent into the surrounding atmosphere. The primary gas–particle separator in pneumatic conveying is a cyclone. Chapter 7 provides details on gas–solid separation.

11.2.3 *Dilute Flow Versus Dense Flow*

The term *dense-phase flow* in contrast to *dilute flow* has been loosely used in the literature. However, there are several criteria commonly employed to define the operating ranges for dense-phase pneumatic conveying, in either vertical or horizontal transport. They include (i) a mass flow ratio of solids to gas that is greater than a certain value, *e.g.*, 15 [Marcus *et al.*, 1990], 20 [Leva, 1959], or 80 [Kunii and Levenspiel, 1969]; (ii) a high volumetric concentration of solids in the pipeline, typically greater than 40 or 50 percent [Chen *et al.*, 1979]; (iii) the conveying of solids by a pipeline that is filled with solids at one or more cross-sectional areas [Konrad *et al.*, 1980]; (iv) particle–particle interaction that is strong and cannot be neglected in comparison with the particle–fluid interaction [Soo, 1989]; and (v) transition demarcated by the minimum pressure on the curve of pressure drop against gas velocity at a constant solids feed rate (see Fig. 11.3 in §11.2.4). Criteria (i) and (ii) are intuitive but without significant analytical justification. Criterion (iii) is only applicable to plug flow and is not specific enough for general use. Criterion (iv) provides general concepts for a dense suspension but lacks quantitative specification. Criterion (v) provides a quantitative means of demarcating the dense and dilute suspension flow. Since the pressure variation due to transition of these suspension flows is closely associated with the particle–particle interactions, Criteria (iv) and (v), in effect, are similar. They are the preferred criteria used in this chapter.

Specifically, Crowe (1982) indicated that the dilute phase can be distinguished from the dense phase on the basis of the effect of the particle–particle collisions on the particle motion. The dilute and the dense gas–particle flows can be directly analogous to, respectively, the free-molecule and continuum flow regimes of a single-phase fluid. In this analogy, the parameter corresponding to the Knudsen number is the Stokes number, which is based on the distance a particle travels between collisions. Consider a particle that has a speed $|\boldsymbol{U} - \boldsymbol{U}_{\mathrm{p}}|$ relative to the gas. The particle travels a distance of $|\boldsymbol{U} - \boldsymbol{U}_{\mathrm{p}}|\tau_{\mathrm{rp}}$ before it stops in a coordinate system moving with the gas flow, where τ_{rp} is the particle relaxation time. If this distance is small compared with the particle–particle collision distance in the same coordinate system, one can regard the particle cloud as dilute. General comparisons of dilute and dense suspensions are given in Table 11.1.

There are significant differences in dynamic behavior (*e.g.*, effects of particle–particle interactions on drag force and mean free path between collision, velocity, and concentration distributions) between vertical and horizontal pneumatic conveying. These differences yield distinctly different operating characteristics; *e.g.*, the onset velocity for the dense-phase flow in horizontal conveying differs from that in vertical conveying. The following discusses flow regime and regime transition in both horizontal and vertical conveying systems.

11.2.4 *Flow Regimes and Regime Transitions*

The flow pattern in a horizontal gas–solid pipe flow varies with the solids concentration. In a horizontal pipe flow at a given gas flow rate, the following flow patterns sequentially occur as the particle loading, defined as the mass flow ratio of particles to gas, increases:

(1) Dilute suspension flow
(2) Sedimentation of particles and dune formation
(3) Stratified flow
(4) Plug flow

Table 11.1. *Comparisons of Dilute and Dense Suspensions.*

	Dilute suspension	Dense suspension
Relative motion between particles	Large	Small
Particle–particle interactions	Weak	Strong
Particle diffusivity	Large	Small
Apparent particle viscosity	Due to gas–particle interaction	Due to particle–particle interaction and gas–particle interaction
Flow regime in applications	Steady, turbulent	Pseudolaminar to unsteady and stratified
Motion above minimum transport velocity	Stable	Unstable
Analogy in molecular system	Rarefied gas flow	Molecular theory of liquids or kinetic theory of gas

Source: S. Soo's *Particulates and Continuum*, Hemisphere, 1989.

(5) Moving bed flow at relatively high gas velocity or solid ripples traveling along the top of a stationary solid layer at relatively low gas velocity

As shown in Fig. 11.2(a), at a very low solids concentration, particles are fully suspended and quite uniformly distributed throughout the pipe, as is characteristic of dilute suspension flow. With increasing particle loading, sedimentation of particles occurs, as shown in Fig. 11.2(b). Particles tend to settle at the bottom of the pipe and may slide over other particles. In this flow, particle dunes are formed and particle flow takes place by way of moving from one dune to another and undergoing alternate acceleration and deceleration. In a low-velocity flow, dunes are formed more readily and the length and height of the dunes are correspondingly larger. An increase in the particle loading yields a stratified flow with a wavy interface, as shown in Fig. 11.2(c). A further increase in the particle loading leads to the formation of a solid plug characterized by the intermittent flow of gas and solids in alternating plugs, as shown in Fig. 11.2(d), and eventually results in a complete blockage of the pipe. As can be seen from Fig. 11.2(e), at a relatively high gas velocity a moving bed flow occurs; at a relatively low gas velocity, the solids travel in the form of ripples at the top portion of the pipe while a large portion of solids remain stationary at the bottom. Thus, dense-phase pneumatic transport through horizontal pipes is usually unstable, with large pressure fluctuations.

The flow regime diagram for a pneumatic conveying system can generally be described in terms of the variation of the pressure drop per unit length with the superficial gas velocity given in Fig. 11.3, in a manner similar to that for a circulating fluidized bed given in Fig. 10.2. The effect of the solids mass flux, shown in Fig. 11.3, indicates that an increase in the solids mass flux increases the pressure drop at a given gas velocity. It is seen in the figure that the pressure drop per unit length decreases and then increases as the gas velocity increases. The minimum pressure drop point marks the dense flow and dilute flow transition for the pneumatic conveying system. For vertical conveying, the choking phenomenon discussed in §10.3.1 may also occur, depending on the pipe size and particle properties [Leung, 1980]. For horizontal conveying, the saltation phenomenon will take place at the minimum pressure drop point for coarse particles, where particles start settling

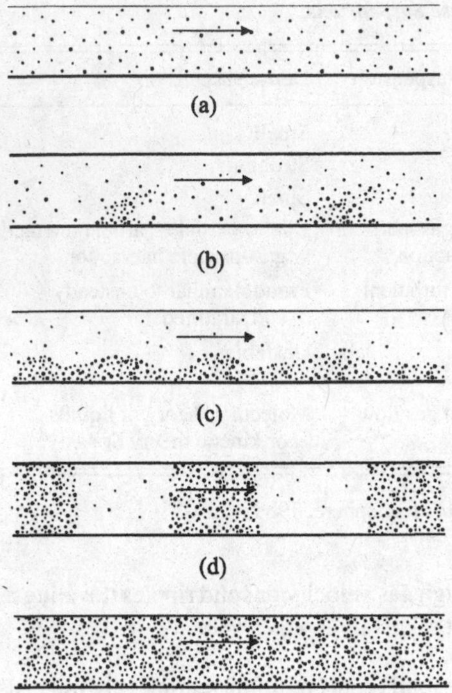

(a) (b) (c) (d) (e)

Figure 11.2. Flow regimes for horizontal conveying of solids: (a) Dilute suspension flow; (b) Sedimentation and dune formation; (c) Stratified flow with wavy interface; (d) Plug flow; (e) Moving bed flow.

at the dilute and dense flow transition. For fine particles, the saltation phenomenon will take place at gas velocity higher than that for the minimum pressure drop point [Marcus *et al.*, 1990]. The flow regimes in pneumatic conveying of fine particles for both vertical and horizontal systems can be further described by the diagrams in log–log scales, as given in Fig. 11.4. The diagrams illustrate that the power-law relationship (or linear relationship in log–log scales) between the pressure drop and the superficial gas velocity holds for dilute suspension flow of fine particles. The diagrams further indicate that, on the basis of visual observations of the flow regimes, dense-phase flow is typically represented by plug flow and moving bed flow. For comparisons the pressure drop for the fixed bed is also indicated in the figure. The unsteady-state plug-dune flow for horizontal conveying and the unsteady-state nonuniform dispersed flow for vertical conveying denoted by broken lines in the diagrams appear to be the regimes which are in need of further study.

11.3 Pressure Drop

For transport of a gas–solid suspension, it is important to be able to evaluate the pressure drop in order to estimate the power consumption. In a pneumatic conveying system, particles are usually fed into the carrying gas and accelerated by the gas flow. The power consumption (or pressure drop per unit pipe length) for the acceleration of particles can be significant compared with that for a fully developed flow condition. Moreover, the pressure

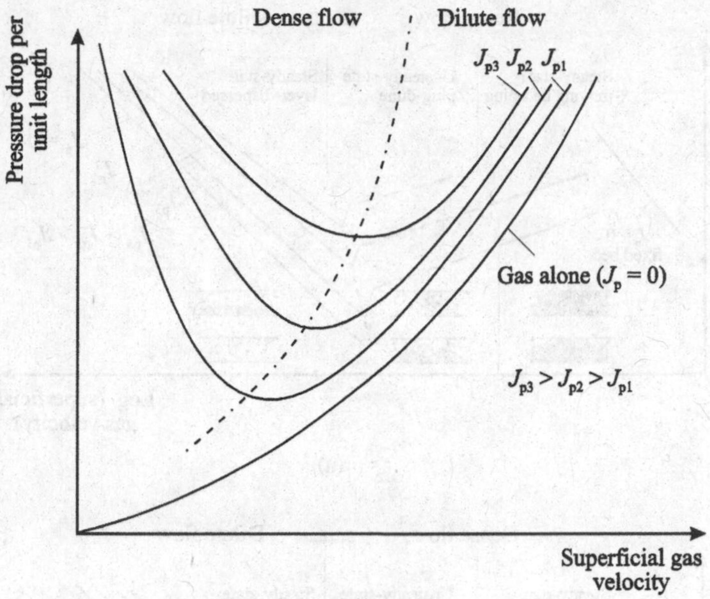

Figure 11.3. General flow regime diagram for a pneumatic conveying system (from Marcus *et al.*, 1990).

drop over a bend or branch is also high because of the deceleration (by the particle–wall collision) and reacceleration of particles. In the following, a general expression of pressure drop in one-dimensional pipe flows is first introduced. A unique phenomenon of drag reduction associated with very dilute suspension flows is then discussed. In addition to pressure drops for solids acceleration, the acceleration length in the developing flow region is also illustrated. The calculation of the pressure drop over a bend is included in this section; however, a detailed discussion of velocity fields of gas–solid flows in a bend is given in §11.5.

11.3.1 *General Pressure Drop in One-Dimensional Flow*

The role of the pressure gradient may be shown in the momentum equation of a gas–solid mixture. Consider a steady pipe flow without mass transfer and with negligible interparticle collisions. From Eq. (5.170), the momentum equations for the gas and particle phases can be given by

$$\nabla \cdot \boldsymbol{J}_{\mathrm{m}} = -\nabla p + \nabla \cdot \boldsymbol{\tau}_{\mathrm{e}} + \boldsymbol{F}_{\mathrm{A}} + \alpha \rho \boldsymbol{g} \tag{11.1}$$

and

$$\nabla \cdot \boldsymbol{J}_{\mathrm{mp}} = -\boldsymbol{F}_{\mathrm{A}} + (1-\alpha)\rho_{\mathrm{p}}\boldsymbol{g} + \boldsymbol{F}_{\mathrm{E}} \tag{11.2}$$

where $\boldsymbol{\tau}_{\mathrm{e}}$ denotes the effective stress tensor per unit volume; $\boldsymbol{F}_{\mathrm{A}}$ accounts for the interfacial interactions between the gas and solids; $\boldsymbol{F}_{\mathrm{E}}$ describes the electrostatic forces of particles; and $\boldsymbol{J}_{\mathrm{m}}$ and $\boldsymbol{J}_{\mathrm{mp}}$ are the momentum fluxes of gas and particle phases, respectively. $\boldsymbol{J}_{\mathrm{m}}$ and $\boldsymbol{J}_{\mathrm{mp}}$ are defined by

$$\boldsymbol{J}_{\mathrm{m}} = \rho\alpha \boldsymbol{U}\boldsymbol{U} - \rho D[(\nabla\alpha)\boldsymbol{U} + \boldsymbol{U}\nabla\alpha] \tag{11.3}$$

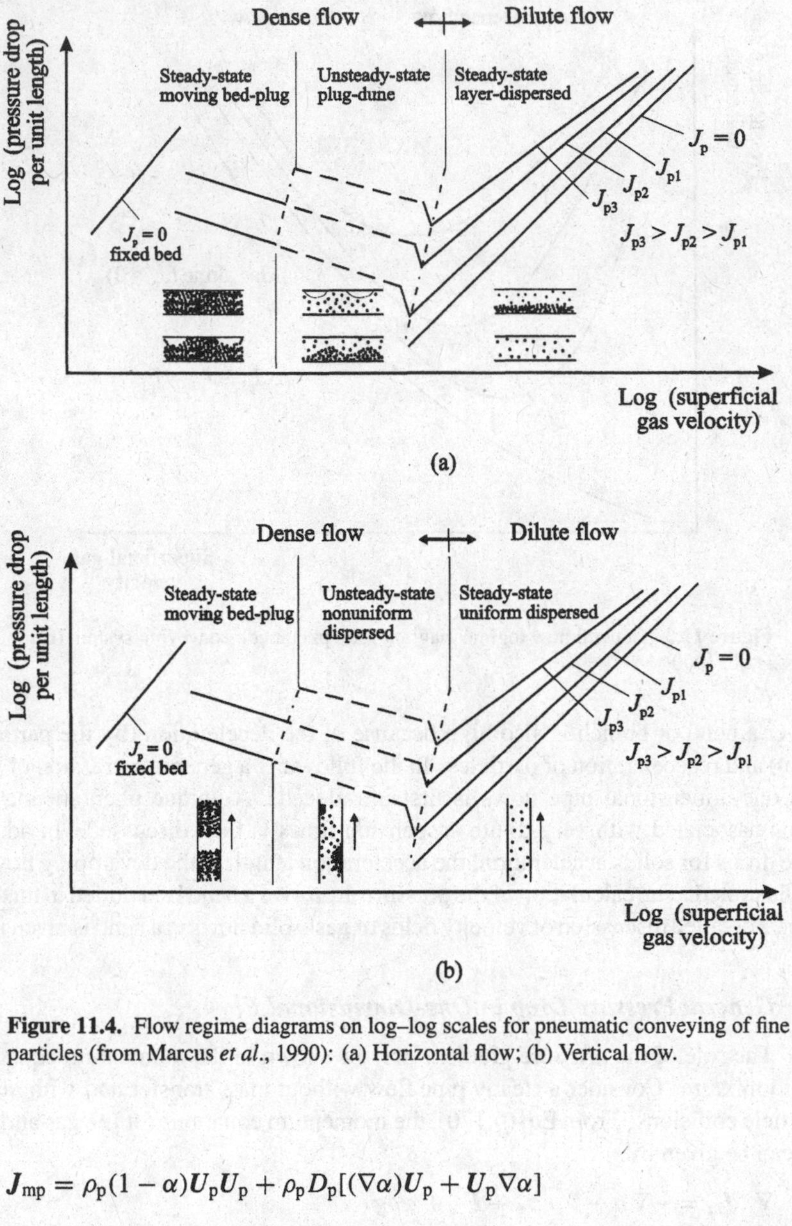

Figure 11.4. Flow regime diagrams on log–log scales for pneumatic conveying of fine particles (from Marcus *et al.*, 1990): (a) Horizontal flow; (b) Vertical flow.

and

$$\boldsymbol{J}_{\mathrm{mp}} = \rho_{\mathrm{p}}(1-\alpha)\boldsymbol{U}_{\mathrm{p}}\boldsymbol{U}_{\mathrm{p}} + \rho_{\mathrm{p}} D_{\mathrm{p}}[(\nabla\alpha)\boldsymbol{U}_{\mathrm{p}} + \boldsymbol{U}_{\mathrm{p}}\nabla\alpha] \tag{11.4}$$

where D and D_{p} represent the diffusivities of gas and particle phases, respectively. The momentum equation of the mixture is obtained by

$$\nabla\cdot(\boldsymbol{J}_{\mathrm{m}} + \boldsymbol{J}_{\mathrm{mp}}) - \nabla\cdot\boldsymbol{\tau}_{\mathrm{e}} - \boldsymbol{F}_{\mathrm{E}} - \alpha\rho\boldsymbol{g} - (1-\alpha)\rho_{\mathrm{p}}\boldsymbol{g} = -\nabla p \tag{11.5}$$

For one-dimensional horizontal pipe flows without electrostatic effects, using Gauss's theorem, Eq. (11.5) yields

$$p_1 - p_2 = J_{\mathrm{m2}} - J_{\mathrm{m1}} + J_{\mathrm{mp2}} - J_{\mathrm{mp1}} - \tau_{\mathrm{e2}} + \tau_{\mathrm{e1}} + (x_2 - x_1)\frac{C_{\mathrm{w}}\tau_{\mathrm{w}}}{A} \tag{11.6}$$

where A is the cross-sectional area, C_w is the perimeter, and x is the axial coordinate. Equation (11.6) shows that the axial pressure drop is balanced by the change of axial viscous force, wall friction, and increase of momentum fluxes of the phases.

11.3.2 *Drag Reduction*

On the basis of Eq. (11.6), it can be deduced that, with the addition of particles into a flow stream, the pressure loss would be higher than that of a single-phase flow. However, experimental data have revealed that under certain operating conditions (typically, the mass loading ranging from 0.5 to 2), the pressure drop of the two-phase mixture is much less than that of a particle-free flow [Halstrom, 1953; Thomas, 1962; Peters and Klinzing, 1972; Kane, 1989]. This unique phenomenon, termed drag reduction, is illustrated in Fig. 11.5 for the acceleration region [Shimizu *et al.*, 1978] and in Fig. 11.6 for the fully developed region [Kane and Pfeffer, 1973]. The phenomenon reflects that it is possible to consume less energy to transport a two-phase flow than to transport a single-phase flow. The effect of drag reduction is significant for small particles (less than 200 μm) in turbulent pipe flow. The decrease in the pressure drop can be up to 30 percent [McCarthy and Olson, 1968; Boyce and Blick; 1969]. It is generally recognized that the effects of particle addition on the drag reduction are mainly through turbulence modulation. The phenomenon of drag reduction is reported to be linked to various factors associated with particle motion near the wall region, including particle spinning, deposition, agglomeration, and electrostatics [Kane, 1989]. These factors ultimately affect the velocity profiles of gas and particles, and particle concentration distributions in the gas–solid flow system. The basic mechanisms of drag reduction are not yet fully known, in part because of a shortage of reliable measurements which quantify these factors. In the following a phenomenological model which accounts for the drag reduction phenomenon from the viewpoint of turbulent intensity attenuation due to the effect of particle addition is introduced.

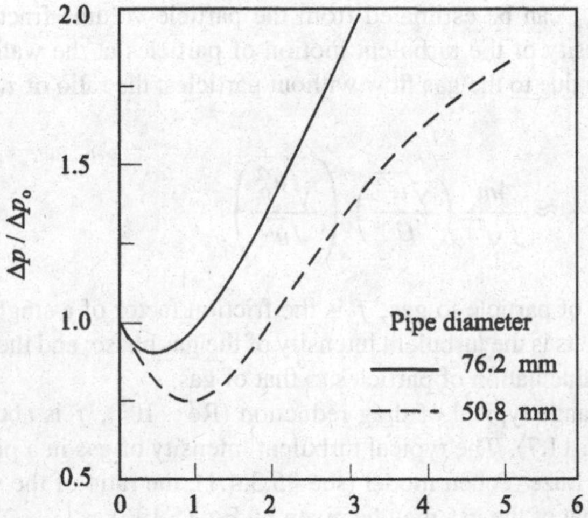

Figure 11.5. Pressure drop ratio as a function of mass flux ratio for 10 μm zinc particles in the acceleration region of Re = 53,000 (after Shimizu *et al.*, 1978).

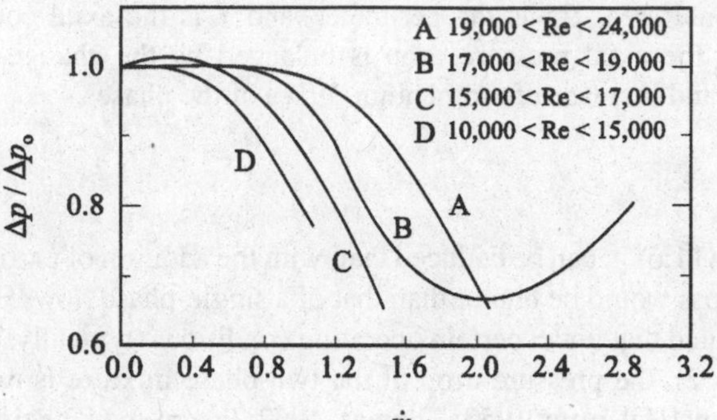

Figure 11.6. Pressure drop ratio as a function of mass flux ratio for 36 μm glass beads in a fully developed region (after Kane and Pfeffer, 1973).

It is recognized that within the range where drag reduction occurs, the solids concentration is so dilute that the averaged distance between particles is usually 10 or more particle diameters. Therefore, under this flow condition, interparticle effects can be neglected. Consider the case of a fully developed horizontal pipe flow with negligible electrostatic effects. From Eq. (11.6), the pressure drop depends only on the wall friction, as given by

$$\frac{\Delta p}{\Delta x} = \frac{C_{\mathrm{w}}(\tau_{\mathrm{gp}} + \tau_{\mathrm{wp}})}{A} \tag{11.7}$$

where τ_{gp} denotes the friction induced by the gas phase in the presence of solid particles; τ_{wp} represents the friction due to collisions of particles with the wall, which can be expressed by [Soo, 1962]

$$\tau_{\mathrm{wp}} = \frac{1}{2\sqrt{\pi}} \alpha_{\mathrm{pw}} \rho_{\mathrm{p}} U_{\mathrm{pw}} \sqrt{u'^2_{\mathrm{pw}}} \tag{11.8}$$

The equation indicates that τ_{wp} can be estimated from the particle volume fraction, the particle velocity, and the intensity of the turbulent motion of particles at the wall. If we approximate τ_{gp} by the friction due to the gas flow without particles, the ratio of τ_{wp} to τ_{gp} can be roughly estimated by

$$\frac{\tau_{\mathrm{wp}}}{\tau_{\mathrm{gp}}} \approx \frac{\frac{1}{2\sqrt{\pi}} \dot{m} \rho U \sqrt{u'^2_{\mathrm{pw}}}}{\frac{f}{8} \rho U^2} \approx \frac{4\dot{m}}{f\sqrt{\pi}} \left(\frac{\sqrt{u'^2}}{U} \right) \left(\frac{\sqrt{u'^2_{\mathrm{p}}}}{\sqrt{u'^2}} \right) \tag{11.9}$$

where $\dot{m}$ is the mass flux ratio of particle to gas; f is the friction factor of a single-phase pipe flow; the first pair of brackets is the turbulent intensity of the gas phase; and the second one is the ratio of the velocity fluctuation of particles to that of gas.

For the Reynolds number range typical of drag reduction ($\mathrm{Re} \sim 10^5$), f is about 0.02 from the Moody chart (see Fig. 11.7). The typical turbulent intensity of gas in a pipe flow is about 5 percent. Using the Hinze–Tchen model (see §5.3.4.1), the ratio of the velocity fluctuation of the particles to that of the gas may be given by Eq. (5.196) as

$$\frac{\sqrt{u'^2_{\mathrm{p}}}}{\sqrt{u'^2}} = \left(1 + \frac{\tau_{\mathrm{S}}}{\tau_{\mathrm{f}}} \right)^{-\frac{1}{2}} \tag{11.10}$$

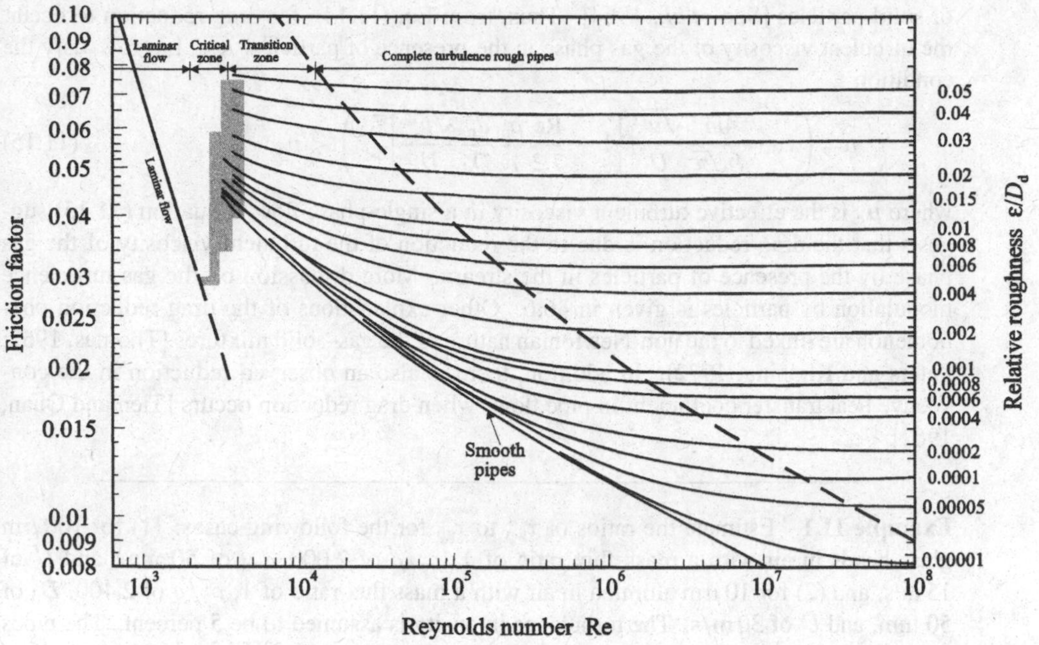

Figure 11.7. Moody chart for pipe friction with smooth and rough walls (from Moody, 1944).

where τ_S is the Stokes relaxation time defined by Eq. (3.39) and τ_f is the characteristic time of the turbulent fluctuation of gas and is given by

$$\tau_f = \frac{l_m}{\sqrt{u'^2}} \tag{11.11}$$

where l_m is the mixing length of turbulence. For a fully developed turbulent pipe flow, l_m near the wall is about 0.4ξ, where ξ is the distance from the wall [Schlichting, 1979]. Thus, we may take $l_m = 0.4d_p$, where d_p is the particle diameter. Substituting Eq. (3.39) and Eq. (11.11) into Eq. (11.10) yields

$$\frac{\sqrt{u_p'^2}}{\sqrt{u'^2}} = \left[1 + \frac{\text{Re}}{7.2}\frac{\rho_p}{\rho}\frac{d_p}{D_d}\frac{\sqrt{u'^2}}{U}\right]^{-\frac{1}{2}} \tag{11.12}$$

where Re is the Reynolds number, based on the pipe diameter D_d and averaged flow velocity U. Hence, Eq. (11.9) becomes

$$\frac{\tau_{wp}}{\tau_{gp}} \approx \frac{4\dot{m}}{f\sqrt{\pi}}\frac{\sqrt{u'^2}}{U}\left[1 + \frac{\text{Re}}{7.2}\frac{\rho_p}{\rho}\frac{d_p}{D_d}\frac{\sqrt{u'^2}}{U}\right]^{-\frac{1}{2}} \tag{11.13}$$

Substitution of Eq. (11.13) into Eq. (11.7) gives

$$\frac{\Delta p}{\Delta x} = \frac{C_w\tau_{gp}}{A}\left\{1 + \frac{4\dot{m}}{f\sqrt{\pi}}\frac{\sqrt{u'^2}}{U}\left(1 + \frac{\text{Re}}{7.2}\frac{\rho_p}{\rho}\frac{d_p}{D_d}\frac{\sqrt{u'^2}}{U}\right)^{-\frac{1}{2}}\right\} \tag{11.14}$$

It is noted from measurements that within the range of particle loading where the drag reduction takes place, the velocity profile of the gas phase is nearly unaltered by the presence of solid particles [Soo *et al.*, 1964]. Thus, from Eq. (11.14), for drag reduction to occur, the turbulent viscosity of the gas phase in the presence of particles, μ_{gp}, has to satisfy the condition

$$\mu_{\mathrm{gp}}\left(1+\frac{4\dot{m}}{f\sqrt{\pi}}\frac{\sqrt{u'^2}}{U}\left[1+\frac{\mathrm{Re}}{7.2}\frac{\rho_{\mathrm{p}}}{\rho}\frac{d_{\mathrm{p}}}{D_{\mathrm{d}}}\frac{\sqrt{u'^2}}{U}\right]^{-\frac{1}{2}}\right) < \mu_{\mathrm{e}} \tag{11.15}$$

where μ_{e} is the effective turbulent viscosity in a single-phase flow. Equation (11.15) suggests that the drag reduction is due to the reduction of the turbulent viscosity of the gas phase by the presence of particles in the stream. More discussion on the gas turbulence modulation by particles is given in §6.6. Other explanations of the drag reduction phenomenon are linked to the non-Newtonian nature of the gas–solid mixtures [Thomas, 1962; Peters and Klinzing, 1972]. In addition, there is also an observed reduction in the convective heat transfer coefficient in pipe flows when drag reduction occurs [Tien and Quan, 1962].

Example 11.1 Estimate the ratios of τ_{wp} to τ_{gp} for the following cases: (1) for 100 μm glass beads in air with a mass flux ratio of 1, ρ_{p}/ρ of 2,000, D_{d} of 50 mm, and U of 15 m/s; and (2) for 10 μm alumina in air with a mass flux ratio of 1, ρ_{p}/ρ of 2,400, D_{d} of 50 mm, and U of 30 m/s. The turbulence intensity is assumed to be 5 percent. The pipes are hydraulically smooth. The air kinematic viscosity is $1.5 \times 10^{-5}\,\mathrm{m^2/s}$.

Solution For case (1), Re can be calculated as

$$\mathrm{Re} = \frac{UD_{\mathrm{d}}}{\nu} = \frac{15 \times 0.05}{1.5 \times 10^{-5}} = 5 \times 10^4 \tag{E11.1}$$

From the Moody chart (Fig. 11.7), the friction factor of smooth pipe for Re of 5×10^4 is 0.021. Thus, the ratio of τ_{wp} to τ_{gp} is obtained from Eq. (11.13) as

$$\begin{aligned}\frac{\tau_{\mathrm{wp}}}{\tau_{\mathrm{gp}}} &\approx \frac{4\dot{m}}{f\sqrt{\pi}}\frac{\sqrt{u'^2}}{U}\left[1+\frac{\mathrm{Re}}{7.2}\frac{\rho_{\mathrm{p}}}{\rho}\frac{d_{\mathrm{p}}}{D_{\mathrm{d}}}\frac{\sqrt{u'^2}}{U}\right]^{-\frac{1}{2}}\\ &= \frac{4\times 1}{0.021\sqrt{\pi}} \times 0.05\left[1+\frac{5\times 10^4}{7.2}\times 2{,}000 \times \frac{1\times 10^{-4}}{5\times 10^{-2}}\times 0.05\right]^{-\frac{1}{2}}\\ &= 0.15\end{aligned} \tag{E11.2}$$

For case (2), Re can be calculated as 10^5, which gives rise to f of 0.018. Thus, on the basis of a similar procedure to case (1), $\tau_{\mathrm{wp}}/\tau_{\mathrm{gp}}$ is obtained as 0.34. Both cases show that τ_{gp} plays a dominant role in the determination of τ_{wp}.

11.3.3 *Pressure Drop and Acceleration Length in Developing Regions*

The pressure drop over the developing regions of gas–solid flows is often significant or even dominant. Typical developing regions of pneumatic conveying systems include the regions at the particle entrances (*i.e.*, solids feeding ports) and regions immediately beyond pipe bends. In the developing regions of pneumatic conveying systems, particles are usually accelerated. The acceleration length of solid particles can be much longer than that of gas as a result of the higher inertia of solid particles.

Consider a one-dimensional gas–solid pipe flow where the electrostatic effect is negligible. The momentum equation of the particle phase can be expressed as [Klinzing, 1981]

$$\frac{dU_p}{dt} = \frac{3}{4} C_D \rho \frac{(U - U_p)^2}{\rho_p d_p} - g \sin \beta - \frac{2 f_p U_p^2}{D_d} \tag{11.16}$$

where the drag coefficient C_D is in general a function of the particle volume fraction; β is the inclined angle of the pipe ($\beta = 0$ for horizontal pipe flows and $\beta = 90°$ for upward pipe flows); and f_p denotes the friction coefficient of particles at the wall. Noting that $dx = U_p\,dt$, the acceleration length L_a is obtained from Eq. (11.16) as

$$L_a = \int_{U_{pi}}^{U_{p\infty}} \left(\frac{3}{4} C_D \rho \frac{(U - U_p)^2}{\rho_p d_p} - g \sin \beta - \frac{2 f_p U_p^2}{D_d} \right)^{-1} U_p\, dU_p \tag{11.17}$$

where U_{pi} and $U_{p\infty}$ represent the particle velocities at the inlet and in the fully developed region, respectively. The pressure drop over the acceleration region can be expressed by

$$\Delta p_a = \int_0^{L_a} \alpha_p \rho_p g \sin \beta\, dx + \int_0^{L_a} \frac{2 f J U dx}{D_d} + \int_0^{L_a} \frac{2 f_p J_p U_p\, dx}{D_d} + J_p (U_{p\infty} - U_{pi}) \tag{11.18}$$

where J and J_p are mass fluxes of gas and particles, respectively.

The acceleration length and the pressure loss in that length can also be estimated by using empirical correlations. Rose and Duckworth (1969) reported that

$$L_a = 6 \left[\frac{D_d J_p}{\rho_p \sqrt{d_p g}} \right]^{\frac{1}{3}} \left(\frac{\rho_p}{\rho} \right)^{\frac{1}{2}} \tag{11.19}$$

and

$$\Delta p_a = 1.12 \left(\frac{\rho U^2}{2} \right) \left(\frac{J_p}{J} \right) \tag{11.20}$$

The gas–solid flow through a bend represents another typical case of developing regions, in which particles are first decelerated by gas impingement and collision with the wall and then accelerated by the gas after the bend turn. A general empirical relation for the estimation of the additional pressure drop due to particles over a bend can be used [Klinzing, 1981]

$$\frac{\Delta p_{Bp}}{\Delta p_L} = a \left(\frac{2 R_B}{D_d} \right)^{-b} \left(\frac{d_p}{D_d} \right)^{c} \tag{11.21}$$

where R_B is the bend radius; Δp_L is the pressure drop over an equivalent length of a straight pipe; and a, b, and c are empirical constants. For coarse particles (1–3 mm) and for $\alpha_p < 5$ percent, Schuchart (1968) suggested that $a = 210$, $b = 1.15$, and $c = 0$. Δp_L in Eq. (11.21) may be determined by [Schuchart, 1968]

$$\Delta p_L = A \left(\frac{\rho U^2}{2} \right) \left(\frac{L_B}{d_p} \right) \frac{\alpha_p}{1 - \alpha_p} \left(1 - \frac{U_p}{U} \right)^2 C_D \tag{11.22}$$

where A is an empirical constant and L_B is the equivalent length of a straight pipe.

It is noted that the total pressure drop over a bend includes contributions from both the gas and particle phases. The pressure drop of gas over a bend, Δp_{Bg}, can be estimated by [Ito, 1959; 1960]

$$\Delta p_{\mathrm{Bg}} = \frac{0.029 + 0.304[\mathrm{Re}(D_{\mathrm{d}}/2R_{\mathrm{B}})^2]^{-0.25}}{\sqrt{2R_{\mathrm{B}}/D_{\mathrm{d}}}}\left(\frac{L_{\mathrm{B}}}{D_{\mathrm{d}}}\right)\left(\frac{\rho U^2}{2}\right) \tag{11.23}$$

The preceding equation is valid for $0.034 < \mathrm{Re}(D_{\mathrm{d}}/2R_{\mathrm{B}})^2 < 300$. When $\mathrm{Re}(D_{\mathrm{d}}/2R_{\mathrm{B}})^2 \leq 0.034$, the bend yields the same pressure drop as for the straight pipe.

Example 11.2 Use Eq. (11.17) to derive a general expression for the acceleration length for dilute gas–solid pipe flows. Assume that the Stokes drag coefficient can be used. The friction coefficient of particles at the wall can be estimated by [Konno and Saito, 1969]

$$f_{\mathrm{p}} = 0.0285\frac{\sqrt{gD_{\mathrm{d}}}}{U_{\mathrm{p}}} \tag{E11.3}$$

Solution For dilute flows, the particle volume fraction $\alpha_{\mathrm{p}} \ll 1$ and, thus, the drag coefficient C_{D} can be approximated to be independent of α_{p}. Substituting Eq. (E11.3) into Eq. (11.17) and noting that $C_{\mathrm{D}} = 24/\mathrm{Re}_{\mathrm{p}}$, we have

$$L_{\mathrm{a}} = \int_{U_{\mathrm{pi}}}^{U_{\mathrm{p}\infty}} \left(\frac{(U - U_{\mathrm{p}})}{\tau_{\mathrm{S}}} - g\sin\beta - 0.057\sqrt{\frac{g}{D_{\mathrm{d}}}}U_{\mathrm{p}}\right)^{-1} U_{\mathrm{p}}\,\mathrm{d}U_{\mathrm{p}} \tag{E11.4}$$

where τ_{S} is the Stokes relaxation time defined by Eq. (3.39). Equation (E11.4) can be integrated directly to become

$$L_{\mathrm{a}} = \int_{U_{\mathrm{pi}}}^{U_{\mathrm{p}\infty}} \frac{U_{\mathrm{p}}\,\mathrm{d}U_{\mathrm{p}}}{A - BU_{\mathrm{p}}} = \frac{A}{B^2}\ln\left(\frac{A - BU_{\mathrm{pi}}}{A - BU_{\mathrm{p}\infty}}\right) - \frac{(U_{\mathrm{p}\infty} - U_{\mathrm{pi}})}{B} \tag{E11.5}$$

where

$$A = \frac{U}{\tau_{\mathrm{S}}} - g\sin\beta; \qquad B = \frac{1}{\tau_{\mathrm{S}}} + 0.057\sqrt{\frac{g}{D_{\mathrm{d}}}} \tag{E11.6}$$

11.4 Critical Transport Velocities

For a horizontal flow, the direction of the gravitational force is perpendicular to that of the main stream of the flow. Gravitational settling of particles occurs when the gas velocity is not sufficiently high. The minimum transport velocity and pickup velocity are defined to account for the transition of the settling phenomenon. The minimum transport velocity is the minimum mean gas velocity required to prevent the accumulation of a layer of stationary or sliding particles on the bottom of a horizontal conveying pipeline. The minimum transport velocity is equivalent to the saltation velocity, below which saltation occurs in the pneumatic conveying system. The gas velocity required to pick up the particles at rest is higher than the saltation velocity because additional interparticle forces or particle–wall forces (such as cohesive and van der Waals forces) must be overcome during the particle pick-up. In this section, simple models for the minimum transport velocity and pick-up velocity are presented.

11.4.1 *Minimum Transport Velocity*

The vertical motion of solids in a horizontal suspension flow is strongly influenced by the ratio of the terminal settling velocity to the friction velocity [Blatch, 1906; Chien and Asce, 1956]. In a circular pipe, the mean stream velocity can be related to the friction velocity by [Taylor, 1954]

$$\frac{U}{U_\mathrm{f}} = 5\log\left(\frac{\rho D_\mathrm{d} U}{\mu}\right) - 3.90 \tag{11.24}$$

where U_f is the friction velocity based on the mixture density and is defined by

$$U_\mathrm{f} = \sqrt{\frac{\tau_\mathrm{w}}{\alpha_\mathrm{p}\rho_\mathrm{p} + (1-\alpha_\mathrm{p})\rho}} \approx \sqrt{\frac{D_\mathrm{d}\Delta p}{4L[\alpha_\mathrm{p}\rho_\mathrm{p} + (1-\alpha_\mathrm{p})\rho]}} \tag{11.25}$$

The friction velocity at the minimum transport condition may be related to the system configuration and the operating condition by a two-step correlation method, *i.e.*, first, to obtain the velocity at the minimum transport condition under infinite dilutions and, second, to correct for the concentration dependence [Thomas, 1962]. The functional dependence of U_f to the solids concentration is given by

$$\frac{U_\mathrm{f}}{U_\mathrm{f0}} = 1 + 2.8\left(\frac{U_\mathrm{pt}}{U_\mathrm{f0}}\right)^{\frac{1}{3}} \alpha_\mathrm{p}^{\frac{1}{2}} \tag{11.26}$$

where U_pt is the particle terminal velocity and U_f0 is the friction velocity at the minimum transport condition and zero solids concentration. Furthermore, U_f0 can be expressed by

$$\frac{U_\mathrm{pt}}{U_\mathrm{f0}} = 4.90\left(\frac{d_\mathrm{p}}{D_\mathrm{d}}\right)\left(\frac{D_\mathrm{d} U_\mathrm{f0}\rho}{\mu}\right)^{0.4}\left(\frac{\rho_\mathrm{p}-\rho}{\rho}\right)^{0.23} \tag{11.27}$$

for $d_\mathrm{p} > 5\mu/\rho U_\mathrm{f0}$, and

$$\frac{U_\mathrm{pt}}{U_\mathrm{f0}} = 0.01\left(\frac{d_\mathrm{p} U_\mathrm{f0}\rho}{\mu}\right)^{2.71} \tag{11.28}$$

for $d_\mathrm{p} < 5\mu/\rho U_\mathrm{f0}$.

Example 11.3 Consider a gas–solid horizontal pipe flow. The pipe diameter is 50 mm. The particle used is 50 μm glass bead with the density of 2,500 kg/m^3. The average particle volume fraction is 0.1 percent. The gas density and kinematic viscosity are 1.2 kg/m^3 and 1.5×10^{-5} m^2/s, respectively. Estimate the minimum transport velocity and power consumption per unit length.

Solution The particle terminal velocity U_pt can be calculated from Eq. (1.7) as

$$U_\mathrm{pt} = \frac{d_\mathrm{p}^2(\rho_\mathrm{p}-\rho)g}{18\rho\nu} = \frac{(50\times10^{-6})^2 \times (2{,}500 - 1.2)\times 9.8}{18\times 1.2\times 1.5\times 10^{-5}} = 0.2\ \mathrm{m/s} \tag{E11.7}$$

The corresponding particle Reynolds number at terminal velocity, *i.e.*, $\mathrm{Re_t}$, is 0.67, which falls within the correct $\mathrm{Re_t}$ range for Eq. (1.7) to be applied to calculate U_pt. The friction velocity at the minimum transport condition and zero solids concentration, U_{f0}, can be obtained from Eq. (11.28) as

$$U_{f0} = \left[100\, U_{pt}\left(\frac{\nu}{d_p}\right)^{2.71}\right]^{\frac{1}{3.71}} = \left[100 \times 0.2\left(\frac{1.5 \times 10^{-5}}{50 \times 10^{-6}}\right)^{2.71}\right]^{\frac{1}{3.71}} = 0.93\ \text{m/s} \quad \text{(E11.8)}$$

The corresponding $d_p U_{f0}/\nu$ is 3.1, which verifies that Eq. (11.28) is a correct equation to use. The friction velocity at the minimum transport condition, U_f, can thus be given by Eq. (11.26) as

$$U_f = U_{f0}\left[1 + 2.8\left(\frac{U_{pt}}{U_{f0}}\right)^{\frac{1}{3}}\sqrt{\alpha_p}\right] = 0.93\left[1 + 2.8\left(\frac{0.2}{0.93}\right)^{\frac{1}{3}} \times \sqrt{0.001}\right] = 0.98\ \text{m/s} \quad \text{(E11.9)}$$

The minimum transport velocity U can be determined from Eq. (11.24) as

$$U = U_f\left[5\log\left(\frac{D_d U}{\nu}\right) - 3.90\right] = 4.9\log\left(3.33 \times 10^3 U\right) - 3.82 \quad \text{(E11.10)}$$

which yields $U = 19.8$ m/s.

The power consumption per unit length (*i.e.*, $\Delta p/L$) is estimated from Eq. (11.25) as

$$\frac{\Delta p}{L} = [\alpha_p \rho_p + (1 - \alpha_p)\rho]\frac{4U_f^2}{D_d} = [0.001 \times 2{,}500 + 1.2]\frac{4 \times 0.98^2}{50 \times 10^{-3}} = 280\ \text{Pa/m} \quad \text{(E11.11)}$$

11.4.2 *Pick-up Velocity*

Particle pick-up is affected by two major factors, *i.e.*, hydrodynamics and particle–wall collisions. When the hydrodynamic forces overcome the gravitational force, the particle can be lifted away from the wall. The bouncing effect of particle–wall collisions can also hinder particles from settling. For fine particles, the particle motion is controlled by gas motion and turbulent diffusion. In this case, the influence of particle–wall collisions is less significant since the particles follow the gas flow after wall collisions. For large particles, the particle motion is dictated by particle inertia and is not strongly influenced by flow turbulence or by the changes in the mean flow. Thus, the motion of large particles is dominated by the particle–wall collision.

The mechanisms of a single particle–wall collision are given in Chapter 2. A particle–wall collision in pneumatic transport systems is a complex process. The bouncing characteristics depend on many parameters, including impact angle, translational and rotational velocities of the particle before collision, physical properties of the wall and particles, and wall roughness and particle shape.

Saltation of solids occurs in the turbulent boundary layer where the wall effects on the particle motion must be accounted for. Such effects include the lift due to the imposed mean shear (Saffman lift, see §3.2.3) and particle rotation (Magnus effect, see §3.2.4), as well as an increase in drag force (Faxen effect). In pneumatic conveying, the motion of a particle in the boundary layer is primarily affected by the shear-induced lift. In addition, the added mass effect and Basset force can be neglected for most cases where the particle

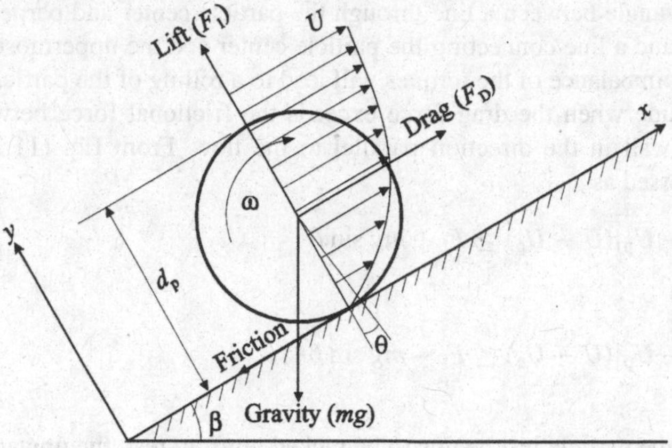

Figure 11.8. Forces acting on the particle on an inclined surface.

acceleration is moderate. In terms of the preceding considerations, the motion of a particle in the wall region of an inclined pipe, as shown in Fig. 11.8, can be expressed as

$$m\frac{\mathrm{d}U_{\mathrm{p}}}{\mathrm{d}t} = \frac{\pi}{8}\rho d_{\mathrm{p}}^2 C_{\mathrm{D}}|\boldsymbol{U} - \boldsymbol{U}_{\mathrm{p}}|(U - U_{\mathrm{p}}) - F_{\mathrm{f}} - mg\sin\beta \tag{11.29}$$

for the x-direction and

$$m\frac{\mathrm{d}V_{\mathrm{p}}}{\mathrm{d}t} = \frac{\pi}{8}\rho d_{\mathrm{p}}^2 C_{\mathrm{D}}|\boldsymbol{U} - \boldsymbol{U}_{\mathrm{p}}|(V - V_{\mathrm{p}}) + 1.62\,\mu d_{\mathrm{p}}^2\sqrt{\frac{1}{\nu}\left|\frac{\partial(U - U_{\mathrm{p}})}{\partial y}\right|}(U - U_{\mathrm{p}}) - mg\cos\beta \tag{11.30}$$

for the y-direction. F_{f} is a friction force, which can be expressed as

$$F_{\mathrm{f}} = f(mg\cos\beta - F_{\mathrm{l}}) \tag{11.31}$$

where F_{l} is the lift force.

A single particle, initially at rest at the bottom wall of the conveying pipe, may undergo various states of motion when the gas velocity increases. Single-particle motion can occur in one or a combination of the following states [Zenz, 1964]:

(1) Rolling, sliding, or bouncing along the bottom of the pipe in the direction of the flow.
(2) Swept up without saltation and obviously rolling and bouncing.
(3) Picked up from rest at the bottom of a pipe and transported in suspension.

The incipient gas velocities for each state have been analyzed by Halow (1973) and are elaborated in the following.

The incipient gas velocity for a particle rolling along the wall can be determined by the torques exerted on the particle. Assuming that the drag force, gravitational force, and lift force are exerted through the central point of the sphere, the balance of the torques about the uppermost point of the particle–wall contact plane (see Fig. 11.8) can be given by

$$F_{\mathrm{D}}\cos\theta + F_{\mathrm{l}}\sin\theta = mg\sin(\theta + \beta) \tag{11.32}$$

where θ is an inclined angle between a line through the particle center and perpendicular to the contact surface and a line connecting the particle center and the uppermost contact point on the wall. Any imbalance of the torques will lead to a rolling of the particle.

The particle will slide when the drag force exceeds the frictional force between the particle and the pipe wall in the direction parallel to the flow. From Eq. (11.29), this condition can be expressed as

$$\frac{\pi}{8}\rho d_{\mathrm{p}}^2 C_{\mathrm{D}}|\boldsymbol{U}-\boldsymbol{U}_{\mathrm{p}}|(U-U_{\mathrm{p}}) \geq F_{\mathrm{f}} + mg\sin\beta \tag{11.33}$$

for uphill sliding and

$$\frac{\pi}{8}\rho d_{\mathrm{p}}^2 C_{\mathrm{D}}|\boldsymbol{U}-\boldsymbol{U}_{\mathrm{p}}|(U-U_{\mathrm{p}}) \leq F_{\mathrm{f}} - mg\sin\beta \tag{11.34}$$

for downhill sliding.

At the point where the particle just begins to be picked up from rest, the upward forces have to be equal to the downward forces so that $\mathrm{d}V_{\mathrm{p}}/\mathrm{d}t = 0$, *i.e.*,

$$\frac{\pi}{8}\rho d_{\mathrm{p}}^2 C_{\mathrm{D}}|\boldsymbol{U}-\boldsymbol{U}_{\mathrm{p}}|(V-V_{\mathrm{p}}) + 1.62\mu d_{\mathrm{p}}^2\sqrt{\frac{1}{\nu}\left|\frac{\partial(U-U_{\mathrm{p}})}{\partial y}\right|}(U-U_{\mathrm{p}}) - mg\cos\beta = 0 \tag{11.35}$$

To develop a criterion to predict the incipient points for rolling, sliding, and pick-up, the gas velocity profile in the boundary layer of the wall region is required. The preceding discussion refers to a flowing gas stream with a single particle in the wall region. When a layer of particles is encountered at the bottom of the pipe, the minimum gas velocity required to pick up a particle from rest and transport it in suspension can be 2.5 times larger than that for the single-particle case as a result of the extra energy needed for overcoming the interparticle forces [Zenz, 1964]. Similar accounts can be extended to other situations involving finite solids concentrations in the wall region. The incipient gas velocities for picking up a particle from rest are of practical importance and have received much attention in theoretical treatment.

11.5 Flows in Bends

The gas–solid flow in a pipe bend is of major interest because it contributes significantly not only to pressure drop effects but also to erosion problems that result from the directional impacts of the solids. Curved pipes or bends provide a large degree of flexibility to gas–solid transport systems by allowing routing and distribution. A basic description of the flow of a suspension through bends is thus considered critical from the viewpoint of transport system design. In the following, the flow field of a single-phase flow in a curved duct is first discussed. The motion of a single particle in such a system is then examined by analyzing its momentum transfer in the bend. The pressure drop of gas–solid flow in the bend may be linked to the particle deceleration induced by the bend.

11.5.1 *Single-Phase Flow in a Curved Pipe*

It is shown in Fig. 11.9 that when a pipe is curved, the distribution of velocity in a bend is appreciably altered by the secondary flow and the highest axial velocity occurs

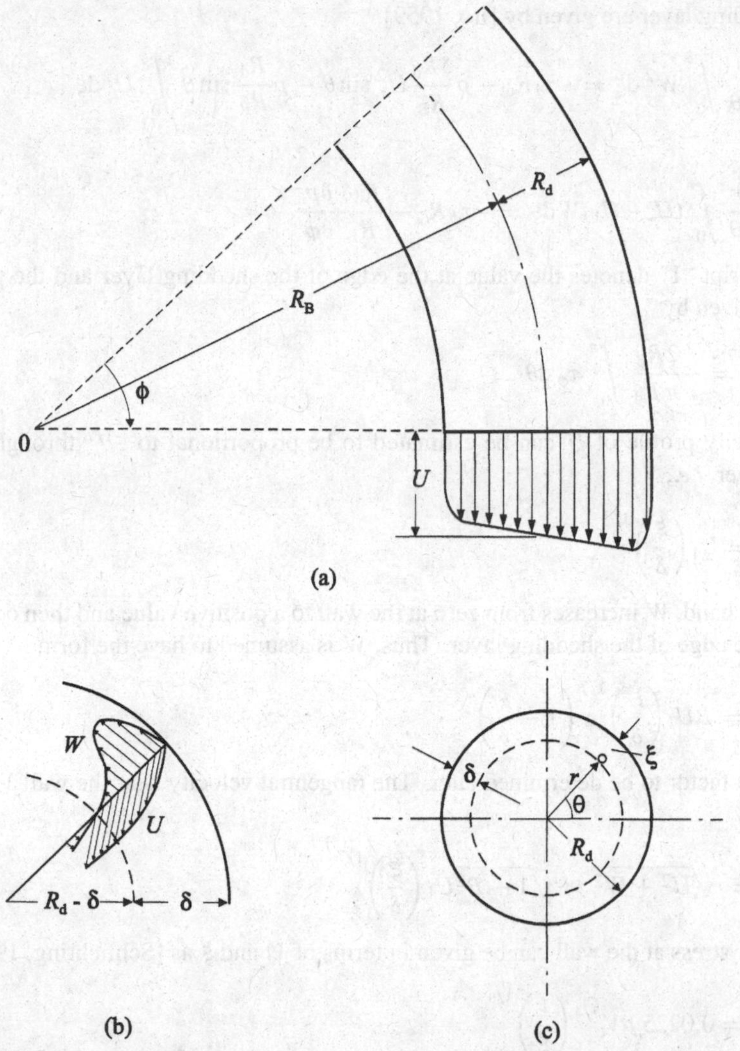

Figure 11.9. Schematic representation of flow in curved pipes: (a) Axial velocity distribution; (b) Velocity distribution in the shedding layer; (c) Notations of pipe cross section.

near the outer wall. The secondary flow takes place primarily in a layer called the shedding layer near the wall. The thickness of the shedding layer δ is defined as being equal to the distance from the wall to the point where the peripheral velocity component W changes its sign, as shown in Fig. 11.9(b).

Let R_B be the radius of curvature of the pipe axis and R_d be the radius of the circular cross section of the pipe. Define U as the axial velocity component and $\xi(=R_d - r)$ as the distance normal to the wall. Denote θ as the angle in the transverse plane with respect to the outward direction of the symmetry line and ϕ as the angle measured in the plane of the curved pipe axis, as shown in Figs. 11.9(a) and (c). Assume that the changes of the flow pattern along the axis of the bend can be neglected. Thus, the momentum integral equations for the shedding layer are given by [Ito, 1959]

$$\rho \frac{\partial}{\partial \theta} \int_0^\delta W^2 \, d\xi = -\tau_\theta R_d + \rho \frac{\delta R_d}{R_B} U_1^2 \sin\theta - \rho \frac{R_d}{R_B} \sin\theta \int_0^\delta U^2 d\xi \tag{11.36}$$

and

$$\rho \frac{\partial}{\partial \theta} \int_0^\delta (U - U_1) W d\xi = -\tau_\phi R_d - \frac{R_d \delta}{R_B} \frac{\partial p}{\partial \phi} \tag{11.37}$$

where subscript "1" denotes the value at the edge of the shedding layer and the pressure gradient is given by

$$\frac{\partial p}{\partial \phi} = -\frac{2R_B}{\pi R_d} \int_0^\pi \tau_\phi \, d\theta \tag{11.38}$$

The velocity profile of U can be estimated to be proportional to $\xi^{1/7}$ throughout the shedding layer, *i.e.*,

$$U = U_1 \left(\frac{\xi}{\delta} \right)^{1/7} \tag{11.39}$$

On the other hand, W increases from zero at the wall to a positive value and then decreases to zero at the edge of the shedding layer. Thus, W is assumed to have the form

$$W = BU_1 \left(\frac{\xi}{\delta} \right)^{1/7} \left(1 - \frac{\xi}{\delta} \right) \tag{11.40}$$

where B is a factor to be determined later. The tangential velocity near the wall V_t is thus expressed by

$$V_t = \sqrt{U^2 + W^2} \approx \sqrt{1 + B^2} U_1 \left(\frac{\xi}{\delta} \right)^{1/7} \tag{11.41}$$

The shear stress at the wall can be given in terms of V_t and ξ as [Schlichting, 1979]

$$\tau_w = 0.0225 \, \rho V_t^{7/4} \left(\frac{\nu}{\xi} \right)^{1/4} \tag{11.42}$$

where ν is the kinematic viscosity. Substitution of Eq. (11.41) into Eq. (11.42) yields

$$\tau_w = 0.0225 \, \rho \left(U_1 \sqrt{1 + B^2} \right)^{7/4} \left(\frac{\nu}{\delta} \right)^{1/4} \tag{11.43}$$

With the relationships of $\tau_\theta / \tau_\phi = W/U$ and $\tau_\theta^2 + \tau_\phi^2 = \tau_w^2$, it gives rise to

$$\tau_\phi = \frac{\tau_w}{\sqrt{1 + B^2}}; \qquad \tau_\theta = \frac{B\tau_w}{\sqrt{1 + B^2}} \tag{11.44}$$

Since the maximum velocity in the secondary flow is lower than the axial velocity at the edge of the shedding layer, B is much smaller than unity. Thus, we have

$$U \approx U_m, \qquad W \approx BU_m, \qquad \tau_\phi \approx \tau_w, \qquad \tau_\theta \approx B\tau_w$$
$$\tau_w \approx 0.03 \, \rho U_m^2 \, \mathrm{Re}^{-1/4} \left(\frac{\delta}{R_d} \right)^{-1/4} \tag{11.45}$$

where U_m is the averaged axial velocity and Re is based on U_m and pipe diameter $2R_d$. As, in arriving at Eq. (11.36) for the momentum balance, the value for each term in the equation is taken to be of the same order of magnitude, B should have the asymptotic form

$$B^2 \approx \frac{R_d}{R_B} \tag{11.46}$$

and the thickness of the shedding layer can be estimated by

$$\frac{\delta}{R_d} \approx \left(\frac{\mathrm{Re} R_d^2}{R_B^2}\right)^{-1/5} \tag{11.47}$$

B and δ can also be determined numerically by solving the momentum integral equations, Eqs. (11.36) and (11.37), with the assumed profiles, Eqs. (11.39) and (11.40), and with the wall shear stress, Eq. (11.44). The axial component in the central core of the fluid is assumed to be in the form

$$U = U_m + Ar\cos\theta \tag{11.48}$$

where the constant A is determined by the condition of continuity of the secondary flow.

Thus, an expression for the friction factor of a curved pipe can be obtained from its definition given below

$$f_B = -\frac{1}{R_B}\frac{\partial p}{\partial \phi}\frac{4R_d}{\rho U_m^2} \tag{11.49}$$

Thus, f_B can be expressed by

$$f_B = C\left(\frac{1}{\mathrm{Re}^2}\frac{R_d}{R_B}\right)^{1/10} \tag{11.50}$$

where C is a constant with a value of about 0.3 [Ito, 1959].

11.5.2 *Particulate Flow in a Bend*

When a dilute gas–solid suspension flow passes through a curved pipe, the particle velocity is lowered by the wall friction, the effects of gravitation, and collision with the wall. The analysis of flows in pipe bends may be represented by three typical arrangements of bends:

(1) A pipe bend in a horizontal plane
(2) A pipe bend in the vertical plane with a horizontal approach flow
(3) A pipe bend in the vertical plane with a vertical approach flow

Without losing generality, in this section we only consider case (3), where the pipe bend is located in the vertical plane with a vertical gas–solid suspension flow at the inlet, as shown in Fig. 11.10. It is assumed that the carried mass and the Basset force are neglected. In addition, the particles slide along the outer surface of the bend by centrifugal force and by the inertia effect of particles. The rebounding effect due to particle collisions with the wall is neglected.

The equation of motion of a single particle in the direction of the pipe axis is expressed by

$$\frac{U_p}{R_d + R_B}\frac{dU_p}{d\phi} = \frac{3}{4}\frac{\rho}{\rho_p}\frac{C_D}{d_p}(U - U_p)^2 - g\cos\phi - f_p\left(\frac{U_p^2}{R_d + R_B} - g\sin\phi\right) \tag{11.51}$$

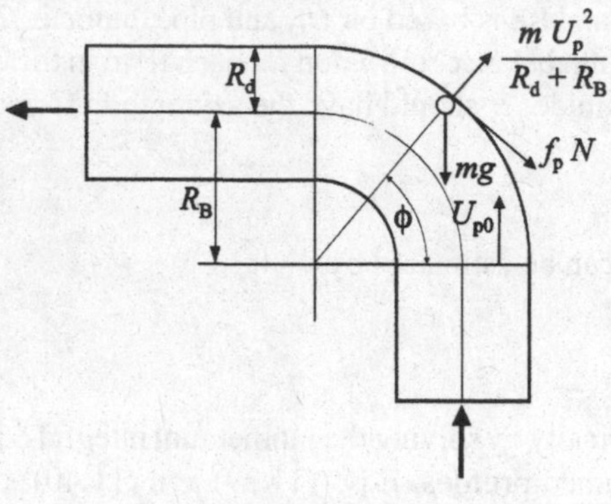

Figure 11.10. Pipe bend in the vertical plane with vertical approach flow.

where f_p is the sliding frictional coefficient between the particles and the pipe wall. The last term in the preceding equation requires the normal force against the wall to be positive within the range of ϕ considered, *i.e.*,

$$U_p^2 \geq (R_d + R_B) g \sin\phi \tag{11.52}$$

Hence, once the local gas velocity and drag coefficient are given, the variation of the particle velocity along the bend can be obtained by numerical integration of Eq. (11.51).

In order to estimate the extent of decrease in the particle velocity due to the pipe curvature, it is assumed that the drag force is negligible in comparison with the sliding force of the particle on the bend surface. With this consideration, Eq. (11.51) is reduced to

$$\frac{dU_p^2}{d\phi} + 2 f_p U_p^2 = -2(R_d + R_B) g (\cos\phi - f_p \sin\phi) \tag{11.53}$$

With the boundary condition of $U_p = U_{p0}$ at $\phi = 0$, the preceding equation gives rise to a solution for U_p as

$$U_p^2 = \left(U_{p0}^2 + \frac{6 f_p (R_d + R_B) g}{1 + 4 f_p^2} \right) e^{-2 f_p \phi} - \frac{2(R_d + R_B) g}{1 + 4 f_p^2} \left(3 f_p \cos\phi + \left(1 - 2 f_p^2\right) \sin\phi \right) \tag{11.54}$$

which yields the expression for the particle velocity at the exit of a right angle bend as

$$U_p^2\Big|_{\phi = \frac{\pi}{2}} = \left(U_{p0}^2 + \frac{6 f_p (R_d + R_B) g}{1 + 4 f_p^2} \right) e^{-f_p \pi} - 2(R_d + R_B) g \frac{1 - 2 f_p^2}{1 + 4 f_p^2} \tag{11.55}$$

11.6 Fully Developed Dilute Pipe Flows

To analyze the dynamic behavior of gas–solid pipe flows, the most common and easiest system to consider is a dilute gas–solid pipe flow which is fully developed and is subject to the effects of electrostatic force and gravitational force. The fully developed flow here refers to the situation where the velocity profiles of both gas and particles are unchanged along the axial direction. The system of this nature was analyzed by Soo and Tung (1971). In this section, the analysis of Soo and Tung (1971) is presented. It is assumed that no particles are deposited on the wall surface of the pipe (or the particle deposition rate is zero). Moreover, the pipe flow is considered to be turbulent, as is true for most flow conditions.

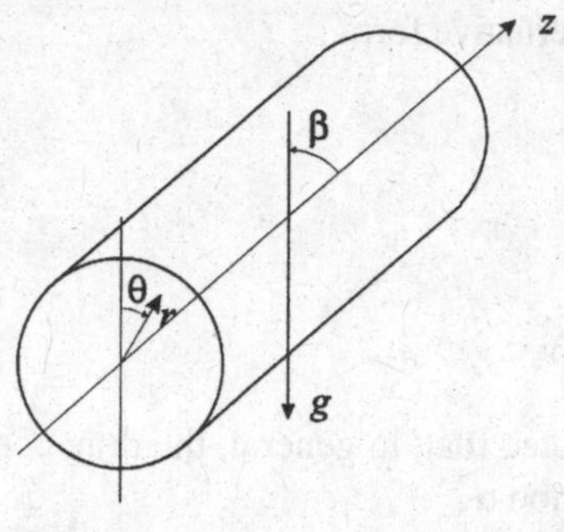

Figure 11.11. General coordinates for the analysis of fully developed dilute pipe flows.

For dilute suspensions, particle–particle interactions can be neglected. The extent of transfer of particles by the gradient in the particle phase density or volume fraction of particles is proportional to the diffusivity of particles D_p. Here D_p accounts for the random motion of particles in the flow field induced by various factors, including the diffusivity of the fluid whether laminar or turbulent, the wake of the particles in their relative motion to the fluid, the Brownian motion of particles, the particle–wall interaction, and the perturbation of the flow field by the particles.

The flow system consists of a circular pipe with an inclined angle of β to the direction of gravitational acceleration $\boldsymbol{g}$, as shown in Fig. 11.11. Thus, for β of 90° the flow is horizontal and for β of 0° or 180° the flow is vertical. In Fig. 11.11, z, r, and θ are the axial, radial, and azimuthal coordinates; U, V, W are the conjugate components of velocity of the fluid; and U_p, V_p, and W_p are those of the particle phase. The particles are assumed to be monodispersed and spherical.

11.6.1 *Basic Equations and Boundary Conditions*

To yield a constitutive relation for the mass flux of particles, it is convenient to begin with the equation of motion of a single particle, which is expressed by

$$\frac{\pi}{6} d_p^3 \rho_p \frac{d\boldsymbol{U}_p}{dt} = \rho C_D \frac{\pi}{8} d_p^2 |\boldsymbol{U} - \boldsymbol{U}_p| (\boldsymbol{U} - \boldsymbol{U}_p) + \boldsymbol{F}_G + \boldsymbol{F}_E \tag{11.56}$$

where $\boldsymbol{F}_G$ is the gravitational force including buoyancy effect and is given by

$$\boldsymbol{F}_G = \frac{\pi}{6} d_p^3 \rho_p \left(1 - \frac{\rho}{\rho_p}\right) \boldsymbol{g} \tag{11.57}$$

and $\boldsymbol{F}_E$ denotes the electrostatic force as

$$\boldsymbol{F}_E = q\boldsymbol{E} = -q\nabla V_E \tag{11.58}$$

where q is the electrostatic charge carried by each particle and V_E is the electric potential. For a steady and fully developed flow in an infinitely long pipe, V_E satisfies the Poisson equation in the form

$$\frac{1}{r}\frac{\partial}{\partial r}\left(r \frac{\partial V_E}{\partial r}\right) + \frac{1}{r^2}\frac{\partial^2 V_E}{\partial \theta^2} = -\frac{q}{m}\frac{(1-\alpha)\rho_p}{\epsilon_0} \tag{11.59}$$

where ϵ_0 is the permittivity in a vacuum.

It is convenient to express Eq. (11.56) in an alternative form

$$\frac{\mathrm{d}\boldsymbol{U}_\mathrm{p}}{\mathrm{d}t} = \frac{\boldsymbol{U} - \boldsymbol{U}_\mathrm{p}}{\tau_\mathrm{rp}} + \left(1 - \frac{\rho}{\rho_\mathrm{p}}\right)\boldsymbol{g} - \frac{q\nabla V_\mathrm{E}}{m} \tag{11.60}$$

where τ_rp is the particle relaxation time, given by

$$\frac{1}{\tau_\mathrm{rp}} = C_\mathrm{D}\frac{3}{4}\frac{\rho}{\rho_\mathrm{p}}\frac{|\boldsymbol{U} - \boldsymbol{U}_\mathrm{p}|}{d_\mathrm{p}} = \frac{3}{4}\frac{\mu}{d_\mathrm{p}^2\rho_\mathrm{p}}C_\mathrm{D}\,\mathrm{Re}_\mathrm{p} \tag{11.61}$$

where Re_p is defined by Eq. (3.2). It should be noted that, in general, the drag coefficient C_D is a function of Re_p and the local volume fraction α.

For a fully developed flow, $\mathrm{d}\boldsymbol{U}_\mathrm{p}/\mathrm{d}t$ is zero and, thus, Eq. (11.60) becomes

$$\frac{\boldsymbol{U} - \boldsymbol{U}_\mathrm{p}}{\tau_\mathrm{rp}} + \left(1 - \frac{\rho}{\rho_\mathrm{p}}\right)\boldsymbol{g} - \frac{q\nabla V_\mathrm{E}}{m} = 0 \tag{11.62}$$

which leads to an expression of the particle mass flux due to relaxation under the gravitational force and electrostatic force as

$$(1-\alpha)\rho_\mathrm{p}(\boldsymbol{U}_\mathrm{p} - \boldsymbol{U}) = (1-\alpha)\rho_\mathrm{p}\tau_\mathrm{rp}\left[\left(1 - \frac{\rho}{\rho_\mathrm{p}}\right)\boldsymbol{g} - \frac{q\nabla V_\mathrm{E}}{m}\right] \tag{11.63}$$

Since a fully developed motion excludes deposition of the solid particle phase under field forces (*i.e.*, zero deposition rate), the flux due to diffusion must be equal to the flux due to relaxation under the field force, *i.e.*,

$$-D_\mathrm{p}\nabla(1-\alpha)\rho_\mathrm{p} + (1-\alpha)\rho_\mathrm{p}(\boldsymbol{U}_\mathrm{p} - \boldsymbol{U}) = 0 \tag{11.64}$$

where D_p is the particle diffusivity. Thus, we obtain a relation for the mass flux of particles in the radial direction

$$\frac{D_\mathrm{p}}{\tau_\mathrm{rp}(1-\alpha)}\frac{\partial\alpha}{\partial r} - \frac{q}{m}\frac{\partial V_\mathrm{E}}{\partial r} - \left(1 - \frac{\rho}{\rho_\mathrm{p}}\right)g\sin\beta\cos\theta = 0 \tag{11.65}$$

and in the azimuthal direction

$$\frac{D_\mathrm{p}}{\tau_\mathrm{rp}(1-\alpha)r}\frac{\partial\alpha}{\partial\theta} - \frac{q}{mr}\frac{\partial V_\mathrm{E}}{\partial\theta} + \left(1 - \frac{\rho}{\rho_\mathrm{p}}\right)g\sin\beta\sin\theta = 0 \tag{11.66}$$

It is seen that Eqs. (11.65) and (11.66) are not independent because both can be integrated to the form

$$\frac{D_\mathrm{p}}{\tau_\mathrm{rp}}\ln(1-\alpha) = C_1 - \frac{q}{m}V_\mathrm{E} - gr\sin\beta\cos\theta\left(1 - \frac{\rho}{\rho_\mathrm{p}}\right) \tag{11.67}$$

where the constant of integration $C_1(\theta) = C_1(r) = C_1$ is determined by the boundary conditions of particle concentration (see Example 11.4 and Problem 11.5). To obtain the velocity profiles of both gas and particle phases, we need to examine further the momentum equations for each phase. A fully developed flow requires that

$$\frac{\partial}{\partial t} = 0; \qquad \frac{\partial}{\partial z} = 0; \qquad V = V_\mathrm{p} = W = W_\mathrm{p} = 0 \tag{11.68}$$

Therefore, we only need to investigate the axial component of the momentum equations. Equation (11.59) indicates that the electrostatic force does not contribute to the momentum transfer in the axial direction. For high Reynolds number and Froude number of the gas phase, the gravity effect does not influence the velocity distribution of that phase. The Froude number here is defined as

$$\mathrm{Fr} = \frac{U_0}{\sqrt{2gR_\mathrm{d}}} \tag{11.69}$$

where R_d is the pipe radius and U_0 is the velocity at the centerline of the pipe. It is assumed that the gas velocity is a function of the radial coordinate only.

The momentum equation for the gas in the axial direction is thus expressed by

$$-\frac{\mathrm{d}p}{\mathrm{d}z} - \alpha\rho g\cos\beta + \frac{1}{r}\frac{\mathrm{d}}{\mathrm{d}r}\left(r\mu_\mathrm{e}\frac{\mathrm{d}U}{\mathrm{d}r}\right) - \frac{(1-\alpha)\rho_\mathrm{p}}{\tau_\mathrm{rp}}(U - U_\mathrm{p}) = 0 \tag{11.70}$$

where μ_e is the effective viscosity of turbulence.

The momentum equation for the particle phase in the axial direction is given by

$$-(1-\alpha)\rho_\mathrm{p}\left(1 - \frac{\rho}{\rho_\mathrm{p}}\right)g\cos\beta + \frac{1}{r}\frac{\partial}{\partial r}(r\tau_\mathrm{pzr}) + \frac{1}{r}\frac{\partial\tau_{\mathrm{pz}\theta}}{\partial\theta} + \frac{(1-\alpha)\rho_\mathrm{p}}{\tau_\mathrm{rp}}(U - U_\mathrm{p}) = 0 \tag{11.71}$$

where τ_p is the shear stress of particles. Since the transport of momentum of the particle phase in the mixture for a dilute suspension is mainly caused by the diffusion of particles through the gas phase, the viscosity of the particle phase is given by

$$\mu_\mathrm{p} \approx A(1-\alpha)\rho_\mathrm{p}D_\mathrm{p} \tag{11.72}$$

where A is a constant with the value of about 1. Therefore, we have

$$\tau_\mathrm{pzr} = (1-\alpha)\rho_\mathrm{p}D_\mathrm{p}\frac{\partial U_\mathrm{p}}{\partial r}; \qquad \tau_{\mathrm{pz}\theta} = (1-\alpha)\rho_\mathrm{p}D_\mathrm{p}\frac{\partial U_\mathrm{p}}{r\partial\theta} \tag{11.73}$$

so that Eq. (11.71) gives

$$\begin{aligned} &-(1-\alpha)\left(1 - \frac{\rho}{\rho_\mathrm{p}}\right)g\cos\beta + \frac{1}{r}\frac{\partial}{\partial r}\left(r(1-\alpha)D_\mathrm{p}\frac{\partial U_\mathrm{p}}{\partial r}\right) \\ &\quad + \frac{1}{r^2}\frac{\partial}{\partial\theta}\left((1-\alpha)D_\mathrm{p}\frac{\partial U_\mathrm{p}}{\partial\theta}\right) + \frac{(1-\alpha)}{\tau_\mathrm{rp}}(U - U_\mathrm{p}) = 0 \end{aligned} \tag{11.74}$$

Now we have four independent equations (Eqs. (11.67), (11.59), (11.70), and (11.74)) for four independent variables: volume fraction α, electric potential V_E, gas velocity U, and particle velocity U_p. We may further assume that U and τ of the gas are not influenced by the presence of the particles so that we may directly use the velocity profile for a single-phase pipe flow. For a fully developed turbulent pipe flow for a single phase, we have, for $r < R_\mathrm{d} - \delta$,

$$\frac{U}{U_0} = \left(1 - \frac{r}{R_\mathrm{d}}\right)^{1/7} \tag{11.75}$$

where the thickness of the laminar sublayer δ is given by

$$\frac{\delta}{R_\mathrm{d}} = \frac{60}{\mathrm{Re}^{7/8}} \tag{11.76}$$

with Re defined by

$$\mathrm{Re} = \frac{U_0 R_\mathrm{d} \rho}{\mu} \tag{11.77}$$

Within the laminar sublayer, *i.e.*, $R_\mathrm{d} - \delta < r < R_\mathrm{d}$, we have

$$\frac{U}{U_0} = 0.0225\left(1 - \frac{r}{R_\mathrm{d}}\right)\mathrm{Re}^{3/4} \tag{11.78}$$

The boundary conditions are given by

(1) Electric potential at the wall

$$V_\mathrm{Ew} = V_\mathrm{Ew}(\theta) \tag{11.79}$$

where $V_\mathrm{Ew}(\theta)$ becomes constant for a conducting pipe.

(2) Slip velocity of particles along the wall

$$U_\mathrm{pw} = -L_\mathrm{p} \left.\frac{\mathrm{d}U_\mathrm{p}}{\mathrm{d}r}\right|_{r=R_\mathrm{d}-\frac{d_\mathrm{p}}{2}} \tag{11.80}$$

where L_p is the length scale of the slip velocity at the wall.

(3) Total flow rates of gas and particles

$$\int_0^{2\pi}\int_0^R \alpha\rho U r \,\mathrm{d}r\,\mathrm{d}\theta = J \tag{11.81}$$

and

$$\int_0^{2\pi}\int_0^R (1-\alpha)\rho_\mathrm{p} U_\mathrm{p} r \,\mathrm{d}r\,\mathrm{d}\theta = J_\mathrm{p} \tag{11.82}$$

where J and J_p are the given total mass flow rates for the gas phase and the particle phase, respectively.

Example 11.4 Consider a dilute gas–solid flow in a horizontal pipe made of electrically conducting materials. The pipe is well grounded. The flow is fully developed. Show that (1) the particle concentration is exponentially distributed and is a function of the vertical distance only and (2) the ratio of particle volume fraction at the top to that at the bottom depends on the density ratio of gas to particle, particle diffusivity, and pipe diameter.

Solution For a horizontal pipe, $\beta = 90°$. Since the pipe is well grounded, the electric potential V_E is zero. Therefore, in terms of Eq. (11.67), the distribution of the particle volume fraction can be expressed by

$$\alpha_\mathrm{p} = \alpha_\mathrm{p0}\exp\left[-\frac{y g \tau_\mathrm{rp}}{D_\mathrm{p}}\left(1 - \frac{\rho}{\rho_\mathrm{p}}\right)\right] \tag{E11.12}$$

where $y(= r\cos\theta)$ is the vertical coordinate and α_p0 represents the particle volume fraction at the centerline of the pipe. Note that the integral constant C_1 in Eq. (11.67) in this case can be related to α_p0 by

$$C_1 = \frac{D_\mathrm{p}}{\tau_\mathrm{rp}}\ln\alpha_\mathrm{p0} \tag{E11.13}$$

Equation (E11.12) shows that the particle concentration is exponentially distributed and varies only with respect to the vertical distance. From Eq. (11.67), we also have

$$\frac{(\alpha_p)_{\theta=\pi}}{(\alpha_p)_{\theta=0}} = \exp\left[\frac{2R_d g}{D_p}\tau_{rp}\left(1-\frac{\rho}{\rho_p}\right)\right] \tag{E11.14}$$

which depends on the density ratio of gas to particle, particle diffusivity, and pipe diameter.

11.6.2 *Characteristic Relations*

To generalize the analysis, various equations and boundary conditions are expressed in dimensionless forms. Consider the following dimensionless quantities

$$r^* = \frac{r}{R_d}, \quad \alpha_p^* = \frac{1-\alpha}{1-\alpha_0}, \quad V_E^* = \frac{qV_E\tau_{rp}}{mD_p}$$
$$U^* = \frac{U}{U_0}, \quad U_p^* = \frac{U_p}{U_0} \tag{11.83}$$

where subscript "0" represents $r=0$. In addition to the Reynolds number and Froude number, defined, respectively, by Eq. (11.77) and Eq. (11.69), three dimensionless numbers of relevance are given in the following:

(1) Electrodiffusion number N_{ED}

$$N_{ED} = \frac{q}{m}\frac{R_d^2}{D_p}\sqrt{\frac{(1-\alpha_0)\rho_p}{\epsilon_0}} \tag{11.84}$$

which is the ratio of the displacement by electrostatic repulsion to that by diffusion.

(2) Diffusion response number N_{DF}

$$N_{DF} = \frac{D_p\tau_{rp}}{R_d^2} \tag{11.85}$$

which represents the ratio of the relaxation time to the diffusion time (R_d^2/D_p).

(3) Momentum transfer number N_m

$$N_m = \frac{U_0\tau_{rp}}{R_d} \tag{11.86}$$

which gives the ratio of the relaxation time to the transport time (R_d/U_0).

Using the preceding definitions, the dimensionless forms of Eqs. (11.59), (11.67), and (11.74) can be given by

$$\frac{1}{r^*}\frac{\partial}{\partial r^*}\left(r^*\frac{\partial V_E^*}{\partial r^*}\right) + \frac{1}{r^{*2}}\frac{\partial^2 V_E^*}{\partial\theta^2} = -N_{ED}^2 N_{DF}\alpha_p^* \tag{11.87}$$

$$\ln\alpha_p^* = -V_E^* - \frac{1}{2}\eta r^*\cos\theta \tag{11.88}$$

$$-\frac{\gamma}{2}\alpha_p^* + \frac{1}{r^*}\frac{\partial}{\partial r^*}\left(r^*\alpha_p^*\frac{\partial U_p^*}{\partial r^*}\right) + \frac{1}{r^{*2}}\frac{\partial}{\partial\theta}\left(\alpha_p^*\frac{\partial U_p^*}{\partial\theta}\right) + \frac{\alpha_p^*}{N_{DF}}(U^*-U_p^*) = 0 \tag{11.89}$$

where η and γ are parameters defined by

$$\gamma = \frac{\mathrm{N_m}\cos\beta}{\mathrm{N_{DF}Fr^2}}\left(1 - \frac{\rho}{\rho_\mathrm{p}}\right) \tag{11.90}$$

and

$$\eta = \frac{\mathrm{N_m^2}\sin\beta}{\mathrm{N_{DF}Fr^2}}\left(1 - \frac{\rho}{\rho_\mathrm{p}}\right) \tag{11.91}$$

The boundary conditions now become

(1) Electric potential at the wall

$$V^*_\mathrm{Ew} = V^*_\mathrm{Ew}(\theta) \tag{11.92}$$

(2) Slip velocity of particles along the wall

$$U^*_\mathrm{pw} = -\mathrm{Kn_p}\left.\frac{\mathrm{d}U^*_\mathrm{p}}{\mathrm{d}r^*}\right|_{r^*=1} \tag{11.93}$$

where $\mathrm{Kn_p}$ is the Knudsen number for particle–fluid interaction, defined by

$$\mathrm{Kn_p} = \frac{L_\mathrm{p}}{R_\mathrm{d}} \tag{11.94}$$

(3) Total flow rate of particles

$$J^* = \frac{J_\mathrm{p}}{\pi R_\mathrm{d}^2(1-\alpha_0)\rho_\mathrm{p}U_0} = \frac{1}{\pi}\int_0^{2\pi}\int_0^1 \alpha^*_\mathrm{p}U^*_\mathrm{p}r^*\,\mathrm{d}r^*\mathrm{d}\theta \tag{11.95}$$

For vertical pipe flows ($\eta = 0$) with negligible electric charge effect ($\mathrm{N_{ED}} = 0$), the particle velocity distributions for various values of $\mathrm{N_{DF}}$, $\mathrm{Kn_p}$, and γ are shown in Fig. 11.12. Figure 11.12(a) illustrates that a small $\mathrm{N_{DF}}$ leads to a particle velocity profile close to the gas velocity profile. As expected, a larger $\mathrm{Kn_p}$ gives rise to a larger slip velocity of particles at the wall and results in a flatter velocity profile as shown in Figs. 11.12(a) and (b). In Fig. 11.12(c), it is interesting to note that as γ increases, the direction of the particle flow may change from concurrent to countercurrent to the gas flow, and, thus, the net characteristic mass flux ratio of particle to gas may become zero (fluidized bed) or negative (countercurrent flow). For a large, negative γ, the velocity for the particle phase is higher than that for the gas phase as a result of the gravity effect (downward flow).

The effects of electrostatic charges of particles can be shown in Fig. 11.13, which illustrates the velocity and concentration profiles in vertical pipe flows ($\eta = 0$) with negligible gravity effect for various values of $\mathrm{N_{ED}}$, $\mathrm{N_{DF}}$, and $\mathrm{Kn_p}$. The influence of the electrostatic charge on the magnitude and the profile of the particle velocity is far less significant than that of the particle concentration. The smaller the diffusion response number $\mathrm{N_{DF}}$, the smaller is the velocity lag at the axis. The Knudsen number is seen again as a key parameter accounting for the velocity slip at the boundary.

The mass flow distribution for various values of η in horizontal pipe flows ($\gamma = 0$) with $\mathrm{N_{DF}}$ of 0.01 and $\mathrm{Kn_p}$ of 0.1 is illustrated in Fig. 11.14, which is compared to the experimental data of Wen (1966). The figure shows that a higher η gives rise to a sharper change in the mass flux profile, namely, a much higher mass flux at the bottom of the horizontal pipe

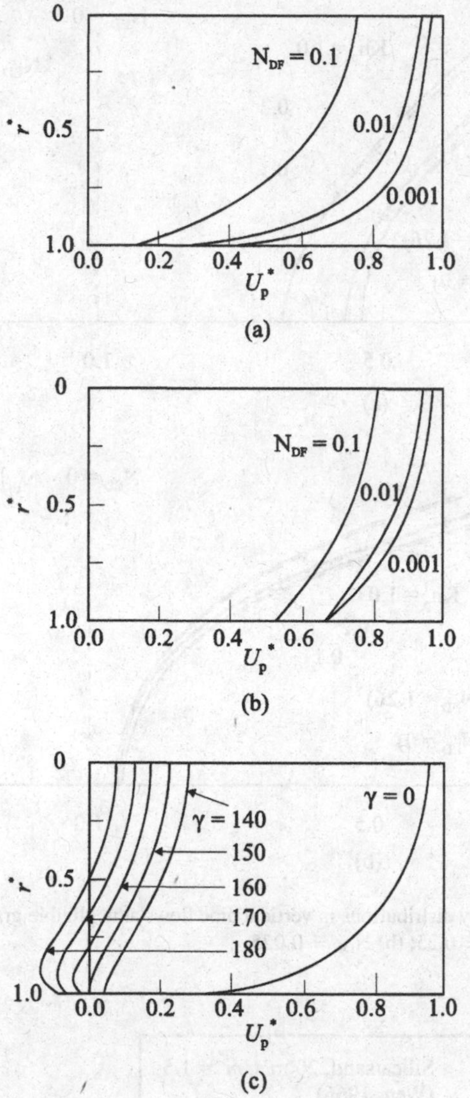

Figure 11.12. Particle velocity distribution in vertical pipe flows with negligible electric charge effect (from Soo, 1990): (a) Effect of N_{DF} at $Kn_p = 0.1$ and $\gamma = 1.0$; (b) Effect of N_{DF} at $Kn_p = 1.0$ and $\gamma = 1.0$; (c) Effect of γ at $Kn_p = 0.1$ and $N_{DF} = 0.01$.

than that at the top. This trend can be understood from Eq. (11.88), which indicates that the particle concentration profile is determined by η in the absence of electrostatic effect. Moreover, in this case, a higher particle mass loading always leads to more nonuniformity in concentration distribution, resulting in a higher value of η.

11.6.3 *Temperature Distributions of Phases*

For a fully developed gas–solid suspension pipe flow where the overall heat transfer rate between the suspension and the pipe wall is dominated by both thermal convection and

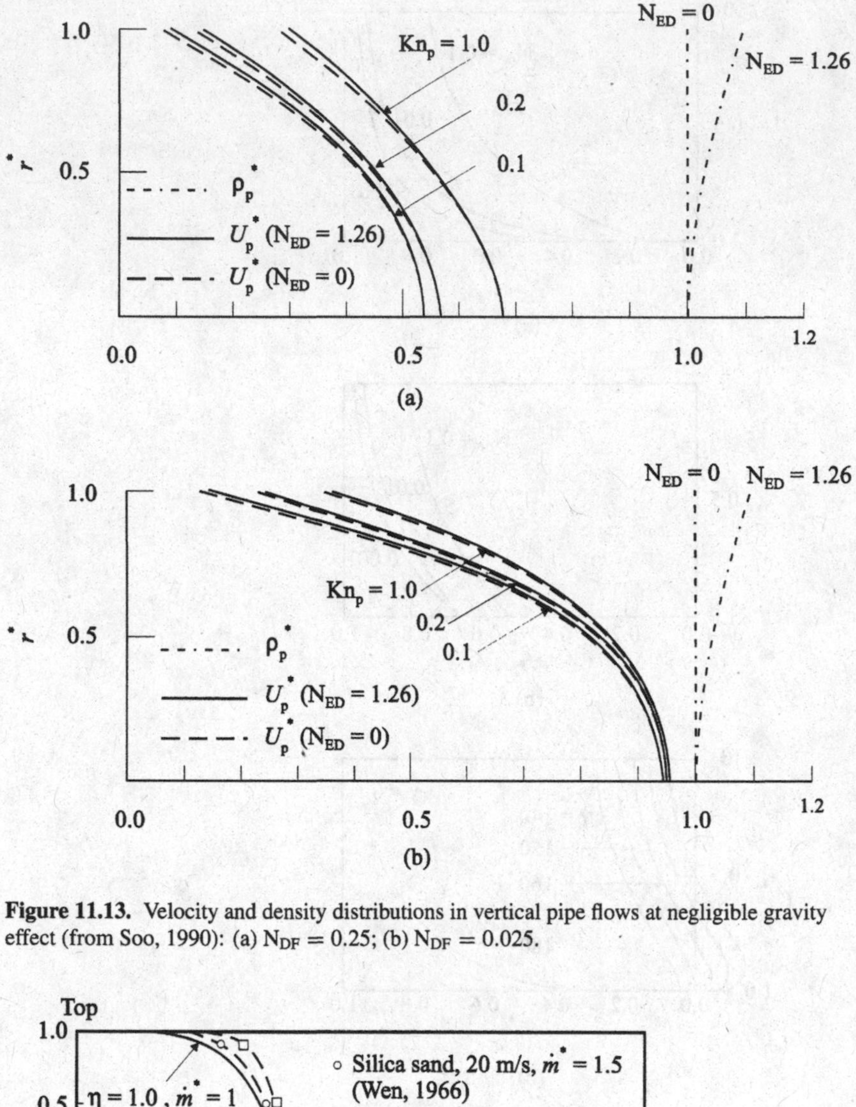

Figure 11.13. Velocity and density distributions in vertical pipe flows at negligible gravity effect (from Soo, 1990): (a) $N_{DF} = 0.25$; (b) $N_{DF} = 0.025$.

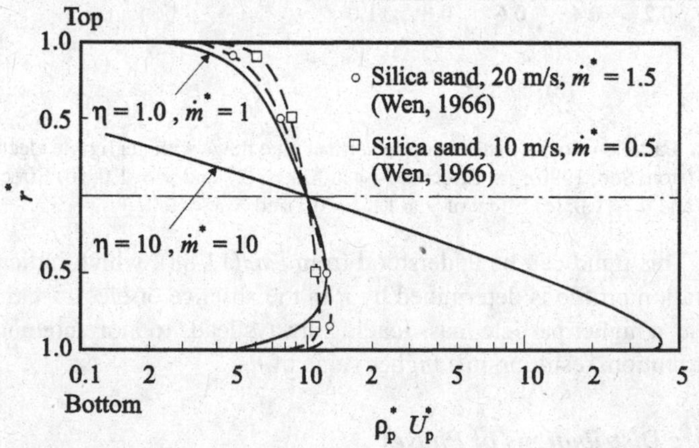

Figure 11.14. Mass flow distributions in horizontal pipe flows at $N_{ED} = 0$, $N_{DF} = 0.01$, and $Kn_p = 0.1$ (from Soo, 1990).

radiation, a thermal analysis of the flow can be made with the following assumptions:

(1) Velocity and volume fraction distributions of phases are independent of the temperature distributions of phases.
(2) Transport properties are constant.
(3) Thermal conductivity of solid particles is very high (Bi $\ll 1$; see Eq. (4.5) for Bi), so that a uniform temperature is maintained in the particle.
(4) Effect of collisions of particles with the pipe wall is negligible.
(5) Particles and wall are perfect gray and gas is transparent for thermal radiation.
(6) Gravitational force is neglected so that the flow is axisymmetric.

The energy equation for a particle can be expressed by

$$mc\frac{\mathrm{d}T_\mathrm{p}}{\mathrm{d}t} = h_\mathrm{p}S_\mathrm{p}(T - T_\mathrm{p}) + \sigma_\mathrm{b}\epsilon_R\left(T_\mathrm{w}^4 - T_\mathrm{p}^4\right) \tag{11.96}$$

where c is the specific heat of particles; h_p is the convective heat transfer coefficient of the particle; S_p is the particle surface area; σ_b is the Stefan–Boltzmann constant; and ϵ_R is the emissivity of the particle. For a steady dilute suspension flow, Eq. (11.96) leads to

$$mcU_\mathrm{p}\frac{\partial T_\mathrm{p}}{\partial z} + h_\mathrm{p}S_\mathrm{p}(T_\mathrm{p} - T) + \sigma_\mathrm{b}\epsilon_\mathrm{R}\left(T_\mathrm{p}^4 - T_\mathrm{w}^4\right) = 0 \tag{11.97}$$

which describes the energy equation for the particle phase.

The energy equation for the gas phase is expressed by

$$\alpha\rho c_\mathrm{p}U\frac{\partial T}{\partial z} = \frac{\alpha}{r}\frac{\partial}{\partial r}\left(rK_\mathrm{e}\frac{\partial T}{\partial r}\right) + nh_\mathrm{p}S_\mathrm{p}(T_\mathrm{p} - T) \tag{11.98}$$

where K_e is the effective thermal conductivity of gas, which includes the effect of eddy diffusion of heat, and n is the particle number density. For given gas and solids velocities and volume fraction distributions, Eqs. (11.97) and (11.98) can be solved numerically with the following boundary conditions:

$$\begin{gathered}\left(\frac{\partial T}{\partial r}\right)_{r=0} = 0, \qquad \left(\frac{\partial T_\mathrm{p}}{\partial r}\right)_{r=0} = 0; \qquad T|_{r=R} = T_\mathrm{w}, \qquad T_\mathrm{p}|_{r=R} = T_\mathrm{w} \\ T|_{z=0} = T_0, \qquad T_\mathrm{p}|_{z=0} = T_0; \qquad T|_{z\to\infty} = T_\mathrm{w}, \qquad T_\mathrm{p}|_{z\to\infty} = T_\mathrm{w}\end{gathered} \tag{11.99}$$

The analytical solution of this problem is difficult because of the highly nonlinear radiation term in Eq. (11.97). However, under some conditions, the term can be simplified. For instance, when the temperature difference between particles and pipe wall is small (*i.e.*, $|T_\mathrm{p} - T_\mathrm{w}| \ll T_\mathrm{w}$), the radiation term in Eq. (11.97) can be linearized to

$$T_\mathrm{w}^4 - T_\mathrm{p}^4 = (T_\mathrm{w} - T_\mathrm{p})(T_\mathrm{w} + T_\mathrm{p})\left(T_\mathrm{w}^2 + T_\mathrm{p}^2\right) \approx 4T_\mathrm{w}^3(T_\mathrm{w} - T_\mathrm{p}) \tag{11.100}$$

Defining nondimensional temperatures

$$T^* = \frac{T - T_\mathrm{w}}{T_0 - T_\mathrm{w}}; \qquad T_\mathrm{p}^* = \frac{T_\mathrm{p} - T_\mathrm{w}}{T_0 - T_\mathrm{w}} \tag{11.101}$$

Eqs. (11.97) and (11.98) can be expressed in dimensionless form as

$$U_\mathrm{p}^*\frac{\partial T_\mathrm{p}^*}{\partial z^*} + \beta_1(T_\mathrm{p}^* - T^*) + \beta_2 T_\mathrm{p}^* = 0 \tag{11.102}$$

where

$$z^* = \frac{z}{R}, \qquad \beta_1 = \frac{Rh_p S_p}{U_0 mc}, \qquad \beta_2 = \frac{4RT_w^3 \sigma_b \epsilon_R}{U_0 mc} \tag{11.103}$$

and

$$U^* \frac{\partial T^*}{\partial z^*} = \frac{1}{r^*}\frac{\partial}{\partial r^*}\left(r^* K^* \frac{\partial T^*}{\partial r^*}\right) + \frac{(1-\alpha)}{\alpha}\beta_3 \left(T_p^* - T^*\right) \tag{11.104}$$

where

$$K^* = \frac{1}{\mathrm{Re}\,\mathrm{Pr}}\frac{K_e}{K}, \qquad \beta_3 = \frac{\rho_p c}{\rho c_p}\beta_1 \tag{11.105}$$

and K^* is a function of r^* whereas Pr is the Prandtl number of the gas and Re is defined by Eq. (11.77). The boundary conditions can be expressed in dimensionless form as

$$\begin{gathered} \left(\frac{\partial T^*}{\partial r^*}\right)_{r^*=0} = 0, \qquad \left(\frac{\partial T_p^*}{\partial r^*}\right)_{r^*=0} = 0; \qquad T^*|_{r^*=1} = 0, \qquad T_p^*|_{r^*=1} = 0 \\ T^*|_{z^*=0} = 1, \qquad T_p^*|_{z^*=0} = 1; \qquad T^*|_{z^*\to\infty} = 0, \qquad T_p^*|_{z^*\to\infty} = 0 \end{gathered} \tag{11.106}$$

It is noted that now the heat transfer of the system is governed by two coupled linear equations, Eqs. (11.102) and (11.104). The last boundary condition in Eq. (11.106) suggests the solution technique for Eqs. (11.102) and (11.104), *i.e.*, separation of variables. Thus, we have

$$T^* = \sum_{n=0}^{\infty} C_n R_n \exp\left(-\lambda_n^2 z^*\right) \tag{11.107}$$

and

$$T_p^* = \sum_{n=0}^{\infty} C_n R_{pn} \exp\left(-\lambda_n^2 z^*\right) \tag{11.108}$$

where $R_n(r^*)$ and $R_{pn}(r^*)$ satisfy

$$R_n = \left(1 + \frac{\beta_2}{\beta_1} - \frac{\lambda_n^2 U_p^*}{\beta_1}\right) R_{pn} \tag{11.109}$$

and

$$\frac{1}{r^*}\frac{d}{dr^*}\left(r^* K^* \frac{dR_n}{dr^*}\right) + \frac{1-\alpha}{\alpha}\beta_3 (R_{pn} - R_n) + \lambda_n^2 U^* R_n = 0 \tag{11.110}$$

with the boundary conditions

$$\left(\frac{dR_n}{dr^*}\right)_{r^*=0} = 0, \qquad \left(\frac{dR_{pn}}{dr^*}\right)_{r^*=0} = 0; \qquad R_n|_{r^*=1} = 0, \qquad R_{pn}|_{r^*=1} = 0 \tag{11.111}$$

Substitution of Eq. (11.109) into Eq. (11.110) yields

$$\frac{1}{r^*}\frac{d}{dr^*}\left(r^* K^* \frac{dR_n}{dr^*}\right) + \frac{1-\alpha}{\alpha}\beta_3 \left[\left(1 + \frac{\beta_2}{\beta_1} - \frac{\lambda_n^2 U_p^*}{\beta_1}\right)^{-1} - 1 + \frac{\lambda_n^2 U^* \alpha}{(1-\alpha)\beta_3}\right] R_n = 0 \tag{11.112}$$

with

$$\left(\frac{\mathrm{d}R_{\mathrm{n}}}{\mathrm{d}r^*}\right)_{r^*=0} = 0, \qquad R_{\mathrm{n}}|_{r^*=1} = 0 \tag{11.113}$$

where all β's are constants and α, K^*, U^*, and U_{p}^* are given functions of r^* (or relations independently solved from the continuity and momentum equations).

The problem imposed by Eq. (11.112) with boundary conditions of Eq. (11.113) is noted as the Sturm–Liouville boundary-value problem [*e.g.*, Derrick and Grossman, 1987], if

$$\frac{\lambda_{\mathrm{n}}^2 U_{\mathrm{p}}^*}{\beta_1 + \beta_2} < 1 \tag{11.114}$$

Since U_{p}^*, β_1, and β_2 are uniquely characteristic for the operating system, the validity of the preceding inequality is largely dependent on the values of λ_{n}^2. As indicated in Eq. (11.107), only the first few successive eigenvalues starting from the smallest one are needed. Therefore, Eq. (11.114) would hold true for the first few eigenvalues. Thus, we have

$$\left(1 + \frac{\beta_2}{\beta_1} - \frac{\lambda_{\mathrm{n}}^2 U_{\mathrm{p}}^*}{\beta_1}\right)^{-1} = \frac{\beta_1}{\beta_1 + \beta_2}\left(1 - \frac{\lambda_{\mathrm{n}}^2 U_{\mathrm{p}}^*}{\beta_1 + \beta_2}\right)^{-1} \approx \frac{\beta_1}{\beta_1 + \beta_2}\left(1 + \frac{\lambda_{\mathrm{n}}^2 U_{\mathrm{p}}^*}{\beta_1 + \beta_2}\right) \tag{11.115}$$

which leads to Eq. (11.110) in the form of the Sturm–Liouville problem

$$\frac{\mathrm{d}}{\mathrm{d}r^*}\left(\zeta(r^*)\frac{\mathrm{d}R_{\mathrm{n}}}{\mathrm{d}r^*}\right) + \left[\eta(r^*) + \lambda_{\mathrm{n}}^2 \xi(r^*)\right] R_{\mathrm{n}} = 0 \tag{11.116}$$

where

$$\begin{aligned} &\zeta(r^*) = r^* K^*, \qquad \eta(r^*) = -\frac{1-\alpha}{\alpha}\frac{\beta_3 \beta_2}{\beta_1 + \beta_2} r^* \\ &\xi(r^*) = \frac{1-\alpha}{\alpha}\frac{\beta_3 \beta_1}{(\beta_1 + \beta_2)^2} r^* U_{\mathrm{p}}^* + r^* U^* \end{aligned} \tag{11.117}$$

with the boundary conditions given in Eq. (11.113).

In principle, eigenfunction R_{n} and its eigenvalue λ_{n} can be obtained numerically by solving Eqs. (11.113) and (11.116). The coefficients are determined in terms of the boundary conditions in Eq. (11.106) and the orthogonality property in the Sturm–Liouville theorems as

$$C_{\mathrm{n}} = \frac{\int_0^1 \xi R_{\mathrm{n}}\,\mathrm{d}r^*}{\int_0^1 \xi R_{\mathrm{n}}^2\,\mathrm{d}r^*} \tag{11.118}$$

In addition, the heat flux to the gas phase at the wall is given by

$$J_{\mathrm{qw}} = K\frac{T_0 - T_{\mathrm{w}}}{R}\left(\frac{\partial T^*}{\partial r^*}\right)_{r^*=1} = K\frac{T_0 - T_{\mathrm{w}}}{R}\sum_{n=0}^{\infty} C_{\mathrm{n}}\left(\frac{\mathrm{d}R_{\mathrm{n}}}{\mathrm{d}r^*}\right)_{r^*=1} \exp\left(-\lambda_{\mathrm{n}}^2 z^*\right) \tag{11.119}$$

The preceding analysis is based on Tien's model of heat transfer by a gas–solid suspension in turbulent pipe flow [Tien, 1961]. However, nonuniform distribution of solids, slip between solids and gas, and effect of thermal radiation were excluded in Tien's work.

Nomenclature

A	Cross-sectional area
A	Constant, defined by Eq. (11.22)
A	Constant, defined by Eq. (11.48)
A	Constant, defined by Eq. (11.72)
B	Factor, defined by Eq. (11.46)
Bi	Biot number
C_D	Drag coefficient
C_w	Perimeter
c	Specific heat of particles
c_p	Specific heat of gas at constant pressure
D	Diffusivity of gas
D_d	Pipe diameter
D_p	Diffusivity of particles
d_p	Particle diameter
$\boldsymbol{E}$	Electric field strength vector
Fr	Froude number, defined by Eq. (11.69)
F_D	Drag force
F_f	Friction force
F_l	Lift force
$\boldsymbol{F}_A$	Interfacial force vector between gas and particles
$\boldsymbol{F}_E$	Electrostatic force vector
$\boldsymbol{F}_G$	Gravitational force vector including buoyancy effect
f	Friction factor
f_B	Friction factor of a bend
f_p	Friction coefficient of particles at wall
g	Gravitational acceleration
h_p	Convective heat transfer coefficient
J	Mass flux of gas
J_p	Mass flux of particles
J_p^*	Total flow rate of particles
$\boldsymbol{J}_m$	Momentum flux tensor of gas
$\boldsymbol{J}_{mp}$	Momentum flux tensor of particles
K	Thermal conductivity of gas
K_e	Effective thermal conductivity of gas
Kn_p	Knudsen number for particle–fluid interaction, defined by Eq. (11.94)
L	Pipe length
L_a	Acceleration length
L_B	Equivalent length of straight pipe
L_p	Length scale of slip velocity at wall
l_m	Mixing length of turbulence
m	Mass of a particle
$\dot{m}$	Mass flux ratio of particle to gas
N	Normal impact force
N_{DF}	Diffusion response number, defined by Eq. (11.85)
N_{ED}	Electrodiffusion number, defined by Eq. (11.84)
N_m	Momentum transfer number, defined by Eq. (11.86)
n	Particle number density
Pr	Prandtl number
p	Pressure
q	Electrostatic charges carried by a particle
R_B	Bend radius
R_d	Pipe radius
Re	Reynolds number
r	Cylindrical coordinate
S_p	Particle surface area
T	Gas temperature
T_0	Temperature at the inlet
T_p	Particle temperature
T_w	Wall temperature
U	Gas velocity component

U	Mean stream velocity
U_0	Gas velocity at pipe centerline
U_1	Gas velocity at the edge of shedding layer
U_f	Friction velocity based on the mixture density
U_{f0}	Friction velocity at minimum transport and zero solids concentration
U_m	Averaged axial velocity of gas
U_p	Particle velocity component
U_{pi}	Particle velocity at inlet
U_{pt}	Particle terminal velocity
U_{pw}	Particle velocity at wall
$U_{p\infty}$	Particle velocity in the fully developed flow
U^*	Dimensionless gas velocity component
U_p^*	Dimensionless particle velocity component
$\boldsymbol{U}$	Gas velocity vector
$\boldsymbol{U}_p$	Particle velocity vector
u'	Gas velocity fluctuation
u'_p	Particle velocity fluctuation
u'_{pw}	Particle velocity fluctuation at wall
V	Gas velocity component
V_E	Electric potential
V_{Ew}	Electric potential at wall
V_p	Particle velocity component
V_t	Tangential velocity near the wall
W	Gas velocity component
W_p	Particle velocity component
x	Axial coordinate
x	Cartesian coordinate
y	Cartesian coordinate

Greek Symbols

α	Volume fraction of gas
α_p	Volume fraction of particles
α_{pw}	Volume fraction of particles at wall
β	Inclined angle of pipe
γ	Parameter, defined by Eq. (11.90)
Δp	Pressure drop
Δp_o	Pressure drop in a single gas flow
Δp_a	Pressure drop over the acceleration region
Δp_{Bg}	Pressure drop of gas over a bend
Δp_{Bp}	Additional pressure drop due to particles over a bend
Δp_L	Pressure drop over an equivalent length of a straight pipe
δ	Shedding layer thickness
ϵ_0	Permittivity in a vacuum
ϵ_R	Particle emissivity
η	Parameter, defined by Eq. (11.91)
θ	Cylindrical coordinate
θ	Inclined angle, defined by Fig. 11.8
θ	Angle in the transverse plane with respect to the outward direction of the symmetry line, defined by Fig. 11.9
μ	Dynamic viscosity of gas
μ_e	Effective turbulent viscosity in a single phase flow
μ_{gp}	Turbulent viscosity of the gas phase in the presence of particles
μ_p	Viscosity of particles
ν	Kinematic viscosity of gas
ξ	Distance normal to wall
ρ	Density of gas
ρ_p	Density of particles
σ_b	Stefan–Boltzmann's constant
$\boldsymbol{\tau}_e$	Effective stress tensor per unit volume

τ_f Characteristic time of turbulent fluctuations of gas
τ_p Shear stress of particles
τ_{gp} Wall shear stress of gas in the presence of particles
τ_{rp} Particle relaxation time
τ_S Stokes relaxation time
τ_w Wall shear stress
τ_{wp} Shear stress due to collisions of particles with the wall
τ_θ Shear stress along θ-axis
τ_ϕ Shear stress along ϕ-axis
ϕ Angle measured in the plane of the curved pipe axis, defined by Fig. 11.9

References

Blatch, N. S. (1906). Discussion: Water Filtration at Washington, D.C. *Trans. Am. Soc. Civil Eng.*, **57**, 400.

Boyce, M. P. and Blick, E. F. (1969). Fluid Flow Phenomena in Dusty Air. ASME Paper No. 69-WA/FE-24.

Chen, T. Y., Walawender, W. P. and Fan, L. T. (1979). Solids Transfer Between Fluidized Beds: A Review. *J. Powder Bulk Solids Tech.*, **3**, 3.

Chien, N. and Asce, A. M. (1956). The Present Status of Research on Sediment Transport. *Trans. Am. Soc. Civil Eng.*, **121**, 833.

Crowe, C. T. (1982). Review: Numerical Models for Dilute Gas–Particle Flows. *Trans. ASME, J. Fluids Eng.*, **104**, 297.

Derrick, W. R. and Grossman, S. I. (1987). *Introduction to Differential Equations with Boundary–Value Problems*, 3rd ed. St. Paul, Minn.: West Publishing.

Haag, A. (1967). Velocity Losses in Pneumatic Conveyer Pipe Bends. *British Chemical Engineering*, **12**, 65.

Halow, J. S. (1973). Incipient Rolling, Sliding, and Suspension of Particles in Horizontal and Inclined Turbulent Flow. *Chem. Eng. Sci.*, **28**, 1.

Haltsrom, E. A. N. (1953). *Design of Experimental Apparatus for the Study of Two-Phase Flow in Circular Straight Pipe*. M.S. Thesis. Princeton University.

Ito, H. (1959). Friction Factors for Turbulent Flow in Curved Pipes. *Trans. ASME, J. Basic Eng.*, **81D**, 123.

Ito, H. (1960). Pressure Loses in Smooth Pipe Bends. *Trans. ASME, J. Basic Eng.*, **82D**, 131.

Kane, R. S. (1989). Drag Reduction by Particle Addition. *Viscous Drag Reduction in Boundary Layers*. Vol. 123. *Progress in Astronautics and Aeronautics*. Ed. Bushnell and Hefner. Washington, D.C.: AIAA.

Kane, R. S. and Pfeffer, R. (1973). Characteristics of Dilute Gas–Solids Suspensions in Drag Reduction Flow. *NASA CR-2267*, 1973.

Klinzing, G. E. (1981). *Gas–Solid Transport*. New York: McGraw–Hill.

Konno, H. and Saito, S. (1969). Pneumatic Conveying of Solids Through Straight Pipes. *J. Chem. Eng. Japan*, **2**, 211.

Konrad, K. (1986). Dense-Phase Pneumatic Conveying: A Review. *Powder Tech.*, **49**, 1.

Konrad, K., Harrison, D., Nedderman, R. M. and Davidson, J. F. (1980). Prediction of the Pressure Drop for Horizontal Dense-Phase Pneumatic Conveying of Particles. *Pneumotransport*, **5**, 225.

Kunii, D. and Levenspiel, O. (1969). *Fluidization Engineering*. New York: Wiley.

Leung, L. S. (1980). Vertical Pneumatic Conveying: A Flow Regime Diagram and a Review of Choking Versus Nonchoking Systems. *Powder Tech.*, **25**, 185.

Leva, M. (1959). *Fluidization*. New York: McGraw-Hill.

Marcus, R. D., Leung, L. S., Klinzing, G. E. and Rizk, F. (1990). *Pneumatic Conveying of Solids*. New York: Chapman & Hall.

McCarthy, H. E. and Olson, J. H. (1968). Turbulent Flow of Gas–Solids Suspensions. *I& EC Fund.*, **7**, 471.

Moody, L. F. (1944). Friction Factors for Pipe Flow. *ASME Trans.*, **66**, 671.

Peters, L. K. and Klinzing, G. E. (1972). Friction in Turbulent Flow of Gas–Solid Systems. *Can. J. Chem. Eng.*, **50**, 441.

Rose, H. E. and Duckworth, R. A. (1969). Transport of Solid Particles in Liquids and Gases. *The Engineer*, **227**, 392.

Roco, M. and Shook, C. (1984). Computational Method for Coal Slurry Pipelines with Heterogeneous Size Distribution. *Powder Tech.*, **39**, 159.

Schlichting, H. (1979). *Boundary Layer Theory*, 7th ed. New York: McGraw-Hill.

Schuchart, P. (1968). Widerstandsgesetze Beim Pneumatischen Transport in Rohrkrümmern. *Chem.-Ing.-Tech.*, **40**, 1060.

Shimizu, A., Echigo, R., Hasegawa, S. and Hishida, M. (1978). Experimental Study on the Pressure Drop and the Entry Length of the Gas–Solid Suspension Flow in a Circular Tube. *Int. J. Multiphase Flow*, **4**, 53.

Soo, S. L. (1962). Boundary Layer Motion of a Gas–Solid Suspension. *Proceedings of Symposium on Interaction Between Fluids and Particles*. Institute of Chemical Engineers, London.

Soo, S. L. (1989). *Particulates and Continuum: Multiphase Fluid Dynamics*. New York: Hemisphere.

Soo, S. L. (1990). *Multiphase Fluid Dynamics*. Beijing: Science Press; Brookfield, Vt.: Gower Technical.

Soo, S. L. and Mei, R. (1987). Dense Suspension and Fluidization. *Proceedings of Powder and Bulk Solids, 1987 Conference*, Rosemont, Ill.

Soo, S. L., Trezek, G. L., Dimick, R. C. and Hohnstreiter, G. F. (1964). Concentration and Mass Flow Distributions in a Gas–Solid Suspension. *I&EC Fund.*, **3**, 98.

Soo, S. L. and Tung, S. K. (1971). Pipe Flow of Suspensions in Turbulent Fluid: Electrostatic and Gravity Effects. *Appl. Sci. Res.*, **24**, 83.

Stoess, H. A. (1983). *Pneumatic Conveying*, 2nd ed. New York: John Wiley & Sons.

Taylor, G. I. (1954). The Dispersion of Matter in Turbulent Flow Through a Pipe. *Proc. R. Soc. London*, **A223**, 446.

Thomas, D. G. (1962). Transport Characteristics of Suspensions. Part IV. Minimum Transport Velocity for Large Particle Size Suspension in Round Horizontal Pipes. *AIChE J.*, **8**, 373.

Tien, C. L. (1961). Heat Transfer by a Turbulently Flowing Fluid–Solids Mixture in a Pipe. *Trans. ASME, J. Heat Transfer*, **83C**, 183.

Tien, C. L. and Quan, V. (1962). Local Heat Transfer Characteristics of Air–Glass and Air–Lead Mixtures in Turbulent Pipe Flow. ASME Paper No. 62–HT-15.

Wen, C. Y. (1966). In *Pneumatic Transportation of Solids*. Ed. Spencer, Joyce, and Farber. Washington, D.C.: Bureau of Mines Information Circular, U.S. Deptartment of Interior.

Williams, O. A. (1983). *Pneumatic and Hydraulic Conveying of Solids*. New York: Marcel Dekker.

Yang, W. C. (1987). Pneumatic Transport in a 10 cm Pipe Horizontal Loop. *Powder Tech.*, **49**, 207.
Zenz, F. A. (1964). Conveyability of Materials of Mixed Particle Size. *I & EC Fund.*, **3**, 65.

Problems

11.1 In a gas–solid horizontal pipe flow, the particle used is 100 μm glass bead with a density of 2,500 kg/m^3. The pipe diameter is 50 mm. The average particle volume fraction in the pipe is 1 percent. The gas density and kinematic viscosity are 1.2 kg/m^3 and 1.5×10^{-5} m^2/s, respectively. Estimate the minimum transport velocity and power consumption per unit length. Repeat this problem for a 10 μm glass bead size.

11.2 For a single-phase flow in a bend, the momentum integral equations for the shedding layer of the secondary flow can be expressed by Eqs. (11.36), (11.37), and (11.38) [Ito, 1959]. On the basis of these equations and the approximations in Eq. (11.45), (a) prove Eq. (11.46) and (b) show that the thickness of the shedding layer can be estimated by Eq. (11.47).

11.3 The pressure drop of a gas flow over a 90° bend may be estimated by either Eq. (11.23) or Eq. (11.50). For the ratios of the pipe radius to the bending radius, R_d/R_B, of 0.1, 0.05, and 0.01, compare the pressure drop predictions using these two equations for the Reynolds numbers ranging from 4,000 to 10^5.

11.4 Consider a pneumatic transport of grains in a sheet metal duct. The sliding frictional coefficient f_p is about 0.36 [Haag, 1967]. To ensure a smooth operation over a 90° bend, the grain particles should always be sliding against the wall during the turn. Estimate the minimum particle velocity required for such a smooth operation. The pipe diameter is 0.1 m. The bend radius is 0.5 m.

11.5 Consider a dilute gas–solid flow in a horizontal rectangular pipe made of electrically conducting materials. The pipe is well grounded. The flow is fully developed. It is assumed that the particle volume fraction distribution in the vertical direction is the same as that in a circular pipe flow. Find out (a) the cross-sectionally averaged particle volume fraction in terms of the particle volume fraction at the centerline and (b) the vertical location at which the particle volume fraction represents the cross-sectionally averaged particle volume fraction.

11.6 Consider a dilute gas–solid flow in a pipe in which the solid particles carry significant electrostatic charges. It is assumed that (a) the flow is fully developed; (b) the gravitational effect is negligible; and (c) the flow and the electrostatic field are axisymmetric. Derive an expression to describe the radial volume fraction distribution of the particles and identify the radial locations where the particle volume fractions are maximum and minimum in the distribution. Also, if the electrostatic charge effects are negligible, derive an expression to describe the radial volume fraction distribution of the particles.

11.7 Consider a fully developed dilute gas–solid flow in a vertical pipe in which solid particles carry significant electrostatic charges. The particle charges vary radially. It is assumed that (a) the flow and the electrostatic field are axisymmetric and (b) the radial charge distribution $C(r)$, defined as $q(r)\alpha_p(r)$, is known. Derive an expression for the radial volume fraction distribution of the particles.

CHAPTER 12

Heat and Mass Transfer Phenomena in Fluidization Systems

12.1 Introduction

Among many gas–solid flow systems involving heat and mass transfer operations, the fluidization system is one of the most frequently encountered. The heat and mass transfer behavior in a gas–solid fluidized bed is important when physical or chemical operations are conducted in the bed, such as drying, coal combustion, polymerization reaction, and chemical synthesis. The fluidized bed is characterized by a high efficiency of heat exchange between particles and the fluidizing gas. The inherent intensive gas–solid contact or rapid mixing between gas and solid phases contributes to this efficient heat transfer. The fluidized bed possesses high heat capacity, and uniform temperature in the bed can generally be maintained. Thus, temperature control in the bed can be effectively carried out. Comparable characteristics are also exhibited for the fluidized bed mass transfer between gas and particles and between the bed and gas or particles.

The heat and mass transfer properties can be represented by heat and mass transfer coefficients, which are commonly given in empirical or semiempirical correlation form. The transfer coefficient is defined in terms of flow models under specific flow conditions and geometric arrangements of the flow system. Thus, when applying the correlations, it is necessary to employ the same flow model to describe the heat and mass transfer coefficients for conditions comparable to those where the correlations were obtained. An accurate characterization of the heat and mass transfer can be made only when the hydrodynamics and underlying mechanism of the transport processes are well understood.

The governing heat transfer modes in gas–solid flow systems include gas–particle heat transfer, particle–particle heat transfer, and suspension–surface heat transfer by conduction, convection, and/or radiation. The basic heat and mass transfer modes of a single particle in a gas medium are introduced in Chapter 4. This chapter deals with the modeling approaches in describing the heat and mass transfer processes in gas–solid flows. In multiparticle systems, as in the fluidization systems with spherical or nearly spherical particles, the conductive heat transfer due to particle collisions is usually negligible. Hence, this chapter is mainly concerned with the heat and mass transfer from suspension to the wall, from suspension to an immersed surface, and from gas to solids for multiparticle systems. The heat and mass transfer mechanisms due to particle convection and gas convection are illustrated. In addition, heat transfer due to radiation is discussed.

12.2 Suspension-to-Surface Heat Transfer

A mechanistic account of suspension-to-surface heat transfer is necessary to quantify the heat transfer behavior accurately and to assess the form of dependency of dimensionless groups in the correlations. In the following, the modes and regimes of suspension-to-surface heat transfer along with the three mechanistic models accounting for this heat transfer behavior are described.

12.2.1 *Heat Transfer Modes and Regimes*

Heat transfer between a fluidized bed and an immersed surface can occur by three modes, namely, particle convection, gas convection, and radiation.

Particle convective heat transfer is due to the convective flow of solid particles from the bulk of the bed to the region adjacent to the heat transfer surface. Solid particles gain heat by thermal conduction from the heated surface (assuming the surface has a higher temperature than the solids). As the particles return to the bulk of the bed, the heat is dissipated. The capacity of moving particles to transfer heat is reflected by the extent of heat conduction by the suspension. Gas convective heat transfer is caused by gas percolating through the bed and also by gas voids in contact with the surface. Radiative heat transfer is due to radiant heat transmitted to fluidized particles or solid surfaces from a heat transfer surface at high temperature. The total heat transfer coefficient, h, can be estimated from the summation of the individual heat transfer coefficients for particle convection, h_{pc}; gas convection, h_{gc}; and radiation, h_r, although their precise relationship may not be additive, *i.e.*,

$$h \approx h_{pc} + h_{gc} + h_r \tag{12.1}$$

The heat transfer coefficients are strongly influenced by the operating conditions of the fluidized bed. Variations in the operating conditions such as bubbling and spouting yield a varied bed structure and hence varied heat transfer coefficients. Understanding the governing mechanisms in heat transfer is important to the development of simplified heat transfer models or equations. The relative importance of the heat transfer modes for the suspension-to-surface heat transfer in gas–solid fluidized beds is illustrated in Fig. 12.1 [Flamant *et al.*, 1992]. The figure indicates that the governing modes of heat transfer mechanism in the fluidized bed depend on both particle size and bed/surface temperature.

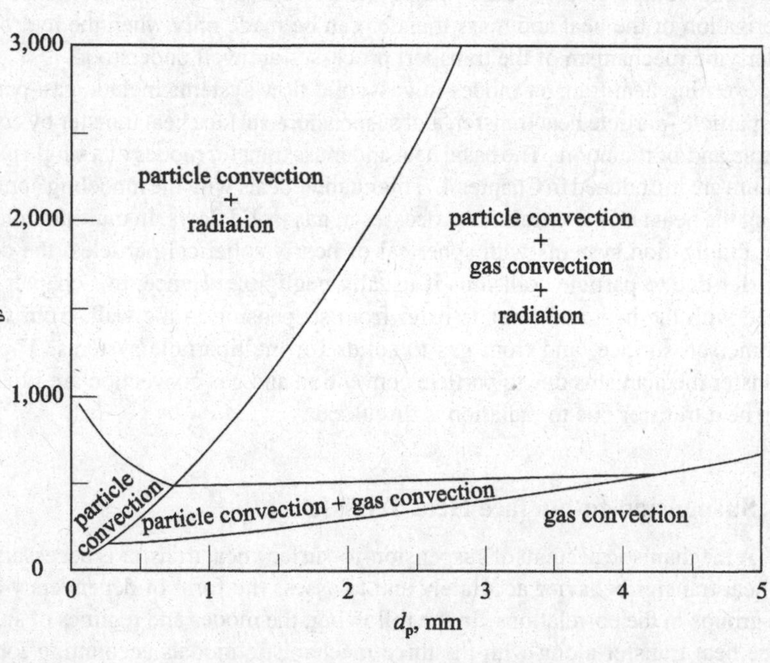

Figure 12.1. Heat transfer diagram for various governing modes (from Flamant *et al.*, 1992).

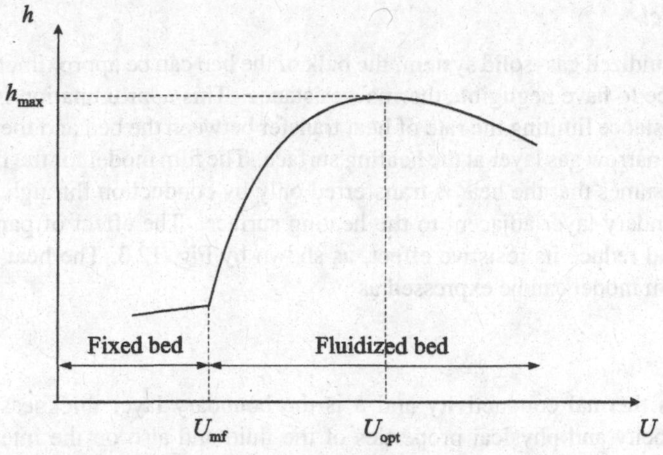

Figure 12.2. Typical dependence of the heat transfer coefficient on gas velocity in dense-phase fluidization systems (from Gel'Perin and Einstein, 1971).

It is seen that particle convection is important for almost all the conditions, except at low bed temperatures in a bed of large particles, where the gas convection becomes important.

In most dense gas–solid fluidization systems, particle circulation (*e.g.*, that induced by the bubbles) is the primary cause of particle convective heat transfer. When particles initially reach the heat transfer surface, the temperature gradient is high and yields high heat transfer rates. As time progresses, the temperature gradient reduces, yielding low heat transfer rates. Thus, the heat transfer rate is high when there is an extensive solids exchange between the bulk and the surface. Increasing particle circulation enhances the heat transfer rate.

Development of a mechanistic model is essential to quantification of the heat transfer phenomena in a fluidized system. Most models that are originally developed for dense-phase fluidized systems are also applicable to other fluidization systems. Figure 12.2 provides basic heat transfer characteristics in dense-phase fluidization systems that must be taken into account by a mechanistic model. The figure shows the variation of heat transfer coefficient with the gas velocity. It is seen that at a low gas velocity where the bed is in a fixed bed state, the heat transfer coefficient is low; with increasing gas velocity, it increases sharply to a maximum value and then decreases. This increasing and decreasing behavior is a result of interplay between the particle convective and gas convective heat transfer which can be explained by mechanistic models given in §12.2.2, §12.2.3, and §12.2.4.

Numerous models have been developed to describe heat transfer behavior in fluidization systems. They can be broadly categorized into the following three models: the film model, the single-particle model, and the emulsion phase/packet model. Each model has its limitations in application. Accordingly, a given model may be more suitable to some fluidization conditions than the others. For example, the film model and the single-particle model are more suitable for a particulate fluidized bed [Gel'Perin and Einstein, 1971] than for a system that contains gas bubbles. The components contributing to heat transfer given by Eq. (12.1) can be evaluated separately by different models. The most challenging aspect of applying these models, however, lies in the determination of the flow and thermal properties of the region directly adjacent to the surface where heat transfer takes place.

12.2.2 *Film Model*

In a well-fluidized gas–solid system, the bulk of the bed can be approximated to be isothermal and hence to have negligible thermal resistance. This approximation indicates that the thermal resistance limiting the rate of heat transfer between the bed and the heating surface lies within a narrow gas layer at the heating surface. The film model for the fluidized bed heat transfer assumes that the heat is transferred only by conduction through the thin gas film or gas boundary layer adjacent to the heating surface. The effect of particles is to erode the film and reduce its resistive effect, as shown by Fig. 12.3. The heat transfer coefficient in the film model can be expressed as

$$h = \frac{K}{\delta} \tag{12.2}$$

where K is the gas thermal conductivity and δ is the boundary layer thickness, which depends on the velocity and physical properties of the fluid and also on the intensity of motion of the solid particles that erode the boundary layer. With an increase in the gas velocity, the particles near the surface move more vigorously but the local concentration of the particles decreases. This interplay results in a maximum in the h–U curve, as shown in Fig. 12.2.

The model based on the concept of pure limiting film resistance involves the steady-state concept of the heat transfer process and omits the essential unsteady nature of the heat transfer phenomena observed in many gas–solid suspension systems. To take into account the unsteady heat transfer behavior and particle convection in fluidized beds, a surface renewal model can be used. The model accounts for the film resistance adjacent to the heat transfer

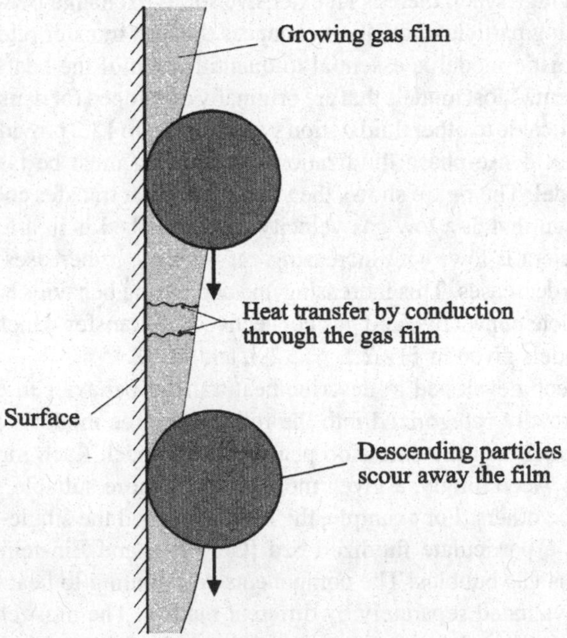

Figure 12.3. Conceptual representation of the thin-film model (from Levenspiel and Walton, 1954).

surface, induced by the particle–gas emulsion phase instead of the pure gas phase. The model is capable of explaining the initial increase in the observed heat transfer with the gas velocity beyond the point of onset of fluidization. Thus, the models suggest different mechanisms useful to describe high heat transfer coefficients in fluidized beds [Botterill, 1975].

The film model discounts the effects of thermophysical properties such as the specific heat of solids and, thus, is not able to predict the particle convective component of heat transfer. Moreover, the measured temperature gradient extends deeper into the bed from the heating surface than that which can be predicted by the film model [Baeyens and Goossens, 1973]. Thus, models based on heat conduction through the gas boundary layer cannot be useful for estimating the contribution of the particle convective component of heat transfer, and the emulsion phase/packet model given in §12.2.4 should be used to describe the temperature gradient from the heating surface to the bed.

12.2.3 *Single-Particle Model*

The single-particle model [Zabrodsky, 1963] postulates that the moving solid particles play an important role in heat transfer by thermal conduction. The model takes into account the thermal conduction through the layer of gas at the heating surface. Thus, allowance is made for the influence of the thermophysical properties of the solid material. In this model, the contributions from gas convection and bed-to-surface thermal radiation are excluded. The high heat transfer coefficient shown in Fig. 12.2 is due to the high temperature gradient during heating of moving solids. The presence of the maximum on the heat transfer curve with gas velocity is a result of the simultaneous effect of the rise in temperature gradient and the fall in concentration of solid particles.

The simplest model of this kind can be represented by one in which an isolated particle surrounded by gas is in contact with or in the vicinity of the heating surface for a certain time, during which the heat transfer between the particle and the heating surface takes place by transient conduction, as shown in Fig. 12.4. In terms of the model, the Fourier equation of thermal conduction can be expressed as

$$\rho_p c \frac{\partial T_p}{\partial t} = K_p \nabla^2 T_p \tag{12.3}$$

for the particle phase and

$$\rho c_p \frac{\partial T}{\partial t} = K \nabla^2 T \tag{12.4}$$

for the gas phase.

The boundary and initial conditions for Eqs. (12.3) and (12.4) are given by the following:

(1) At the symmetric plane

$$\frac{\partial T_p}{\partial n} = 0; \qquad \frac{\partial T}{\partial n} = 0 \tag{12.5}$$

where $\partial T_p/\partial n$ and $\partial T/\partial n$ are the normal derivatives with respect to the temperature symmetric plane.

(2) At the heating surface

$$T_p = T_s; \qquad T = T_s \tag{12.6}$$

For the case without particle–surface contact, only $T = T_s$ should be imposed.

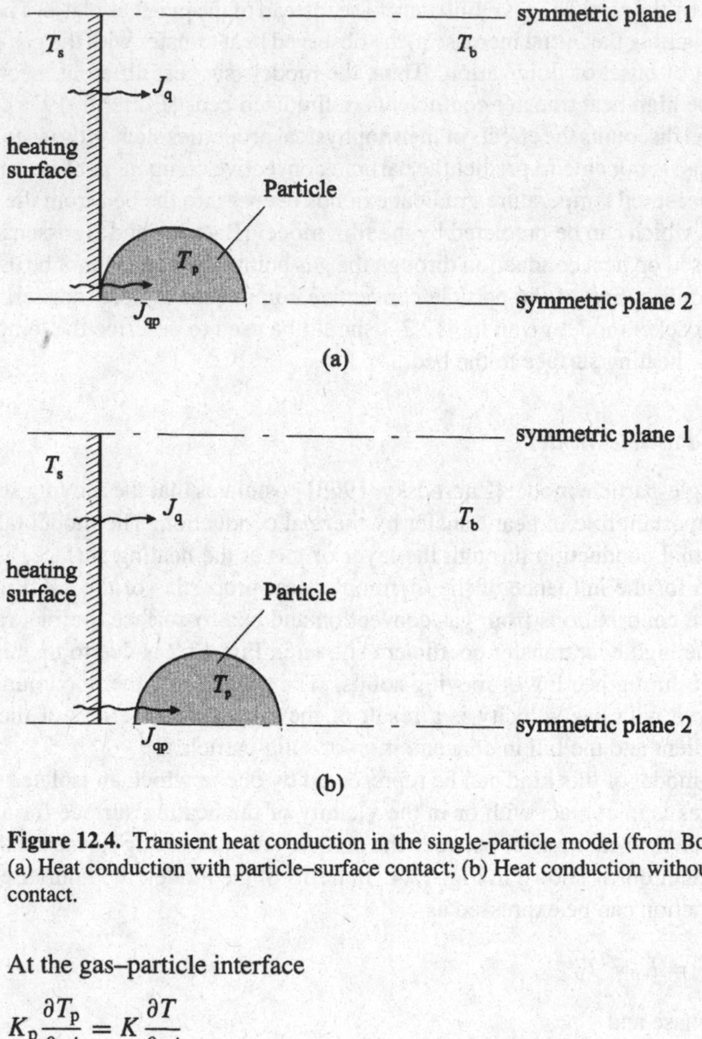

Figure 12.4. Transient heat conduction in the single-particle model (from Botterill, 1975): (a) Heat conduction with particle–surface contact; (b) Heat conduction without particle–surface contact.

(3) At the gas–particle interface

$$K_p \frac{\partial T_p}{\partial n'} = K \frac{\partial T}{\partial n'} \tag{12.7}$$

where $\partial T_p/\partial n'$ and $\partial T/\partial n'$ are the normal derivatives with respect to the gas–particle interface.

(4) Far inside the bed

$$T = T_b \tag{12.8}$$

(5) At $t = 0$ (initial condition)

$$T = T_p = T_b \tag{12.9}$$

Sample isotherm simulation results for a glass sphere of 200 μm in contact with a heat transfer surface surrounded by static air are shown in Fig. 12.5 for two contact times, *i.e.*, 1.2 ms and 52.4 ms [Botterill and Williams, 1963]. The initial temperature difference between the sphere and the surface is 10°C. It is seen that at the instant of contact heat begins to flow around the upper surface of the sphere and significant heat transfer takes place at

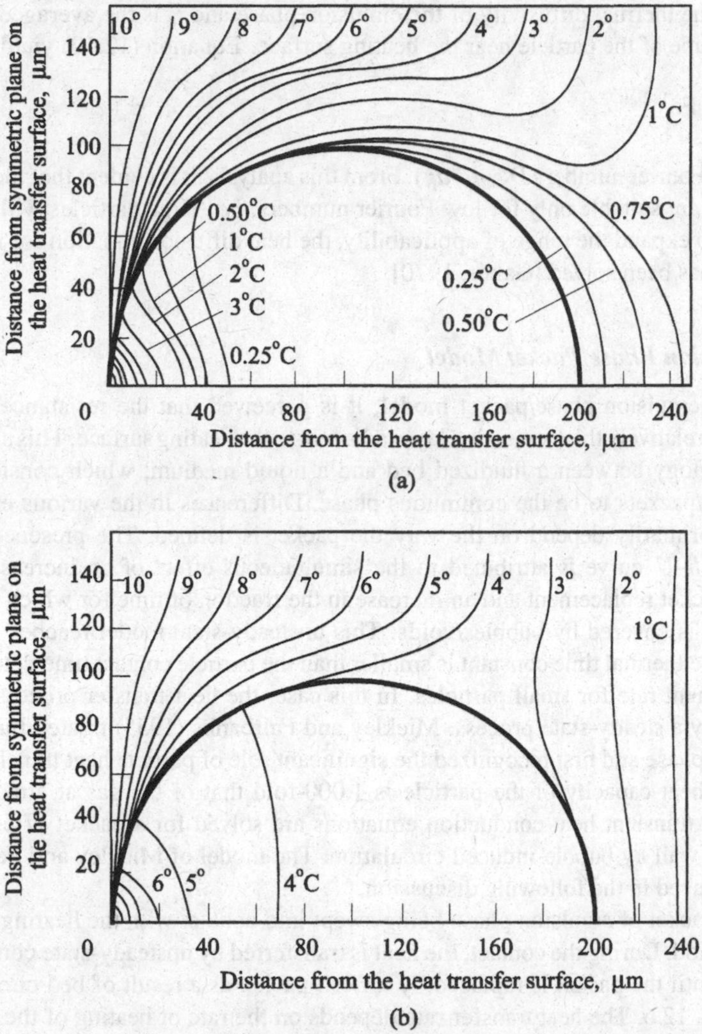

Figure 12.5. Isotherms for a glass sphere of 200 μm in contact with a heat transfer surface surrounded by air at an initial temperature difference of 10°C at contact time: (a) 1.2 ms; (b) 52.4 ms (from Botterill and Williams, 1963).

the point of contact between the heat transfer surface and the sphere. Compared to that in air, the temperature rise in the sphere is relatively slow as a result of larger heat capacity but increases significantly as the contact time increases. The model can be extended from a single particle to a single layer of particles at the surface. The model requires precise information on the position and the residence time of the particle near the surface, and this requirement may limit its usage. It is applicable only when the heat from the heat transfer surface does not penetrate beyond the single particle layer. The depth of penetration into the bed (δ_{em}) can be estimated from the temperature gradient at the heating surface. In terms of Eq. (4.26), we have

$$\delta_{em} \propto (D_{tem} t_c)^{1/2} \qquad (12.10)$$

where D_{tem} is the thermal diffusivity of the emulsion phase and t_c is the averaged contact (or residence) time of the particle near the heating surface. Equation (12.10) yields

$$\frac{\delta_{em}}{d_p} \propto Fo^{1/2} \tag{12.11}$$

where Fo is the Fourier number $(D_{tem}t_c/d_p^2)$. From this analysis, it is evident that the single-particle models are suitable only for low Fourier numbers, *i.e.*, large particles with a short contact time. To expand the range of applicability, the heat diffusion equation for multiple particle layers has been solved [Gabor, 1970].

12.2.4 *Emulsion Phase/Packet Model*

In the emulsion phase/packet model, it is perceived that the resistance to heat transfer lies in a relatively thick emulsion layer adjacent to the heating surface. This approach employs an analogy between a fluidized bed and a liquid medium, which considers the emulsion phase/packets to be the continuous phase. Differences in the various emulsion phase models primarily depend on the way the packet is defined. The presence of the maxima in the h–U curve is attributed to the simultaneous effect of an increase in the frequency of packet replacement and an increase in the fraction of time for which the heat transfer surface is covered by bubbles/voids. This unsteady-state model reaches its limit when the particle thermal time constant is smaller than the particle contact time determined by the replacement rate for small particles. In this case, the heat transfer process can be approximated by a steady-state process. Mickley and Fairbanks (1955) treated the packet as a continuum phase and first recognized the significant role of particle heat transfer since the volumetric heat capacity of the particle is 1,000-fold that of the gas at atmospheric conditions. The transient heat conduction equations are solved for a packet of emulsion swept up to the wall by bubble-induced circulation. The model of Mickley and Fairbanks (1955) is introduced in the following discussion.

Consider a packet of emulsion phase being swept into contact with the heating surface for a certain period. During the contact, the heat is transferred by unsteady-state conduction at the surface until the packet is replaced by a fresh packet as a result of bed circulation, as shown in Fig. 12.6. The heat transfer rate depends on the rate of heating of the packets (or emulsion phase) and on the frequency of their replacement at the surface. To simplify the model, the packet of particles and interstitial gas can be regarded as having the uniform thermal properties of the quiescent bed. The simplest case is represented by the problem of one-dimensional unsteady thermal conduction in a semiinfinite medium. Thus, the governing equation with the boundary conditions and initial condition can be imposed as

$$\frac{\partial T}{\partial t} = \frac{K_{em}}{c\rho_{em}} \frac{\partial^2 T}{\partial x^2} \qquad 0 \le x \le \infty \tag{12.12}$$

$$\begin{aligned} T &= T_s && \text{at } x = 0 \\ T &= T_\infty && \text{at } x = \infty \\ T &= T_\infty && \text{at } t = 0 \end{aligned} \tag{12.13}$$

where K_{em} and ρ_{em} are the conductivity and density of the emulsion phase, respectively.

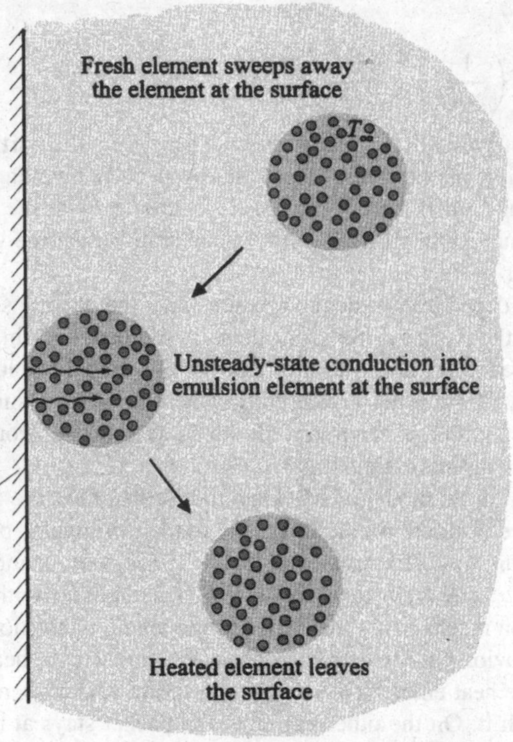

Figure 12.6. Conceptual representation of the emulsion-contact model of Mickley and Fairbanks (1955).

The solution can be obtained as

$$T = T_s + (T_\infty - T_s)\,\mathrm{erf}\left(x\sqrt{\frac{c\rho_{em}}{4K_{em}t}}\right) \tag{12.14}$$

where $\mathrm{erf}(x)$ is the error function of argument x. Therefore, the instantaneous heat flux is expressed by

$$J_q = -K_{em}\left.\frac{\partial T}{\partial x}\right|_{x=0} = \frac{T_s - T_\infty}{\sqrt{\pi t}}\sqrt{K_{em}c\rho_{em}} \tag{12.15}$$

Consequently, the instantaneous local heat transfer coefficient is obtained as

$$h_i = \sqrt{\frac{K_{em}c\rho_{em}}{\pi t}} \tag{12.16}$$

Furthermore, the area-averaged local heat transfer coefficient can be expressed by

$$h = \frac{1}{A_m}\int_0^{A_m}\left(\int_0^\infty h_i\psi(\tau)\,d\tau\right)dA = \sqrt{K_{em}c\rho_{em}S} \tag{12.17}$$

where A_m is the area of the packet in contact with the heating surface, $\psi(\tau)$ represents the frequency of occurrence in time of the packet of age τ, and S is the area mean stirring factor, defined as

$$S^{1/2} = \frac{1}{A_m} \int_0^{A_m} \left(\frac{1}{\sqrt{\pi}} \int_0^{\infty} \frac{\psi(\tau)}{\sqrt{\tau}} \, d\tau \right) dA \tag{12.18}$$

The preceding model successfully explains the role played by the particles in the heat transfer processes occurring in the dense-phase fluidized bed at voidage $\alpha < 0.7$. But it predicts very large values when the contact time of particles with the heating surface decreases because the nonuniformity of the solids concentration near the wall is not taken into account in this model.

As a result of the time-dependent voidage variations near the heating surface, the thermophysical properties of the packet differ from those in the bed, and this difference has not been included in the packet model. The limitation of this model lies in not taking into account the nonuniformity of the solids concentration near the heating surface. Thus, the packet model under this condition is accurate only for large values of Fourier number, in general agreement with the discussion in §4.3.3.

An important variation of the model of Mickley and Fairbanks is the film-penetration model developed for more accurate prediction. This model, originally proposed for the case of a single-phase liquid flow [Toor and Marchello, 1958], was extended to gas–solid suspensions by Yoshida *et al.* (1969) by treating packets as a continuum with finite thickness (δ_{em}). The film-penetration mechanism, analogous to the single-phase flow, includes two extremes of emulsion behavior. On one extreme, the packet contacts the heating surface for a short time so that all the heat entering the packet is used to heat it (penetration theory) while none passes through it. On the other extreme, the packet stays at the surface long enough to achieve steady-state and simply provides resistance for heat conduction.

In the heat transfer process a thin layer of emulsion of thickness δ_{em} is in contact with a heat transfer surface and after a time t_c, is replaced by a fresh element of emulsion from the bulk of the suspension, as shown in Fig. 12.7. Mathematically, the governing equation can be expressed by Eq. (12.12), with x, which is the distance from the heat transfer surface, confined to

$$0 \le x \le \delta_{em} \tag{12.19}$$

The governing equation must be solved with two boundary conditions and an initial condition which are identical to Eq. (12.13) except

$$T = T_\infty \quad \text{at } x = \delta_{em} \tag{12.20}$$

For convenience, we define the thermal diffusivity of the emulsion phase as

$$D_{tem} = \frac{K_{em}}{\rho_{em} c} \approx \frac{K_{em}}{\rho_p c (1 - \alpha_{mf})} \tag{12.21}$$

The solution for the instantaneous local heat transfer coefficient can be obtained as [Yoshida *et al.*, 1969]

$$h_i = \frac{K_{em}}{\sqrt{\pi D_{tem} t}} \left[1 + 2 \sum_{n=1}^{\infty} \exp\left(-\frac{\delta_{em}^2 n^2}{D_{tem} t} \right) \right]; \quad \pi \le \frac{\delta_{em}^2}{D_{tem} t} < \infty \tag{12.22}$$

$$h_i = \frac{K_{em}}{\delta_{em}} \left[1 + 2 \sum_{n=1}^{\infty} \exp\left(-\frac{\pi^2 D_{tem} t n^2}{\delta_{em}^2} \right) \right]; \quad 0 < \frac{\delta_{em}^2}{D_{tem} t} \le \pi \tag{12.23}$$

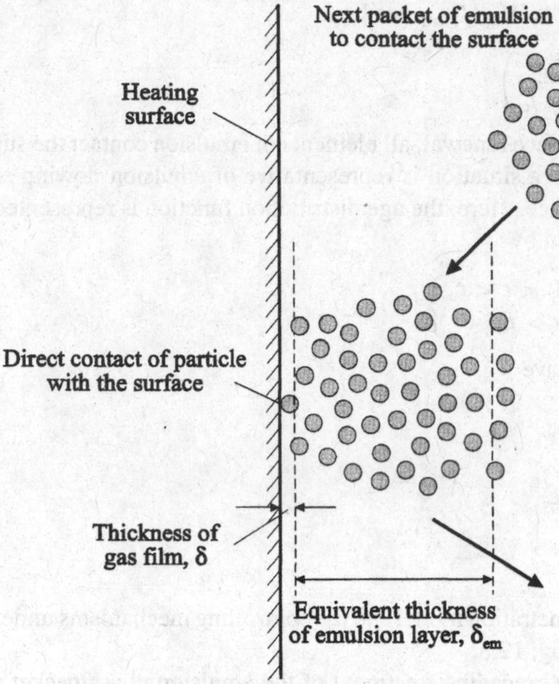

Figure 12.7. Conceptual representation of the film-penetration model for suspension-to-surface heat transfer (from Yoshida *et al.*, 1969).

where the dimensionless group $D_{tem}t/\delta_{em}^2$ characterizes the intensity of renewal or the magnitude of contact time. Generally, when $t_c \gg \delta_{em}^2/D_{tem}$, the film theory holds, and when $t_c \ll \delta_{em}^2/D_{tem}$, the penetration theory is valid. The time-averaged heat transfer coefficient h is obtained as

$$h = \int_0^\infty h_i \mathrm{I}(t)\,\mathrm{d}t \tag{12.24}$$

where $\mathrm{I}(t)$ is the age distribution function, representing the fraction of surface occupied by packets of age between t and $t + \mathrm{d}t$. Thus, the age distribution function of emulsion elements on the surface needs to be defined. Two commonly used age distribution functions for random surface renewal and for uniform surface renewal are discussed in the following.

The mode of random surface renewal exists for a surface that is in the main body of a suspension and is continually contacted by rising bubbles. The age distribution of elements on the surface is represented by that of a continuous-stirred tank reactor (CSTR), as given by

$$\mathrm{I}(t) = \frac{1}{t_c}\exp\left(-\frac{t}{t_c}\right) \tag{12.25}$$

Therefore, we obtain the following expressions for the average heat transfer coefficient for two extreme cases: For rapid renewal when $t_c \ll \delta_{em}^2/D_{tem}$, we have

$$h = \frac{K_{em}}{\sqrt{D_{tem}t_c}}\left[1 + 2\sum_{n=1}^{\infty}\exp\left(-\frac{2\delta_{em}n}{\sqrt{D_{tem}t_c}}\right)\right] \tag{12.26}$$

and for slow renewal when $t_c \gg \delta_{em}^2/D_{tem}$, we have

$$h = \frac{K_{em}}{\delta_{em}}\left(1 + \frac{\delta_{em}^2}{3D_{tem}t_c}\right) \tag{12.27}$$

In the mode of uniform surface renewal, all elements of emulsion contact the surface for the same length of time; such a situation is representative of emulsion flowing smoothly past a small heat transfer surface. Here, the age distribution function is represented by that of a plug flow reactor, as given by

$$\mathrm{I}(t) = \begin{Bmatrix} 1/t_c & \text{for } 0 < t < t_c \\ 0 & \text{for } t > t_c \end{Bmatrix} \tag{12.28}$$

Thus, for rapid renewal, we have

$$h = \frac{2K_{em}}{\sqrt{\pi D_{tem} t_c}} \tag{12.29}$$

and for slow renewal, we have

$$h = \frac{K_{em}}{\delta_{em}} \tag{12.30}$$

Different forms of the film-penetration model and the controlling mechanisms under various conditions are illustrated in Fig. 12.8.

It should be noted that the preceding treatment of the emulsion phase/packet model is suitable for a system with homogeneous emulsion phase (*i.e.*, particulate fluidization). The model needs to be modified when applied to the fluidized bed with a discrete bubble phase.

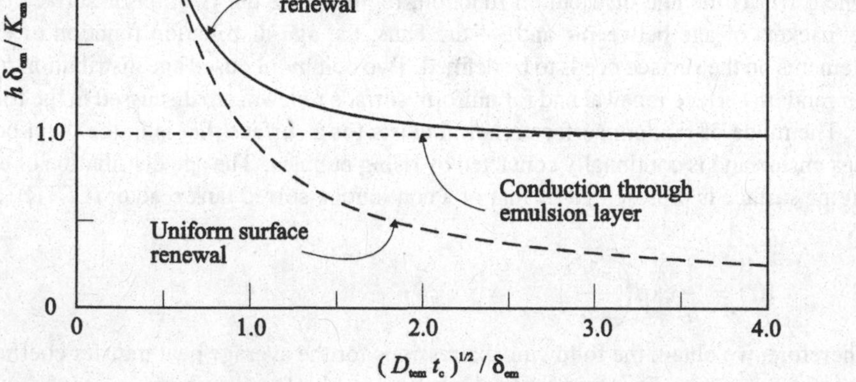

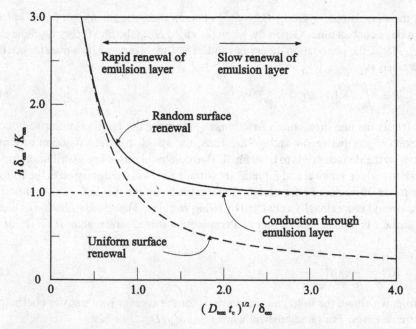

Figure 12.8. Heat transfer coefficient as a function of $(D_{tem}t_c)^{1/2}/\delta_{em}$ for random and uniform surface renewal (from Toor and Marchello, 1958; Yoshida *et al.*, 1969).

Modifications should account for the time fraction of the heating surface covered by the bubbles/voids and changes in the pseudothermal properties.

Example 12.1 In fluidization of microspherical catalysts, the particle and gas properties are given as [Yoshida *et al.*, 1969] $d_p = 152\ \mu m$, $\rho_p = 1{,}540\ kg/m^3$, $U_{mf} = 0.02$ m/s, $\alpha_{mf} = 0.505$, $c = 920$ J/kg · K, $K_p = 0.139$ W/m · K, and $K = 2.62 \times 10^{-2}$ W/m · K. The following conditions are also given: fluidizing velocity $U = 0.1$ m/s and $Nu_p = 2.33$.

(a) Assuming the film model applies, find the film thickness.
(b) Using the emulsion phase model and assuming random surface renewal, determine the mean surface renewal time with rapid replacement and the effective thickness of the emulsion layer with slow replacement.
(c) Calculate the mean surface renewal time, assuming uniform surface renewal.

Solution

(a) The value for h can be calculated from Nu_p as

$$h = \frac{K\,Nu_p}{d_p} = \frac{2.33 \times 2.62 \times 10^{-2}}{152 \times 10^{-6}} = 402\ W/m^2 \cdot K \tag{E12.1}$$

and assuming the film model, we have

$$\delta = \frac{K}{h} = \frac{2.62 \times 10^{-2}}{402} = 65.2\ \mu m \tag{E12.2}$$

(b) Assuming random surface renewal, for rapid renewal, we have (from Eq. (12.26))

$$h = \frac{K_{em}}{\sqrt{D_{tem} t_c}} \tag{E12.3}$$

Combining this equation with Eq. (12.21) yields

$$\begin{aligned} t_c &= \frac{K_{em}\rho_p c(1-\alpha_{mf})}{h^2} = \frac{[K\alpha_{mf} + K_p(1-\alpha_{mf})]\rho_p c(1-\alpha_{mf})}{h^2} \\ &= \frac{[2.62 \times 10^{-2} \times 0.505 + 0.139 \times (1-0.505)] \times 1{,}540 \times 920 \times (1-0.505)}{402^2} \\ &= 0.36\ s \end{aligned} \tag{E12.4}$$

For slow renewal, from Eq. (12.27), we have

$$\begin{aligned} \delta_{em} &= \frac{K_{em}}{h} = \frac{K\alpha_{mf} + K_p(1-\alpha_{mf})}{h} = \frac{2.62 \times 10^{-2} \times 0.505 + 0.139 \times (1-0.505)}{402} \\ &= 204\ \mu m \end{aligned} \tag{E12.5}$$

(c) Assuming uniform surface renewal, from Eq. (12.29), we have

$$\begin{aligned} t_c &= \frac{4K_{em}\rho_p c(1-\alpha_{mf})}{\pi h^2} = \frac{4[K\alpha_{mf} + K_p(1-\alpha_{mf})]\rho_p c(1-\alpha_{mf})}{\pi h^2} \\ &= \frac{4 \times [2.62 \times 10^{-2} \times 0.505 + 0.139 \times (1-0.505)] \times 1{,}540 \times 920 \times (1-0.505)}{3.14 \times 402^2} \\ &= 0.45\ s \end{aligned} \tag{E12.6}$$

12.3 Heat Transfer in Dense-Phase Fluidized Beds

As noted, most of the heat transfer models and correlations for gas–solid fluidization systems were originally developed for dense-phase fluidized beds (see Chapter 9). In the following, the behavior of heat transfer coefficients between the suspension (or bed) and the particle, between the suspension (or bed) and the gas, and between the suspension (or bed) and the wall or heat transfer surface are discussed.

12.3.1 *Particle-to-Gas and Bed-to-Gas Heat Transfer*

The particle-to-gas heat transfer can be obtained by unsteady-state experiments which measure the time required for cold particles of temperature T_{po}, mass M, and surface area S_p to reach the bed temperature when they are introduced into the bed. Assuming that the bed is well mixed and the bulk solids temperature in the bed is the same as the gas temperature, the heat balance gives

$$cM\,\mathrm{d}T_p = h_{gp}S_p(T_b - T_p)\,\mathrm{d}t \tag{12.31}$$

where h_{gp} is the local particle-to-gas heat transfer coefficient. Integrating the preceding equation yields

$$\frac{T_p - T_{po}}{T_b - T_{po}} = 1 - \exp\left(-\frac{h_{gp}S_p t}{Mc}\right) \tag{12.32}$$

The particle-to-gas heat transfer coefficient in dense-phase fluidization systems can be determined from the correlation [Kunii and Levenspiel, 1991]

$$\mathrm{Nu}_{gp} = \frac{h_{gp}d_p}{K} \approx 2 + (0.6 \sim 1.8)\,\mathrm{Re}_{pf}^{1/2}\,\mathrm{Pr}^{1/3} \tag{12.33}$$

where $\mathrm{Re}_{pf} = d_pU\rho/\mu$; $\mathrm{Pr} = c_p\mu/K$; and U is the superficial gas velocity. Equation (12.33) indicates that the values of the particle-to-gas heat transfer coefficient in a fluidized bed lie between those for a fixed bed with large isometric particles (with a factor of 1.8 in the second term; Ranz, 1952) and those with a factor of 0.6 in the second term of the equation. It is noted that the equation with the factor of 0.6 coincides with that for the single-particle heat transfer coefficient, *i.e.*, Nu_p, obtained with Re_p based on the relative velocity of particle to gas as given in Eq. (4.40).

For suspension-to-gas (or bed-to-gas) heat transfer in a well-mixed bed of particles, the heat balance over the bed under low Biot number (*i.e.*, negligible internal thermal resistance) and, if the gas flow is assumed to be a plug flow, steady temperature conditions can be expressed as

$$c_pU\rho\,\mathrm{d}T = -h_{bg}S_B(T - T_b)\,\mathrm{d}H \tag{12.34}$$

where $\mathrm{d}T$ is the change in gas temperature when flow is through a layer of bed with height $\mathrm{d}H$; h_{bg} is the suspension-to-gas heat transfer coefficient; and S_B is the surface area of particles per unit bed volume as given by

$$S_B = \frac{6(1-\alpha)}{d_p} \tag{12.35}$$

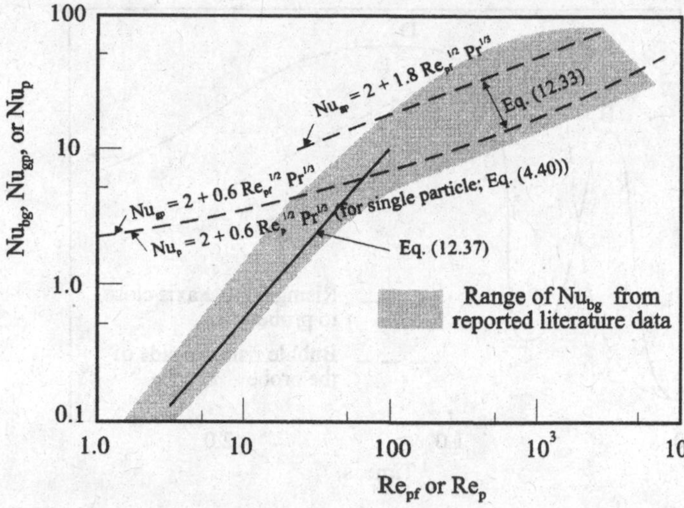

Figure 12.9. Particle-to-gas and bed-to-gas heat transfer coefficients under various flow conditions (from Kunii and Levenspiel, 1991).

Integrating Eq. (12.34) gives

$$\ln\left(\frac{T - T_\mathrm{b}}{T_\mathrm{i} - T_\mathrm{b}}\right) = -\left(\frac{h_{\mathrm{bg}} S_\mathrm{B}}{U \rho c_\mathrm{p}}\right) H \tag{12.36}$$

where T_i is the gas temperature at the inlet.

The range of data for the bed-to-gas heat transfer coefficient reported in the literature, which were primarily determined on the basis of Eq. (12.36), is shown in the shadowed area in Fig. 12.9. It can be seen from the figure that, under high Reynolds numbers ($\mathrm{Re}_{\mathrm{pf}} > 100$), values of h_{bg} are very close to those of h_{gp} determined by Eq. (12.33) as the plug flow assumption for the gas phase in the bed is realistic. However, this is not the case under low Reynolds numbers ($\mathrm{Re}_{\mathrm{pf}} < 100$). Values of $\mathrm{Nu}_{\mathrm{bg}}$ under low Reynolds numbers, as in fine particle fluidization, are smaller than $\mathrm{Nu}_{\mathrm{gp}}$ in terms of Eq. (12.33) and are much smaller than the value of 2 for an isolated spherical particle in a stationary condition expressed by Eq. (4.15). $\mathrm{Nu}_{\mathrm{bg}}$ under this Reynolds number range follows the correlation given by

$$\mathrm{Nu}_{\mathrm{bg}} = \frac{h_{\mathrm{bg}} d_\mathrm{p}}{K} = 0.03\,\mathrm{Re}_{\mathrm{pf}}^{1.3} \tag{12.37}$$

It should be mentioned that this deviation is more model-dependent than mechanistic because the real gas–solid contact is much poorer than that portrayed by plug flow, on which Eq. (12.36) is based [Kunii and Levenspiel, 1991]. The deviation can also be related to the effects of the particle boundary layer reduction due to particle collision and the generation of turbulence by bubble motion and particle collision [Brodkey *et al.*, 1991].

12.3.2 *Bed-to-Surface Heat Transfer*

For the bed-to-surface heat transfer in a dense-phase fluidized bed, the particle circulation induced by bubble motion plays an important role. This can be seen in a study of heat transfer properties around a single bubble rising in a gas–solid suspension conducted

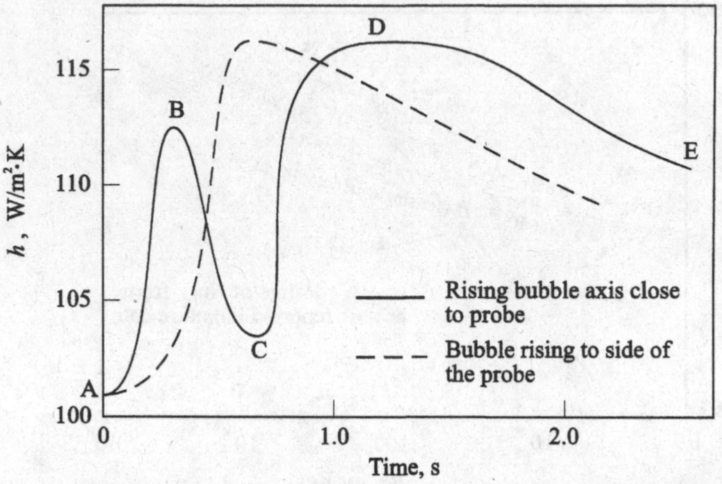

Figure 12.10. Probe-to-bed heat transfer coefficient variations in a fluidized bed (from Tuot and Clift, 1973).

by Tuot and Clift (1973). Using a sensitive probe of low heat capacity with fast response time, they observed that the heat transfer coefficient increased as the bubble rose toward the probe (points A to B on the solid line in Fig. 12.10). The increase results from some particle movement close to the probe surface as the bubble approaches from beneath. As the bubble envelopes the probe, the heat transfer coefficient decreases (point C on the solid line) because of the lower thermal conductivity and heat capacity of the gas. Further rising of the bubble leads to a peak in the signal behind the bubble (point D) which is the effect of the high concentration of particles carried by the wake passing the probe. The relatively slow decay beyond point D to a new steady value is due to the turbulence in the medium. The dashed line in Fig. 12.10 shows the heat transfer signal caused by a bubble rising to the side of the probe and the maximum is again the effect of the high concentration of particles carried in the wake of the bubble. Thus, it is evident that the bubble wake plays a significant role in particle circulation and hence the heat transfer in gas–solid fluidization.

The surface heat transfer is contributed to by three major components, as given in Eq. (12.1), *i.e.*, the particle convective component, the gas convective component, and radiation. In gas–solid fluidization systems, radiation may be neglected when the bed temperature is lower than 400°C. The significance of particle convection and gas convection depends mostly on the types of particles used. As a rule of thumb, particle convection is the dominant mechanism for small particles ($d_p < 400\ \mu$m); it usually plays a key role for Group A particles (see §9.2.1). Gas convection becomes dominant for large particles ($d_p > 1{,}500\ \mu$m) and for high-pressure or high-velocity fluidizations; it usually plays a key role for Group D particles [Maskaev and Baskakov, 1974]. For Group B particles, both components are significant.

12.3.2.1 *Particle Convective Component*

Particle convection, caused by the particle motion within the bed, is concerned with the heat transfer from a surface when it is in contact with the particulate emulsion phase instead of the void/bubble phase. Thus, the heat transfer coefficient of particle convection can be defined as

$$h_{pc} = \frac{(1-\alpha_b)}{\text{particle convective heat resistance}} \tag{12.38}$$

where the particle convective heat resistance can be further divided into the following two series resistances: (1) average packet (particulate phase) resistance, $1/h_p$, and (2) film resistance, $1/h_f$. Thus, Eq. (12.38) can be expressed by

$$h_{pc} = \frac{(1-\alpha_b)}{1/h_p + 1/h_f} \tag{12.39}$$

where bubble fraction α_b can be calculated from Eq. (9.43) and h_p can be calculated from

$$h_p = \frac{1}{t_c}\int_0^{t_c} h_i \, dt \tag{12.40}$$

where h_i is the instantaneous heat transfer coefficient averaged over the contact area.

Considering the thermal diffusion through an emulsion packet and assuming that the properties of the emulsion phase are the same as those of minimum fluidization, h_i can be expressed by [Mickley *et al.*, 1961]

$$h_i = \left(\frac{K_{em}\rho_p(1-\alpha_{mf})c}{\pi t}\right)^{1/2} \tag{12.41}$$

Substituting Eq. (12.41) into Eq. (12.40) yields

$$h_p = \frac{2}{\sqrt{\pi}}\left(\frac{K_{em}\rho_p(1-\alpha_{mf})c}{t_c}\right)^{1/2} \tag{12.42}$$

The surface–emulsion phase contact time t_c can be estimated by assuming that the time fraction for the surface to be covered by bubbles equals the bubble volume fraction in the bed

$$t_c = \frac{1-\alpha_b}{f_b} \tag{12.43}$$

where f_b is the bubble frequency at the surface. Equations (12.42) and (12.43) yield

$$h_p = \frac{2}{\sqrt{\pi}}\left(\frac{K_{em}\rho_p(1-\alpha_{mf})cf_b}{1-\alpha_b}\right)^{1/2} \tag{12.44}$$

Furthermore, K_{em} can be expressed by [Ranz, 1952]

$$K_{em} = K_e + 0.1\rho c_p d_p U_{mf} \tag{12.45}$$

where K_e is the effective thermal conductivity of a fixed bed with stationary gas. In the fixed bed, heat transfers through the gas and solid via parallel paths, as seen in Fig. 12.11(a). For this case, the effective conductivity is given by

$$K_e = \alpha_{mf}K + (1-\alpha_{mf})K_p \tag{12.46}$$

To account for the actual geometry and small contact region between the adjacent particle in the fixed bed with stationary gas, as shown in Fig. 12.11(b), Eq. (12.46) was modified by Kunii and Smith (1960) to

$$K_e = \alpha_{mf}K + (1-\alpha_{mf})K_p\left(\frac{1}{\varphi_b(K_p/K)+2/3}\right) \tag{12.47}$$

where φ_b is the ratio of an equivalent thickness of gas film to particle diameter.

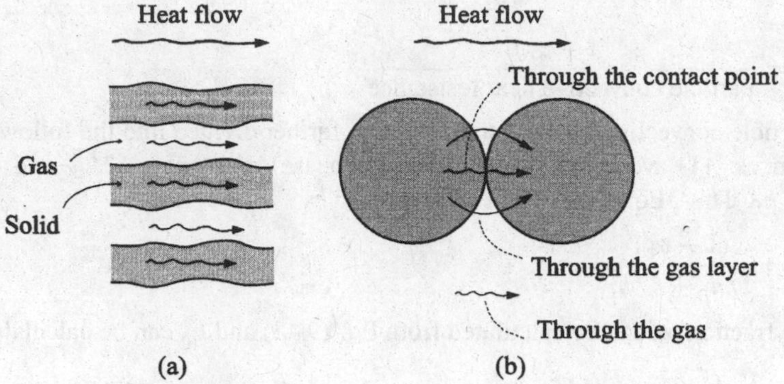

Figure 12.11. Conceptual representation of models for evaluating the effective thermal conductivity of a fixed bed (from Kunii and Levenspiel, 1991): (a) Simple parallel path model; (b) Modified model of Kunii and Smith (1960).

For film resistance, the film heat transfer coefficient can be expressed by

$$h_f = \frac{\xi K}{d_p} \tag{12.48}$$

where ξ is a factor ranging from 4 to 10 [Xavier and Davidson, 1985]. Thus, the particle convective heat transfer component h_{pc} can be calculated from Eqs. (12.39), (12.44), and (12.48).

12.3.2.2 *Gas Convective Component*

The gas convective component is concerned with the surface heat transfer due to gas percolating through the particulate phase and the surface heat transfer due to gas bubbles. For small particles, though the gas convective component is small in the bulk (*i.e.*, in-bed region), it can be important in the freeboard region. The heat transfer coefficient of gas convection in general varies with the geometry of the heat transfer surface. However, it can be approximated without treating specific surface geometries, as suggested by Baskakov *et al.* (1974) for $0.16\text{ mm} \leq d_p \leq 4\text{ mm}$

$$\frac{h_{gc} d_p}{K} = 0.009\,\mathrm{Ar}^{1/2}\,\mathrm{Pr}^{1/3} \tag{12.49}$$

or as proposed by Denloye and Botterill (1978) for $10^3 < \mathrm{Ar} < 2 \times 10^6$ and for operating pressure up to 1 MPa

$$\frac{h_{gc}\sqrt{d_p}}{K} = 0.86\,\mathrm{Ar}^{0.39} \tag{12.50}$$

where the coefficient 0.86 has the unit of meter$^{-1/2}$

The heat transfer rate for the gas convective component can be regarded as comparable to that at incipient fluidizing conditions. Thus, assuming $h_{gc} = h_{mf}$, Xavier and Davidson (1985) simulated the system by considering a pseudofluid with the apparent thermal conductivity K_a of the gas–solid medium flowing at the same superficial velocity and the same inlet and outlet temperatures as the gas. Therefore, the heat conduction of the fluid flowing through a cylindrical column can be expressed by Eq. (4.43) as

$$\frac{\partial T}{\partial z} = \frac{K_a}{\rho c_p U}\left(\frac{\partial^2 T}{\partial r^2} + \frac{1}{r}\frac{\partial T}{\partial r}\right) \tag{12.51}$$

Here the axial conduction is considered negligible in comparison with the convection due to the flowing gas. This assumption may be invalid for fluidizations with very low gas velocities.

For a vertical heating surface, the boundary conditions can be imposed as

$$\begin{aligned} 0 < r < \frac{D}{2}, \quad & z = 0, \quad T = T_1 \\ 0 < z < L, \quad & r = \frac{D}{2}, \quad T = T_s \end{aligned} \tag{12.52}$$

If T_2 is the mean temperature of the gas leaving the bed at $z = L$, the temperature distribution in the bed can be obtained as

$$\frac{T_2 - T_s}{T_1 - T_s} = 4\sum_{n=1}^{\infty} \frac{1}{\lambda_{n^2}} \exp\left(-4\lambda_{n^2}\frac{K_a L}{\rho c_p U D^2}\right) \tag{12.53}$$

where λ_n are eigenvalues of the eigenequation $J_0(\lambda) = 0$, with the first three being $\lambda_1 = 2.450$, $\lambda_2 = 5.520$, and $\lambda_3 = 8.645$.

12.3.2.3 *Radiative Component*

Radiative heat transfer plays an important part in many fluidized bed processes operated at high temperatures, such as coal combustion and gasification. When treating a fluidized bed as a whole "solid" gray body, the radiative heat transfer coefficient h_r between the fluidized bed at temperature T_b and a heating surface at temperature T_s is defined as

$$h_r = \frac{J_r}{T_b - T_s} = \sigma_b \epsilon_{bs}(T_b^2 + T_s^2)(T_b + T_s) \tag{12.54}$$

where J_r is the radiant heat flux, σ_b is the Stefan–Boltzmann constant, and ϵ_{bs} is the generalized emissivity, which depends on the shape, material property, and emissivity of the radiating and receiving bodies [Baskakov, 1985]. For two large, parallel, and perfect gray planes, we have

$$\epsilon_{bs} = (1/\epsilon_b + 1/\epsilon_s - 1)^{-1} \tag{12.55}$$

The effective bed-to-surface emissivity is larger than the particle emissivity, ϵ_p, because of the multiple surface reflections. For the same reason, an irregular surface has a higher emissivity than a smooth surface of the same material. It can be recognized that the importance of radiative heat transfer significantly increases with the temperature. Increasing particle size also increases the relative radiative heat transfer [Baskakov *et al.*, 1973].

The general bed emissivity for bed-to-surface radiation may be estimated from the emissivity of an individual particle as [Grace, 1982]

$$\epsilon_{bs} \approx 0.5(1 + \epsilon_p) \tag{12.56}$$

Another expression for ϵ_{bs} is suggested by Baskakov (1985)

$$\epsilon_{bs} = \frac{1}{1/\epsilon_b + 1/\epsilon_s - 1} \approx \epsilon_b \epsilon_s \tag{12.57}$$

Equation (12.57) is valid when ϵ_b and ϵ_s are greater than 0.8.

The radiant heat transfer coefficient becomes important at temperatures above 600°C, but it is difficult to predict. In general, depending on particle size, h_r increases from being approximately 8–12 percent of the overall heat transfer coefficient at 600°C to being 20–30 percent of h at 800°C [Baskakov *et al.*, 1973].

Example 12.2 An exothermic chemical reaction occurs in a gas–solid fluidized bed reactor. To maintain the reaction temperature at 350 K, 50 MJ/h of heat of reaction must be removed from this reactor. Determine the heat transfer area needed in this reactor with the following data: $U = 0.25$ m/s; $\rho = 0.716$ kg/m^3; $c_p = 2{,}185$ J/kg · K; $K = 0.034$ W/m · K; $\mu = 1.09 \times 10^{-5}$ kg/m · s; $\rho_p = 2{,}600$ kg/m^3; $c = 840$ J/kg · K; $K_p = 1.9$ W/m · K; $d_p =$ 0.5 mm; $\alpha_{mf} = 0.47$; $\alpha_b = 0.6$; $f_b = 3$; $U_{mf} = 0.192$ m/s; $\varphi_b = 0.85$, $\xi = 6$. The cooling water temperature is 293 K.

Solution To determine the heat transfer area, the overall heat transfer coefficient given in the following needs to be calculated first from Eq. (12.1). Radiative heat transfer would be low for the temperature of 350 K and, thus, it is neglected.

(1) The particle convective component h_{pc} may be calculated from Eq. (12.39). The heat transfer coefficients of film can be calculated from Eq. (12.48)

$$h_f = \frac{\xi K}{d_p} = \frac{6 \times 0.034}{0.5 \times 10^{-3}} = 408\ \mathrm{W/m^2 \cdot K} \tag{E12.7}$$

Similarly, the heat transfer coefficients of the packet can be calculated from Eqs. (12.42), (12.45), and (12.47) and can be obtained as 759 W/m^2 · K. Thus, h_{pc} is found to be 106 W/m^2 · K.

(2) The gas convective component can be calculated by Eq. (12.50) as

$$\begin{aligned} h_{gc} &= 0.86\,\mathrm{Ar}^{0.39}\, K / \sqrt{d_p} = 0.86 \times (19.2 \times 10^3)^{0.39} \times 0.034/\sqrt{0.5 \times 10^{-3}} \\ &= 61.2\ \mathrm{W/m^2 \cdot K} \end{aligned} \tag{E12.8}$$

After adding h_p to h_{gc}, the surface to bed heat transfer coefficient becomes 167 W/m^2· K. Thus, the heat transfer area can be determined as

$$A = \frac{J_q}{(T_b - T_w)h} = \frac{50 \times 10^6/3{,}600}{(350 - 293) \times 167} = 1.46\,\mathrm{m^2} \tag{E12.9}$$

12.3.3 *Effect of Operating Conditions*

A unique feature of the dense-phase fluidized bed is the existence of a maximum convective heat transfer coefficient h_{max} when the radiative heat transfer is negligible. This feature is distinct for fluidized beds with small particles. For beds with coarse particles, the heat transfer coefficient is relatively insensitive to the gas flow rate once the maximum value is reached.

For a given system, h_{max} varies mainly with particle and gas properties. For coarse particle fluidization at $U > U_{mf}$, the heat transfer is dominated by gas convection. Thus, h_{max} can be evaluated from Eq. (12.50). On the other hand, h_{max} in a fine particle bed can be reasonably evaluated from the equations for h_{pc}. In general, h_{max} is a complicated function

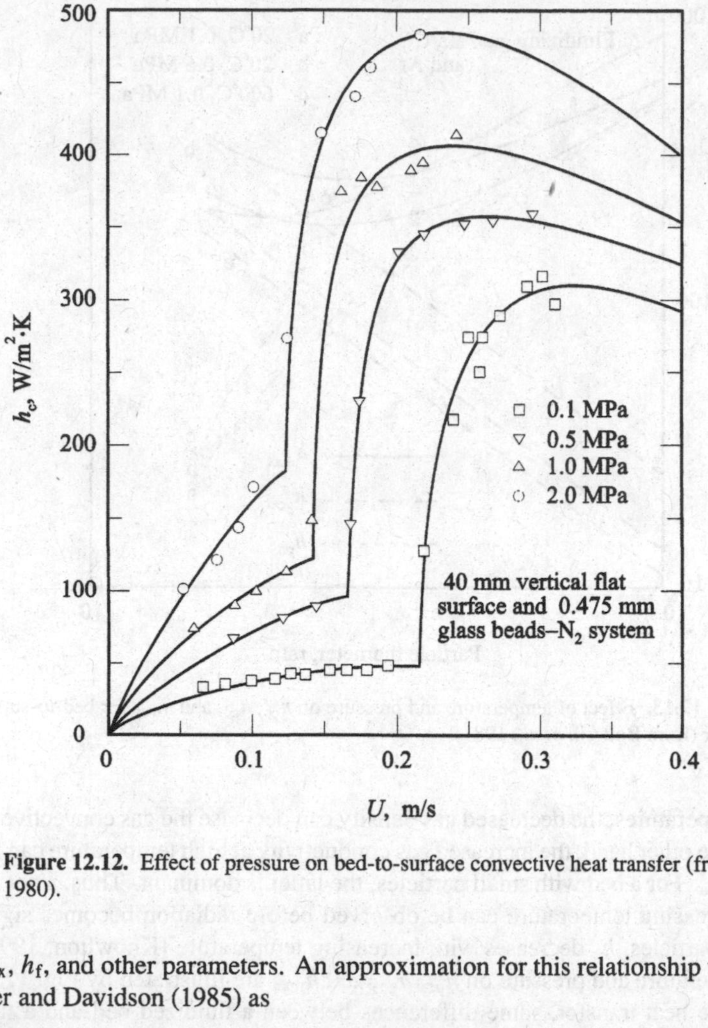

Figure 12.12. Effect of pressure on bed-to-surface convective heat transfer (from Xavier *et al.*, 1980).

of $h_{pc,max}$, h_f, and other parameters. An approximation for this relationship was suggested by Xavier and Davidson (1985) as

$$\frac{h_{max}}{h_{pc,max}} = \left(\frac{h_f}{h_f + 2h_{pc,max}} + \frac{h_{gc}}{h_{pc,max}} \right)^{0.84} \tag{12.58}$$

The convective heat transfer h_c $(=h_{pc} + h_{gc})$ is influenced by both the pressure and the temperature. An increase in pressure increases the gas density, yielding a lower U_{mf}. Thus, a pressurized operation enhances the convective heat transfer, as illustrated in Fig. 12.12. Under vacuum operation, h_c is lower than under ambient pressure operation as a result of lower gas density and a decrease in K_e with decreasing pressure [Bock and Molerus, 1985]. For fluidized beds with small particles, increasing pressure enhances solids mixing and hence particle convection [Borodulya *et al.*, 1982]. For large particles, the heat transfer is dominated by gas convection, which is enhanced with an increase in gas density through the elevation of pressure. In general, the convective heat transfer significantly increases with pressure for Group D particles; however, the pressure effect decreases with decreasing particle size. For Group A and Group B particles, the increase in h_c with pressure is small.

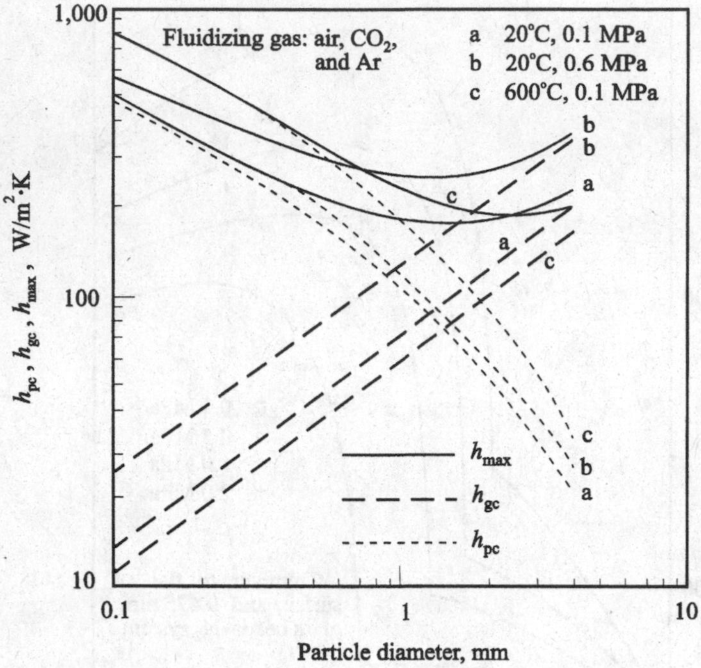

Figure 12.13. Effect of temperature and pressure on h_{pc}, h_{gc}, and h_{max} for bed-to-surface heat transfer (from Botterill *et al.*, 1981).

At high temperatures, the decreased gas density can decrease the gas convective component h_{gc}. On the other hand, the increased gas conductivity at high temperature can increase h_{gc}, K_e, and h_{pc}. For a bed with small particles, the latter is dominant. Thus, a net increase in h_c with increasing temperature can be observed before radiation becomes significant. For Group D particles, h_c decreases with increasing temperature [Knowlton, 1992]. The effects of temperature and pressure on h_{pc}, h_{gc}, and h_{max} are illustrated by Fig. 12.13.

For radiative heat transfer, some differences between a fluidized bed and a fixed bed exist [Kovenskii, 1980]. When a bed starts to fluidize from the fixed state, the radiative heat transfer rapidly increases by 10–20 percent; it remains constant as the gas velocity increases further in the bubbling bed.

In fluidized bed heat transfer, it is a common practice to use internals such as water-cooled tubes in the bed. It is important to know the effect of immersed objects on the local fluidization behavior and local heat transfer coefficient. Without loss of generality, we consider a case in which a horizontal water-cooled tube is placed in a hot fluidized bed [Botterill *et al.*, 1984]. The interference of the immersed tube with the particle circulation pattern leads to an increased average particle residence time at the heat transfer surface. The common formation of stagnant regions of particles on the top or nose of the horizonal tube is evidence of such interference. Furthermore, at high gas flow rates, gas packets are frequently present at the upper part of the horizontal tube. Thus, there is a significant variation in the heat transfer coefficient around the tube. As illustrated in Fig. 12.14, for fluidization in a bed of an immersed horizontal tube, the heat transfer coefficient near the nose is considerably smaller than that around the sides, indicating the reduction of convective heat transfer by the stagnation of gas

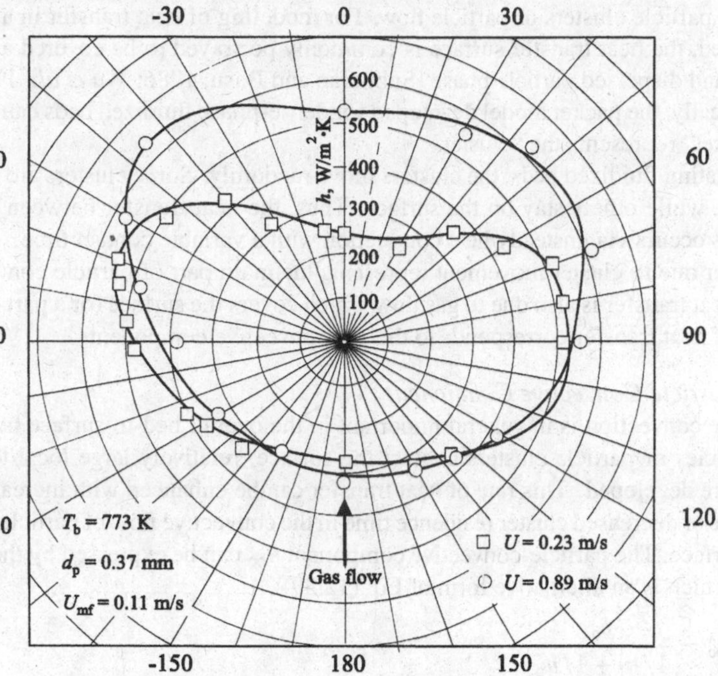

Figure 12.14. Variation of local heat transfer coefficient around a horizontal tube immersed in a fluidized bed of alumina (from Botterill *et al.*, 1984).

near the nose. For large particles for which the interphase gas convective component of heat transfer is significant, the circumferential variations would be somewhat similar with peaks at the sides, but the variation is less than that with small particles, since the particle convection is no longer dominant. The overall heat transfer coefficient as a result of orientation usually does not vary much; that for a horizontal tube is slightly lower than that for a vertical tube.

It is noted that most of the models and correlations that are developed are based on bubbling fluidization. However, most of them can be extended to the turbulent regime with reasonable error margins. The overall heat transfer coefficient in the turbulent regime is a result of two counteracting effects, the vigorous gas–solid movement, which enhances the heat transfer and the low particle concentration, which reduces the heat transfer.

12.4 Heat Transfer in Circulating Fluidized Beds

The mechanism of heat transfer in circulating fluidized beds is described in this section. Effects of the operating variables on the local and overall heat transfer coefficients are discussed.

12.4.1 *Mechanism and Modeling*

The suspension-to-wall surface heat transfer mechanism in a circulating fluidized bed (see Chapter 10) comprises various modes, including conduction due to particle clusters on the surface or particles falling along the walls, thermal radiation, and convection due to gas flow or particle clusters or particle flow. For modeling of heat transfer in a circulating fluidized bed, the heat transfer surface is commonly portrayed to be covered alternatively by cluster and dispersed particle phase [Subbarao and Basu, 1986; Wu *et al.*, 1990]. Thus, mechanistically, the packet model developed for dense-phase fluidized beds can be applied, if the "packet" represents the "cluster."

In circulating fluidized beds, the clusters move randomly. Some clusters are swept from the surface, while others stay on the surface. Thus, the heat transfer between the surface and clusters occurs via unsteady heat conduction with a variable contact time. This part of heat transfer due to cluster movement represents the main part of particle convective heat transfer. Heat transfer is also due to gas flow which covers the surface (or a part of surface). This part of heat transfer corresponds to the gas convective component.

12.4.1.1 *Particle Convective Component*

The particle convection is in general important in the overall bed-to-surface heat transfer. When particles or particle clusters contact the surface, relatively large local temperature gradients are developed. This rate of heat transfer can be enhanced with increased surface renewal rate or decreased cluster residence time in the convective flow of particles in contact with the surface. The particle-convective component h_{pc} can be expressed by the following equation, which is an alternative form of Eq. (12.39):

$$h_{pc} = \frac{\delta_c}{1/h_f + 1/h_p} \tag{12.59}$$

Thus, h_{pc} is determined from the wall (film) resistance, $1/h_f$, in series with the transient conduction resistance of a homogeneous semiinfinite medium, $1/h_p$. By analogy with Eq. (12.48), h_f can be expressed by [Gloski *et al.*, 1984]

$$h_f = \frac{K_f}{\delta^* d_p} \tag{12.60}$$

where δ^* is the dimensionless effective gas layer thickness between wall and cluster (ratio of gas layer thickness to particle size), which is mainly a function of cross-sectionally averaged particle volume fraction [Lints and Glicksman, 1993]. The range of δ^* is exemplified in Fig. 12.15.

Similar to Eq. (12.42), h_p can be expressed as [Lints and Glicksman, 1993]

$$h_p = \left(\frac{K_c \rho_p (1-\alpha_c) c_{pc}}{\pi t_c}\right)^{1/2} \tag{12.61}$$

For a very short cluster contact time or very large fluidizing particle, h_f dominates the particle convection and thus Eq. (12.59) yields

$$h_{pc} = \frac{\delta_c K_f}{\delta^* d_p} \tag{12.62}$$

12.4.1.2 *Gas Convective Component*

In practice, h_{gc} may be evaluated by one of the following approaches:

(1) Extended from Eq. (12.50) for h_{gc} in dense-phase fluidized beds
(2) Approximated as that for dilute-phase pneumatic transport [Wen and Miller, 1961; Basu and Nag, 1987]
(3) Estimated by the convective coefficient of single-phase gas flow [Sleicher and Rouse, 1975]

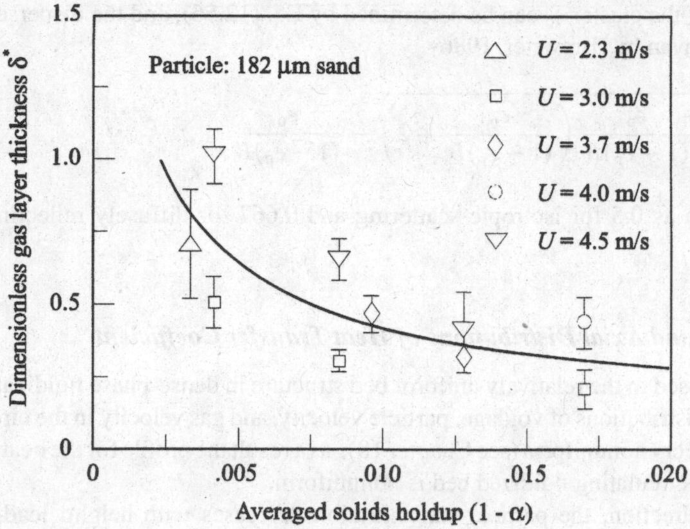

Figure 12.15. Variation of gas layer thickness, δ^*, between the cluster and wall with averaged solids holdup (from Lints and Glicksman, 1993).

For high particle concentration on the surface with large dimension, any one of these approaches is reasonable because of the small value of h_{gc}. For low particle concentrations and high temperatures, using these approaches may cause discrepancies in h_{gc}.

12.4.1.3 *Radiative Component*

To understand the radiative heat transfer in a circulating fluidized bed, the bed can be regarded as a pseudogray body. The radiative heat transfer coefficient is [Wu *et al.*, 1989]

$$h_r = \frac{\sigma_b\left(T_b^4 - T_s^4\right)}{\left(\dfrac{1}{\epsilon_{sus}} + \dfrac{1}{\epsilon_s} - 1\right)(T_b - T_s)} \tag{12.63}$$

where ϵ_{sus} is the emissivity of the suspension.

An alternative treatment for radiative heat transfer in a circulating fluidized bed is to consider the radiation from the clusters (h_{cr}) and from the dispersed phase (*i.e.*, the remaining aspect of gas–solid suspension except clusters, h_{dr}), separately [Basu, 1990]

$$h_r = \alpha_c h_{cr} + (1 - \alpha_c) h_{dr} \tag{12.64}$$

where α_c is the volume fraction of clusters in the bed. These two components can be defined by

$$h_{cr} = \frac{\sigma_b\left(T_b^4 - T_s^4\right)}{\left(\dfrac{1}{\epsilon_c} + \dfrac{1}{\epsilon_s} - 1\right)(T_b - T_s)} \tag{12.65}$$

$$h_{dr} = \frac{\sigma_b\left(T_b^4 - T_s^4\right)}{\left(\dfrac{1}{\epsilon_d} + \dfrac{1}{\epsilon_s} - 1\right)(T_b - T_s)} \tag{12.66}$$

The emissivity of the cluster ϵ_c can be determined by Eq. (12.56), and the dispersed phase emissivity ϵ_d is given by [Brewster, 1986]

$$\epsilon_d = \sqrt{\frac{\epsilon_p}{(1-\epsilon_p)B}\left(\frac{\epsilon_p}{(1-\epsilon_p)B}+2\right)} - \frac{\epsilon_p}{(1-\epsilon_p)B} \tag{12.67}$$

where B is taken as 0.5 for isotropic scattering and 0.667 for diffusely reflecting particles.

12.4.2 *Radial and Axial Distributions of Heat Transfer Coefficient*

As opposed to the relatively uniform bed structure in dense-phase fluidization, the radial and axial distributions of voidage, particle velocity, and gas velocity in the circulating fluidized bed are very nonuniform (see Chapter 10); as a result the profile for the heat transfer coefficient in the circulating fluidized bed is nonuniform.

In the axial direction, the particle concentration decreases with height, leading to a decrease in the cross section averaged heat transfer coefficient. In addition, the influence of the solids circulation rate is significant at lower bed sections but less significant at upper bed sections, as illustrated in Fig. 12.16. In the radial direction, the situation is more complicated as a result of the uneven radial distribution of the particle concentration as well as the opposite solids flow directions in the wall and center regions. In general, the coefficient is relatively low and approximately constant in the center region. The coefficient increases sharply toward the wall region. Three representative radial profiles of the heat transfer coefficient with various particle holdups reported by Bi *et al.* (1989) are shown in

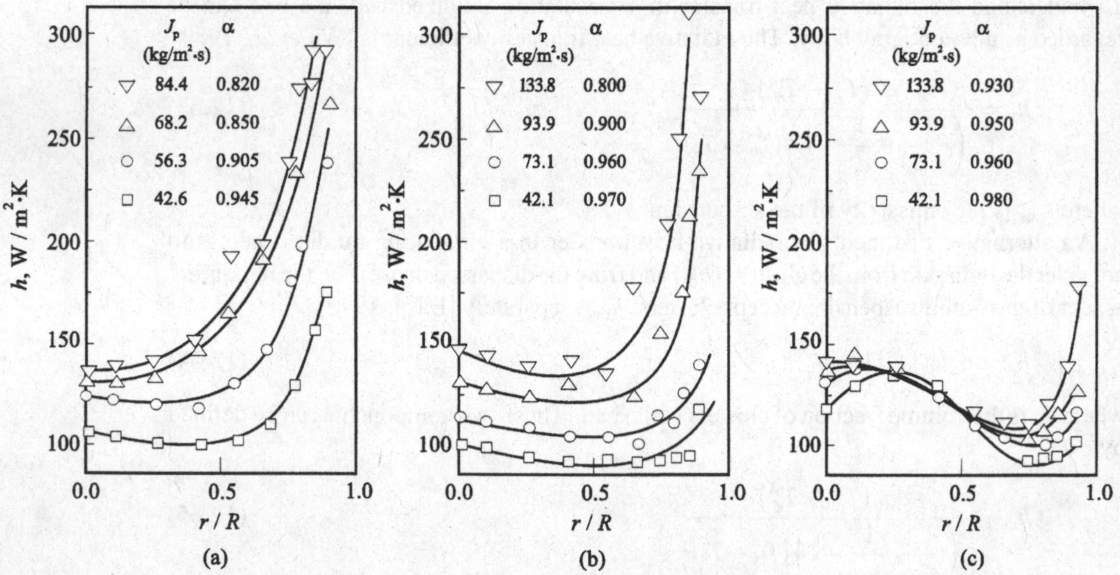

Figure 12.16. Radial distributions of overall heat transfer coefficient in a circulating fluidized bed of $d_p = 280\ \mu$m and $\rho_p = 706$ kg/m^3 (from Bi *et al.*, 1989): (a) $U = 3.7$ m/s, $H = 1.25$ m; (b) $U = 6.0$ m/s, $H = 1.25$ m; (c) $U = 6.0$ m/s, $H = 6.5$ m.

Fig. 12.16 as described in the following:

(1) When the particle holdup is high, the contribution of h_{pc} plays a dominant role and h_{gc} is less important. The radial distribution of the heat transfer coefficient is nearly parabolic, as shown in Fig. 12.16(a). Such a heat transfer profile is similar to the solids concentration profile described in Chapter 10.

(2) As the gas velocity increases, the solids holdup decreases and, thus, h_{gc} begins to become as important as h_{pc}. In the center region of the riser, h_{gc} is dominant, and its influence decreases with an increase in the solids holdup along the radial direction toward the wall. In the near-wall region, h_{pc} dominates the heat transfer. The contribution of h_{pc} decreases with a decrease in the particle concentration toward the bed center. As a result, a minimum value of h appears at r/R of about 0.5–0.8, as indicated in Fig. 12.16(b).

(3) With further decrease in the particle concentration at $\alpha > 0.93$, h_{gc} becomes dominant except at a region very close to the wall. Thus, the heat transfer coefficient decreases with increasing r/R in most parts of the riser, as shown in Fig. 12.16(c), exemplifying the same trend as the radial profile of the gas velocity. In the region near the wall h_{pc} increases sharply, apparently as the effect of relatively high solids concentration in that region.

12.4.3 *Effect of Operating Parameters*

The overall heat transfer coefficient can be influenced by the suspension density, solids circulation rate, gas velocity, particle properties, bed temperature, pressure, and dimensions of the heating surface. The overall heat transfer coefficient increases with suspension density [Wu *et al.*, 1989] and with particle circulation rate. The increase in gas velocity appears to have two counteracting effects, namely, enhancing h_{gc} through an increased gas convection effect while reducing particle convective heat transfer through reduced particle concentration, as illustrated in Fig. 12.17. When h_{pc} dominates (in the near-wall region with high particle concentration), h decreases with increasing U. On the other hand, h increases with U if h_{gc} is important (*e.g.*, in the central region where the particle concentration is small). Another reason for the decrease of h in the near-wall region is the reduced particle downward velocity caused by increasing U, which results in prolonged particle–surface contact.

In general, small/light particles can enhance heat transfer. The cluster formation in small/light particle systems contributes to the enhancement of h_{pc}. Also the gas film resistance can be reduced by fluidizing with small particles [Wu *et al.*, 1987]. When the temperature is lower than 400°C, the effect of bed temperature on the heat transfer coefficient is due to the change of gas properties, while h_r is negligible. At higher temperatures, h increases with temperature, mainly because of the sharp increase of radiative heat transfer.

Measurements of heat transfer in circulating fluidized beds require use of very small heat transfer probes, in order to reduce the interference to the flow field. The dimensions of the heat transfer surface may significantly affect the heat transfer coefficient at any radial position in the riser. All the treatment of circulating fluidized bed heat transfer described is based on a small dimension for the heat transfer surface. The heat transfer coefficient decreases asymptotically with an increase in the vertical dimension of the heat transfer surface [Bi *et al.*, 1990]. It can be stated that the large dimensions of the heat transfer surface

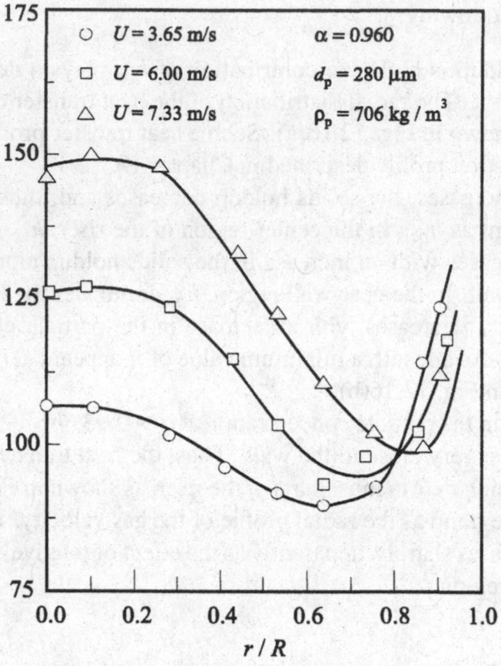

Figure 12.17. Effect of gas velocity on the radial distribution of the heat transfer coefficient in a circulating fluidized bed (from Bi *et al.*, 1989).

can prolong the residence time of particles or particle clusters on the surface, resulting in lower renewal frequency and hence a low apparent heat transfer coefficient.

12.5 Heat Transfer in Spouted Beds

This section discusses the behavior of heat transfer between gas and particles, and between the bed and the surface in a spouted bed.

12.5.1 *Gas-to-Particle Heat Transfer*

The heat transfer behavior in a spouted bed (see §9.8) is different from that in dense-phase and circulating fluidized bed systems as a result of the inherent differences in their flow structures. The spouted bed is represented by a flow structure that can be characterized by two regions: the annulus and the central spouting region (see Chapter 9). The heat transfers in these two regions are usually modeled separately. For the central spouting region, the correlation of Rowe and Claxton (1965) can be used for $\mathrm{Re_{pf}} > 1{,}000$

$$\mathrm{Nu_{gp}} = \frac{2}{1-(1-\alpha_{\mathrm{cs}})^{1/3}} + \frac{2}{3\alpha_{\mathrm{cs}}}\mathrm{Pr}^{1/3}\,\mathrm{Re_{pf}^{0.55}} \tag{12.68}$$

where α_{cs} is the gas volume fraction in the central spouting region. In the annulus, the heat transfer can be described by using the correlations for fixed beds, for example, Littman and Sliva's (1971) equation for $\mathrm{Re_{pf}} < 100$

$$\mathrm{Nu_{gp}} = 0.42 + 0.35\,\mathrm{Re_{pf}^{0.8}} \tag{12.69}$$

Substituting the corresponding values for the spouted bed into Eqs. (12.32), (12.68), and (12.69) reveals that the distance required for the gas to travel to achieve a thermal equilibrium with the solids in the annulus region is on the order of magnitude of centimeters, while this distance in the spout region is one or two orders of magnitude larger.

The importance of the intraparticle heat transfer resistance is evident for particles with relatively short contacting time in the bed and those with a large Biot number. Thus, for a shallow spouted bed, the overall heat transfer rate and thermal efficiency are controlled by the intraparticle temperature gradient. This gradient effect is most likely to be important when particles enter the lowest part of the spout and come in contact with the gas at high temperature, while it is negligible when the particles are slowly flowing through the annulus. Thus, in the annulus, unlike the spout, thermal equilibrium between gas and particles can usually be achieved even in a shallow bed.

12.5.2 *Bed-to-Surface Heat Transfer*

Compared to the fluidized bed, a spouted bed with immersed heat exchangers is less frequently encountered. Thus, the bed-to-surface heat transfer in a spouted bed mainly is related to bed-to-wall heat transfer. The bed-to-immersed-object heat transfer coefficient reaches a maximum at the spout–annulus interface and increases with the particle diameter [Epstein and Grace, 1997].

Since the solid particles in the spouted bed are well mixed, their average temperature in different parts of the annulus can be considered to be the same, just as in the case of a fluidized bed. The maximum value of the heat transfer coefficient in the h–U plot is also similar to that in a dense-phase fluidized bed [Mathur and Epstein, 1974].

12.6 Mass Transfer in Multiparticle Gas–Solid Systems

Values of the mass transfer coefficient for multiphase systems depend on the model which defines this coefficient. Hence, mass transfer coefficients should be used in conjunction with the corresponding model equations.

12.6.1 *Mass Transfer in Dense-Phase Fluidized Beds*

In a dense-phase fluidized bed (see Chapter 9), mass transfer can take place between the particle and the gas, between the bubble and the emulsion, and between the bed and the surface. These processes are discussed in the following.

12.6.1.1 *Gas-to-Particle Mass Transfer*

Under fairly low gas velocity conditions where U is close to U_{mf} or the bed is in particulate fluidization, the plug flow assumption for the gas phase can be reasonably made [Wen and Fane, 1982]. Considering the sublimation of species A from the solid phase to the gas phase, the mass balance on the concentration for species A in the gas phase, C_A, over the incremental height dH can be expressed as

$$U\frac{\mathrm{d}C_A}{\mathrm{d}H} = k_f S_B\left(C_A^s - C_A\right) \tag{12.70}$$

where C_A^s is the saturation concentration of species A on the solid surface and k_f is the particle-to-gas mass transfer coefficient.

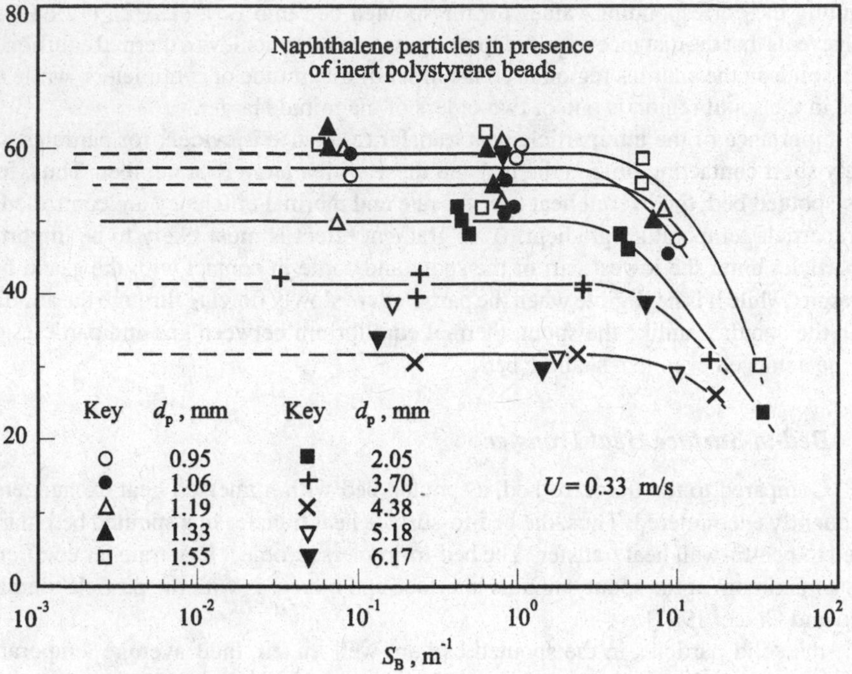

Figure 12.18. Dependence of mass transfer coefficient on naphthalene-specific interfacial area and particle size (after de Kok *et al.*, 1986).

Assuming k_f does not change with height, de Kok *et al.* (1986) studied the dependence of the overall mass transfer coefficient on the specific interfacial area. Figure 12.18 depicts the dependence, which shows that k_f is constant up to $S_B \cong 2\,\text{m}^{-1}$ and then decreases. Integrating Eq. (12.70) yields

$$k_f = \frac{U}{S_B H_f} \ln\left(\frac{C_A^s - C_{A,in}}{C_A^s - C_{A,out}}\right) \tag{12.71}$$

Equation (12.71) provides a means to quantify k_f on the basis of experimentally measured inlet and outlet concentrations of species A in the bed under low gas velocity conditions. Wen and Fane (1982) proposed the following empirical correlations for overall mass transfer coefficients in gas–solid fluidized beds, using the experimental data of Kato *et al.* (1970) based on Eq. (12.71):

$$\text{Sh} = 0.43\left[\text{Re}_{pf}\left(\frac{d_p}{H_e}\right)^{0.6}\right]^{0.97} \text{Sc}^{0.33} \quad \text{for } 0.5 \le \text{Re}_{pf}\left(\frac{d_p}{H_e}\right)^{0.6} \le 80 \tag{12.72}$$

and

$$\text{Sh} = 12.5\left[\text{Re}_{pf}\left(\frac{d_p}{H_e}\right)^{0.6}\right]^{0.2} \text{Sc}^{0.33} \quad \text{for } 80 \le \text{Re}_{pf}\left(\frac{d_p}{H_e}\right)^{0.6} \le 1{,}000 \tag{12.73}$$

where $\text{Sh} = k_f d_p/D_m$, $\text{Sc} = \mu/(\rho D_m)$, $H_e = x_s H_f$, and x_s is the volume fraction of active solids taking part in the mass transfer. The definition for Sh given assumes that the logarithmic mean fraction of the inert or nondiffusing component is unity. When the bed is made up of all active particles, $x_s = 1$.

Under relatively high gas velocity, *e.g.*, bubbling conditions, the axial mixing effect needs to be added to Eq. (12.70). Thus, we have

$$-D_a \frac{d^2 C_A}{dH^2} + U \frac{dC_A}{dH} = k_f' S_B \left(C_A^s - C_A\right) \tag{12.74}$$

where D_a is the axial dispersion coefficient, which can be evaluated empirically. Clearly, in the formulation given by Eq. (12.74), the mass transfer resistance surrounding the bubble phase is neglected.

12.6.1.2 *Bubble-to-Emulsion Mass Transfer*

A realistic treatment of mass transfer between the gas and solid phases requires consideration of the bed structure comprising the bubble phase and the emulsion phase (see §9.4). Considering bubbles containing species A passing through a fluidized bed where species A is in depletion, the mass transfer, or the mass interchange coefficient from the bubble phase to the emulsion phase, K_{be}, can be related to the difference in the concentration of species A in the bubble phase, $C_{A,b}$, and that in the emulsion phase, $C_{A,e}$, by [Kunii and Levenspiel, 1968]

$$-\frac{1}{V_b}\frac{dN_{A,b}}{dt} = -\frac{dC_{A,b}}{dt} = -U_b \frac{dC_{A,b}}{dH} = K_{be}(C_{A,b} - C_{A,e}) \tag{12.75}$$

The mass transfer resistance between the bubble and emulsion phases can be described in terms of bubble-to-cloud and cloud-to-emulsion resistances in series. Thus, we have

$$K_{be}(C_{A,b} - C_{A,e}) = K_{bc}(C_{A,b} - C_{A,c}) = K_{ce}(C_{A,c} - C_{A,e}) \tag{12.76}$$

where K_{bc} and K_{ce} are bubble-to-cloud and cloud-to-emulsion interchange coefficients, respectively. K_{be} relates to K_{bc} and K_{ce} by the equation

$$\frac{1}{K_{be}} = \frac{1}{K_{bc}} + \frac{1}{K_{ce}} \tag{12.77}$$

Define k_{be} according to

$$-U_b V_b \frac{dC_{A,b}}{dH} = A_{be} k_{be} (C_{A,b} - C_{A,e}) \tag{12.78}$$

where A_{be} is the bubble–emulsion interfacial area. Combining Eqs. (12.75) and (12.78) yields

$$K_{be} = k_{be}\frac{A_{be}}{V_b} = \frac{6}{d_b} k_{be} \tag{12.79}$$

By experimentally measuring the concentration of species A in the gas phase and emulsion phase and the inlet gas concentration in a gas–solid system without chemical reaction or adsorption, K_{be} can be determined on the basis of Eq. (12.75).

K_{be} can also be accounted for mechanically from its relationship with K_{bc} and K_{ce} given in Eq. (12.77). Following the analysis of Davidson and Harrison (1963) for K_{bc} and that of Kunii and Levenspiel (1968) for K_{ce} for bubbling fluidized beds based on the bubble configuration described in §9.4.2.1, a material balance on a single rising bubble gives

$$-\frac{dN_{A,b}}{dt} = -V_b \frac{dC_{A,b}}{dt} = (Q + k_{bc} S_{bc})(C_{A,b} - C_{A,c}) \tag{12.80}$$

where Q is the volumetric flow rate of circulating cloud gas illustrated in Fig. 9.7(a) and S_{bc} is the interfacial area between the bubble and its cloud ($\approx \pi d_b^2$). The gas flow velocity through a bubble is assumed to be $3U_{mf}$ (see §9.4.5). Thus, Q can be expressed as

$$Q = \frac{3\pi}{4} U_{mf} d_b^2 \tag{12.81}$$

For the mass transfer coefficient between the bubble and its cloud due to diffusion, k_{bc}, Davidson and Harrison (1963) also derived the following expression by assuming a spherical cap bubble with $\theta_w = 50°$ (Problem 12.5)

$$k_{bc} = 0.975 D_m^{1/2} d_b^{-1/4} g^{1/4} \tag{12.82}$$

where D_m is the molecular diffusion coefficient of gas. Combining Eqs. (12.75) and (12.80) yields

$$K_{bc} = \frac{Q + k_{bc} S_{bc}}{V_b} \tag{12.83}$$

From Eqs. (12.81) and (12.82), K_{bc} in Eq. (12.83) can be expressed as

$$K_{bc} = 4.5\left(\frac{U_{mf}}{d_b}\right) + 5.85\left(\frac{D_m^{1/2} g^{1/4}}{d_b^{5/4}}\right) \tag{12.84}$$

Davidson and Harrison (1963) expressed the total interchange coefficient for mass transfer from the bubble to the emulsion, K_{be}, by using Eq. (12.84), which is reasonable for very fast bubbles with negligible cloud. For bubbles with a large cloud, the cloud–emulsion mass transfer coefficient, K_{ce}, should also be considered, as indicated in Eq. (12.77).

For the cloud–emulsion mass transfer, the mass balance over the cloud–emulsion interface gives

$$-\frac{dN_{A,c}}{dt} = S_{ce} k_{ce} (C_{A,c} - C_{A,e}) \tag{12.85}$$

where k_{ce} is the cloud–emulsion mass transfer coefficient and S_{ce} is the interfacial area between the cloud and the emulsion. Following the analogy of bubble contact by liquid, the penetration model can be applied, leading to the following expression for k_{ce} (Problem 12.6)

$$k_{ce} \cong \left(\frac{4 D_{em} \alpha_{mf}}{\pi t_e}\right)^{1/2} \tag{12.86}$$

where D_{em} is the effective diffusion coefficient in the emulsion phase and t_e is the surface renewal time in the penetration model for bubble in contact with the emulsion phase. For a bubble with cloud, t_e can be approximated by

$$t_e = \frac{d_c}{U_{b\infty}} \cong \frac{d_b}{U_{b\infty}} \tag{12.87}$$

Since the mass transfer from bubble-to-cloud and to-emulsion is a continuous process, Eq. (12.75) can also be written as

$$-\frac{1}{V_b}\frac{dN_{A,c}}{dt} = K_{ce}(C_{A,c} - C_{A,e}) \tag{12.88}$$

Combining Eqs. (12.85) and (12.88) yields

$$K_{ce} = \frac{S_{ce} k_{ce}}{V_b} \tag{12.89}$$

For a thin-cloud bubble, S_{ce}/V_b can be approximated by $6/d_b$. From Eqs. (12.86), (12.87), and (9.37) with $d_{b\infty}$ replaced with d_b, Eq. (12.89) becomes

$$K_{ce} = 5.71\left(\frac{D_{em}\alpha_{mf}(gd_b)^{1/2}}{d_b^3}\right)^{1/2} \tag{12.90}$$

Example 12.3 Determine the bubble-to-emulsion phase interchange coefficient for mass transfer in the following gas–solid fluidization system operated at 400°C and the ambient pressure: $d_p = 510\ \mu$m, $\rho_p = 2{,}242$ kg/m³, shape factor = 0.84, $U = 0.5$ m/s, $M_p = 50$ kg, $D = 0.3$ m, and $D_m = 3\times10^{-5}$ m²/s. The fluidizing gas is air.

Solution The minimum fluidization velocity can be obtained by Eq. (9.14) as $U_{mf} = 0.113$ m. The bed voidage at minimum fluidization can be determined by the Ergun equation (Eq. (9.11)) as $\alpha_{mf} = 0.422$. The expanded bed height can be obtained from Eq. (9.47) as $H_f = 0.638$ m. The average bubble size in the bed can be calculated by Eq. (E9.1) as 0.091 m. D_{em} can be approximated by [Kunii and Levenspiel, 1991]

$$D_{em} \cong D_m \tag{E12.10}$$

The bubble-to-cloud interchange coefficient is obtained by Eq. (12.84)

$$K_{bc} = 4.5\left(\frac{0.113}{0.091}\right) + 5.85\left(\frac{(3\times10^{-5})^{1/2}9.8^{1/4}}{0.091^{5/4}}\right) = 6.72\ \text{s}^{-1} \tag{E12.11}$$

The cloud-to-emulsion interchange coefficient is calculated by Eq. (12.90)

$$K_{ce} = 5.71\left(\frac{3\times10^{-5}\times0.422\times(9.8\times0.091)^{1/2}}{0.091^3}\right)^{1/2} = 0.72\ \text{s}^{-1} \tag{E12.12}$$

Therefore, Eq. (12.77) gives the gas-to-emulsion phase interchange coefficient for mass transfer

$$K_{be} = \left(\frac{1}{\dfrac{1}{6.72}+\dfrac{1}{0.72}}\right) = 0.65\ \text{s}^{-1} \tag{E12.13}$$

12.6.1.3 *Bed-to-Surface Mass Transfer*

Mass transfer from a surface to the gas–solid suspension is significantly higher than that for particle-free conditions. This increase is due to a possible reduction of the mass transfer boundary layer and an increase in the interstitial gas velocity when particles are present. The effect on heat transfer is even more significant as solid particles also act as heat carriers.

For gas–solid fluidized beds, Wen and Fane (1982) suggested that the determination of the bed-to-surface mass transfer coefficient can be conducted by using the corresponding heat transfer correlations, replacing the Nusselt number with the Sherwood number, and replacing the Prandtl number by $\mathrm{Sc}(c_p\rho)/(c\rho_p)/(1-\alpha)$. Few experimental results on bed-to-surface mass transfer are available, especially for gas–solid fluidized beds operated at relatively high gas velocities.

12.6.2 *Mass Transfer in Circulating Fluidized Beds*

The effectiveness of the gas–solid mass transfer in a circulating fluidized bed (see Chapter 10) can be reflected by the contact efficiency, which is a measure of the extent to which the particles are exposed to the gas stream. As noted in Chapter 10, fine particles tend to form clusters, which yield contact resistance of the main gas stream with inner particles in the cluster. The contact efficiency was evaluated by using hot gas as a tracer [Dry *et al.*, 1987] and using the ozone decomposition reaction with iron oxide catalyst as particles [Jiang *et al.*, 1991]. It was found that the contact efficiency decreases as the particle concentration in the bed increases. At lower gas velocities, the contact efficiency is lower as a result of lower turbulence levels, allowing a greater extent of aggregate formation. The contact efficiency increases with the gas velocity, but the rate of increase falls with the gas velocity.

Kwauk *et al.* (1986) measured the gas–solid mass transfer coefficient using CCl_4 as a gas tracer in a circulating fluidized bed of Group B particles. The mass transfer coefficient in their study is defined in terms of plug flow for the gas phase and complete mixing for the solid phase. They reported an increase in the mass transfer coefficient with increasing gas velocity within the fast fluidization regime. Halder and Basu (1988) used the sublimation of naphthalene particles to study gas–solid mass transfer in a circulating fluidized bed. They indicated that the mass transfer coefficient from particle to bed varies significantly with the slip velocity between the gas and the solid phase, U_s $(=[(U/\alpha) - U_p/(1-\alpha)])$. A smaller overall slip velocity yields a smaller gas–solid mass transfer coefficient. They proposed a modified form of the Froessling equation (Eq. (4.96)):

$$\frac{k_f d_p}{D_m} = 2\alpha + 0.69 \left(\frac{U_s d_p \rho}{\mu\alpha}\right)^{0.5} \mathrm{Sc}^{0.33} \tag{12.91}$$

where α is the bed voidage.

Nomenclature

A_{be}	Bubble–emulsion interfacial area
A_m	Area of packet in contact with the heating surface
Ar	Archimedes number
B	Constant, defined by Eq. (12.67)
C_A	Concentration of species A
C_A^s	Saturation concentration of species A
$C_{A,b}$	Concentration of species A in the bubble
$C_{A,c}$	Concentration of species A in the cloud
$C_{A,e}$	Concentration of species A in the emulsion phase
$C_{A,in}$	Inlet concentration of species A
$C_{A,out}$	Outlet concentration of species A
c	Specific heat of particles
c_p	Specific heat at constant pressure
c_{pc}	Specific heat of clusters
D_a	Axial dispersion coefficient of gas
D_m	Molecular diffusion coefficient of gas
D_{em}	Effective diffusion coefficient in the emulsion phase
D_{tem}	Thermal diffusivity of the emulsion phase

d_b	Volume diameter of bubble
d_c	Diameter of bubble cloud
Fo	Fourier number
f_b	Bubble frequency at surface
g	Gravitational acceleration
H	Height
H_e	Effective bed height
H_f	Expansion bed height
h	Heat transfer coefficient
h	Bed-to-surface heat transfer coefficient
h_{bg}	Bed-to-gas heat transfer coefficient
h_c	Convective bed-to-surface heat transfer coefficient
h_{cr}	Radiative heat transfer coefficient of clusters
h_{dr}	Radiative heat transfer coefficient of the dispersed phase
h_f	Gas film heat transfer coefficient
h_{gc}	Gas convective component of h_c
h_{gp}	Gas-to-particle heat transfer coefficient of the bed
h_i	Instantaneous heat transfer coefficient
h_{max}	Maximum value of h
h_p	Average heat transfer coefficient between the particulate phase and surface in the absence of gas film resistance
h_{pc}	Particle convective component of h_c
$h_{pc,max}$	Maximum value of h_{pc}
h_r	Radiative heat transfer coefficient
I(t)	Age distribution function in the film-penetration model
J_q	Heat flux
J_r	Radiant heat flux
K	Thermal conductivity of gas
K_a	Apparent thermal conductivity of a gas–solid suspension flowing through an empty bed
K_{bc}	Bubble-to-cloud interchange coefficient for mass transfer
K_{be}	Bubble-to-emulsion interchange coefficient for mass transfer
K_c	Thermal conductivity of clusters
K_{ce}	Cloud-to-emulsion interchange coefficient for mass transfer
K_e	Effective thermal conductivity of a fixed bed with stagnant gas
K_{em}	Thermal conductivity of the emulsion phase
K_f	Thermal conductivity of gas film
K_p	Thermal conductivity of particles
k_{bc}	Bubble-to-cloud mass transfer coefficient
k_{be}	Bubble-to-emulsion mass transfer coefficient
k_{ce}	Cloud-to-emulsion mass transfer coefficient
k_f	Gas-to-solid mass transfer coefficient
k'_f	Gas-to-solid mass transfer coefficient, defined by Eq. (12.74)
M	Mass of particles
M_p	Total mass of particles in the bed
$N_{A,b}$	Number of moles of species A in the bubble
$N_{A,c}$	Number of moles of species A in the cloud
Nu_{bg}	Bed-to-gas Nusselt number
Nu_{gp}	Particle-to-gas Nusselt number
Nu_p	Particle-to-gas Nusselt number for a single particle
n	Coordinate normal to the temperature symmetric plane
n'	Coordinate normal to the gas–particle interface

Pr	Prandtl number
Q	Volumetric flow rate in and out of a single bubble
r	Radial coordinate
R	Radius of the bed
Re_p	Particle Reynolds number based on particle diameter and relative velocity
Re_{pf}	Particle Reynolds number based on particle diameter and superficial gas velocity
S	Area mean stirring factor
S_B	Surface area of solids per unit volume of the bed
S_{bc}	Interfacial area between the bubble and cloud
S_{ce}	Interfacial area between the cloud and emulsion
S_p	Surface area of particles
Sc	Schmidt number
Sh	Gas-to-particle Sherwood number
t_c	Contacting time of clusters or the particulate phase and surface
t_e	Surface renewal time in the penetration model for bubbles in contact with the emulsion phase
T_1	Gas temperature at the inlet of the bed
T_2	Gas temperature at the outlet of the bed
T_∞	Temperature of the emulsion phase when $x \to \infty$
T_b	Bed temperature
T_p	Temperature of particles
T_s	Temperature of the heating surface
U	Superficial gas velocity
U_b	Bubble rise velocity
$U_{b\infty}$	Rise velocity of an isolated bubble
U_p	Superficial particle velocity
U_s	Slip velocity between the gas and solid phases
U_{mf}	Minimum fluidization velocity
U_{opt}	Superficial gas velocity at $h = h_{max}$
V_b	Bubble volume
x	Distance from surface
x_s	Volume fraction of solids taking part in mass transfer
z	Axial coordinate

Greek Symbols

α	Bed voidage
α_b	Volume fraction of bubbles in the bed
α_c	Volume fraction of clusters in the bed
α_{cs}	Volume fraction in the central spouting region
α_{mf}	Bed voidage at minimum fluidization
δ	Boundary layer thickness
δ^*	Dimensionless gas layer thickness
δ_c	Time-averaged fraction of wall area covered by clusters
δ_{em}	Layer thickness of emulsion on the surface
ϵ_{bs}	General bed emissivity for bed-to-surface radiation
ϵ_b	Emissivity of bed suspension
ϵ_c	Emissivity of clusters
ϵ_d	Emissivity of dispersed phase
ϵ_p	Emissivity of particle surface
ϵ_s	Emissivity of heat transfer surface
ϵ_{sus}	Emissivity of suspension phase
ξ	Parameter, defined by Eq. (12.48)

ρ_{em}	Density of emulsion phase	$\psi(t_c)$	Age distribution function in the packet model
ρ_p	Density of particles		
σ_b	Stefan–Boltzmann's constant	φ_b	Ratio of the equivalent thickness of gas film to particle diameter

References

Baeyens, J. and Goossens, W. R. A. (1973). Some Aspects of Heat Transfer Between a Vertical Wall and a Gas Fluidized Bed. *Powder Tech.*, **8**, 91.

Baskakov, A. P. (1985). Heat Transfer in Fluidized Beds: Radiative Heat Transfer in Fluidized Beds. In *Fluidization*, 2nd ed. Ed. Davidson, Clift and Harrison. London: Academic Press.

Baskakov, A. P., Berg, B. V., Vitt, O. K., Filippovsky, N. F., Kirakosyan, V. A., Goldobin, J. M. and Maskaev, V. K. (1973). Heat Transfer to Objects Immersed in Fluidized Beds. *Powder Tech.*, **8**, 273.

Baskakov, A. P., Vitt, O. K., Kirakosyan, V. A., Maskaev, V. K. and Filippovsky, N. F. (1974). Investigation of Heat Transfer Coefficient Pulsations and of the Mechanism of Heat Transfer from a Surface Immersed into a Fluidized Bed. *Proceedings of International Symposium on Fluidization Applications*. Toulouse: Cepadues-Editions.

Basu, P. (1990). Heat Transfer in High Temperature Fast Fluidized Beds. *Chem. Eng. Sci.*, **45**, 3123.

Basu, P. and Nag, P. K. (1987). An Investigation into Heat Transfer in Circulating Fluidized Beds. *Int. J. Heat & Mass Transfer*, **30**, 2399.

Bi, H.-T., Jin, Y., Yu, Z. Q. and Bai, D.-R. (1989). The Radial Distribution of Heat Transfer Coefficients in Fast Fluidized Bed. In *Fluidization VI*. Ed. Grace, Shemilt and Bergougnou. New York: Engineering Foundation.

Bi, H.-T., Jin, Y., Yu, Z.-Q. and Bai, D.-R. (1990). An Investigation on Heat Transfer Circulating Fluidized Bed. In *Circulating Fluidized Bed Technology III*. Ed. Basu, Horio and Hasatani. Oxford: Pergamon Press.

Bock, H.-J. and Molerus, O. (1985). Influence of Hydrodynamics on Heat Transfer in Fluidized Beds. In *Fluidization*. Ed. Grace and Matsen. New York: Plenum.

Borodulya, V. A., Ganzha, V. L. and Kovensky, V. I. (1982). *Nauka I Technika*. Minsk, USSR.

Botterill, J. M. S. (1975). *Fluid-Bed Heat Transfer*. London: Academic Press.

Botterill, J. M. S., Teoman, Y. and Yüregir, K. R. (1981). Temperature Effects on the Heat Transfer Behaviour of Gas Fluidized Beds. *AIChE Symp. Ser.*, **77**(208), 330.

Botterill, J. M. S., Teoman, Y. and Yüregir, K. R. (1984). Factors Affecting Heat Transfer Between Gas–Fluidized Beds and Immersed Surfaces. *Powder Tech.*, **39**, 177.

Botterill, J. M. S. and Williams, J. R. (1963). The Mechanism of Heat Transfer to Gas–Fluidized Beds. *Trans. Instn. Chem. Engrs.*, **41**, 217.

Brodkey, R. S., Kim, D. S. and Sidner, W. (1991). Fluid to Particle Heat Transfer in a Fluidized Bed and to Single Particles. *Int. J. Heat & Mass Transfer*, **34**, 2327.

Brewster, M. Q. (1986). Effective Absorptivity and Emissivity of Particulate Medium with Applications to a Fluidized Bed. *Trans. ASME, J. Heat Transfer*, **108**, 710.

Davidson, J. F. and Harrison, D. (1963). *Fluidized Particles*. Cambridge: Cambridge University Press.

de Kok, J. J., Stark, N. L. and van Swaaij, W. P. M. (1986). The Influence of Solids Specific Interfacial Area on Gas–Solid Mass Transfer in Gas Fluidized Beds. In *Fluidization V*. Ed. Ostergaard and Sorensen. New York: Engineering Foundation.

Denloye, A. O. O. and Botterill, J. M. S. (1978). Bed to Surface Heat Transfer in a Fluidized Bed of Large Particles. *Powder Tech.*, **19**, 197.

Dry, R. J., Christensen, I. N. and White, C. C. (1987). Gas–Solids Contact Efficiency in a High Velocity Fluidized Bed. *Powder Tech.*, **52**, 243.

Epstein, N. and Grace, J. R. (1997). Spouting of Particulate Solids. In *Handbook of Powder Science and Technology*, 2nd ed. Ed. Fayed and Otten. New York: Chapman & Hall.

Flamant, G., Fatah, N. and Flitris, Y. (1992). Wall-to-Bed Heat Transfer in Gas–Solid Fluidized Beds: Prediction of Heat Transfer Regimes. *Powder Tech.*, **69**, 223.

Gabor, J. D. (1970). Wall-to-Bed Heat Transfer in Fluidized and Packed Beds. *Chem. Eng. Prog. Symp. Ser.*, **66**(105), 76.

Gel'Perin, N. I. and Einstein, V. G. (1971). Heat Transfer in Fluidized Beds. In *Fluidization*. Ed. Davidson and Harrison. New York: Academic Press.

Gloski, D., Glicksman, L. and Decker, N. (1984). Thermal Resistance at a Surface in Contact with Fluidized Bed Particles. *Int. J. Heat & Mass Transfer*, **27**, 599.

Grace, J. (1982). Fluidized-Bed Heat Transfer. In *Handbook of Multiphase Systems*. Ed. G. Hetsroni. New York: McGraw-Hill.

Halder, P. K. and Basu, P. (1988). Mass Transfer from a Coarse Particle to a Fast Bed of Fine Solids. *AIChE Symp. Ser.*, **84**(262), 58.

Higbie, R. (1935). The Rate of Absorption of a Pure Gas into a Still Liquid During Short Periods of Exposure. *Trans. AIChE*, **31**, 365.

Jiang, P., Inokuchi, K., Jean, R.-H., Bi, H.-T. and Fan, L.-S. (1991). Ozone Decomposition in a Catalytic Circulating Fluidized Bed Reactor. In *Circulating Fluidized Bed Technology III*. Ed. Basu, Horio, and Hasatani. Oxford: Pergamon Press.

Kato, K., Kubota, H. and Wen, C. Y. (1970). Mass Transfer in Fixed and Fluidized Beds. *Inst. Chem. Eng. Symp. Ser.*, **66**(105), 86.

Knowlton, T. M. (1992). Pressure and Temperature Effects in Fluid–Particle System. In *Fluidization VII*. Ed. Potter and Nicklin. New York: Engineering Foundation.

Kovenskii, V. I. (1980). Calculation of Emittance of a Disperse System. *J. Eng. Physics*, **38**, 602.

Kunii, D. and Levenspiel, O. (1968). Bubbling Bed Model for Kinetic Processes in Fluidized Beds. *I&EC Proc. Des. Dev.*, **7**, 481.

Kunii, D. and Levenspiel, O. (1991). *Fluidization Engineering*, 2nd ed. Boston: Butterworth-Heinemann.

Kunii, D. and Smith, J. M. (1960). Heat Transfer Characteristics of Porous Rocks. *AIChE J.*, **6**, 71.

Kwauk, M., Wang, N., Li, Y., Chen, B. and Shen, Z. (1986). Fast Fluidization at ICM. In *Circulating Fluidized Bed Technology*. Ed. P. Basu. Toronto: Pergamon Press.

Levenspiel, O. and Walton, J. S. (1954). Bed–Wall Heat Transfer in Fluidized Systems. *Chem. Eng. Prog. Symp. Ser.*, **50**(9), 1.

Lints, M. C. and Glicksman, L. R. (1993). Parameters Governing Particle-to-Wall Heat Transfer in a Circulating Fluidized Bed. In *Circulating Fluidized Bed Technology IV*. Ed. A. A. Avidan. New York: AIChE Publications.

Littman, H. and Sliva, D. E. (1971). Gas–Particle Heat Transfer Coefficient in Packed Beds at Low Reynolds Number. In *Heat Transfer 1970, Paris-Versailles*, CT 1.4. Amsterdam: Elsevier.

Maskaev, V. K. and Baskakov, A. P. (1974). Features of External Heat Transfer in a Fluidized Bed of Coarse Particles. *Int. Chem. Eng.*, **14**, 80.

Mathur, K. B. and Epstein, N. (1974). *Spouted Beds*. New York: Academic Press.

Mickley, H. S. and Fairbanks, D. F. (1955). Mechanism of Heat Transfer to Fluidized Beds. *AIChE J.*, **1**, 374.

Mickley, H. S., Fairbanks, D. F. and Hawthorn, R. D. (1961). The Relation Between the Transfer Coefficient and Thermal Fluctuations in Fluidized Bed Heat Transfer. *Chem. Eng. Prog. Symp. Ser.*, **57**(32), 51.

Ranz, W. E. (1952). Friction and Transfer Coefficients for Single Particles and Packed Beds. *Chem. Eng. Prog.*, **48**, 247.

Rowe, P. N. and Claxton, K. T. (1965). Heat and Mass Transfer from a Single Sphere to a Fluid Flowing Through an Array. *Trans. Instn. Chem. Engrs.*, **43**, T321.

Sleicher, C. A. and Rouse, M. W. (1975). A Convective Correlation for Heat Transfer to Constant and Variable Property Fluids in Turbulent Pipe Flow. *Int. J. Heat & Mass Transfer*, **18**, 677.

Subbarao, D. and Basu, P. (1986). A Model for Heat Transfer in Circulating Fluidized Beds. *Int. J. Heat & Mass Transfer*, **29**, 487.

Toor, R. L. and Marchello, J. M. (1958). Film-Penetration Model for Mass and Heat Transfer. *AIChE J.*, **4**, 97.

Tuot, J. and Clift, R. (1973). Heat Transfer Around Single Bubbles in a Two-Dimensional Fluidized Bed. *Chem. Eng. Prog. Symp. Ser.*, **69**(128), 78.

Wen, C. Y. and Fane, A. G. (1982). Fluidized-Bed Mass Transfer. In *Handbook of Multiphase System*. Ed. G. Hetsroni. Washington, D.C.: Hemisphere.

Wen, C. Y. and Miller, E. N. (1961). Heat Transfer in Solid–Gas Transport Lines. *I&EC*, **53**, 51.

Wender, L. and Cooper, G. T. (1958). Heat Transfer between Fluidized-Solids, Beds, and Boundary Surfaces: Correlation of Data. *AIChE J.*, **4**, 15.

Wu, R. L., Grace, J. R., Lim, C. J. and Brereton, C. M. H. (1989). Suspension-to-Surface Heat Transfer in a Circulating Fluidized Bed Combustor. *AIChE J.*, **35**, 1685.

Wu, R. L., Grace, J. R. and Lim, C. J. (1990). A Model for Heat Transfer in Circulating Fluidized Beds. *Chem. Eng. Sci.*, **45**, 3389.

Wu, R. L., Lim, C. J., Chaouki, J. and Grace, J. R. (1987). Heat Transfer from a Circulating Fluidized Bed to Membrane Waterwall Surfaces. *AIChE J.*, **33**, 1888.

Xavier, A. M., King, D. F., Davidson, J. F. and Harrison, D. (1980). Surface–Bed Heat Transfer in a Fluidized Bed at High Pressure. In *Fluidization*. Ed. Grace and Matsen. New York: Plenum.

Xavier, A. M. and Davidson, J. F. (1985). Heat Transfer in Fluidized Beds: Convective Heat Transfer in Fluidized Beds. In *Fluidization*, 2nd ed. Ed. Davidson, Clift, and Harrison. London: Academic Press.

Yoshida, K., Kunii, D. and Levenspiel, O. (1969). Heat Transfer Mechanisms Between Wall Surface and Fluidized Bed. *Int. J. Heat & Mass Transfer*, **12**, 529.

Zabrodsky, S.S. (1963). Heat Transfer Between Solid Particles and a Gas in a Nonuniformly Aggregated Fluidized Bed. *Int. J. Heat & Mass Transfer*, **6**, 23 and 991.

Problems

12.1 Hot air at 473 K passes through a dense-phase fluidized bed from the bottom. The bed is maintained at a constant temperature of 273 K. The expanded bed height is 3 m. Determine the bed-to-gas heat transfer coefficient and gas temperature at different heights of the bed if the superficial gas velocity is 0.5 m/s. Also consider the following conditions: $\alpha = 0.5$, $d_p = 0.5\,\text{mm}$, $k_g = 0.026\,\text{W/m} \cdot \text{K}$, $c_p = 1{,}005\,\text{J/kg} \cdot \text{K}$.

12.2 The bed-to-surface heat transfer coefficient may vary with the radial position in the bed. Wender and Cooper (1958) correlated the data of coefficients for heat transfer to immersed vertical tubes and proposed the following empirical correlation for $0.01 < \mathrm{Re_p} < 100$

$$\frac{hd_p}{K} = 3.51 \times 10^{-4} C_R (1-\alpha) \left(\frac{c_p \rho}{K}\right)^{0.43} \left(\frac{d_p \rho U}{\mu}\right)^{0.23} \left(\frac{c}{c_p}\right)^{0.8} \left(\frac{\rho_p}{\rho}\right)^{0.66} \qquad \text{(P12.1)}$$

where C_R is a parameter determined by Fig. P12.1. Calculate the cross-sectional averaged heat transfer coefficient for the conditions given in Example 9.2(b). Additional conditions provided are $c = 840$ J/kg · K, $c_p = 1{,}005$ J/kg · K, $K_p = 1.9$ W/m · K, $K = 0.0262$ W/m · K, and $t_c = 0.8$ s. Also, compare the results with those obtained by the approach given in §12.3.2.

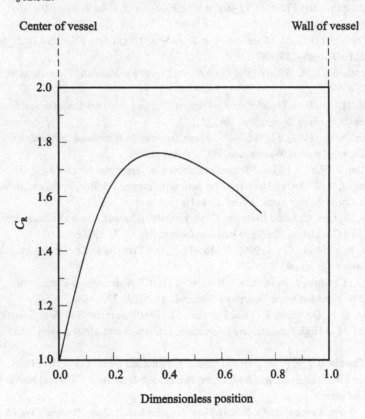

Figure P12.1. Value of C_R for nonaxial location of immersed vertical tubes (from Wender and Cooper, 1958).

12.3 Derive the axial profile of the particle convective heat transfer coefficient in a circulating fluidized bed of fine particles using the information given in §10.4.1. It can be assumed in the derivation that particles in the bed are all in a cluster form.

12.4 Derive the radial profile of the particle convective heat transfer coefficient in a circulating fluidized bed of fine particles using the information given in §10.4.2, and simplify the expression for the center region and the near-wall region, respectively.

12.5 The mass transfer coefficient between a bubble and its cloud due to diffusion, k_{bc}, given in Eq. (12.82) for a gas–solid fluidized bed can be derived on the basis of the mass transfer analogy of gas bubble in liquids [Davidson and Harrison, 1963]. Consider gas circulation inside the bubble and the diffusion of species A from the bubble to the cloud through the bubble surface to follow the radial gradient of the gas concentration, C_A. The stream function, ψ, for gas circulation near the bubble surface can be approximated by

$$\psi = R_b V_s y \sin\theta \tag{P12.2}$$

where R_b is the bubble radius; y is the distance from the bubble surface in the direction normal to the surface; θ is the angle between a radius and the vertical axis of the bubble; and V_s is the emulsion phase velocity at the bubble surface relative to the rise velocity of the bubble. V_s is given by

$$V_s = \sqrt{2gz} \tag{P12.3}$$

where z is the vertical distance from the nose of the bubble.

(a) Assuming that the diffusion along the circulation streamlines and the convection in the radial direction are negligible, derive the following equation based on the mass balance of species A in the bubble

$$D_m R_b \left(\frac{\partial^2 C_A}{\partial y^2}\right)_\theta = V_s \left(\frac{\partial C_A}{\partial \theta}\right)_\psi \tag{P12.4}$$

(b) Convert Eq. (P12.4) into the following equation

$$D_m \left(\frac{\partial^2 C_A}{\partial \psi^2}\right)_\Phi = \left(\frac{\partial C_A}{\partial \Phi}\right)_\psi \tag{P12.5}$$

by expressing y in terms of ψ and θ in terms of Φ, where Φ is a parameter defined as

$$\frac{d\Phi}{d\theta} = R_b^3 V_s \sin^2\theta \tag{P12.6}$$

Specify the boundary conditions for Eq. (P12.5).

(c) Solve Eq. (P12.5) to obtain the concentration distribution of species A near the bubble surface and express the rate of mass transfer across the frontal surface of the bubble.

(d) Obtain k_{bc} for the case of $\theta_w = 50°$, based on the following definition for k_{bc}

$$N_A = k_{bc}\pi d_b^2 (C_{A,b} - C_{A,c}) \tag{P12.7}$$

where N_A is the mass transfer rate of species A.

12.6 Derive the interchange coefficient for mass transfer between a cloud and an emulsion (Eq. (12.86)) considering Higbie's penetration model [Higbie, 1935] for mass transfer. The Higbie penetration model assumes that the exposure time is the same for all the interface elements moving along the periphery of the bubble.

APPENDIX

Summary of Scalar, Vector, and Tensor Notations

The major notations of scalars, vectors, and tensors and their operations presented in the text are summarized in Tables A1 through A5. Table A1 gives the basic definitions of vector and second-order tensor. Table A2 describes the basic algebraic operations with vector and second-order tensor. Tables A3 through A5 present the differential operations with scalar, vector, and tensor in Cartesian, cylindrical, and spherical coordinates, respectively. It is noted that in these tables, the product of quantities with the same subscripts, *e.g.*, $a_i b_i$, represents the Einstein summation and δ_{ij} refers to the Kronecker delta. The boldface symbols represent vectors and tensors.

Table A1. *Definitions of Vector and Second-Order Tensor*

Vector

$$\boldsymbol{a} = a_1\boldsymbol{e}_1 + a_2\boldsymbol{e}_2 + a_3\boldsymbol{e}_3 = a_i\boldsymbol{e}_i$$

Second-order tensor

$$\boldsymbol{\tau} = \begin{pmatrix} \tau_{11} & \tau_{12} & \tau_{13} \\ \tau_{21} & \tau_{22} & \tau_{23} \\ \tau_{31} & \tau_{32} & \tau_{33} \end{pmatrix} = \tau_{ij}\boldsymbol{e}_i\boldsymbol{e}_j$$

Unit tensor

$$\boldsymbol{I} = \begin{pmatrix} 1 & 0 & 0 \\ 0 & 1 & 0 \\ 0 & 0 & 1 \end{pmatrix} = \delta_{ij}\boldsymbol{e}_i\boldsymbol{e}_j = \boldsymbol{e}_i\boldsymbol{e}_i$$

Trace

$$\mathrm{tr}\,\boldsymbol{\tau} = \tau_{11} + \tau_{22} + \tau_{33} = \tau_{ii}$$

Table A2. *Algebraic Operations with Vector and Second-Order Tensor*

Scalar product of two vectors

$$\boldsymbol{a} \cdot \boldsymbol{b} = a_1 b_1 + a_2 b_2 + a_3 b_3 = a_i b_i$$

Vector product of two vectors

$$\boldsymbol{a} \times \boldsymbol{b} = \begin{vmatrix} \boldsymbol{e}_1 & \boldsymbol{e}_2 & \boldsymbol{e}_3 \\ a_1 & a_2 & a_3 \\ b_1 & b_2 & b_3 \end{vmatrix} = (a_2 b_3 - a_3 b_2)\boldsymbol{e}_1 + (a_3 b_1 - a_1 b_3)\boldsymbol{e}_2 + (a_1 b_2 - a_2 b_1)\boldsymbol{e}_3$$

Dyadic product of two vectors

$$\begin{aligned}\boldsymbol{ab} &= a_1 b_1 \boldsymbol{e}_1\boldsymbol{e}_1 + a_1 b_2 \boldsymbol{e}_1\boldsymbol{e}_2 + a_1 b_3 \boldsymbol{e}_1\boldsymbol{e}_3 + a_2 b_1 \boldsymbol{e}_2\boldsymbol{e}_1 + a_2 b_2 \boldsymbol{e}_2\boldsymbol{e}_2 + a_2 b_3 \boldsymbol{e}_2\boldsymbol{e}_3 + a_3 b_1 \boldsymbol{e}_3\boldsymbol{e}_1 \\ &\quad + a_3 b_2 \boldsymbol{e}_3\boldsymbol{e}_2 + a_3 b_3 \boldsymbol{e}_3\boldsymbol{e}_3 \\ &= a_i b_j \boldsymbol{e}_i\boldsymbol{e}_j\end{aligned}$$

Table A2. (*cont.*)

Scalar product of two tensors

$$\boldsymbol{\tau}:\boldsymbol{\sigma} = \tau_{11}\sigma_{11} + \tau_{12}\sigma_{21} + \tau_{13}\sigma_{31} + \tau_{21}\sigma_{12} + \tau_{22}\sigma_{22} + \tau_{23}\sigma_{32} + \tau_{31}\sigma_{13} + \tau_{32}\sigma_{23} + \tau_{33}\sigma_{33} = \tau_{\mathrm{ij}}\sigma_{\mathrm{ji}}$$

Vector product of a vector with a tensor

$$\boldsymbol{a}\cdot\boldsymbol{\tau} = (a_1\tau_{11} + a_2\tau_{21} + a_3\tau_{31})\boldsymbol{e}_1 + (a_1\tau_{12} + a_2\tau_{22} + a_3\tau_{32})\boldsymbol{e}_2 + (a_1\tau_{13} + a_2\tau_{23} + a_3\tau_{33})\boldsymbol{e}_3 = a_{\mathrm{j}}\tau_{\mathrm{ji}}\boldsymbol{e}_{\mathrm{i}}$$

Vector product of a tensor with a vector

$$\boldsymbol{\tau}\cdot\boldsymbol{a} = (\tau_{11}a_1 + \tau_{12}a_2 + \tau_{13}a_3)\boldsymbol{e}_1 + (\tau_{21}a_1 + \tau_{22}a_2 + \tau_{23}a_3)\boldsymbol{e}_2 + (\tau_{31}a_1 + \tau_{32}a_2 + \tau_{33}a_3)\boldsymbol{e}_3 = \tau_{\mathrm{ij}}a_{\mathrm{j}}\boldsymbol{e}_{\mathrm{i}}$$

Table A3. *Differential Operations in Cartesian Coordinates* (x, y, z)

Gradient of a scalar

$$\nabla s = \frac{\partial s}{\partial x}\boldsymbol{e}_{\mathrm{x}} + \frac{\partial s}{\partial y}\boldsymbol{e}_{\mathrm{y}} + \frac{\partial s}{\partial z}\boldsymbol{e}_{\mathrm{z}}$$

Divergence of a vector

$$\nabla\cdot\boldsymbol{a} = \frac{\partial a_{\mathrm{x}}}{\partial x} + \frac{\partial a_{\mathrm{y}}}{\partial y} + \frac{\partial a_{\mathrm{z}}}{\partial z}$$

Laplacian of a scalar

$$\nabla^2 s = \frac{\partial^2 s}{\partial x^2} + \frac{\partial^2 s}{\partial y^2} + \frac{\partial^2 s}{\partial z^2}$$

Laplacian of a vector

$$\nabla^2\boldsymbol{a} = \left(\frac{\partial^2 a_{\mathrm{x}}}{\partial x^2} + \frac{\partial^2 a_{\mathrm{x}}}{\partial y^2} + \frac{\partial^2 a_{\mathrm{x}}}{\partial z^2}\right)\boldsymbol{e}_{\mathrm{x}} + \left(\frac{\partial^2 a_{\mathrm{y}}}{\partial x^2} + \frac{\partial^2 a_{\mathrm{y}}}{\partial y^2} + \frac{\partial^2 a_{\mathrm{y}}}{\partial z^2}\right)\boldsymbol{e}_{\mathrm{y}} + \left(\frac{\partial^2 a_{\mathrm{z}}}{\partial x^2} + \frac{\partial^2 a_{\mathrm{z}}}{\partial y^2} + \frac{\partial^2 a_{\mathrm{z}}}{\partial z^2}\right)\boldsymbol{e}_{\mathrm{z}}$$

Divergence of a tensor

$$\nabla\cdot\boldsymbol{\tau} = \left(\frac{\partial \tau_{\mathrm{xx}}}{\partial x} + \frac{\partial \tau_{\mathrm{yx}}}{\partial y} + \frac{\partial \tau_{\mathrm{zx}}}{\partial z}\right)\boldsymbol{e}_{\mathrm{x}} + \left(\frac{\partial \tau_{\mathrm{xy}}}{\partial x} + \frac{\partial \tau_{\mathrm{yy}}}{\partial y} + \frac{\partial \tau_{\mathrm{zy}}}{\partial z}\right)\boldsymbol{e}_{\mathrm{y}} + \left(\frac{\partial \tau_{\mathrm{xz}}}{\partial x} + \frac{\partial \tau_{\mathrm{yz}}}{\partial y} + \frac{\partial \tau_{\mathrm{zz}}}{\partial z}\right)\boldsymbol{e}_{\mathrm{z}}$$

Gradient of a vector

$$\nabla\boldsymbol{a} = \frac{\partial a_{\mathrm{x}}}{\partial x}\boldsymbol{e}_{\mathrm{x}}\boldsymbol{e}_{\mathrm{x}} + \frac{\partial a_{\mathrm{y}}}{\partial x}\boldsymbol{e}_{\mathrm{x}}\boldsymbol{e}_{\mathrm{y}} + \frac{\partial a_{\mathrm{z}}}{\partial x}\boldsymbol{e}_{\mathrm{x}}\boldsymbol{e}_{\mathrm{z}} + \frac{\partial a_{\mathrm{x}}}{\partial y}\boldsymbol{e}_{\mathrm{y}}\boldsymbol{e}_{\mathrm{x}} + \frac{\partial a_{\mathrm{y}}}{\partial y}\boldsymbol{e}_{\mathrm{y}}\boldsymbol{e}_{\mathrm{y}} + \frac{\partial a_{\mathrm{z}}}{\partial y}\boldsymbol{e}_{\mathrm{y}}\boldsymbol{e}_{\mathrm{z}} + \frac{\partial a_{\mathrm{x}}}{\partial z}\boldsymbol{e}_{\mathrm{z}}\boldsymbol{e}_{\mathrm{x}} + \frac{\partial a_{\mathrm{y}}}{\partial z}\boldsymbol{e}_{\mathrm{z}}\boldsymbol{e}_{\mathrm{y}} + \frac{\partial a_{\mathrm{z}}}{\partial z}\boldsymbol{e}_{\mathrm{z}}\boldsymbol{e}_{\mathrm{z}}$$

Substantial derivatives following the gas flow

$$\frac{\mathrm{D}s}{\mathrm{D}t} = \frac{\partial s}{\partial t} + \boldsymbol{U}\cdot\nabla s, \qquad \frac{\mathrm{D}\boldsymbol{a}}{\mathrm{D}t} = \frac{\partial \boldsymbol{a}}{\partial t} + \boldsymbol{U}\cdot\nabla\boldsymbol{a}$$

Substantial derivatives following the particle flow

$$\frac{\mathrm{D}s}{\mathrm{D}t} = \frac{\partial s}{\partial t} + \boldsymbol{U}_{\mathrm{p}}\cdot\nabla s, \qquad \frac{\mathrm{D}\boldsymbol{a}}{\mathrm{D}t} = \frac{\partial \boldsymbol{a}}{\partial t} + \boldsymbol{U}_{\mathrm{p}}\cdot\nabla\boldsymbol{a}$$

Table A4. *Differential Operations in Cylindrical Coordinates* (r, θ, z)

$$\nabla s = \frac{\partial s}{\partial r}\boldsymbol{e}_r + \frac{1}{r}\frac{\partial s}{\partial \theta}\boldsymbol{e}_\theta + \frac{\partial s}{\partial z}\boldsymbol{e}_z$$

$$\nabla \cdot \boldsymbol{a} = \frac{1}{r}\frac{\partial}{\partial r}(r a_r) + \frac{1}{r}\frac{\partial a_\theta}{\partial \theta} + \frac{\partial a_z}{\partial z}$$

$$\nabla^2 s = \frac{1}{r}\frac{\partial}{\partial r}\left(r\frac{\partial s}{\partial r}\right) + \frac{1}{r^2}\frac{\partial^2 s}{\partial \theta^2} + \frac{\partial^2 s}{\partial z^2}$$

$$\nabla^2 \boldsymbol{a} = \left[\frac{\partial}{\partial r}\left(\frac{1}{r}\frac{\partial}{\partial r}(r a_r)\right) + \frac{1}{r^2}\frac{\partial^2 a_r}{\partial \theta^2} + \frac{\partial^2 a_r}{\partial z^2} - \frac{2}{r^2}\frac{\partial a_\theta}{\partial \theta}\right]\boldsymbol{e}_r$$
$$+ \left[\frac{\partial}{\partial r}\left(\frac{1}{r}\frac{\partial}{\partial r}(r a_\theta)\right) + \frac{1}{r^2}\frac{\partial^2 a_\theta}{\partial \theta^2} + \frac{\partial^2 a_\theta}{\partial z^2} + \frac{2}{r^2}\frac{\partial a_r}{\partial \theta}\right]\boldsymbol{e}_\theta$$
$$+ \left[\frac{1}{r}\frac{\partial}{\partial r}\left(r\frac{\partial a_z}{\partial r}\right) + \frac{1}{r^2}\frac{\partial^2 a_z}{\partial \theta^2} + \frac{\partial^2 a_z}{\partial z^2}\right]\boldsymbol{e}_z$$

$$\nabla \cdot \boldsymbol{\tau} = \left[\frac{1}{r}\frac{\partial}{\partial r}(r\tau_{rr}) + \frac{1}{r}\frac{\partial \tau_{\theta r}}{\partial \theta} + \frac{\partial \tau_{zr}}{\partial z} - \frac{\tau_{\theta\theta}}{r}\right]\boldsymbol{e}_r$$
$$+ \left[\frac{1}{r^2}\frac{\partial}{\partial r}(r^2\tau_{r\theta}) + \frac{1}{r}\frac{\partial \tau_{\theta\theta}}{\partial \theta} + \frac{\partial \tau_{z\theta}}{\partial z} + \frac{\tau_{\theta r} - \tau_{r\theta}}{r}\right]\boldsymbol{e}_\theta$$
$$+ \left[\frac{1}{r}\frac{\partial}{\partial r}(r\tau_{rz}) + \frac{1}{r}\frac{\partial \tau_{\theta z}}{\partial \theta} + \frac{\partial \tau_{zz}}{\partial z}\right]\boldsymbol{e}_z$$

$$\nabla \boldsymbol{a} = \frac{\partial a_r}{\partial r}\boldsymbol{e}_r\boldsymbol{e}_r + \frac{\partial a_\theta}{\partial r}\boldsymbol{e}_r\boldsymbol{e}_\theta + \frac{\partial a_z}{\partial r}\boldsymbol{e}_r\boldsymbol{e}_z$$
$$+ \left(\frac{1}{r}\frac{\partial a_r}{\partial \theta} - \frac{a_\theta}{r}\right)\boldsymbol{e}_\theta\boldsymbol{e}_r + \left(\frac{1}{r}\frac{\partial a_\theta}{\partial \theta} + \frac{a_r}{r}\right)\boldsymbol{e}_\theta\boldsymbol{e}_\theta + \frac{1}{r}\frac{\partial a_z}{\partial \theta}\boldsymbol{e}_\theta\boldsymbol{e}_z$$
$$+ \frac{\partial a_r}{\partial z}\boldsymbol{e}_z\boldsymbol{e}_r + \frac{\partial a_\theta}{\partial z}\boldsymbol{e}_z\boldsymbol{e}_\theta + \frac{\partial a_z}{\partial z}\boldsymbol{e}_z\boldsymbol{e}_z$$

Table A5. *Differential Operations in Spherical Coordinates* (R, θ, φ)

$$\nabla s = \frac{\partial s}{\partial R}\boldsymbol{e}_R + \frac{1}{R}\frac{\partial s}{\partial \theta}\boldsymbol{e}_\theta + \frac{1}{R\sin\theta}\frac{\partial s}{\partial \varphi}\boldsymbol{e}_\varphi$$

$$\nabla \cdot \boldsymbol{a} = \frac{1}{R^2}\frac{\partial}{\partial R}(R^2 a_R) + \frac{1}{R\sin\theta}\frac{\partial}{\partial \theta}(a_\theta \sin\theta) + \frac{1}{R\sin\theta}\frac{\partial a_\varphi}{\partial \varphi}$$

$$\nabla^2 s = \frac{1}{R^2}\frac{\partial}{\partial R}\left(R^2\frac{\partial s}{\partial R}\right) + \frac{1}{R^2\sin\theta}\frac{\partial}{\partial \theta}\left(\sin\theta\frac{\partial s}{\partial \theta}\right) + \frac{1}{R^2\sin^2\theta}\frac{\partial^2 s}{\partial \varphi^2}$$

$$\nabla^2 \boldsymbol{a} = \left[\frac{\partial}{\partial R}\left(\frac{1}{R^2}\frac{\partial}{\partial R}(R^2 a_R)\right) + \frac{1}{R^2\sin\theta}\frac{\partial}{\partial \theta}\left(\sin\theta\frac{\partial a_R}{\partial \theta}\right) + \frac{1}{R^2\sin^2\theta}\frac{\partial^2 a_R}{\partial \varphi^2}\right.$$
$$\left. - \frac{2}{R^2\sin\theta}\frac{\partial}{\partial \theta}(a_\theta\sin\theta) - \frac{2}{R^2\sin\theta}\frac{\partial a_\varphi}{\partial \varphi}\right]\boldsymbol{e}_R$$

Table A5. (*cont.*)

$$+\left[\frac{1}{R^2}\frac{\partial}{\partial R}\left(R^2\frac{\partial a_\theta}{\partial R}\right)+\frac{1}{R^2}\frac{\partial}{\partial\theta}\left(\frac{1}{\sin\theta}\frac{\partial}{\partial\theta}(a_\theta\sin\theta)\right)+\frac{1}{R^2\sin^2\theta}\frac{\partial^2 a_\theta}{\partial\varphi^2}\right.$$
$$\left.+\frac{2}{R^2}\frac{\partial a_R}{\partial\theta}-\frac{2\cot\theta}{R^2\sin\theta}\frac{\partial a_\varphi}{\partial\varphi}\right]\boldsymbol{e}_\theta$$
$$+\left[\frac{1}{R^2}\frac{\partial}{\partial R}\left(R^2\frac{\partial a_\varphi}{\partial R}\right)+\frac{1}{R^2}\frac{\partial}{\partial\theta}\left(\frac{1}{\sin\theta}\frac{\partial}{\partial\theta}(a_\varphi\sin\theta)\right)+\frac{1}{R^2\sin^2\theta}\frac{\partial^2 a_\varphi}{\partial\varphi^2}\right.$$
$$\left.+\frac{2}{R^2\sin\theta}\frac{\partial a_R}{\partial\varphi}+\frac{2\cot\theta}{R^2\sin\theta}\frac{\partial a_\theta}{\partial\varphi}\right]\boldsymbol{e}_\varphi$$

$$\nabla\cdot\boldsymbol{\tau}=\left[\frac{1}{R^2}\frac{\partial}{\partial R}(R^2\tau_{RR})+\frac{1}{R\sin\theta}\frac{\partial}{\partial\theta}(\tau_{\theta R}\sin\theta)+\frac{1}{R\sin\theta}\frac{\partial\tau_{\varphi R}}{\partial\varphi}-\frac{\tau_{\theta\theta}+\tau_{\varphi\varphi}}{R}\right]\boldsymbol{e}_R$$
$$+\left[\frac{1}{R^3}\frac{\partial}{\partial R}(R^3\tau_{R\theta})+\frac{1}{R\sin\theta}\frac{\partial}{\partial\theta}(\tau_{\theta\theta}\sin\theta)+\frac{1}{R\sin\theta}\frac{\partial\tau_{\varphi\theta}}{\partial\varphi}+\frac{(\tau_{\theta R}-\tau_{R\theta})-\tau_{\varphi\varphi}\cot\theta}{R}\right]\boldsymbol{e}_\theta$$
$$+\left[\frac{1}{R^3}\frac{\partial}{\partial R}(R^3\tau_{R\varphi})+\frac{1}{R\sin\theta}\frac{\partial}{\partial\theta}(\tau_{\theta\varphi}\sin\theta)+\frac{1}{R\sin\theta}\frac{\partial\tau_{\varphi\varphi}}{\partial\varphi}+\frac{(\tau_{\varphi R}-\tau_{R\varphi})+\tau_{\varphi\theta}\cot\theta}{R}\right]\boldsymbol{e}_\varphi$$

$$\nabla\boldsymbol{a}=\frac{\partial a_R}{\partial R}\boldsymbol{e}_R\boldsymbol{e}_R+\frac{\partial a_\theta}{\partial R}\boldsymbol{e}_R\boldsymbol{e}_\theta+\frac{\partial a_\varphi}{\partial R}\boldsymbol{e}_R\boldsymbol{e}_\varphi$$
$$+\left(\frac{1}{R}\frac{\partial a_R}{\partial\theta}-\frac{a_\theta}{R}\right)\boldsymbol{e}_\theta\boldsymbol{e}_R+\left(\frac{1}{R}\frac{\partial a_\theta}{\partial\theta}+\frac{a_R}{R}\right)\boldsymbol{e}_\theta\boldsymbol{e}_\theta+\frac{1}{R}\frac{\partial a_\varphi}{\partial\theta}\boldsymbol{e}_\theta\boldsymbol{e}_\varphi$$
$$+\left(\frac{1}{R\sin\theta}\frac{\partial a_R}{\partial\varphi}-\frac{a_\varphi}{R}\right)\boldsymbol{e}_\varphi\boldsymbol{e}_R+\left(\frac{1}{R\sin\theta}\frac{\partial a_\theta}{\partial\varphi}-\frac{a_\varphi}{R}\cot\theta\right)\boldsymbol{e}_\varphi\boldsymbol{e}_\theta$$
$$+\left(\frac{1}{R\sin\theta}\frac{\partial a_\varphi}{\partial\varphi}+\frac{a_R}{R}+\frac{a_\theta}{R}\cot\theta\right)\boldsymbol{e}_\varphi\boldsymbol{e}_\varphi$$

Index

abrasion
  definition and effects of, 252–254
  in particle flows, 246
absorption, cross section for, 143
acceleration, of particles, 87, 88, 92, 93
acceleration number, 108
acicular particles, 4
acoustic fluidized beds, 377
acoustic waves, 244, 259–265
activated carbon, 25, 29
adsorption
  chemical, 25
  in dense-phase fluidized beds, 372
  physical, 25
aeration, in standpipes, 365, 366
age distribution function, 510
agglomeration, of particles, 190, 445
air slides, solids transport by, 461
Allen's equation, 7
alloys
  as ferromagnetic materials, 38
  as poor electric conductors, 33
  yield strength of, 31
alumina, properties of, 30, 31, 35, 37
aluminum, thermal properties of, 35
aluminum alloys, properties of, 30, 31
American Society for Testing and Materials. *See* ASTM
Amontons's law of sliding friction, 64, 70, 76
angle of repose. *See* flow properties of powders
angular momentum, conservation of, 49
arc discharge, from particle charges, 87
Archimedes number, 373–374, 379, 409, 427, 532
arching, at hopper outlets, 333, 342, 343, 346
argon, use in gas adsorption, 13
arithmetic mean diameter, of particulate systems, 23
ASTM Standard sieves, 5, 10, 11
attrition. *See also* abrasion; fragmentation
  mechanisms of, 252–254
  in particle collisions, 46, 190, 191
averages, 182–189
  ensemble average, 211
  intrinsic average, 182–183
  minimum control volume of phase average, 183–187
  phase average, 182–183
averaging theorems, 182–189
axial dispersion coefficient, 529
axial distribution, of heat transfer coefficient, 524, 538
axial flow cyclone, 299
Babinet's principle, 147
baffles, in dense-phase fluidized beds, 376, 377, 398
bag filters, 315–319, 320
Bagnold number, 210
barium titanate, permittivity of, 37
Basset, Boussinesq, and Oseen (BBO) equation, 107–108, 198–199, 259, 260
Basset coefficient, 108, 123
Basset force, 87, 88–95, 107, 125, 199, 231, 261, 476, 481
  carried mass and, 92–95
bed-to-gas heat transfer, in dense-phase fluidized beds, 512–513
bed-to-surface heat transfer, 503, 513–518, 519, 520
  coefficient for, 538
  in spouted beds, 527
bed-to-surface mass transfer, 531
Beer–Lambert law, 14
belt conveyors, solids transport by, 461
beryllia, thermal properties of, 35
Biot number, 131, 132, 159, 512, 527
blackbody
  definition of, 34
  monochromatic-emissive power of, 144
blowers, in circulating fluidized beds, 423, 437–438
body forces. *See* field forces
Boltzmann constant, 114, 116, 171
boron, elastic properties of, 30
boron nitride, thermal properties of, 35
Bose–Einstein statistics, 170
Bourger–Beer relation, 152
Boussinesq formulation, 176, 194, 198
brass, yield strength of, 31
breakage, of particles, 46, 80
brittle erosion, 245–247, 250–251
brittle fracture, of particles, 3, 30–32
bronze, yield strength of, 31
Brunauer–Emmett–Teller (BET) isotherm, 26–28
Brunauer–Emmett–Teller (BET) method, for surface area analysis, 13, 26
bubble column, 371, 384
bubble Reynolds number, 391
bubbles
  breakup of, 375, 377, 389, 390
  in circulating fluidized beds, 433
  circulation induced by, 506
  cloudless, 382, 383
  cloud-type, 381, 382, 383, 417
  coalescence of, 375, 376, 387, 388–389, 390, 401
  configurations of, 382–385
  in dense-phase fluidized bed, 375, 376, 377, 385–389, 390, 401, 417, 513–514

bubbles (*cont.*)
  effect on heat transfer, 500, 501, 510, 513–514, 515
  effect on mass transfer, 538–539
  flow around, 385–388
  formation of, 381
  fraction of, 396, 515
  frequency of, 515
  invisible gas flow through, 393
  jet formation from, 388, 389–392
  onset of bubbling, 381–382
  phase flow of, 392
  pressure distribution around, 387
  rise velocity of, 374, 381, 383, 389–392, 396, 399, 418, 539
  size of, 373, 375–376, 381, 390–391, 392, 396
  stability of, 390
  wake angle of, 383–384
  wake of, 381, 383, 385, 387, 514
bubble-to-cloud interchange coefficient, 529, 531
bubble-to-cloud mass transfer, in dense-phase fluidized beds, 529–531, 538
bubbling fluidization, in dense-phase fluidized beds, 373, 374, 375, 381–396
bucket elevators, solids transport by, 461
Buckingham pi theorem, 230
buoyancy force, 93, 107
Burke and Plummer's theory, 166, 226, 228–230

cake filters, 314, 315, 316
calcium hydroxide, specific surface area of, 27–28
capillaric model, 224–225, 226
carbide (sintered), properties of, 30, 35
Carman correlation, 282, 283
carried mass, Basset force and, 92–95, 108
carried mass coefficient, 108
cascade impactor, particle sizing using, 17
catalytic cracking
  circulating fluidized bed use in, 421
  dense–phase fluidized bed use in, 381
  standpipe use in, 350, 351
cement clinker, mass distribution of, 21
center of compression, 53–56
centrifugal fluidized beds, 378, 415–416
centrifugal sedimentation, particle sizing by, 10, 12
centrifugal separators, 298, 300
centrifugation, gas–solid separation by, 297
ceramics
  as elastic-brittle particles, 29
  pipe walls of, 244–245
  yield strength of, 31
channeling, in dense-phase fluidized beds, 374, 375, 376
charge generation, 111–123
  corona charging, 115, 117–118
  diffusion charging, 115, 116–117
  field charging, 115–116
  by ion collection, 115–119
  in ionized gases, 117–118
  saturation charge, 116
  by surface contact, 111–115
  by thermionic emission, 119
charge transfer, 111–123
  basic mechanism of, 114
  by collision, 119–123
  momentum transfer and, 87–129
  by single impact, 119–120
charge transfer coefficient, 119
choking
  in circulating fluidized beds, 425, 428, 432, 460
  in pneumatic conveying, 465
circulating fluidized beds (CFB), 421–460
  applications of, 421
  axial profile of voidage in, 438–440, 459
  blowers in, 423, 437–438
  choking in, 425, 428, 432, 460
  clusters and clustering in, 425, 445–446, 447, 453, 522, 532
  compressors in, 423, 437–438
  computational fluid dynamics for, 451–453
  core-annular flow model for, 448, 459
  core-annular flow structure in, 446, 447
  cyclones in, 422, 431
  dense-phase fluidization regime in, 423, 424, 425, 427
  deposition coefficient for, 449, 450
  dilute transport regime of, 421, 443
  downcomer in, 421, 422–423, 431, 432
  entrainment in, 440, 448, 449
  entrance geometry of, 440, 441
  exit geometry of, 440, 441
  fast fluidization regime in, 423, 424, 425, 429, 447–453
  flow regimes and transitions in, 423–438
  gas kinetic theory applied to, 452
  granular flow in, 452
  granular temperature in, 452
  heat transfer coefficient in, 524–525
  heat transfer in, 521–526
  hydrodynamics of, 422, 434, 440–442, 448, 451
  intermittency index for, 445–447
  intermittent solids flow in, 445–447
  internals for, 421
  J-valve in, 423
  L-valve in, 423, 431, 435
  mass transfer in, 532
  maximum gas velocities in, 428–429
  maximum stable bubble size in, 460
  minimum gas velocities in, 428–429, 437
  nonmechanical valves in, 422, 423, 429
  operating conditions for, 421, 525–526
  particle–particle collision in, 444, 452, 525
  particle turbulent diffusion coefficient for, 449
  particle turbulent diffusion in, 448
  particle–wall collisions in, 444, 525
  pressure balance in, 429–430
  pressure drop in, 423, 425, 430–431, 433
  radial profile of voidage in, 440–442, 524
  reactor applications of, 421
  risers in, 421, 422, 423, 430–431, 434–436, 440, 442, 448
  seal pots in, 423
  solids circulation in, 524, 525

solids flow control devices in, 422–423, 431, 433–437
solids flow structure in, 443–447
solids holdup in, 442–443
solids mixing in, 443–453
S-shaped profiles of voidage in, 439, 440
suspension-to-wall heat transfer in, 521–522
system configuration in, 422–423
tangential gas injection in, 422
transport velocity in, 425, 459
two-phase flow models applied to, 451–453
voidage in, 438–442, 445
V-valve in, 423
wall region of, 423, 424, 442, 444
wavy solids layer in, 444–445
cloud-to-emulsion interchange coefficient, 529, 530, 531, 539
clusters and clustering
in circulating fluidized beds, 425, 445–446, 447, 453, 522, 532
in dense-phase fluidized beds, 401, 521–522, 523–524
role in heat transfer, 521–522, 523–524
coal
as elastic-brittle solids, 29
Rosin–Rammler distribution, 20, 21
coal combustion
circulating fluidized bed use in, 421
dense-phase fluidized bed use in, 372
heat transfer in, 130, 499, 517
particle changes in, 253
particle tracking in, 166
Sauter's diameter applied to, 6
cobalt, as ferromagnetic material, 38
coefficient of diffusion, 198
coefficient of kinetic friction, 64, 75, 76, 248, 252
coefficient of restitution, 47, 48, 78, 80–83
coefficient of self-diffusion, 173–174
coefficient of thermal expansion, of particles, 32–33, 35
coefficient of viscosity, 173
collinear impact, of spheres, 47–48
collision
charge transfer by, 119–123
due to wake attraction, 93–95, 128
of elastic spheres, 49–50, 72–78, 133–138
frequency of, 46, 172–173
heat conduction in, 133–138
of inelastic spheres, 78–83, 248
interparticle kinetic theory modeling of, 164, 166, 197
interparticle particle viscosity due to, 202–204
kinetic theory applied to, 210
speed of, 249
collisional deformation, of spheres, 49
collisional force, 49, 73, 87, 108, 203
erosion from, 248
collisional heat transfer coefficient, 136, 137
collisional pair distribution function, 215–217
collision mechanics, of solids, 46–86
compatibility requirement, 341
compressors, in circulating fluidized beds, 423, 437–438
conical hoppers
design diagram for, 345
stress analysis of, 339
conservation of angular momentum, 49, 266
contact
axisymmetric, 51, 52
elastic, 49–50
frictional, 63–71, 76
frictionless, 59–63
inelastic, 49, 78–83
nonsliding, 66–69
normal, 59–63
oblique, 46, 63–64
plastic, 78–80
quasi-static, 51
torsionless, 51
contact efficiency, effect on gas–solid mass transfer, 532
contact potential, 114
contact theories, collisional mechanics and, 46
continuity equation, 169, 175, 179, 195, 214
continuity waves
in dense-phase fluidized beds, 382
generation of, 280–282
continuum modeling. *See also* Eulerian continuum approach
of multiphase flows, 164, 182–205
convection
heat transfer by, 130, 138–142
mass transfer by, 156–157
copper
as isotropic material, 29
thermal properties of, 33, 35
copper alloys, elastic properties of, 30
copper ball, restitution coefficient of, 82–83
core-annular flow model, 448, 459
core (funnel) flow hoppers, 333, 334
corona charging, 115, 117–118
Coulomb force, 119, 310
Coulomb powders, 336–337, 347
Coulomb's law, 104
Coulter principle, particle sizing using, 17
cracking mechanism, 246
crack initiation and growth, 32
creeping flows. *See* Stokes flows
crystalline particles, 4
cyclones, 297. *See also* rotary flow dust separators; tangential flow cyclones
in circulating fluidized beds, 422, 431
collection efficiency of, 303–309
cut-off size of, 307–308
in dense-phase fluidized beds, 376
flow field in, 300–303
geometric configuration of, 304, 305
principles of, 297–309
reverse-flow cyclone, 298
uniform cyclone, 298
cyclone scrubbers, 323, 325
cylindrical column, stresses in, 338–339

Darcy's law, 164, 166, 222–224, 226, 386, 419
Davidson–Harrison model, for flow around a bubble in fluidized beds, 385, 387, 417, 460, 529, 539
Davies and Taylor equation, 391
DeBroucker's mean diameter, 24, 40, 379
Debye shielding distance, 117
deformation
  elastic, 72–78
  of particles, 3, 24, 28–32, 46
  plastic. *See* plastic deformation
dendritic particles, 4
dense-phase fluidized beds, 371–420, 421
  acoustic fluidized beds, 377
  applications of, 371, 372
  baffles in, 376, 377, 398, 399
  bed-collapsing technique applied to, 418
  bed expansion in, 378, 380, 392–396
  bed voidage in, 371, 388, 398, 399–400, 408, 419
  bubble behavior in. *See* bubbles
  bubbling fluidization in, 374, 375, 381–396
  centrifugal fluidized beds, 378
  channeling in, 374, 375, 376
  components of, 376–378
  convective heat transfer coefficient of, 518
  cyclones in, 376
  diplegs in, 376, 377
  elutriation in, 400–402
  emulsion phase in, 392–393, 394–395
  entrainment in, 371, 400–402
  flow around bubbles in, 385–388
  fluidization regimes in, 374–376
  force fields for, 377
  freeboard of, 376, 377, 384, 400–401, 440
  gas flow division in, 392–396
  heat exchangers in, 376
  heat transfer in, 501, 508, 512–518
  hydrodynamics of, 373, 399–400, 418
  instability of, 381
  internals in, 376, 398, 399
  interparticle forces in, 372
  jetting in, 381, 388–389
  lean-phase fluidization in, 371, 374, 396, 397
  magnetofluidized beds, 378
  mass transfer in, 527–532
  minimum bubbling velocity in, 382, 416
  minimum fluidization in, 378–380, 385
  minimum fluidization velocity in, 378, 395, 407
  particle classifications in, 371–376
  particle interaction in, 381
  particulate fluidization in, 373, 374–375, 380–381
  reactor applications, 371
  regime classifications in, 371–376
  regime transition in, 396–398
  slugging in, 374, 375–376, 403–406
  solids ejection in, 401
  solids mixing in, 371, 381, 384
  spouting in, 372, 374, 376, 388, 406–409
  transition velocity in, 398
  transport disengagement height (TDH), 401, 402, 403
  turbulent fluidization in, 374, 375, 396–400
  two-phase theory of fluidization applied to, 375, 393, 405, 420
  vibrofluidized beds, 377
  wall effect in, 403, 418
dense-phase transport modeling, 211–213
density, of gas–solid mixtures, 244, 254–257
density functions, 18
deposition coefficient, 449, 450
depth filters, 297, 314–325
deterministic trajectory method, for multiphase flow, 165, 205, 206–208
detonation combustion, acoustic waves from, 259
Deutsch equation, 312, 313
diamagnetic solids, 37, 38, 39
dielectric materials, electrical properties of, 37, 105
diffraction
  effect on thermal radiation, 143
  effects on scattering, 150
  from large spheres, 147–148
diffusion
  heat transfer by. *See* thermal diffusivity
  mass transfer by, 156–157
diffusion charging, 115, 116–117
diffusivity, thermal. *See* thermal diffusivity
dilute transport, 374, 421, 522
dimensional analysis, similarity and, 230–235
diplegs, in dense-phase fluidized beds, 376, 377
direct numerical simulation (DNS), 174, 288
discharge flow, 364
discrete-vortex simulation (DVS), 174, 288
dispersion forces, 25
displacement
  from boundary point forces, 58
  radial displacement, 58
  vertical displacement, 58, 60
displacement–strain relationships, 52, 58
Doppler effect, 16
downcomer, in circulating fluidized beds, 421, 422–423, 431, 432, 433–437
drag coefficient, 7, 40, 88, 94, 108, 123, 268, 287, 484, 494
drag force, 87–88, 96, 190, 247, 261, 267, 276–280, 318, 353, 464, 477, 478
drag reduction, 469–472
drying, in dense-phase fluidized beds, 372
ductile erosion, 245–247, 249
ductile fracture, of particles, 3, 30–32
dust explosion, from particle charges, 87
dust removal, 297
dynamic diameter, of particles, 6–9, 13
dynamic waves, 282–284
  in dense-phase fluidized beds, 382, 417

efficiency factors
  for absorption, 143
  for extinction, 143, 144
  for scattering, 143, 144, 151
Einstein summation, 175, 219, 540
elastic-brittle particles, 29
elastic collision, 46
elastic constants, 29
elastic contact theory, of solids, 49–50

elastic deformation, 72–78, 245
  of particles, 3, 28, 29–30, 31
elasticity, 50
  restoration of, 80
elasticity modulus, 382
elastic–plastic impact model, 80, 81, 83
elastomeric deformation, 28
elastoplastic particles, 29–30, 31
electrical conductivity, of particles, 35–36
electric force, 101, 105
electric porcelain, yield strength of, 31
electric wind, in electrostatic precipitation, 297, 311–312
electrification. *See* charge generation
electron microscopy, of particles, 6, 10
electrophoresis, 108
electrostatic ball probe theory, 111, 119, 121–123
electrostatic forces, 25, 47, 87, 104, 107, 108, 111, 307, 308
electrostatic precipitation and precipitators, 105
  collection efficiency in, 312–314
  cylinder-type precipitators, 311
  gas-solid separation by, 297, 309–314, 315
  migration velocity and electric wind in, 311–312
  particle charging in, 310
  particle discharging in, 310
  plate-type precipitators, 311
  principles of, 310–311
elutriation, in dense-phase fluidized beds, 400–402
elutriators, in settling chambers, 321, 322
emulsion phase flow, 392, 394, 395
emulsion phase/packet model, for heat transfer, 501, 503, 506–511
energy equation, 169–170, 196, 207–208, 213, 214–215, 268
entrainment
  in dense-phase fluidized beds, 371, 400–402, 440
  of fluid, 97
  model for, 448, 451
ϵ-equation, 178, 201
equation of equilibrium, 50, 51, 52, 341
equation of motion, 106, 108–110, 232, 481
equation of state, for gas–solid mixtures, 244, 254–257
equilibrium speed of sound, in gas–solid mixtures, 265
equivalent diameters, of nonspherical particles, 4–10
Ergun equation, 164, 225–230, 234, 379
  application to suspensions, 167
  flow-rate effects, 166, 225
  fluid-property effects on, 225–226
  particle-property effects on, 229–230
  voidage of packing effects on, 226
erosion. *See also* flow properties of powders
  abrasive, 246, 252–254
  brittle, 245–247
  ductile, 245–247
  locations of, 247–252
  particle properties affecting, 3
  of wall surfaces, 46, 244–254
Eulerian continuum approach, 164–165, 182, 183, 206, 323
Eulerian integral length scale, 285

fabric filters, 318
  collection efficiency of, 319–320
  diffusion collection in, 319
  inertial impaction collection in, 319
  interception collection in, 319–320
  materials used in, 315
far-field diffraction. *See* Fraunhofer diffraction
fast fluidization regime, 421
  in circulating fluidized beds, 423, 424, 425, 429, 432–438, 447–453
  gas velocities for, 428–429
  mass transfer in, 532
  operating constraints of, 432–438
fatigue, of particles, 30–32
fatigue cracking, 245
Faxen effect, 476
Feret's diameter, 5–6
Fermi–Dirac distribution function, 114
Fermi energy level, 111, 114, 115
ferrimagnetic and ferroelectric materials, 37, 39
fibrous particles, 4
Fick's equation, 156, 158
field charging, 115–116
field forces, 105–107. *See also* electric force; gravitational force; magnetic force
film model, for heat transfer, 501, 502–503
film-penetration model, for heat transfer, 508–510
filtration and filters
  cake filters, 314, 315
  collection efficiency of, 319–320
  fabric filters, 315, 318, 319–320
  mechanisms and types of, 314–315
  pressure drops in, 314, 315–319
Fischer–Tropsch synthesis, circulating fluidized bed use in, 421
fixed beds, 371, 374, 379
flaky particles, 4
flowability of powders, 337, 342–346
  arching effects on, 333, 342, 343, 346
  Jenike's model for, 342
  particle shape effects on, 3
  piping effects on, 333, 342
  segregation effects on, 342
flow properties of powders
  angle of fall, 342
  angle of repose, 342
  compressibility, 342
  dispersibility, 342
fluid catalytic cracking (FCC) particles, in fluidized beds, 381, 397, 421, 426, 428, 434, 440, 443, 459
fluid element, of continuum, 164
fluidization
  regimes of, in dense-phase fluidized beds, 374–376
  two-phase theory of, 381
fluidized beds. *See also* circulating fluidized beds; dense-phase fluidized beds; lean-phase fluidization
  bubbles in, 510–511. *See also* bubbles

fluidized beds (*cont.*)
  dense-phase type. *See* dense-phase fluidized beds
  Ergun equation applied to, 164
  flow around bubbles in, 385
  gas–particle heat transfer in, 499
  heat transfer coefficients in, 503
  heat transfer in, 136, 499–532
  mass transfer in, 499–532
  particle interaction in, 256
  particle properties affecting, 3
  pressure drops in, 364
  scaling relationships for, 232–235
  speed of sound in, 263
  stability of, 270
  suspension–surface heat transfer in, 499, 500
  temperature control in, 499
flux equation, 157
fountain height, in spouted beds, 408
Fourier equation, 156, 503
Fourier number, 131, 132, 134, 135, 137, 142, 159, 236, 506, 508
Fourier's law, 170
fracture, of particles, 3, 25, 30–32
fragmentation, definition and effects of, 252–254
Fraunhofer diffraction, particle sizing by, 10, 13–15, 16, 18
freeboard, in dense-phase fluidized beds, 376, 384
Freundlich isotherm, 25
friction, in powder flows, 341, 342
frictional force, 64
Froessling equation, 158, 532
Froude number, 231, 235, 236, 380, 409, 484, 485, 487
frozen speed of sound, in gas–solid mixtures, 265

gas adsorption, use in particle sizing, 10, 13
gas convection, heat transfer by, 500, 501
gas convective component, in heat transfer, 514, 516, 522–523
gas distributors, for dense-phase fluidized beds, 376
gas kinetic theory, granular flow and, 452
gas–solid mixture momentum equation, 451
gas–solid separation, 297–332
  in circulating fluidized beds, 422
  by electrostatic precipitation, 297, 309–314
  by filtration, 297, 314–320
  by gravity settling, 297, 320–323
  by rotating flow, 297–309
  by wet scrubbing, 297, 321, 323–324
gas-to-emulsion phase interchange coefficient, 531
gas-to-particle heat transfer
  in fluidized beds, 499, 526–527
  in spouted beds, 526–527
gas-to-particle mass transfer, in dense-phase fluidized beds, 527–529
gas turbulent diffusion coefficient, 453
Gaussian distribution, 19, 21, 209, 210
Gaussian quadrature formula, scattering integral approximation by, 153–154
Gauss's law, 112, 115
Gauss theorem, 168, 189, 272, 468
Geldart's powder classification, 372, 427
glass beads, elastic displacement in, 62–63
glass-ceramics, elastic properties of, 30
glass powder, mass distribution of, 21
glasses
  index of refraction, 40
  thermal properties of, 35
gliding, of particles, 249, 251
gradient-related forces, 95–97. *See also* pressure gradient force; radiometric force; Saffman force
grain inertia flows, kinetic theory modeling of, 210
grains, pneumatic transport of, 461, 498
granular materials
  flow of, 220
  hopper flows of, 333
granular particles, 4
granular temperature, 222
graphite
  properties, 30, 31, 35
  in suspension coolants, heat transfer by, 130
Grashof number, 158, 159
gravitational force, 92, 101, 105, 107, 108
gravitational sedimentation, particle sizing by, 10, 12
gravity chutes, solids transport by, 461
gravity settling and settlers
  collection efficiency of, 321
  gas–solid separation by, 297, 315
  particle residence time in, 321, 323
  particle settling time in, 323
  pick-up velocity in, 321

Haar–von Karman hypothesis, 347
Hagen–Poiseuille equation, 223, 224–225, 226
Hamaker's constant, 102–103, 123, 128
heat conduction, 3. *See also* thermal diffusivity
heat diffusion equation, 506
heat exchanger
  in dense-phase fluidized beds, 371, 372, 376, 377
  downcomer use as, 423
heat loss, by colliding spheres, 46, 80, 121
heat transfer, 130–163, 499–539
  in annulus region, 526–527
  bed-to-gas type, 512–513
  bed-to-surface type, 503, 513–518, 519, 520
  in circulating fluidized beds, 521–526
  in collisions, 133–138
  by convection, 130, 138–142, 500
  in dense-phase fluidized beds, 501, 508, 512–518
  emulsion phase/packet model for, 501, 503, 506–511
  film model for, 501, 502–503
  film-penetration model for, 508–510
  gas convective component in, 514, 516, 522–523
  gas-to-particle type, 526–527
  mass transfer compared to, 130, 157–159
  modes and regimes of, 500–501
  particle convective component in, 514–516, 518, 522
  by radiation, 499, 500, 523–524
  single-particle model for, 501, 503
  of single spheres, 131–133

in spouted beds, 526–527
by thermal conduction, 503
unsteady behavior of, 502
heat transfer coefficient(s), 136, 137, 148, 268, 499
axial distribution of, 524–525
for dense-phase fluidized beds, 512
in emulsion phase/packet model, 507
in film model, 502
gas velocity effects on, 501
of particle convection, 514–515, 522
radial distribution of, 524–525
in turbulent regime, 521
heat transfer probes, 514, 525
Hertzian contact theory, 46, 49, 50, 59–63, 64, 72, 74, 76, 78–79, 81, 133, 246, 248
Higbie's penetration model, 509, 530, 539
Hinze's model, 285
Hinze–Tchen model, 165, 197–201, 470
Hooke's law, 29, 51, 58, 334, 341. *See also* elasticity
hoppers and hopper flows, 333–350
aerating jets for, 333
core (funnel) flow hoppers, 333, 334
flow patterns in, 333
flow-promoting devices for, 333
hopper–standpipe–discharge flow, 354–357
mass flow hoppers, 333, 334, 345
moving bed flows in, 346–350
powder flowability in. *See* flowability of powders
powder mechanics in, 333–346
stress distribution in. *See* stress distribution
stress failure in. *See* stress failure
vibrators for, 333
horizontal flow settling chamber, 321, 322
Howard settling chamber, 321, 322
hydrodynamics
of circulating fluidized beds, 422
of dense-phase fluidized beds, 372
fundamental equations for, 213, 214, 215
hydrogenation, heat transfer in, 130

impact. *See also* collision
collinear, 47–48
damage due to, 245
directional, 244, 251
duration of, 46, 72
dynamic, 46
of elastic bodies, 46, 72
force of, 46
of particles, 244, 245, 247
planar, 48
random, 244
single, charge transfer by, 119–120
stereomechanical, 46, 47–49
velocity of, 48, 72, 81, 82
impact angle, 252
impact number, 249
impulse, definition of, 46
impulse–momentum law, 46
incident radiation, on dielectric spheres, 146, 147
inclined standpipe, 364
index of refraction, 14, 40
inelastic collision, 46
inelastic contact, of solids, 49
instability analysis, 244, 270–288
of batch-solid fluidization system, 284–285
intermittency index, for circulating fluidized beds, 445–447
internal energy, of gas–solid mixtures, 244, 257–258
internals, in fluidized beds, 376, 398, 399, 412, 520
intraparticle heat transfer resistance, 527
Invar, thermal properties of, 35
ion collection, charging by, 115
irregular particles, 4
isentropic change of state, in gas–solid mixtures, 258–259

Jacobian determinant, 168, 236
Janssen's model, for stress distribution, 333, 338–339, 419
Jenike's model, for powder flowability, 342
Jenike translatory shear cell, for powder flowability, 341, 342
jetting
in dense-phase fluidized beds, 376, 381, 388–389
jet formation in, 388–389
jet size in, 389–392
J-valve
in circulating fluidized beds, 423
in standpipe systems, 365

Kelvin effect, 44
k–ϵ–k_p model, 165, 197, 201–202
k–ϵ model, for turbulent flow, 156, 165, 175, 176, 177–178, 179, 180, 201
k-equation, 178, 181, 201
kinetic energy, loss due to collisions, 46, 48, 76, 80, 259
kinetic theory, transport coefficients and, 170–174, 197, 200–201
kinetic theory modeling
collisional flux of fluctuation energy, 219, 222
collisional pair distribution function, 215–217
collisional stress tensor, 217, 219, 220, 221
for collision-dominated dense suspensions, 210–222
constitutive relations, 217–222
dense-phase transport modeling, 211–213
energy dissipation rate, 219, 221
hydrodynamic equations, 213–215
for interparticle collisions, 164, 166
Kirchoff's law, 34, 152
Knudsen diffusion, 156
Knudsen number, 464, 488, 494
Kolmogorov dissipative scale, 447
Kozeny theory, 166, 226–228
Kronecker delta, 169, 239, 540
krypton, use in gas adsorption, 13
Kutta–Joukowsky lift formula, 100

Lagrangian correlation coefficient, 198, 200, 237
Lagrangian trajectory model, 107, 164, 165–166, 167, 205–210, 323
deterministic trajectory method, 165
stochastic trajectory method, 165

laminar flow, description of, 165
laminar transport coefficients, 173, 179
Langevin model, 210
Langmuir isotherm, 25–26
large-eddy simulation (LES), 174, 288
laser Doppler phase shift, particle sizing by, 16–17
laser Doppler velocimetry (LDV) measurements, on cyclones, 302
lean-phase fluidization, 371, 374, 396, 397, 421
Lifshitz–van der Waals constant, 102–103, 124
lift force, 95, 100, 101
lifting, of particles, 251
light wavelength, 40
log-normal distribution, 19, 20, 21
London's constant, 101, 126
loopseal, in standpipe systems, 365
Lorentz constant, 33, 41
Lorentz force, 105, 124
Lorenz–Mie theory, 14
Loschmidt number, 255
Love's stress function, 52, 85
L-valve
  in circulating fluidized beds, 423, 431–432, 435
  in standpipe systems, 365, 366

macropores, 13
magnetic force, 101, 105, 108
magnetic properties, of solids, 3, 37–38
magnetofluidized beds, 105, 378, 416, 417
Magnus effect, 97, 476
Magnus force, 87, 97–101, 108, 124, 190
maleic anhydride production, circulating fluidized bed use in, 421
Martin's diameter, 5–6, 40
mass flow hoppers, 333, 334, 345
mass interchange coefficient, 529, 539
mass transfer, 156–159
  bubble-to-cloud type, 529–531
  in circulating fluidized beds, 532
  in dense-phase fluidized beds, 527–531
  by diffusion and convection, 156–157
  in gas–solid systems, 527–532
  gas-to-particle type, 527–529
  heat transfer compared to, 130, 157–159
  in multiphase systems, 527–532
mass transfer boundary layer, 531
mass transfer coefficient, 159, 499, 527, 528, 530
material densities, 38–39
Maxwell–Boltzmann velocity distribution, 165, 170–172, 210, 215, 216
  transport coefficients based on, 174
mesopores, 13
metals
  electrical properties of, 35–36
  yield strength of, 31
mica, properties of, 31, 37
Mickley–Fairbanks model for heat transfer, 506–508
microorganism cultivation, dense-phase fluidized bed use in, 372
micropores, 13
microscopy, in particle sizing, 10, 11–12
microslip contact, tangential displacement from, 68
Mie scattering theory, 13, 16, 143, 144–148
migration velocity, in electrostatic precipitation, 311–312
Mindlin's theory for frictional contact, 46, 49, 50, 64, 74
minimum fluidization, 378–380, 385
mixing length model, for turbulent flows, 165, 167, 175, 176–177
modular particles, 4
Mohr circle, for plane stresses, 334–335, 337, 340, 341, 342, 344
Mohr–Coulomb failure criterion, for granular materials, 336–337, 342, 347
molecular diffusion coefficient, of gas, 530
molecular diffusivity, 157
momentum equation, 169, 175, 179, 195, 207, 214, 271, 317
monochromatic emissivity, of particles, 34, 41
Monte Carlo method, 153, 165, 206, 209, 323
Moody chart for pipe friction, 470, 471, 472
moon dust, Rosin–Rammler distribution of, 20–21
multiple-tray settling chamber, 321

Navier–Stokes equations, 89, 97, 107, 109, 165, 198, 301, 316
  time-averaged, 174, 175–176
Newton's equation for drag coefficient, 7, 88
Newton's law of acceleration, 35, 203
nickel, as ferromagnetic material, 38
nonmechanical valves, 365, 366
nonspherical particles, equivalent diameters, 4–10
Nusselt number, 132, 133, 138, 139, 141, 157, 159, 268, 512, 513, 526, 531, 533
nylon, properties of, 30, 31, 37

oblique, sliding contact, 64–66
optical microscopy, in particle sizing, 6, 10, 11
optical properties, of solids, 3, 39–40
oscillation, of spherical particles, 259–261
Oseen approximation, 98, 99
Oseen drag, 97, 100
overflow weir, in dense-phase fluidized beds, 419

packed beds. *See also* fixed beds
  equations for flows through, 222–230
packet resistance, for heat conduction, 508
packets, clusters as, 522
paramagnetic solids, 37, 38, 39
particle(s)
  acceleration of, 87, 88, 92, 93
  agglomeration of, 46
  attrition of, 46
  circulation of, effects on heat transfer, 501, 513, 514
  deformation of. *See* deformation
  in dense-phase fluidized beds, 371–376
  density, 38
  diameter of, 255
  eddy interaction with, 287
  electrification of, 46
  fluid interactions of, 87–101
  forces controlling motions of, 87

Geldart's classification of, 372
interactions of, 87, 269, 381, 452, 464, 465, 513
interparticle forces affecting, 101–107, 373
intraparticle heat transfer, 499
motion of single, 87, 107–110
number density, 17, 18, 20, 23, 173, 185, 203, 206, 211
rotation of, 87, 97
shape effect, 3
size and properties of, 3–45
sizing methods for, 10–17
trajectories of, 109–111
turbulent kinetic energy of, 285
viscosity from interparticle collisions, 202–204
wall collisions of, 76–78, 82, 87, 205
particle clouds, thermal radiation with, 130, 150–156
particle convective heat transfer, 501, 503, 514–516, 518, 522
particle convective heat transfer coefficient, axial and radial profiles of, 538
particle Reynolds number, 87, 125, 149, 160, 237, 268, 350, 379, 410, 453, 534
particle terminal velocity, 7–8
particle-to-gas heat transfer, in dense-phase fluidized beds, 512–513
particle-to-gas mass transfer coefficient, 527
particle-turbulence interaction, 244, 285–288
particulate fluidization, in dense-phase fluidized beds, 373, 374–375
particulate fluidized beds, heat transfer in, 501, 510
Peclet number, 139, 181
permittivity, of particles, 36–37, 105
perturbation method, for fluidized-bed instability studies, 270
phase average, 182–183
Phase Doppler Particle Analyzer (PDPA), 17
phonons, heat conduction by, 33
photophoresis, 96, 108
physical adsorption, of particles, 3, 25
pick-up velocity
in gravity settling chambers, 321
in horizontal pneumatic conveying, 474, 476–478
piezoelectric materials, properties and uses of, 37
pipes, stratified, wave motion in, 270–280
pipe walls
erosion of, 244–254
materials used in, 244–245
planar impact, of spheres, 48–49
Planck intensity function, 152
Planck's formula, 34, 142
plastic deformation, 121, 245
onset of, 78–80
of particles, 3, 28, 29–30, 47, 49, 80, 81
plastics
fatigue wear of, 246
pipe walls of, 244–245
Plexiglas, index of refraction, 40
plugs
in pneumatic conveying, 464, 466
square-nosed, 404
pneumatic conveying methods, 448, 461–498
acceleration length in, 472–473
advantages and disadvantages of, 461
choking in, 465
classifications of, 461
critical transport velocities in, 474–475
dilute suspension flow in, 464, 466, 467
dilute transport regime in, 421
drag reduction in, 469–472
flow characteristics in, 464–466
flow regimes and transitions in, 464–466
fully developed dilute pipe flows in, 482–493
horizontal transport in, 461, 466, 468, 474
inclination angle effects on, 477
minimum transport velocity in, 474, 475–476, 498
momentum integral equations for, 479–480
moving bed flow in, 465–466
negative- and positive-pressure systems in, 461
operational modes of, 461, 462–463
particle sedimentation in, 464, 465, 466
particle–wall forces in, 474, 476, 477
pick-up velocity in, 474, 476–478
pipe-bend flows in, 467, 473, 474, 478–482
plug flow in, 464, 466
pressure drop in, 466–472
saltation in, 465, 474
shedding layer in, 479–480, 498
single-phase flow in curved pipes in, 478–481
slip velocity of particles in, 488
stratified flow in, 464, 466
temperature distributions of phases in, 489–493
variables affecting, 461
vertical transport in, 461, 466, 468
wall region of, 469, 470
Poisson contraction, 29
Poisson distribution, 172
Poisson equation, 115, 483
Poisson's ratio, 29, 30, 42, 51, 63, 79, 84, 246, 291, 335, 368
polydispersed particle systems, density functions affecting, 3
polyethylene
elastic properties of, 30, 76, 79
permittivity of, 37
thermal properties of, 35
polymerization, heat transfer in, 130, 499
polymers
in triboelectric series, 113–114
use in fabric filters, 315
yield strength of, 31
polystyrene
plastic deformation of, 29
yield strength of, 31
polyvinyl chloride, permittivity of, 37
porcelain, permittivity of, 37
pores, size classification of, 13
powder(s). *See also* Coulomb powders
flowability of. *See* flowability of powders
flow properties of, 342
motion onset in. *See* stress failure
shear testing of, 341, 342
sieve classification of, 10

powder mechanics
angle of friction, 342
bulk density, 342
cohesion, 342
in hopper flows, 333–346
major principal stress, 342
unconfined yield strength, 342
Prandtl number, 138, 139, 157, 158, 181, 492, 494, 531, 534
pressure, of gas–solid mixtures, 244, 254–257
pressure diffusion, 156
pressure distribution, 58
in frictionless contact, 61, 69
pressure drops
in circulating fluidized beds, 423, 425, 433
in fluidized beds, 364, 378–379
pressure gradient, force due to, 95, 96, 107
pressure waves. *See also* acoustic waves; shock waves
through gas–solid mixtures, 244, 259–269
projected area diameter, 5–6
pseudocontinuum model, 141–142, 150, 186, 338, 350
instability analysis using, 244
PVC–acrylic alloy, yield strength of, 31

radiant heating, of particles, 148–150
radiant heat transfer coefficient, 518
radiation, gas absorption of, 130
radiation transport equation, 151–153
radiative heat transfer, 499, 500, 514, 517–518, 523–524. *See also* thermal radiation
radiative heat transfer coefficient, 523
radiometric force, 96
random surface renewal, heat transfer and, 509
Rankine's combined vortex model, 300–301
rare earth elements, as ferromagnetic materials, 38
Rayleigh number, 158, 160
Rayleigh scattering, 14, 143, 144
reactors
circulating fluidized bed use in, 421
dense-phase fluidized bed use in, 371, 419
reflection
from diffuse spheres, 147
effect on thermal radiation, 143, 150
from specularly reflecting spheres, 146–147
refraction, effect on thermal radiation, 143, 150
regime classifications, in dense-phase fluidized beds, 371–376
relaxation time
of particle, 207, 231
Stokes relaxation time, 93, 108, 126, 199, 239, 265, 285, 306, 330, 471, 497
retardation effects, 101, 102
Reynolds decomposition concept, application to turbulent flow analysis, 174, 175, 193
Reynolds number, 6, 88, 97, 100, 101, 107, 160, 181, 231, 232, 234, 302, 329, 470, 471, 484, 487, 513
heat transfer and, 130, 139
Reynolds stress, 175, 176, 177, 194, 195, 452
Richardson–Dushman equation, 119
Richardson–Zaki correlation, 352, 367, 370, 448
risers
circular, model for, 448
in circulating fluidized beds, 421, 422, 423, 429, 434–436, 440, 442
unstable operation, 428, 432, 433, 437
Rosin–Rammler distribution, 19, 20–21, 22–23, 41
rotary flow dust separators, 297
axial flow cyclones, 299
mechanism and types of, 297–300
tangential flow cyclones, 297, 298, 302, 303–305
tornado dust collectors, 299
rubber
abrasive wear of, 246
thermal properties of, 35

Saffman force, 87, 95–97, 108, 124, 125, 128, 190, 251, 476
saltation, in pneumatic conveying, 465, 474
saturation charge, 116
Sauter's diameter, 6, 13, 23, 24
scaling relationships
for fluidized beds, 232–235
for pneumatic transport, 230–231
scanning electron microscopy (SEM), for particle sizing, 10, 11, 12
scattering
cross section for, 143
diffraction effects on, 150
efficiency factor for, 143, 144
geometric scattering, 143
independent vs. dependent scattering, 150–151
from large spheres, 146–148
Mie scattering theory, 143, 144–148
by multiple particles, 150
by particle clouds, 155
of particles, 249, 251
radiative transport by, 143
Rayleigh scattering, 14, 143
by single particles, 143–148, 150
scattering integral, approximation of, 153–154
Schmidt number, 157, 158, 160, 534
Schrödinger's wave equation, 165, 171
screw conveyors, solids transport by, 461
seal pots, in circulating fluidized beds, 423
Sedigraph particle analyzer, 8
sedimentation, particle sizing by, 10, 11, 12–13
settling chambers, gas–solid separation by, 297
shape coefficients, 3
shape factors, 3
shear strain, 51
shear stress, 51, 56
shear testing
Jenike translatory cell for, 341, 342
of powders, 341, 345
rotational split-level shear cell for, 342
Sherwood number, 157, 158, 160, 531, 534
shock waves
force generated by, 96, 108, 244
propagation of, 108, 265–269
sieve standards, 4–5
sieving, for particle sizing, 10, 18

silica glass, elastic properties of, 30, 62
silo design, 344
silver, as electric conductor, 33
similarity, dimensional analysis and, 230–235
single-particle model, for heat transfer, 501, 503
single-phase flows, modeling of, 167–181
skeletal density, 38, 39
slip velocity
  cluster concept and, 447
  between particles, 451, 532
  between phases, 279, 281
  in standpipes, 363
slugging
  in circulating fluidized beds, 425
  continuous, 404–406
  in dense-phase fluidized beds, 374, 375–376, 392, 403–406
soda-lime glass, thermal properties of, 35
solids circulation
  in circulating fluidized beds, 524, 525
solids mixing
  in circulating fluidized beds, 443–453
  in dense-phase fluidized beds, 371, 381, 384, 401
specific heat
  of gas–solid mixtures, 244, 257–258
  of particles, 32–33
  ratio of, 257
speed of sound, 244, 261–265
  equilibrium speed of sound, 265
  frozen, 265
  in gas–solid mixtures, 244, 261–265
sphere(s), 4
  elastic, 69–71, 104, 133–138
  frictional, 46, 49, 50, 64, 74–78
  frictionless, 59–63
  motion of, 87
  oscillation of, 259–261
  perfect elastic spheres, 48, 49
  perfect inelastic spheres, 48
  rotation of, 97–101
spouted beds
  in dense-phase fluidization, 372, 374, 376, 388, 406–409, 419
  heat transfer in, 526–527
spouting
  effect on heat transfer coefficients, 500
  fountain height, 408
  maximum spoutable bed depth in, 408
  minimum spouting velocity, 407
  onset of, 407
  spout diameter in, 406, 408
spray chamber scrubbers, 323–324
square-nosed slugs, formation in dense-phase fluidized beds, 404
standpipes and standpipe flows, 333, 350–354
  aeration of, 365, 366
  downcomer use as, 423
  inclined standpipes, 364–366
  leakage of flow of gas in, 359–361
  multiplicity of steady flows in, 357–359
  nonmechanical valves in, 364–366
  overflow standpipes, 361–364
  stress distributions in, 337–340
  system types, 361–366
  underflow standpipes, 361–364
static friction coefficient, 64, 66, 68
steel
  elastic properties of, 30
  thermal properties of, 35
  yield strength of, 31
steel ball, restitution coefficient of, 82
Stefan–Boltzmann constant, 161, 491, 495, 517, 535
Stefan–Boltzmann's law, 34, 42, 45, 142
Stefan flow, 158, 159, 160
Stefan–Maxwell equation, 157
stereomechanical impact, 46, 47–49
stochastic trajectory method, for multiphase flow, 165, 205, 208–210
Stokes drag force, 89, 93, 95, 97, 100, 108, 109, 125, 260, 304, 321, 474
Stokes expansion, 98, 99
Stokes flows, 7, 88–90, 94, 108, 139, 140, 261, 323
Stokes number, 231, 450, 453, 464
Stokes relaxation time, 93, 108, 126, 199, 239, 265, 285, 306, 330, 471, 497
Stokes stream function, 97
straight capillaric model, 224–225, 226
strain(s)
  deformation caused by, 28–29, 46, 49
  normal, 51
  shear, 51
stratified flows
  in horizontal pipes, 273–276
  wave motions in, 270–280
stream function, 90, 91–92, 97, 317
  for linear motion of spheres, 89–92
stress(es)
  from boundary point forces, 56–58
  collisional, 104–105
  deformation caused by, 28–29, 46, 49
  general relations of, 50–52
  normal, 55
  shear, 51, 56
  yield, 78, 79
stress distribution
  equation of equilibrium, 341
  in hoppers, 337–340
  Janssen's model (cylindrical column), 333, 338–339
  in steady hopper flow, 340–342
  Walker's analysis, 339–340
  Walters' analysis, 339–340
stress failure
  active failure, 337, 340
  incipient failure, 336
  Mohr–Coulomb failure criterion, 336–337
  passive failure, 337, 340
  of powders, 333
Strouhal number, wake shedding frequency and, 286
Sturm–Liouville boundary-value problem, 493
surface abrasion model, 253
surface area, determination in particles, 13
surface diameter, of particles, 6
surface mean diameter, of particulate systems, 23

surface renewal model, for heat transfer, 502
suspension density, effects on heat transfer coefficient, 525
suspension-surface heat transfer. *See also* bed-to-surface heat transfer
in fluidized beds, 499–511, 521–522
suspension-to-gas heat transfer. *See also* bed-to-gas heat transfer
in dense-phase fluidized beds, 512

tangential displacements, 63–64
tangential flow cyclones, 297, 298, 300, 302, 303
collection efficiency of, 303–307
tangential tractions
in nonsliding contact, 66–69
in sliding contact, 64–66
Taylor expansion, 121, 173
tensile strength, of particles, 29–30, 31
thermal conductivity, 33, 174
thermal convection
heat transfer by, 130
in a pseudocontinuum flow, 141–142
of spheres, 139
thermal diffusivity, 133, 157
thermal properties, of particles, 3, 13, 25, 32–35
thermal radiation
heat transfer by, 130, 142–152
of particle clouds, 130, 150–156
of particles, 3, 13, 33–35
in pseudocontinuum flows, 130
through isothermal and diffuse scattering media, 154–156
thermionic emission, charging by, 119
thermionic work function, 114
thermodynamic properties
equation of state, 244, 254–257
of gas–solid mixtures, 244, 254–259
specific heat, 257–258
specific heat ratio, 257
speed of sound, 244, 261–265
thermophoresis effect, 96, 97, 108, 323
time-averaging
application to turbulent flow analysis, 174
of Navier–Stokes equations, 174, 175–176
tornado dust collector, 299
torsion, of elastic spheres in contact, 69–71, 75
traction, distribution of, 68, 69
trajectory modeling of multiphase flows, 205–210. *See also* Lagrangian trajectory model
deterministic trajectory model, 165, 205, 206–208
stochastic trajectory model, 165, 205, 208–210
transmission, effect on thermal radiation, 143, 150
transmission electron microscopy (TEM), for particle sizing, 10, 11, 12
transport coefficients
based on Maxwell–Boltzmann distribution, 174
for diffusivity, 173–174
for eddy diffusivity, 194
kinetic theory and, 170–174
for particle viscosity, 173
for single-phase flows, 167
for thermal conductivity, 174
turbulence models and, 196–204
for turbulent viscosity, 177
for viscosity, 173
transport disengagement height (TDH), in dense-phase fluidized beds, 401, 402, 403, 410
transport theorem, 167–168
continuity equation in, 169, 175, 179, 195
energy equation in, 169–170, 196, 207–208, 213, 214–215
momentum equation in, 169, 175, 179, 195, 207, 214, 271
transport velocity, in circulating fluidized beds, 425
Tresca criterion, 78, 79
triboelectric charging, 112
triboelectric series, 112, 113–114
tungsten, elastic properties of, 30
turbulence
in cyclone separators, 302
modulation of, 286, 287
particle interaction with, 285–288
turbulent dissipation rate, transport equation of, 177, 178
turbulent flow modeling, 174–179
Hinze–Tchen model, 165, 197
k–ϵ–k_p model, 165, 197, 201–202
k–ϵ model, 165, 175, 176, 177–179
mixing length model, 165, 175, 176–177
transport coefficients and, 196–204
turbulent fluidization, in dense-phase fluidized beds, 374, 375, 396–400
turbulent kinetic energy
equation for dissipation rate of, 179
equation of, 179
of particles, 285
transport equation of, 177–178
turbulent mixing, 304, 312
turbulent viscosity, 181
two-phase theory of fluidization, 375, 393, 405, 420, 451
Tyler Standard sieves, 4, 10, 11

unsteady heat transfer behavior, 502

Van der Waals adsorption, 24, 25
Van der Waals forces, 87, 101–103, 108, 124, 128, 191, 315, 373, 445, 474
macroscopic approach to, 102–104
microscopic approach to, 101–102
venturi scrubbers, 323, 324, 326
vibratory conveyors, solids transport by, 461
vibrofluidized beds, 377
viscoelastic deformation, 28
viscosity
coefficient of, 173
effective, definition of, 179
of gas–solid suspensions, 384
Volta potential, 114
volume averaging, 165, 182–183
volume-averaging theorems, 187–189, 190
volume diameter, of particles, 6
volume fraction, of particles, 268, 269, 272
volume mean diameter, of particulate systems, 23

volume–time-averaged equations, 193–196, 206
Von Karman's constant, 180, 239
Von Mises criterion, 78
vortex shedding, 286, 287
V-valve, in circulating fluidized beds, 423

wake
  from bubbles, 381, 387
  shedding of, 286, 385
  in slugging bed, 404
wake angle, of bubbles, 383–384
wake attraction, particle collision due to, 93–95, 128
Walters's analysis, of stress distribution, 339–340
walls
  effect on dense-phase fluidized beds, 403, 418
  particle collisions with, 76–78, 82, 87, 307
  role in heat transfer, 521–522, 527
wall slugs, formation in dense-phase fluidized beds, 404
Walters's analysis, of stress distribution, 339–340
wave equation, 279
waves. *See also* pressure waves
  in circulating fluidized beds, 444
  continuity waves, 280–282
  in dense-phase fluidized beds, 382
  dynamic waves, 282–284
  motion of, 270–280
  speed of, 281, 283
  wave equation, 279
Weber number, 385
Wen–Yu equation, 379
wet scrubbing and scrubbers
  cocurrent-flow mode in, 324, 328
  countercurrent-flow mode in, 324, 327
  cross-flow mode in, 324
  cyclone scrubbers, 323
  gas–solid separation by, 297, 321
  mechanisms and types of, 323–324
  modeling of, 324–328
  spray chamber scrubbers, 323–324
  venturi scrubbers, 323, 324, 326
Wiedemann–Franz law, 33
Wien's displacement law, 34, 143
wood, elastic properties of, 30

yield strength, of particles, 29–30, 31
yield stress, 245
Young's modulus, 29, 30, 41, 51, 63, 83